Electromagnetic Fields

Electromagnetic Fields

Theory and Applications

Ahmad Shahid Khan and
Saurabh Kumar Mukerji

CRC Press
Taylor & Francis Group
Boca Raton London New York

CRC Press is an imprint of the
Taylor & Francis Group, an informa business

First edition published 2021
by CRC Press
6000 Broken Sound Parkway NW, Suite 300, Boca Raton, FL 33487-2742

and by CRC Press
2 Park Square, Milton Park, Abingdon, Oxon, OX14 4RN

ISBN: 978-0-367-49430-8 (hbk)
ISBN: 978-1-003-04613-4 (ebk)

Typeset in Times
by Deanta Global Publishing Services, Chennai, India

Visit the eResources: http://routledge.com/9780367494308

Contents

Preface

Electromagnetics is one of the most fundamental subjects included in the electrical and electronics engineering curriculum. In general, this subject is taught at the undergraduate level. Some of the institutions also give application-oriented courses on electromagnetics at the postgraduate level, whereas others include research facilities. In spite of its importance due to widespread applications in various devices and systems, this subject has never gained the favor it rightly deserves. In recent years, more areas in which electromagnetics is deeply involved have been identified. The fruitful development of these areas is waiting for people well versed in field theory.

Electromagnetic field theory revolves around a number of quantities that can neither be seen nor easily measured. Most of these quantities require imagination and can be described only through mathematical language. This aspect makes this subject a bit repulsive and students in general avoid specializing in areas related to field theory. In such a scenario any further development in the areas involving field theory appears to be remote. In fact, this scenario led to many surveys conducted at global and regional levels. The reports of these surveys were the basic motivating force to write the present text. Effort has been made to accommodate some of the feasible suggestions of these surveys.

This text contains all the relevant topics essential for the understanding of electromagnetics. It tries to present the fundamental concepts in a simplified way. It tries to make mathematical complexities reader-friendly and supplements mathematical expressions with examples to provide critical insight.

This book contains 22 chapters and 4 appendices. The first two chapters include introduction and applications, and the next two chapters dwell through some required mathematical tools. Chapters 5 to 15 present the basics of electromagnetic field theory. The last seven chapters deal with some of the applications of electromagnetics. All appendices are made available at the eResources: http://routledge.com/9780367494308. The contents of these chapters are briefly given below.

Chapter 1 highlights the essence of field theory, includes brief history of development in the area and includes a world scenario vis-à-vis the teaching methods and student's response.

Chapter 2 classifies the fields and highlights the involvement of electromagnetic field theory in different disciplines ranging from machines to microwaves, computers to communication, transmission lines to television, radio waves to radars, and many other areas.

Chapter 3 introduces coordinate systems and vector algebra. These are the basic tools for the understanding and interpretation of different field phenomena.

Chapter 4 briefly describes gradient, divergence and curl operations, introduces Del and Laplacian operators, line and surface integrals, and various theorems required for the understanding and interpretation of field phenomena.

Chapter 5 introduces Coulomb's law and electric field intensity and obtains relations for electric field intensity for various charge distributions.

Chapter 6 introduces Gauss's law, electric flux, and electric flux density. It also includes relations for electric flux density for various charge distributions.

Chapter 7 introduces potential difference, potential, field due to a dipole, and the energy stored in electrostatic field. It also includes relations for potential fields and configurations for equipotential surfaces for various charge distributions.

Chapter 8 discusses the current and current density, different type of electrical materials, polarization, continuity equation, relaxation time, and boundary conditions for electrostatic fields. This chapter introduces two basic circuit elements, resistance and capacitance.

Chapter 9 includes the boundary value problems related to the Laplacian field. This chapter deals with one-, two-, and three-dimensional field problems belonging to cartesian, cylindrical, and spherical coordinate systems.

Chapter 10 includes the boundary value problems related to the Poisson's field. This deals with one and two dimensional field problems in cartesian, cylindrical, and spherical coordinate systems.

Chapter 11 introduces the steady magnetic fields, the related field quantities (viz. H, Φ, and B) and the governing laws viz. Biot-Savart's Law and Ampère's circuital law. It also introduces the scalar and vector magnetic potentials.

Chapter 12 discusses the forces due to various current distributions, torque, dipole moment, magnetization, magnetic susceptibility, relative permeability, properties of magnetic materials, and the boundary conditions for magnetostatic fields.

Chapter 13 deals with the magnetic circuits and introduces inductance. This also discusses energy stored in the magnetic field.

Chapter 14 discusses two boundary value problems related to the magnetostatic field.

Chapter 15 deals with Faraday's law of electromagnetic induction and various aspects of time varying field including the displacement current, Maxwell's equations, retarded potentials, and the Poynting theorem. This chapter also describes the impact of relativistic effect on field relations.

Chapter 16 describes various aspects of the plane wave propagation in lossless and lossy media. This also explains the meaning of attenuation, propagation, phase velocity, characteristic impedance, polarization, depth of penetration, and the surface impedance.

Chapter 17 discusses the reflection and refraction of plane waves from perfect conductor and perfect dielectric for both normal and oblique incidences. Cases of reflection from imperfect dielectric and refraction phenomenon in the ionosphere have also been discussed.

Chapter 18 deals with the theory of transmission lines. The study includes the equations governing line behavior, characterizing parameters, applications of line sections at UHF and above, and stub matching. It includes a brief overview of various transmission structures. It also describes the Smith chart and its applications in transmission lines.

Chapter 19 describes the theory of rectangular and circular waveguides and that of different cavity resonators. It includes the mathematical aspects of modes, characterizing parameters, field patterns, and methods of their excitation.

Chapter 20 deals with radiation mechanism, including the related mathematical aspects. It describes the far and near fields, radiation patterns, and the effect of lengths and ground thereupon.

Chapter 21 deals with the genesis, importance, and applications of eddy currents. It also includes the solutions of some one-, two-, and three-dimensional eddy current problems.

Chapter 22 discusses different aspects of the electromagnetic compatibility including the sources, types of electromagnetic interference, coupling mechanisms and EMC control plans.

Each chapter contains solved examples, unsolved problems, and descriptive questions. References are also given for further reading. Since Chapters 3 and 4 require almost the same references, all of these are given at the end of Chapter 4. Similarly, Chapters 5 to 15 require almost the same references and all of these are given at the end of Chapter 15. In other chapters these are given at their end.

This book also has a supplemental solutions manual. Chapterwise PowerPoint slides are also provided. Both of these are to be made available by the publisher on their website. In addition, this book is provided with a subject index to facilitate the search of desired topics.

The authors deeply condole the loss of Ms Nazia Shahid, daughter of A. S. Khan, the first author, who departed during the course of preparation of this text.

Finally, the authors thankfully acknowledge the help and encouragement received from the members of their families and colleagues.

A. S. Khan and S. K. Mukerji

Authors

Ahmad Shahid Khan obtained his B.Sc. Engg. M.Sc. Engg. and Ph.D. degrees in 1968, 1971, and 1980 respectively from Aligarh Muslim University (AMU), Aligarh. He joined AMU in March 1972 and served in various capacities including as Professor and Chairman, Department of Electronics Engineering, Registrar AMU, and as Estate Officer (Gazetted). After retirement in December 2006 from AMU, he served at IIMT Meerut, VCTM Aligarh, and JCET Rampur, all in the capacity of Director. He also served as Professor in KIET Ghaziabad and IMS Ghaziabad. All these engineering institutes are affiliated to U. P. Technical (now APJ Abul Kalam) University, Lucknow, India. He also served as a visiting Professor at MEC (Wakf) Palla, Nuh, affiliated to MDU Rohtak, Haryana. He has nearly 45 years of teaching, research, and administrative experience.

Dr. Khan is a Fellow of the Institution of Electronics and Telecommunication Engineers (India), life member of the Institution of Engineers (India), life member of the Indian Society for Technical Education, and life member of the Systems Society of India. He has attended many international and national conferences and refresher/orientation courses in the emerging areas of electronics and telecommunication. His area of interest is mainly related with electromagnetics, antennas and wave propagation, microwaves, and radar systems. He has published 23 papers mainly related to electromagnetics. His Ph.D. work was related to the application of electromagnetic fields to electrical machines. He is a recipient of Pandit Madan Mohan Malviya Memorial Gold Medal for one of his research papers published in the *Journal of Institution of Engineers India*.

Dr. Khan has edited a book entitled *A Guide to Laboratory Practices in Electronics & Communication Engineering* published in 1998. Its revised version was published in 2002. Both of these were published by the Department of Electronics Engineering, AMU Aligarh. Dr. Khan revised the book entitled *Antennas for All Applications* by Kraus and Marhefka. His name was added as a coauthor in its Indian third edition, published by Tata McGraw Hill, New Delhi, in 2006. The fourth edition of this book appeared in April 2010 with a new title, *Antennas and Wave Propagation*. Dr. Khan added six new chapters in this edition. Its nineteenth reprint appeared in December 2017. The fifth edition of this book, with learning objective mode, was published in 2019. His book entitled *Microwave Engineering: Concepts and Fundamentals* was published by CRC Press, Taylor & Francis Group, in March 2014. Another book entitled *Electromagnetics for Electrical Machines*, for which Dr. Khan is a coauthor, was also published by CRC Press in March 2015.

Saurabh Kumar Mukerji obtained his B.Sc. Engg. degree in Electrical Engineering from AMU Aligarh in 1958, and M. Tech. and Ph.D. degrees in Electrical Engineering from I.I.T. Bombay in 1963 and 1968 respectively. He has more than 50 years of teaching and research experience.

During this period, he served in various capacities including as Professor and Chairman, Department of Electrical Engineering, Aligarh Muslim University,

Aligarh (India). He also worked at Madhav Engineering College, Gwalior (India), S.R.M.S. College of Engineering and Technology, Bareilly (India), and M.M.U., Melaka (Malaysia), in various capacities including as Professor. He has also served as Senior Professor at Mangalayatan University, Aligarh (India). He has more than forty research papers in journals and conference proceedings, both national and international, to his credit. He has taught electrical machines and electromagnetic fields to graduate and post-graduate classes for more than twenty years. He has also taught network theory, optimization techniques, electrical machine design, antenna and wave propagation, microwave engineering, electrical measurements and measuring instruments, etc. His main area of research is electromagnetic field applications to electrical machines.

Professor Mukerji has been the reviewer of research papers for the *Journal of the Institution of Engineers* (India), *Progress in Electromagnetic Research* and *Journal of Electromagnetic Waves and Applications* (PIER and JEMWA) M.I.T., USA, and the *Journal of Engineering Science and Technology* (JESTEC), Malaysia. He has served as an Executive Editor with *Thomson George Publishing House*, Malaysia. A book entitled *Electromagnetics for Electrical Machines*, for which Dr. Mukerji is the main author, was published by CRC Press in March 2015.

1 Introduction

1.1 INTRODUCTION

A region of empty or occupied space wherein certain physical states occur is referred to as a field. A field may contain some solid, liquid, or gaseous material. A field has a manifestation of energy in some form, which may correspond to the fall of bodies, transfer of heat, flow of fluids, movement of electric charges, or attraction or repulsion of magnets. Engineers study the way a magnetic flux distributes in an air-gap, the current flows in a thick conductor, the stresses develop in a mechanical device, and vortices and eddies form in the water gushing out of a dam. Any electrical or electronic engineer must always be concerned about the characteristic structures of electrical fields, their nature and sources, the quantities involved, and the utility vis-à-vis the operations of electrical and electronic devices.

Throughout history, people have tried to solve the puzzle of life, their surroundings, and the universe. The giants of electrical sciences, particularly in the area of electromagnetics, include Coulomb, Gauss, Ampère, Hennery, Volta, Faraday, Kirchhoff, Galvani, Oersted, Fleming, Maxwell, Marconi, Bose, and many others. What they sowed we are reaping today, and what we cultivate today future generations will reap. One has to understand what these giants have done and how their successors utilized achievements for the benefit of humanity. Section 1.2 will give a brief account of the history of electromagnetic study.

1.2 HISTORICAL PERSPECTIVE

The historical perspective is divided into three stages: (1) the conceptual stage, (2) the era of basic laws, and (3) the era of inventions.

1.2.1 Conceptual Stage

The concept of static electricity is not a new one. The ability of rubbed amber to attract light particles was recorded in 600 BCE by the Greek thinker Thales of Miletus. For nearly 2,000 years, this concept remained almost confined to history. Then in 1296 CE, a magnetic compass was reportedly brought to Venice from the court of Kublai Khan by Marco Polo. In 1600 an English scientist, William Gilbert, demonstrated the effect on a compass of a metalized sphere (similar to earth), and published *De Magnete*. The word "electricity" was first used by Sir Thomas Browne in 1646 and the first electrostatic machine was reportedly built by Otto Von Guericke in 1650. In 1733 Charles Du Fay discovered two kinds of charges, which he called positive and negative. He also noted the attraction and repulsion of unlike and like charged particles. In 1735 conducting and dielectric properties were demonstrated by Stephen Grey. In 1745 the Leyden jar, the first capacitor, was invented by E. J. G. Von Kleist and P. V. Musschenbrock independently. In 1746 Benjamin Franklin classified electricity into negative (excess of electrons) and positive (deficiency of electrons). He also demonstrated the electrical nature of lightning and invented the lightning conductor in 1752. In 1787, M. Lammond invented the telegraph. In 1790 Alessandro Volta found that the chemistry acting on two dissimilar metals generates electricity, and in 1800 he invented the voltaic pile battery.

1.2.2 Era of Basic Laws

The amount of force exerted between unlike and like charges was first measured by Charles Augustine de Coulomb. The result, now known as Coulomb's law, was first published in 1785. In 1820 Hans Christian Oersted demonstrated that electric current affects a compass needle. Biot-Savart's law was also proposed in 1820. In 1822 André-Marie Ampère formulated the law which describes the force between current carrying conductors, known as Ampère's (force) law. In 1826 George Simon Ohm formulated another law relating voltage, current, and resistance, known as Ohm's law. Michael Faraday formulated his law of electromagnetic induction in 1831. He presented the idea of electric fields at the Royal Society of London. He also studied the effect of currents on magnets and magnets inducing electric currents. In 1835 Carl Friedrich Gauss formulated the mathematical theory now known as Gauss's law.

1.2.3 Era of Inventions

Dynamo, the first practical generator, was invented by Hypolite Pixii in 1832. In 1934 André-Marie Ampère invented the galvanometer. The invention of the magnetic telegraph in 1838 is attributed to Samuel Morse and Charles Wheatstone, respectively. The first long distance telegram was exchanged by Samuel Morse and Alfred Viol in 1844. Mehlon Loomis invented radio telegraphy in 1864. The same year, James Clerk Maxwell predicted the existence of electromagnetic waves. His *Treatise on Electricity and Magnetism* was published in 1873. Maxwell's equations have become the governing laws for all electromagnetic phenomena at the macro level inside material media that are described by constitutive relations. In 1873 DC Electric Motor was invented by Zénobe Gramme and in 1876 Alexander Graham Bell invented the microphone. Graham Bell also invented the telephone, which was patented that same year. The gramophone and incandescent electric lamps were invented by Thomas Alva Edison in 1878 and 1879, respectively.

Subsequent years saw the invention of a number of important devices as well as discoveries of phenomena, including cathode ray tubes by William Crookes (1878), the AC transformer by William Stanley (1885), the induction motor by Nikola Tesla (1888), the existence of electromagnetic waves by Heinrich Hertz (1888), the transmission of radio waves by J. C. Bose (1894), magnetic tape recorders by Valdemar Poulsen (1899), and the loudspeaker by Horace Short (1900). The radiotelegraph was invented by Guglielmo Marconi in 1900.

From the beginning of the twentieth century, the rate of discoveries and inventions multiplied enormously based on the knowledge thus far acquired. As a result, in 1900 Marconi successfully established communication between U.S. and British battleships that were thirty-eight miles apart. On December 12, 1901, Marconi received the first transatlantic radio telegraphic signals. Inventions in 1902 included the synchronous motor by Ernest Danielson, the photo-electric-cell by Arthur Korn, and the radio telephone by Valdemar Poulsen and Reginald Fessenden. The subsequent years saw the discoveries of the electrocardiograph (ECG or EKG) by Willem Einthoven (1903), the electrostatic precipitator by Frederick Gardner Cottrell (1905), the vacuum tube triode by Lee De Forest (1906), the radio receiver by Ernst Alexanderson and Reginald Fessenden (1913), and the electrical method for recording sound by L. Guest and H. O. Merriman (1920).

One of the most fascinating and useful devices, radar, was invented by A. H. Taylor and L. C. Young (1922). Another useful device, television, was developed over time by Philo Farnsworth (1923), C. Francis Jenkins (1925), John Baird (1926), and P. T. Farnsworth (1927). Subsequent years witnessed the development of many other useful tools such as the electroencephalograph (EEG) by Hans Berger (1929) and the radio telescope by Karl Jansky and Grote Reber (1931). In 1943 Zoltán Bay sent ultra-short radio waves to the moon. Robert Dicke and Robert Beringer used the word microwave in an astronomical context in 1946. Further developments included mobile telephone service by AT & T and Southwestern Bell (1946), the first-generation computer with vacuum tube technology by John Mauchly and John Eckert (1946), the transistor by W. Shockley, J. Bardeen, and W. Brattain (1947), optical fiber by N. S. Kapany (1952), the integrated circuit by Jack Kilby and Robert Noyce (1948), the communications satellite by Kenneth Masterman Smith (1958), the microprocessor by Robert Noyce and Gordon Moore, the computerized tomography scanner (CAT-SCAN) by Sir Godfrey Newbold Hounsfield, magnetic resonance imaging (MRI) by Raymond V. Damadian (1971), and the mobile phone by Bell Labs in 1977. The idea of maglev, a contraction of magnetic levitation was put forward in 1988, along with the Global Positioning System (GPS) by the U.S. Department of Defense in 1993.

1.3 SPHERE OF ELECTROMAGNETICS

The word "field" is used in various contexts. It is used for land under cultivation, for playgrounds, and for areas of studies, such as "field of medicine", "field of engineering", and so on. Thus, the word "field" covers a variety of areas and actions which need not be confined to living creatures; these actions may also belong to non-living entities such as tiny particles.

1.3.1 Tiny Charged Particles

As all matter in the universe is composed of tiny particles called atoms which in itself is composed of a number of subatomic particles viz. electrons, protons, and neutrons. Electrons and protons are called charge particles since these are assumed to possess an electrical property called charge. The charge content of an electron [$(1.60210 \pm 0.00007) \times 10^{-19}$ C] is assumed to be negative whereas that of a proton is of opposite polarity (thus positive) but of the same order. The (rest) mass of an electron (m) is estimated to be [$(9.1091 \pm 0.0004) \times 10^{-31}$ kg], whereas its radius (at rest) is said to be 3.8×10^{-15} meters. The study of activities of these tiny particles has not only revolutionized the world, it influenced almost every sphere of human life.

1.3.2 Behaviour of Tiny Charged Particles

The tiny charged particles referred to above may behave in a variety of ways. When these charges are stationary, their field of influence is called an *electrostatic field.* The moving charges (with constant speed) constitute steady electric current, which results in a *magnetostatic field.* The static charges give an effect termed "static electricity," which does not have any magnetic effect. A steady (dc) current flowing in a conductor produces a magnetic field without any electric field. However, if time-varying or alternating current (ac) flows in a conductor both electric and magnetic fields are produced, such a field is referred to as an *electromagnetic* or *time-varying field.* The charged particles act differently in free space or vacuum and in the presence of matter in the media. It is the behavior of these particles that leads to the general categorization of materials as *conductors*, *semiconductors*, and *insulators* (or dielectrics). Thus the fields resulting due to different modes of behavior of charges in different materials need to be thoroughly and systematically studied.

1.3.3 Role of Maxwell Equations

Electromagnetics relates space, time, spatial and temporal frequencies, spatial vectors, complex vectors (or phasors), power, and distributions in three-dimensional space and time. It is based on one of the most fundamental sets of equations, known as Maxwell's equations. These equations have applications in almost all natural sciences and modern technologies, including that of electromagnetics and its offshoots. Therefore, the study of electromagnetics mainly revolves around Maxwell's equations. In addition, operations of all of the devices listed in Section 1.2.3 involve field phenomena in one way or the other. To understand the essence of electromagnetics, its involvement in some of the most commonly known devices is discussed briefly in Chapter 2, which is entirely devoted to applications of electromagnetics.

1.4 TEACHING OF ELECTROMAGNETICS

Electromagnetics (EM) is not a new subject introduced at the B.Tech. level. At the senior secondary or its equivalent level, which is generally the entry point to B.Tech., almost all the terminology, laws, theorems, and equations related to the field phenomenon are briefly covered in physics courses. The new entrants are fully aware of the concepts of charge, current, voltage, power, force, frequency, and torque, etc. They are well aware of Coulomb's law, Gauss's law, Biot-Savart's law, Ampère's circuital law, and Faraday's law of electromagnetic induction. In addition, before their entry to B.Tech. courses, students become well-versed in vector algebra, vector calculus, integral and differential calculus, coordinate and solid geometry, and many other related areas of mathematics. With such a solid background, the study of electromagnetics in fact repeats much of what B.Tech. students have already learned.

Thus, the subject at this level is to be presented in somewhat more elaborated fashion, with new perspectives and with more emphasis on practical utility.

1.4.1 Importance of Electromagnetics

There is no dispute about the importance of electromagnetics. It is one of the most fundamental subjects included in all undergraduate courses in electrical and electronics engineering. In some institutions, an advanced course on electromagnetics is also given at the postgraduate level. Although some academic institutions also include research facilities, an insignificant number of students take advantage of this. A review of the curriculum of any engineering institution reveals that due recognition is given to its importance.

1.4.2 Acceptability of the Subject

From the students' point of view, this subject has not been very popular. The majority of students go through this subject as a necessary evil. The usual complaints against the subject are that it is too mathematical, too abstract, and, at least at the beginning level, devoid of any practical utility. A dislike developed for the subject at an early stage usually persists in students' minds and, when options are available, few students opt to specialize in this area. This is an unfortunate situation and can be changed only by adopting a new approach to teaching, which can make the subject both interesting and informative so that the significance and applications of the ideas developed can be clearly brought out.

The issue of the teaching of electromagnetic fields was deliberated at Indian Institute of Technology Kanpur (India) in 1978. The aspects of discussion included when, what, and how to teach it, in order that the subject may secure, both in coverage and student response, the place it rightly deserves. Since then, almost four decades have passed, more areas of its applications have been identified, and now its relevance and significance vis-à-vis electrical and electronics disciplines have been better appreciated. However, dislike for this subject still persists and innovative and elaborative efforts to make the subject popular or at least dispel the dislike for it are needed.

1.4.3 Present Scenario

In order to understand worldwide teaching of electromagnetics and the response of students toward the subject, a survey was conducted in 1990 in thirteen countries, including Australia, Belgium, Brazil, China, Egypt, Germany, India, Israel, Japan, Lithuania, South Africa, Switzerland, and the United Kingdom. This survey was based on some common questions related to (i) the number of students at graduation and postgraduation levels, (ii) the curriculum and beginning point, (iii) the usual methods of instruction used, (iv) the number of teaching hours per week, (v) the length of the session, (vi) the textbooks used, (vii) any novel or innovative techniques employed, (viii) student responses, and (ix) the usefulness of new teaching aids. The response to these queries showed a universal decline in student interest in electromagnetics. It was noted that most courses used traditional teaching methods, whereas students are more interested in application-oriented courses that use computers and software simulators.

Another survey encompassing some North American and Canadian universities was conducted in 2005. In its report, the committee spelled the following prevailing approaches for the teaching of electromagnetics:

- The traditional model starts with a review of vector analysis. This is followed by static fields and some basic time-varying fields. Thereafter, the transmission lines, plane waves, and other concepts are given.
- The second model starts with the time and frequency domain analysis of transmission line theory. This is followed by a review of vectors, static fields, and time-varying fields.
- The third model starts with the general time-varying fields. This is followed by Maxwell's equations and then static fields are introduced.

This survey revealed that in spite of a background in physics and mathematics, students in general found electromagnetics courses very challenging and demanding. The report suggested using simulation and visualization tools, including some popular commercial computer-aided EM software produced by faculty at various universities. It further suggested the importance of discussion of practical applications along with abstract theory. The committee also said that proper motivation of students by teachers may greatly change the scenario. The survey observed that most of the initiatives reported in the literature fit into one of the following categories:

- Use of modeling, simulation, and visualization software to improve comprehension.
- Alternatives to vector calculus for analysis of EM fields.
- Development of virtual laboratory environments.
- Incorporation of active (experiential) learning in the classroom.
- Incorporation of problem-based learning activities integrated with other courses.
- Incorporation of electromagnetic compatibility (EMC) issues into its curriculum.
- Development of assessment tools such as concept inventories.
- Web-based instruction.

The report also suggested that the IEEE Societies viz. Antennas and Propagation Society (APS), Microwave Theory and Techniques (MTT) Society, Electromagnetic compatibility (EMC) society, etc. should support projects to develop innovative and interesting teaching tools.

1.4.4 Features of the Present Text

- This book is based on the traditional model as followed in many institutions in the developing world.
- It includes all of the essential topics relating to the theory of electromagnetics.

- It tries to present the field concepts in such a way that the reader, through personal study, can understand the subject. These concepts are aided with appropriate illustrations, mathematical relations, and a good number of solved examples.
- It tries to address mathematical complexities to make them reader-friendly.
- It includes chapters on some important applications of field theory.
- It includes two detailed chapters on electrostatic boundary value problems involving Laplacian and Poissonion fields, one chapter on magnetostatic boundary value problems, one chapter on eddy currents, and one on electromagnetic compatibility.
- It does not include MATLAB®-based problems and experimental methods.

DESCRIPTIVE QUESTIONS

Q1.1 List the basic laws related to the electromagnetic field phenomenon.

Q1.2 List ten commonly used electrical and electronic devices wherein the electromagnetic field phenomenon is involved.

Q1.3 How does the behavior of tiny charged particles result in different forms of fields?

Q1.4 Why is it essential for students of the electrical sciences to study electromagnetics?

Q1.5 Present your opinion about positive and negative aspects of books on electromagnetics that you have read.

FURTHER READING

H. Roussel, M. Hèlier, "Difficulties in teaching electromagnetism: an eight year experience at Pierre and Marie Curie University," *Advanced Electromagnetics*, Vol. 1, No. 1, May 2012.

J. Leppävirta, H. Kettunen, A. Sihvola, "Complex problem exercises in developing engineering student's conceptual and procedural knowledge of electromagnetics," *IEEE Transactions on Education*, Vol. 54, No. 1, Feb. 2011.

X. M. Zhang, S. L. Zheng, Y. Du, X. F. Ye, K. S. Chen, "Perspective of electromagnetics education," in *Progress in Electromagnetics Research Symposium Proceedings*, Xi'an, China, March 22–26, pp. 131–134, 2010.

M. Lumori, E. Kim, "Engaging students in applied electromagnetics at the University of San Diego," *IEEE Transactions on Education*, Vol. 53, No. 3, Aug. 2010.

J. Bunting, R. Cheville, "VECTOR: a hands-on approach that makes electromagnetics relevant to students," *IEEE Transactions on Education*, Vol. 52, No. 3, August 2009.

L. Sevgi, "A new electromagnetic engineering program and teaching via virtual tools," *Progress in Electromagnetics Research B*, Vol. 6, pp. 205–224, 2008.

Y. J. Dori, E. Hult, L. Breslow, J. Belcher, "How much have they retained? Making unseen concepts seen in a freshman electromagnetism," Course at MIT, *Journal of Science Education and Technology*, Vol. 16, No. 4, August 2007.

L. Ding, R. Chabay, B. Sherwood, R. Beichner, "Evaluating an electricity and magnetism assessment tool: brief electricity and magnetism assessment," *Physical review special Topics – Physics Education Research*, Vol. 2, 2006.

M. Popović, D. D. Giannacopoulos, "Assessment-based use of CAD tools in electromagnetic field courses," *IEEE Transactions*, Vol. 41, No. 5, pp. 1824–1827, May 2005.

S. Rengarajan, D. Kelley, C. Furse, L. Shafai, "Electromagnetics education in North America," in *Proceedings of the URSI General Assembly*, New Delhi, India, Oct. 2005.

J. Mur, A. Usón, J. Letosa, M. Samplón, S. J.Artal, "Teaching electricity and magnetism in electrical engineering curriculum: applied methods and trends," in *Proceedings of the International Conference in Engineering Education*, Gainesville, FL, Oct. 16–21, 2004.

A. Norman, M. Mani, "A new approach in teaching electromagnetism: how to teach EM to all levels from freshman to graduate and advanced-level students," in *Proceedings of the American Society for Engineering Education Annual Conference*, Iowa State University, 2003.

D. L. Evans, D. Gray, S. Krause, J. Martin, C. Midkiff, B. M. Notaros, M. Pavelich, D. Rancour, T. Reed-Rhoads, P. Steif, R. Streveler, K. Wage, "Progress on concept inventory assessment tools," in *Proceedings of the 33rd ASEE/IEEE FIE*, Boulder, CO, Nov. 5–8, 2003, pp. T4G.1–T4G.8.

A, Fagen, C. Crouch, E. Mazur, "Peer instruction: results from a range of classrooms," *The Physics Teacher*, Vol. 40, Apr. 2002.

M. F. Iskander, "Technology-based electromagnetic education," *IEEE Transactions on Microwave Theory and Techniques*, Vol. 50, No. 3, pp. 1015–1020, Mar. 2002.

B. M. Notaros, "Concept inventory assessment instruments for electromagnetics education," in *IEEE Antennas Propagation Society International Symposium Digest*, San Antonio, TX, June 16–21, 2002, Vol. 1, pp. 684–687.

D. Ioan, I. Munteanu, "Symbolic computation with Maple V for undergraduate electromagnetics education," *IEEE Transactions on Education*, Vol. 44, p. 217, 2001.

B. Beker, D. W. Bailey, G. J. Cokkinides, "An application-enhanced approach to introductory electromagnetics," *IEEE Transactions on Education*, Vol. 41, No. 1, pp. 31–36, Feb. 1998.

K. F. Warnick, R. H. Selfridge, D. V. Arnold, "Teaching electromagnetic field theory using differential forms," *IEEE Transactions on Education*, Vol. 40, No. 1, pp. 53–68, Feb. 1997.

O. de los Santos Vidal, R. Jameson, M. Iskander, "Interaction and simulation-based multimedia modules for electromagnetics education," in *FIE'96 Proceedings*, 1996.

Fred J. Rosenbaum, "Teaching electromagnetics around the world: a survey," *IEEE Transactions on Education*, Vol. 33, No. 1, pp. 22–34, February 1990.

Course organized by IIT Kanpur on "Electromagnetics, Computer Techniques and Recent Trends," Dec. 11–22, 1978.

2 Field Applications

2.1 INTRODUCTION

In the modern era, electricity has now become the backbone of the global economy. The generation, transportation, conversion, and many other aspects of electric power are closely related to electromagnetics. The devices, methods, and systems involving electrical energy require electromagnetic theory for a better understanding of their design and operation. In this era of information technology, the transmission of information with or without wires, a sound knowledge of electromagnetic field is essential. In electromagnetics, both electric and magnetic fields are defined in terms of the forces they produce. Thus, a strong grasp of field theory is essential for proper use of these fields to create forces, motion, and effects to do useful work, and to transmit noise free signals loaded with information.

In electromagnetic (EM) field theory, there is no dearth of examples of practical applications. As no electrical or electronic device or system can be conceived without the presence of charges or currents, the involvement of field concept is imminent. A lightning stroke from cloud to the ground, a digital wristwatch, a coffee maker, the remote car door opener, the power generating plants, the propulsion motors of a naval aircraft, the communication satellite, the mobile handset; all involve electromagnetic phenomena. All of the devices referred to in Section 1.2.3 fall under the domain of electromagnetics. The true meaning of electrical components involved in circuit theory viz. resistance (R), inductance (L), capacitance (C), and conductance (G) emanate from the field theory. Many of the terms viz. current, potential, potential difference, power flow, energy stored, force exerted, and the torque developed cannot be fully understood without considering their field aspects. Field concepts are deeply rooted in the phenomena of radiation, propagation, reflection, refraction, scattering, and polarization of EM waves. Without field theory, the entire structure of the communication system will crumble to the ground. Its applications are so widespread that it directly or indirectly encompasses almost every sphere of human life; therefore, it is extremely important to thoroughly study and understand it properly.

Field theory revolves around Maxwell's four magical equations. Without these, the human race still would be in the Dark Ages, as there would have been no electrical power, no electronic communication, and no computers. In order to understand the essence of these equations, it is necessary to briefly describe some of the commonly known practical applications. Section 2.2 includes the classification of fields, and Sections 2.3, 2.4, and 2.5 describe some of the usual applications that most people come across in daily life. These discussions hopefully will help the reader in appreciating the importance of electromagnetics, and may even help to develop a liking for this subject.

2.2 CLASSIFICATION OF FIELDS

The common field categories include electrostatic, magnetostatic, and electromagnetic. These are briefly described below.

2.2.1 Electrostatic Field

The theory which describes physical phenomena related to the interaction between stationary electric charges or charge distributions in a finite space which has stationary boundaries is called **electrostatics**. For a long time, electrostatics, under the name electricity, was considered an independent physical theory of its own, alongside other physical theories such as magnetism, mechanics, optics and thermodynamics.

Electromagnetic Field Theory **by Bo Thidé**

Physicist and philosopher Pierre Duhem (1861–1916) once wrote:

> The whole theory of electrostatics constitutes a group of abstract ideas and general propositions, formulated in the clear and concise language of geometry and algebra, and connected with one another by the rules of strict logic. This whole fully satisfies the reason of a French physicist and his taste for clarity, simplicity and order.

An electric (including electrostatic) field is a force field that acts upon material bodies by virtue of their property of charge. It is analogous to a gravitational field, which is also a force field that acts upon material bodies by virtue of their property of mass. The electrostatic (or static electric) fields may result due to the presence of stationary (or quasi-stationary) point charges, cluster of discrete charges, line charge, surface charge, or volume charge distribution. This field exerts a force on other (moving or stationary) charges. Although this electrostatically induced force appears to be weak, the magnitude of this force in a hydrogen atom between an electron and a proton is nearly 36 times (Ref: https://en.wikipedia.org/.../Electrostatics) stronger than the gravitational force acting between them. This category of field dwells around electric field intensity, electric flux density, electric potential, etc. As long as the field remains electrostatic in nature, these quantities have no relation with those involved in other categories, viz. magnetostatic and electromagnetic fields.

2.2.2 Magnetostatic Field

While electrostatics deals with static electric charges, magnetostatics deals with stationary electric currents, i.e., electric charges moving with constant speeds, and the interaction between these currents.

Electromagnetic Field Theory **by Bo Thidé**

A magnetic (or magnetostatic) field is a force field that acts upon charges in motion. The source of the steady magnetic field may be a permanent magnet, a steady (time invariant, i.e., DC) current or an electric field linearly changing with time. The current distribution may be in the form of line current, surface current density, or volume current density distribution. It is the magnetic analog of electrostatics, where stationary charges are distributed. In the magnetostatic field, the magnetization need not be static; the equations of magnetostatics can be used to predict fast magnetic switching events that occur on time scales of nanoseconds or even less. Magnetostatics is a good approximation when the currents do not alternate rapidly. In this category, the field quantities include magnetic field intensity, magnetic flux density, magnetic scalar, vector potentials, etc. As long as the field remains magnetostatic, these quantities have no relation with those involved in the other two categories, viz. electrostatic and electromagnetic fields.

2.2.3 Electromagnetic Field

When the field is dynamic or varies with time, the independence of electric and magnetic fields is lost and the quantities involved in the two fields become interdependent. Thus, an electric field can be produced by a changing magnetic field and a magnetic field can be produced by a changing electric field. Such a field is referred to as an electromagnetic field. This category of field is conceivable only with the involvement of time, both periodic and aperiodic.

2.3 APPLICATIONS OF AN ELECTROSTATIC FIELD

The use of electrostatic forces in industry is quite common. These are involved in (i) the operation of miniature electrostatic motors used in sensors and control devices; (ii) electrostatic generators used for high voltage low current applications such as electron microscope, ion implantation, etc. (iii) electrostatic precipitators employed for separating suspended particles from a gas; (iv) electrostatic coating with paint or other materials; (v) electrostatic imaging in the photostat process; (vi) X–Y recorder for holding paper; (vii) highly sensitive charge-coupled device (CCD) cameras; (viii) non-impact printing (e.g., in inkjet printers); and (ix) electrophoresis (separation of charged colloidal particles by electric field) used in biology, etc. Some of the above referenced applications are described in the following subsections.

2.3.1 Electrostatic Generators

The imbalance of surface charges on an object exhibits attractive or repulsive forces and yields static electricity. This static electricity can be generated by touching and then separating the two differing surfaces due to the phenomena of contact electrification and the triboelectric effect. Rubbing of two non-conductive objects generates a good amount of static electricity. This is not only the friction even the placement of two non-conductive surfaces one over the other can make them charged. The charging of objects through simple contact takes longer than through the rubbing due to the rough texture of most surfaces. Rubbing of objects increases the amount of adhesive contact between the surfaces. Usually insulators (e.g., rubber, plastic, glass, and pith) are good at both generating and holding surface charges. Conductive objects rarely generate charge imbalance except when a metal surface is impacted by solid or liquid non-conductors. The charge that is transferred during contact electrification is stored on the surface of each object. Static electric generators rely on this effect. These generators produce very high voltage at very low current and thus can be used for classroom physics demonstrations. The presence of electric current does not detract from the electrostatic forces, nor from the sparking from the corona discharge or other phenomena and both phenomena can co-exist in the same system.

2.3.2 Electrostatic Filters

These are used for removing fine particles from exhaust gases. The particles are charged, separated from the rest of the gas by a strong electric field, and finally attracted to a pollutant-collecting electrode.

2.3.3 Photocopiers

The modern copier uses a photosensitive material such as selenium. Selenium is normally a dielectric, but when illuminated, it becomes conductive. A photosensitive plate is first charged over its surface and then illuminated by an image of the document. The dark places containing letters or figures are charged and the rest of the plate is discharged by light. An image of the document is thus obtained. This is followed by a process of obtaining a copy of the image on a sheet of paper, which requires that the charge image (i.e. the surface charge) must remain on the plate for a sufficiently long time. This time can be determined by using the Ohms law ($\boldsymbol{J} = \sigma \boldsymbol{E}$), Continuity equation ($\nabla \bullet \boldsymbol{J} = -\partial \rho / \partial t$), and Maxwell's first equation $\left(\nabla \bullet \boldsymbol{D} = \nabla \bullet (\varepsilon \boldsymbol{E}) = \rho\right)$. In these relations, ρ is the volume charge density, ε is the permittivity, and σ is the conductivity of dark selenium. The assumption that $\rho = \rho_0$ at $t = 0$ these relations yield to another relation $\rho = \rho_0\, e^{-(\sigma/\varepsilon)t} = \rho_0\, e^{-t/\tau}$. The quantity ($\tau = \varepsilon/\sigma$) is called the *charge transfer time constant* or relaxation time constant, and is similar to the *RC* time constant of an *RC* circuit.

2.3.4 Electrostatic Separators

Electrostatic separators are used in industry for the purification of food and ores and sorting of reusable wastes according to the size and weight. In all these processes, particles are charged, and then separated by an electric force, or by a combination of an electric force and some other force.

2.3.5 Production of Ions

Electric charges in vacuum or gases are propelled by the electric field of stationary charges. In these media, the point form of Ohm's law does not hold. In rarefied gases, the paths of accelerated ions between two successive collisions are relatively long, so that they can acquire considerable kinetic energy. As a consequence, various new effects can be produced. The best known is probably a chain production of new pairs of ions by collisions of high-velocity ions with neutral molecules, which may result either in a corona (ionized layer around charged bodies), or in breakdown of gas (discharge), depending on the structure geometry and voltage.

2.3.6 Two- or Four-Point Probe Instruments

These are commonly used in semiconductor labs for measuring the resistivity of a material. The two-point probe instruments have only two probes, which are used both for injecting the current in the material and for measuring the voltage between these two points. The four-point probe instruments contain two current and two voltage probes. In these, an electric field is created over the surface of the specimen by two current probes and then the two voltage probes are used to measure the voltage between two convenient points. In this way, the contact between the current probes and the material, which is very difficult to control, practically does not influence the results. The contact between the probes and the material two-point probes is not well-defined and does not yield accurate results.

2.4 APPLICATIONS OF A MAGNETOSTATIC FIELD

Magnetostatics is widely used in applications of micro-magnetics such as models of magnetic recording devices. Magnetostatics is used in non-destructive testing, magnetic levitation of high-speed vehicles, magneto-hydro-dynamic (MHD) pumps and generators and in controlled thermonuclear fusion, etc. The development of the motors, transformers, microphones, compasses, telephone bell ringers, television focusing controls, advertising displays, memory stores, magnetic separators, etc., which play an important role in our everyday life involve magnetostatic phenomena. Some of the above applications are discussed in the following subsections, whereas some others are described in the category of electromagnetic field application.

2.4.1 Non-Destructive Testing

A magneto-inductive device may be used for non-destructive testing. It can detect the defects in the metallic ropes. The device contains two permanent magnets which are placed a distance apart but closer to the rope. The magnetic flux between two magnets passes through the rope. Any defect in the rope may result either in the reduction of the main flux or in the modification of its path. These defects may include localized fault (LF) or loss of metallic area (LMA). The quantum of reduction or path modification will depend on the magnetic characteristic of the rope. These changes in flux can be observed by the flux sensors attached to the rope at different locations. The magneto-inductive instruments designed on this principle can be classified depending on the way the flux is measured. In the case of LF the leakage flux is measured, whereas in LMA the main flux is measured.

2.4.2 Magnetic Levitation

Magnetic levitation (maglev) is a system of transportation that suspends guides and propels vehicles, predominantly trains, using magnetic levitation from a very large number of magnets for lift and propulsion. This method has the potential to be faster, quieter, and smoother than wheeled mass transit systems. In an evacuated tunnel, this technology has the potential to provide speed exceeding 6400 km/h. In an unevacuated tube, the power needed for levitation is usually not a large percentage and most of the power is used to overcome air drag as with any other high-speed train. The highest recorded speed of a maglev train is 581 kilometers per hour, achieved in Japan in 2003.

The two notable types of maglev technologies include electromagnetic suspension (EMS) and electrodynamic suspension (EDS). In EMS, electronically controlled electromagnets in the train attract it to a magnetically conductive (usually steel) track. EDS uses superconducting electromagnets or strong permanent magnets which create a magnetic field that induces currents in nearby metallic conductors when there is relative movement which pushes and pulls the train towards the designed levitation position on the guideway.

2.4.3 Magnetic Separator

Magnetic separation is a process in which magnetically susceptible material is extracted from a mixture using a magnetic force. This technique is useful in mining iron as it is attracted to a magnet. It is also used in electromagnetic cranes that separate magnetic material from scraps. Magnetic separators that used permanent magnets generate fields of low intensity only. High-intensity magnetic separators which employ electromagnets are found more effective in the collection of very fine paramagnetic particles

2.4.4 Magnetic Storage

Magnetic storage and magnetic recording refer to the storage of data on a magnetized medium. Magnetic storage uses different patterns of magnetization in a magnetizable material to store data and is a form of non-volatile memory. This stored information can be accessed by using read/write head(s). Magnetic storage media, primarily hard disks are widely used to store computer data and audio and video signals. Other examples of magnetic storage media include floppy disks, magnetic recording tape, and magnetic stripes on credit cards.

2.4.5 MHD Generator

The magneto-hydro-dynamic (MHD) generator transforms thermal and kinetic energy directly into electricity. Unlike electric generators, these have no moving parts and operate at high temperatures. Like a conventional generator, these rely on moving a conductor through a magnetic field to generate electric current. It uses hot conductive plasma as the moving conductor. The Lorentz Force Law $\{\boldsymbol{F} = Q\,(\upsilon \times \boldsymbol{B})\}$

describes the effects of a charged particle moving in a constant magnetic field. In this relation $\boldsymbol{F}$ is the force acting on the particle, Q is the charge of the particle, υ is the velocity of the particle, and $\boldsymbol{B}$ is the magnetic field. According to the right-hand rule, vector $\boldsymbol{F}$ is perpendicular to both υ and $\boldsymbol{B}$.

2.5 FIELD APPLICATIONS OF ELECTROMAGNETIC FIELDS

The applications of electromagnetic (or time-varying) field are quite widespread in both the electrical and electronic engineering. These applications can further be divided in accordance with the speed of time variation. These may be referred to as the applications involving (i) slow time-varying fields and (ii) fast time-varying fields. The devices employed in these applications are sometimes called low and high-frequency devices. The devices such as transformers, relays, circuit breakers, and most of the measuring instruments fall under the category of slow time-varying fields. The operation of these devices is deeply related to the field phenomena. Ampère's, Faraday's, and Lenz's laws of electromagnetic field form the basis of the design and analysis of rotating electrical machines. The field phenomena also form the basis of tape recording. Similarly there are many applications which fall under the category of fast time-varying fields. The radiation and propagation of electromagnetic waves, working on all high-frequency devices and systems, are some such examples. Some of these applications are discussed below. It is to be noted that in some of the devices and systems electrostatic, magnetostatic, and electromagnetic fields are simultaneously employed.

2.5.1 Electrical Machines

Electric machines are used for manufacturing processes, water supplies, data equipment, and a score of other systems. In a modern home the electrically operated machines may include computer disk drives, DVD players, and large motors for appliances and space conditioning. A modern hybrid electric vehicle uses electric motors for propulsion, power steering, cooling, and many other functions. In addition, industrial automation and robotics rely on electric machines. According to a conservative estimate, electrical machines consume about 70% of the world's electricity.

The rotating electric machines, transformers, inductors, and other devices require proper magnetic field arrangement. In micro-electro-mechanical systems (MEMS) wherein size scales down to nanometers both magnetic and electric fields are used for motion control. Electrical motors, generators, and actuators are referred to as energy conversion devices. These conversions between electrical and mechanical energy take place in the coupling fields. Thus there is inescapable involvement of field theory in the design and operation of all rotating electrical machines.

2.5.2 Transformers

National and international electricity grids are enabled by transformers, which convert voltage and current to the desired levels. Transformers enable the use of long-range high voltage power transmission. They provide low-voltage electricity for electronic devices and home appliances. Their design and operation require a clear understanding of magnetics, including effects such as eddy current and hysteresis loss that are related to fundamental laws of Ampère and Faraday.

2.5.3 Transmission Lines

The power transmission grids supply enormous energy to cities and towns around the globe. The transmission towers carry millions of meters long lines having up to a million volts and hundreds of amperes on each conductor. EM theory is an important tool used for the design and operation of these lines and many other devices (viz. circuit breakers, relays etc.) connected therein.

2.5.4 Circuit Theory

Many circuit theory laws (viz. Ohm's law, Kirchhoff's voltage, and current laws, etc.) are derived from the laws of electromagnetic theory. When the clock rates of computer signals are increased, the electrical signals in computer circuits and chips become more electromagnetic. Thus, these signals require a basic understanding of electromagnetics for their manipulation.

2.5.5 Power Electronics

Power electronic circuits are used in computer power supplies, automotive systems, alternative energy production, motor controllers, efficient lighting, portable electronics, and in many other applications. These circuits use silicon switching devices such as transistors and diodes to manage energy flow and also involve high-frequency magnetic components, including transformers and inductors for energy storage. Magnetic components are often the largest and most expensive components in power converters. Thus a thorough understanding of magnetic design is fundamental to their application. In a power converter circuit design, EM theory plays another role. Fast switching of large currents and voltages radiates EM energy that interacts with nearby parts. The resulting noise and interference are difficult to manage. The concepts of coupling capacitance, mutual inductance, and signal transmission play important roles here. All these aspects can only be understood with a proper background in EM theory.

2.5.6 Electron Tubes

The operations of microwave tubes (viz. klystron, magnetron, traveling wave tube, backward wave-oscillator, and many other cross-field devices), cathode ray tubes (CRTs), TV picture tubes, computer monitors, and radar displays are all dependent on electromagnetics for their operational understanding. In most of these tubes, it is basically the behavior of electrons that needs to be studied. These electrons may come under the influence of electrostatic, magnetostatic, and electromagnetic fields after emerging from the cathode, and may be required for various purposes including accelerating, controlling, and focusing of electron beams.

2.5.7 Semiconductor Devices

The semiconductor (SC) devices, viz. SC diodes, bipolar junction transistors (BJTs), heterojunction bipolar transistors (HBTs), field effect transistors (FETs), metal-oxide-semiconductor field effect transistors (MOSFETs), metal semiconductor field effect transistors (MESFETs), high electron mobility transistors (HEMTs), optical devices, viz. light-emitting diodes (LEDs) and photo diodes (PDs), and various types of microwave SC devices (viz. tunnel, Gunn, p-type, intrinsic, and n-type (PIN), IMPact avalanche transit time (IMPATT), TRApped plasma avalanche trigger transit-time (TRAPATT), BARrier injected transit time (BARITT) diodes, etc. need the electromagnetic field theory to explain their working and behavior.

2.5.8 Signal Processing

No field is more fundamental to high technology than electromagnetics. In the case of signal processing, the vast majority of signals processed in high-tech systems and components are electromagnetic waves. It is therefore essential to know how to model signal propagation in the physical medium of interest, be it optical fiber, coaxial cable, twisted pair wires, or in the air. Its knowledge is employed by communication system designers in the design of algorithms and architectures for transmitting data reliably over a noisy channel.

2.5.9 Antennas

The antenna theory, analysis, and design are totally based on electromagnetic field theory. The radiation pattern, radiation resistance, gain, directivity, effective aperture, polarization, and many other antenna parameters are defined in terms of field quantities. Thus no antenna can be conceived without the involvement of electromagnetics. In the case of antennas arrays, the effect of mutual coupling between elements cannot be accounted without thorough understanding of field phenomena. Antennas are the key component in radars, satellites, navigation, and all other wireless systems.

2.5.10 Wave Propagation

The plane wave propagation is the most widely used mean of communication over long distances. They provide the only means of studying extraterrestrial objects and form the basis of radar and radio astronomy. The knowledge of the reflection and refraction at the interfaces between media with different refractive indices is of utmost importance for understanding the wave behavior. The concept of reflection and refraction is practically involved in the bending of waves in the lower atmosphere and the reflection by the ionosphere. The fiber optic communication and certain beyond the horizon communication systems make use of reflections at dielectric-dielectric interfaces.

The concepts of phase and group velocities find applications in radars for detection and ranging, in interferometers for direction finding and in surface acoustic wave (SAW) devices. The significance of polarization can be brought out by examples such as the use of vertical versus horizontal polarization in ground wave, line of sight (LOS) propagation, and the study of ionosphere by Faraday's rotation of satellite signals. Propagation in conducting and lossy media and the significance of the skin depth are understood through the examples of extremely-low-frequency (ELF) wave propagation for communication in mines and submarines whose depth below the surface is small compared to the skin depth of seawater at these frequencies.

The propagation of unguided waves through troposphere, stratosphere, and ionosphere requires study and thorough understanding of field phenomena for their behavior. Similarly, the guided waves between parallel planes, parallel wires, coaxial-cables, and waveguides of various shapes and sizes require an understanding of Maxwell's equations and the related electromagnetic field phenomena. The basic characteristic equation, which governs the mode of propagation of light waves through optical fibers, too, is in essence derived from Maxwell's equations. The understanding of the radiation pattern of any antenna basically requires an understanding of electromagnetics vis-à-vis its beam formation, radiation resistance, gain, and directivity. Besides, microwave components and planar lines (viz. striplines, microstrips, coplanar lines, slot lines, fin lines, etc.), microwave integrated circuits (MICs), multi-element antenna arrays cannot be fully understood without understanding the very nature of electromagnetic field involved in these devices.

2.5.11 Radar Systems

The term *radar* is a kind of acronym for **RA**dio, **D**etection, **A**nd **R**anging. Radars operate over a very wide range of frequencies. Over the horizon (OTH) radar uses short waves (3–30 MHz), most of the other radars operate in the microwave range (3–30 GHz) whereas some others employ millimetric waves or MMW (30–300 GHz). Radars operate on the ground, in the air, in sea, and in space. Radars are used for (i) military and civilian applications; (ii) detection, ranging, and tracking of the targets; (iii) remote sensing for exploring minerals and water resources, mapping of the land: for agriculture, crop assessment, routes for trains and buses, for calamity assessment (viz. damages caused due to earthquakes, landslides, forest fires, floods, etc.), for locating the crashes of planes, etc., (iv) guiding/navigating the aircrafts, ships, and space vehicles; and (v) assessing the age, movement, and size of navigation buoys and velocity, direction, and height of sea waves.

Owing to some inherent attractive features (viz. large bandwidth, higher spatial resolution, low probability of interception and interference, small size and weight of antenna and equipment, more ruggedness, greater reliability of the system, low-voltage supplies, MMW radar technology is gaining ground. From the military aspects, in surveillance and weaponry systems, these features lead to better performance, better precision, and better maneuverability. MMW systems, however, suffer from atmospheric attenuation, which is found to be maximum at 22,183 and 325 GHz due to H_2O and at 19 and 60 GHz due to O_2. The low attenuation windows are available at 35, 94, 140, and 240 GHz. The MMW radars are used for short ranges for target detection and tracking. These are normally used up to 10 Km and up to 50 Km and more with a special design.

Radar is needed for landing and takeoff, enroute guidance, air surface detection equipment (ASDE), and for aircraft taxiing and parking. The type of radars needed for enroute navigation include height finders, direction finders, navigation, communication, and instrumentation. Radars are also needed for enroute safety for weather avoidance, weather following, terrain avoidance, terrain following, collision avoidance, and for identification of friends and foes (Beacons/IFF).

One can easily conclude that radar technology is a vast area vis-à-vis its frequency range and applications. All of the radar systems discussed here involve antennas and the propagation of radio waves; therefore, none of these can be fully understood without a thorough understanding of the electromagnetics, which is the soul of wave propagation and antenna theory.

2.5.12 Navigation

Navigation is the art of directing/guiding the movement of an aircraft, ship, or space vehicle from one point to another along a desired path. The radio navigation is based on the use of electromagnetic waves to find the location of the vehicles. It may include enroute navigation, direction finding, and height finding. The instrumentation and microwave landing systems also involve electromagnetic waves and sophisticated antenna systems. The navigation systems include basic r/l direction finder, automatic direction finder (ADF), Ballini-Tosi system or goniometer, long wave ranging, very high omni ranging (VOR), differential distance ranging using hyperbolic systems, viz. long-range air navigation (LORAN), decca, omega, distance measuring equipment (DME), tactical air navigation (TACAN), and VORTAC (VOR + TACAN).

2.5.13 Space Exploration

The human race is physically confined to the inner neighborhood of the solar system. It has no other medium but electromagnetic fields by which to know anything about the greater universe. The development of giant optical and radio telescopes is the great triumph of modern technology. These make it possible to receive electromagnetic waves from the outermost limits of the cosmos and to make images and analyze the substance of the various objects inhabiting it. An intimate knowledge of electromagnetics is necessary, not only to design and build the instruments but also to interpret their findings.

2.5.14 Wireless Technology

Electromagnetic waves are the basis of all wireless systems. These waves carry energy and information from one point in space to another. These waves are also a good tool for probing something from a distance. Radars used for detection and ranging also use radio waves which is a subset of electromagnetic waves. Wireless electromagnetic waves provide access to Internet, video and audio communications, intelligent utility control, entertainment, and many other services at any time, anywhere. Besides electrical and computer engineering, wireless technology plays a role in many other engineering disciplines, including mechanical engineering, chemical and material engineering, environmental and civil engineering, and biomedical engineering.

All wireless communication systems contain antennas for transmission and reception of signals which operate on the principles of electromagnetics. These systems may include mobile communication, pagers, Wi-Fi, personal communication, communication satellites, remote sensing satellites, etc. The remote sensing provides the necessary information to monitor the status of the global environment, information vital to the pursuit of the sustainable development of human society. The use of laws of electromagnetics can be extended into the realms of remote sensing and subsurface sensing.

2.5.15 High-Frequency Devices

At frequencies beyond a few GHz electronic devices can no longer be treated as simple lumped components. When wavelength starts diminishing to the point where it is comparable to integrated circuit dimensions, electromagnetic phenomena called *transmission line effects* become critical. These effects include conductor loss, dielectric loss, and radiation loss. At this stage, optimal matching among parts of the circuit and to limit signal attenuation caused by transmission line effects becomes a necessity. This requires the solid foundation of electromagnetics.

2.5.16 Integrated Circuits

Electrical signals move from one part of an IC to another according to the laws of electromagnetics. Unwanted coupling of electrical signals among different parts of an IC can be explained and solved, only by using the fundamental knowledge of electromagnetics. If the components in a system are electromagnetically incompatible the system will not function properly.

Circuits comprising microchips may suffer from crosstalk. The crosstalk in a computer chip may be due to the high clock rate of the chip. The high clock rate makes the inductive and capacitive coupling between non-contact lines significant. One can visualize the leakage of electromagnetic energy over to the other lines even though only one line is excited in the circuit. Analysis of such systems may require computational electromagnetics to model the small length scale physics in a microchip.

2.5.17 Computers

Due to the increase of embedded microprocessor clock frequencies well above 20 MHz, the circuit boards radiate electromagnetic energy. This radiated wave is capable of inducing noise in linear traces and loops of neighboring circuit boards, which act like small antennas. If these radiations are strong enough, the induced noise can cause these neighboring circuits to malfunction.

Besides the plethora of nearby wireless RF devices may also interfere with the operation of an embedded system. The sources of such interference include an increasing number of satellites, broadcast radio and TV stations, cellular telephones, cordless telephones, wireless remote control devices, wireless car keys, wireless Internet, and even pill capsules that wirelessly send back pictures. In addition, there are a large number of nearby radiating devices, such as switching dc power supplies, digital audio and video devices that employ the latest digital signal processing techniques, and personal computing devices. This spectrally rich electromagnetic environment demands that a computer engineer must know how to design "electromagnetically compatible" (EMC) systems that perform well even in the presence of unintended electromagnetic radiation from nearby electronic equipment. In addition, these systems are so designed that these do not themselves pollute the electromagnetic spectrum further. The designer, therefore, must have a solid understanding of electric fields, magnetic fields, wave propagation, signal-coupling mechanisms, filtering, shielding, and grounding techniques. A deep understanding of electromagnetics, Maxwell's equations, wave propagation, transmission lines, and waveguides becomes essential for the efficient embedded and digital system design.

2.5.18 Digital Subscriber Lines

Digital subscriber lines (DSLs) for broadband services use copper telephone-line twisted pair at or near its fundamental data-carrying limits. Such high-performance transmission requires a fundamental understanding of the physical channel, and in particular, electromagnetic field theory. A twisted pair line can be divided into a series of incrementally small circuits. Each of such circuits is characterized by fundamental passive circuit elements of resistance, inductance, capacitance, and conductance. These elements often vary as a function of frequency. The electromagnetic theory in terms of the basic Maxwell's equations is the basis of construction of these incremental circuits and their cascade. These segments allow calculation of various transfer functions and impedances and then characterize the achievable data rates of the DSL. Thus, electromagnetic theory is fundamental to the understanding and design of DSL systems.

Electromagnetic theory has recently been used for modeling of simple isolated transmission lines by a binder of copper twisted pairs using vector and or matrix generalizations wherein telephone lines are regarded as big antennas, radiating into one another and receiving each other's signals. The modeling of this "crosstalk" is important for understanding the limits of transmission of all the lines within the binder and their mutual effects upon one another. Electromagnetic theory allows such characterization and the calculation of the impact of the various transmission lines upon one another.

2.5.19 Optical Fibers

In optical networks signals travel at the speed of light and can transmit at data rates in excess of 10 gigabits/s. These signals suffer from dispersion as light waves of different wavelengths travel at different speeds. This results in spreading the information over a longer time period as the optical signal travels through a long span of fiber. Because of dispersion, a clean signal at 0 Km may become almost unrecognizable at 120 Km.

It is reported that a chipset called the Smart Clock-Data Recovery (CDR) compensates for dispersion in a 12.5 gigabit per second optical link. This algorithm incorporates electromagnetic properties of signal propagation along with advanced statistical techniques. It is implemented with high-frequency mixed-signal ICs. The basic requirement for CDR include (i) adequate input matching to reduce reflections as analog signals enter the device; (ii) modeling of inductors to obtain a high-quality, low phase-noise, voltage controlled oscillator for the clock recovery unit (CRU); (iii) modeling of the interconnect and terminations in order to guarantee synchronized transmission of the 12.5 GHz clock signal from the CRU to the analog–digital converter; and (iv) design of the output buffers and on-package traces to provide a balanced 32-bit differential, 1.56 gigabit per second interface to the digital chip. The design of all the above stages requires a good knowledge of electromagnetics.

Moore's law accurately predicted the rate of increase in chip density (hence, computing power). Similarly, Snell's law relates angles of incidence and refraction of waves as they move in different media. This happens with optical and electrical signals as they course through fibers, cables, wires; onto and off boards and chips and through packages. Advanced high-tech products such as the Smart CDR exploit both of these laws in order to achieve system performance that would otherwise be impossible. These laws, particularly that of Snell, fall into the domain of electromagnetics.

2.5.20 Optical Communication

Optical communication system, which operates in the terra-hertz (THz) range, has now become a key tool for high-speed Internet. As light follows the theory of electromagnetic waves, classical electromagnetics plays a crucial role in understanding of the systems of guided wave phenomena. These systems involve four important areas of devices, including generation, modulation, propagation, and detection of light. The wave nature of light plays a vital role in all these devices and, thus, it is absolutely necessary to have a knowledge of electromagnetics.

2.5.20.1 Light Generation

Light is generated by using semiconductor lasers, light-emitting diodes (LEDs), and erbium-doped fiber optical amplifiers. The laser structure requires a waveguide or cavity in which light is confined in the form of optical resonator modes. The phenomenon of wave propagation in a waveguide or cavity is governed by Maxwell's equations. Such solutions have to satisfy the boundary conditions specified by the laser cavity. The design of high-extraction efficiency LEDs also requires a good understanding of geometric optics.

2.5.20.2 Light Modulation

Light can be modulated through electro-optical modulators, electro-absorption modulators, and optical phase modulators. The refractive index or absorption coefficient of the materials plays a pivotal role in the formation of these devices. These parameters are controlled by an applied electric field or voltage. These devices usually require bulk or dielectric waveguide geometry, which in turn involves Maxwell's equations.

2.5.20.3 Light Propagation

The propagation of light requires optical fibers and optical dielectric waveguides and thus use of Maxwell's equations becomes imminent.

2.5.20.4 Light Detection

The modulated light carrying the transmitted data illuminates the active region of the semiconductor photodetectors, which in turn converts this light into photocurrents. The normal incidence and waveguide geometry of photodetectors require a good understanding of electromagnetic wave theory.

2.5.21 Biomedical Optical Imaging

The light interacts with biological tissues and cells. This interaction can provide helpful diagnostic information about the structure and function of these tissues and cells. Optical biomedical imaging relies on detecting differences in the properties of light after such interactions. It is therefore essential to understand the properties of light, its propagation through tissue, scattering and absorption effects, and change of its polarization. All of these properties of light fall in the domain of electromagnetics.

The constituents of biological tissue, such as hemoglobin in blood, melanin in skin, and ubiquitous water, have wavelength-dependent optical properties over the visible and near-infrared electromagnetic spectrum. The spectroscopic wavelength-content of light, therefore, provides a new dimension of diagnostic information. This led to the development of novel optical imaging technologies.

Optical coherence tomography (OCT) is one such biomedical imaging technology. It relies on the principle of optical ranging in tissue, and is the optical analog to ultrasound imaging. As wavelength of light is smaller than that of sound, OCT enables high-resolution imaging that can identify individual cells in tissue to depths of several millimeters. It can be used as a form of "optical biopsy", and can eliminate the need for removing tissue for examination. The study of EM has direct relevance to understanding how light interacts with tissues, and novel technology for medical and biological imaging can be developed based on EM principles.

2.5.22 Magnetic Resonance Imaging

Magnetic resonance imaging (MRI) is a good example of an application of electromagnetics. The image formation process uses three magnetic fields to interact with a nuclear spin system for signal generation, detection, and spatial information encoding.

Certain nuclei such as those with odd atomic weights and/or odd atomic number have an intrinsic angular momentum called spin. These spins have an associated magnetic dipole vector. Under thermal equilibrium conditions, these magnetic dipoles are oriented in random directions and macroscopic magnetism cannot be detected.

The image of an object can be created in terms of its magnetic dipole vectors. MRI first uses a strong magnetic field ($\boldsymbol{B}_0$) to create a non-zero bulk magnetization for the object. It then uses a short-lived, oscillating field ($\boldsymbol{B}_i$), or RF pulse. This second field oscillates in the RF range to tip the bulk magnetization away from the direction of the $\boldsymbol{B}_0$ field. Although magnetic dipole vectors behave quantum-mechanically, the bulk magnetization vector can be accurately described by classical electromagnetic theory. This tipped bulk magnetization vector precesses about the $\boldsymbol{B}_0$ field, thus induces an electrical signal in the receiver coil placed near the object according to Faraday's law of induction. MRI further imposes on the $\boldsymbol{B}_0$ field a linear gradient field so that the frequency and/or phase of the MR signals become linearly dependent on the spatial origin of the signal. The spatially encoded MRI signals can be easily processed using the Fourier transform or Radon transform to generate the desired image. MRI is a vibrant field with many opportunities for new technology development, which demands a good background in electromagnetics.

2.5.23 Nanotechnology

As reported in the literature a new phenomenon termed as flow-limited field-injection electrostatic spraying (FFESS) was used to produce patterns, films, nanoparticles, nanofibers, and nanowires for various cutting-edge scientific applications. In this process, the uniform drops of a given material with precisely controlled size and charge were produced. The charges were injected by inserting a charge injection needle into the material in the liquid phase and invoked either field ionization or field emission. On this charged material the electrical forces were applied to disrupt the liquid at the charged surface. The electromagnetic forces (like the Lorentz force) were used to manipulate their trajectories onto a substrate. As a result nanoparticles with the diameter of one 10,000th of a human hair were produced. This process also produced nanofibers of biodegradable materials, copper, and silver. Fibers of 100 nm diameters were produced to serve as electron emitters in a kind of flat-panel (field-emission) display.

Development of an advanced compact electromagnetic railgun to accelerate 3 mm × 6 mm frozen hydrogen pellets to a velocity in excess of 3 km/s, that is, much faster than any high-speed bullet has also been reported. In this, the armature used to accelerate the hydrogen pellet is high-density plasma produced by electrically breaking down hydrogen gas. These hypervelocity hydrogen pellets serve to refuel a magnetic confinement fusion device to replenish burnt fuel, which consists of mixtures of hydrogen isotopes.

2.6 CONCLUSION

In view of the developments discussed here, it can be concluded that electromagnetics is an important field with far-reaching impact and influence on many areas of research, including the newly emerging area of nanotechnology. Developments of MEMS, nanostructures, high-speed chips, and biosensors are related to electromagnetics. Furthermore, in view of the applications discussed, it will not be a tall claim that electromagnetic theory is one of the supreme accomplishments of the human intellect. Its usefulness in science and engineering makes it an indispensable tool in virtually any area of technology or physical research. These are sufficient reasons for it to be included in the curriculum, particularly that of electrical and electronics engineering, and it must receive due attention from students and researchers alike.

It is to be noted that the quantities and relations given in some of the subsections above are discussed at length in the following chapters.

DESCRIPTIVE QUESTIONS

Q2.1 Discuss the classification of electromagnetic fields in view of the nature of sources.

Q2.2 Discuss the practical applications of electrostatic and magnetostatic fields.

Q2.3 Discuss the practical applications of electromagnetic fields.

Q2.4 Describe the application of a magnetostatic field to Maglev and MHD generators.

Q2.5 Discuss the relevance of field theory with the working of an MRI.

Q2.6 Discuss the involvement of field theory in the area of communication technology.

FURTHER READING

C. Aldo, F. Fabio and V. Bruno, "3D magnetostatic analysis of magneto-inductive devices for non-destructive-testing (NDT)", in *8th EMF Conference*, 2009. http://www.polito.it/cadema.

A. J. Mestel, "Magnetic levitation of liquid metals", *Journal of Fluid Mechanics* Vol. 117, p. 27, 2006. Bibcode:1982JFM...117...27M.

J. R. Hull, "Attractive levitation for high-speed ground transport with large guide way clearance and alternating-gradient stabilization", *IEEE Transactions on Magnetics* Vol. 25, No. 5, p. 3272, 1989. Bibcode:1989ITM....25.3272H. doi:10.1109/20.42275.

E. R. Laithwaite, "Linear electric machines—a personal view", *Proceedings of IEEE* Vol. 63, No. 2, p. 250, 1975. doi:10.1109/PROC.1975.9734.

M. Tsuchiya, H. Ohsaki, "Characteristics of electromagnetic force of EMS-type maglev vehicle using bulk superconductors", *IEEE Transactions on Magnetics* Vol. 36, No. 5, pp. 3683–3685, Sept. 2000. doi:10.1109/20.908940.

R. Goodall, "The theory of electromagnetic levitation", *Physics in Technology* Vol. 16, No. 5, pp. 207–213, Sept. 1985. doi:10.1088/0305-4624/16/5/I02.

Marc T. Thompson, R. D. Thornton, "Flux-canceling electrodynamic maglev suspension: part II test results and scaling laws", *IEEE Transactions on Magnetics* Vol. 35, No. 3, May 1999.

Richard F. Post, "MagLev: a new approach", *Scientific American*, pp. 64–69, Jan. 2000.

G. Areq, "Maglev—a super fast train", *The Journal of the Acoustical Society of America* Vol. 108, No. 5, p. 2527, Nov. 2000.

J. P. Eckert, "A survey of digital computer memory systems", *Proceedings of IRE*, Oct. 1953.

D. R. Edwin, *Milestones in Computer Science and Information Technology*, Greenwood Press, Westport, CT, 2003, p. 164. ISBN 1-57356-521-0.

Jay W. Forrester, "Digital information in three dimensions using magnetic cores", *Journal of Applied Physics* Vol. 22, pp. 44–48, 1951.

H. K. Messerle, *Magneto-Hydrodynamic Power Generation*, Part of the UNESCO Energy Engineering Series, John Wiley, Chichester, 1994.

R. J. Rosa, *Magneto-Hydrodynamic Energy Conversion*, Hemisphere Publishing, Washington, DC, 1987.

G. J. Womac, *MHD Power Generation*, Chapman and Hall, London, 1969.

H. C. Bazett, "An analysis of the time-relations of electrocardiograms", *Heart* Vol. 7, pp. 353–370, 1920.

J. W. Hurst, "Naming of the waves in the ECG, with a brief account of their genesis", *Circulation* Vol. 98, No. 18, pp. 1937–1942, 1998. doi:10.1161/01.CIR.98.18.1937. PMID 9799216.

G. T. Herman, *Fundamentals of Computerized Tomography: Image Reconstruction from Projection*, 2nd Ed, Springer, 2009.

J. K. Udupa, G. T. Herman, *3D Imaging in Medicine*, 2nd ed., CRC Press, 2000.

W. H. Oldendorf, "The quest for an image of brain: a brief historical and technical review of brain imaging techniques", *Neurology* Vol. 28, No. 6, pp. 517–533, June 1978.

F. Natterer, *The Mathematics of Computerized Tomography (Classics in Applied Mathematics)*, Society for Industrial Mathematics. ISBN 0898714931.

3 Coordinate Systems and Vector Algebra

3.1 INTRODUCTION

Mankind is gifted by nature with certain inherent qualities called *senses*. These senses help us in seeing our surroundings, hearing sounds, recognizing smells, identifying tastes, and feeling temperature, hardness, and softness. Vision is the most important of all of these senses. It helps us in recognizing shapes, sizes, colors and shades, etc. of objects. Our capability to resolve the details of these aspects is quite limited. The necessity to enhance these limits has led to the development of a number of tools. In spite of the great advances in science, many aspects are yet to be circumvented by human senses or tools. These limitations, therefore, create a demand for the adoption of a tool of entirely different nature; this tool is known as *mathematics*. Mathematics is more than a physical tool; it is an aid to extend our faculty of reasoning to guide our thought and permit us to make brief, precise, and unambiguous statements.

In electromagnetics there are a number of quantities (e.g. charge, flux, flux density, field intensity, etc.), which can neither be seen nor are conveniently measurable. Their integrated effects (viz. current, voltage, power, etc.), however, lead to the real physical actions (e.g. force, torque, motion, heating, transportation of energy and information to distant places with or without physical connections, etc.). Mathematical tools are more frequently employed to tackle the field problems or to solve its puzzles, and these are briefly described in this and the subsequent chapter. Since our basic aim is to understand electromagnetic field phenomena involved in electrical and electronic devices and systems and not to study mathematics, these tools need only to be covered briefly. For a detailed study, the reader will find references included at the end of the next chapter. This chapter deals with the coordinate systems and vector algebra and the next chapter discusses various aspects of vector calculus.

3.2 COORDINATE SYSTEMS

Geometry is a branch of mathematics that deals with the measurement, property, and relations of lines, angles, surfaces, and volumes. A practical device or part thereof in which the field distribution is to be studied for its behavior requires representation of its shape and size in such a way that all its physical aspects fully conform to the geometry. The system that deals with geometrical aspects of an object is referred to as a *coordinate system*. The necessity to enhance these limits has led to the development of number of tools. A coordinate system helps us to visualize relative positions of independent points or those belong to a line, a surface or a volume. Thus, while introducing the coordinate systems, it is presumed that a reader fully understands the meanings of point, line, surface, and volume, and also is aware of normal, tangent, and geometrical properties. Furthermore, as and when any point on, or a segment in terms of line, area, or volume of, the system is to be represented (or identified), it should fit in well in the coordinate system selected. The following subsections describe the types of coordinate systems.

3.2.1 Types of Coordinate Systems

There are many types of coordinate systems; each can be divided into two broad categories referred to as two-dimensional (2D) and three-dimensional (3D) systems.

3.2.1.1 Two-Dimensional Coordinate Systems

These include (i) Cartesian, (ii) polar, (iii) bipolar, (iv) parabolic, (v) hyperbolic, and (vi) elliptic coordinate systems. All of these are orthogonal systems wherein the coordinate axes involved are perpendicular to each other.

3.2.1.2 Three-Dimensional Coordinate Systems

The three-dimensional coordinate systems can further be classified into (i) parallel coordinate systems wherein a point is visualized in n-dimensional space as a polyline connecting points on n vertical lines; (ii) curvilinear coordinate systems that are generalized coordinate systems based on intersection of curves; (iii) circular (polar) coordinate systems that represent a point in a plane by an angle and a distance from the origin; (iv) Plücker coordinate systems, in which lines are represented in 3D Euclidean space using a six-tuple of numbers as homogeneous coordinates; (v) generalized coordinate systems that are used in a Lagrangian treatment of mechanics; (vi) canonical coordinate systems that are used in Hamiltonian treatment of mechanics; and (vii) orthogonal coordinate systems wherein the coordinate surfaces meet at right angles. This last category is commonly used in the study of field theory.

3.2.2 Orthogonal Coordinate Systems

The orthogonal systems include (i) Cartesian, (ii) cylindrical, (iii) spherical, (iv) parabolic cylindrical, (v) paraboloidal, (vi) oblate spheroidal, (vii) prolate spheroidal, (viii) ellipsoidal, (ix) elliptic cylindrical, (x) toroidal, (xi) bispherical, (xii) bipolar cylindrical, (xiii) conical, (xiv) flat-ring cyclide, (xv) flat-disk cyclide, (xvi) cap-cyclide, (xvii) bi-cyclide, (xviii) concave bi-sinusoidal single-centered, (xix) concave bi-sinusoidal double-centered, (xx) convex inverted-sinusoidal spherically aligned, and (xxi) quasi-random-intersection Cartesian coordinates.

3.2.2.1 Commonly Used Orthogonal Systems

Out of this long list of orthogonal systems, Cartesian, cylindrical, and (iii) spherical coordinate systems are the most commonly used coordinate systems.

A Cartesian system uses three numbers for representing distances from 3D flat surfaces, namely, $x = 0$, $y = 0$, and $z = 0$ surfaces. In a cylindrical system, the position of a point in space is represented by a distance from the axis ρ, an angle φ, and height from the origin along the axis z. Lastly, in a spherical system a point in space is represented with two angles θ and φ, and a distance r from the origin. Figure 3.1 shows the location of a point (P) in these three coordinate systems. Note that the two

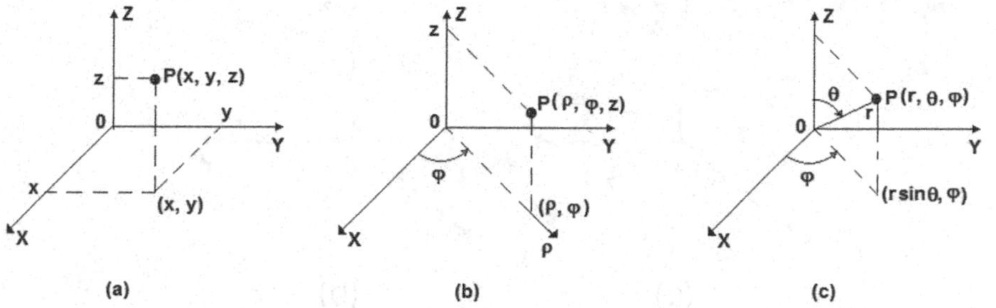

FIGURE 3.1 Point P in (a) Cartesian (b) cylindrical and (c) spherical systems.

angles θ and φ, are defined with reference to the three axes of the Cartesian system of space coordinates.

The meaning of orthogonal system can be further understood through Figure 3.2. Figure 3.2a illustrates a Cartesian system having three orthogonal coordinate axes (X, Y, Z), wherein the directed line-segments 0X, 0Y, and 0Z are perpendicular to each other. Figure 3.2b shows a cylindrical system with three orthogonal coordinates (ρ, φ, z) and Figure 3.2c shows a spherical system having three orthogonal coordinates (r, θ, φ).

As per the prevalent convention in texts, rectangular coordinate axes are represented by capital letters (X, Y, Z) whereas the location of a point along (X, Y, Z) is identified by small letters (x, y, z). Coordinate axes for cylindrical are represented by (ρ, φ, z) and for spherical by (r, θ, φ), instead of (P, Φ, Z) and (R, Θ, Φ) respectively. Similar to (x, y, z), (ρ, φ, z), and (r, θ, φ) are the locations of a point along (P, Φ, Z) and (R, Θ, Φ) axes respectively. This convention is used throughout this text.

3.2.2.2 Right- and Left-Handed Systems

Orthogonal systems can further be classified as right-handed and left-handed. These are illustrated in Figure 3.3.

Figure 3.3a shows a right-handed system with the sequence of coordinates (X, Y, Z). If a (right-handed) screw placed along the z-axis is rotated in the direction from x-axis to y-axis, it moves in the positive direction of the z-axis. The left-handed case is shown in Figure 3.3b. If (a right-handed) screw placed along the z-axis is rotated in

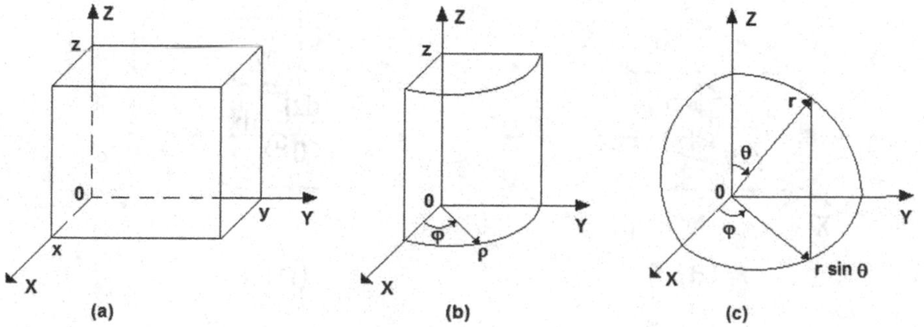

FIGURE 3.2 (a) Cartesian (b) cylindrical and (c) spherical coordinate systems.

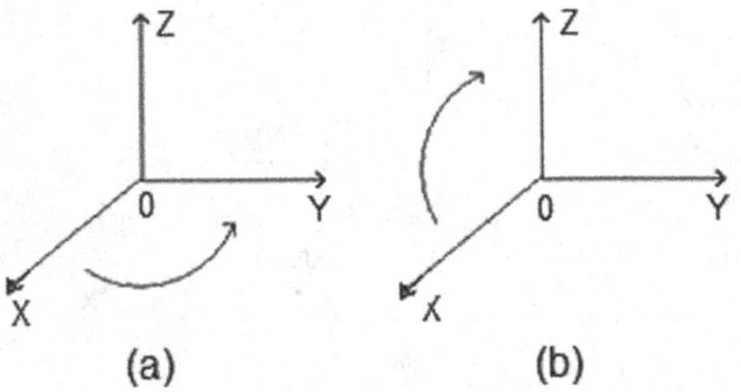

FIGURE 3.3 (a) Right-handed and (b) left-handed coordinate systems.

the direction from x- to y-axis, it moves in the negative direction of the z-axis. While dealing with the vectors the sequences of coordinates to be maintained for the three right-handed systems are (a) Cartesian system (x, y, z), (b) cylindrical system (ρ, φ, z), and (c) spherical system (r, θ, φ).

In this text, only the right-handed Cartesian, cylindrical, and spherical coordinate systems are used throughout. These systems, therefore, are discussed in detail in the following sections.

3.2.3 Cartesian Coordinate System

This is the simplest of the three orthogonal coordinate systems and is illustrated in Figure 3.4. It is commonly used and can be easily visualized. This system is also called rectangular coordinate system of space coordinates. It has three orthogonal coordinates, X, Y, and Z. While dealing with the vectors this sequence is to be maintained. Figure 3.4a shows apoint P located in this system. The values of x, y, and z along the coordinates X, Y, and Z may assume any value between $-\infty$ to $+\infty$.

3.2.3.1 Elemental Volume

Figure 3.4b shows the elemental volume (dv) along with its sides (dx, dy, and dz). Its value in the Cartesian coordinate system is:

$$dv = dx \cdot dy \cdot dz \tag{3.1}$$

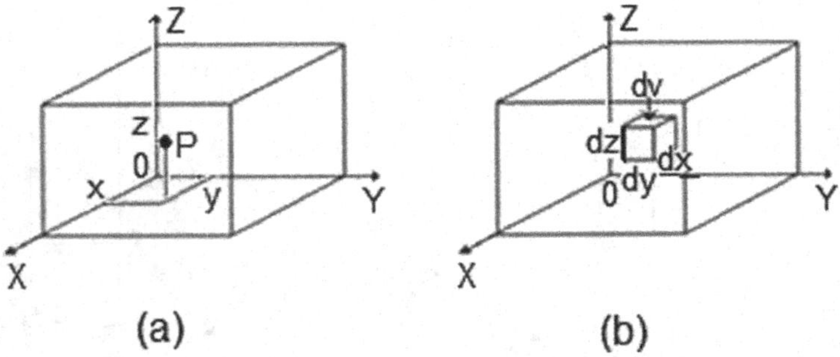

FIGURE 3.4 (a) point P located at x, y, z (b) elemental volume.

3.2.3.2 Shapes of Constant Coordinate Surfaces

Figures 3.5a, b and c illustrate constant x, constant y, and constant z surfaces, respectively. The shapes of all these surfaces are rectangular planes like loafs of bread cut along different axes.

3.2.3.3 Applications

This coordinate system is used for the configurations involving mutually perpendicular flat surfaces such as the flat current sheets, rectangular waveguides, planar lines, etc.

3.2.4 Cylindrical Coordinates System

The visualization of this coordinate system requires little more effort than that in case of the Cartesian system. It has three orthogonal coordinates (ρ, φ, and z). While dealing with the vectors this sequence is to be retained as such. As shown in Figure 3.6a a point P at an arbitrary location in this system may involve values of ρ, φ, and z. In this system ρ may vary from 0 to ∞, φ from 0 to 2π and z from $-\infty$ to $+\infty$.

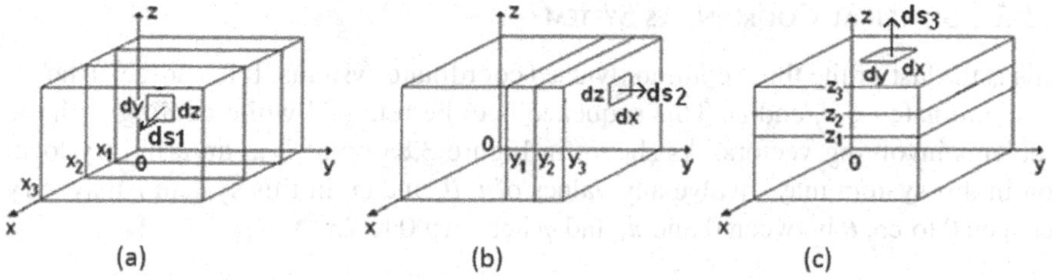

FIGURE 3.5 (a) constant x (b) constant y and (c) constant z surfaces in a Cartesian system.

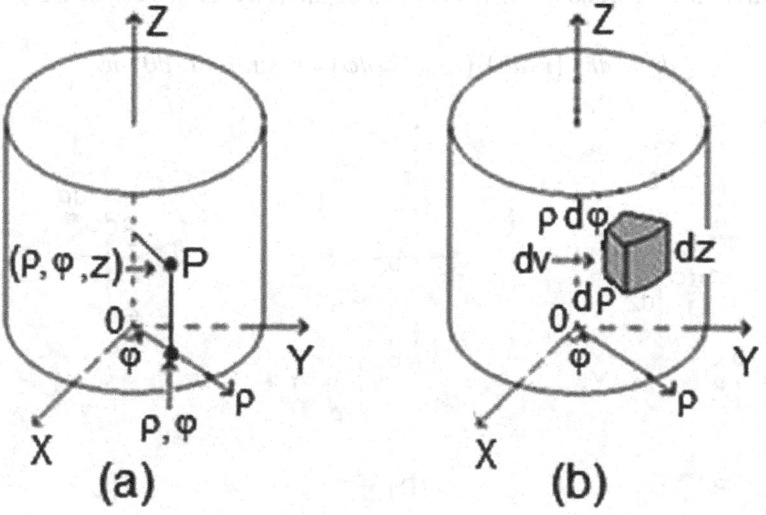

FIGURE 3.6 (a) point P located at ρ, φ, z (b) elemental volume.

3.2.4.1 Elemental Volume

Figure 3.6b shows the shape of elemental volume (*dv*) along with its sides ($d\rho$, $\rho{\cdot}d\varphi$, and dz). Its value in the cylindrical coordinate system is:

$$dv = (d\rho)\cdot(\rho\cdot d\varphi)\cdot(dz) = \rho\cdot d\rho\cdot d\varphi\cdot dz \tag{3.2}$$

3.2.4.2 Shapes of Constant Coordinate Surfaces

Figure 3.7a illustrates constant ρ surfaces. These are like that of outer surfaces of coaxial circular pipes of different radii, each with infinite axial length. Figure 3.7b illustrates constant φ surfaces. These semi-infinite flat surfaces are like the pages of a book each emanating from the z-axis. Figure 3.7c illustrates constant z surfaces. These are flat circular coin like surfaces placed one over the other all parallel to the $z = 0$ plane.

3.2.4.3 Applications

Using this coordinate system the problem of a coaxial cable, a solenoid, a cylindrical rotor of an electrical machine, a cylindrical slow wave structure, and such similar physical devices (involving shapes of circular cylinders) can be tackled.

3.2.5 Spherical Coordinates System

This is the last of the three commonly used coordinate systems. It has three orthogonal coordinates r, θ, and φ. This sequence is to be retained while dealing with the problems involving vectors. As shown in Figure 3.8a point P at an arbitrary location in this system may involve any values of r, θ, and φ. In this system r may vary between 0 to ∞, θ between 0 and π, and φ between 0 to 2π.

3.2.5.1 Elemental Volume

Figure 3.8b shows the shape of elemental volume (*dv*) along with its three orthogonal sides $\left(dr,\, r\sin\theta\cdot d\varphi,\text{ and } rd\theta\right)$. Its value in the spherical coordinate system is given as:

$$dv = (dr)\cdot(r\cdot d\theta)\cdot(r\cdot\sin\theta\cdot d\varphi) = r^2\sin\theta\cdot dr\cdot d\theta\cdot d\varphi \tag{3.3}$$

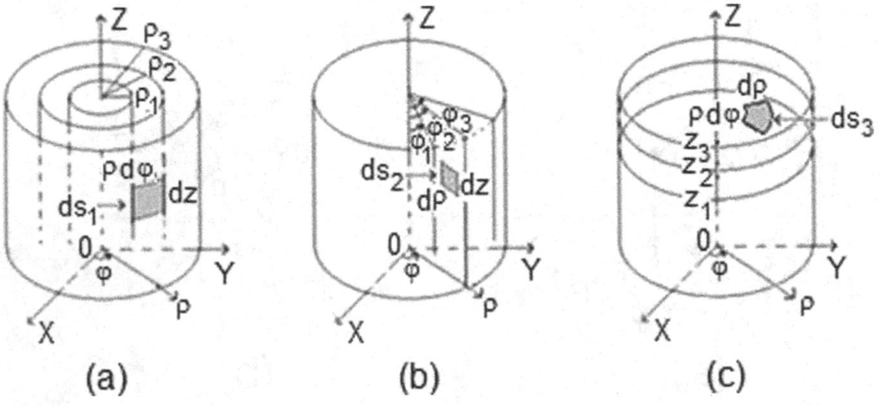

FIGURE 3.7 (a) Constant ρ (b) constant φ and (c) constant z surfaces in a cylindrical system.

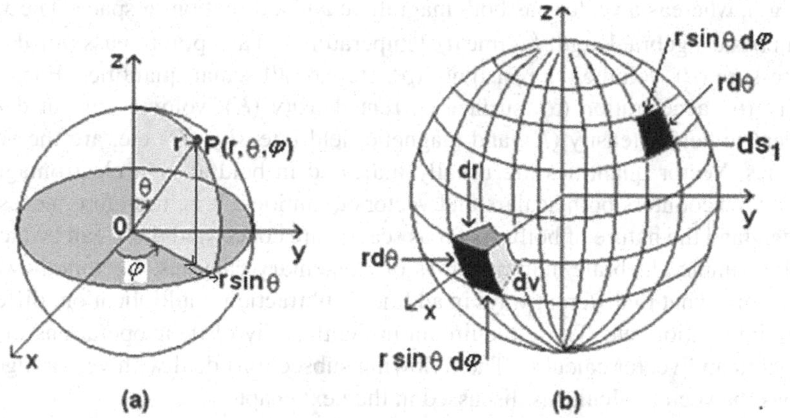

FIGURE 3.8 (a) Point P located at r, θ, φ (b) elemental volume.

3.2.5.2 Shapes of Constant Coordinate Surfaces

Figure 3.9a shows constant r surfaces. These are the surfaces of concentric spheres of different radii. Figure 3.9b illustrates constant θ surfaces. These are the family of surfaces of cones with a common vertex, where the vertex angle θ varies from zero to π radian. Figure 3.9c shows constant φ surfaces. These semi-infinite flat surfaces are like the pages of a book, each of semi-circular shape and emanating from the z-axis.

3.2.5.3 Applications

This coordinate system is suitable for dealing with the problems wherein the spherical geometry gets involved. Such a situation arises in the radiation of an electromagnetic wave from a short dipole or a point source.

In certain problems, e.g. in linear antenna, a mixed system of coordinates may also be used.

3.3 SCALARS AND VECTORS

We limit our study to those quantities which can be classified either as a scalar or a vector. The term scalar refers to a quantity which only has a magnitude (with +

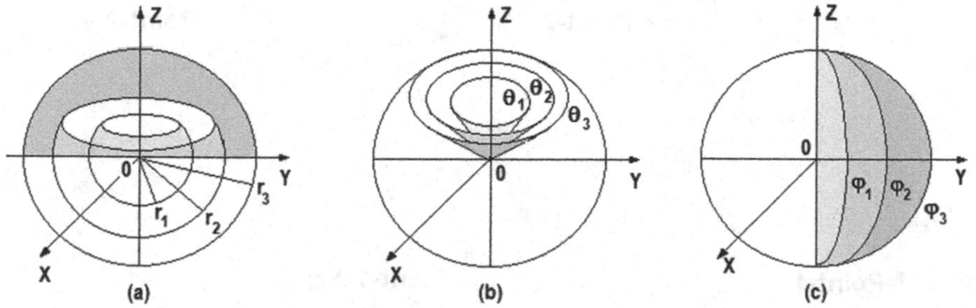

FIGURE 3.9 (a) Constant r (b) constant θ and (c) constant φ surfaces in a spherical system.

or – sign), whereas a vector has both magnitude and a direction in space. The x, y, z used in basic algebra, length ℓ, time (t), temperature (T) at a point, mass (m), density (d), pressure (P), volume (v), resistivity (ρ), etc. are all scalar quantities. Force ($\boldsymbol{F}$), velocity ($\boldsymbol{v}$), acceleration (α), surface current density ($\boldsymbol{K}$), volume current density ($\boldsymbol{J}$), electric field intensity ($\boldsymbol{E}$), and magnetic field intensity ($\boldsymbol{H}$) etc. are the vector quantities. Vector quantities are usually indicated in boldface. In electromagnetic fields one encounters both scalars and vectors quantities. It is, therefore, necessary to understand the nature of both. As far as scalars are concerned these can be tackled by using simple algebraic manipulations or elementary calculus. Vectors, however, behave somewhat in different ways in addition, subtraction, multiplication, differentiation, integration, etc. These require an in-depth study of their operations in vector algebra and vector calculus. The following subsections deal with vector algebra, whereas the vector calculus is discussed in the next chapter.

3.3.1 Vector Representation

A vector is geometrically represented by a straight-line segment with an arrowhead. Its length represents the magnitude and its arrowhead indicated the direction. Figure 3.10 shows a two-line segment drawn between points 1 and 2. One of these is directed from point 1 to point 2 and is marked as $\boldsymbol{R}_{12}$ whereas the second directed from point 2 to point 1 marked as $\boldsymbol{R}_{21}$. These directed line-segments ($\boldsymbol{R}_{12}$ and $\boldsymbol{R}_{21}$) can be referred to as vectors.

The magnitudes of $\boldsymbol{R}_{12}$ and $\boldsymbol{R}_{21}$ can be written as $|\boldsymbol{R}_{12}|$ (or R_{12}) and $|\boldsymbol{R}_{21}|$ (or R_{21}), respectively. The orientation of these vectors can be given by $\boldsymbol{a}_{R12}$ and $\boldsymbol{a}_{R21}$, respectively, which are referred to as unit vectors. These two unit vectors are defined as:

$$\left.\boldsymbol{a}_{R12} = \frac{\boldsymbol{R}_{12}}{|\boldsymbol{R}_{12}|} = \frac{\boldsymbol{R}_{12}}{R_{12}} \ (\text{a}), \quad \boldsymbol{a}_{R21} = \frac{\boldsymbol{R}_{21}}{|\boldsymbol{R}_{21}|} = \frac{\boldsymbol{R}_{21}}{R_{21}} \ (\text{b})\right\} \tag{3.4}$$

Inspection of $\boldsymbol{R}_{12}$ and $\boldsymbol{R}_{21}$, in Figure 3.10 reveals that

$$\left.\boldsymbol{R}_{12} = -\boldsymbol{R}_{21} \ (\text{c}), \quad \boldsymbol{a}_{R12} = -\boldsymbol{a}_{R21} \ (\text{d}), \text{ and } \boldsymbol{R}_{12} = \boldsymbol{R}_{21} \ (\text{e})\right\} \tag{3.4}$$

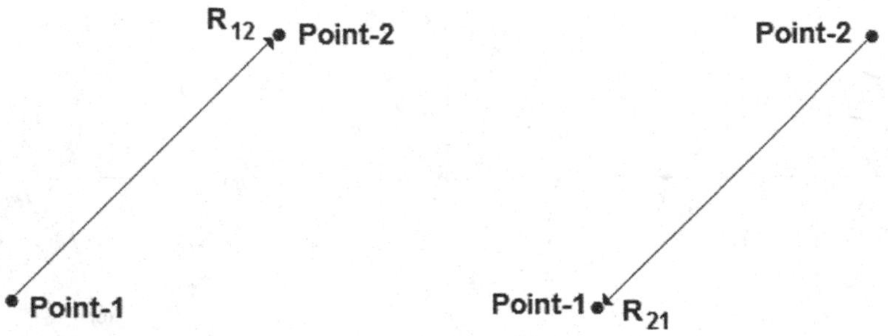

FIGURE 3.10 Representation of (a) $\mathbf{R}_{12}$ and (b) $\mathbf{R}_{21}$ vectors.

3.3.2 Vector Algebra

Figure 3.11a illustrates a vector which is represented by the line segment 1–2 and named as ***A***. Similarly, Figure 3.11b shows another vector named as ***B***, which is represented by the line segment 3–4. The addition and subtraction of these vectors are given in the subsections 3.3.2.1 and 3.3.2.2, respectively.

3.3.2.1 Vector Addition

For adding vectors ***A*** and ***B*** geometrically, join point 2 (the last tip or head of ***A***) and point 3 (the first tip or tail of ***B***). This step is shown in Figure 3.12a. The line segment (with arrowhead) 1–4 in Figure 3.12b is the resultant vector ***A*** + ***B***.

Mere graphical representation of the sum vector is not of much value unless it is translated into proper mathematical (algebraic) language. This translation is possible by using Figure 3.13. Let θ be the angle between the directions of orientation of vectors ***A*** and ***B***. This figure illustrates the projections of vector ***B*** in parallel and perpendicular planes of vector ***A***. The respective magnitudes of these projections are $\boldsymbol{B}\cos\theta$ and $\boldsymbol{B}\sin\theta$. In this figure resultant sum vector $\boldsymbol{R}_s$ can be written as:

$$\boldsymbol{R}_s = \boldsymbol{A} + \boldsymbol{B} = A\boldsymbol{a}_A + B\boldsymbol{a}_B \tag{3.5a}$$

Where the symbols $\boldsymbol{a}_A$ and $\boldsymbol{a}_B$ indicate unit vectors in the directions of vectors ***A*** and ***B*** respectively. While, $A = |\boldsymbol{A}|$ and $B = |\boldsymbol{B}|$ are the magnitudes of these vectors. The

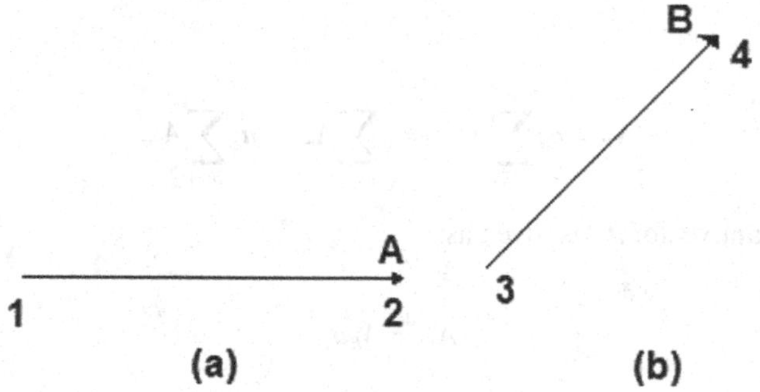

FIGURE 3.11 Representation of (a) vector **A** and (b) vector **B**.

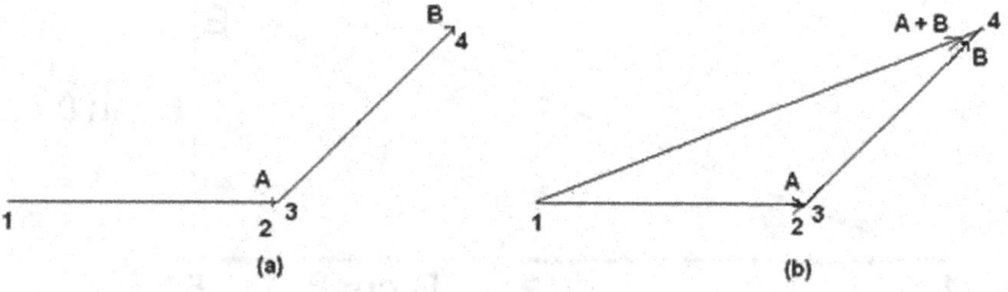

FIGURE 3.12 (a) Point 2 of **A** is joined to point 3 of **B** (b) resultant vector shown by line 1–4.

magnitude of the resultant vector $\boldsymbol{R}_s$ is found from Figure 3.13 using the Pythagorean Theorem, (also known as Pythagoras' theorem), as given below:

$$R_s = |\boldsymbol{R}_s| = \sqrt{(A + B \cdot \cos\theta)^2 + (B \cdot \sin\theta)^2} \tag{3.5b}$$

The unit vector $\boldsymbol{a}_{Rs}$ in the direction of the resultant vector R_s can be written as:

$$\boldsymbol{a}_{Rs} \overset{\text{def}}{=} \frac{\boldsymbol{R}_s}{R_s} = \frac{A\boldsymbol{a}_A + B\boldsymbol{a}_B}{\sqrt{(A + B \cdot \cos\theta)^2 + (B \cdot \sin\theta)^2}} \tag{3.5c}$$

The graphic method is rather inconvenient if three or more vectors are to be added, even if all the vectors are coplanar. A better method calls for expressing each vector in terms of its Cartesian components with appropriate unit vectors, thus let the vector $\boldsymbol{A}_m$ be written as:

$$\boldsymbol{A}_m = A_{mx}\boldsymbol{a}_x + A_{my}\boldsymbol{a}_y + A_{mz}\boldsymbol{a}_z \tag{3.6a}$$

Therefore, the vector sum is obtained as follows:

$$\boldsymbol{A}_R \overset{\text{def}}{=} \sum_{m=1}^{n} \boldsymbol{A}_m = \sum_{m=1}^{n} \left(A_{mx}\boldsymbol{a}_x + A_{my}\boldsymbol{a}_y + A_{mz}\boldsymbol{a}_z\right)$$

or

$$\boldsymbol{A}_R = \boldsymbol{a}_x \sum_{m=1}^{n} A_{mx} + \boldsymbol{a}_y \sum_{m=1}^{n} A_{my} + \boldsymbol{a}_z \sum_{m=1}^{n} A_{mz} \tag{3.6b}$$

Let the resultant vector $\boldsymbol{A}_R$ is given as:

$$\boldsymbol{A}_R \overset{\text{def}}{=} A_R \boldsymbol{a}_R \tag{3.7a}$$

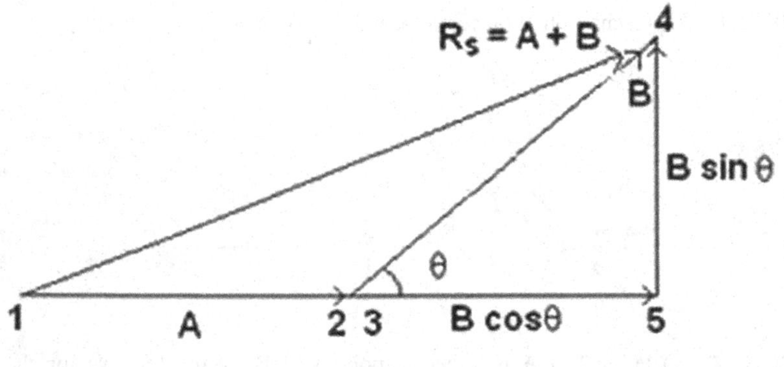

FIGURE 3.13 Representation of the vector sum.

where,

$$A_R \stackrel{\text{def}}{=} |\boldsymbol{A}_R| = \sqrt{\left(\sum_{m=1}^{n} A_{mx}\right)^2 + \left(\sum_{m=1}^{n} A_{my}\right)^2 + \left(\sum_{m=1}^{n} A_{mz}\right)^2} \tag{3.7b}$$

and

$$\boldsymbol{a}_R = \frac{\boldsymbol{A}_R}{A_R} \tag{3.7c}$$

This method permits us to add some vectors and subtract some other vectors in one go; all we need to do is to change the sign of Cartesian scalar components for vectors to be subtracted.

3.3.2.2 Vector Subtraction

For subtracting vector ***B*** from vector ***A*** first the orientation of vector ***B*** is to be reversed to represent (–***B***). Add the vectors ***A*** and (–***B***) by joining them as shown in Figure 3.14a wherein point 2 and 3 are again joined together. The resultant vector (***A***–***B***), called the difference vector ($\boldsymbol{R}_d$) shown by line 1–4 in Figure 3.14b can be written as:

$$\boldsymbol{R}_d = \boldsymbol{A} - \boldsymbol{B} = A\boldsymbol{a}_A - B\boldsymbol{a}_B \tag{3.8a}$$

Where the symbols $\boldsymbol{a}_A$ and $\boldsymbol{a}_B$ again indicate unit vectors in the directions of vectors ***A*** and ***B***, respectively.

The translation of the difference vector is possible using Figure 3.15. This figure illustrates the projections of vector –***B*** in parallel and perpendicular planes of vector ***A***. The respective values of these projections are ($-B\cos\theta$) and ($-B\sin\theta$). In this figure resultant difference vector $\boldsymbol{R}_d$ can be written as:

$$\boldsymbol{R}_d = (A - B\cos\theta)\boldsymbol{a}_A - (B\sin\theta)\boldsymbol{a}_B \tag{3.8b}$$

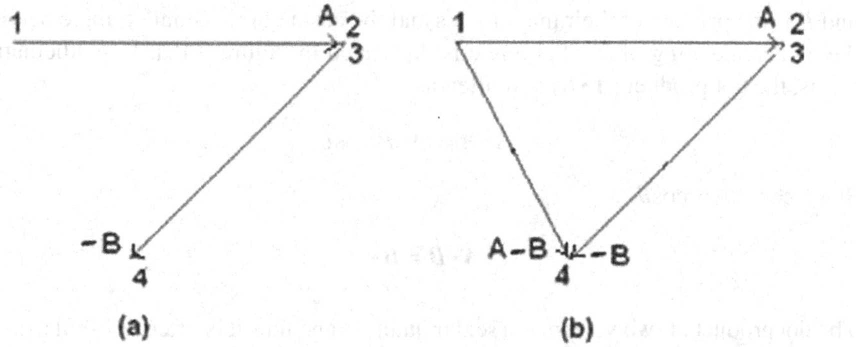

FIGURE 3.14 (a) point 2 of A joined to point 3 of –B, (b) resultant vector shown by line 1–4.

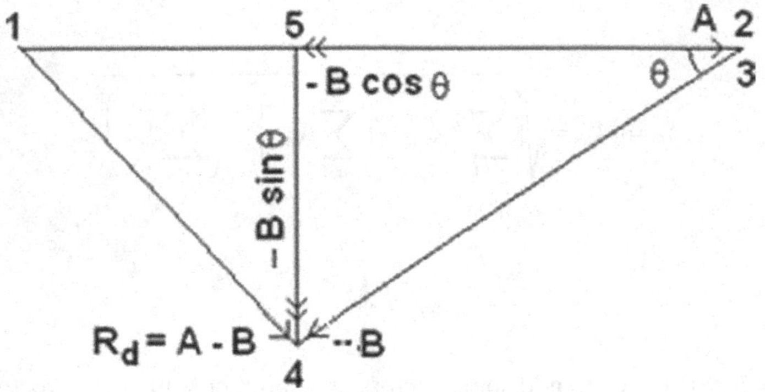

FIGURE 3.15 Representation of the difference vector.

Its magnitude is

$$|R_d| = \left[(A - B\cos\theta)^2 + (B\sin\theta)^2 \right]^{1/2} \tag{3.8c}$$

The unit vector $\boldsymbol{a}_{Rd}$, which represents the direction of its orientation is:

$$\boldsymbol{a}_{Rd} = \frac{\boldsymbol{R}_d}{R_d} \tag{3.8d}$$

This method of addition and subtraction can be used for obtaining sum and difference vectors for any number of vectors.

3.3.2.3 Multiplication of Vectors

The multiplication of two vectors can be defined in two ways, referred to as the scalar product and the vector product.

3.3.2.3.1 Scalar Product

The scalar product is also called a *dot product* as it is written with a dot (•) sign between the two (say ***A*** and ***B***) vectors, and is read as "***A*** dot ***B***". The dot product of ***A*** and ***B*** is the product of their magnitudes and the cosine of the smaller angle between them. The meaning of smaller angle is illustrated in Figure 3.16a. In mathematical terms, the dot product ($\boldsymbol{A} \bullet \boldsymbol{B}$) is written as:

$$\boldsymbol{A} \bullet \boldsymbol{B} = |\boldsymbol{A}|\,|\boldsymbol{B}|\cos\theta \tag{3.9a}$$

Since $\cos(-\theta) = \cos\theta$

$$\boldsymbol{A} \bullet \boldsymbol{B} = \boldsymbol{B} \bullet \boldsymbol{A} \tag{3.9b}$$

The dot product of two vectors is a scalar quantity and thus it is often referred to as the scalar product. As shown in Equation 3.9b, this product obeys the *commutative law.*

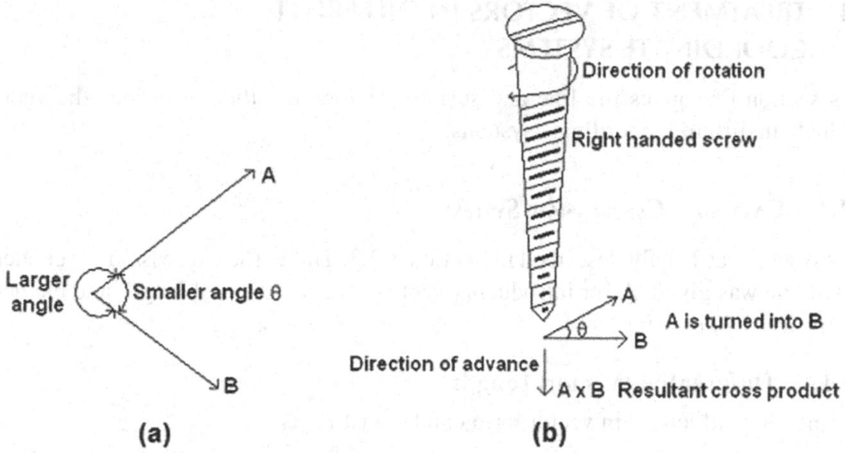

FIGURE 3.16 (a) Smaller and larger angles, (b) illustration of cross-product.

By taking suitable vectors (***A***, ***B***, and ***C***) it can be proved that the dot product also obeys *distributive law* i.e.

$$\boldsymbol{C}\bullet(\boldsymbol{A}+\boldsymbol{B})=\boldsymbol{C}\bullet\boldsymbol{A}+\boldsymbol{C}\bullet\boldsymbol{B} \tag{3.9c}$$

3.3.2.3.2 *Vector Product*

This is also called *cross-product* as it is written with a cross (×) sign between the two (say ***A*** and ***B***) vectors and is read as "***A*** cross ***B***". The *magnitude* of this product is equal to the product of magnitudes of ***A*** and ***B*** and sine of the smaller angle (say θ) between ***A*** and ***B***. Since the result of this product is a vector it is to be multiplied by a unit vector (say $\boldsymbol{a}_N$). The direction of this unit vector is perpendicular to the plane containing vectors ***A*** and ***B***. This perpendicular is in the direction of advance of a right-handed screw as it is turned from ***A*** toward ***B***. The vectors ***A*** and ***B*** and the direction of ***A*** × ***B*** are shown in Figure 3.16b. In mathematical terms:

$$\boldsymbol{A}\times\boldsymbol{B}=|\boldsymbol{A}|\cdot|\boldsymbol{B}|\cdot\sin\theta\cdot\boldsymbol{a}_N \tag{3.10a}$$

If the order of vectors in cross-product is reversed, the direction of the unit vector will also get reversed. That is:

$$\boldsymbol{B}\times\boldsymbol{A}=|\boldsymbol{A}|\cdot|\boldsymbol{B}|\cdot\sin\theta\cdot(-\boldsymbol{a}_N)=-|\boldsymbol{A}|\cdot|\boldsymbol{B}|\cdot\sin\theta\cdot\boldsymbol{a}_N=-\boldsymbol{A}\times\boldsymbol{B} \tag{3.10b}$$

Thus, the cross-product does not follow the commutative law, i.e.

$$\boldsymbol{A}\times\boldsymbol{B}\neq\boldsymbol{B}\times\boldsymbol{A} \tag{3.10c}$$

3.4 TREATMENT OF VECTORS IN DIFFERENT COORDINATE SYSTEMS

This section first gives the line and surface vectors and then describes the vector products in different coordinate systems.

3.4.1 Cartesian Coordinate System

This system was briefly described in Section 3.2.3. There, the expression for elemental volume was given. After introducing vectors, the differential length and elemental surfaces can now be given.

3.4.1.1 Differential (Vector) Length

The incremental length in vector form can be written as:

$$d\boldsymbol{\ell} = dx\boldsymbol{a}_x + dy\boldsymbol{a}_y + dz\boldsymbol{a}_z \tag{3.11}$$

3.4.1.2 Differential (Vector) Surfaces

The three elemental vector surfaces with the sides involved therein are shown in Figure 3.17.

The elemental **vector** surfaces, named as $\boldsymbol{ds}_x$, $\boldsymbol{ds}_y$, $\boldsymbol{ds}_z$, in the vector form can be written as:

$$\boldsymbol{ds}_x = dy \cdot dz \cdot \boldsymbol{a}_x \ (\text{a}), \quad \boldsymbol{ds}_y = dz \cdot dx \cdot \boldsymbol{a}_y \ (\text{b}), \quad \boldsymbol{ds}_z = dx \cdot dy \cdot \boldsymbol{a}_z \ (\text{c})\} \tag{3.12}$$

where $\boldsymbol{a}_x$, $\boldsymbol{a}_y$, and $\boldsymbol{a}_z$ are the unit vectors taken to be directed along with the x-, y-, and z-axis, respectively. This is the only system of space coordinates where all the unit vectors, viz. $\boldsymbol{a}_x$, $\boldsymbol{a}_y$, and $\boldsymbol{a}_z$, are constant quantities.

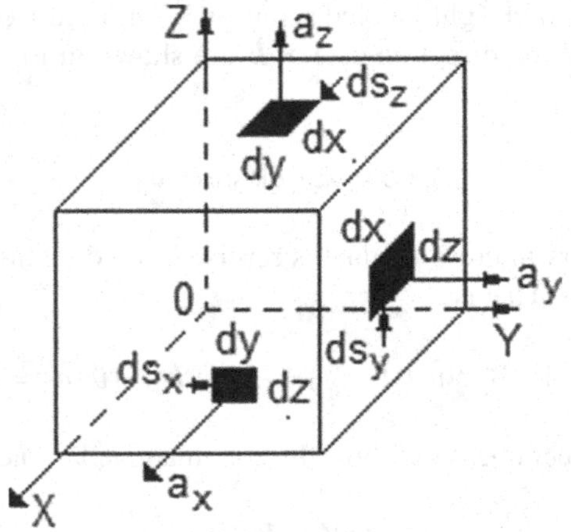

FIGURE 3.17 Differential (vector) surfaces in a Cartesian coordinate system.

3.4.1.3 Vector Representation

Earlier two vectors represented by straight line-segments were shown in Figure 3.10. The same vectors are revisited again in Figure 3.18 wherein points 1 and 2 are assigned with coordinates (x_1, y_1, z_1) and (x_2, y_2, z_2), respectively. In this figure one of the two vectors is oriented (i.e. directed) from point 1 to point 2, whereas the other from point 2 to point 1.

Vectors $\boldsymbol{R}_{12}$ and $\boldsymbol{R}_{21}$ can be written in terms of the coordinates of the two points as:

$$\boldsymbol{R}_{12} = (x_2 - x_1)\boldsymbol{a}_x + (y_2 - y_1)\boldsymbol{a}_y + (z_2 - z_1)\boldsymbol{a}_z \quad (3.13a)$$

$$\boldsymbol{R}_{21} = (x_1 - x_2)\boldsymbol{a}_x + (y_1 - y_2)\boldsymbol{a}_y + (z_1 - z_2)\boldsymbol{a}_z \quad (3.13b)$$

In these relations $\boldsymbol{a}_x$, $\boldsymbol{a}_y$, and $\boldsymbol{a}_z$ are the unit vectors along the positive x-, y-, and z-axes.

The magnitudes $|\boldsymbol{R}_{12}|$ and $|\boldsymbol{R}_{12}|$ of the above vectors can be written as:

$$|\boldsymbol{R}_{12}| = \left[(x_2 - x_1)^2 + (y_2 - y_1)^2 + (z_2 - z_1)^2\right]^{1/2} \quad (3.13c)$$

$$|\boldsymbol{R}_{21}| = \left[(x_1 - x_2)^2 + (y_1 - y_2)^2 + (z_1 - z_2)^2\right]^{1/2} \quad (3.13d)$$

As noted earlier the direction (or orientation) of vector $\boldsymbol{R}_{12}$ and $\boldsymbol{R}_{21}$ can be specified by a unit vectors $\boldsymbol{a}_{R12}$ and $\boldsymbol{a}_{R21}$. These unit vectors are defined as:

$$\left.\boldsymbol{a}_{R12} = \frac{\boldsymbol{R}_{12}}{|\boldsymbol{R}_{12}|} \;(\mathrm{e}) \qquad \boldsymbol{a}_{R21} = \frac{\boldsymbol{R}_{21}}{|\boldsymbol{R}_{21}|} \;(\mathrm{f})\right\} \quad (3.13)$$

As given earlier by Equation 3.4, the inspection of Equations 3.13a and 3.13c reveals that:

$$\boldsymbol{R}_{21} = -\boldsymbol{R}_{12} \quad (3.13g)$$

In Equation 3.13a or Equation 3.13b vector $\boldsymbol{A}$ in Cartesian coordinates can now be written as:

$$\boldsymbol{A} = A_x\boldsymbol{a}_x + A_y\boldsymbol{a}_y + A_z\boldsymbol{a}_z \quad (3.14)$$

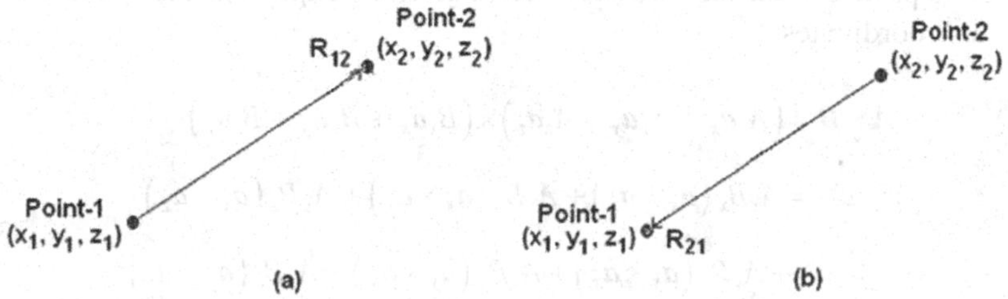

FIGURE 3.18 Vectors (a) R_{12} and (b) R_{21}.

where A_x, A_y, A_z are the magnitudes and $\boldsymbol{a}_x$, $\boldsymbol{a}_y$, and $\boldsymbol{a}_z$ are the unit vectors along the positive direction of x-, y-, and z-axes, respectively.

3.4.1.4 Dot Product of Two Vectors

Let the two vectors $\boldsymbol{A}$ and $\boldsymbol{B}$ in Cartesian coordinate system are:

$$\boldsymbol{A} = A_x\boldsymbol{a}_x + A_y\boldsymbol{a}_y + A_z\boldsymbol{a}_z \quad (\mathrm{a}), \quad \boldsymbol{B} = B_x\boldsymbol{a}_x + B_y\boldsymbol{a}_y + B_z\boldsymbol{a}_z \quad (\mathrm{b})\} \tag{3.15}$$

The dot product of these vectors can be written as:

$$\begin{aligned} \boldsymbol{A}\bullet\boldsymbol{B} &= \left(A_x\boldsymbol{a}_x + A_y\boldsymbol{a}_y + A_z\boldsymbol{a}_z\right)\bullet\left(B_x\boldsymbol{a}_x + B_y\boldsymbol{a}_y + B_z\boldsymbol{a}_z\right) \\ &= A_xB_x\left(\boldsymbol{a}_x\bullet\boldsymbol{a}_x\right) + A_xB_y\left(\boldsymbol{a}_x\bullet\boldsymbol{a}_y\right) + A_xB_z\left(\boldsymbol{a}_x\bullet\boldsymbol{a}_z\right) \\ &\quad + A_yB_x\left(\boldsymbol{a}_y\bullet\boldsymbol{a}_x\right) + A_yB_y\left(\boldsymbol{a}_y\bullet\boldsymbol{a}_y\right) + A_yB_z\left(\boldsymbol{a}_y\bullet\boldsymbol{a}_z\right) \\ &\quad + A_zB_x\left(\boldsymbol{a}_z\bullet\boldsymbol{a}_x\right) + A_zB_y\left(\boldsymbol{a}_z\bullet\boldsymbol{a}_y\right) + A_zB_z\left(\boldsymbol{a}_z\bullet\boldsymbol{a}_z\right) \end{aligned} \tag{3.16}$$

In Equation 3.16, this dot product yields sum of the nine scalar terms with each term has the product of two unit vectors. In Equation 3.9a, the cosine of the angle between the two vectors in the product of like unit vectors yields *one* whereas the product of unlike unit vectors yields *zero*. Thus:

$$\boldsymbol{a}_x\bullet\boldsymbol{a}_x = \boldsymbol{a}_y\bullet\boldsymbol{a}_y = \boldsymbol{a}_z\bullet\boldsymbol{a}_z = 1 \tag{3.17a}$$

$$\boldsymbol{a}_x\bullet\boldsymbol{a}_y = \boldsymbol{a}_y\bullet\boldsymbol{a}_x = \boldsymbol{a}_y\bullet\boldsymbol{a}_z = \boldsymbol{a}_z\bullet\boldsymbol{a}_y = \boldsymbol{a}_z\bullet\boldsymbol{a}_x = \boldsymbol{a}_x\bullet\boldsymbol{a}_z = 0 \tag{3.17b}$$

In Equations 3.17a and 3.17b, Equation 3.16 reduces to:

$$\boldsymbol{A}\bullet\boldsymbol{B} = A_xB_x + A_yB_y + A_zB_z \tag{3.18a}$$

In Equation 3.18a, if B is replaced by A, the resulting relation is:

$$\boldsymbol{A}\bullet\boldsymbol{A} = A_xA_x + A_yA_y + A_zA_z = A_x^2 + A_y^2 + A_z^2 \stackrel{\text{def}}{=} A^2 \tag{3.18b}$$

3.4.1.5 Cross-Product of Two Vectors

For a cross-product, consider the two vectors given by Equations 3.15a and 3.15b in Cartesian coordinates:

$$\begin{aligned} \boldsymbol{A}\times\boldsymbol{B} &= \left(A_x\boldsymbol{a}_x + A_y\boldsymbol{a}_y + A_z\boldsymbol{a}_z\right)\times\left(B_x\boldsymbol{a}_x + B_y\boldsymbol{a}_y + B_z\boldsymbol{a}_z\right) \\ &= A_xB_x\left(\boldsymbol{a}_x\times\boldsymbol{a}_x\right) + A_xB_y\left(\boldsymbol{a}_x\times\boldsymbol{a}_y\right) + A_xB_z\left(\boldsymbol{a}_x\times\boldsymbol{a}_z\right) \\ &\quad + A_yB_x\left(\boldsymbol{a}_y\times\boldsymbol{a}_x\right) + A_yB_y\left(\boldsymbol{a}_y\times\boldsymbol{a}_y\right) + A_yB_z\left(\boldsymbol{a}_y\times\boldsymbol{a}_z\right) \\ &\quad + A_zB_x\left(\boldsymbol{a}_z\times\boldsymbol{a}_x\right) + A_zB_y\left(\boldsymbol{a}_z\times\boldsymbol{a}_y\right) + A_zB_z\left(\boldsymbol{a}_z\times\boldsymbol{a}_z\right) \end{aligned} \tag{3.19}$$

The cross-product given by Equation 3.19 is the sum of nine terms. Each of these terms contains the product of two scalar terms (magnitudes of two components) and cross-product of two unit vectors. In Equation 3.10a the product of like unit vectors yields *zero*, whereas the product of unlike unit vectors yields a *new unit vector*, which is perpendicular to the two unit vectors involved in the product. The sign of resulting unit vector depends on the sequence of involved unit vectors. Thus:

$$\boldsymbol{a_x} \times \boldsymbol{a_x} = \boldsymbol{a_y} \times \boldsymbol{a_y} = \boldsymbol{a_z} \times \boldsymbol{a_z} = 0 \tag{3.20a}$$

$$\begin{cases} \boldsymbol{a_x} \times \boldsymbol{a_y} = \boldsymbol{a_z}; \ \boldsymbol{a_y} \times \boldsymbol{a_x} = -\boldsymbol{a_z} \\ \boldsymbol{a_y} \times \boldsymbol{a_z} = \boldsymbol{a_x}; \ \boldsymbol{a_z} \times \boldsymbol{a_y} = -\boldsymbol{a_x} \\ \boldsymbol{a_z} \times \boldsymbol{a_x} = \boldsymbol{a_y}; \ \boldsymbol{a_x} \times \boldsymbol{a_z} = -\boldsymbol{a_y} \end{cases} \tag{3.20b}$$

In these vector products, wherever the sequence of unit vectors is in accordance with the right-handed coordinate system, the result is positive. The reversal of sequence leads to a negative outcome. In Equations 3.20a and 3.20b, Equation 3.19 reduces to:

$$\begin{aligned} \boldsymbol{A} \times \boldsymbol{B} &= A_x B_x (0) + A_x B_y (\boldsymbol{a_z}) + A_x B_z (-\boldsymbol{a_y}) + A_y B_x (-\boldsymbol{a_z}) + A_y B_y (0) \\ &\quad + A_y B_z (\boldsymbol{a_x}) + A_z B_x (\boldsymbol{a_y}) + A_z B_y (-\boldsymbol{a_x}) + A_z B_z (0) \\ &= A_x B_y \boldsymbol{a_z} - A_x B_z \boldsymbol{a_y} - A_y B_x \boldsymbol{a_z} + A_y B_z \boldsymbol{a_x} + A_z B_x \boldsymbol{a_y} - A_z B_y \boldsymbol{a_x} \\ &= (A_y B_z - A_z B_y)\boldsymbol{a_x} + (A_z B_x - A_x B_z)\boldsymbol{a_y} + (A_x B_y - A_y B_x)\boldsymbol{a_z} \end{aligned} \tag{3.21a}$$

Using determinant of a matrix, Equation 3.21a can be written in the following compact form:

$$\boldsymbol{A} \times \boldsymbol{B} = \begin{vmatrix} \boldsymbol{a_x} & \boldsymbol{a_y} & \boldsymbol{a_z} \\ A_x & A_y & A_z \\ B_x & B_y & B_z \end{vmatrix} \tag{3.21b}$$

3.4.2 Cylindrical Coordinates System

This system was given in Section 3.2.4 along with the expression for elemental volume. This section gives the differential length and elemental surfaces in vector form.

3.4.2.1 Differential (Vector) Length

In the system of cylindrical space coordinates, the incremental length in vector form can be written as:

$$d\boldsymbol{\ell} = d\rho \cdot \boldsymbol{a_\rho} + \rho \cdot d\varphi \cdot \boldsymbol{a_\varphi} + dz \cdot \boldsymbol{a_z} \tag{3.22}$$

3.4.2.2 Differential (Vector) Surfaces

The three elemental vector surfaces $\boldsymbol{ds}_\rho$, $\boldsymbol{ds}_\varphi$, and $\boldsymbol{ds}_z$ with the sides involved therein are shown in Figure 3.19. These vector surfaces in mathematical terms can be written as:

$$\boldsymbol{ds}_\rho = \rho \cdot d\varphi \cdot dz \cdot \boldsymbol{a}_\rho \ (\text{a}), \quad \boldsymbol{ds}_\varphi = dz \cdot d\rho \cdot \boldsymbol{a}_\varphi \ (\text{b}), \quad \boldsymbol{ds}_z = d\rho \cdot \rho \cdot d\varphi \cdot \boldsymbol{a}_z \ (\text{c})\} \tag{3.23}$$

where $\boldsymbol{a}_\rho$, $\boldsymbol{a}_\varphi$, and $\boldsymbol{a}_z$ are unit vectors with unit length and directed along positive (ρ, φ, and z) axes, each has the dimension of length.

It is to be noted that ρ and z have dimension of length, whereas φ being in radians (or degrees), becomes length only when it is multiplied by ρ. Thus, in the expressions of elemental volume or surfaces wherever $d\varphi$ appears it is to be multiplied by ρ.

3.4.2.3 Vector Representation

In cylindrical coordinates, a vector $\boldsymbol{A}$ can be written as:

$$\boldsymbol{A} = A_\rho \boldsymbol{a}_\rho + A_\varphi \boldsymbol{a}_\varphi + A_z \boldsymbol{a}_z \tag{3.24}$$

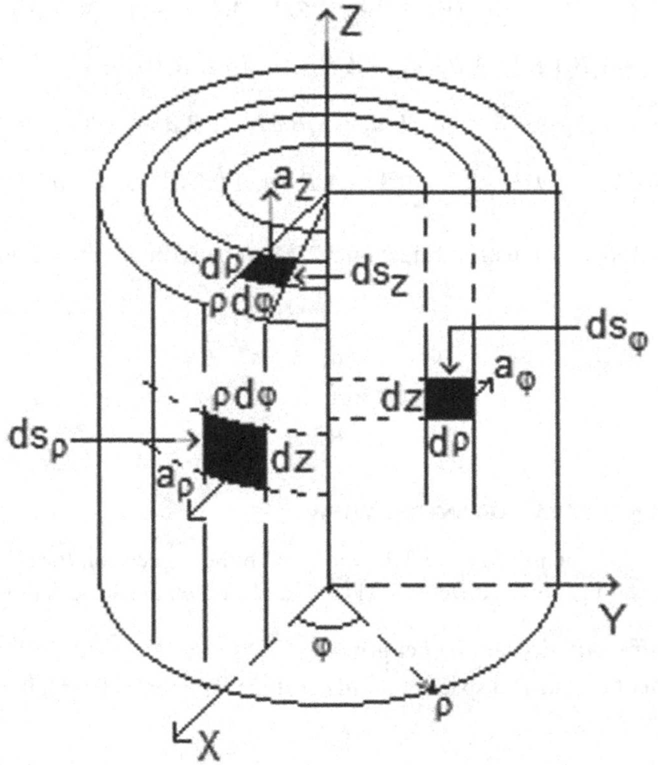

FIGURE 3.19 Constant ρ, constant φ, and constant z surfaces in a cylindrical system.

where A_ρ, A_φ, A_z are the magnitudes and $\boldsymbol{a}_\rho$, $\boldsymbol{a}_\varphi$, $\boldsymbol{a}_z$ are the unit vectors along ρ-, φ-, and z-axes.

It is to be noted that this is the same vector, which is given by Equation 3.14 and shown by Figure 3.18a. However in this case the coordinates of points 1 and 2 are to be specified in terms of ρ, $\rho\varphi$, z instead of x, y, z.

3.4.2.4 Dot Product of Two Vectors

Let the two vectors $\boldsymbol{A}$ and $\boldsymbol{B}$ in cylindrical coordinates are:

$$\boldsymbol{A} = A_\rho \boldsymbol{a}_\rho + A_\varphi \boldsymbol{a}_\varphi + A_z \boldsymbol{a}_z \;\; (\mathrm{a}), \quad \boldsymbol{B} = B_\rho \boldsymbol{a}_\rho + B_\varphi \boldsymbol{a}_\varphi + B_z \boldsymbol{a}_z \;\; (\mathrm{b})\} \tag{3.25}$$

The dot product of these vectors can be written as:

$$\begin{aligned} \boldsymbol{A}\bullet\boldsymbol{B} &= \left(A_\rho \boldsymbol{a}_\rho + A_\varphi \boldsymbol{a}_\varphi + A_z \boldsymbol{a}_z\right)\bullet\left(B_\rho \boldsymbol{a}_\rho + B_\varphi \boldsymbol{a}_\varphi + B_z \boldsymbol{a}_z\right) \\ &= A_\rho B_\rho \left(\boldsymbol{a}_\rho \bullet \boldsymbol{a}_\rho\right) + A_\rho B_\varphi \left(\boldsymbol{a}_\rho \bullet \boldsymbol{a}_\varphi\right) + A_\rho B_z \left(\boldsymbol{a}_\rho \bullet \boldsymbol{a}_z\right) \\ &\quad + A_\varphi B_\rho \left(\boldsymbol{a}_\varphi \bullet \boldsymbol{a}_\rho\right) + A_\varphi B_\varphi \left(\boldsymbol{a}_\varphi \bullet \boldsymbol{a}_\varphi\right) + A_\varphi B_z \left(\boldsymbol{a}_\varphi \bullet \boldsymbol{a}_z\right) \\ &\quad + A_z B_\rho \left(\boldsymbol{a}_z \bullet \boldsymbol{a}_\rho\right) + A_z B_\varphi \left(\boldsymbol{a}_z \bullet \boldsymbol{a}_\varphi\right) + A_z B_z \left(\boldsymbol{a}_z \bullet \boldsymbol{a}_z\right) \end{aligned} \tag{3.26}$$

In Equation 3.26, this dot product yields sum of the nine scalar terms with each term has the scalar product of two unit vectors. In Equation 3.9a, the cosine of the angle between the two vectors the product of like unit vectors yields *one*, whereas the product of unlike unit vectors yields *zero*. Thus, *for a given point* the scalar product between various unit vectors are given as follows:

$$\boldsymbol{a}_\rho \bullet \boldsymbol{a}_\rho = \boldsymbol{a}_\varphi \bullet \boldsymbol{a}_\varphi = \boldsymbol{a}_z \bullet \boldsymbol{a}_z = 1 \tag{3.27a}$$

$$\boldsymbol{a}_\rho \bullet \boldsymbol{a}_\varphi = \boldsymbol{a}_\varphi \bullet \boldsymbol{a}_\rho = \boldsymbol{a}_\varphi \bullet \boldsymbol{a}_z = \boldsymbol{a}_z \bullet \boldsymbol{a}_\varphi = \boldsymbol{a}_z \bullet \boldsymbol{a}_\rho = \boldsymbol{a}_\rho \bullet \boldsymbol{a}_z = 0 \tag{3.27b}$$

In Equations 3.27a and 3.27b, Equation 3.26 reduces to:

$$\boldsymbol{A}\bullet\boldsymbol{B} = A_\rho B_\rho + A_\varphi B_\varphi + A_z B_z \tag{3.28a}$$

In Equation 3.28a if $\boldsymbol{B}$ is replaced by $\boldsymbol{A}$ the resulting relation is:

$$\boldsymbol{A}\bullet\boldsymbol{B} = \boldsymbol{A}\bullet\boldsymbol{A} = A_\rho A_\rho + A_\varphi A_\varphi + A_z A_z = A_\rho^2 + A_\varphi^2 + A_z^2 \overset{\text{def}}{=} A^2 \tag{3.28b}$$

3.4.2.5 Cross-Product of Two Vectors

The cross-product of two vectors given by Equations 3.25a and 3.25b is:

$$\begin{aligned} \boldsymbol{A}\times\boldsymbol{B} &= \left(A_\rho \boldsymbol{a}_\rho + A_\varphi \boldsymbol{a}_\varphi + A_z \boldsymbol{a}_z\right)\times\left(B_\rho \boldsymbol{a}_\rho + B_\varphi \boldsymbol{a}_\varphi + B_z \boldsymbol{a}_z\right) \\ &= A_\rho B_\rho \left(\boldsymbol{a}_\rho \times \boldsymbol{a}_\rho\right) + A_\rho B_\varphi \left(\boldsymbol{a}_\rho \times \boldsymbol{a}_\varphi\right) + A_\rho B_z \left(\boldsymbol{a}_\rho \times \boldsymbol{a}_z\right) \\ &\quad + A_\varphi B_\rho \left(\boldsymbol{a}_\varphi \times \boldsymbol{a}_\rho\right) + A_\varphi B_\varphi \left(\boldsymbol{a}_\varphi \times \boldsymbol{a}_\varphi\right) + A_\varphi B_z \left(\boldsymbol{a}_\varphi \times \boldsymbol{a}_z\right) \\ &\quad + A_z B_\rho \left(\boldsymbol{a}_z \times \boldsymbol{a}_\rho\right) + A_z B_\varphi \left(\boldsymbol{a}_z \times \boldsymbol{a}_\varphi\right) + A_z B_z \left(\boldsymbol{a}_z \times \boldsymbol{a}_z\right) \end{aligned} \tag{3.29}$$

This cross-product given by Equation 3.29 is the sum of nine terms. Each of these terms contains the product of two scalar terms (magnitudes of components of two vectors) and cross-product of two unit vectors. In Equation 3.10a the cross-product of like unit vectors yields *zero* whereas the product of unlike unit vectors yields a *new unit vector*, which is perpendicular to the two unit vectors involved in the product. Thus:

$$\boldsymbol{a}_\rho \times \boldsymbol{a}_\rho = \boldsymbol{a}_\varphi \times \boldsymbol{a}_\varphi = \boldsymbol{a}_z \times \boldsymbol{a}_z = \mathbf{0} \tag{3.30a}$$

$$\begin{cases} \boldsymbol{a}_\rho \times \boldsymbol{a}_\varphi = \boldsymbol{a}_z;\ \boldsymbol{a}_\varphi \times \boldsymbol{a}_\rho = -\boldsymbol{a}_z \\ \boldsymbol{a}_\varphi \times \boldsymbol{a}_z = \boldsymbol{a}_\rho;\ \boldsymbol{a}_z \times \boldsymbol{a}_\varphi = -\boldsymbol{a}_\rho \\ \boldsymbol{a}_z \times \boldsymbol{a}_\rho = \boldsymbol{a}_\varphi;\ \boldsymbol{a}_\rho \times \boldsymbol{a}_z = -\boldsymbol{a}_\varphi \end{cases} \tag{3.30b}$$

In these vector products, wherever the sequence of unit vectors is in accordance with the right-handed coordinate system the result is positive. The reversal of sequence leads to a negative outcome. In Equations 3.30a and 3.30b, Equation 3.29 reduces to:

$$\begin{aligned} \boldsymbol{A} \times \boldsymbol{B} &= A_\rho B_\rho (0) + A_\rho B_\varphi (\boldsymbol{a}_z) + A_\rho B_z (-\boldsymbol{a}_\varphi) + A_\varphi B_\rho (-\boldsymbol{a}_z) + A_\varphi B_\varphi (0) \\ &\quad + A_\varphi B_z (\boldsymbol{a}_\rho) + A_z B_\rho (\boldsymbol{a}_\varphi) + A_z B_\varphi (-\boldsymbol{a}_\rho) + A_z B_z (0) \\ &= A_\rho B_\varphi \boldsymbol{a}_z - A_\rho B_z \boldsymbol{a}_\varphi - A_\varphi B_\rho \boldsymbol{a}_z + A_\varphi B_z \boldsymbol{a}_\rho + A_z B_\rho \boldsymbol{a}_\varphi - A_z B_\varphi \boldsymbol{a}_\rho \\ &= (A_\varphi B_z - A_z B_\varphi)\boldsymbol{a}_\rho + (A_z B_\rho - A_\rho B_z)\boldsymbol{a}_\varphi + (A_\rho B_\varphi - A_\varphi B_\rho)\boldsymbol{a}_z \end{aligned} \tag{3.31a}$$

Equation 3.24a can also be written using the notation for determinant, in the following form:

$$\boldsymbol{A} \times \boldsymbol{B} = \begin{vmatrix} \boldsymbol{a}_\rho & \boldsymbol{a}_\varphi & \boldsymbol{a}_z \\ A_\rho & A_\varphi & A_z \\ B_\rho & B_\varphi & B_z \end{vmatrix} \tag{3.31b}$$

3.4.3 Spherical Coordinates System

This system was given in Section 3.2.5 along with the elemental volume and some applications. Here the differential (vector) length and elemental (vector) surfaces are given.

3.4.3.1 Differential (Vector) Length

In this case the incremental length in vector form can be written as:

$$\boldsymbol{d\ell} = dr \cdot \boldsymbol{a}_r + r \cdot d\theta \cdot \boldsymbol{a}_\theta + r \sin\theta \cdot d\varphi \cdot \boldsymbol{a}_\varphi \tag{3.32}$$

3.4.3.2 Differential (Vector) Surfaces

These elemental surfaces in vector form are:

$$\boldsymbol{ds}_r = r^2 \sin\theta \cdot d\theta \cdot d\varphi \cdot \boldsymbol{a}_r \tag{3.33a}$$

$$ds_{\theta} = r\sin\theta \cdot dr \cdot d\varphi \cdot \boldsymbol{a_{\theta}} \tag{3.33b}$$

$$ds_{\varphi} = r \cdot dr \cdot d\theta \cdot \boldsymbol{a_{\varphi}} \tag{3.33c}$$

One can imagine ds_r as a white or black patch on the outer surface of a football. A small patch ds_{θ} on the conical surfaces is also shown. Lastly, ds_{φ} is a seed-like patch on a flat semi-circular surface of a watermelon. All of these patches are shown in Figure 3.20.

It is to be noted that both the angles θ and φ are in radians (or degrees). These will assume units of length after getting multiplied by r. Thus, in the expressions of elemental volume or elemental surfaces both $d\theta$ and $d\varphi$ are to be multiplied by r wherever they appear. Furthermore, in Figure 3.6a r and φ do not lie in the same plane. Thus, the projection of point P located at r, θ becomes $r\sin\theta$ in the φ plane. As evident from the expressions of dv, ds_r, and ds_{θ}, wherever r and $d\varphi$ appear together their product is to be further multiplied by $\sin\theta$.

3.4.3.3 Vector Representation

In spherical coordinates a vector A can be written as:

$$\boldsymbol{A} = A_r\boldsymbol{a_r} + A_{\theta}\boldsymbol{a_{\theta}} + A_{\varphi}\boldsymbol{a_{\varphi}} \tag{3.34}$$

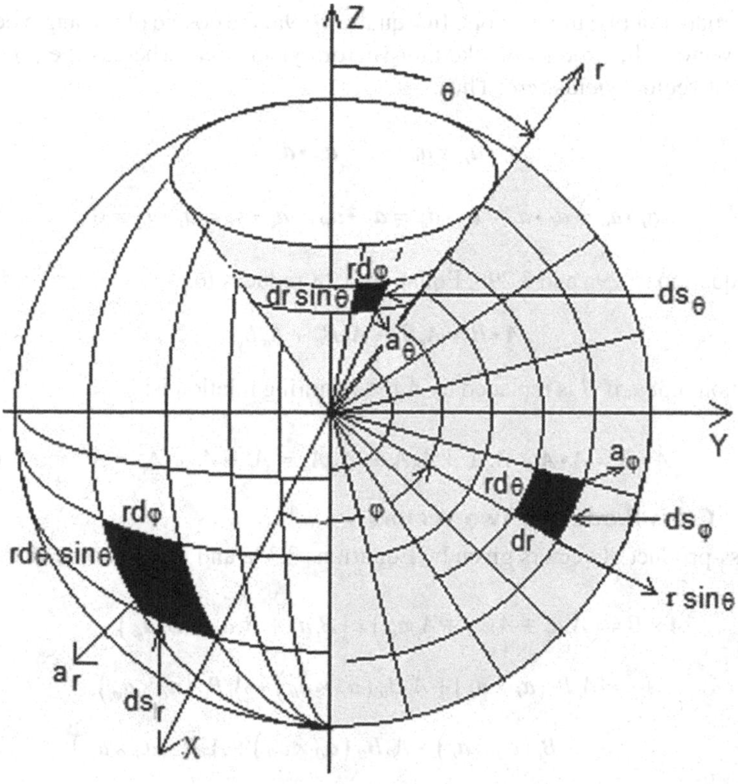

FIGURE 3.20 Constant r, constant θ, and constant φ surfaces in a spherical system.

where A_r, A_θ, and A_φ are the scalar components of the vector A and $\boldsymbol{a}_r$, $\boldsymbol{a}_\theta$, and $\boldsymbol{a}_\varphi$ are the unit vectors along the r, θ, and φ axes. This is again to be noted that this is the same vector which is given by Equation 3.15a and shown by Figure 3.18a. In this case the coordinates of points 1 and 2 are to be specified in terms of r, $r\theta$, $r\varphi sin\theta$ instead of x, y, z.

3.4.3.4 Dot Product of Two Vectors

Let the two vectors $\boldsymbol{A}$ and $\boldsymbol{B}$ in spherical coordinates be:

$$\boldsymbol{A} = A_r\boldsymbol{a}_r + A_\theta\boldsymbol{a}_\theta + A_\varphi\boldsymbol{a}_\varphi \quad (\mathrm{a}), \quad \boldsymbol{B} = B_r\boldsymbol{a}_r + B_\theta\boldsymbol{a}_\theta + B_\varphi\boldsymbol{a}_\varphi \quad (\mathrm{b})\Big\} \tag{3.35}$$

The dot product of these vectors can be written as:

$$\begin{aligned} \boldsymbol{A}\bullet\boldsymbol{B} &= \left(A_r\boldsymbol{a}_r + A_\theta\boldsymbol{a}_\theta + A_\varphi\boldsymbol{a}_\varphi\right)\bullet\left(B_r\boldsymbol{a}_r + B_\theta\boldsymbol{a}_\theta + B_\varphi\boldsymbol{a}_\varphi\right) \\ &= A_rB_r\left(\boldsymbol{a}_r\bullet\boldsymbol{a}_r\right) + A_rB_\theta\left(\boldsymbol{a}_r\bullet\boldsymbol{a}_\theta\right) + A_rB_\varphi\left(\boldsymbol{a}_r\bullet\boldsymbol{a}_\varphi\right) \\ &+ A_\theta B_r\left(\boldsymbol{a}_\theta\bullet\boldsymbol{a}_r\right) + A_\theta B_\theta\left(\boldsymbol{a}_\theta\bullet\boldsymbol{a}_\theta\right) + A_\theta B_\varphi\left(\boldsymbol{a}_\theta\bullet\boldsymbol{a}_\varphi\right) \\ &+ A_\varphi B_r\left(\boldsymbol{a}_\varphi\bullet\boldsymbol{a}_r\right) + A_\varphi B_\theta\left(\boldsymbol{a}_\varphi\bullet\boldsymbol{a}_\theta\right) + A_\varphi B_\varphi\left(\boldsymbol{a}_\varphi\bullet\boldsymbol{a}_\varphi\right) \end{aligned} \tag{3.36}$$

In Equation 3.36, this dot product yields sum of the nine scalar terms with each term has the product of two unit vectors. In Equation 3.9a, the cosine of the angle between the two vectors the product of like unit vectors yields *one*, whereas the product of unlike unit vectors yields *zero*. Thus:

$$\boldsymbol{a}_r\bullet\boldsymbol{a}_r = \boldsymbol{a}_\theta\bullet\boldsymbol{a}_\theta = \boldsymbol{a}_\varphi\bullet\boldsymbol{a}_\varphi = 1 \tag{3.37a}$$

$$\boldsymbol{a}_r\bullet\boldsymbol{a}_\theta = \boldsymbol{a}_\theta\bullet\boldsymbol{a}_r = \boldsymbol{a}_\theta\bullet\boldsymbol{a}_\varphi = \boldsymbol{a}_\varphi\bullet\boldsymbol{a}_\theta = \boldsymbol{a}_\varphi\bullet\boldsymbol{a}_r = \boldsymbol{a}_r\bullet\boldsymbol{a}_\varphi = 0 \tag{3.37b}$$

Using Equations 3.29a and 3.29b, Equation 3.28 reduces to:

$$\boldsymbol{A}\bullet\boldsymbol{B} = A_rB_r + A_\theta B_\theta + A_\varphi B_\varphi \tag{3.38a}$$

In Equation 3.38a, if B is replaced by A the resulting relation is:

$$\boldsymbol{A}\bullet\boldsymbol{B} = \boldsymbol{A}\bullet\boldsymbol{A} = A_rA_r + A_\theta A_\theta + A_\varphi A_\varphi = A_r^2 + A_\theta^2 + A_\varphi^2 = A^2 \tag{3.38b}$$

3.4.3.5 Cross-Product of Two Vectors

The cross-product of vectors given by Equations 3.35a and 3.35b can be written as:

$$\begin{aligned} \boldsymbol{A}\times\boldsymbol{B} &= \left(A_r\boldsymbol{a}_r + A_\theta\boldsymbol{a}_\theta + A_\varphi\boldsymbol{a}_\varphi\right)\times\left(B_r\boldsymbol{a}_r + B_\theta\boldsymbol{a}_\theta + B_\varphi\boldsymbol{a}_\varphi\right) \\ &= A_rB_r\left(\boldsymbol{a}_r\times\boldsymbol{a}_r\right) + A_rB_\theta\left(\boldsymbol{a}_r\times\boldsymbol{a}_\theta\right) + A_rB_\varphi\left(\boldsymbol{a}_r\times\boldsymbol{a}_\varphi\right) \\ &+ A_\theta B_r\left(\boldsymbol{a}_\theta\times\boldsymbol{a}_r\right) + A_\theta B_\theta\left(\boldsymbol{a}_\theta\times\boldsymbol{a}_\theta\right) + A_\theta B_\varphi\left(\boldsymbol{a}_\theta\times\boldsymbol{a}_\varphi\right) \\ &+ A_\varphi B_r\left(\boldsymbol{a}_\varphi\times\boldsymbol{a}_r\right) + A_\varphi B_\theta\left(\boldsymbol{a}_\varphi\times\boldsymbol{a}_\theta\right) + A_\varphi B_\varphi\left(\boldsymbol{a}_\varphi\times\boldsymbol{a}_\varphi\right) \end{aligned} \tag{3.39}$$

This cross-product is the sum of nine terms. Each of these terms contains the product of two scalar terms (magnitude of components involved) and cross-product of two unit vectors. In Equation 3.10a the product of like unit vectors yields *zero*, whereas that of unlike unit vectors yields a *new unit vector* which is perpendicular to the two unit vectors involved in the product. Thus:

$$\boldsymbol{a}_r \times \boldsymbol{a}_r = \boldsymbol{a}_\theta \times \boldsymbol{a}_\theta = \boldsymbol{a}_\varphi \times \boldsymbol{a}_\varphi = \boldsymbol{0} \quad (3.40a)$$

$$\begin{cases} \boldsymbol{a}_r \times \boldsymbol{a}_\theta = \boldsymbol{a}_\varphi; & \boldsymbol{a}_\theta \times \boldsymbol{a}_r = -\boldsymbol{a}_\varphi \\ \boldsymbol{a}_\theta \times \boldsymbol{a}_\varphi = \boldsymbol{a}_r; & \boldsymbol{a}_\varphi \times \boldsymbol{a}_\theta = -\boldsymbol{a}_r \\ \boldsymbol{a}_\varphi \times \boldsymbol{a}_r = \boldsymbol{a}_\theta; & \boldsymbol{a}_r \times \boldsymbol{a}_\varphi = -\boldsymbol{a}_\theta \end{cases} \quad (3.40b)$$

In these vector products, wherever the sequence of unit vectors is in accordance with the right-handed coordinate system the result is positive. The reversal of sequence leads to a negative outcome. Thus using Equations 3.40a and 3.40b, Equation 4.39 reduces to:

$$\begin{aligned} \boldsymbol{A} \times \boldsymbol{B} &= A_r B_r (0) + A_r B_\theta (\boldsymbol{a}_\varphi) + A_r B_\varphi (-\boldsymbol{a}_\theta) + A_\theta B_r (-\boldsymbol{a}_\varphi) + A_0 B_\theta (0) \\ &\quad + A_\theta B_\varphi (\boldsymbol{a}_r) + A_\varphi B_r (\boldsymbol{a}_\theta) + A_\varphi B_\theta (-\boldsymbol{a}_r) + A_\varphi B_\varphi (0) \\ &= A_r B_\theta \boldsymbol{a}_\varphi - A_r B_\varphi \boldsymbol{a}_\theta - A_\theta B_r \boldsymbol{a}_\varphi + A_\theta B_\varphi \boldsymbol{a}_r + A_\varphi B_r \boldsymbol{a}_\theta - A_\varphi B_\theta \boldsymbol{a}_r \\ &= (A_\theta B_\varphi - A_\varphi B_\theta) \boldsymbol{a}_r + (A_\varphi B_r - A_r B_\varphi) \boldsymbol{a}_\theta + (A_r B_\theta - A_\theta B_r) \boldsymbol{a}_\varphi \end{aligned} \quad (3.41a)$$

Equation 3.41a can also be written in the following form:

$$\boldsymbol{A} \times \boldsymbol{B} = \begin{vmatrix} \boldsymbol{a}_r & \boldsymbol{a}_\theta & \boldsymbol{a}_\varphi \\ A_r & A_\theta & A_\varphi \\ B_r & B_\theta & B_\varphi \end{vmatrix} \quad (3.41b)$$

In Sections 3.4.1, 3.4.2, and 3.4.3 parameters x, y, z, ρ, and r are generally taken in meters or centimeters, while angles θ and φ are taken in radians unless specified otherwise.

3.5 COORDINATE TRANSFORMATION

As noted earlier, all three discussed coordinate systems are orthogonal systems since x-, y-, and z-coordinates in rectangular, ρ-, φ-, and z-coordinates in cylindrical, and r-, θ-, and φ-coordinates in spherical systems are perpendicular to each other. Also, as long as the sequence of coordinates (x, y, z), (ρ, φ, z), and (r, θ, φ) in respective systems is maintained all these can be referred to as right-handed. If this sequence in any of these is altered to the form of, e.g. (x, z, y), (ρ, z, φ), and (r, φ, θ) that particular system will be termed as a left-handed system.

3.5.1 Relation between Coordinates

The cylindrical coordinate system is defined with reference to the rectangular Cartesian system of space coordinates. In Figure 3.2b (or Figure 3.6a) it can be noted that the coordinates of Cartesian and cylindrical systems bear definite relations. Similarly, Figure 3.2c (or 3.8a) shows that the coordinates of Cartesian and spherical systems are also related. These relations are as given below.

3.5.1.1 Cartesian and Cylindrical

In Figure 3.6a the cylindrical system is superimposed over the rectangular system. The relations between the coordinates of two systems are summarized in Table 3.1.

3.5.1.2 Cartesian and Spherical

The spherical coordinate system is defined with reference to the Cartesian system of space coordinates. In Figure 3.8a the spherical system is superimposed over the rectangular Cartesian system. The relations between the coordinates of two systems are summarized in Table 3.2.

In order to simplify practical problems coordinate transformation sometimes becomes imminent. This transformation is a bilateral process, i.e. rectangular coordinates can be transformed to cylindrical or spherical coordinates and vice versa. The process of transformation from cylindrical to spherical or vice versa is a bit

TABLE 3.1
Relationship between Cartesian and Cylindrical Coordinates

$x = \rho \cdot \cos\varphi$	$y = \rho \cdot \sin\varphi$	$z = z$
$\rho = (x^2 + y^2)^{1/2}$	$\varphi = \tan^{-1}(y/x)$	$z = z$
$\sin\varphi = \frac{y}{\rho} = \frac{y}{\sqrt{x^2 + y^2}}$	$\cos\varphi = \frac{x}{\rho} = \frac{x}{\sqrt{x^2 + y^2}}$	$\tan\varphi = \frac{y}{x}$

TABLE 3.2
Relationship between Cartesian and Spherical Coordinates

$x = r \cdot \sin\theta \cdot \cos\varphi$	$y = r \cdot \sin\theta \cdot \sin\varphi$	$z = r \cdot \cos\theta$
$r = \sqrt{x^2 + y^2 + z^2}$	$\theta = \cos^{-1}\left[\frac{z}{\sqrt{x^2 + y^2 + z^2}}\right]$	$\varphi = \tan^{-1}\left(\frac{y}{x}\right)$
$\cos\theta = \frac{z}{\sqrt{x^2 + y^2 + z^2}}$	$\sin\theta = \frac{r'}{r} = \frac{\sqrt{x^2 + y^2}}{\sqrt{x^2 + y^2 + z^2}}$	$r' = \sqrt{x^2 + y^2}$
$\sin\varphi = \frac{y}{r'} = \frac{y}{\sqrt{x^2 + y^2}}$	$\cos\varphi = \frac{x}{r'} = \frac{x}{\sqrt{x^2 + y^2}}$	$\tan\varphi = \frac{y}{x}$

more involved. If required, it can be carried out by using a rectangular system as the base. Accordingly, the spherical coordinates can be first converted to rectangular coordinates and the resulting vector can then be converted to cylindrical coordinates. A reverse process can be adopted to convert cylindrical to spherical via rectangular system of space coordinates. The two commonly used transformations are discussed below.

3.5.2 Transformation between Rectangular and Cylindrical Coordinates

Figure 3.21 superimposes a cylindrical system over a rectangular coordinate system. It illustrates angles between different unit vectors belonging to rectangular and cylindrical systems. In these angles the dot products of unit vectors are given in Table 3.3.

The definition of dot product makes the transformation of coordinates much easier. If a particular component of a given vector A is to be obtained that vector is to be simply multiplied by the corresponding unit vector. Thus:

$$\left.\begin{aligned} A_x &= \boldsymbol{A} \bullet \boldsymbol{a}_x = \left(A_x\boldsymbol{a}_x + A_y\boldsymbol{a}_y + A_z\boldsymbol{a}_z\right) \bullet \boldsymbol{a}_x \quad &(a) \\ A_y &= \boldsymbol{A} \bullet \boldsymbol{a}_y = \left(A_x\boldsymbol{a}_x + A_y\boldsymbol{a}_y + A_z\boldsymbol{a}_z\right) \bullet \boldsymbol{a}_y \quad &(b) \\ A_z &= \boldsymbol{A} \bullet \boldsymbol{a}_z = \left(A_x\boldsymbol{a}_x + A_y\boldsymbol{a}_y + A_z\boldsymbol{a}_z\right) \bullet \boldsymbol{a}_z \quad &(c) \end{aligned}\right\} \quad (3.42)$$

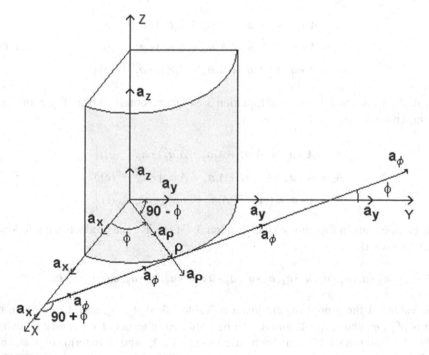

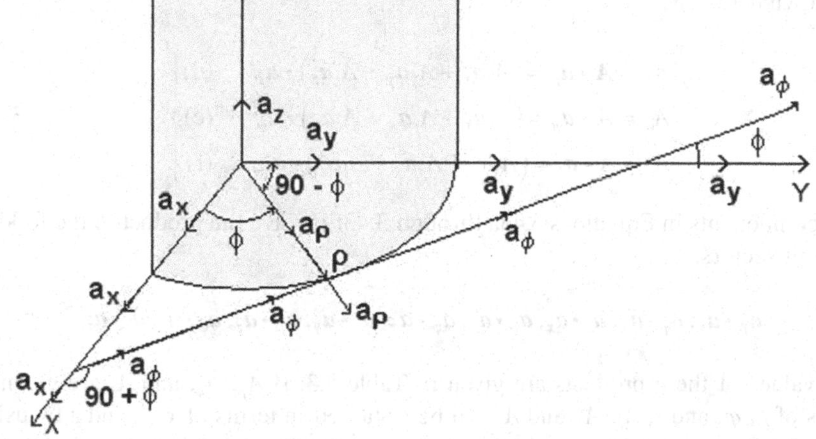

FIGURE 3.21 Angles between different unit vectors.

TABLE 3.3
Unit Vectors, Angles, and Dot Products

Unit vectors	Angle between unit vectors	Dot product
$\boldsymbol{a}_x$ & $\boldsymbol{a}_\rho$	φ	$\boldsymbol{a}_x \bullet \boldsymbol{a}_\rho = \cos\varphi$
$\boldsymbol{a}_x$ & $\boldsymbol{a}_\varphi$	$90+\varphi$	$\boldsymbol{a}_x \bullet \boldsymbol{a}_\varphi = -\sin\varphi$
$\boldsymbol{a}_x$ & $\boldsymbol{a}_z$	90	$\boldsymbol{a}_x \bullet \boldsymbol{a}_z = 0$
$\boldsymbol{a}_y$ & $\boldsymbol{a}_\rho$	$90-\varphi$	$\boldsymbol{a}_y \bullet \boldsymbol{a}_\rho = \sin\varphi$
$\boldsymbol{a}_y$ & $\boldsymbol{a}_\varphi$	φ	$\boldsymbol{a}_y \bullet \boldsymbol{a}_\varphi = \cos\varphi$
$\boldsymbol{a}_y$ & $\boldsymbol{a}_z$	90	$\boldsymbol{a}_y \bullet \boldsymbol{a}_z = 0$
$\boldsymbol{a}_z$ & $\boldsymbol{a}_\rho$	90	$\boldsymbol{a}_z \bullet \boldsymbol{a}_\rho = 0$
$\boldsymbol{a}_z$ & $\boldsymbol{a}_\varphi$	90	$\boldsymbol{a}_z \bullet \boldsymbol{a}_\varphi = 0$
$\boldsymbol{a}_x$ & $\boldsymbol{a}_z$	0	$\boldsymbol{a}_z \bullet \boldsymbol{a}_z = 1$

Vector A, given by Equations 3.14, 3.24, and 3.34, is the same physical entity in all the three coordinate systems. Vector A, given by Equations 3.14 and 3.24, is reproduced here:

$$\boldsymbol{A} = A_x\boldsymbol{a}_x + A_y\boldsymbol{a}_y + A_z\boldsymbol{a}_z \quad (3.14), \qquad \boldsymbol{A} = \left(A_\rho\boldsymbol{a}_\rho + A_\varphi\boldsymbol{a}_\varphi + A_z\boldsymbol{a}_z\right) \quad (3.24)\}$$

A_x, A_y, and A_z in Equation 3.14 can be obtained from Equation 3.24 as shown here:

$$\left.\begin{aligned} A_x &= \boldsymbol{A} \bullet \boldsymbol{a}_x = \left(A_\rho\boldsymbol{a}_\rho + A_\varphi\boldsymbol{a}_\varphi + A_Z\boldsymbol{a}_z\right) \bullet \boldsymbol{a}_x \quad &(a)\\ A_y &= \boldsymbol{A} \bullet \boldsymbol{a}_y = \left(A_\rho\boldsymbol{a}_\rho + A_\varphi\boldsymbol{a}_\varphi + A_Z\boldsymbol{a}_z\right) \bullet \boldsymbol{a}_y \quad &(b)\\ A_Z &= \boldsymbol{A} \bullet \boldsymbol{a}_z = \left(A_\rho\boldsymbol{a}_\rho + A_\varphi\boldsymbol{a}_\varphi + A_Z\boldsymbol{a}_z\right) \bullet \boldsymbol{a}_z \quad &(c) \end{aligned}\right\} \tag{3.43}$$

Similarly A_ρ, A_φ, and A_z involved Equation 3.24 can be obtained from Equation 3.14 as shown here:

$$\left.\begin{aligned} A_\rho &= \boldsymbol{A} \bullet \boldsymbol{a}_\rho = \left(A_x\boldsymbol{a}_x + A_y\boldsymbol{a}_y + A_z\boldsymbol{a}_z\right) \bullet \boldsymbol{a}_\rho \quad &(d)\\ A_\varphi &= \boldsymbol{A} \bullet \boldsymbol{a}_\varphi = \left(A_x\boldsymbol{a}_x + A_y\boldsymbol{a}_y + A_z\boldsymbol{a}_z\right) \bullet \boldsymbol{a}_\varphi \quad &(e)\\ A_z &= \boldsymbol{A} \bullet \boldsymbol{a}_z = \left(A_x\boldsymbol{a}_x + A_y\boldsymbol{a}_y + A_z\boldsymbol{a}_z\right) \bullet \boldsymbol{a}_z \quad &(f) \end{aligned}\right\} \tag{3.43}$$

The components in Equations 3.43a through 3.43f involve the product of the following unit vectors:

$$\boldsymbol{a}_\rho \bullet \boldsymbol{a}_x,\ \boldsymbol{a}_\varphi \bullet \boldsymbol{a}_x,\ \boldsymbol{a}_z \bullet \boldsymbol{a}_x,\ \boldsymbol{a}_\rho \bullet \boldsymbol{a}_y,\ \boldsymbol{a}_\varphi \bullet \boldsymbol{a}_y,\ \boldsymbol{a}_z \bullet \boldsymbol{a}_y,\ \boldsymbol{a}_\rho \bullet \boldsymbol{a}_z,\ \boldsymbol{a}_\varphi \bullet \boldsymbol{a}_z,\ \boldsymbol{a}_z \bullet \boldsymbol{a}_z$$

The values of these products are given in Table 3.3. If A_ρ, A_φ, and A_z are given in terms of ρ, φ, and z, A_x, A_y, and A_z can be obtained in terms of x, y, and z by using Table 3.1. The same table can also be used to get A_ρ, A_φ, and A_z in terms of ρ, φ, and z when A_x, A_y, and A_z are given in terms of x, y, and z.

3.5.3 Transformation between Rectangular and Spherical Coordinates

Figure 3.22 illustrates angles between different unit vectors belonging to rectangular and spherical coordinate systems. In these angles the dot products of unit vectors are given in Table 3.4.

Note: $\boldsymbol{a}_{xy}$ is the unit vector along the projection of r on the x–y-plane, which makes angles θ and $90-\theta$ with $\boldsymbol{a}_\theta$ and $\boldsymbol{a}_r$, respectively.

Vector A in rectangular and spherical systems is given here:

$$\left.\boldsymbol{A} = A_x\boldsymbol{a}_x + A_y\boldsymbol{a}_y + A_z\boldsymbol{a}_z \quad (3.14), \qquad \boldsymbol{A} = A_r\boldsymbol{a}_r + A_\theta\boldsymbol{a}_\theta + A_\varphi\boldsymbol{a}_\varphi \quad (3.34)\right\}$$

A_x, A_y, and A_z in Equation 3.14 can be obtained from Equation 3.34 as here:

$$\left.\begin{aligned} A_x &= \boldsymbol{A}\bullet\boldsymbol{a}_x = \left(A_r\boldsymbol{a}_r + A_\theta\boldsymbol{a}_\theta + A_\varphi\boldsymbol{a}_\varphi\right)\bullet\boldsymbol{a}_x \quad &(a)\\ A_y &= \boldsymbol{A}\bullet\boldsymbol{a}_y = \left(A_r\boldsymbol{a}_r + A_\theta\boldsymbol{a}_\theta + A_\varphi\boldsymbol{a}_\varphi\right)\bullet\boldsymbol{a}_y \quad &(b)\\ A_Z &= \boldsymbol{A}\bullet\boldsymbol{a}_z = \left(A_r\boldsymbol{a}_r + A_\theta\boldsymbol{a}_\theta + A_\varphi\boldsymbol{a}_\varphi\right)\bullet\boldsymbol{a}_z \quad &(c)\end{aligned}\right\} \quad (3.44)$$

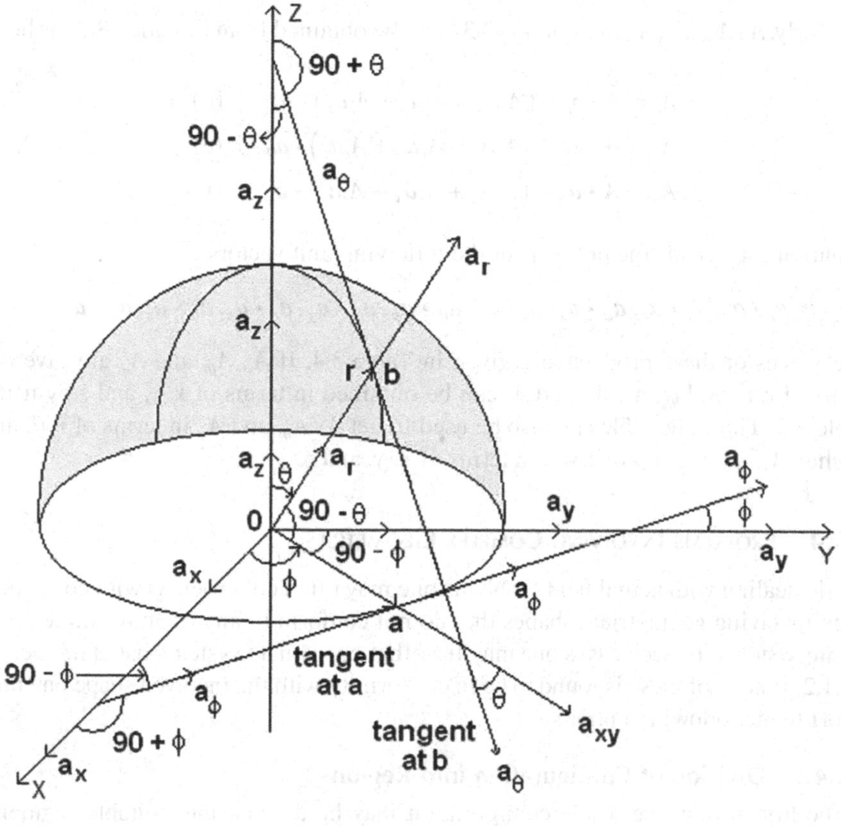

FIGURE 3.22 Angles between different unit vectors.

TABLE 3.4
Unit vectors, Angles, and Dot Products

Unit vectors	Angle between unit vectors	Dot product
$\boldsymbol{a}_{xy}$ & $\boldsymbol{a}_r$	$\angle 90-\theta$	$\boldsymbol{a}_{xy}\bullet\boldsymbol{a}_r=\sin\theta$
$\boldsymbol{a}_{xy}$ & $\boldsymbol{a}_\theta$	$\angle\theta$	$\boldsymbol{a}_{xy}\bullet\boldsymbol{a}_\theta=\cos\theta$
$\boldsymbol{a}_x$ & $\boldsymbol{a}_r$	$\angle(\boldsymbol{a}_r\ \&\boldsymbol{a}_{xy})\bullet\angle(\boldsymbol{a}_{xy}\ \&\boldsymbol{a}_x)=\angle 90-\theta\bullet\angle\varphi$	$\boldsymbol{a}_x\bullet\boldsymbol{a}_r=\sin\theta\cos\varphi$
$\boldsymbol{a}_x$ & $\boldsymbol{a}_\theta$	$\angle(\boldsymbol{a}_\theta\ \&\boldsymbol{a}_{xy})\bullet\angle(\boldsymbol{a}_{xy}\ \&\boldsymbol{a}_x)=\angle\theta\bullet\angle\varphi$	$\boldsymbol{a}_x\bullet\boldsymbol{a}_\theta=\cos\theta\cos\varphi$
$\boldsymbol{a}_x$ & $\boldsymbol{a}_\varphi$	$\angle 90+\theta$	$\boldsymbol{a}_x\bullet\boldsymbol{a}_\varphi=-\sin\theta$
$\boldsymbol{a}_y$ & $\boldsymbol{a}_r$	$\angle(\boldsymbol{a}_r\ \&\ \boldsymbol{a}_{xy})\bullet\angle(\boldsymbol{a}_{xy}\ \&\ \boldsymbol{a}_y)=\angle 90-\theta\bullet\angle 90-\varphi$	$\boldsymbol{a}_y\bullet\boldsymbol{a}_r=\sin\theta\ \sin\varphi$
$\boldsymbol{a}_y$ & $\boldsymbol{a}_\theta$	$\angle(\boldsymbol{a}_\theta\ \&\ \boldsymbol{a}_{xy})\bullet\angle(\boldsymbol{a}_{xy}\ \&\ \boldsymbol{a}_y)=\angle\theta\bullet\angle 90-\varphi$	$\boldsymbol{a}_y\bullet\boldsymbol{a}_\theta=\cos\theta\ \sin\varphi$
$\boldsymbol{a}_y$ & $\boldsymbol{a}_\varphi$	$\angle\varphi$	$\boldsymbol{a}_y\bullet\boldsymbol{a}_\varphi=\cos\varphi$
$\boldsymbol{a}_z$ & $\boldsymbol{a}_r$	$\angle\theta$	$\boldsymbol{a}_z\bullet\boldsymbol{a}_r=\cos\theta$
$\boldsymbol{a}_z$ & $\boldsymbol{a}_\theta$	$\angle 90+\theta$	$\boldsymbol{a}_z\bullet\boldsymbol{a}_\theta=-\sin\theta$
$\boldsymbol{a}_z$ & $\boldsymbol{a}_\varphi$	$\angle 90$	$\boldsymbol{a}_z\bullet\boldsymbol{a}_\varphi=0$

Similarly, A_r, A_θ, and A_φ in Equation 3.34 can be obtained from Equation 3.14 as here:

$$\left.\begin{aligned} A_r &= \boldsymbol{A}\bullet\boldsymbol{a}_r=\left(A_x\boldsymbol{a}_x+A_y\boldsymbol{a}_y+A_z\boldsymbol{a}_z\right)\bullet\boldsymbol{a}_r && \text{(d)}\\ A_\theta &= \boldsymbol{A}\bullet\boldsymbol{a}_\theta=\left(A_x\boldsymbol{a}_x+A_y\boldsymbol{a}_y+A_z\boldsymbol{a}_z\right)\bullet\boldsymbol{a}_\theta && \text{(e)}\\ A_\varphi &= \boldsymbol{A}\bullet\boldsymbol{a}_\varphi=\left(A_x\boldsymbol{a}_x+A_y\boldsymbol{a}_y+A_z\boldsymbol{a}_z\right)\bullet\boldsymbol{a}_\varphi && \text{(f)}\end{aligned}\right\} \tag{3.44}$$

Equation 3.44 yields the products of the following unit vectors

$$\boldsymbol{a}_r\bullet\boldsymbol{a}_x,\ \boldsymbol{a}_\theta\bullet\boldsymbol{a}_x,\ \boldsymbol{a}_\varphi\bullet\boldsymbol{a}_x,\ \boldsymbol{a}_r\bullet\boldsymbol{a}_y,\ \boldsymbol{a}_\theta\bullet\boldsymbol{a}_y,\ \boldsymbol{a}_\varphi\bullet\boldsymbol{a}_y,\ \boldsymbol{a}_r\bullet\boldsymbol{a}_z,\ \boldsymbol{a}_\theta\bullet\boldsymbol{a}_z,\ \boldsymbol{a}_\varphi\bullet\boldsymbol{a}_z$$

The values of these products are given in Table 3.4. If A_r, A_θ, and A_φ are given in terms of r, θ, and φ, A_x, A_y, and A_z can be obtained in terms of x, y, and z by using Table 3.2. The same table can also be used to get A_r, A_θ, and A_φ in terms of r, θ, and φ when A_x, A_y, and A_z are given in terms of x, y, and z.

3.5.4 Problems Involving Complex Geometries

While dealing with actual field problems one may often come across with configurations involving geometrical shapes that do not conform to any of above three coordinate systems. In such cases one may try other coordinate systems listed in Section 3.2.1.2. If none of these is found to be in conformity with the involved shape one may resort to the following options.

3.5.4.1 Division of Configuration into Regions

In the first option, the whole configuration may be divided into suitable segments (or regions) and solutions are independently obtained for each segment. Latter these solutions are matched at the boundaries between adjacent regions. The problems wherein this option is better suited may arise in the configurations which involve either typical geometrical shapes (e.g. electrical machines) or different materials (e.g. dielectric waveguides and step-index optical fibers, etc.).

3.5.4.2 Use of More than One Coordinate System

In this option, more than one coordinate system may be used. Latter the solutions obtained may be transformed to one of the more suitable coordinate systems. The examples for this option may include problems of eddy current in transformer tank, field due to bundle conductors, annular slots in flat sheets, helix antenna, combination of a helix and a horn or of a helix and a corner reflector, etc. None of these configurations can be tackled with a single coordinate system.

Example 3.1

Find $|\mathbf{R}_{12}|$ and $\mathbf{a}_{R12}$ if vector $\mathbf{R}_{12}$ extends from $(x_1 = 1, y_1 = 2, z_1 = 3)$ to $(x_2 = -3, y_2 = 3, z_2 = 1)$.

SOLUTION

In view of Equations 3.13a, c, and e we get:

$$\mathbf{R}_{12} = (-3-1)\,\mathbf{a}_x + (3-2)\,\mathbf{a}_y + (1-3)\,\mathbf{a}_z = -4\mathbf{a}_x + \mathbf{a}_y - 2\mathbf{a}_z$$

$$|\mathbf{R}_{12}| = \sqrt{(-4)^2 + (1)^2 + (-2)^2} = \sqrt{21} = 4.5826$$

$$\mathbf{a}_{R12} = \frac{\mathbf{R}_{12}}{|\mathbf{R}_{12}|} = \frac{-4\mathbf{a}_x + \mathbf{a}_y - 2\mathbf{a}_z}{4.5826} = -0.87287\mathbf{a}_x + 0.2182\mathbf{a}_y \quad 0.4364\mathbf{a}_z$$

Example 3.2

Write $\mathbf{R}_{12}$ in terms of its components if the respective coordinates of point 1 and 2 lie at: $\left(\rho = 1, \varphi = \frac{\pi}{6}, z = 3\right)$ and $\left(\rho = 2, \varphi = \frac{\pi}{3}, z = 4\right)$ in a cylindrical spherical system.

SOLUTION

In view of Equation 3.24:

$$\mathbf{R}_{12} = (2-1)\,\mathbf{a}_\rho + \left(2\times\frac{\pi}{3} - 1\times\frac{\pi}{6}\right)\mathbf{a}_\varphi + (4-3)\,\mathbf{a}_z = \mathbf{a}_\rho + \left(\frac{\pi}{2}\right)\mathbf{a}_\varphi + \mathbf{a}_z$$

Example 3.3

Write $\mathbf{R}_{12}$ in terms of its components if the respective coordinates of point 1 and 2 are at $\left(r_1 = 1, \theta_1 = \frac{\pi}{6}, \varphi_1 = \frac{\pi}{4}\right)$ and $\left(r_2 = 2, \theta_2 = \frac{\pi}{3}, \varphi_2 = \frac{\pi}{3}\right)$ in a spherical system.

SOLUTION

In view of Equation 3.34:

$$\boldsymbol{R}_{12} = (2-1)\mathbf{a}_r + \left(2\times\frac{\pi}{3} - 1\times\frac{\pi}{6}\right)\mathbf{a}_\theta + \left(2\times\frac{\pi}{3}\sin\frac{\pi}{3} - 1\times\frac{\pi}{4}\sin\frac{\pi}{6}\right)\mathbf{a}_\varphi$$

$$= \mathbf{a}_r + \left(\frac{\pi}{2}\right)\mathbf{a}_\theta - \left\{\left(\frac{2\pi}{3}\right)\left(\frac{\sqrt{3}}{2}\right) - \frac{\pi}{4}\frac{1}{2}\right\}\mathbf{a}_\varphi = \mathbf{a}_r + \left(\frac{\pi}{2}\right)\mathbf{a}_\theta + \left\{\frac{\pi}{\sqrt{3}} - \frac{\pi}{8}\right\}\mathbf{a}_\varphi$$

$$= \mathbf{a}_r + (1.571)\mathbf{a}_\theta + (1.8138 - 0.3927)\mathbf{a}_\varphi = \mathbf{a}_r + 1.571\mathbf{a}_\theta + 1.4211\mathbf{a}_\varphi$$

Example 3.4

Find $\boldsymbol{A} + \boldsymbol{B}$ **and** $\boldsymbol{A} - \boldsymbol{B}$ if $\boldsymbol{A} = 5\mathbf{a}_x + 6\mathbf{a}_y - \mathbf{a}_z$ and $\boldsymbol{B} = 3\mathbf{a}_x - 2\mathbf{a}_y + 3\mathbf{a}_z$.

SOLUTION

$$\boldsymbol{A} + \boldsymbol{B} = (5\mathbf{a}_x + 6\mathbf{a}_y - \mathbf{a}_z) + (3\mathbf{a}_x - 2\mathbf{a}_y + 3\mathbf{a}_z) = 8\mathbf{a}_x + 4\mathbf{a}_y + 2\mathbf{a}_z$$

$$\boldsymbol{A} - \boldsymbol{B} = (5\mathbf{a}_x + 6\mathbf{a}_y - \mathbf{a}_z) - (3\mathbf{a}_x - 2\mathbf{a}_y + 3\mathbf{a}_z) = 2\mathbf{a}_x + 8\mathbf{a}_y - 4\mathbf{a}_z$$

Example 3.5

Find scalar product $\boldsymbol{A}\bullet\boldsymbol{B}$ if $\boldsymbol{A} = 3\mathbf{a}_x + 2\mathbf{a}_y - \mathbf{a}_z$ and $\boldsymbol{B} = -2\mathbf{a}_x + 3\mathbf{a}_y + 4\mathbf{a}_z$. Also, find the angle between these vectors.

SOLUTION

On comparing the components of vectors $\boldsymbol{A}$ and $\boldsymbol{B}$ with those given by Equations 3.15a and 3.15b, we get:

$$A_x = 3,\ A_y = 2,\ A_z = -1,\ B_x = -2,\ B_y = 3 \text{ and } B_z = 4$$

Substitution of these components in Equation 3.18a gives:

$$\boldsymbol{A}\bullet\boldsymbol{B} = A_xB_x + A_yB_y + A_zB_z = -6 + 6 - 4 = -4$$

Using Equation 3.9a $(\boldsymbol{A}\bullet\boldsymbol{B} = |\boldsymbol{A}|\,|\boldsymbol{B}|\cos\theta)$, where:

$$|\boldsymbol{A}| = \sqrt{\{(3)^2 + (2)^2 + (-1)^2\}} = \sqrt{14},\quad |\boldsymbol{B}| = \sqrt{\{(-2)^2 + (3)^2 + (4)^2\}} = \sqrt{29}$$

$$\boldsymbol{A}\bullet\boldsymbol{B} = \sqrt{14}\sqrt{29}\cdot\cos\theta = -4 \quad\text{or}\quad \cos\theta = \frac{-4}{\sqrt{14}\sqrt{29}} = -0.1985$$

Thus, $\theta = \cos^{-1}(-0.1985) = 101.45°$.

Example 3.6

Find the vector product ($\boldsymbol{A} \times \boldsymbol{B}$) if $\boldsymbol{A} = 2\boldsymbol{a_x} + 2\boldsymbol{a_y} + 4\boldsymbol{a_z}$ and $\boldsymbol{B} = -2\boldsymbol{a_x} - 3\boldsymbol{a_y} + 3\boldsymbol{a_z}$.

SOLUTION

The components of $\boldsymbol{A}$ and $\boldsymbol{B}$ are: $A_x = 2$, $A_y = 2$, $A_z = 4$, $B_x = -2$, $B_y = -3$, $B_z = 3$.
In view of Equation 3.21a:

$$\boldsymbol{A} \times \boldsymbol{B} = \left(A_yB_z - A_zB_y\right)\boldsymbol{a_x} + \left(A_zB_x - A_xB_z\right)\boldsymbol{a_y} + \left(A_xB_y - A_yB_x\right)\boldsymbol{a_z}$$

$$= \left(2\cdot 3 - 4\cdot(-3)\right)\boldsymbol{a_x} + \left(4\cdot(-2) - 2\cdot 3\right)\boldsymbol{a_y} + \left(2\cdot(-3) - 2\cdot(-2)\right)\boldsymbol{a_z}$$

$$= \left(6 + 12\right)\boldsymbol{a_x} + \left(-8 - 6\right)\boldsymbol{a_y} + \left(-6 + 4\right)\boldsymbol{a_z}$$

$$= 18\boldsymbol{a_x} - 14\boldsymbol{a_y} - 2\boldsymbol{a_z}$$

Example 3.7

Transform vector $\boldsymbol{A} = x\boldsymbol{a_x} + y\boldsymbol{a_y} - z\boldsymbol{a_z}$ from a Cartesian to a cylindrical coordinate system.

SOLUTION

In view of Equation 3.43d to 3.43f:

$$A_\rho = (x\boldsymbol{a_x} + y\boldsymbol{a_y} - z\boldsymbol{a_z})\bullet \boldsymbol{a_\rho} = x\boldsymbol{a_x} \bullet \boldsymbol{a_\rho} + y\boldsymbol{a_y} \bullet \boldsymbol{a_\rho} - z\boldsymbol{a_z} \bullet \boldsymbol{a_\rho}$$

$$A_\varphi = (x\boldsymbol{a_x} + y\boldsymbol{a_y} - z\boldsymbol{a_z})\bullet \boldsymbol{a_\varphi} = x\boldsymbol{a_x} \bullet \boldsymbol{a_\varphi} + y\boldsymbol{a_y} \bullet \boldsymbol{a_\varphi} - z\boldsymbol{a_z} \bullet \boldsymbol{a_\varphi}$$

$$A_z = (x\boldsymbol{a_x} + y\boldsymbol{a_y} - z\boldsymbol{a_z})\bullet \boldsymbol{a_z} = x\boldsymbol{a_x} \bullet \boldsymbol{a_z} + y\boldsymbol{a_y} \bullet \boldsymbol{a_z} - z\boldsymbol{a_z} \bullet \boldsymbol{a_z}$$

From Table 3.3 the above components can be written as:

$$A_\rho = x\cos\varphi + y\sin\varphi, \quad A_\varphi = -x\sin\varphi + y\cos\varphi \text{ and } A_z = -z$$

Substitution of $x = \rho\cos\varphi$, $y = \rho\sin\varphi$ *and* $z = z$ from Table 3.1 gives:

$$A_\rho = \rho\left(\cos^2\varphi + \sin^2\varphi\right) = \rho, \quad A_\varphi = -\rho\cos\varphi\sin\varphi + \rho\sin\varphi\cos\varphi = 0, \quad A_z = -z$$

Thus, in CCS $A = \rho a_\rho - z a_z$.

Example 3.8

Transform vector $\boldsymbol{A} = \rho\boldsymbol{a_\rho} - \rho\boldsymbol{a_\varphi} + z\boldsymbol{a_z}$ from a cylindrical to a rectangular coordinate system.

SOLUTION

In view of Equation 3.43a and 3.43c:

$$A_x = (\rho \mathbf{a}_\rho - \rho \mathbf{a}_\varphi + z\mathbf{a}_z) \bullet \mathbf{a}_x = \rho \mathbf{a}_\rho \bullet \mathbf{a}_x - \rho \mathbf{a}_\varphi \bullet \mathbf{a}_x + z\mathbf{a}_z \bullet \mathbf{a}_x$$

$$A_y = (\rho \mathbf{a}_\rho - \rho \mathbf{a}_\varphi + z\mathbf{a}_z) \bullet \mathbf{a}_y = \rho \mathbf{a}_\rho \bullet \mathbf{a}_y - \rho \mathbf{a}_\varphi \bullet \mathbf{a}_y + z\mathbf{a}_z \bullet \mathbf{a}_y$$

$$A_z = (\rho \mathbf{a}_\rho - \rho \mathbf{a}_\varphi + z\mathbf{a}_z) \bullet \mathbf{a}_z = \rho \mathbf{a}_\rho \bullet \mathbf{a}_z - \rho \mathbf{a}_\varphi \bullet \mathbf{a}_z + z\mathbf{a}_z \bullet \mathbf{a}_z$$

From Table 3.3 the above components can be written as:

$$A_x = \rho\cos\varphi + \rho\sin\varphi \quad A_y = \rho\sin\varphi - \rho\cos\varphi \quad A_z = z$$

Substitution of $\rho = (x^2 + y^2)^{1/2}$, $\varphi = \tan^{-1}\left(\frac{y}{x}\right)$ {from which $\sin\varphi = \frac{y}{\sqrt{x^2 + y^2}}$ and $\cos\varphi = \frac{x}{\sqrt{x^2 + y^2}}$ and $z = z$ from Table 3.1} gives:

$$A_x = \rho\{\cos\varphi + \sin\varphi\} = (x^2 + y^2)^{1/2}\left\{\frac{x}{(x^2+y^2)} + \frac{y}{(x^2+y^2)}\right\}$$

$$= (y + x)(x^2 + y^2)^{-1/2}$$

$$A_y = \rho\{\sin\varphi - \cos\varphi\} = (x^2 + y^2)^{1/2}\left\{\frac{y}{(x^2+y^2)} - \frac{x}{(x^2+y^2)}\right\}$$

$$= (y - x)(x^2 + y^2)^{-1/2}$$

$$A_z = z$$

Thus, in RCS $A = (y + x)(x^2 + y^2)^{-\frac{1}{2}} \mathbf{a}_x + (y - x)(x^2 + y^2)^{-\frac{1}{2}} \mathbf{a}_y + z\mathbf{a}_z$.

Example 3.9

Transform vector $\mathbf{A} = x\mathbf{a}_x - y\mathbf{a}_y + z\mathbf{a}_z$ from a Cartesian to a spherical coordinate system.

SOLUTION

In view of Equation 3.44d to 3.44f:

$$A_r = (x\mathbf{a}_x - y\mathbf{a}_y + z\mathbf{a}_z) \bullet \mathbf{a}_r = (x\mathbf{a}_x \bullet \mathbf{a}_r - y\mathbf{a}_y \bullet \mathbf{a}_r + z\mathbf{a}_z \bullet \mathbf{a}_r)$$

$$A_\theta = (x\mathbf{a}_x - y\mathbf{a}_y + z\mathbf{a}_z) \bullet \mathbf{a}_\theta = (x\mathbf{a}_x \bullet \mathbf{a}_\theta - y\mathbf{a}_y \bullet \mathbf{a}_\theta + z\mathbf{a}_z \bullet \mathbf{a}_\theta)$$

$$A_\varphi = (x\mathbf{a}_x - y\mathbf{a}_y + z\mathbf{a}_z) \bullet \mathbf{a}_\varphi = (x\mathbf{a}_x \bullet \mathbf{a}_\varphi - y\mathbf{a}_y \bullet \mathbf{a}_\varphi + z\mathbf{a}_z \bullet \mathbf{a}_\varphi)$$

On substitution of involved dot products from Table 3.4, we get:

$$A_r = x\sin\theta\cos\varphi - y\sin\theta\sin\varphi + z\cos\theta$$

$$A_\theta = x\cos\theta\cos\varphi - y\cos\theta\sin\varphi - z\sin\theta$$

$$A_\varphi = -x\sin\theta - y\cos\varphi$$

Substituting from Table 3.2:

$$x = r\cdot\sin\theta\cdot\cos\varphi,\ y = r\cdot\sin\theta\cdot\sin\varphi \text{ and } z = r\cdot\cos\theta$$

$$A_r = r\cdot\sin\theta\cdot\cos\varphi\sin\theta\cos\varphi - r\cdot\sin\theta\cdot\sin\varphi\sin\theta\sin\varphi + r\cdot\cos\theta\cos\theta$$

$$A_\theta = r\cdot\sin\theta\cdot\cos\varphi\cos\theta\cos\varphi - r\cdot\sin\theta\cdot\sin\varphi\cos\theta\sin\varphi - r\cdot\cos\theta\sin\theta$$

$$A_\varphi = -r\cdot\sin\theta\cdot\cos\varphi\sin\theta - r\cdot\sin\theta\cdot\sin\varphi\cos\varphi$$

$$A_r = r\left(\sin^2\theta\cos^2\varphi - \sin^2\theta\sin^2\varphi + \cos^2\theta\right)$$

$$A_\theta = r\left(\sin\theta\cos\theta\cos^2\varphi - \sin\theta\cos\theta\sin^2\varphi - \cos\theta\sin\theta\right)$$

$$A_\varphi = -r\left(\sin^2\theta\cos\varphi + \sin\theta\cdot\sin\varphi\cos\varphi\right)$$

Thus:

$$\begin{aligned}\mathbf{A} = r\Big[&\left(\sin^2\theta\cos^2\varphi - \sin^2\theta\sin^2\varphi + \cos^2\theta\right)\mathbf{a}_r \\ &+\left(\sin\theta\cos\theta\cos^2\varphi - \sin\theta\cos\theta\sin^2\varphi - \cos\theta\sin\theta\right)\mathbf{a}_\theta \\ &-\left(\sin^2\theta\cos\varphi + \sin\theta\cdot\sin\varphi\cos\varphi\right)\mathbf{a}_\varphi\Big]\end{aligned}$$

Example 3.10

Transform vector $\boldsymbol{A} = r\mathbf{a}_r + r\mathbf{a}_\theta - r\mathbf{a}_\varphi$ from a spherical to a rectangular coordinate system.

SOLUTION

In view of Equation 3.44a to 3.44c:

$$A_x = (r\mathbf{a}_r + r\mathbf{a}_\theta - r\mathbf{a}_\varphi)\bullet\mathbf{a}_x = r\mathbf{a}_r\bullet\mathbf{a}_x + r\mathbf{a}_\theta\bullet\mathbf{a}_x - r\mathbf{a}_\varphi\bullet\mathbf{a}_x$$

$$A_y = (r\mathbf{a}_r + r\mathbf{a}_\theta - r\mathbf{a}_\varphi)\bullet\mathbf{a}_y = r\mathbf{a}_r\bullet\mathbf{a}_y + r\mathbf{a}_\theta\bullet\mathbf{a}_y - r\mathbf{a}_\varphi\bullet\mathbf{a}_y$$

$$A_z = (r\mathbf{a}_r + r\mathbf{a}_\theta - r\mathbf{a}_\varphi)\bullet\mathbf{a}_z = r\mathbf{a}_r\bullet\mathbf{a}_z + r\mathbf{a}_\theta\bullet\mathbf{a}_z - r\mathbf{a}_\varphi\bullet\mathbf{a}_z$$

From Table 3.4 we get:

$$A_x = r\sin\theta\cos\varphi + r\cos\theta\cos\varphi + r\sin\theta$$

$$A_y = r\sin\theta\sin\varphi + r\cos\theta\sin\varphi - r\cos\varphi$$

$$A_z = r\cos\theta - r\sin\theta$$

$$A_x = r\left(\sin\theta\cos\varphi + \cos\theta\cos\varphi + \sin\varphi\right)$$

$$A_y = r\left(\sin\theta\sin\varphi + \cos\theta\sin\varphi - \cos\varphi\right)$$

$$A_z = r\left(\cos\theta - \sin\theta\right)$$

In Table 3.2:

$r = \sqrt{x^2 + y^2 + z^2}$, $\cos\theta = \dfrac{z}{\sqrt{x^2 + y^2 + z^2}}$, $\sin\theta = \dfrac{\sqrt{x^2 + y^2}}{\sqrt{x^2 + y^2 + z^2}}$, $\cos\varphi = \dfrac{x}{\sqrt{x^2 + y^2}}$ and $\sin\varphi = \dfrac{y}{\sqrt{x^2 + y^2}}$

Values of A_x, A_y, and A_z can be obtained in terms of the Cartesian coordinates (x, y, and z).

PROBLEMS

P3.1 Write $\boldsymbol{R}_{12}$, $|\boldsymbol{R}_{12}|$ and $\boldsymbol{a}_{R12}$ for the vector that extends from $\left(x_1 = 2, y_1 = 2, z_1 = 4\right)$ to $\left(x_2 = -1, y_2 = 3, z_2 = 1\right)$.

P3.2 Write $\boldsymbol{R}_{12}$, $|\boldsymbol{R}_{12}|$ and $\boldsymbol{a}_{R12}$ for the vector that extends from $\left(\rho_1 = 2, \varphi_1 = \dfrac{\pi}{4}, z_1 = 2\right)$ to $\left(\rho_2 = 1, \varphi_2 = \dfrac{\pi}{3}, z_2 = 4\right)$.

P3.3 Write $\boldsymbol{R}_{21}$, $|\boldsymbol{R}_{21}|$ and $\boldsymbol{a}_{R21}$ for the vector that extends from $\left(r_2 = 3, \theta_2 = \dfrac{\pi}{3}, \varphi_2 = \dfrac{\pi}{3}\right)$ to $\left(r_1 = 2, \theta_1 = \dfrac{\pi}{6}, \varphi_1 = \pi/4\right)$.

P3.4 Find $\boldsymbol{B} + \boldsymbol{A}$ and $\boldsymbol{B} - \boldsymbol{A}$ if $\boldsymbol{A} = 2\boldsymbol{a}_x + 3\boldsymbol{a}_y - 7\boldsymbol{a}_z$ and $\boldsymbol{B} = 3\boldsymbol{a}_x - 2\boldsymbol{a}_y + 3\boldsymbol{a}_z$.

P3.5 Find scalar product $\boldsymbol{B} \bullet \boldsymbol{A}$ if $\boldsymbol{A} = 5\boldsymbol{a}_x + 2\boldsymbol{a}_y - 2\boldsymbol{a}_z$ and $\boldsymbol{B} = -3\boldsymbol{a}_x + \boldsymbol{a}_y + 3\boldsymbol{a}_z$. Also find the angle between these vectors.

P3.6 Find the cross-products ($\boldsymbol{B} \times \boldsymbol{A}$) if $\boldsymbol{A} = 5\boldsymbol{a}_x + 3\boldsymbol{a}_y - 4\boldsymbol{a}_z$ and $\boldsymbol{B} = -3\boldsymbol{a}_x - 3\boldsymbol{a}_y + 4\boldsymbol{a}_z$.

P3.7 Transform vector $\boldsymbol{A} = 3\boldsymbol{a}_x + 5\boldsymbol{a}_y - 2\boldsymbol{a}_z$ from a Cartesian to a cylindrical coordinate system.

P3.8 Transform vector $\boldsymbol{A} = 5\boldsymbol{a}_\rho - 2\boldsymbol{a}_\varphi + 3\boldsymbol{a}_z$ from a cylindrical to a Cartesian coordinate system.

P3.9 Transform vector $\boldsymbol{A} = 2\boldsymbol{a}_x - 3\boldsymbol{a}_y - 2\boldsymbol{a}_z$ from a Cartesian to a spherical coordinate system.

P3.10 Transform vector $\boldsymbol{A} = 2\boldsymbol{a}_r - 3\boldsymbol{a}_\theta + 2\boldsymbol{a}_\varphi$ from a spherical to a Cartesian coordinate system.

DESCRIPTIVE QUESTIONS

Q3.1 Illustrate the incremental volumes in Cartesian, cylindrical, and spherical coordinate systems. Write their mathematical relations in terms of the related parameters.

Q3.2 Illustrate different elemental surfaces in Cartesian, cylindrical, and spherical coordinate systems. Give their mathematical relations in terms of related parameters.

Q3.3 Illustrate: (a) constant ρ and constant φ surfaces in cylindrical coordinates and (b) constant r, constant φ and constant θ surfaces in spherical coordinates.

Q3.4 Illustrate the addition and subtraction of two vectors $\boldsymbol{A}\angle 30°$ and $\boldsymbol{B}\angle 60°$.

Note 1: Summary of commonly used relations is given in Appendix A3 in the eResources.

Note 2: References for further reading are given at the end of Chapter 4.

4 Vector Calculus

4.1 INTRODUCTION

In Chapter 3, the commonly used coordinate systems and basics of vector algebra were discussed briefly. This chapter deals with another mathematical tool, referred to as vector calculus. It includes the physical description, and mathematical formulation of gradient, divergence, and curl operations. It also describes the vector operator Del, scalar operator Laplacian operating on "scalar", and vector functions for different coordinate systems. Lastly, the line and surface integrals and some commonly used theorems (viz. Divergence, Stokes', and Green's theorems) involved in vector calculus have also been briefly described. The following text describes all the essential aspects of vector calculus required for the study of electromagnetics.

4.2 GRADIENT OPERATION

For any continuously differentiable scalar function V, the vector function "*grad V*" is the gradient of the scalar function V:

$$\text{grad}\, V \stackrel{\text{def}}{=} \nabla V \tag{4.1}$$

In a Cartesian system of coordinates, the gradient of the scalar function V can be given as:

$$\nabla V = \frac{\partial V}{\partial x} \cdot \boldsymbol{a}_x + \frac{\partial V}{\partial y} \cdot \boldsymbol{a}_y + \frac{\partial V}{\partial z} \cdot \boldsymbol{a}_z \tag{4.1a}$$

where the symbol ∇ is the vector differential operator called *Del*. In the Cartesian coordinate system, it is given as follows:

$$\nabla = \boldsymbol{a}_x \cdot \frac{\partial}{\partial x} + \boldsymbol{a}_y \cdot \frac{\partial}{\partial y} + \boldsymbol{a}_z \cdot \frac{\partial}{\partial z} \tag{4.1b}$$

4.2.1 Physical Description

The gradient at a point is a fancy word for slope, or the rate of change of a scalar function with respect to the space coordinates at that point. The gradient of a scalar function is a vector that points in the direction of the greatest rate of increase of the function. The magnitude of this vector gives the value for the greatest rate of increase of the function. The magnitude is zero at a local maximum or a local minimum. The term gradient (or grad) typically refers to the (first) derivative of a scalar function of one or more space variables, with appropriate unit vectors inserted, as given in Equation 4.1a. The gradient of a constant is a null vector.

An ordinary derivative of a function of a single variable gives the rate of change of the function with the variable. For example, the derivative dF/dx of a scalar function F of the variable x tells how much the function F at a given value of x, and changes for a small change in x. It is the slope of the curve F plotted against x. If this scalar function F has multiple variables (say α, β, γ), it will have partial derivatives $\left(\partial F/\partial\alpha, \partial F/\partial\beta, \partial F/\partial\gamma\right)$, with suitable multiplying factors indicating slopes in different directions in the functional space. Each slope indicates the scalar component of a vector, called the gradient of the scalar function F. Thus, in an n-dimensional functional space, a function with n variables will have a gradient with n components. $F(x)$ has one variable and an ordinary derivative; while $F(x,y,z)$ has three variables and three partial derivatives. The gradient of a multivariable function has a component for each direction. The gradient of a scalar function at any given point indicates the magnitude and the direction of the greatest rate of increase in the value of the function at this point.

In vector calculus, gradient is defined only for scalar functions. The gradient of a scalar function is a vector function. As the gradient has a derivative for each variable of a scalar function, in the Cartesian coordinate it can be written as:

$$\textbf{grad}\, F(x,y,z) \stackrel{\text{def}}{=} \nabla F(x,y,z) = \frac{\partial F}{\partial x}\boldsymbol{a}_x + \frac{\partial F}{\partial y}\boldsymbol{a}_y + \frac{\partial F}{\partial z}\boldsymbol{a}_z \tag{4.2}$$

The coefficients of the three unit vectors at a given point indicate the rate of change of the value of the scalar function in the respective directions. These directions being mutually perpendicular, the resulting variation is given as:

$$|\nabla F| = \sqrt{\left(\frac{\partial F}{\partial x}\right)^2 + \left(\frac{\partial F}{\partial y}\right)^2 + \left(\frac{\partial F}{\partial z}\right)^2} \tag{4.2a}$$

Thus, the magnitude of the vector gives the maximum variation of the function at the given point. The direction of this maximum variation is the direction of this vector. This is indicated by the unit vector $\boldsymbol{a}_{\max}$ as shown here:

$$\boldsymbol{a}_{\max} = \frac{\nabla F}{|\nabla F|} \tag{4.2b}$$

The gradient of a scalar function may also be written as:

$$\textbf{grad}\, F(x,y,z) \stackrel{\text{def}}{=} \left(\frac{\partial F}{\partial x}, \frac{\partial F}{\partial y}, \frac{\partial F}{\partial z}\right) \tag{4.3}$$

In this discussion, the gradient comes out to be the generalization of the usual concept of derivative to the function of several variables. A differentiable function of several variables $\{f(x_1, \ldots, x_n)\}$ may be referred to as n-dimensional scalar field. Its gradient involving first order n partial derivatives of such a field is a vector and the resulting vector function is referred to as vector field. Each partial derivative gives its slop in the respective direction. The magnitude and the direction of the gradient vector at any point give the magnitude and the direction of the greatest rate of increase of the scalar function. The components of the gradient are (usually) the nonconstant coefficients of the equation of the tangent space to the graph.

Many examples can be cited to enable proper understanding of the gradient. Some of these are presented in the following paragraphs.

Consider a room with space variation of temperature. This temperature is assumed to be time-invariant and is given by a scalar field T. The value of this scalar field at a space location (x,y,z) in the room is given by $T(x,y,z)$. The gradient of T at any point in the room will show the magnitude and the direction of the maximum rate of temperature rise at that point.

Consider a surface located above sea level. Its height at a point (x,y) is given as $h(x,y)$. The gradient of h at a point is a vector pointing in the direction of the steepest slope at that point. The steepness of the slope at that point is given by the magnitude of the gradient vector.

The gradient is not only meant just for the direction of greatest change it can also be used to measure how a scalar field changes in other directions by taking a dot product with the unit vector in that direction. Suppose that the steepest slope on a hill is 30%. If a road goes directly up the hill, then the steepest slope on the road will also be 30%. The road runs around the hill at an angle, then its slope will depend on this angle. If the angle between the road and the uphill direction, projected onto the horizontal plane, is 60°, then the steepest slope along the road will be 15%, which is the product of 30% and cos60°. In mathematical terms if the hill height-function h is differentiable, then the scalar product of the gradient of h with a unit vector gives the slope of the hill in the direction of the vector. More precisely, when h is differentiable, the dot product of gradient of h with a given unit vector is equal to the directional derivative of h in the direction of that unit vector.

4.2.2 Mathematical Formulation

Imagine a physical configuration (Figure 4.1a) wherein a large tank is provided with two distantly located holes marked as A and B. Each of these holes is connected to a pipe to fill and drain some incompressible fluid. If the diameters of these holes

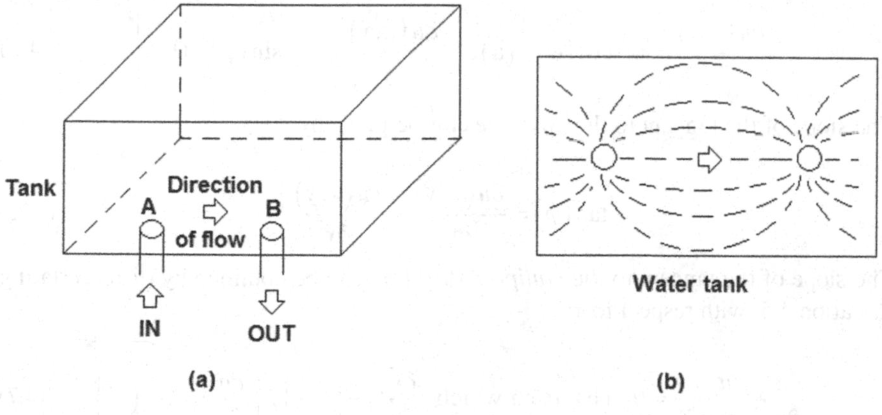

FIGURE 4.1 (a) A water tank with in and out pipes (b) orthogonal lines.

are much smaller in comparison to the dimensions of tank, holes *A* and *B* can be regarded as points. When fluid is fed in the tank from hole *A* it will flow in all the possible directions inside the tank. In hole *B* the fluid will exit in all the possible directions. Thus, *A* may be referred to as the source point and *B* as the sink point. Take a hypothetical case wherein this flowing fluid assumes some discrete paths to travel from *A* to *B* (inside the tank). These paths may be represented by lines drawn between points *A* and *B*. *A* set of such (dotted) lines is shown in Figure 4.1b. This figure also shows another set of (firm) lines. The lines (also referred to as trajectories) of these two sets are orthogonal to each other.

To understand the meaning of the orthogonal trajectories let us assume that the velocity vector $\upsilon(x,y)$ for the flow of fluid is defined at each point of the x–y-plane. If this vector depends solely on the position of the point in the plane, but independent of the time, the motion is called stationary or steady-state. Let us further assume that the motion of fluid under consideration falls under this category. It can also be assumed that there exists a potential of velocities that is a function $\boldsymbol{u}(x,y)$, such that the projections of vector $\upsilon(x,y)$ on the coordinate axes, $\upsilon_x(x,y)$ and $\upsilon_y(x,y)$, are its partial derivatives with respect to x and y. That is:

$$\left.\nabla u = \upsilon \quad (a), \qquad \frac{\partial u}{\partial x} = \upsilon_x \quad (b), \qquad \frac{\partial u}{\partial y} = \upsilon_y \quad (c)\right\} \tag{4.4}$$

The lines satisfying the following equation are called *equipotential lines* or *lines of equipotential* if these satisfy the relation:

$$u(x,y) = C \tag{4.5}$$

where, C is a constant, indicating the value of the potential.

The lines, whose tangents at all points coincide with the vector $\upsilon(x,y)$, are called *flow lines* and they yield the trajectories of moving points.

To show that the flow lines are the orthogonal trajectories of a family of equipotential lines let φ be an angle formed by the velocity vector υ with the x-axis. Then in Equation 4.5:

$$\left.\frac{\partial u(x,y)}{\partial x} = |\upsilon| \cdot \cos(\varphi) \quad (a), \qquad \frac{\partial u(x,y)}{\partial y} = |\upsilon| \cdot \sin(\varphi) \quad (b),\right\} \tag{4.6}$$

The slope of the tangent to the *flow line* can be given by:

$$\tan(\varphi) = \frac{\partial u(x,y)}{\partial y} \Big/ \frac{\partial u(x,y)}{\partial x} \tag{4.7a}$$

The slope of the tangent to the *equipotential line* can be obtained by differentiating Equation 4.5, with respect to x:

$$\left.\frac{\partial u}{\partial x} + \frac{\partial u}{\partial y} \cdot \frac{\partial y}{\partial x} = 0 \quad (b) \text{ from which } \frac{\partial y}{\partial x} = -\left(\frac{\partial u}{\partial x}\right) \Big/ \left(\frac{\partial u}{\partial y}\right) = 0 \quad (c)\right\} \tag{4.7}$$

Thus in the magnitude and sign, the slope of the tangent to the equipotential line is the inverse of the slope to the flow line. This illustrates that the *equipotential* and the *flow lines* are mutually orthogonal.

The analogy of fluid flow can be extended to electric and magnetic fields. In both of these cases, the line of force of field serves as orthogonal trajectories of the family of equipotential lines. In order to correlate the concept of the gradient with the electric field, consider Figure 4.2 which illustrates two points *A* and *B* whereat two equal and opposite point charges are located. This also shows the ***E*** lines emerging from the positive charge at *A* and sinking into the negative charge at *B*. These ***E*** lines shown by the broken lines are the same which earlier referred to as flow lines. Another set of firm orthogonal lines represents the *equipotential lines* in the region of ***E*** field. These too are the same which earlier represented the equipotential lines in case of fluid flow.

It needs to be noted that this emergence and sinking of ***E*** lines is a three-dimensional phenomenon whereas ***E*** lines shown in the figure represent a two-dimensional view only. The *equipotential lines* shown in Figure 4.2 are in fact surfaces which may be referred to as *equipotential surfaces* (see Chapter 7). The equipotential surfaces can be defined as surfaces on which each point has the same electrostatic potential. For a three-dimensional view the figure is to be rotated about an axis passing through the points *A* and *B*. The two sets of orthogonal surfaces of revolution can be identified as a family of equipotential surfaces and a family of ***E*** lines. The former is a set of open surfaces, each with infinite area, and the latter is a set of closed surfaces, each with finite area. Thus Figure 4.2 can be treated as the cross-sectional view cut across a plane passing through the axis of revolution. The component of the electric field parallel to any of these equipotential surfaces must be zero since the change in the potential between all points on this surface is zero. This implies that the direction of the electric field is perpendicular to the equipotential surfaces.

In Figure 4.2 the incremental potential difference ΔV between two points (*C* and *D*) located in the region of vector field ***E*** and separated by an incremental vector distance $\Delta \boldsymbol{L}$ can be written in terms of dot product as $\Delta V = -\boldsymbol{E} \cdot \Delta \boldsymbol{L}$. The length

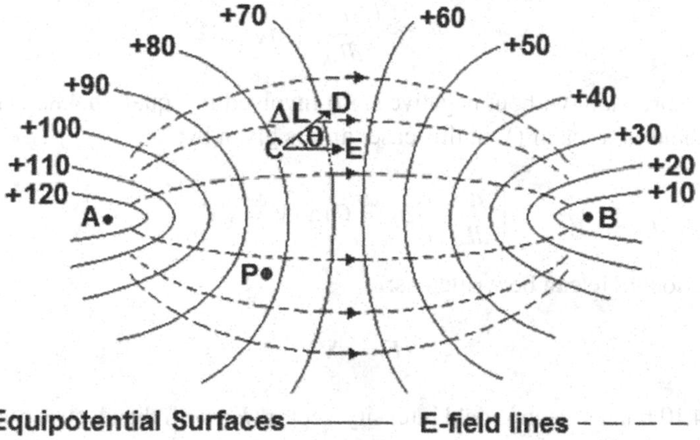

FIGURE 4.2 Flow lines (or E field) and equipotentials.

$\boldsymbol{\Delta L}$ shown in Figure 4.2 is to be assigned a direction to make it a vector quantity. If points C and D are separated by distance $\boldsymbol{L}$ the potential difference V can be evaluated by integrating $\boldsymbol{E}$ over this length $\boldsymbol{L}$. For integration $\boldsymbol{\Delta L}$ can be replaced by $\boldsymbol{d\ell}$. The resulting equation is:

$$V = -\int_0^L \boldsymbol{E} \bullet \boldsymbol{d\ell} \tag{4.8a}$$

If θ is the angle between the orientations of $\boldsymbol{E}$ field and $\boldsymbol{\Delta L}$, the incremental potential difference ΔV can be written as:

$$\Delta V = -\boldsymbol{E} \bullet \boldsymbol{\Delta L} = -E \cdot \Delta L \cdot \cos\theta \tag{4.8b}$$

This equation can be rearranged in the following form:

$$\lim_{\Delta L \to 0} \frac{\Delta V}{\Delta L} = \frac{dV}{dL} = -E \cdot \cos\theta \tag{4.8c}$$

The ratio dV/dL is maximum for $\theta = 0$ and can be written as:

$$\left.\frac{dV}{dL}\right|_{\max} = -E \tag{4.8d}$$

In Equation 4.8d, the term on the LHS represents the maximum rate of change of potential with the displacement. Since $\boldsymbol{E}$ at any point lying on an equipotential surface is a vector quantity it needs to be assigned a direction. This can be done by rearranging this equation and by replacing the incremental length dL by an incremental length dL_N directed normal to the equipotential surface where the field point is lying, and by multiplying the resulting expression by a unit vector $\boldsymbol{a}_N$ which is along the incremental length dL_N ($\boldsymbol{a}_N = \boldsymbol{dL}_N/dL_N$). This unit vector $\boldsymbol{a}_N$ is chosen along the direction of increasing V. Thus:

$$\boldsymbol{E} = -\left(\frac{dV}{dL_N}\right) \cdot a_N \tag{4.8e}$$

The RHS expression (without negative sign) involved in Equation 4.8e is referred to as the gradient of a scalar (V in this case) and is given by:

$$\left(\frac{dV}{dL_N}\right) \cdot \boldsymbol{a}_N \stackrel{\text{def}}{=} \text{Grad}\, V \stackrel{\text{def}}{=} \nabla V. \tag{4.9}$$

Thus Equation 4.11e can be written as:

$$\boldsymbol{E} = -\nabla V \tag{4.10}$$

Equation 4.10 relates electric field intensity vector ($\boldsymbol{E}$), with the electric potential (V). The magnitude of this vector at a point is equal to the maximum rate of change of the scalar with space variables and its direction lies along that change.

4.2.3 Information Imparted by the Gradient Relation

Equation 4.10 indicates that the electric field intensity at any point is the negative of potential gradient at that point and the direction of electric field intensity is the direction of its gradient. This equation also indicates that the field that is expressible as the gradient of scalar potential is conservative in the sense that the work done in moving a point charge with constant velocity straight from point "A" to point "B" depends exclusively upon the difference of potential between the two points and if the energy radiation is neglected it is independent of the path. Further since the total work done in moving around a closed path is equal to zero the field is not turbulent. As pointed out earlier, this is essentially an approximation since a point charge cannot be moved around a closed path without acceleration and consequential power radiation. Electrostatic lines of flow (also known as lines of force or flux lines) are directed lines with arrowheads. These lines do not close upon themselves. They originate from a positive point charge known as *source* and terminate in a negative point charge known as *sink*. The potential field is produced by a source; in the case the field is present in a source-free region, it must be due to the source present elsewhere outside the region within finite distances. In general, if divergence is taken of potential gradient, it may yield a nonzero value. The divergence of potential gradient in a source-free homogeneous region however is zero. In a later chapter on time-varying electromagnetic fields, we shall come across electric lines of force closing upon themselves. These lines neither originate from a positive point charge nor terminate in a negative point charge.

4.2.4 Gradient Expressions in Different Coordinate Systems

Since V may be a function of more than one coordinate $\boldsymbol{a}_N$ (in Equation 4.8e) may be replaced by $\boldsymbol{a}_x, \boldsymbol{a}_y, \boldsymbol{a}_z$ in a Cartesian system, $\boldsymbol{a}_\rho, \boldsymbol{a}_\varphi, \boldsymbol{a}_z$ in a cylindrical system and $\boldsymbol{a}_r, \boldsymbol{a}_\theta, \boldsymbol{a}_\varphi$ in a spherical system. The resulting expressions of gradient in these systems are:

$$\left.\begin{aligned}
\nabla V &= \frac{\partial V}{\partial x}\cdot\boldsymbol{a}_x + \frac{\partial V}{\partial y}\cdot\boldsymbol{a}_y + \frac{\partial V}{\partial z}\cdot\boldsymbol{a}_z && \text{(Cartesian)} \quad \text{(a)}\\
\nabla V &= \frac{\partial V}{\partial \rho}\cdot\boldsymbol{a}_\rho + \left(\frac{1}{\rho}\right)\frac{\partial V}{\partial \varphi}\cdot\boldsymbol{a}_\varphi + \frac{\partial V}{\partial z}\cdot\boldsymbol{a}_z && \text{(Cylindrical)} \quad \text{(b)}\\
\nabla V &= \frac{\partial V}{\partial r}\cdot\boldsymbol{a}_r + \left(\frac{1}{r}\right)\cdot\frac{\partial V}{\partial \theta}\cdot\boldsymbol{a}_\theta + \left(\frac{1}{r\cdot\sin\theta}\right)\cdot\frac{\partial V}{\partial \varphi}\cdot\boldsymbol{a}_\varphi && \text{(Spherical)} \quad \text{(c)}
\end{aligned}\right\} \quad (4.11)$$

These equations may also be rewritten as follows:

$$\left.\begin{aligned}
\nabla V &= \boldsymbol{a}_x\cdot\frac{\partial V}{\partial x} + \boldsymbol{a}_y\cdot\frac{\partial V}{\partial y} + \boldsymbol{a}_Z\cdot\frac{\partial V}{\partial z} && \text{(Cartesian)} \quad \text{(a)}\\
\nabla V &= \boldsymbol{a}_\rho\cdot\frac{\partial V}{\partial \rho} + \boldsymbol{a}_\varphi\cdot\left(\frac{1}{\rho}\right)\frac{\partial V}{\partial \varphi} + \boldsymbol{a}_Z\cdot\frac{\partial V}{\partial z} && \text{(Cylindrical)} \quad \text{(b)}\\
\nabla V &= \boldsymbol{a}_r\cdot\frac{\partial V}{\partial r} + \boldsymbol{a}_\theta\cdot\left(\frac{1}{r}\right)\cdot\frac{\partial V}{\partial \theta} + \boldsymbol{a}_\varphi\cdot\left(\frac{1}{r\cdot\sin\theta}\right)\cdot\frac{\partial V}{\partial \varphi} && \text{(Spherical)} \quad \text{(c)}
\end{aligned}\right\} \quad (4.12)$$

4.2.5 Vector Differential Operator Del

In these equations, the expressions for the Del operator for different coordinate systems are as follows:

$$\left.\begin{aligned}
&\nabla \equiv \boldsymbol{a}_x \cdot \frac{\partial}{\partial x} + \boldsymbol{a}_y \cdot \frac{\partial}{\partial y} + \boldsymbol{a}_z \cdot \frac{\partial}{\partial z} && \text{(Cartesian)} \quad \text{(a)} \\
&\nabla \equiv \boldsymbol{a}_\rho \cdot \frac{\partial}{\partial \rho} + \boldsymbol{a}_\varphi \cdot \left(\frac{1}{\rho}\right) \cdot \frac{\partial}{\partial \varphi} + \boldsymbol{a}_z \cdot \frac{\partial}{\partial z} && \text{(Cylindrical)} \quad \text{(b)} \\
&\nabla \equiv \boldsymbol{a}_r \cdot \frac{\partial}{\partial r} + \boldsymbol{a}_\theta \cdot \left(\frac{1}{r}\right) \cdot \frac{\partial}{\partial \theta} + \boldsymbol{a}_\varphi \cdot \left(\frac{1}{r \cdot \sin\theta}\right) \cdot \frac{\partial}{\partial \varphi} && \text{(Spherical)} \quad \text{(c)}
\end{aligned}\right\} \quad (4.13)$$

4.2.6 Applications

The gradient operation has applications in many fields of science and engineering including in the study of fluid flow, heat flow, and that of electromagnetics. Its application greatly simplifies the problems of electromagnetics.

4.3 DIVERGENCE OPERATION

The dot product of vector operator ∇ and any vector (say $\boldsymbol{F}$) gives a scalar function and is called the divergence of that vector, provided that the vector operator ∇ operates on the vector $\boldsymbol{F}$. It is written as $\nabla \bullet \boldsymbol{F}$ or div$\boldsymbol{F}$. In Equation 3.9b, it was noted that dot product obeys the commutative law (i.e. $\boldsymbol{A} \bullet \boldsymbol{B} = \boldsymbol{B} \bullet \boldsymbol{A}$). This however is not true here (i.e. $\boldsymbol{F} \bullet \nabla \neq \nabla \bullet \boldsymbol{F}$) as ∇ is not a simple vector but a differential vector operator. Note that $\boldsymbol{F} \bullet \nabla$ is a scalar differential operator, while $\nabla \bullet \boldsymbol{F}$ is a scalar mathematical expression.

4.3.1 Physical Description

To further elaborate the meaning of divergence operation, consider a vector field $\boldsymbol{F}$ depicting the flow of a fluid. A two dimensional the velocity of some fluid flow is shown in Figure 4.3. Figure 4.3a shows the fluid exploding outward from the origin whereas Figure 4.3b shows the imploding (contracting) fluid to the origin. The expansion (or contraction) of fluid flow with velocity field $\boldsymbol{F}$ can be accounted by the divergence (or div) of $\boldsymbol{F}$. The divergence of the vector field is taken to be positive for the expanding or outward flow and negative for contracting or inward flow.

Consider that an imaginary spherical region centered at the origin is submerged in some fluid. In case of expansion the fluid will flow out of the sphere in all the possible directions. This flow has positive divergence everywhere inside the sphere. Since the above vector field has positive divergence everywhere, the flow out of the sphere will be positive even if we move the sphere away from the origin (i.e. location 1 to location 2) as shown in Figure 4.4a. If the vector field represents the velocity of fluid flow, then irrespective of the position of the sphere, more fluid is flowing out of the sphere than entering into the sphere. This is so even if the sphere is made smaller and smaller till it is reduced to a point. This suggests that the vector field has positive divergence everywhere. This figure also shows that the arrows continue to get longer

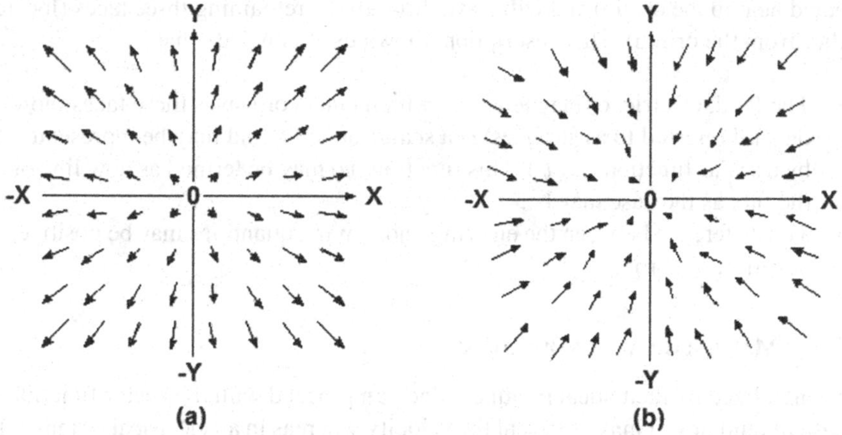

FIGURE 4.3 Fluid flow (a) expanding, (b) contracting.

as one moves away from the origin so that the fluid is flowing faster when it leaves the sphere than when it enters. Moreover, since the arrows are radiating outward, the fluid is always entering the sphere over less than half its surface and is leaving the sphere over greater than half of its surface. Hence, the flow out of the sphere is always greater than the flow into the sphere. The divergence of a vector field simply measures how much the flow is expanding at a given point. It does not indicate the direction of expansion. Thus the divergence is a scalar.

Figure 4.4b illustrates that without affecting the nature of divergence the infinitesimal sphere of Figure 4.4a can conveniently be replaced by an infinitesimal cube. This cube with its six tiny sides (or infinitesimally small faces) allows a deeper look into the flow process. It can now easily be visualized that the fluid entering through a particular face will leave through the face which is located opposite to it. If this flow is taken to be a three-dimensional phenomenon and its direction is assumed to be along the positive coordinates then the fluid will enter through three of the faces

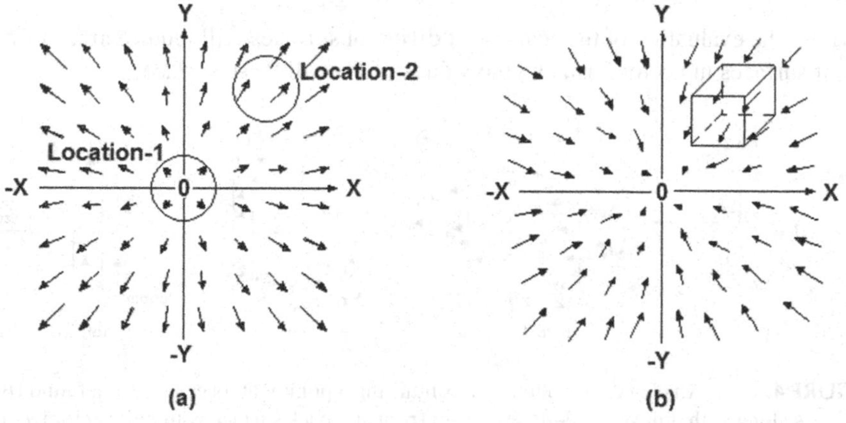

FIGURE 4.4 (a) Movement of sphere, (b) location of cube.

(located near to the origin) and will leave through the remaining three faces (located farther from the origin). This description allows us to conclude that:

- The fluid, electric, or magnetic flux which enters or leaves these faces (now onward referred to as surfaces) is a scalar quantity and may be represented by a scalar function (say ψ). This function (ψ) may be termed as flow line or the flux as the case may be.
- The difference between the entering and leaving quantities may be positive, negative, or zero.

4.3.2 Mathematical Formulation

For generalized mathematical treatment one can proceed with the vector function $\boldsymbol{F}$. In case of fluid flow it may represent the velocity whereas in an electrical or magnetic field it may represent the electric or magnetic flux densities.

4.3.2.1 Field Configuration

Consider Figure 4.5a, which shows an elemental volume Δv (=$\Delta x \Delta y \Delta z$) containing a point P at its center (located at x_o, y_o and z_o). This volume is enclosed by six elemental surfaces, which can be represented as vector surfaces by assigning unit vectors as shown in Figures 4.5b, c, and d. Assume that this volume is located in some region containing a vector field $\boldsymbol{F}$. It can safely be assumed that at the location of this elemental volume, $\boldsymbol{F}$ is changing in the increasing order along the positive directions of three (Cartesian) coordinates x, y, and z. Thus the value of $\boldsymbol{F}$ could be different at all its surfaces. If this vector field $\boldsymbol{F}$ is multiplied by the areas of vector surface through the dot product the resultant will be a scalar quantity say ψ. This scalar quantity ψ may be termed as flux of the vector $\boldsymbol{F}$ to represent the quantity of flow of a fluid, gas, or that of electric flux. To calculate the values of ψ at different elemental surfaces let $\boldsymbol{F}_o$ be the value of the vector $\boldsymbol{F}$ at a point P (which is located in the center of the elemental volume Δv (or dv) with coordinates: x_o, y_o, and z_o). Let the components of $\boldsymbol{F}_o$ be given as:

$$\boldsymbol{F}_o = F_{xo}\boldsymbol{a}_x + F_{yo}\boldsymbol{a}_y + F_{zo}\boldsymbol{a}_z \tag{4.14}$$

Further, the evaluation of the scalar ψ at different surfaces will require areas of different surfaces in vector form and the values of $\boldsymbol{F}$ at all these surfaces.

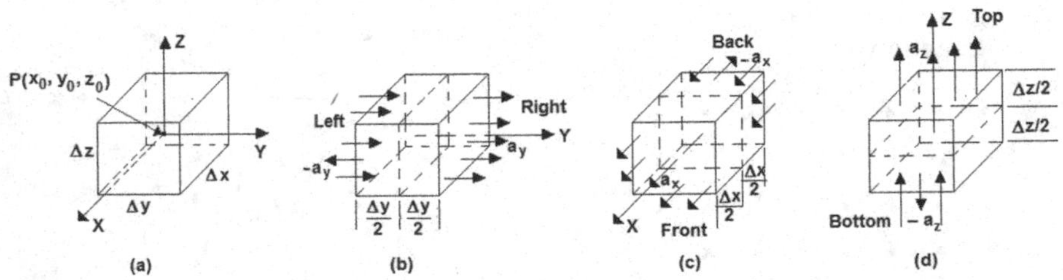

FIGURE 4.5 (a)An elemental volume Δv containing a point P at its center; (b) left and right surfaces along with unit vectors $-a_y$ and a_y; (c) front and back surface with unit vectors a_x and $-a_x$; (d) bottom and top surfaces with unit vectors $-a_z$ and a_z.

4.3.2.2 Vector Surfaces

The areas of six elemental surfaces enclosing the volume Δv can be written, in vector form as:

$$\left.\begin{array}{llll} \Delta \boldsymbol{s}_{\text{left}} = \Delta z \Delta x(-\boldsymbol{a}_y) & \text{(a)}, & \Delta \boldsymbol{s}_{\text{right}} = \Delta z \Delta x(\boldsymbol{a}_y) & \text{(b)}, \\ \Delta \boldsymbol{s}_{\text{front}} = \Delta y \Delta z(\boldsymbol{a}_x) & \text{(c)}, & \Delta \boldsymbol{s}_{\text{back}} = \Delta y \Delta z(-\boldsymbol{a}_x) & \text{(d)}, \\ \Delta \boldsymbol{s}_{\text{bottom}} = \Delta x \Delta y(-\boldsymbol{a}_z) & \text{(e)}, & \Delta \boldsymbol{s}_{\text{top}} = \Delta x \Delta y(\boldsymbol{a}_z) & \text{(f)} \end{array}\right\} \quad (4.15)$$

4.3.2.3 Flux Densities at Different Surfaces

By taking the first two terms of Taylor's series, the values of F at six different surfaces can be written as:

$$\left.\begin{array}{ll} F_{\text{front}} \approx F_{xo} + \left(\dfrac{\Delta x}{2}\right) \times \text{rate of change of } F_x \text{ with } x \approx F_{xo} + \dfrac{\Delta x}{2} \cdot \dfrac{\partial F_x}{\partial x} & \text{(a)} \\ F_{\text{back}} \approx F_{xo} - \left(\dfrac{\Delta x}{2}\right) \times \text{rate of change of } F_x \text{ with } x \approx F_{xo} - \dfrac{\Delta x}{2} \cdot \dfrac{\partial F_x}{\partial x} & \text{(b)} \\ F_{\text{left}} \approx F_{yo} - \left(\dfrac{\Delta y}{2}\right) \times \text{rate of change of } F_y \text{ with } y \approx F_{yo} - \dfrac{\Delta y}{2} \cdot \dfrac{\partial F_y}{\partial y} & \text{(c)} \\ F_{\text{right}} \approx F_{yo} + \left(\dfrac{\Delta y}{2}\right) \times \text{rate of change of } F_y \text{ with } y \approx F_{yo} + \dfrac{\Delta y}{2} \cdot \dfrac{\partial F_y}{\partial y} & \text{(d)} \\ F_{\text{bottom}} \approx F_{zo} - \left(\dfrac{\Delta z}{2}\right) \times \text{rate of change of } F_z \text{ with } z \approx F_{zo} - \dfrac{\Delta z}{2} \cdot \dfrac{\partial F_z}{\partial z} & \text{(e)} \\ F_{\text{top}} \approx F_{zo} + \left(\dfrac{\Delta z}{2}\right) \times \text{rate of change of } F_z \text{ with } z \approx F_{zo} + \dfrac{\Delta z}{2} \cdot \dfrac{\partial F_z}{\partial z} & \text{(f)} \end{array}\right\} \quad (4.16)$$

4.3.2.4 Flux Crossing through Different Surfaces

The dot product of the values of $\boldsymbol{F}$ from Equation 4.16 by the corresponding vector surfaces given by Equation 4.15 will result in small fluxes emerging from different elemental surfaces. If we designate these surfaces as front (*ft*), back (*bk*), right (*rt*), left (*lt*), top (*tp*), and bottom (*bm*) surfaces, these values are:

$$\left.\begin{array}{ll} \Delta\psi_{ft} \approx \left(F_{xo} + \dfrac{\Delta x}{2} \cdot \dfrac{\partial F_x}{\partial x}\right)\boldsymbol{a}_x \bullet \Delta y \Delta z(\boldsymbol{a}_x) \approx \left(F_{xo} + \dfrac{\Delta x}{2} \cdot \dfrac{\partial F_x}{\partial x}\right)\Delta y \Delta z & \text{(a)} \\ \Delta\psi_{bk} \approx \left(F_{xo} - \dfrac{\Delta x}{2} \cdot \dfrac{\partial F_x}{\partial x}\right)\boldsymbol{a}_x \bullet \Delta y \Delta z(-\boldsymbol{a}_x) \approx \left(-F_{xo} + \dfrac{\Delta x}{2} \cdot \dfrac{\partial F_x}{\partial x}\right)\Delta y \Delta z & \text{(b)} \\ \Delta\psi_{lt} \approx \left(F_{yo} - \dfrac{\Delta y}{2} \cdot \dfrac{\partial F_y}{\partial y}\right)\boldsymbol{a}_y \bullet \Delta z \Delta x(-\boldsymbol{a}_y) \approx \left(-F_{yo} + \dfrac{\Delta y}{2} \cdot \dfrac{\partial F_y}{\partial y}\right)\Delta z \Delta x & \text{(c)} \\ \Delta\psi_{rt} \approx \left(F_{yo} + \dfrac{\Delta y}{2} \cdot \dfrac{\partial F_y}{\partial y}\right)\boldsymbol{a}_y \bullet \Delta z \Delta x(\boldsymbol{a}_y) \approx \left(F_{yo} + \dfrac{\Delta y}{2} \cdot \dfrac{\partial F_y}{\partial y}\right)\Delta z \Delta x & \text{(d)} \\ \Delta\psi_{bm} \approx \left(F_{zo} - \dfrac{\Delta z}{2} \cdot \dfrac{\partial F_z}{\partial z}\right)\boldsymbol{a}_z \bullet \Delta x \Delta y(-\boldsymbol{a}_z) \approx \left(-F_{zo} + \dfrac{\Delta z}{2} \cdot \dfrac{\partial F_z}{\partial z}\right)\Delta x \Delta y & \text{(e)} \\ \Delta\psi_{tp} \approx \left(F_{zo} + \dfrac{\Delta z}{2} \cdot \dfrac{\partial F_z}{\partial z}\right)\boldsymbol{a}_z \bullet \Delta x \Delta y(\boldsymbol{a}_z) \approx \left(F_{zo} + \dfrac{\Delta z}{2} \cdot \dfrac{\partial F_z}{\partial z}\right)\Delta x \Delta y & \text{(f)} \end{array}\right\} \quad (4.17)$$

It is to be noted that in Equations 4.17a to 4.17f all unit vectors associated with the components of $\boldsymbol{F}$ are taken as positive as flux density is assumed to be increasing along the positive directions of the coordinates. In elemental surfaces the unit vectors may have positive or negative signs as given in Equations 4.15a to 4.15f and shown in Figure 4.5b, c, and d.

4.3.2.5 Total Flux

Addition of fluxes at front and back, right and left, and top and bottom surfaces yields:

$$\left.\begin{aligned} \Delta\psi_{(ft+bk)} &\approx \left(\frac{\partial F_x}{\partial x}\right)\cdot \Delta x \Delta y \Delta z \qquad \text{(a)} \\ \Delta\psi(rt+lt) &\approx \left(\frac{\partial F_y}{\partial y}\right)\cdot \Delta x \Delta y \Delta z \qquad \text{(b)} \\ \Delta\psi(tp+bm) &\approx \left(\frac{\partial F_z}{\partial z}\right)\cdot \Delta x \Delta y \Delta z \qquad \text{(c)} \end{aligned}\right\} \tag{4.18}$$

The values of fluxes given by Equations 4.18a, 4.18b, and 4.18c are the differences of fluxes leaving and entering the opposite elemental surfaces in their surface vectors. If all these fluxes are added together the net flux ($\Delta\psi$) is obtained to be:

$$\Delta\psi \approx \left(\frac{\partial F_x}{\partial x}+\frac{\partial F_y}{\partial y}+\frac{\partial F_z}{\partial z}\right)\cdot \Delta x\,\Delta y\,\Delta z \stackrel{\text{def}}{=} \left(\frac{\partial F_x}{\partial x}+\frac{\partial F_y}{\partial y}+\frac{\partial F_z}{\partial z}\right)\cdot \Delta v \tag{4.19a}$$

From Equation 4.19a:

$$\left(\frac{\partial F_x}{\partial x}+\frac{\partial F_y}{\partial y}+\frac{\partial F_z}{\partial z}\right) \approx \frac{\Delta\psi}{\Delta v} \tag{4.19b}$$

On integrating both sides of Equation 4.19a we get:

$$\int d\psi = \psi = \iiint_v \left(\frac{\partial F_x}{\partial x}+\frac{\partial F_y}{\partial y}+\frac{\partial F_z}{\partial z}\right)\cdot dv \tag{4.19c}$$

4.3.2.6 Mathematical Expressions

The left-hand side of Equation 4.19b represents the dot product of vector operator ∇ and vector $\boldsymbol{F}$ in Cartesian coordinates which by definition is the divergence of vector $\boldsymbol{F}$ and can be written as:

$$\left(\frac{\partial F_x}{\partial x}+\frac{\partial F_y}{\partial y}+\frac{\partial F_z}{\partial z}\right) = \nabla \bullet \boldsymbol{F} \stackrel{\text{def}}{=} \text{divergance of } \boldsymbol{F} = \text{div}\,\boldsymbol{F} \tag{4.20a}$$

In Equation 4.20a, Equation 4.19b leads to the general definition for the divergence of any vector in any system of space coordinates at any point P as:

$$\boxed{\nabla \bullet \boldsymbol{F} \overset{\text{def}}{=} \lim_{\Delta v \to 0} \left(\frac{\Delta \psi}{\Delta v} \right) = \lim_{\Delta v \to 0} \left(\frac{\oiint_s \boldsymbol{F} \bullet \mathbf{ds}}{\Delta v} \right)} \qquad (4.20b)$$

where s is the total surface of the elemental volume Δv around the point P. The integration over this closed surface s gives the total flux emerging from the elemental volume Δv.

On integrating both sides of this equation over a volume v bounded by its surface s, this equation leads to the *divergence theorem* given as:

$$\psi = \boxed{\oiint_s \boldsymbol{F} \bullet \mathbf{ds} = \iiint_v (\nabla \bullet \boldsymbol{F}) \cdot dv} \qquad (4.20c)$$

4.3.2.7 Physical Interpretation of Mathematical Relation

The scalar quantity on the right-hand side of Equation 4.20b may be positive, negative, or zero. In the beginning it was assumed that the flux density $\boldsymbol{F}$ is increasing along the positive x-, y-, and z-axes. Figures 4.5b, c and d show the entry of flux from left, back, and bottom surfaces of the incremental volume and its emergence from its right, front, and top surfaces respectively. This leads to the following three situations:

- When these emerging fluxes are more than the entering fluxes the net value of flux will be positive. Thus one can imagine that there is some (positive) *source* present inside the elemental volume, which adds some more flux to the entering flux.
- When the right-hand side of Equation 4.20b is negative there is some *sink* (i.e. negative source) present inside the elemental volume which subtracts some flux from the flux leaving the volume from three different sides and makes the emerging flux from three opposing surfaces less than the entering fluxes.
- In case the right-hand side of Equation 4.20b is zero there is neither a source nor a sink and the same flux which enters the elemental volume emerges.

4.3.3 Solenoidal Field

As noted above the right-hand side of Equation 4.20b may become zero when there is neither a source nor a sink. Thus, when the divergence of a vector field is zero, such a vector field is called a *solenoidal field*.

4.3.4 Divergence Expressions in Different Coordinates

Expressions for divergence in different coordinate systems are:

$$\left.\begin{aligned}
\nabla \bullet \boldsymbol{F} &\equiv \frac{\partial F_x}{\partial x} + \frac{\partial F_y}{\partial y} + \frac{\partial F_z}{\partial z} && \text{(Cartesian)} && \text{(a)}\\
\nabla \bullet \boldsymbol{F} &\equiv \frac{1}{\rho}\cdot\frac{\partial\left(\rho\cdot F_\rho\right)}{\partial\rho} + \frac{1}{\rho}\cdot\frac{\partial F_\varphi}{\partial\varphi} + \frac{\partial F_Z}{\partial z} && \text{(Cylindrical)} && \text{(b)}\\
\nabla \bullet \boldsymbol{F} &\equiv \frac{1}{r^2}\cdot\frac{\partial\left(r^2\cdot F_r\right)}{\partial r} + \frac{1}{r\cdot\sin\theta}\cdot\frac{\partial\left(\sin\theta\cdot F_\theta\right)}{\partial\theta} + \frac{1}{r\cdot\sin\theta}\cdot\frac{\partial\left(F_\varphi\right)}{\partial\varphi} && \text{(Spherical)} && \text{(c)}
\end{aligned}\right\} \tag{4.21}$$

4.3.5 Applications

The divergence operation has applications in many fields of science and engineering including that of electromagnetics. Some of its applications will appear in the subsequent chapters.

4.4 CURL OPERATION

The cross product of vector operator ∇ and any vector ***F*** gives a vector called curl of ***F***. The curl of a field is represented by a vector at every point in that field. The length and direction of this vector characterize the rotation at that point. The direction of the curl is along the axis of rotation, determined by the right-hand rule and the magnitude of the curl is the magnitude of rotation. If the vector field represents the velocity of flow of a fluid, then the curl is the circulation density of the fluid. The curl is a form of vector differentiation for vector fields. Sometimes alternative term *rotational* (or rot) is also used for curl. Thus, curl of a vector field ***F*** is denoted as curl ***F***, $\nabla \times \boldsymbol{F}$, or rot ***F***.

Figure 4.6 shows that the curl of a vector field ***F*** at a point is defined in terms of its projection onto various lines through the point. If ***n*** is any unit vector, the projection of curl ***F*** onto ***n*** is defined to be the limiting value of a closed line integral in a plane orthogonal to ***n*** as the path (*C*) used in the integral becomes infinitesimally close to the point, divided by the area (*A*) enclosed.

4.4.1 Physical Description

Imagine a large tank filled with some incompressible fluid. This fluid is somehow made to rotate (or flow) in anticlockwise direction inside the tank. Furthermore, a tiny ball (or paddle wheel) is assumed to be placed in this tank with its center fixed at a point. If the surface of this ball is rough, the flowing fluid will force this ball to rotate. Figure 4.7 depicts some of the parameters involved or to be reckoned with. The vector field of this flowing fluid describes the velocity field (v). The orientation of rotation axis (n) (in accordance with the right-hand rule), points in the direction of

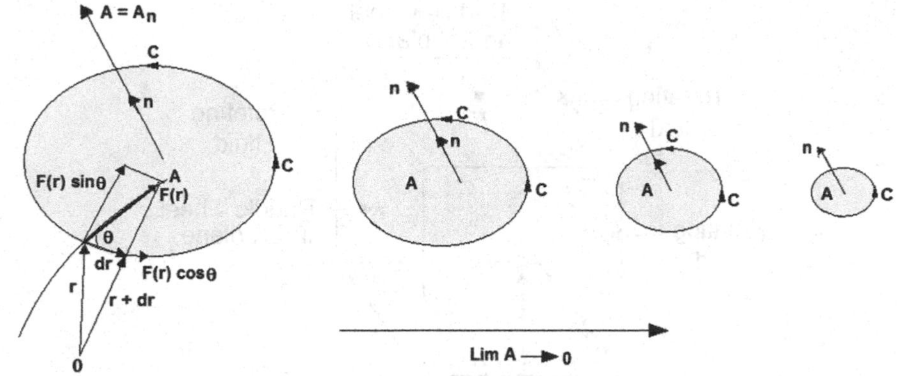

FIGURE 4.6 Normal and tangential components of field **F** at position r on a closed curve C in a plane, enclosing a planar vector area $A = A_n$.

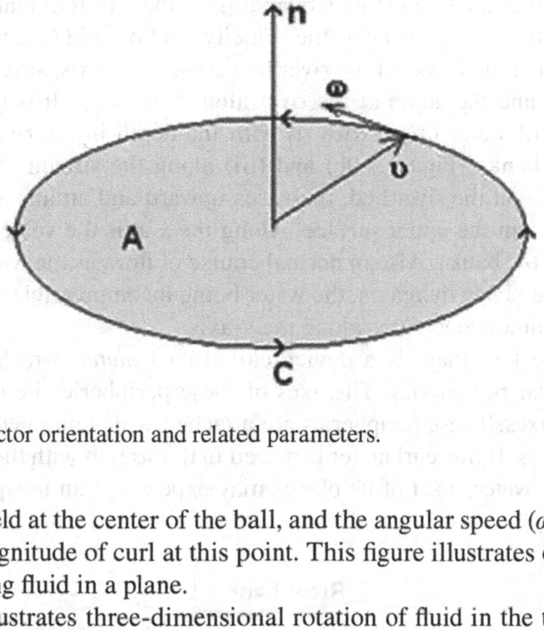

FIGURE 4.7 Vector orientation and related parameters.

the curl of the field at the center of the ball, and the angular speed (ω) of the rotation is half of the magnitude of curl at this point. This figure illustrates only two dimensional the rotating fluid in a plane.

Figure 4.8 illustrates three-dimensional rotation of fluid in the tank. The visual inspection of this figure reveals that if a paddle wheel is placed anywhere (say) in the x–y plane it will have the tendency to rotate clockwise. In the right-hand rule (in right-handed coordinate system) the curl will be in the negative z-direction. Since curl is a vector, its direction is normal (perpendicular) to the surface with the vector field and its magnitude is the circulation per unit area. By convention the counterclockwise circulation will give a curl pointing out of the page. Similarly, if this paddle wheel is placed in y–z or z–x plane, it will rotate around the x- or y-axis as per its location.

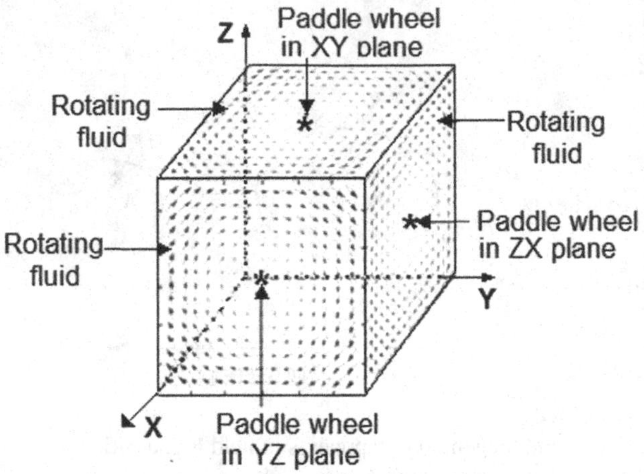

FIGURE 4.8 Three-dimensional rotating fluid.

Figure 4.9 further adds to the understanding of the curl (circulation or rotation). This figure depicts a vector field υ as the velocity field of fluid (e.g. water) flowing in a river. Assume that the banks of the river lie across the *x*-axis, stream of water flow along the *y*-axis, and the depth of the river along the *z*-axis. It is generally known that the velocity of water flow varies (i) with the depth in the river (Figure 4.9a), (ii) between the banks (Figure 4.9b) and (iii) along the stream. Along the *z*-axis the velocity is zero at the river bed, increases upward and attains maxima at about one-third depth from the upper surface. Along the *x*-axis the velocity is maximum midway between the banks. Also in normal course of flow, as the width and/or depth of the river progressively increases, the water being incompressible fluid the velocity of water flow gradually decreases along the *y*-axis.

Let us assume that there is a device called *curl meter*, which contains three orthogonal circular peripheries. The axes of these peripheries lie perpendicular to the *x*-, *y*-, and *z*-axes. These peripheries are attached with a number of infinitesimal turbines like blades. If this curl meter is placed in the stream with the plane of blades along the flow of water, a set of its blades may experience an unequal force due to

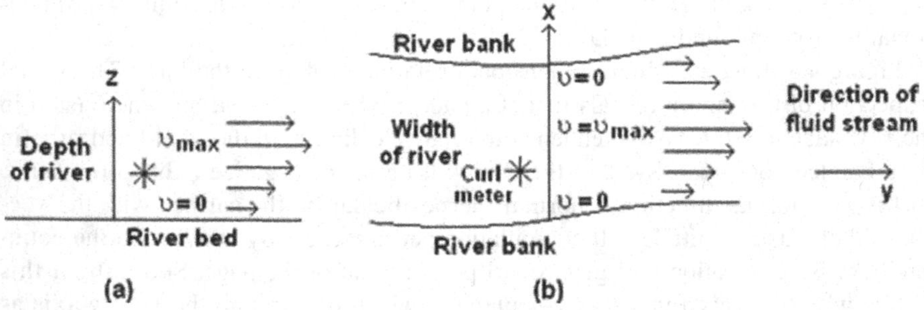

FIGURE 4.9 A curl meter located (a) at some depth lying along the *z*-axis and (b) along the *x*-axis.

variation of velocity along the z-axis provided it is not located at the location of maximum velocity. This curl meter will rotate and the direction of rotation will depend on its location. The upper blades experience more force than the lower ones it will rotate in a clockwise direction and if the lower blades experience stronger force than the upper ones it will rotate in an anticlockwise direction. Similarly, when this curl meter is located along the x- or y-axis, another set of its blades may experience a similar situation and the curl meter will rotate in a clockwise or anticlockwise direction along the corresponding axis. Thus the blades of this curl meter may simultaneously rotate in accordance with its location and the forces exerted. It is so located that its blades experience equal forces there will be no rotation. The rotation of blades is an indicator of the curl of the velocity vector for the water flow at various locations in the river. In the following subsection, the mathematical formulation of curl operation is given in a Cartesian coordinate system for easy visualization.

4.4.2 Mathematical Formulation

Figure 4.10a illustrates three elemental close loops drawn in constant x, y, and z planes as visualized separately in different planes. In Figure 4.10b all these loops are overlapped together. The incremental lengths of the sides of these loops are also shown in these figures. The centers of these loops lie at the same location, i.e. at x_0, y_0, and z_0. For mathematical formulation of the curl operation, one can start with any vector field $\boldsymbol{F}$ without assigning any physical meaning to it. This three-dimensional vector field is assumed to be increasing along the positive directions of the coordinate axes. In its nature of variation it may assume different values at different geometrical locations. Assume that its value at x_0, y_0, and z_0 be given by:

$$\boldsymbol{F}_0 = F_{x0}\boldsymbol{a}_x + F_{y0}\boldsymbol{a}_y + F_{z0}\boldsymbol{a}_z \tag{4.22}$$

The close loop in the x-, y-, and z-plane has incremental lengths of (Δy,Δz), (Δx,Δz), and (Δx,Δy) and their sides are marked as (1-2, 2-3, 3-4, 4-1), (5-6, 6-7, 7-8, 8-5), and (9-10, 10-11, 11-12, 12-9), respectively.

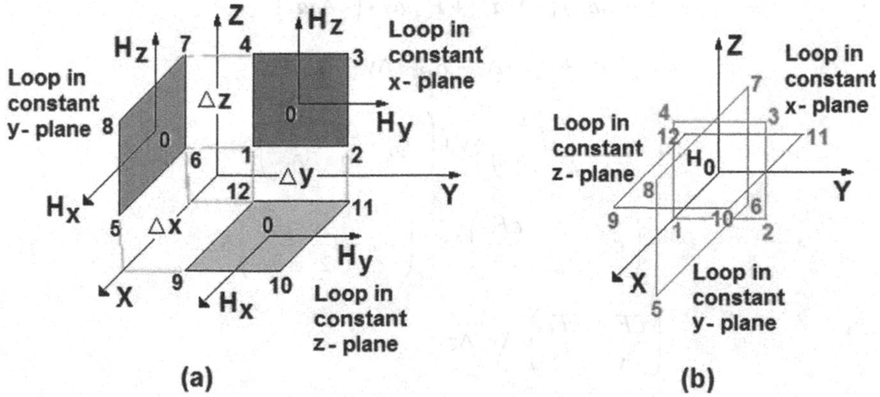

FIGURE 4.10 Elemental close loops in constant x, y, and z planes, (a) as seen separately in different planes and (b) overlapped together.

4.4.2.1 Values of F at Different Loop Segments

Consider the loop (1-2-3-4-1) placed on a constant x plane. The approximate values of $\boldsymbol{F}$ at different loop segments can be obtained by considering the rate of change of relevant component of $\boldsymbol{F}$ in the directions of segments with reference to $\boldsymbol{F}_0$. The values obtained are as follows:

$$\left.\begin{aligned} F_{y12} &\approx F_{y0} - \frac{\Delta z}{2}\cdot\frac{\partial F_y}{\partial z} \quad \text{(a)}, \qquad F_{z23} \approx F_{z0} + \frac{\Delta y}{2}\cdot\frac{\partial F_z}{\partial y} \quad \text{(b)},\\ F_{y34} &\approx F_{y0} + \frac{\Delta z}{2}\cdot\frac{\partial F_y}{\partial z} \quad \text{(c)}, \qquad F_{z41} \approx F_{z0} - \frac{\Delta y}{2}\cdot\frac{\partial F_z}{\partial y} \quad \text{(d)} \end{aligned}\right\} \tag{4.23}$$

4.4.2.2 Lengths of Different Loop Segments

The lengths of different segments of loops in vectorial forms are:

$$\left.\begin{aligned} &\text{From 1 to 2} \quad \Delta y\boldsymbol{a}_y \quad \text{(a)}, \qquad \text{From 2 to 3} \quad \Delta z\boldsymbol{a}_z \quad \text{(b)},\\ &\text{From 3 to 4} \quad -\Delta y\boldsymbol{a}_y \quad \text{(c)}, \qquad \text{From 4 to 1} \quad -\Delta z\boldsymbol{a}_z \quad \text{(d)} \end{aligned}\right\} \tag{4.24}$$

4.4.2.3 Contour Integration of F

Equations 4.23a to 4.23d give the values of components of $\boldsymbol{F}$ in scalar form in the directions of different segments. If these are multiplied by the appropriate unit vectors these values assume vector forms. These resulting values can further be multiplied (through dot product) by the vectorial lengths of the corresponding segments given by Equations(4.24a to 4.24d. The sum of such products can be approximately equated to a closed integral around the loop segment 1-2-3-4-1 shown in Figure 4.10a and b. Thus:

$$\begin{aligned} \oint_{1,2,3,4,1} \boldsymbol{F}\bullet d\boldsymbol{\ell} &\approx F_{y12}\boldsymbol{a}_y \bullet \Delta y\boldsymbol{a}_y + F_{z23}\boldsymbol{a}_z \bullet \Delta z\boldsymbol{a}_z \\ &\quad + F_{y34}\boldsymbol{a}_y \bullet \left(-\Delta y\boldsymbol{a}_y\right) + F_{z41}\boldsymbol{a}_z \bullet \left(-\Delta z\boldsymbol{a}_z\right) \\ &\approx F_{y12}\cdot\Delta y + F_{z23}\cdot\Delta z - F_{y34}\cdot\Delta y - F_{z41}\cdot\Delta z \\ &\approx \left(F_{y0} - \frac{\Delta z}{2}\cdot\frac{\partial F_y}{\partial z}\right)\Delta y + \left(F_{z0} + \frac{\Delta y}{2}\cdot\frac{\partial F_z}{\partial y}\right)\Delta z \\ &\quad - \left(F_{y0} + \frac{\Delta z}{2}\cdot\frac{\partial F_y}{\partial z}\right)\Delta y - \left(F_{z0} - \frac{\Delta y}{2}\cdot\frac{\partial F_z}{\partial y}\right)\Delta z \\ &\approx \left(\frac{\partial F_z}{\partial y} - \frac{\partial F_y}{\partial z}\right)\cdot\Delta y\cdot\Delta z \end{aligned}$$

or

$$\oint_{1,2,3,4,1} \boldsymbol{F} \bullet d\boldsymbol{\ell} \approx \left(\frac{\partial F_z}{\partial y} - \frac{\partial F_y}{\partial z} \right) \cdot \Delta y \cdot \Delta z \tag{4.25a}$$

Similarly, for loops (5-6-7-8-5) and (9-10-11-12-9) (see Figure 4.10a and b), we get:

$$\left.\begin{aligned} &\oint_{5,6,7,8,5} \boldsymbol{F} \bullet d\boldsymbol{\ell} \approx \left(\frac{\partial H_x}{\partial z} - \frac{\partial H_Z}{\partial x} \right) \cdot \Delta x \cdot \Delta z \quad \text{(b)} \\ &\oint_{9,10,11,12,9} \boldsymbol{F} \bullet d\boldsymbol{\ell} \approx \left(\frac{\partial H_y}{\partial x} - \frac{\partial H_x}{\partial y} \right) \cdot \Delta x \cdot \Delta y \quad \text{(c)} \end{aligned}\right\} \tag{4.25}$$

If Equation 4.25a is divided by $(\Delta y \cdot \Delta z)$ with the limits of $(\Delta y \cdot \Delta z)$ tending to zero, it will yield:

$$\lim_{(\Delta y . \Delta z) \to 0} \left[\frac{\oint_{1,2,3,4,1} \boldsymbol{F} \bullet d\boldsymbol{\ell}}{(\Delta y \cdot \Delta z)} \right] = \left(\frac{\partial H_z}{\partial y} - \frac{\partial H_y}{\partial z} \right) \tag{4.26a}$$

Similarly, if Equation 4.25b is divided by $(\Delta x \cdot \Delta y)$ and Equation 4.25c is divided by $(\Delta x \cdot \Delta y)$ with the limits of $(\Delta x \cdot \Delta z)$ and $(\Delta x \cdot \Delta y)$ tending to zero, it will yield:

$$\left.\begin{aligned} &\lim_{(\Delta x . \Delta z) \to 0} \left[\frac{\oint_{5,6,7,8,5} \boldsymbol{F} \bullet d\boldsymbol{\ell}}{(\Delta x \cdot \Delta z)} \right] = \left(\frac{\partial F_x}{\partial z} - \frac{\partial F_z}{\partial x} \right) \quad \text{(b)} \\ &\lim_{(\Delta x . \Delta y) \to 0} \left[\frac{\oint_{9,10,11,12,9} \boldsymbol{F} \bullet d\boldsymbol{\ell}}{(\Delta x \cdot \Delta y)} \right] = \left(\frac{\partial F_y}{\partial x} - \frac{\partial F_x}{\partial y} \right) \quad \text{(c)} \end{aligned}\right\} \tag{4.26}$$

Noting that the contour integrations are made on mutually perpendicular planes let each of these equations is multiplied by unit vectors corresponding to respective planes and then added up. These result in:

$$\lim_{\Delta S_n \to 0} \left(\frac{\oint_c \boldsymbol{F} \bullet d\boldsymbol{\ell}}{\Delta S_n} \right) \boldsymbol{a}_n = \left(\frac{\partial H_z}{\partial y} - \frac{\partial H_y}{\partial z} \right) \boldsymbol{a}_x + \left(\frac{\partial F_x}{\partial z} - \frac{\partial F_z}{\partial x} \right) \boldsymbol{a}_y + \left(\frac{\partial F_y}{\partial x} - \frac{\partial F_x}{\partial y} \right) \boldsymbol{a}_z \tag{4.27}$$

The symbol ΔS_n indicates the elemental surface area bounded by the contour c and $\boldsymbol{a}_n$ is the unit vector normal to this surface.

4.4.2.4 Definition of Curl

The curl of a vector (say $\boldsymbol{F}$) is defined as:

$$\nabla\times\boldsymbol{F}\overset{\text{def}}{=}\lim_{\Delta S_n\to 0}\left(\frac{\oint_c \boldsymbol{F}\cdot d\boldsymbol{\ell}}{\Delta S_n}\right)\boldsymbol{a}_n \tag{4.28}$$

Since the contour integrations in deriving Equation 4.28 are made in a Cartesian system of coordinates, the RHS of this equation gives the expression for the curl of the vector $\boldsymbol{F}$ in the Cartesian coordinates:

$$\nabla\times\boldsymbol{F}=\left(\frac{\partial F_z}{\partial y}-\frac{\partial F_y}{\partial z}\right)\boldsymbol{a}_x+\left(\frac{\partial F_x}{\partial z}-\frac{\partial F_z}{\partial x}\right)\boldsymbol{a}_y+\left(\frac{\partial F_y}{\partial x}-\frac{\partial F_x}{\partial y}\right)\boldsymbol{a}_z \tag{4.29}$$

4.4.2.5 Some Salient Features of Curl

- In a vector field describing the linear velocities of each part of a (rigid) rotating disk, the curl has the same value at all points.
- Faraday's law (discussed in Chapter 15) and Ampère's law for time-invariant fields (discussed in Chapter 11) can be compactly expressed by using the curl notation.
- Faraday's law states that the curl of electric field is equal to the negative of the time rate of change of the magnetic field.
- Ampère's law for time-varying fields (discussed in Chapter 15) relates the curl of the magnetic field to the current density and rate of change of the electric field of the flux density.
- A field having curl is not conservative. Thus in a field with curl (like a whirlpool) one with zero mass can get a free ride by moving in the direction of the twist. This free trip for moving with the current in a circle will require no energy. Any movement against the current one has to use energy. A body with nonzero mass will be subjected to the centrifugal force of rotation and shall be thrown away.
- A conservative field has zero curl. The gravity and electrostatic fields are examples of conservative field. Thus if a rock is lifted and then allowed to fall the energy obtained from fall is the same as it is put in to lift the rock. Theoretically no energy is gained or lost in this transition, provided the wind friction is neglected.

4.4.3 Rotational Field

As noted earlier, the curl of a vector $\boldsymbol{F}$ can be written as curl $\boldsymbol{F}$, $\nabla\times\boldsymbol{F}$, or rot $\boldsymbol{F}$. The word rot is the shortened version of the word rotation. This term indicates that a vector field may be rotational or irrotational. The rotational of irrotational can thus be written as:

$$\nabla\times\boldsymbol{F}=\boldsymbol{0}\ (\text{Irrotational})\ (\text{a})\quad \nabla\times\boldsymbol{F}\neq\boldsymbol{0}\ (\text{rotational})\ (\text{b})\} \tag{4.30}$$

4.4.4 Curl Expressions in Different Coordinate Systems

Since $\boldsymbol{F}$ may be a function of more than one coordinate $\boldsymbol{a}_N$ (in Equation 4.28) may be replaced by $\boldsymbol{a}_x$, $\boldsymbol{a}_y$, $\boldsymbol{a}_z$ in Cartesian, $\boldsymbol{a}_\rho$, $\boldsymbol{a}_\varphi$, $\boldsymbol{a}_z$ in cylindrical and $\boldsymbol{a}_r$, $\boldsymbol{a}_\theta$, $\boldsymbol{a}_\varphi$ in spherical systems. The resulting expressions for curl in these systems are:

$$\nabla\times\boldsymbol{F}=\left[\frac{\partial F_z}{\partial y}-\frac{\partial F_y}{\partial z}\right]\boldsymbol{a}_x+\left[\frac{\partial F_x}{\partial z}-\frac{\partial F_z}{\partial x}\right]\boldsymbol{a}_y+\left[\frac{\partial F_y}{\partial x}-\frac{\partial F_x}{\partial y}\right]\boldsymbol{a}_z \quad (\text{Cartesian}) \tag{4.31a}$$

$$\begin{aligned}\nabla\times\boldsymbol{F}&=\left[\left(\frac{1}{\rho}\right)\cdot\frac{\partial F_z}{\partial\varphi}-\frac{\partial F_\varphi}{\partial z}\right]\boldsymbol{a}_\rho+\left[\frac{\partial F_\rho}{\partial z}-\frac{\partial F_z}{\partial\rho}\right]\boldsymbol{a}_\varphi\\&+\left[\left(\frac{1}{\rho}\right)\cdot\left\{\frac{\partial(\rho F_\varphi)}{\partial\rho}-\frac{\partial F_\rho}{\partial\varphi}\right\}\right]\boldsymbol{a}_z \quad (\text{Cylindrical})\end{aligned} \tag{4.31b}$$

$$\begin{aligned}\nabla\times\boldsymbol{F}&=\left[\left(\frac{1}{r\cdot\sin\theta}\right)\cdot\left\{\frac{\partial(\sin\theta\cdot F_\varphi)}{\partial\theta}-\frac{\partial F_\theta}{\partial\varphi}\right\}\right]\boldsymbol{a}_r\\&+\left[\left(\frac{1}{r}\right)\cdot\left\{\left(\frac{1}{\sin\theta}\right)\cdot\frac{\partial F_r}{\partial\varphi}-\frac{\partial(r\cdot F_\varphi)}{\partial r}\right\}\right]\boldsymbol{a}_\theta\\&+\left[\left(\frac{1}{r}\right)\cdot\left\{\frac{\partial(r\cdot F_\theta)}{\partial r}-\frac{\partial F_r}{\partial\theta}\right\}\right]\boldsymbol{a}_\varphi \quad (\text{Spherical})\end{aligned} \tag{4.31c}$$

These equations can also be written in another form by using the notation of determinant:

$$\nabla\times\boldsymbol{F}=\begin{vmatrix}\boldsymbol{a}_x & \boldsymbol{a}_y & \boldsymbol{a}_z\\ \dfrac{\partial}{\partial x} & \dfrac{\partial}{\partial y} & \dfrac{\partial}{\partial z}\\ F_x & F_y & F_z\end{vmatrix} \quad (\text{Cartesian}) \tag{4.32a}$$

$$\nabla\times\boldsymbol{F}=\frac{1}{\rho}\cdot\begin{vmatrix}\boldsymbol{a}_\rho & \rho\cdot\boldsymbol{a}_\varphi & \boldsymbol{a}_z\\ \dfrac{\partial}{\partial\rho} & \dfrac{\partial}{\partial\varphi} & \dfrac{\partial}{\partial z}\\ F_\rho & \rho\cdot F_\varphi & F_z\end{vmatrix} \quad (\text{Cylindrical}) \tag{4.32b}$$

$$\nabla\times\boldsymbol{F}=\frac{1}{r^2\cdot\sin\theta}\cdot\begin{vmatrix}\boldsymbol{a}_r & r\cdot\boldsymbol{a}_\theta & r\cdot\sin\theta\cdot\boldsymbol{a}_\varphi\\ \dfrac{\partial}{\partial r} & \dfrac{\partial}{\partial\theta} & \dfrac{\partial}{\partial\varphi}\\ F_r & r\cdot F_\theta & r\cdot\sin\theta\cdot F_\varphi\end{vmatrix} \quad (\text{Spherical}) \tag{4.32c}$$

4.4.5 Applications

The curl operation is widely used in many fields of science and engineering including that of electromagnetics. Its applications will appear in many subsequent chapters.

4.5 LAPLACIAN OPERATOR AND ITS APPLICATIONS

This section describes the meaning and applications of the Laplacian operator.

4.5.1 Laplacian Operator

The Laplacian operator is a scalar differential operator. It is denoted by the square of |∇|or the scalar (or dot) product of two Del operators, i.e.:

$$\nabla \bullet \nabla = \nabla^2 \tag{4.33}$$

Its expression in a Cartesian system of space coordinates is given as:

$$\nabla^2 = \left(\boldsymbol{a}_x \frac{\partial}{\partial x} + \boldsymbol{a}_y \frac{\partial}{\partial y} + \boldsymbol{a}_z \frac{\partial}{\partial z}\right) \bullet \left(\boldsymbol{a}_x \frac{\partial}{\partial x} + \boldsymbol{a}_y \frac{\partial}{\partial y} + \boldsymbol{a}_z \frac{\partial}{\partial z}\right) \equiv \frac{\partial^2}{\partial x^2} + \frac{\partial^2}{\partial y^2} + \frac{\partial^2}{\partial z^2} \tag{4.33}$$

4.5.1.1 Laplacian for Scalar Operand

The general definition of Laplacian for scalar operand V in any system of coordinates is given as:

$$\nabla^2 V \overset{\text{def}}{=} \nabla \bullet (\nabla V) \tag{4.34}$$

The expressions for the Laplacian of a scalar function V in different coordinate systems can be written as:

$$\nabla^2 V = \frac{\partial^2 V}{\partial x^2} + \frac{\partial^2 V}{\partial y^2} + \frac{\partial^2 V}{\partial z^2} \quad \text{(Cartesian)} \tag{4.35a}$$

$$\nabla^2 V = \frac{1}{\rho} \cdot \frac{\partial}{\partial \rho}\left(\rho \cdot \frac{\partial V}{\partial \rho}\right) + \frac{1}{\rho^2} \cdot \frac{\partial^2 V}{\partial \varphi^2} + \frac{\partial^2 V}{\partial z^2} \quad \text{(Cylindrical)} \tag{4.35b}$$

$$\nabla^2 V = \frac{1}{r^2} \cdot \frac{\partial}{\partial r}\left(r^2 \cdot \frac{\partial V}{\partial r}\right) + \frac{1}{r^2 \sin\theta} \cdot \frac{\partial}{\partial \theta}\left(\sin\theta \cdot \frac{\partial V}{\partial \theta}\right) + \frac{1}{r^2 \sin^2\theta} \cdot \frac{\partial^2 V}{\partial \varphi^2} \quad \text{(Spherical)} \tag{4.35c}$$

4.5.1.2 Laplacian for Vector Operand

Since the gradient of a vector is not defined, the Laplacian of a vector needs to be defined afresh. The general definition of Laplacian for vector operand $\boldsymbol{A}$ in any system of coordinates is:

$$\nabla^2 \boldsymbol{A} \overset{\text{def}}{=} \nabla(\nabla \bullet \boldsymbol{A}) - \nabla \times (\nabla \times \boldsymbol{A}) \tag{4.36}$$

Thus, for *vector operand* its "del square" is the gradient of divergence minus curl of curl of the vector operand. To appreciate the advantage of this definition, let the vector be expressed in terms of its Cartesian components as:

$$\boldsymbol{A} \overset{\text{def}}{=} A_x \cdot \boldsymbol{a}_x + A_y \cdot \boldsymbol{a}_y + A_z \cdot \boldsymbol{a}_z \tag{4.37a}$$

Thus,

$$\nabla \bullet \boldsymbol{A} = \frac{\partial A_x}{\partial x} + \frac{\partial A_y}{\partial y} + \frac{\partial A_z}{\partial z} \tag{4.37b}$$

The components of Equation 4.36 are worked out here:

$$\begin{aligned} \nabla(\nabla \bullet \boldsymbol{A}) = \boldsymbol{a}_x \frac{\partial}{\partial x}\left(\frac{\partial A_x}{\partial x} + \frac{\partial A_y}{\partial y} + \frac{\partial A_z}{\partial z}\right) + \boldsymbol{a}_y \frac{\partial}{\partial y}\left(\frac{\partial A_x}{\partial x} + \frac{\partial A_y}{\partial y} + \frac{\partial A_z}{\partial z}\right) \\ + \boldsymbol{a}_z \frac{\partial}{\partial z}\left(\frac{\partial A_x}{\partial x} + \frac{\partial A_y}{\partial y} + \frac{\partial A_z}{\partial z}\right) \end{aligned} \tag{4.38a}$$

Further in Equation 4.31a in a Cartesian system:

$$\nabla \times \boldsymbol{A} = \boldsymbol{a}_x \left(\frac{\partial A_z}{\partial y} - \frac{\partial A_y}{\partial z}\right) + \boldsymbol{a}_y \left(\frac{\partial A_x}{\partial z} - \frac{\partial A_z}{\partial x}\right) + \boldsymbol{a}_z \left(\frac{\partial A_y}{\partial x} - \frac{\partial A_x}{\partial y}\right) \tag{4.38b}$$

$$\begin{aligned} \nabla \times (\nabla \times \boldsymbol{A}) = \boldsymbol{a}_x \left\{\frac{\partial}{\partial y}\left(\frac{\partial A_y}{\partial x} - \frac{\partial A_x}{\partial y}\right) - \frac{\partial}{\partial z}\left(\frac{\partial A_x}{\partial z} - \frac{\partial A_z}{\partial x}\right)\right\} \\ + \boldsymbol{a}_y \left\{\frac{\partial}{\partial z}\left(\frac{\partial A_z}{\partial y} - \frac{\partial A_y}{\partial z}\right) - \frac{\partial}{\partial x}\left(\frac{\partial A_y}{\partial x} - \frac{\partial A_x}{\partial y}\right)\right\} \\ + \boldsymbol{a}_z \left\{\frac{\partial}{\partial x}\left(\frac{\partial A_x}{\partial z} - \frac{\partial A_z}{\partial x}\right) - \frac{\partial}{\partial y}\left(\frac{\partial A_z}{\partial y} - \frac{\partial A_y}{\partial z}\right)\right\} \end{aligned} \tag{4.38c}$$

From Equations 4.38a and 4.38c, we get:

$$\nabla(\nabla \bullet \boldsymbol{A})-\nabla\times(\nabla\times\boldsymbol{A})$$

$$=\boldsymbol{a}_x\left[\frac{\partial}{\partial x}\left(\frac{\partial A_x}{\partial x}+\frac{\partial A_y}{\partial y}+\frac{\partial A_z}{\partial z}\right)-\left\{\frac{\partial}{\partial y}\left(\frac{\partial A_y}{\partial x}-\frac{\partial A_x}{\partial y}\right)-\frac{\partial}{\partial z}\left(\frac{\partial A_x}{\partial z}-\frac{\partial A_z}{\partial x}\right)\right\}\right]$$

$$+\boldsymbol{a}_y\left[\frac{\partial}{\partial y}\left(\frac{\partial A_x}{\partial x}+\frac{\partial A_y}{\partial y}+\frac{\partial A_z}{\partial z}\right)-\left\{\frac{\partial}{\partial z}\left(\frac{\partial A_z}{\partial y}-\frac{\partial A_y}{\partial z}\right)-\frac{\partial}{\partial x}\left(\frac{\partial A_y}{\partial x}-\frac{\partial A_x}{\partial y}\right)\right\}\right]$$

$$+\boldsymbol{a}_z\left[\frac{\partial}{\partial z}\left(\frac{\partial A_x}{\partial x}+\frac{\partial A_y}{\partial y}+\frac{\partial A_z}{\partial z}\right)-\left\{\frac{\partial}{\partial x}\left(\frac{\partial A_x}{\partial z}-\frac{\partial A_z}{\partial x}\right)-\frac{\partial}{\partial y}\left(\frac{\partial A_z}{\partial y}-\frac{\partial A_y}{\partial z}\right)\right\}\right]$$

or

$$\nabla(\nabla \bullet \boldsymbol{A})-\nabla\times(\nabla\times\boldsymbol{A})$$

$$=\boldsymbol{a}_x\left[\frac{\partial^2 A_x}{\partial x^2}+\frac{\partial^2 A_x}{\partial y^2}+\frac{\partial^2 A_x}{\partial z^2}\right]+\boldsymbol{a}_y\left[\frac{\partial^2 A_y}{\partial x^2}+\frac{\partial^2 A_y}{\partial y^2}+\frac{\partial^2 A_y}{\partial z^2}\right]$$

$$+\boldsymbol{a}_z\left[\frac{\partial^2 A_z}{\partial x^2}+\frac{\partial^2 A_z}{\partial y^2}+\frac{\partial^2 A_z}{\partial z^2}\right]$$

Thus:

$$\nabla^2\boldsymbol{A}=\boldsymbol{a}_x\nabla^2 A_x+\boldsymbol{a}_y\nabla^2 A_y+\boldsymbol{a}_z\nabla^2 A_z \tag{4.38d}$$

or

$$\nabla^2\boldsymbol{A}=\boldsymbol{a}_x\left\{\nabla\bullet(\nabla A_x)\right\}+\boldsymbol{a}_y\left\{\nabla\bullet(\nabla A_y)\right\}+\boldsymbol{a}_z\left\{\nabla\bullet(\nabla A_z)\right\} \tag{4.38e}$$

Therefore, in Cartesian coordinates the Laplacian of a vector is the sum of the Laplacian of each scalar component with appropriate unit vectors multiplied; the Laplacian of each scalar component being the divergence of its gradient.

In Equations 4.31 and 4.36, the Laplacian operator may interact with any scalar (say V) or vector (say **A**). Such operation may result in four types of equations. These are referred to as *Laplace's*, *Poisson's*, *wave*, and *diffusion (or eddy current) equations*. These are discussed in Chapters 9, 10, 16, and 22, respectively.

Equation 4.38 ultimately leads to Equation 4.36.

4.5.2 Defined and Undefined Vector Operations

The vector operations discussed in Sections 4.2, 4.3, and 4.4 can be divided into the defined and undefined categories. The defined operations include gradient of a scalar, divergence, and curl of a vector, whereas the undefined operations include gradient of a vector and divergence and curl of a scalar. Thus $\nabla(\nabla \bullet \mathbf{V})$, $\nabla \bullet (\nabla S) \equiv \nabla^2 S$, $\nabla \bullet (\nabla \times \mathbf{V}) \equiv 0$, $\nabla \times (\nabla S) \equiv 0$, and $\nabla \times (\nabla \times \mathbf{V}) \equiv \nabla(\nabla \bullet \mathbf{V}) - \nabla^2 \mathbf{V}$ fall under the category of defined relations. Similarly, $\nabla(\nabla S)$, $\nabla(\nabla \times \mathbf{V})$, $\{\nabla \bullet (\nabla \cdot V)\}$, and $\{\nabla \times (\nabla \bullet V)\}$ are the undefined relations.

4.5.3 Vector Identities

The vector identities are helpful in simplifying the field analysis. Some of the commonly used identities obtained in gradient, divergence, and curl operations are listed below. In these identities, V and W are scalars and **A**, **B**, and **C** are vectors:

$$\nabla(V+W) \equiv \nabla V + \nabla W \tag{4.39a}$$

$$\nabla(VW) \equiv V(\nabla W) + W(\nabla V) \tag{4.39b}$$

$$\nabla(\boldsymbol{A} \bullet \boldsymbol{B}) \equiv (\boldsymbol{A} \bullet \nabla)\boldsymbol{B} + (\boldsymbol{B} \bullet \nabla)\boldsymbol{A} + \boldsymbol{A} \times (\nabla \times \boldsymbol{B}) + \boldsymbol{B} \times (\nabla \times \boldsymbol{A}) \tag{4.39c}$$

$$\nabla \bullet (\boldsymbol{A}+\boldsymbol{B}) \equiv \nabla \bullet \boldsymbol{A} + \nabla \bullet \boldsymbol{B} \tag{4.40a}$$

$$\nabla \bullet (V\boldsymbol{A}) \equiv \boldsymbol{A} \bullet (\nabla V) + V(\nabla \bullet \boldsymbol{A}) \tag{4.40b}$$

$$\nabla \bullet (\boldsymbol{A} \times \boldsymbol{B}) \equiv \boldsymbol{B} \bullet (\nabla \bullet \boldsymbol{A}) - \boldsymbol{A} \bullet (\nabla \times \boldsymbol{B}) \tag{4.40c}$$

$$\nabla \bullet (\nabla S) \equiv \nabla^2 S \tag{4.40d}$$

$$\nabla \bullet (\nabla \times V) \equiv 0 \tag{4.40e}$$

$$\nabla \times (\boldsymbol{A}+\boldsymbol{B}) \equiv \nabla \times \boldsymbol{A} + \nabla \times \boldsymbol{B} \tag{4.41a}$$

$$\nabla \times (V\boldsymbol{A}) \equiv (\nabla V) \times \boldsymbol{A} + V(\nabla \times \boldsymbol{A}) \tag{4.41b}$$

$$\nabla \times (\boldsymbol{A} \times \boldsymbol{B}) \equiv \boldsymbol{A}(\nabla \bullet \boldsymbol{B}) - \boldsymbol{B}(\nabla \bullet \boldsymbol{A}) + (\boldsymbol{B} \bullet \nabla)\boldsymbol{A} - (\boldsymbol{A} \bullet \nabla)\boldsymbol{B} \tag{4.41c}$$

$$\nabla \times (\nabla \times \boldsymbol{A}) \equiv \nabla(\nabla \bullet \boldsymbol{A}) - \nabla^2 \boldsymbol{A} \tag{4.41d}$$

$$\nabla \times (\nabla V) \equiv 0 \tag{4.41e}$$

4.6 INTEGRALS

The integrals are part of the field analysis. These can be classified as (i) definite and indefinite integrals, (ii) open and closed integrals, and (iii) line, surface, and volume integrals. This section briefly describes the line and surface integrals involving vectors. As before, the field vector F is used in both, the line and surface integrals.

4.6.1 Line Integral

Figure 4.11 shows a continuous curve *AB* described in the sense *A* to *B*. This curve, divided into n infinitesimal vector elements termed as $\boldsymbol{d\ell}_1, \boldsymbol{d\ell}_2, \ldots, \boldsymbol{d\ell}_n$, is located in a vector field $\boldsymbol{F}$ in space. Let the values of vector $\boldsymbol{F}$ at the centers of these elemental vector lengths be given as $\boldsymbol{F}_1, \boldsymbol{F}_2 \ldots$ and $\boldsymbol{F}_n$, respectively. The scalar products of the components of $\boldsymbol{F}$ and $\boldsymbol{d\ell}$ can be written as $\boldsymbol{F}_1 \bullet \boldsymbol{d\ell}_1$, $\boldsymbol{F}_2 \bullet \boldsymbol{d\ell}_2$, …, and $\boldsymbol{F}_n \bullet \boldsymbol{d\ell}_n$. The sum *S* of these scalar products along the entire length of the line can be written as:

$$S = \sum_{A}^{B} \boldsymbol{F}_n\big|_{\boldsymbol{d\ell}_n} \bullet \boldsymbol{d\ell}_n \tag{4.42a}$$

If the number of line segments is quite large (i.e. $n \to \infty$) the summation can be replaced by integration *I* and Equation 4.42a becomes:

$$I = \int_{A}^{B} \boldsymbol{F}\big|_{\boldsymbol{d\ell}} \bullet \boldsymbol{d\ell} \tag{4.42b}$$

Equation 4.42b is called the line integral of $\boldsymbol{F}$ along the curve *AB*, from *A* to *B*.

4.6.1.1 Line Integral in Different Coordinates

In the representation of vectors given by Equations 3.14, 3.24, and 3.34 and the expressions of incremental lengths given by Equations 3.11, 3.22, and 3.32 the line integral in different coordinate systems can be written as:

$$\int_A^B \boldsymbol{F} \bullet \boldsymbol{d\ell} = \begin{cases} \int_A^B \left[F_x dx + F_y dy + F_z dz\right] & \text{(Cartesian)} \quad \text{(a)} \\ \int_A^B \left[F_\rho d\rho + F_\varphi(\rho \cdot d\varphi) + F_Z dz\right] & \text{(Cylindrical)} \quad \text{(b)} \\ \int_A^B \left[F_r dr + F_\theta(r \cdot d\theta) + F_\varphi(r \sin\theta \cdot d\varphi)\right] & \text{(Spherical)} \quad \text{(c)} \end{cases} \tag{4.43}$$

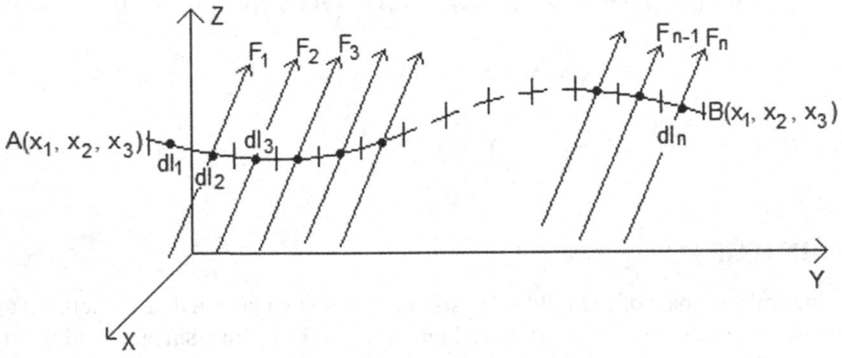

FIGURE 4.11 Line A-B, its segments, and F field.

If the coordinates of points A and B in Figure 4.11 are given as (x_1,y_1,z_1) and (x_2,y_2,z_2), respectively, the RHS of Equations 4.43a, 4.43b, and 4.43c can be rewritten as:

$$\left.\begin{aligned}
&\int_A^B \left[F_x dx + F_y dy + F_z dz\right] = \int_{x_1}^{x_2} F_x dx + \int_{y_1}^{y_2} F_y dy + \int_{z_1}^{z_2} F_z dz && \text{(a)}\\
&\int_A^B \left[F_\rho d\rho + F_\varphi(\rho.d\varphi) + F_Z dz\right] = \int_{\rho_1}^{\rho_2} F_\rho d\rho + \int_{\varphi_1}^{\varphi_2} F_\varphi(\rho.d\varphi) + \int_{z_1}^{z_2} F_z dz && \text{(b)}\\
&\int_A^B \left[F_r dr + F_\theta r d\theta + F_\varphi r\sin\theta d\varphi\right] = \int_{r_1}^{r_2} F_r dr + \int_{\theta_1}^{\theta_2} F_\theta r d\theta + \int_{\varphi_1}^{\varphi_2} F_\varphi r\sin\theta d\varphi && \text{(c)}
\end{aligned}\right\} \tag{4.44}$$

4.6.1.2 Evaluation of Line Integrals

The segments of Equation 4.44 give the expressions for line integrals in different coordinates. In these integrals A and B are known limits of integration, the integration must result in definite quantities. This cannot be obtained by integrating a component of a vector (say F_x) over only one variable (say x). This is because F_x could be a function of all three variables. Let us assume F_x a function of x, y, and z can be written as $F_x = f_1(x,y,z)$. Similarly $F_y = f_2(x,y,z)$ and $F_z = f_3(x,y,z)$. Thus, Equation 4.44a can be written as:

$$\int_A^B F \bullet d\ell = \int_{x_1}^{x_2} f_1(x,y,z)dx + \int_{y_1}^{y_2} f_2(x,y,z)dy + \int_{z_1}^{z_2} f_2(x,y,z)dz \tag{4.45a}$$

If some parametric relations between x, y, and z are known, through which y and z in the first integral, x and z in the second integral, and x and y in the third can be replaced only by x, y, and z, respectively, the integrals involved in Equation 4.45a can be transformed to:

$$\int_A^B F \bullet d\ell = \int_{x_1}^{x_2} f_1(x \text{ alone})dx + \int_{y_1}^{y_2} f_2(y \text{ alone})dy + \int_{z_1}^{z_2} f_2(z \text{ alone})dz \tag{4.45b}$$

Let us consider that points B and A are connected through a straight line. The equations for this line from B to A can be written in terms of the coordinates of B and A (viz. x_A, x_B, y_A, y_B, z_A, z_B, or x_1, x_2, y_1, y_2, z_1, and z_2):

$$\left.\begin{aligned}
y - y_B &= \frac{y_A - y_B}{x_A - x_B}(x - x_B) \quad \text{or} \quad y - y_2 = \frac{y_1 - y_2}{x_1 - x_2}(x - x_2) && \text{(a)}\\
z - z_B &= \frac{z_A - z_B}{y_A - y_B}(y - y_B) \quad \text{or} \quad z - z_2 = \frac{z_1 - z_2}{y_1 - y_2}(y - y_2) && \text{(b)}\\
x - x_B &= \frac{x_A - x_B}{z_A - z_B}(z - z_B) \quad \text{or} \quad x - x_2 = \frac{x_1 - x_2}{z_1 - z_2}(z - z_2) && \text{(c)}
\end{aligned}\right\} \tag{4.46}$$

For A to B the signs of these relations will get change. In these, Equation 4.46a relates y to x, Equation 4.46b relates z to y, and Equation 4.46c relates x to z. These relations can be used to get the integral of the form given by Equation 4.45b. A similar exercise will be required to transform Equations 4.44b and 4.44c. The method is further evident from the given examples of line integrals.

4.6.1.3 Applications

Line integrals arise in many field applications including the one described in Chapter 7 in relation to the work done, potential difference, and absolute potential.

4.6.2 Surface Integral

This section deals with the surface integrals involving vector quantities commonly encountered in field problems.

4.6.2.1 Surface and Its Types

In mathematics, a *surface* is a geometrical shape that resembles a deformed plane. A surface is referred to as *closed* if it has no boundary curves - a boundary curve being defined as a curve which separates the inside of a surface from the outside. A sphere, a cube, solid cylinder, and a torus are all closed surfaces. A cube is closed because it has edges but no boundary. A surface is said to be *open* if it has one or more boundary curves. A disk, a sheet of paper, a cylinder open at one end (but having no volume) is an open surface.

4.6.2.2 Evaluation of Surface Integral

Figure 4.12 shows a surface s through which a vector field $\boldsymbol{F}$ emerges in some specified direction. This surface is divided into n elemental vector surfaces, which can be identified as n vector surfaces $\Delta s_1, \Delta s_2, \ldots \Delta s_n$. If vector $\boldsymbol{F}$ assumes values $\boldsymbol{F}_1, \boldsymbol{F}_2, \ldots, \boldsymbol{F}_n$ at $\Delta s_1, \Delta s_2, \ldots \Delta s_n$ respectively the dot product $(\boldsymbol{F} \bullet \boldsymbol{ds})$ can be evaluated simply by summing these products $\boldsymbol{F}_1 \bullet \Delta s_1$, $\boldsymbol{F}_1 \bullet \Delta s_2$ etc. This sum can be written as:

$$\sum_{i=1}^{i=n} \boldsymbol{F}_i \bullet \Delta s_i \tag{4.47}$$

If the elemental surfaces are infinitesimally small (i.e. $\Delta s_i \to 0$) and n is very large $(n \to \infty)$ Equation 4.47 reduces to the surface integral. It can thus be written as:

$$\lim_{n\to\infty,\, \Delta s_i \to 0} \sum_{i=1}^{i=n} \boldsymbol{F}_i \bullet \Delta s_i = \iint_s \boldsymbol{F} \bullet \mathbf{ds} \tag{4.48a}$$

Here $\boldsymbol{F}$ is a function of three space variables, two variables along the elemental surface (where integration is to be performed) and one indicating the location of the elemental surface. This third variable may be accounted if Equation 4.48a is written as:

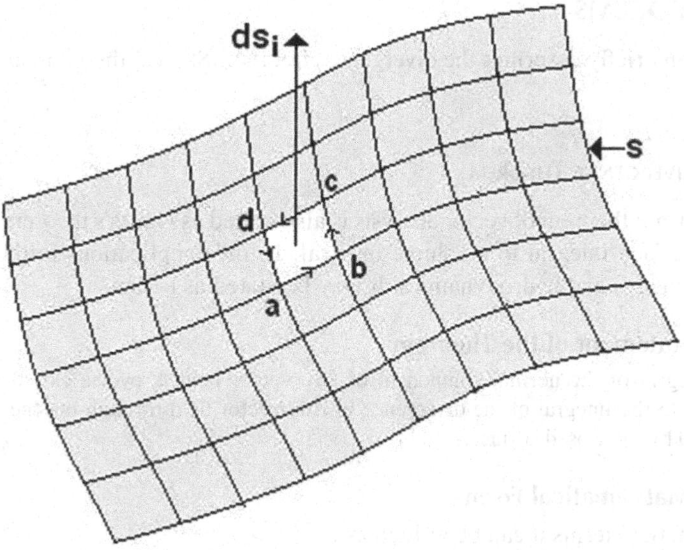

FIGURE 4.12 Surface s, its segments, and surface vector ds$_i$.

$$\lim_{n\to\infty,\,\Delta s_i\to 0}\sum_{i=1}^{i=n} F_i\Big|_{s_i}\bullet\Delta s_i=\iint_s F\Big|_s\bullet \mathbf{ds} \tag{4.48b}$$

Equation 4.48b can be written in terms of components as below

$$\left.\begin{aligned}
&\int_{z_1}^{z_2}\int_{y_1}^{y_2} F\Big|_x\bullet ds_x+\int_{z_1}^{z_2}\int_{x_1}^{x_2} F\Big|_y\bullet ds_y+\int_{y_1}^{y_2}\int_{x_1}^{x_2} F\Big|_z\bullet ds_z \quad \text{(Cartesian)} \quad \text{(a)}\\
&\int_{z_1}^{z_2}\int_{\varphi_1}^{\varphi_2} F\Big|_\rho\bullet ds_\rho+\int_{z_1}^{z_2}\int_{\rho_1}^{\rho_2} F\Big|_\varphi\bullet ds_\varphi+\int_{\varphi_1}^{\varphi_2}\int_{\rho_1}^{\rho_2} F\Big|_z\bullet ds_z \quad \text{(Cylindrical)} \quad \text{(b)}\\
&\int_{\varphi_1}^{\varphi_2}\int_{\theta_1}^{\theta_2} F\Big|_r\bullet ds_r+\int_{\varphi_1}^{\varphi_2}\int_{r_1}^{r_2} F\Big|_\theta\bullet ds_\theta+\int_{\theta_1}^{\theta_2}\int_{r_1}^{r_2} F\Big|_\varphi\bullet ds_\varphi \quad \text{(Spherical)} \quad \text{(c)}
\end{aligned}\right\} \tag{4.49}$$

Equation 4.48 is called the surface integral of ***F*** over the surface ***s***. The sign of elemental vector surface ***ds*** could be arbitrarily defined if the surface ***s*** is an open surface. If the surface ***s*** is a closed surface, the positive of the elemental vector surface ***ds*** must indicate the direction away from the closed surface, thus the outward normal is taken as positive. This outward normal will depend on which face the sign of vector surface is taken as positive. In the above relations vector ***F*** may represent any vector field quantity defined over the surface.

4.6.2.3 Applications

Such integrals may arise in many field applications including the one described in Chapter 11 in relation to the evaluation of the magnetic flux.

4.7 THEOREMS

This section briefly describes the divergence theorem, Stokes' theorem, and Green's theorem.

4.7.1 Divergence Theorem

This important theorem of vector analysis is also called as Gauss's theorem. It relates a closed surface integral to a volume integral. It finds applications in the fields of electromagnetics and hydrodynamics. It may be stated as below.

4.7.1.1 Statement of the Theorem

The integral of the normal component of any vector field **A** over a closed surface is equal to the integral of the divergence of this vector field through-out the volume enclosed by the closed surface.

4.7.1.2 Mathematical Form

In mathematical terms it can be written as

$$\oiint_s A\big|_s \bullet \boldsymbol{ds} \equiv \iiint_v (\nabla \bullet A)\, dv \tag{4.50}$$

where, s is the closed surface of the volume v.

On the LHS of this equation, the vector $\boldsymbol{A}$ must be taken on the surface $\boldsymbol{s}$; on the RHS of this equation, the divergence of the vector $\boldsymbol{A}$ should be taken at every point inside the volume.

4.7.2 Stokes' Theorem

This theorem relates a closed line integral to the surface integral. This finds frequent applications in electromagnetics and many other fields of engineering. Its statement and mathematical form are given below.

4.7.2.1 Statement of the Theorem

The integral of the curl of a vector field **A** taken over any surface s is equal to the closed line integral of **A** around the periphery c of the surfaces.

A corollary of this theorem states that: "The integral of the curl of a vector field **A** taken over any closed surface S is identically equal to zero."

4.7.2.2 Mathematical Form

In mathematical terms the above statements of Stokes' theorem can be written as:

$$\iint_s (\nabla \times A)\big|_s \bullet \boldsymbol{ds} = \oint_c A\big|_c \bullet d\ell \tag{4.51a}$$

and

$$\oiint_s (\nabla \times \boldsymbol{A})\Big|_s \bullet \boldsymbol{ds} \equiv 0 \tag{4.51b}$$

The RHS of this equation involves a line integral around the closed boundary c of an arbitrary open surface s. On the LHS of this equation, the surface integral is to be evaluated over any arbitrary surface with the same closed boundary c. For a given closed boundary, there will be infinitely different surfaces of various shapes and sizes. One such surface ($\boldsymbol{ds_i}$) and the closed path (a-b-c-d-a) are shown in Figure 4.12.

4.7.3 Green's Theorems

Let $\boldsymbol{A}$ be the product of scalar function u and the gradient of another scalar function w.

$$\boldsymbol{A} = u\nabla w \tag{4.52}$$

In Equation 4.52 the expression of divergence in the Cartesian system can be manipulated as below:

$$\nabla \bullet \boldsymbol{A} = \frac{\partial A_x}{\partial x} + \frac{\partial A_y}{\partial y} + \frac{\partial A_z}{\partial z} = \frac{\partial}{\partial x}\left(u\frac{\partial w}{\partial x}\right) + \frac{\partial}{\partial y}\left(u\frac{\partial w}{\partial y}\right) + \frac{\partial}{\partial z}\left(u\frac{\partial w}{\partial z}\right)$$

$$= u\left(\frac{\partial^2 w}{\partial x^2} + \frac{\partial^2 w}{\partial y^2} + \frac{\partial^2 w}{\partial z^2}\right) + \frac{\partial u}{\partial x}\frac{\partial w}{\partial x} + \frac{\partial u}{\partial y}\frac{\partial w}{\partial y} + \frac{\partial u}{\partial z}\frac{\partial w}{\partial z}$$

or,

$$\boxed{\nabla \bullet (u\nabla w) = u\nabla^2 w + \nabla u \bullet \nabla w} \tag{4.53}$$

4.7.3.1 First Form of Green's Theorem

Integration of Equation 4.52, over a volume v, and the application of Gauss' (or divergence) theorem gives the *first form of Green's theorem*, i.e.:

$$\boxed{\iiint_v \left(u\nabla^2 w + \nabla u \bullet \nabla w\right) dv = \oiint_S (u\nabla w) \bullet \boldsymbol{ds}} \tag{4.54}$$

4.7.3.2 Second Form of Green's Theorem

In Equation 4.52 interchange u and w to get:

$$\iiint_v \left(w\nabla^2 u + \nabla w \bullet \nabla u\right) dv = \oiint_S (w\nabla u) \bullet \boldsymbol{ds} \tag{4.55}$$

Subtraction of Equation 4.55 from Equation 4.54 gives:

$$\boxed{\iiint_v \left(u\nabla^2 w - w\nabla^2 u\right) dv = \oiint_S (u\nabla w - w\nabla u) \bullet \boldsymbol{ds}} \tag{4.56}$$

This transformation is referred to as the *second form of Green's theorem.*

These transformations are of extreme importance in the fields of electromagnetics and hydrodynamics.

4.8 OTHER USEFUL TOOLS

In Chapter 3 the coordinate systems, vector algebra, and the coordinate transformation were described in brief. Similarly, Sections 4.2 to 4.4 describe the gradient, divergence, and curl operations. The Del operator involved in the gradient, divergence, and curl operations is discussed in Subsection 4.2.5. In addition to these, the field analysis ultimately leads to a set of equations referred to as Maxwell's equations. These equations too involve divergence and curl operations and thus Del operator. The manipulation of Maxwell's equations gives another set of equations, referred to as Laplace, Poisson, wave, and eddy current equations. This latter set of equations involve the Laplacian operator given in Section 4.5. Both Del and Laplacian operators involve partial derivatives. As long the field problems remain two- or three-dimensional, the involved (Del and Laplacian) operators retain their partial derivative forms. In a one-dimensional case the partial derivatives are replaced by the complete derivatives.

In addition, there are many other areas of mathematics, which are required to handle practical field problems. The exponential, logarithmic, trigonometric, and hyperbolic functions, differentiation, and integration find very frequent applications in the problems related to the field theory. Similarly, involvement of summations of series, Fourier series, and Fourier integrals is quite imminent, particularly in the boundary value problems. In some cases, wherein harmonic fields are encountered, the use of complex numbers becomes imminent. Some of the problems require the use of linear algebraic equations and matrices whereas others require special functions (viz. Bessel's and Legendre functions). The reader will find the involvement of the above tools at one stage or the other in the ensuing text. Some of these, however, are more frequently used than others. A very brief description about the involvement of some of the tools is added to mentally prepare the reader to face the situation with confidence. In Chapter 1 it was noted that by the time a student is given the basic course on electromagnetics he/she is not only conversant but fully equipped with all the mathematical tools cited above. Lastly, in cases where closed form solutions are not available numerical techniques are used. Appendix A3 in the eResources includes some most commonly used relations belonging to the useful tools referred to above and Appendix A4 in the eResources briefly discusses some special functions.

Example 4.1

Find the gradient of: $V = 5(x+1)^2(y-2)^2(z+4)$ at $x = y = z = 1$.

SOLUTION

Given $V = 5(x+1)^2(y-2)^2(z+4)$.

In view of Equation 4.11a,

$$\nabla V = \frac{\partial V}{\partial x}\cdot \mathbf{a}_x + \frac{\partial V}{\partial y}\cdot \mathbf{a}_y + \frac{\partial V}{\partial z}\cdot \mathbf{a}_z$$

$$= 10(x+1)(y-2)^2(z+4)\mathbf{a}_x + 10(x+1)^2(y-2)(z+4)\mathbf{a}_y$$

$$+ 5(x+1)^2(y-2)^2\,\mathbf{a}_z$$

$$\nabla \mathbf{V}\big]_{x=y=z=1} = 100\mathbf{a}_x - 200\mathbf{a}_y + 20\mathbf{a}_z$$

Example 4.2

Find the gradient of $V = 3\rho^3\left(2\cos^2\varphi - \sin^2\varphi\right)z$ at $\rho = 1, \varphi = \frac{\pi}{4}, z = 2.$

SOLUTION

Given $V = 3\rho^3\left(2\cos^2\varphi - \sin^2\varphi\right)z$

Since $\sin^2\varphi = 1 - \cos^2\varphi,\ 2\cos^2\varphi - \sin^2\varphi = 3\cos^2\varphi - 1$

Further, $\cos^2\varphi = \dfrac{(\cos 2\varphi + 1)}{2}\quad 3\cos^2\varphi - 1 = \dfrac{3\cos 2\varphi + 1}{2}$

Thus $V = 3\rho^3(1.5\cos 2\varphi + 0.5)z$.

In view of Equation 4.11b:

$$\nabla V = \frac{\partial V}{\partial \rho}\cdot \mathbf{a}_\rho + \left(\frac{1}{\rho}\right)\frac{\partial V}{\partial \varphi}\cdot \mathbf{a}_\varphi + \frac{\partial V}{\partial z}\cdot \mathbf{a}_z$$

$$= 9\rho^2(1.5\cos 2\varphi + 0.5)z\,a_\rho - 9\rho^2\sin 2\varphi z\ a_\varphi$$

$$+ 3\rho^3(1.5\cos 2\varphi + 0.5)a_z$$

$$\nabla \mathbf{V}\big]_{\rho=1,\ \varphi=\frac{\pi}{4},\ z=2} = 9a_\rho - 18a_\varphi + 1.5a_z$$

Example 4.3

Find the gradient of $V = r^2\cos\theta\sin\varphi$ at $r = 2, \theta = \pi/4, \varphi = \pi/3$.

SOLUTION

Given $V = r^2\cos\theta\sin\varphi$.

In view of Equation 4.11c,

$$\nabla V = \frac{\partial V}{\partial r}\cdot \mathbf{a}_r + \left(\frac{1}{r}\right)\cdot\frac{\partial V}{\partial \theta}\cdot \mathbf{a}_\theta + \left(\frac{1}{r\cdot\sin\theta}\right)\cdot\frac{\partial V}{\partial \varphi}\cdot \mathbf{a}_\varphi$$

$$= 2r\cos\theta\sin\varphi\,\mathbf{a}_r - r\sin\theta\sin\varphi\,\mathbf{a}_\theta + r\cot\theta\cos\varphi\,\mathbf{a}_\varphi$$

$$\nabla V\Big]_{r=2,\,\theta=\frac{\pi}{4},\,\varphi=\frac{\pi}{3}} = 4\cdot\left(\frac{1}{\sqrt{2}}\right)\cdot\left(\frac{\sqrt{3}}{2}\right)\mathbf{a}_r - 2\cdot\left(\frac{1}{\sqrt{2}}\right)\cdot\left(\frac{\sqrt{3}}{2}\right)\mathbf{a}_\theta + 2\cdot 1\cdot(1/2)\,\mathbf{a}_\varphi$$

$$= 2\cdot\left(\frac{\sqrt{3}}{\sqrt{2}}\right)\mathbf{a}_r - \left(\frac{\sqrt{3}}{\sqrt{2}}\right)\mathbf{a}_\theta + \mathbf{a}_\varphi = 2.45\mathbf{a}_r - 1.225\mathbf{a}_\theta + \mathbf{a}_\varphi$$

Example 4.4

Evaluate the divergence of vector $\boldsymbol{D}$ (C/m²) at $x = 0.5$ if $\boldsymbol{D} = \left(6\mathbf{a}_x + 8\mathbf{a}_y + 10\mathbf{a}_z\right)$ $\mathbf{e}^{-3x}\mathbf{e}^{-4y}\mathbf{e}^{-5z}$.

SOLUTION

Given $\boldsymbol{D} = \left(6\mathbf{a}_x + 8\mathbf{a}_y + 10\mathbf{a}_z\right)e^{-3x}e^{-4y}e^{-5z}$

$$D_x = 6e^{-3x}e^{-4y}e^{-5z}\mathbf{a}_x \qquad D_y = 8e^{-3x}e^{-4y}e^{-5z}\mathbf{a}_y \qquad D_z = 10e^{-3x}e^{-4y}e^{-5z}\mathbf{a}_z$$

$$\frac{\partial D_x}{\partial x} = -18e^{-3x}e^{-4y}e^{-5z} \qquad \frac{\partial D_y}{\partial y} = -32e^{-3x}e^{-4y}e^{-5z} \qquad \frac{\partial D_z}{\partial z} = -50e^{-3x}e^{-4y}e^{-5z}$$

In view of Equation 4.21a,

$$\nabla\bullet\boldsymbol{D} \equiv \frac{\partial D_x}{\partial x} + \frac{\partial D_y}{\partial y} + \frac{\partial D_z}{\partial z}$$

$$= (-18-32-50)e^{-3x}e^{-4y}e^{-5z} = -100e^{-3x}e^{-4y}e^{-5z}$$

$$(\nabla\bullet\boldsymbol{D})\Big|_{x=0.5,\,y=0,\,z=0} = -100e^{-1.5} = -22.313$$

Example 4.5

Evaluate the divergence of vector $\boldsymbol{D}$ (C/m²) at $x = 0.5$ if $\boldsymbol{D} = (1/\rho^2)\mathbf{a}_\rho$.

SOLUTION

Given $\boldsymbol{D} = \dfrac{1}{\rho^2}\mathbf{a}_\rho = D_\rho a_\rho$ thus $D_\rho = \dfrac{1}{\rho^2}$, $D_\varphi = D_z = 0$

$$\rho D_\rho = \frac{1}{\rho} \qquad \frac{\partial(\rho\cdot D_\rho)}{\partial\rho} = -\frac{1}{\rho^2} \qquad \frac{1}{\rho}\cdot\frac{\partial(\rho\cdot D_\rho)}{\partial\rho} = -\frac{1}{\rho^3}$$

In view of Equation 4.21b,

$$\nabla\bullet\boldsymbol{D} \equiv \frac{1}{\rho}\cdot\frac{\partial(\rho\cdot D_\rho)}{\partial\rho} + \frac{1}{\rho}\cdot\frac{\partial D_\varphi}{\partial\varphi} + \frac{\partial D_z}{\partial z} \equiv -\frac{1}{\rho^3}$$

and

$$\nabla\bullet\boldsymbol{D} = -\frac{1}{\rho^3}\Big]_{\rho=x=0.5} = -8$$

Example 4.6

Evaluate the divergence of vector $\boldsymbol{D}$ (C/m²) at $x = 0.5$ if $= (1/r)\boldsymbol{a_r}$.

SOLUTION

Given $\boldsymbol{D} = \frac{1}{r} a_r = D_r a_r$ thus $D_r = \frac{1}{r}$, $D_\theta = D_\varphi = 0$

$$r^2 D_r = r, \frac{\partial\left(r^2 \cdot D_r\right)}{\partial r} = 1, \frac{1}{r^2} \cdot \frac{\partial\left(r^2 \cdot D_r\right)}{\partial r} = \frac{1}{r^2}$$

In view of Equation 4.21c,

$$\nabla \bullet \boldsymbol{D} \equiv \frac{1}{r^2} \cdot \frac{\partial\left(r^2 \cdot D_r\right)}{\partial r} + \frac{1}{r \cdot \sin\theta} \cdot \frac{\partial\left(\sin\theta \cdot D_\theta\right)}{\partial \theta} + \frac{1}{r \cdot \sin\theta} \cdot \frac{\partial\left(D_\varphi\right)}{\partial \varphi} = \frac{1}{r^2}$$

and

$$\nabla \bullet \boldsymbol{D} = \frac{1}{r^2}\Bigg]_{r=x=0.5} = 4$$

Example 4.7

Evaluate the divergence of vector $= 5e^{2x} \cos y\boldsymbol{a_x} \quad 4e^{4x} \sin y\boldsymbol{a_y} + 3ze^{5x}\boldsymbol{a_z}$ at $x = y = z = 0$.

SOLUTION

Given $\boldsymbol{F} = 5e^{2x} \cos y\boldsymbol{a_x} \quad 4e^{4x} \sin y\boldsymbol{a_y} + 3ze^{5x}\boldsymbol{a_z}$

$$F_x = 5e^{2x} \cos y \qquad \frac{\partial F_x}{\partial x} = 10e^{2x} \cos y \qquad \left.\frac{\partial F_x}{\partial x}\right|_{x=0,\, y=0,\, z=0} = 10$$

$$F_y = -4e^{4x} siny \qquad \frac{\partial F_y}{\partial y} = -4e^{4x} \cos y \qquad \left.\frac{\partial F_y}{\partial y}\right|_{x=0,\, y=0,\, z=0} = -4$$

$$F_z = 3ze^{5x} \qquad \frac{\partial F_z}{\partial z} = 3e^{5x} \qquad \left.\frac{\partial F_z}{\partial z}\right|_{x=0,\, y-0,\, z=0} = 3$$

In view of Equation 4.21a:

$$\nabla \bullet \boldsymbol{F} = \frac{\partial F_x}{\partial x} + \frac{\partial F_y}{\partial y} + \frac{\partial F_z}{\partial z} = 10 - 4 + 3 = 9$$

Example 4.8

Find the curl of vector $\boldsymbol{F} = 2xy\boldsymbol{a_x} + \left(z^2 - y^2\right)\boldsymbol{a_y} + 3yz\boldsymbol{a_z}$.

SOLUTION

Given $\boldsymbol{F} = 2xy\boldsymbol{a_x} + \left(z^2 - y^2\right)\boldsymbol{a_y} + 3yz\boldsymbol{a_z}$

$$F_x = 2xy \qquad \frac{\partial F_x}{\partial z} = 0 \qquad \frac{\partial F_x}{\partial y} = 2x$$

$$F_y = z^2 - y^2 \qquad \frac{\partial F_y}{\partial z} = 2z \qquad \frac{\partial F_y}{\partial x} = 0$$

$$F_z = 3yz \qquad \frac{\partial F_z}{\partial y} = 3z \qquad \frac{\partial F_z}{\partial x} = 0$$

In view of Equation 4.31a and the above values of derivatives:

$$\nabla \times \boldsymbol{F} = \left[\frac{\partial F_z}{\partial y} - \frac{\partial F_y}{\partial z}\right]\boldsymbol{a}_x + \left[\frac{\partial F_x}{\partial z} - \frac{\partial F_z}{\partial x}\right]\boldsymbol{a}_y + \left[\frac{\partial F_y}{\partial x} - \frac{\partial F_x}{\partial y}\right]\boldsymbol{a}_z$$

$$= [3z - 2z]\boldsymbol{a}_x + [0 - 0]\boldsymbol{a}_y + [0 - 2x]\boldsymbol{a}_z = z\boldsymbol{a}_x - 2x\boldsymbol{a}_z$$

Example 4.9

Find the curl of vector $\boldsymbol{F}$, if $\boldsymbol{F} = \left(2 - \frac{1}{\rho^2}\right)\sin\varphi\boldsymbol{a}_\rho + \left(2 + \frac{1}{\rho^2}\right)\cos\varphi\boldsymbol{a}_\varphi + 3\rho z\boldsymbol{a_z}$.

SOLUTION

Given $\boldsymbol{F} = \left(2 - \frac{1}{\rho^2}\right)\sin\varphi\boldsymbol{a}_\rho + \left(2 + \frac{1}{\rho^2}\right)\cos\varphi\, \boldsymbol{a}_\varphi + 3\rho z\boldsymbol{a_z}$

$$F_\rho = \left(2 - \frac{1}{\rho^2}\right)\sin\varphi \qquad \frac{\partial F_\rho}{\partial z} = 0 \qquad \frac{\partial F_\rho}{\partial \varphi} = \left(2 - \frac{1}{\rho^2}\right)\cos\varphi$$

$$F_\varphi = \left(2 + \frac{1}{\rho^2}\right) \qquad \frac{\partial F_\varphi}{\partial z} = 0 \qquad \frac{\partial(\rho F_\varphi)}{\partial \rho} = \left(2 - \frac{1}{\rho^2}\right)$$

$$F_z = 3\rho z \qquad \frac{\partial F_z}{\partial \varphi} = 0 \qquad \frac{\partial F_z}{\partial \rho} = 3z$$

In view of Equation 4.31b and the values of above derivatives:

$$\nabla \times \boldsymbol{F} = \left[\left(\frac{1}{\rho}\right)\cdot\frac{\partial F_z}{\partial \varphi} - \frac{\partial F_\varphi}{\partial z}\right]\boldsymbol{a}_\rho + \left[\frac{\partial F_\rho}{\partial z} - \frac{\partial F_z}{\partial \rho}\right]\boldsymbol{a}_\varphi + \left[\left(\frac{1}{\rho}\right)\cdot\left\{\frac{\partial(\rho F_\varphi)}{\partial \rho} - \frac{\partial F_\rho}{\partial \varphi}\right\}\right]\boldsymbol{a}_z$$

$$= [0 - 0]\boldsymbol{a}_\rho + [0 - 3z]\boldsymbol{a}_\varphi + \left[\left(\frac{1}{\rho}\right)\cdot\left\{\left(2 - \frac{1}{\rho^2}\right) - \left(2 - \frac{1}{\rho^2}\right)\cos\varphi\right\}\right]\boldsymbol{a}_z$$

$$= -3z\boldsymbol{a}_\varphi + \left(\frac{2}{\rho} - \frac{1}{\rho^3}\right)(1 - \cos\varphi)\,\boldsymbol{a}_z$$

Example 4.10

Find the curl of vector $\boldsymbol{F}$, if $\boldsymbol{F} = r\sin\varphi\,\boldsymbol{a}_r + r^2\cos\theta\cos\varphi\,\boldsymbol{a}_\theta + 3\sin\theta\sin\varphi\,\boldsymbol{a}_\varphi$.

SOLUTION

Given $\boldsymbol{F} = r\sin\varphi\,\boldsymbol{a}_r + r^2\cos\theta\cos\varphi\,\boldsymbol{a}_\theta + 3\sin\theta\sin\varphi\,\boldsymbol{a}_\varphi$

$$F_r = r\sin\varphi \qquad \frac{\partial F_r}{\partial\theta} = 0 \qquad \frac{1}{r}\left(\frac{1}{\sin\theta}\right)\cdot\frac{\partial F_r}{\partial\varphi} = \frac{\cos\varphi}{\sin\theta}$$

$$F_\theta = r^2\cos\theta\cos\varphi \qquad \frac{1}{r\cdot\sin\theta}\frac{\partial F_\theta}{\partial\varphi} = -r\cot\theta\sin\varphi \qquad \frac{1}{r}\frac{\partial(r\cdot F_\theta)}{\partial r} = 3r\cos\theta\cos\varphi$$

$$F_\varphi = 3\sin\theta\sin\varphi \qquad \frac{1}{r}\frac{\partial(r\cdot F_\varphi)}{\partial r} = \frac{3}{r}\sin\theta\sin\varphi \qquad \frac{1}{r\cdot\sin\theta}\frac{\partial(\sin\theta\cdot F_\varphi)}{\partial\theta} = \frac{6}{r}\cos\theta\sin\varphi$$

In view of Equation 4.31c and the above evaluated terms:

$$\nabla\times\boldsymbol{F} = \left[\left(\frac{1}{r\cdot\sin\theta}\right)\cdot\left\{\frac{\partial(\sin\theta\cdot F_\varphi)}{\partial\theta} - \frac{\partial F_\theta}{\partial\varphi}\right\}\right]\boldsymbol{a}_r$$

$$+\left[\left(\frac{1}{r}\right)\cdot\left\{\left(\frac{1}{\sin\theta}\right)\cdot\frac{\partial F_r}{\partial\varphi} - \frac{\partial(r\cdot F_\varphi)}{\partial r}\right\}\right]\boldsymbol{a}_\theta + \left[\left(\frac{1}{r}\right)\cdot\left\{\frac{\partial(r\cdot F_\theta)}{\partial r} - \frac{\partial F_r}{\partial\theta}\right\}\right]\boldsymbol{a}_\varphi$$

$$= \left[\frac{6}{r}\cos\theta\sin\varphi - r\cot\theta\sin\varphi\right]\boldsymbol{a}_r + \left[\frac{\cos\varphi}{\sin\theta} - 3r\cos\theta\cos\varphi\right]\boldsymbol{a}_\theta$$

$$+\left[3r\cos\theta\cos\varphi\right]\boldsymbol{a}_\varphi$$

Example 4.11

Evaluate the line integral between points A ($x = 0.8$, $y = 0.6$, $z = 1$) and B ($x = 1$, $y = 0$, $z = 1$), along the path defined as $x^2 + y^2 = 1$ and $z = 1$ if the vector field is given as $\boldsymbol{F} = y\boldsymbol{a}_x + x\boldsymbol{a}_y + 2\boldsymbol{a}_z$.

SOLUTION

Since $\boldsymbol{F} = y\boldsymbol{a}_x + x\boldsymbol{a}_y + 2\boldsymbol{a}_z = F_x\boldsymbol{a}_x + F_y\boldsymbol{a}_y + F_z\boldsymbol{a}_z$ or $F_x = y$, $F_y = x$ and $F_z = 2$, in view of Equation 4.44a:

$$I = \int_{x_1}^{x_2} F_x dx + \int_{y_1}^{y_2} F_y dy + \int_{z_1}^{z_2} F_z dz = \int_{x_1}^{x_2} y dx + \int_{y_1}^{y_2} x dy + \int_{z_1}^{z_2} 2dz$$

$$= \int_{0.8}^{1} y dx + \int_{0.6}^{0} x dy + \int_{1}^{1} 2dz = \int_{0.8}^{1} y dx + \int_{0.6}^{0} x dy + 0$$

Since $x^2 + y^2 = 1\, x = \sqrt{1-y^2}$ and $y = \sqrt{1-x^2}$

$$\text{Thus } I = \int_{0.8}^{1} \sqrt{(1\ \ x^2)}dx + \int_{0.6}^{0} \sqrt{(1\ \ y^2)}dy$$

$$\text{Since } \int \sqrt{(a^2 - x^2)}\, dx = \frac{x}{2}\sqrt{(a^2 - x^2)} + \frac{a^2}{2}\sin^{-1}\frac{x}{a} + C$$

On substituting $a = 1$ we get $\int \sqrt{(1 - x^2)}\, dx = \frac{x}{2}\sqrt{(1 - x^2)} + \frac{1}{2}\sin^{-1} x + C$

Thus:

$$I = \left[\frac{x}{2}\sqrt{(1 - x^2)} + \frac{1}{2}\sin^{-1} x\right]_{0.8}^{1} + \left[\frac{y}{2}\sqrt{(1 - y^2)} + \frac{1}{2}\sin^{-1} y\right]_{0.6}^{0}$$

$$= \frac{1}{2}\big[(0 + 1.571 - 0.48 - 0.927) + (0 + 0 - 0.48 - 0.644)\big] = -0.48$$

Example 4.12

Evaluate the line integral between points A ($\rho = 1$, $\varphi = 0°$, $z = 0$) and B ($\rho = 1$, $\varphi = 90°$, $z = 5$), if vector field $\boldsymbol{F} = \frac{1}{\rho}\boldsymbol{a}_\rho - \frac{1}{\rho}\boldsymbol{a}_\varphi + 2\boldsymbol{a}_z$, and the paths followed is specified by $z = 10\varphi$, $\rho = 1$.

SOLUTION

Since $\boldsymbol{F} = \frac{1}{\rho}\boldsymbol{a}_\rho - \frac{1}{\rho}\boldsymbol{a}_\varphi + 2\boldsymbol{a}_z$ thus $F_\rho = \frac{1}{\rho}$, $F_\varphi = -\frac{1}{\rho}$, $F_z = 2$.

In view of Equation 4.44b, the line integral (say I) is:

$$I = \int_{\rho_1}^{\rho_2} F_\rho d\rho + \int_{\varphi_1}^{\varphi_2} F_\varphi (\rho \cdot d\varphi) + \int_{z_1}^{z_2} F_z dz = \int_{1}^{1} \frac{1}{\rho} d\rho - \int_{0}^{90} \frac{1}{\rho}(\rho \cdot d\varphi) + \int_{0}^{5} 2dz$$

$$= \rho\Big|_1^1 - \varphi\Big|_0^{90} + 2z\Big|_0^5 = 0 - \frac{\pi}{2} + 10 = 8.43$$

Example 4.13

Points A ($r = 1$, $\theta = \pi/4$, $\varphi = 0°$) and B ($r = 0.5$, $\theta = \pi/3$, $\varphi = \pi/2$) are located in a non-uniform field $\boldsymbol{F} = \left(\frac{1}{r^2}\right)\boldsymbol{a}_r - \left(\frac{1}{r\sin\theta}\right)\boldsymbol{a}_\theta - \left(\frac{1}{r\sin\theta}\right)\boldsymbol{a}_\varphi$. Evaluate the line integral from A to B along the path proceeding in the direction $\boldsymbol{a}_\varphi$, $\boldsymbol{a}_\theta$, and $-\boldsymbol{a}_r$ and in that sequence.

SOLUTION

$$\text{Since } \boldsymbol{F} = \left(\frac{1}{r^2}\right)\boldsymbol{a}_r - \left(\frac{1}{r\sin\theta}\right)\boldsymbol{a}_\theta - \left(\frac{1}{r\sin\theta}\right)\boldsymbol{a}_\varphi = F_r\boldsymbol{a}_r - F_\theta\boldsymbol{a}_\theta - F_\varphi\boldsymbol{a}_\varphi,$$

Thus $F_r = \left(\frac{1}{r^2}\right)$, $F_\theta = \frac{1}{r\sin\theta}$ and $F_\varphi = \frac{1}{r\sin\theta}$

In view of Equation 4.44c:

$$I = \int_{r_1}^{r_2} F_r dr + \int_{\theta_1}^{\theta_2} F_\theta r d\theta + \int_{\varphi_1}^{\varphi_2} F_\varphi r\sin\theta d\varphi = \int_{1}^{0.5} \frac{1}{r^2} dr - \int_{\frac{\pi}{4}}^{\frac{\pi}{3}} \frac{1}{r\sin\theta} r d\theta - \int_{0}^{\pi/2} \frac{1}{r\sin\theta} r\sin\theta d\varphi$$

$$= \int_{1}^{0.5} \frac{1}{r^2} dr - \int_{\frac{\pi}{4}}^{\frac{\pi}{3}} \frac{d\theta}{\sin\theta} - \int_{0}^{\pi/2} d\varphi = -\frac{1}{r}\Big|_1^{0.5} - \log\tan\left(\frac{\theta}{2}\right)\Big|_{\pi/4}^{\pi/3} - \varphi\Big|_0^{\pi/2}$$

$$= -1 - \log\tan\left(\frac{\pi}{6}\right) + \log\tan\left(\frac{\pi}{8}\right) - \frac{\pi}{2}$$

$$= -1 - \log(0.5236) + \log(0.3929) - \frac{\pi}{2}$$

$$= -1 - \log\left(\frac{0.5236}{0.3929}\right) - \frac{\pi}{2} = -1 - \log(1.33) - \frac{\pi}{2} = -1 - 0.1247 - \frac{\pi}{2} = -2.695$$

Example 4.14

Evaluate the surface integral over the surface of a cylindrical segment of radius ρ = 2 cm extending from 0 to 1 cm along the z-axis and $\frac{\pi}{2}$ to π over φ axis. The surface is located in the field $\boldsymbol{F} = 5\frac{\cos(\varphi/2)}{\rho}\boldsymbol{a}_\rho$.

SOLUTION

Given $\boldsymbol{F} = 5\frac{\cos(\varphi/2)}{\rho}\boldsymbol{a}_\rho$, $\rho = 2$ cm, $0 \le z \le 1$ cm and $\frac{\pi}{2} \le \varphi \le \pi$

In view of Equations 4.49b and 3.23a, surface integral (say I) can be written as:

$$I = \iint_s \boldsymbol{F} \cdot \boldsymbol{ds} = \int_{\frac{\pi}{2}}^{\pi}\int_{0}^{1} 5\frac{\cos(\varphi/2)}{\rho}\boldsymbol{a}_\rho \cdot \rho \cdot d\varphi \cdot dz \cdot \boldsymbol{a}_\rho = 5\int_{\frac{\pi}{2}}^{\pi}\int_{0}^{1} \cos(\varphi/2) d\varphi \cdot dz$$

$$= 5\int_{\frac{\pi}{2}}^{\pi} \cos(\varphi/2) d\varphi = 2.5\sin(\varphi/2)\Big|_{\frac{\pi}{2}}^{\pi}$$

$$= 2.5\left(1 - \frac{1}{\sqrt{2}}\right) = 0.73$$

Example 4.15

Evaluate both sides of the divergence theorem in a region $-3 \le x \le 3, -2 \le y \le 2,$ $-5 \le z \le 5$ if $\boldsymbol{D} = z\boldsymbol{a_z}$.

SOLUTION

In view of Equation 4.50, the mathematical form of divergence theorem is written as:

$$\oiint_s \boldsymbol{D}|_s \bullet \boldsymbol{ds} \equiv \iiint_v (\nabla \bullet \boldsymbol{D}) dv$$

The left-hand side (LHS) and right-hand-side (RHS) of this equation can be evaluated as here:

$$\text{LHS} = \oiint_s \boldsymbol{D}|_s \bullet \boldsymbol{ds} = \int_{y=-2}^{2} \int_{x=-3}^{3} z\boldsymbol{a_z} \bullet dxdy\boldsymbol{a_z} = 24z$$

The flux emerging from $z = -5$ $24z = 24 \times 5 = 120$
The flux emerging from $z = 5$ $24z = 24 \times 5 = 120$
Total flux = 120 + 120 = 240

$$\text{RHS} = \iiint_v (\nabla \bullet \boldsymbol{D}) dv = \int_{z=-5}^{5} \int_{y=-2}^{2} \int_{x=-3}^{3} \left(\frac{\partial D_x}{\partial x} + \frac{\partial D_y}{\partial y} + \frac{\partial D_z}{\partial z} \right) dxdydz$$

Since $\boldsymbol{D} = z\boldsymbol{a_z}$ $D_z = z$ $D_x = D_y = 0$ $\dfrac{\partial D_z}{\partial z} = 1$

$$\text{RHS} = \int_{z=-5}^{5} \int_{y=-2}^{2} \int_{x=-3}^{3} \left(\frac{\partial D_z}{\partial z} \right) dxdydz = \int_{-5}^{5} \int_{-2}^{2} \int_{-3}^{3} 1 \cdot dxdydz = 240$$

Example 4.16

Prove that the vector $\boldsymbol{F} = x^2yz\boldsymbol{a_x} + xy^2z\boldsymbol{a_y} + xyz^2\boldsymbol{a_z}$ satisfies Stokes' theorem over a surface lying in the $z = 1$ plane and bounded by the four points (0, 0), (1, 0), (1, 1), (0, 1).

SOLUTION

In view of Equation 4.51a, Stokes' theorem is given as:

$$\iint_s (\nabla \times \boldsymbol{F})|_s \bullet \boldsymbol{ds} = \oint_c \boldsymbol{F}|_c \bullet \boldsymbol{d\ell}$$

This relation involves $\nabla \times \boldsymbol{F}$ given by Equation 4.31a, surface vector $\boldsymbol{ds}$ given by Equation 3.12c and elemental length $d\ell$ given by Equation 3.11, all in rectangular coordinate system. These are:

$$\nabla \times \boldsymbol{F} = \left[\frac{\partial F_z}{\partial y} - \frac{\partial F_y}{\partial z}\right]\mathbf{a}_x + \left[\frac{\partial F_x}{\partial z} - \frac{\partial F_z}{\partial x}\right]\mathbf{a}_y + \left[\frac{\partial F_y}{\partial x} - \frac{\partial F_x}{\partial y}\right]\mathbf{a}_z$$

$$\boldsymbol{ds} = dxdy\mathbf{a}_z \qquad \boldsymbol{d\ell} = dx\mathbf{a}_x + dy\mathbf{a}_y + dz\mathbf{a}_z$$

The LHS and RHS of Stokes' theorem are:

$$\text{LHS} = \iint_s (\nabla \times \boldsymbol{F})\big|_s \bullet \boldsymbol{ds} \quad \text{and} \quad \text{RHS} = \oint_C \boldsymbol{F}\big|_C \bullet \boldsymbol{d\ell}$$

Since $\boldsymbol{F} = x^2yz\mathbf{a}_x + xy^2z\mathbf{a}_y + xyz^2\mathbf{a}_z = F_x\mathbf{a}_x + F_y\mathbf{a}_y + F_z\mathbf{a}_z$,

$$F_x = x^2yz \quad F_y = xy^2z \quad F_z = xyz^2$$

$$\nabla \times \boldsymbol{F} = \left[\frac{\partial(xyz^2)}{\partial y} - \frac{\partial(xy^2z)}{\partial z}\right]\mathbf{a}_x + \left[\frac{\partial(x^2yz)}{\partial z} - \frac{\partial(xyz^2)}{\partial x}\right]\mathbf{a}_y + \left[\frac{\partial(xy^2z)}{\partial x} - \frac{\partial(x^2yz)}{\partial y}\right]\mathbf{a}_z$$

$$= (xz^2 - xy^2)\mathbf{a}_x + (x^2y - yz^2)\mathbf{a}_y + (y^2z - x^2z)\mathbf{a}_z$$

$$(\nabla \times \boldsymbol{F})\big|_{z=1} = (x - xy^2)\mathbf{a}_x + (x^2y - y)\mathbf{a}_y + (y^2 - x^2)\mathbf{a}_z$$

$$\iint_s (\nabla \times \boldsymbol{F})\big|_{z=1} \bullet \boldsymbol{ds} = \iint_s \left[(x - xy^2)\mathbf{a}_x + (x^2y - y)\mathbf{a}_y + (y^2 - x^2)\mathbf{a}_z\right] \bullet dxdy\mathbf{a}_z$$

$$= \iint_s (y^2 - x^2)dxdy = \int_{x=0}^{1}\int_{y=0}^{1}(y^2 - x^2)dxdy$$

$$= \int_{y=0}^{1}\left(y^2x - \frac{x^3}{3}\right)\Bigg|_{x=0}^{x=1} dy = \int_{y=0}^{1}\left(y^2 - \frac{1}{3}\right)dy = \left(\frac{y^3}{3} - \frac{y}{3}\right)_{y=0}^{y=1} = 0$$

On the RHS, the integral is to be carried out around the close loop lying in the $z = 1$, $\boldsymbol{d\ell} = dx\mathbf{a}_x + dy\mathbf{a}_y$ and the value of $\boldsymbol{F}$ is:

$$\boldsymbol{F}\big|_{z=1} = (x^2yz\mathbf{a}_x + xy^2z\mathbf{a}_y + xyz^2\mathbf{a}_z)_{z=1} = (x^2y\mathbf{a}_x + xy^2\mathbf{a}_y + xy\mathbf{a}_z)$$

$$= F_x\mathbf{a}_x + F_y\mathbf{a}_y + F_z\mathbf{a}_z$$

Thus $F_x = x^2y$, $F_y = xy^2$, $F_z = xy$

$$\boldsymbol{F} \bullet \boldsymbol{d\ell} = (F_x\mathbf{a}_x + F_y\mathbf{a}_y + F_z\mathbf{a}_z) \bullet (dx\mathbf{a}_x + dy\mathbf{a}_y) = F_xdx + F_ydy$$

The sequence of integrals [(0, 0), (1, 0), (1, 1), (0, 1)] is shown in the figure.

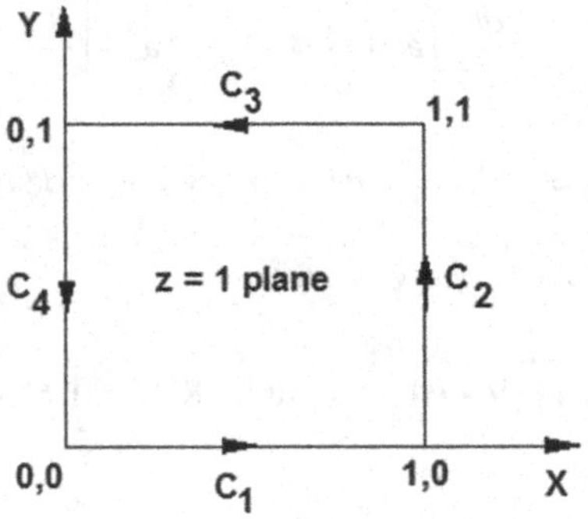

Thus:

$$\text{RHS} = \oint_C \boldsymbol{F}\big|_c \bullet d\boldsymbol{\ell} = \int_{C1} F_x dx + \int_{C2} F_y dy + \int_{C3} F_x dx + \int_{C4} F_y dy$$

$$= \int_0^1 \left(x^2 y\right)_{y=0} dx + \int_0^1 \left(xy^2\right)_{x=1} dy + \int_1^0 \left(x^2 y\right)_{y=1} (-dx) + \int_1^0 \left(xy^2\right)_{x=0} (-dy)$$

$$= 0 + \int_0^1 -y^2 dy + \int_1^0 x^2 dx + 0 = -\frac{y^3}{3}\bigg|_0^1 + \frac{x^3}{3}\bigg|_0^1 = -\frac{1}{3} + \frac{1}{3} = 0$$

Thus, LHS = RHS and the given vector satisfies Stokes' theorem.

PROBLEMS

P4.1 Find the gradient of the scalar field $V = 5x^2y^3z^4$ at $x = 1$, $y = 2$, $z = -1$.

P4.2 Find the gradient of $V = 5\rho^3\left(3\cos^2\varphi - 2\sin^2\varphi\right)z^2$ at $\rho = 2, \varphi = \pi/2, z = 1$.

P4.3 Find the gradient of the scalar field $V = r^2\left(2\cos^2\theta + 3\sin^2\varphi\right)$ at $r = 1$, $\theta = \dfrac{\pi}{3}, \varphi = \pi/6$.

P4.4 Evaluate the divergence of vector $\boldsymbol{F} = 3e^x \sin y\, \boldsymbol{a_x} \quad 4e^{2x} \sin y\, \boldsymbol{a_y} + 3e^{3z}\, \boldsymbol{a_z}$ at $x = y = z = 0$.

P4.5 Evaluate the divergence of vector $\boldsymbol{F} = \rho^3 \boldsymbol{a}_\rho + \rho^2 \boldsymbol{a}_\varphi - z^2\, \boldsymbol{a}_z$ at $\rho = 1$, $\varphi = \dfrac{\pi}{4}, z = 1$.

P4.6 Evaluate the divergence of vector $\boldsymbol{F} = r\boldsymbol{a}_r + \left(\dfrac{1}{r}\right)\boldsymbol{a}_\theta - \left(\dfrac{3}{r}\right)\boldsymbol{a}_\varphi$ at $r = 1$, $\theta = \pi/4, \varphi = \pi/6$.

P4.7 Evaluate the divergence of vector $\boldsymbol{F} = \rho^2\, \boldsymbol{a}_\rho + \rho\, \boldsymbol{a}_\varphi - 3z\, \boldsymbol{a}_z$ at $\rho = 2$, $\varphi = \pi/2, z = 1$.

P4.8 Find the divergence of the given field at the specified point: (a) $\boldsymbol{D} = 4x^3y^3z^2\boldsymbol{a_x} + 3x^3y^2z^3\boldsymbol{a_y} + 2x^2y^3z^3\boldsymbol{a_z}$ at (1,2,3) (b) $\boldsymbol{D} = z\sin\varphi\, \boldsymbol{a}_\rho +$

$z\cos\varphi\, \boldsymbol{a}_{\varphi} + \rho\sin\varphi\, \boldsymbol{a}_z$ at $\left(1, \frac{\pi}{3}, 2\right)$ (c) $\boldsymbol{D} = \frac{\sin\theta}{r}\boldsymbol{a}_r + \frac{\cos\theta(\ln r)}{r}\boldsymbol{a}_{\theta}$ at $\left(2, \frac{\pi}{3}, \frac{\pi}{2}\right)$.

P4.9 Evaluate the divergence of vector $\boldsymbol{F} = r^2\boldsymbol{a}_r + (1/r)\boldsymbol{a}_{\theta} - 3r\,\boldsymbol{a}_{\varphi}$ at $r = 1$, $\theta = \pi/6, \varphi = \pi/3$.

P4.10 Find the curl of a vector $\boldsymbol{F} = 2xz\,\boldsymbol{a}_x + (x^2 - y^2)\boldsymbol{a}_y + 3xy\,\boldsymbol{a}_z$.

P4.11 Find $\nabla \times H$ if $H = y^2 z a_x + 2(x+1)yz a_y - (x+1)z^2 a_z$.

P4.12 Find the curl of a vector $\boldsymbol{F} = \left(\frac{1}{\rho^2}\right)\sin\varphi\,\boldsymbol{a}_{\rho} + \left(\frac{1}{\rho}\right)\cos\varphi\,\boldsymbol{a}_{\varphi} + 3z\sin\varphi\,\boldsymbol{a}_z$.

P4.13 Find $\nabla \times \boldsymbol{H}$ if $\boldsymbol{H} = 5\rho\cos\varphi\boldsymbol{a}_{\rho} - 4\rho\sin\varphi\boldsymbol{a}_{\varphi} + 2z\boldsymbol{a}_z$.

P4.14 Find the curl of a vector $\boldsymbol{F} = r^2\sin\varphi\boldsymbol{a}_r + r\cos\theta\cos\varphi\,\boldsymbol{a}_{\theta} + 3r\sin\theta\sin\varphi\,\boldsymbol{a}_{\varphi}$.

P4.15 Find $\nabla \times H$ if $H = 4r\cos\theta\boldsymbol{a}_r - 2r\sin\theta\boldsymbol{a}_{\theta}$.

P4.16 Evaluate the line integral between points A (1,2,3) and B (2,4,5), if the path is identified by $y = x^2$ and the vector field is given as $\boldsymbol{F} = y\boldsymbol{a}_x + x\boldsymbol{a}_y + z\boldsymbol{a}_z$.

P4.17 Evaluate the line integral between points A (1,2,3) and B (3,3,0), if the path is specified by $x = y^2$ and the vector field $\boldsymbol{F} = y\boldsymbol{a}_x + x\boldsymbol{a}_y + z\boldsymbol{a}_z$.

P4.18 Evaluate the line integral between points $A(\rho = 1, \varphi = 0, z = 0)$ and $B(\rho = 2, \varphi = \pi/2, z = 5)$ if vector field $\boldsymbol{F} = \frac{1}{\rho}\boldsymbol{a}_{\rho} + \frac{z}{\rho}\boldsymbol{a}_{\varphi} + 5\boldsymbol{a}_z$, and the paths to be followed is identified by $z = 5\varphi/8$.

P4.19 Evaluate the line integral in the field $\boldsymbol{F} = r^2\boldsymbol{a}_r + \frac{1}{r\sin\theta}\boldsymbol{a}_{\theta} - \frac{1}{2r\sin\theta}\boldsymbol{a}_{\varphi}$ from $A(r = 1, \theta = \pi/6, \varphi = 0)$ to $B(r = 0.5, \theta = \pi/3, \varphi = \pi/2)$ by moving in the sequence $\boldsymbol{a}_{\varphi}, \boldsymbol{a}_{\theta}, -\boldsymbol{a}_r$.

P4.20 Evaluate the surface integral of $\nabla \times \boldsymbol{F}$ over the surface of a sphere of radius 'a' if vector $\boldsymbol{F} = r^2\sin\theta\,\boldsymbol{a}_r + r^2\sin\theta\cos\varphi\,\boldsymbol{a}_{\theta}$.

P4.21 Prove that vector $\boldsymbol{F} = \rho^2\cos^2\varphi\,\boldsymbol{a}_{\rho}$ satisfies divergence theorem over a closed surface specified by $\rho = 4$ and $0 < z < 1$.

P4.22 Prove that the field $\boldsymbol{F} = 2\rho^2\sin\varphi\,\boldsymbol{a}_{\varphi}$ satisfies the divergence theorem in the region bounded by $\rho = 3, 0.1\pi \le \varphi \le 0.2\pi, 0 \le z \le 5$.

P4.23 Evaluate both sides of the Stokes Theorem for vector $\boldsymbol{F} = 10x^2 y\boldsymbol{a}_x + 15xy^2\boldsymbol{a}_y + x^2y^2z^2\boldsymbol{a}_z$ over a surface lying in the $z = 3$ plane and bounded by the four points (0, 0), (0, 1), (1, 1), (1, 0).

P4.24 Evaluate both sides of the Stokes theorem for $\boldsymbol{A} = 2\rho^2(z+1)\sin^2\varphi\boldsymbol{a}_{\varphi}$ which exists over the portion of a cylindrical surface (Figure 4.3) defined by $\rho = 4, \frac{\pi}{4} \le \varphi \le \frac{\pi}{2}, 1 < z < 2$.

P4.25 Using the expression for the gradient of the scalar function *V*, find the expression for the Laplacian of *V* in the cylindrical system of space coordinates.

P4.26 Using the definition of the Laplacian for vector operand (Equation 4.36), find the expression for the Laplacian of vector ***F*** in the cylindrical system of space coordinates.

P4.27 Using the expression for the Del operator in the gradient of a scalar function, find the expression for the divergence of vector ***A*** in the cylindrical coordinate system.

DESCRIPTIVE QUESTIONS

Q4.1 What do you understand by gradient? How does it differ from the derivative? Give some examples of its application.

Q4.2 With the necessary illustrations explain the meaning of divergence. Give its physical interpretation. Describe the condition under which a vector field is referred to as a solenoidal field.

Q4.3 Explain the meaning of curl. Why it is also termed a rotation? Describe the condition under which a vector field is referred to as rotational field.

Q4.4 Using the expression for the gradient of the scalar function *V*, find the expression for the Laplacian of *V* in the spherical system of space coordinates.

Q4.5 Explain the meaning of line integral. Give its expressions in different coordinate systems.

Q4.6 Describe the procedure for the evaluation of line integral.

Q4.7 Describe the procedure for the evaluation of surface integral.

Q4.8 State and explain divergence theorem and its applications in field problems.

Q4.9 State and explain Stokes' theorem and its applications in field problems.

Q4.10 List various mathematical tools which are commonly required to solve field problems.

FURTHER READING

W. H. Hayt, Jr., J. A. Buck, *Engineering Electromagnetics*, 6th Ed, McGraw-Hill, New York, 2001.

M. N. O. Sadiku, *Elements of Electromagnetics*, 3rd Ed, Oxford University Press, New York, 2001.

N. N. Rao, *Elements of Engineering Electromagnetics*, 6th Ed, Prentice-Hall, Upper Saddle River, NJ, 2004.

M. F. Iskander, *Electromagnetic Fields and Waves*. Waveland Press, Prospect Hills, IL, 2000.

F. T. Ulaby, *Fundamentals of Applied Electromagnetics*. Prentice-Hall, Upper Saddle River, NJ, 2001.

U. S. Inan, A. S. Inan, *Engineering Electromagnetics*. Addison-Wesley-Longman, Menlo Park, CA, 1999.

D. K. Cheng, *Field and Wave Electromagnetics*, 2nd Ed. Addison-Wesley, Reading, MA, 1989.

J. D. Kraus, D. A. Fleisch, *Electromagnetics with Applications*, 5th Ed, McGraw-Hill, New York, 1999.

K. R. Demarest, *Engineering Electromagnetics.* Prentice-Hall, Upper Saddle River, NJ, 1998.

S. M. Wentworth, *Fundamentals of Electromagnetics with Engineering Applications.* Wiley, New York, 2005.

C. R. Paul, *Electromagnetics for Engineers with Applications.* Wiley, New York, 2004.

K. E. Lonngren, S. V. Savov, R. J. Jost, *Fundamentals of Electromagnetics with MATLAB*, 2nd Ed, SciTech, Raleigh, NC, 2007.

C. T. A. Johnk, *Engineering Electromagnetic Fields and Waves*, 2nd Ed. Wiley, New York, 1988.

N. Ida, *Engineering Electromagnetics*, 2nd Ed, Springer, New York, 2004.

D. J. Griffiths, *Introduction to Electrodynamics*, 3rd Ed, Pearson-Addison-Wesley, Upper Saddle River, NJ, 1999.

B. M. Notaroš, *Electromagnetics.* Pearson Prentice-Hall, Upper Saddle River, NJ, 2010.

L. A. Pipes, *Applied Mathematics*, 2nd Ed, McGraw Hill Book Co., New York, 1958.

Sunil Bhooshan, *Fundamentals of Engineering Electromagnetics.* Oxford University Press, 2012.

N. Piskunov, *Differential and Integral Calculus*, Vol. II, Mir Publishers, Moscow, 1974.

B. M. Budak, S. V. Fomin, *Multiple Integrals, Field Theory and Series.* Mir Publishers, Moscow, 1978.

5 Electric Field Intensity

5.1 INTRODUCTION

In Chapter 1, it was noted that electrons and protons possess a property called charge, later referred to as an electrical charge. The presence of stationary electrical charges in a region causes an effect commonly called static electricity or an electrostatic field. This static electricity was known to the Greeks as early as 600 BCE. In 1600 CE, Dr. Gilbert, a physician, explored electricity as a real and useful entity for the first time. It was also known that the charges of the same polarity repel and of different polarity attract each other.

A point charge exerts force on another point charge or charges located in its vicinity. A force is any physical cause that is capable of modifying the motion of a body. Charles Coulomb, a colonel in the French army, was the first person to evaluate this force due to charges. His experimental result, published in 1785, is now referred to as Coulomb's law. This chapter describes Coulomb's law of force and a related quantity called the electric field intensity.

5.2 COULOMB'S LAW AND ELECTRIC FIELD INTENSITY

This section deals first with Coulomb's law and its validity conditions. It also describes the similarity between electric and gravitational forces. It discusses Coulomb's electric force in vector form and its computation. Finally, it introduces electric field (or force) lines and the electric field intensity.

5.2.1 Coulomb's Law

Coulomb's law states:

> The repulsive force (F_1 or F_2) between two (point) charges (Q_1 and Q_2) located "R" distance apart in free space is directly proportional to the product of the two charges and inversely proportional to the square of the distance between the two charges.

In mathematical terms the magnitude of this force in the free space is given as:

$$|F_1| = k \cdot \frac{|Q_1 \cdot Q_2|}{R^2} \tag{5.1}$$

Where $|F_1|$ indicates the magnitude of the force F_1 acting on the charge Q_1, and k is the constant of proportionality. Its value in rationalized MKS unit can be written as:

$$k = \frac{1}{4\pi\varepsilon_o} \cong 9 \times 10^9 \text{ Nm}^2/\text{C}^2 \tag{5.2}$$

The term ε_o $\{\varepsilon_0 = 8.854 \times 10^{-12} \cong \frac{10^{-9}}{36\pi}$ Farad per meter (F/m)$\}$ is called the permittivity of free space (or vacuum). Thus the magnitude of force $\boldsymbol{F}_1$ acting on the charge Q_1 {in Newton (N)} given by Equation 5.1 can be rewritten as:

$$|\boldsymbol{F}_1| = \frac{1}{4\pi\varepsilon_o} \cdot \frac{|Q_1.Q_2|}{R^2} \tag{5.3}$$

The force $\boldsymbol{F}_2$ acting on the charge Q_2 is always equal in magnitude of the force acting on the charge Q_1. Thus,

$$|\boldsymbol{F}_2| = |\boldsymbol{F}_1| \tag{5.3a}$$

Coulomb's law of force given by Equation 5.3 spells that the magnitude of the force between any two charges is directly proportional to the product of charge contents of the two charges and inversely proportional to the square of the distance between them. Also in view of the given constant of proportionality, this force is inversely proportional to the media constant i.e. its permittivity.

5.2.2 Coulomb's Constant

The constant of proportionality k given by Equation 5.2 is sometimes referred to as Coulomb's constant. The quantity ke^2 i.e. the product of Coulomb's constant (k) and the square of electron charge $\{(e = -1.6 \times 10^{-19}\ \text{C})^2\}$ is another important quantity which is often used to describe the electric forces in atoms and nuclei.

5.2.3 Validity of Coulomb's Law

The relation of Equation 5.3 is true under the conditions illustrated by Figure 5.1.

This figure indicates that charges Q_1 and Q_2 must be concentrated at points 1 and 2. These points may be considered as two infinitesimally small spheres both with radius r (where r tends to zero). The separation (R) between the two charges is much greater than r (i.e. $R >> r$). Also the distance $\Re$ of the nearest point on the boundary from the two charges is much larger than the separation distance R (i.e. $\Re >> R$), thus the boundary of the region where the two charges are located must be far removed from the points where the charges are placed. Further, the charges must be stationary and time invariant. The medium containing the two charges must be linear, homogeneous, isotropic, and stationary with time invariant permittivity. Thus a point charge can be visualized as charge content of some specified value which is not only located at but confined to a point. In case of deviation of any of the above conditions the force given by Equation 5.3 will not represent the correct value of force. It is to be noted that electric sources are inherently monopole or point charges and thus Equation 5.3 can safely be applied to evaluate the force between the two point charges given by Coulomb's law.

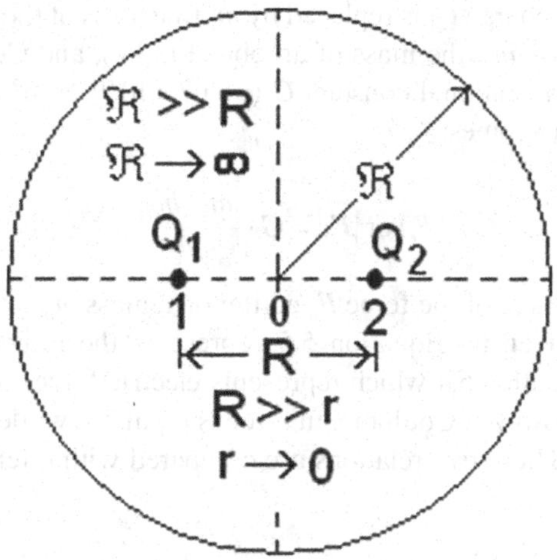

FIGURE 5.1 Illustration of conditions for validity of Coulomb's law.

In view of the above discussion, Coulomb's law is applicable only in the case of point charges or infinitesimally small volume. Charges of non-point form too exert force on other charges, but such cases cannot be easily tackled through Coulomb's law. As illustrated in Figure 5.2, if charges occupy some volume with non zero dimensions, then identification of the distance between the charges may create problems.

5.2.4 Similarity between Electric and Gravitational Forces

The relation of force given by Equation 5.1 can be rewritten by replacing distance R by d.

$$|F_1| = |F_2| = k \cdot \frac{|Q_1 \cdot Q_2|}{d^2} \tag{5.4}$$

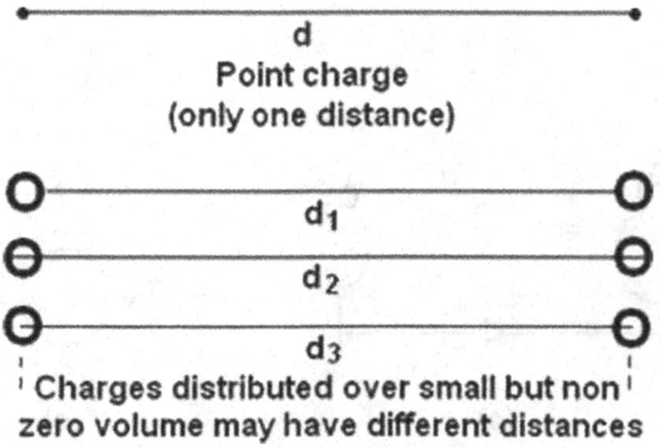

FIGURE 5.2 Distance identification problem.

If in Equation 5.4 the charge Q_1 is replaced by m_e (the mass of the earth in kg), charge Q_2 is replaced by mass m_0 (the mass of an object in kg), and Coulomb's constant k is replaced by the Gravitational constant G (=6.67 × 10^{-11} N-m²/kg² or m³/kg.s²) the resulting expression becomes:

$$|\boldsymbol{F}_e| \text{ or } |\boldsymbol{F}_o| = G \cdot \frac{|m_e \cdot m_o|}{d^2} \tag{5.5}$$

This gives the magnitude of the force $\boldsymbol{F}_o$ acting on a mass m_o.

The expression given by Equation 5.5 represents the gravitational force. It is similar to that of Equation 5.4 which represents electric force. Both forces obey the inverse square law, however, Coulombian force is repulsive while Newtonian force is attractive in nature. These two relations are compared with reference to Figure 5.3a and b.

5.2.5 Coulomb's Force in Vector Form

Since force is a vector quantity, this force vector will lie along a line oriented from the force exerting charge to the charge on which force is exerted. It can be expressed as

$$\boldsymbol{F}_{12} = \frac{1}{4\pi\varepsilon_o} \cdot \frac{Q_1.Q_2}{R_{12}^{\ 2}} \boldsymbol{a}_{R12} \tag{5.6a}$$

Where $\boldsymbol{F}_{12}$ is the vector force exerted by Q_1 on Q_2, $\boldsymbol{R}_{12}$ is the vector distance directed from point 1 to point 2, $|R_{12}|$ is the magnitude of $\boldsymbol{R}_{12}$, and $\boldsymbol{a}_{R12}$ (=$\boldsymbol{R}_{12}/|R_{12}|$) is the unit vector which is also directed from point 1 to 2.

Similarly the force exerted by Q_2 on Q_1 is given by

$$\boldsymbol{F}_{21} = \frac{1}{4\pi\varepsilon_o} \cdot \frac{Q_1.Q_2}{R_{21}^{\ 2}} \boldsymbol{a}_{R21} \tag{5.6b}$$

Where $\boldsymbol{R}_{21}$ is the vector distance, $|R_{21}|$ is its magnitude, and $\boldsymbol{a}_{R21}$ (=$\boldsymbol{R}_{21}/|R_{21}|$) is the unit vector directed from point 2 to point 1.

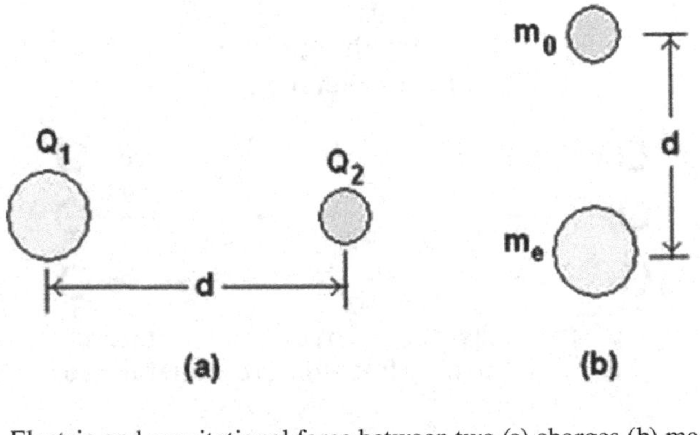

FIGURE 5.3 Electric and gravitational force between two (a) charges (b) masses.

The comparison of Equations 5.6a and 5.6b gives

$$\boldsymbol{a}_{R21} = -\boldsymbol{a}_{R12} \tag{5.6c}$$

If the two charges are of opposite sign, their product will be negative; since the magnitude of the force cannot be negative; this reverses the direction of the force acting on the two charges. It may thus be concluded that like charges repel and unlike charges attract each other.

In Equations 5.6a and b, if Q_1 or Q_2 is multiplied by a factor n the resulting force will also get multiplied by n. Thus, *Coulomb's law is linear in nature.*

5.2.6 Charge Accumulation

In connection with the electricity, the terms power, voltage, and current are understood even by a layperson. The power is measured in watts (W), the voltage in volts (V), and the current in amperes (A). The (instantaneous) electric power (w) is the product of voltage (v) and current (i) (i.e. $w = vi$). If a device consumes 100 W of electric power and operates at 100 V it will draw a current of 1A. As 1A current transports 1C of charge/sec through the conductor, thus 1C is the charge transported through the device in 1 second. Thus the force between two charge concentrations of 1C each, at points 1 m apart, obtained by applying Coulomb's law (Equation 5.3) comes out to be 9×10^9 Newtons or about 1.01 million tons, which is tremendously large. Since the two charges are of the same polarity this force is repulsive. The two charges will move apart from each other even if they have to rip themselves out of solid steel to do so. It may be noted that nature never allows concentration of 1C charge at one point and there is never much departure from electrical neutrality at a given point in a conductor.

5.2.7 Computation of Force

Let $\boldsymbol{F}_{12}$ be the force exerted by charge Q_1 {located at point 1 with coordinates (x_1, y_1, z_1)}on charge Q_2{located at point 2 with coordinates (x_2, y_2, z_2)}. It can be evaluated by using Equation 5.6a and the following steps.

1. As a thumb rule, subtract the coordinates of the charge which exerts force (point 1 or the source point) from that on which the force is exerted (point 2 or the field point), to get the distance vector. Thus:

$$\boldsymbol{R}_{12} = (x_2 - x_1)\boldsymbol{a}_x + (y_2 - y_1)\boldsymbol{a}_y + (z_2 - z_1)\boldsymbol{a}_z \tag{5.7a}$$

2. Identify the magnitudes of x, y, and z components of $\boldsymbol{R}_{12}$, i.e.:

$$(x_2 - x_1), (y_2 - y_1) \text{ and } (z_2 - z_1)$$

3. Sum the squares of the identified terms as:

$$\left\{(x_2 - x_1)^2 + (y_2 - y_1)^2 + (z_2 - z_1)^2\right\}$$

4. Take the square-root of this sum to get the magnitude of R_{12}:

$$|R_{12}| = \sqrt{\left\{(x_2 - x_1)^2 + (y_2 - y_1)^2 + (z_2 - z_1)^2\right\}} \tag{5.7b}$$

5. Evaluate the unit vector a_{R12}:

$$a_{R12} = R_{12} / |R_{12}| \tag{5.7c}$$

6. Substitute Equations 5.7a, b, and c into Equation 5.6a to get the vector force $\mathbf{F}_{12}$.
7. In order to evaluate $\mathbf{F}_{21}$ evaluate R_{21} $\{= (x_1 - x_2)a_x + (y_1 - y_2)a_y + (z_1 - z_2)a_z\}$
8. Evaluate $|R_{21}|$ and a_{R21} and substitute these into Equation 5.6b.
9. In the case of more charges, the force exerted by each charge is to be separately evaluated.
10. Estimate the total force by summing all these separately evaluated forces.

5.2.8 Magnitude of Force

To further understand the electric force, consider two spheres, each having a volume of 1 cm^3. These spheres are assumed to be of a good conducting material (say copper). Copper has one electron in the outer-most orbit of its atom which is fairly free to move around in the solid material. The density of metallic copper is about 9 gm/cm^3 and one mole of copper is 63.5 grams. Thus 1 cm^3 of copper contains about one-seventh of a mole or about 8.5×10^{22} copper atoms. With one mobile electron (having a charge of 1.6×10^{-19} C) per atom there will be about 13,600 C of potentially mobile charge in 1 cm^3 volume. If enough electrons are removed from the two spheres these spheres will be left with enough net positive charge on each of them. If one of the spheres is suspended over the other (Figure 5.3), the force (0.088 Newtons) to lift one of the spheres would be equal to its weight.

As shown in Figure 5.4, the repulsion is likely to cause the net charge to reside at points of spheres most distant from each other. The force of repulsion is set equal to

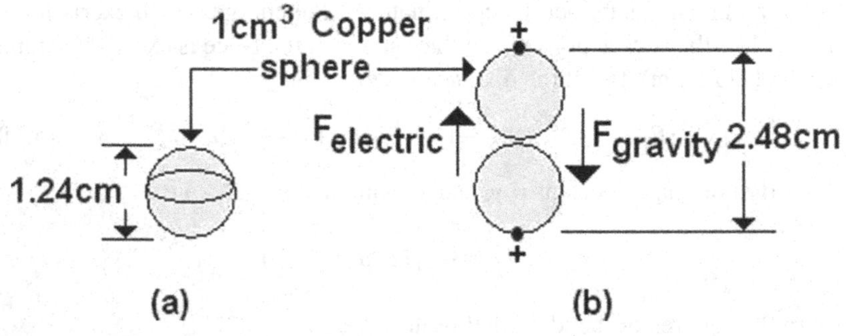

FIGURE 5.4 Force between two copper spheres.

the weight of a sphere. The diameter of a 1 cm^3 sphere is 1.24 cm, so the force can be treated as that between two point charges 2.48 cm apart (i.e., twice the sphere diameter apart). Using Coulomb's law, this requires a charge of 7.8×10^{-8} C. Compared to the total mobile charge of 13,600 C, this amounts to removing just one valence electron out of every 5.7 trillion (5.7×10^{12}) from each copper sphere. The final result is that removal of just one out of roughly six trillion of free electrons from each copper sphere would cause enough electric repulsion on the top sphere to lift it, overcoming the gravitational pull of the entire Earth.

5.2.9 Electric Flux (or Force) Lines

Figures 5.5a and b illustrate a positive charge and a negative charge. Both of these charges result in electric flux (or force) lines (only truncated line segments are shown). As can be seen, these lines uniformly emerge (radially outward in all directions) from an isolated positive charge and uniformly merge (radially inward from all directions) into an isolated negative charge. These are in fact imaginary lines along which a charge exerts force on another charge (or charges). Alternatively it can be stated that a force exerting charge transmits its force to another charge (or charges) along these imaginary lines referred to as electric flux (or force) lines.

These lines are like tentacles of an octopus or legs of a spider through which they catch their prey. The number of these force lines depends on the quantity of charge. More is the charge more lines emerge (or enter) and more intense is the force exerted on another charge or charges.

When both the charges are simultaneously present, electric flux lines originate from positive charge and end up at negative charge. A positive charge exerts force out and a negative charge exerts force in, both equally and symmetrically in all the directions. As shown in Figure 5.6, if the two charges are of opposite polarity, the force between these charges is attractive, and if they are of the same polarity the force between them is repulsive. Note that most flux lines are curved lines of various shapes and sizes.

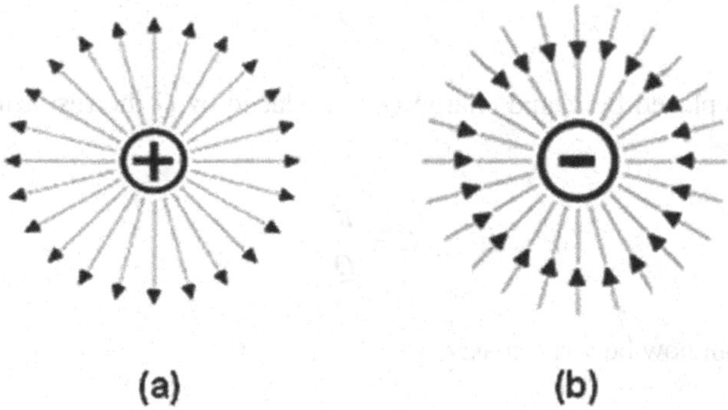

FIGURE 5.5 Electric field (or force) lines due to (a) positive charge (b) negative charge.

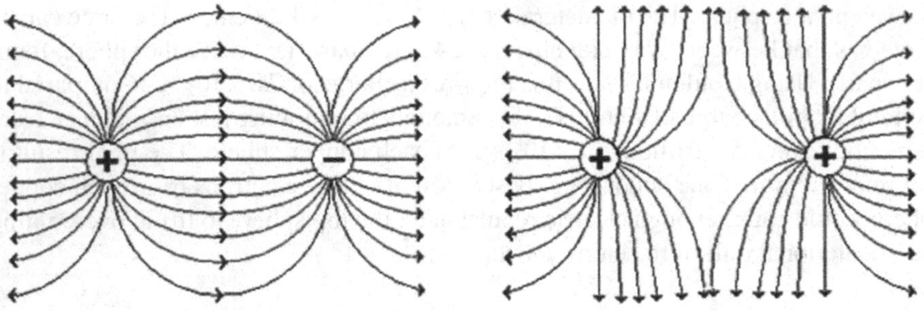

FIGURE 5.6 Electric field lines due to two (a) unlike charges placed between two large parallel conducting plates, (b) like charges placed in a rectangular conducting box. The plates and the box are not shown, flux lines are perpendicular to the conducting surfaces.

5.2.10 Electric Field Intensity

In Equation 5.6a, if charge Q_2 is replaced by Q_t (referred to as a test charge), the force $\boldsymbol{F}_t$ on Q_t due to Q_1 can be written as:

$$\boldsymbol{F}_{1t} = \frac{1}{4\pi\varepsilon_o} \cdot \frac{Q_1 \cdot Q_t}{R_{1t}{}^2} \boldsymbol{a}_{R1t} \stackrel{\text{def}}{=} \boldsymbol{F}_t \tag{5.8}$$

If both the sides of Equation 5.8 are divided by Q_t it reduces to:

$$\frac{\boldsymbol{F}_t}{Q_t} = \frac{1}{4\pi\varepsilon_o} \cdot \frac{Q_1}{R_{1t}^2} \boldsymbol{a}_{R1t} \tag{5.9a}$$

The ratio $\boldsymbol{F}_t/Q_t$ denoted by $\boldsymbol{E}$ can be written as:

$$\boldsymbol{E} \stackrel{\text{def}}{=} \frac{\boldsymbol{F}_t}{Q_t} \tag{5.9b}$$

Now if $\boldsymbol{F}_t$ is replaced by $\boldsymbol{F}$ and charge Q_t is replaced by Q the resulting expression becomes:

$$\boldsymbol{E} \stackrel{\text{def}}{=} \frac{\boldsymbol{F}}{Q} \tag{5.9c}$$

The force $\boldsymbol{F}$ can now be written as:

$$\boldsymbol{F} = Q\boldsymbol{E} \tag{5.9d}$$

5.3 FIELD DUE TO POINT CHARGES

This section first formulates the electric field due to a single point charge. The formulation is later extended to the discrete charge distribution. It also describes the analogy between gravitational and electric fields.

5.3.1 Field Due to a Single Point Charge

The vector quantity $\boldsymbol{E}$ is referred to as "Electric Field Intensity". It has the unit of Volt/meter (V/m). In view of Equations 5.9a and 5.9c, the expression of $\boldsymbol{E}$ due to a point charge Q becomes:

$$\boldsymbol{E} = \frac{1}{4\pi\varepsilon_o} \cdot \frac{Q}{R^2} \boldsymbol{a}_R \tag{5.10a}$$

Equation 5.10a can also be written in the following form:

$$\boldsymbol{E} = \frac{1}{4\pi\epsilon_0} \cdot \frac{Q}{R_{12}^2} \cdot \boldsymbol{a}_{R12} \tag{5.10b}$$

In Equations 5.10a and b the unit vector $\boldsymbol{a}_R$ or $\boldsymbol{a}_{R12}$ is directed from the source point where the charge is located to the field point where the electric field is measured. The symbol R or R_{12} stands for the distance between these two points. These relations give the electric field intensity, or simply electric field due to a point charge.

Since the measured electric field depends upon a reference frame, a more general definition of the electric field emerges from the Lorentz force law. The electric field is defined as the electromagnetic force (emf) per unit charge (i.e. $\boldsymbol{E}=\boldsymbol{F}/Q$) in the rest frame of the charge. Here $\boldsymbol{F}$ is in Newtons (N), charge is in Coulombs (C), and $\boldsymbol{E}$ is in N/C or V/m. The Lorentz force law spells that a charge that is moving relative to the source will experience part of the force as a magnetic force. Mathematically this can be written as:

$$\boldsymbol{F} = Q\boldsymbol{E} + Q(\boldsymbol{U} \times \boldsymbol{B}) \tag{5.11}$$

This relation is composed of two parts. The first one represents the electric force whereas the second represents the magnetic force. The Lorentz force law is described in Chapter 13.

In the above derivation a test charge Q_t is presumed to be located in the field of another charge Q. Figure 5.7a shows such a configuration wherein distance R is replaced by d. Figure 5.7b shows another configuration in which this test charge Q_t is located at a point in the field of two like or unlike charges (Q_1, and Q_2). For this case the net force exerted on Q_t can be evaluated simply by superimposing the forces due to the individual charges Q_1, and Q_2. Flux lines shown in these figures are those due to individual charges, flux lines due to the combined effects of all the charges present will, however, be different (see Figure 5.6).

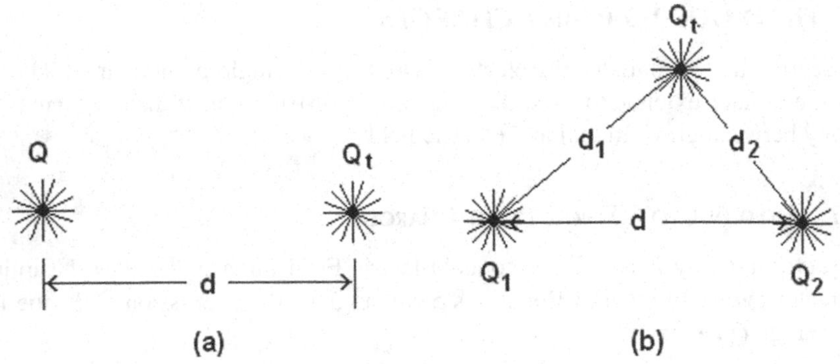

FIGURE 5.7 A test charge in the field of (a) another point charge (b) two point charges.

5.3.2 Field due to Discrete Charges

If charges of (different) specified values exist at different discrete locations in some space it may be called discrete charge distribution. These charges may be distributed along a line, over a surface or in a volume, each of arbitrary shape and size. Also these charges may have different polarities and magnitudes. Such charges may be denoted by Q_n wherein suffix n indicates the nth charge, one of those charges involved in the distribution. Figure 5.8 illustrates such discrete charge distribution along with the fields of individual charges.

In this case the net force can also be evaluated simply by superimposing the forces due to individual charges involved. Thus the electric field at a point due to n

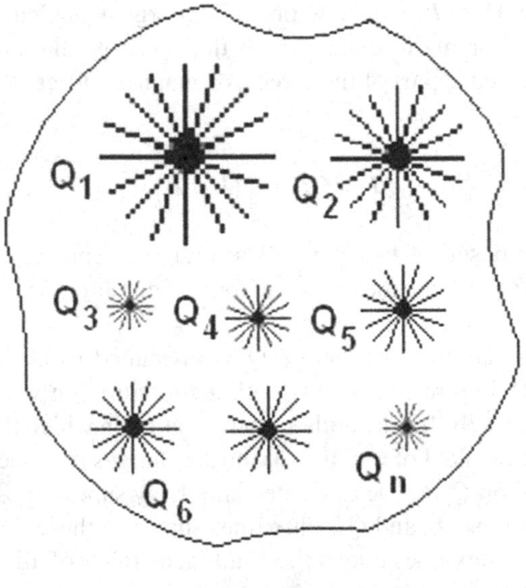

FIGURE 5.8 Discrete charge distribution and fields of individual charges.

discretely distributed point charges is the vector sum of the fields due to individual charges. It can be written as

$$\boldsymbol{E} = \frac{1}{4\pi\varepsilon_o}\cdot\left[\frac{Q_1}{R_1^2}\boldsymbol{a}_{R1} + \frac{Q_2}{R_2^2}\boldsymbol{a}_{R2} + \ldots + \frac{Q_n}{R_n^2}\boldsymbol{a}_{Rn}\right] = \frac{1}{4\pi\varepsilon_o}\cdot\sum_{m=1}^{n}\frac{Q_m}{R_m^2}\boldsymbol{a}_{Rm} \quad (5.12)$$

5.3.3 Analogy between Gravitational and Electric Fields

The expression for force given by Equation 5.9d is similar to that of the force of gravitational field. Correspondingly in Equation 5.5 if (Gm_e/d^2) is replaced by a parameter g, where g represents the acceleration due to gravity of the earth, the resulting expression for the magnitude of force acting on a mass m_o due to the gravitational field becomes:

$$F = m_o \cdot g \quad (5.13)$$

Since G (=6.67×10^{-11} Nm²/kg² or m³/kg.s²) is the universal gravitational constant, me (=5.98×10^{24} kg) is the mass of earth and d (=6.38×10^{6} m) is the radius of the earth. Thus:

$$g = G\frac{m_e}{d^2} = \left[\left[\frac{6.67\times10^{-11})\times(5.98\times10^{24}}{\left(6.38\times10^{6}\right)^2}\right]\right] \cong 9.8 \text{ m/s}^2 \quad (5.14)$$

As can be seen from Equation 5.9d the electric force is linearly proportional to the charge (Q) and in view of Equation 5.13 the Gravitational force is linearly proportional to the mass (m_0) of the object.

Besides the above charge distributions, wherein the field at a particular location is to be evaluated due to single point charge, two point charges or n discrete point charges of arbitrary values and polarities; the other charge configurations of interest include the continuous or piecewise continuous distribution of charges (a) along a line, (b) over a surface, or (c) in a volume. These cases require evaluation of line, surface, or volume integrals. All these integrals have to confirm to the appropriate limits imposed by the geometrical shapes and sizes. These cases are discussed in the subsequent three sections.

5.4 FIELD DUE TO LINE CHARGE

If charges are assumed to be continuously located one after another along a line the distribution is called line charge distribution. As an example, an electron-beam (with negligible diameter) in a Cathode Ray Tube (CRT) can be considered a line charge. Figure 5.9 shows some of the line charge distributions. These charge distributions along a straight line, a circular segment, and an arbitrary shaped line are shown in Figures 5.9a, b, and c respectively.

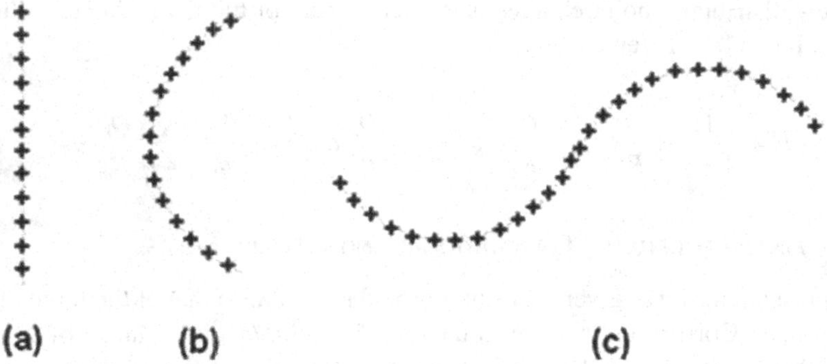

FIGURE 5.9 Charges along (a) a straight line (b) a circular segment (c) an arbitrary shaped line.

5.4.1 Forms of Line Charge Distribution

This distribution can be classified in terms of uniform, non-uniform, finite, or infinite line charges. The charge distribution is termed as uniform if the charge contents at each point along the line are the same, as in Figure 5.9. If the charge concentration differs from point to point along the distribution it is termed as non-uniform. The terms finite and infinite have their usual meaning in accordance with the length. Sometimes these usual meanings get modified in view of the distance from the field point or the location at which the field is to be computed. A short line from a closely located point may be considered as infinite whereas a long line from a very far distance may appear to be a finite line or even a point. In the following subsections the distribution is assumed to be uniform for both infinite and finite line charges.

5.4.2 Line Charge Density and the Total Charge

The line charge density ρ_L measured in C/m is given by:

$$\rho_L \stackrel{\text{def}}{=} \lim_{\Delta\ell\to 0} \frac{\Delta Q}{\Delta \ell} = \frac{dQ}{d\ell} \tag{5.15a}$$

For a given line charge density the total charge along a line can be obtained as:

$$Q = \int_{\ell} \rho_L \cdot d\ell \tag{5.15b}$$

5.4.3 Field Configurations due to Different Types of Line Charges

The configurations due to an infinite and finite line charges are described below.

5.4.3.1 Field due to an Infinite Line Charge

Figure 5.10a illustrates the two-dimensional view of the flux lines emerging out from a positive point charge. Consider a uniform line charge placed along the z-axis.

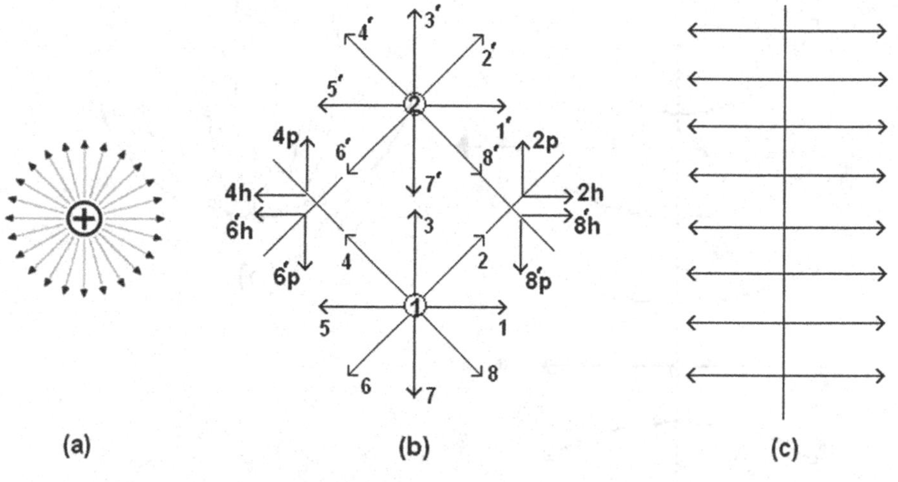

FIGURE 5.10 Field lines due to (a) an individual point charge (b) two point charges and cancellation of components (c) the entire infinite line charge in the final form.

This can be divided into a large number of charges each of infinitesimally small length that may be treated as point charges. Since point charges are located one over the other along a straight line, every point charge will result in flux lines shown in Figure 5.10a. Figure 5.10b shows two such point charges (1 and 2) placed along the z-axis.

As given in Figure 5.10b point charge-1 yields in lines 1 to 8 and point charge-2 in lines 1′ to 8′. Consider lines 1 to 5 due to charge-1 and 5′ to 8′ and 1′ of charge-2. In these lines, 1, 5, 5′, and 1′ are oriented along ρ-axis. Since the lines 3 and 7′ are oppositely oriented, fields due to these lines will get canceled. Also the field line-2 is inclined upward it can be resolved into perpendicular and horizontal components 2p and 2h respectively. Similarly the field line 8′ is inclined downward and it can also be resolved into 8′p and 8′h. For uniform charge, 2p and 8′p are equal and get canceled due to their opposite orientations. Owing to their same direction, the horizontal components 2h and 8′h get added to yield a ρ component. The study of line 4 and 6′also leads to the same conclusion. Thus for every (upward) inclined line due to charge-1, there exists a (downward) inclined line due to charge-2 which will yield ρ component. If a third charge is placed just above charge-2, a similar cancelation will take place. Thus if there are infinite charges along a line these will yield only ρ components. Figure 5.10c shows the final form of the field distribution due to a uniform line charge. It shows that an infinitely long straight line with uniform line charge density has no variation in the magnitude of $\boldsymbol{E}$ with z or φ and contains only ρ component of $\boldsymbol{E}$. Since the direction of the vector $\boldsymbol{E}$ changes with φ, this vector is a function of φ as well.

5.4.3.2 Field due to a Finite Line Charge

Figure 5.11 shows the field lines emerging out of a finite line charge. The line charge density has been assumed to be an even function of the z-coordinates. It shows that

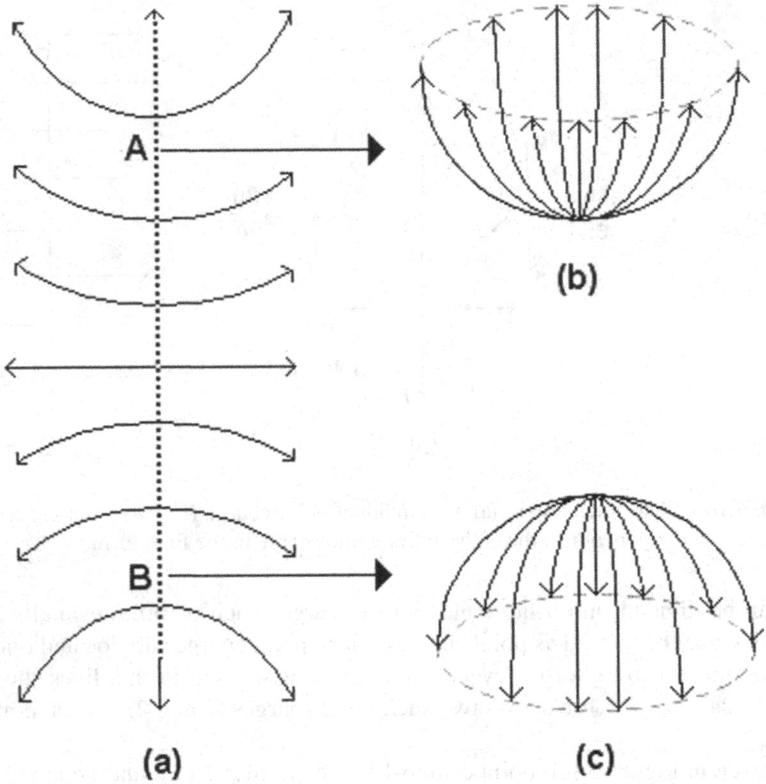

FIGURE 5.11 Field due to finite line charge.

the field lines start twisting upward (or downward) at the locations away from the center of the line. This is due to imperfect cancelation of perpendicular components due to inclined lines described above. The field lines become completely upward or downward at the uppermost and the lowermost tips of the line (of finite length). In this case the magnitude of the field variation is still independent of φ (due to symmetry) but varies with ρ and z. The field $\boldsymbol{E}$ has no φ-component. The z-component vanishes at $\rho=0$ and also ast $z=0$ plane. Figure 5.11 also illustrates three-dimensional field as it would appear at points A and B respectively marked on the line.

5.4.4 E Field due to Different Line Charge Distributions

Let ρ_L be the uniform charge density distributed over a line extended from $-\infty$ to $+\infty$ along the z-axis. In view of Equation 5.15a the charge contents in length $d\ell$ is $dQ\left(=\rho_L d\ell\right)$. This charge will result in the incremental electric field $d\boldsymbol{E}$ which can be written as:

$$d\boldsymbol{E} = \frac{dQ}{4\pi\varepsilon_o \rho^2}\boldsymbol{a}_\rho = \frac{\rho_L\, d\ell}{4\pi\varepsilon_o \rho^2}\boldsymbol{a}_\rho \tag{5.16a}$$

The total electric field $\boldsymbol{E}$ can now be written as:

$$\boldsymbol{E} = \int_L \frac{\rho_L \, d\ell}{4\pi\varepsilon_o \rho^2} \boldsymbol{a}_\rho \tag{5.16b}$$

In the following two subsections this integral is evaluated for obtaining field at a point P when it is located at (i) y-axis and (ii) (ρ,φ,z). Since line charge lies along the z-axis, in both the cases $d\ell$ is replaced by dz and thus $dQ=\rho_L dz$.

5.4.4.1 Field due to Uniform Line Charge at Point P on *y*-Axis

Figure 5.12 illustrates electric field components due to a uniform line charge along with the coordinate system. For convenience, the point P at which the field is to be evaluated is taken at the y-axis. As y and ρ axes are taken to be the same in view of Figure 5.12:

$$\left. dE_y = dE_\rho = dE \cdot \sin\theta \quad (\text{a}) \qquad \text{where, } \sin\theta = \frac{y}{R} = \frac{\rho}{R} \quad (\text{b}) \right\} \tag{5.17}$$

From Figure 5.12 vector $\boldsymbol{R}$, its magnitude $|\boldsymbol{R}|$ and the unit vector $\boldsymbol{a}_R$ can also be written as:

$$\left. \begin{array}{ll} \boldsymbol{R} = y\boldsymbol{a}_y - L\boldsymbol{a}_z = \rho\boldsymbol{a}_\rho - L\boldsymbol{a}_z & (\text{a}) \\ |\boldsymbol{R}| = \sqrt{\rho^2 + L^2} \quad (\text{b}) \quad \boldsymbol{a}_R = \dfrac{\boldsymbol{R}}{|\boldsymbol{R}|} = \dfrac{\rho\boldsymbol{a}_\rho - L\boldsymbol{a}_z}{\sqrt{\rho^2 + L^2}} & (\text{c}) \end{array} \right\} \tag{5.18}$$

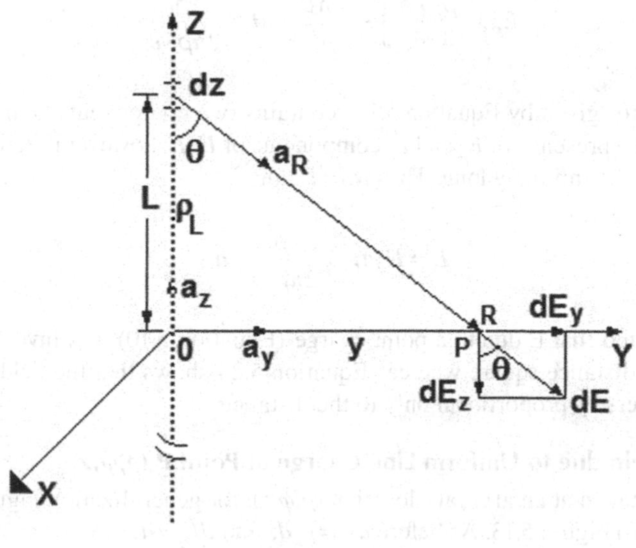

FIGURE 5.12 Electric field at point P located on the y-axis due to an infinite line charge.

In view of Equations 5.16, 5.17, and 5.18b:

$$dE_\rho = \frac{\rho_L dL}{4\pi\varepsilon_o R^2}\sin\theta = \frac{\rho_L dL}{4\pi\varepsilon_o R^2}\cdot\frac{\rho}{R} = \frac{\rho_L dL}{4\pi\varepsilon_o}\cdot\frac{\rho}{R^3} = \frac{\rho_L dL}{4\pi\varepsilon_o}\cdot\frac{\rho}{\left(\rho^2+L^2\right)^{3/2}} \tag{5.19}$$

An expression for E_ρ obtained by integrating dE_ρ over $-\infty$ to $+\infty$ is:

$$E_\rho = \int_{-\infty}^{+\infty}\frac{\rho_L}{4\pi\varepsilon_o}\cdot\frac{\rho}{\left(\rho^2+L^2\right)^{3/2}}\cdot dL = \frac{\rho_L\cdot\rho}{4\pi\varepsilon_o}\int_{-\infty}^{+\infty}\frac{dL}{\left(\rho^2+L^2\right)^{3/2}} \tag{5.20}$$

For simplification, substitute:

$$\left.\begin{array}{ll} L = \rho\cdot\cot\theta \quad (a) & dL = -\rho\cdot\operatorname{cosec}^2\theta\cdot d\theta \quad (b) \\ \rho^2+L^2 = \rho^2+\rho^2\cdot\cot^2\theta = \rho^2\cdot\operatorname{cosec}^2\theta \quad (c) & \\ \left(\rho^2+L^2\right)^{3/2} = \rho^3\cdot\operatorname{cosec}^3\theta \quad (d) & \dfrac{dL}{\left(\rho^2+L^2\right)^{3/2}} = -\dfrac{\sin\theta}{\rho^2}d\theta \quad (e) \end{array}\right\} \tag{5.21}$$

The limits of integration evaluated in view of the above substitution are:

$$\theta = \pi, \quad \text{for } L = -\infty \quad (a) \qquad \theta = 0, \quad \text{for } L = +\infty \quad (b)\} \tag{5.22a}$$

In view of Equation 5.21, Equation 5.20 reduces to:

$$E_\rho = \frac{\rho_L\cdot\rho}{4\pi\varepsilon_o}\int_{\pi}^{0} -\frac{\sin\theta}{\rho^2}\cdot d\theta = \frac{\rho_L}{2\pi\rho\cdot\varepsilon_o} \tag{5.23}$$

The unit vector given by Equation 5.18c contains two components (a and a_z) which amounts to the presence of E_ρ and E_z components of $\boldsymbol{E}$. E_z however is zero as the line is assumed to be infinitely long. Thus $|\boldsymbol{E}| = E_\rho$ and:

$$\boldsymbol{E} = E_\rho \boldsymbol{a}_\rho = \frac{\rho_L}{2\pi\rho\cdot\varepsilon_o}\boldsymbol{a}_\rho \tag{5.24}$$

It may be noted that E due to a point charge (Equation 5.10) was inversely proportional to the distance square whereas Equation 5.24 shows that the field due to line charge is inversely proportional only to the distance.

5.4.4.2 Field due to Uniform Line Charge at Point P (ρ,φ,z)

If point P is taken at an arbitrary location (ρ,φ,z), for generalization Figure 5.12 can be modified to Figure 5.13. As before $dQ = \rho_L dz$ and $dE_y = dE_\rho$.

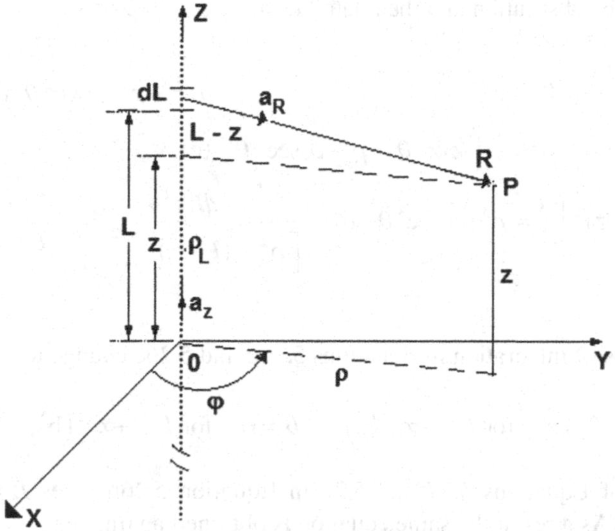

FIGURE 5.13 Electric field at an arbitrarily located point due to an infinite line charge.

The values of $\boldsymbol{R}$, its magnitude $|\boldsymbol{R}|$, and the unit vector $\boldsymbol{a_R}$ can be given as:

$$\left.\begin{aligned} &\boldsymbol{R} = \rho \boldsymbol{a}_\rho - (L-z)\boldsymbol{a_Z} \quad \text{(a)} \qquad |\boldsymbol{R}| = \sqrt{\rho^2 + (L-z)^2} \quad \text{(b)} \\ &\boldsymbol{a}_R = \frac{\boldsymbol{R}}{|\boldsymbol{R}|} = \frac{\rho \cdot \boldsymbol{a}_\rho - (L-Z)\boldsymbol{a}_z}{\sqrt{\rho^2 + (L-Z)^2}} \quad \text{(c)} \end{aligned}\right\} \quad (5.25)$$

Equation 5.25c shows that the unit vector $\boldsymbol{a_R}$ contains two unit vectors $\boldsymbol{a}_\rho$ and $\boldsymbol{a}_z$. Thus at first glance $\boldsymbol{E}$ may have E_ρ and E_z components.

In view of the above relations and Figure 4.13, the expression for the field becomes:

$$\boldsymbol{E} = \int_{L=-\infty}^{L=+\infty} \frac{\rho_L dL}{4\pi\varepsilon_o |\boldsymbol{R}|^2} \boldsymbol{a}_R = \int_{L=-\infty}^{L=+\infty} \frac{\rho_L \left[\rho \cdot \boldsymbol{a}_\rho - (L-z)\cdot \boldsymbol{a}_z\right]}{4\pi\varepsilon_o \left[\rho^2 + (L-z)^2\right]^{3/2}} dL \overset{\text{def}}{=} E_\rho \boldsymbol{a}_\rho + E_z \boldsymbol{a}_z \quad (5.26a)$$

Expressions for E_ρ and E_z can separately be written from Equation 5.26a:

$$E_\rho = \frac{\rho_L}{4\pi\varepsilon_o} \cdot \int_{L=-\infty}^{L=+\infty} \frac{\rho}{\left[\rho^2 + (L-z)^2\right]^{3/2}} dL \quad (5.26b)$$

$$E_z = \frac{\rho_L}{4\pi\varepsilon_o} \cdot \int_{L=-\infty}^{L=+\infty} \frac{-(L-z)}{\left[\rho^2 + (L-z)^2\right]^{3/2}} dL \quad (5.26c)$$

In this case the substitution and the related terms are:

$$\left.\begin{array}{ll} L-Z \overset{\text{def}}{=} \rho\cdot\cot\theta \quad \text{(a)} & dL = -\rho\cdot\operatorname{cosec}^2\theta\cdot d\theta \quad \text{(b)} \\ \rho^2+(L-z)^2 = \rho^2+\rho^2\cdot\cot^2\theta = \rho^2\cdot\operatorname{cosec}^2\theta \quad \text{(c)} & \\ \left[\rho^2+(L-z)^2\right]^{3/2} = \rho^3\cdot\operatorname{cosec}^3\theta \quad \text{(d)} & \dfrac{dL}{\left[\rho^2+(L-z)^2\right]^{3/2}} = -\dfrac{\sin\theta}{\theta^2}d\theta \quad \text{(e)} \end{array}\right\} \tag{5.27}$$

Also the limits of integration in Equations 5.26b and 5.26c change to:

$$\theta = \pi,\ \text{ for } L = -\infty\ \ \text{(a)},\quad \theta = 0,\ \text{ for } L = +\infty\ \ \text{(b)}\} \tag{5.28}$$

Substitution of Equations 5.27 and 5.28 in Equation 5.26b gives $E_z=0$ and thus $\boldsymbol{E}=E_\rho \boldsymbol{a}_\rho$ only. As a result the same equation is obtained again, i.e.:

$$\boldsymbol{E} = \frac{\rho_L}{2\pi\varepsilon_o\cdot\rho}\boldsymbol{a}_\rho \tag{5.29}$$

Figures 5.12 and 5.13 represent the simple cases of line charge distributions wherein the line charges were aligned along one of the coordinate axes (z). Also the line charges were taken to be infinitely long and thus $\boldsymbol{E}$ has only the ρ component in both cases. This section explores some slightly difficult cases, as described below. The method of evaluation for these cases will still remain the same but with the following considerations.

5.4.4.3 Field due to Finite and Asymmetrical Line Charges

Figure 5.14 illustrates three typical cases of line charge distributions. Figure 5.14a shows a line charge lying along the z-axis, Figure 5.14b along the y-axis, and Figure 5.14c along the x-axis. In all these cases the line charges are not only finite

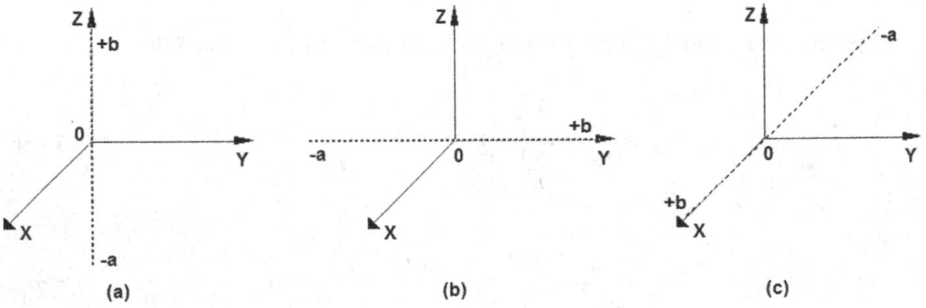

FIGURE 5.14 Line charge of finite length along (a) z-axis (b) y-axis (c) x-axis.

but asymmetrical vis-à-vis the origin of the coordinate system. The field due to any of these line charges can be evaluated by considering the following aspects:

- The field emerging out of these line charges will be symmetrical along the center of lines and not along the origin of the coordinate system.
- The field will be symmetrical along the periphery of the circle drawn with any of these line charges at its center.
- The field at the upper and lower tips of these lines will lie along the coordinate with which the line charge is aligned.
- The field between the upper and lower tips and the centers of these line charges will start twisting in the forward and backward directions as explained in Subsection 5.3.3.2.

5.4.4.4 Field due to an Infinite Line Charge Lying in the *x–y* Plane

Figure 5.15a shows an infinite line charge lying in the *x–y* plane. The field at any point on the *z*-axis can be evaluated by considering Figure 5.15b wherein the field components on the *z*-axis due to symmetrically located elemental segments of the line charge are shown. In view of this figure, only the *z*-component of $\boldsymbol{E}$ will survive and thus $\boldsymbol{E}=E_z\boldsymbol{a}_z$ only and no other component of $\boldsymbol{E}$ will be present. If the line charge is finite and symmetrically located vis-à-vis the origin of the coordinate axis, $\boldsymbol{E}=E_z\boldsymbol{a}_z$ on the *z*-axis, but at the two end tips of the line (lying in the *x–y* plane) $\boldsymbol{E}=E_{xy}\boldsymbol{a}_{xy}$. In the case that the line charge is finite but asymmetrical vis-à-vis the origin, the situation will get complicated and will have components along all the axes. The field will still be symmetrical at the center of the line and along the peripheral direction.

5.4.4.5 Field due to Two Infinite Line Charges in the *x–y* Plane

Figure 5.16a shows two infinite line charges lying in the *x–y* plane. The field at any point on the *z*-axis or elsewhere can be separately evaluated by considering Figure 5.16b and c. These fields can then be summed to get the net field.

To deal with such typical cases for the computation of E field at any desired point draw the line diagram representing the coordinate system, the location of line charge

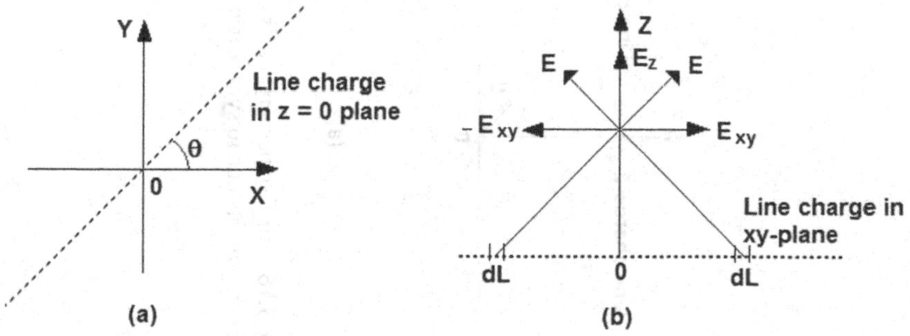

FIGURE 5.15 (a) Infinite line charge located in the *x–y* plane. (b) Field components on the *z*-axis.

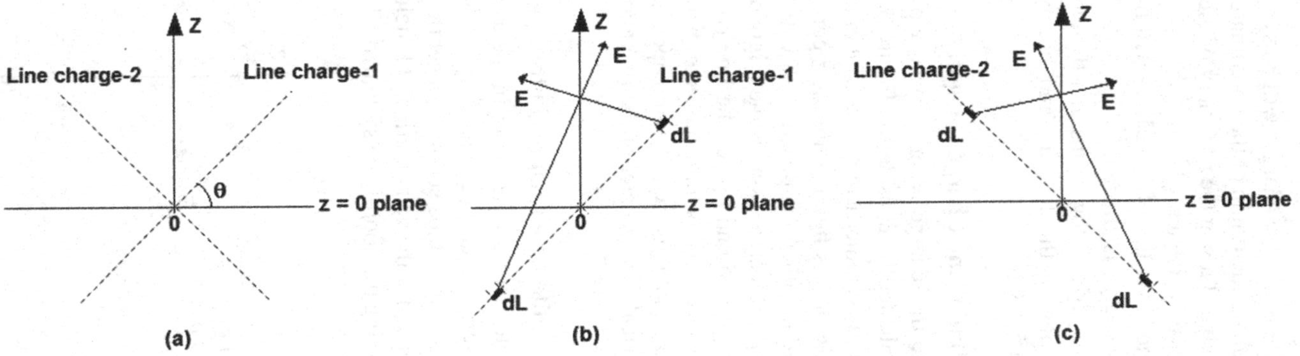

FIGURE 5.16 (a) Two perpendicular line charges lying in the $z=0$ plane. (b) Field components due to symmetrically located segments of line charge-1. (c) Field components due to symmetrically located segments of line charge-2.

(source point) therein and the location of the point at which the field is desired (field point). Using a little bit of imagination the following conclusions can be easily drawn.

- If the charge density along a line is positive the field will be emerging-out and if it is negative the field will be merging-in the line. This aspect can, however, be ignored as the sign of ρ_L itself accounts for this mathematically.
- From every infinite line charge the field will always be in the radial direction at every location of the line.
- If a line charge lies in an arbitrary direction, the perpendicular drawn to this line may be referred to as radial direction.
- In the case of the line charge of infinite or finite lengths, there will be no variation of field in the peripheral direction. This periphery corresponds to a cylinder wherein the line charge is taken at its central axis.
- In the case of finite line charge, the emerging (or merging) field will have only the radial component in its center and the tangential component at the two ends of the line. The field will have both radial and tangential components in between the center and the end. The radial component will dominate near the center and the tangential nearer to the ends.

5.5 SURFACE CHARGE DISTRIBUTION

If the charges are distributed over a surface (of any shape), or a sheet is formed due to continuous distribution of charges, such a distribution may be referred to as surface charge distribution. An analogy of such a sheet may be drawn with the Milky Way where the stars are so densely populated that the entire region appears to be a sheet of stars (or light). Figure 5.17 illustrates a few practical cases of surface charge distributions. Figure 5.17a shows a parallel plate capacitor wherein charges are distributed over the surfaces of two parallel (conducting) planes. Figure 5.17b shows a circular cylinder containing charges over its outer surface. Lastly Figure 5.17c illustrates a charged spherical surface.

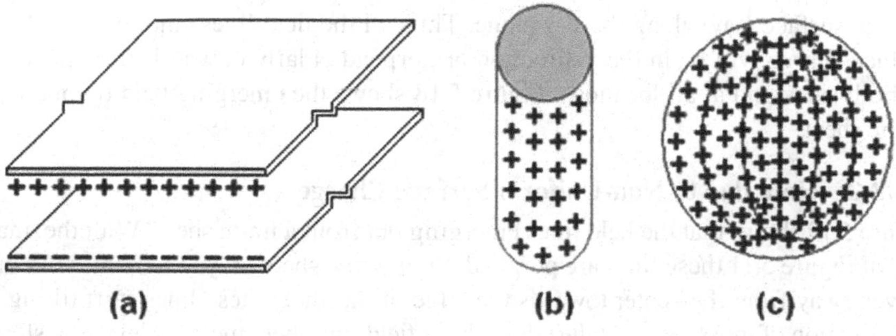

FIGURE 5.17 Configuration of charged (a) parallel plates (b) cylinder (c) sphere.

5.5.1 Forms of Surface Charge Distribution

The surface charge distribution can also be uniform or non-uniform. The charge distribution is termed as uniform if the charge contents at each point over the entire surface are the same. If the charge concentration over the surface differs from point to point it is termed as non-uniform. This distribution can also be of finite or infinitein extents. As in the case of line charge, this classification depends not only on the extent of the surface charge but also on its distance from the location of the field point or the point at which the field is to be computed. A short finite surface from a very closely located field point may be considered as an infinite surface charge whereas a large surface from a very far distance may appear to be a finite surface or even a point charge. In the following subsections the distribution is assumed to be uniform for both infinite and finite surface charges.

5.5.2 Surface Charge Density and the Total Charge

The surface charge density "ρ_s" measured in C/m^2 is given by $Q = \iint_s \rho_s \cdot ds$

$$\rho_s \stackrel{\text{def}}{=} \lim_{\Delta s \to 0} \frac{\Delta Q}{\Delta s} = \frac{dQ}{ds} \text{ C/m}^2 \tag{5.30a}$$

For a given surface charge densitythe total charge over a surface can be obtained as:

$$Q = \iint_s \rho_s \cdot ds \tag{5.30b}$$

5.5.3 Field Configurations for Different Surface Charge Distributions

This subsection deals with the fields due to uniform finite and infinite surface charge distributions.

5.5.3.1 Field due to Uniform Surface Charge

Figure 5.18 shows flux lines due to a uniform surface charge distribution over a flat infinite surface (say) along the x–y plane. Thus all the field lines emerging out from surface charge will be in the z-direction or perpendicularly outward from the sheet on both sides and at all locations. Figure 5.18 shows the emerging field on one face of the sheet.

5.5.3.2 Field due to Non-Uniform Surface Charge

Figure 5.19 shows that the field lines emerging out from a finite sheet. With the analogy of Figure 5.11 these lines are perpendicular to the sheet only at its center. As one moves away from the center towards the edge of the sheet these lines start tilting in the direction of movement. At the edges these field lines become parallel to the sheet. This tilting effect is shown by field lines at "A–A" and "B–B".

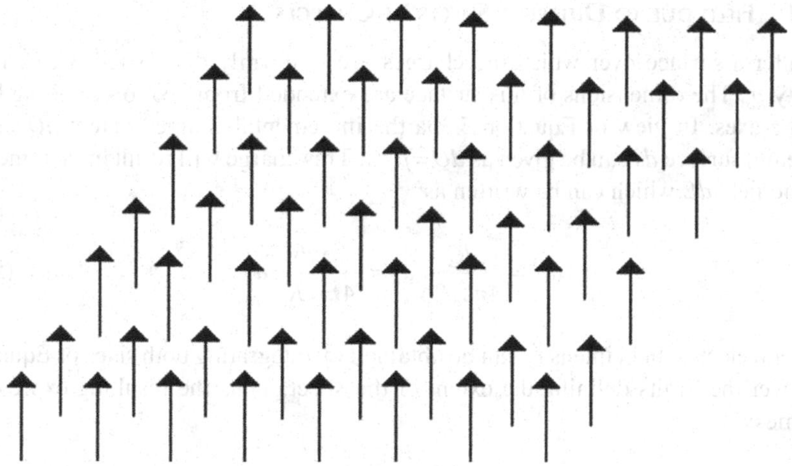

FIGURE 5.18 Field due to an infinite sheet of charge.

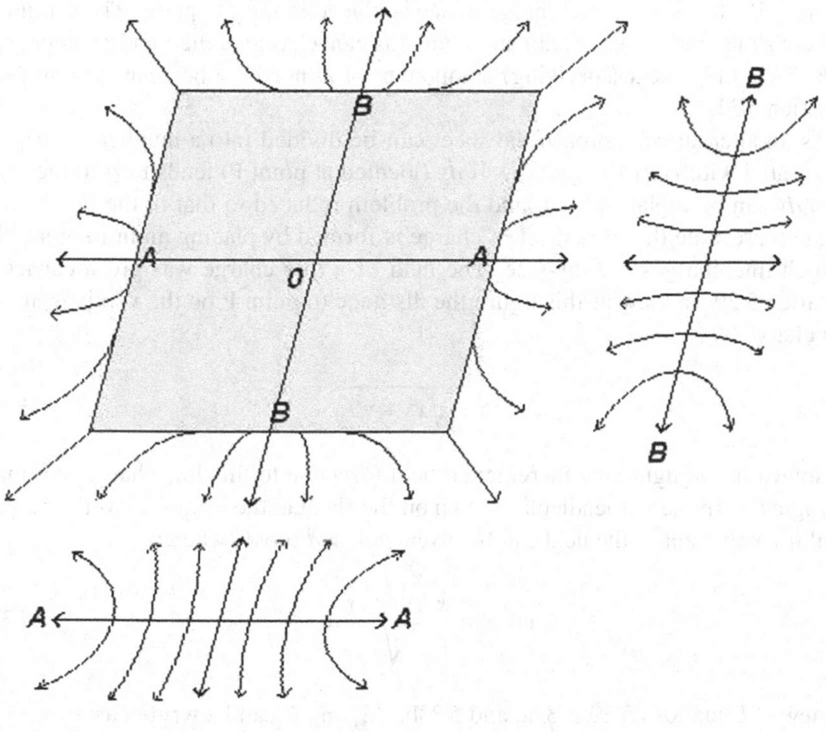

FIGURE 5.19 Field due to a finite sheet of charge.

5.5.4 Field due to Different Sheets of Charges

Consider a surface over which the charges are uniformly distributed with charge density ρ_s. The dimensions of this surface are extended from $-\infty$ to $+\infty$ along both y and z axes. In view of Equation 5.30a the incremental charge content dQ on an elemental surface ds can be given as $dQ = \rho_s ds$. This charge will result in incremental electric field dE which can be written as:

$$dE = \frac{dQ}{4\pi\varepsilon_o R^2}\boldsymbol{a}_R = \frac{\rho_s ds}{4\pi\varepsilon_o R^2}\boldsymbol{a}_R \tag{5.31}$$

The total electric field intensity can be obtained by integrating both sides of Equation 5.31 over the limits defining the extent of the sheet. Thus the resulting expression becomes:

$$\boldsymbol{E} = \iint_s \frac{\rho_s ds}{4\pi\varepsilon_o R^2}\boldsymbol{a}_R \tag{5.32}$$

5.5.4.1 Field due to a Single Sheet

Figure 5.20 shows a sheet of charge which is placed in the y–z plane. The symmetry of sheet along the y and z coordinates demands cancelation of the y and z components of $\boldsymbol{E}$. Thus only the x (surviving) component of $\boldsymbol{E}$ needs to be evaluated by using Equation 5.32.

As an alternative approach the sheet can be divided into a number of strips of differential widths dy (or dz). Now if dy (located at point P) tends to zero the product $\rho_s dy$ can be replaced by ρ_L and the problem reduced to that of the line charge. One can presume that this sheet of charge is formed by placing an infinite number of such line charges side-by-side. The field of a line charge was given earlier by Equation 5.29. In view of this figure the distance to point P on the x-axis from this line charge is:

$$r = \sqrt{x^2 + y^2} \tag{5.33a}$$

As shown in the figure the incremental field ($d\boldsymbol{E}$) due to this line charge will make an angle θ with the perpendicular drawn on the sheet at the origin. Thus the perpendicular component of the field can be given as $E_x = dE\cos\theta$, where:

$$\cos\theta = \frac{x}{r} = \frac{x}{\sqrt{x^2 + y^2}} \tag{5.33b}$$

In view of Equations 5.29, 5.33a, and 5.33b, dE_x and E_x can be written as:

$$dE_x = \frac{\rho_L}{2\pi\varepsilon_o \cdot \boldsymbol{r}}\cos\theta = \frac{\rho_{s} \cdot dy}{2\pi\varepsilon_o \cdot \sqrt{x^2 + y^2}}\cos\theta = \frac{\rho_s}{2\pi\varepsilon_o} \cdot \frac{x}{x^2 + y^2} \cdot dy \tag{5.34}$$

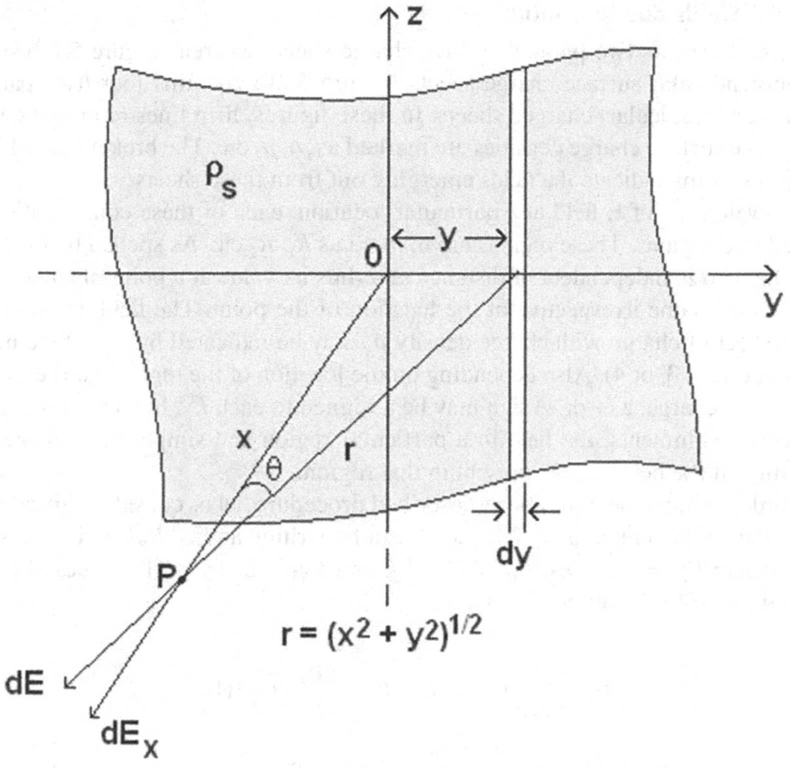

FIGURE 5.20 Electric field components due to an infinite sheet of charge.

$$E_x = \frac{\rho_s}{2\pi\varepsilon_o} \cdot \int_{-\infty}^{+\infty} \frac{x}{x^2 + y^2} \cdot dy = \frac{\rho_s}{2\pi\varepsilon_o} \cdot \left[\tan^{-1}(y/x)\right]_{y=-\infty}^{y=+\infty} = \frac{\rho_s}{2\varepsilon_o} \tag{5.35}$$

Since, $\boldsymbol{E}=E_x\boldsymbol{a}_x$, where the unit vector $\boldsymbol{a}_x$ is outward from the sheet of charge it may assume positive or negative sign depending on the location of the point P. If it lies on the positive x-axis, $\boldsymbol{E}=E_x\boldsymbol{a}_x$, and if P lies on the negative x-axis, $\boldsymbol{E}=-E_x\boldsymbol{a}_x$. Thus the relation for $\boldsymbol{E}$ in general can be written as:

$$\boldsymbol{E} = \frac{\rho_s}{2\varepsilon_o}\boldsymbol{a}_N \tag{5.36}$$

Where $\boldsymbol{a}_N$ is the unit vector which is outward normal to the sheet of charge. Here it can be noted that the field $\boldsymbol{E}$ is no more a function of the distance from the sheet. Stars of the Milky Way appear at the same location if observed from different parts of the world. As noted earlier a surface charge may be considered finite or infinite in accordance with the distance to the location at which the field is to be computed. The next section shows that the relation given by Equation 5.36 can be used to evaluate the field due to more than one sheet of charges.

5.5.4.2 Fields due to Multiple Sheets

Figure 5.21a shows two parallel surface charge sheets whereas Figure 5.21b shows two perpendicular surface charge sheets. Figure 5.21c contains four (two parallel and two perpendicular) charged sheets. In these figures, firm lines represent charge sheets. The surface charge densities are marked as ρ_1, ρ_2 etc. The broken/dotted lines of various forms indicate the fields emerging out from these sheets.

For evaluation of E field at a particular location, each of these configurations is divided into regions. These regions are marked as R_1, R_2 etc. As spelled by Equation 5.36, the field is independent of distance and thus its value at a point in a particular region is the same irrespective of the location of the point. The field due to a particular sheet of charge with charge density ρ_n may be indicated by E_n (where n may take value 1, 2, 3, or 4). Also depending on the location of the region with respect to the sheet of charge, a (+ or –) sign may be assigned to each E_n. In view of the above symbolic assignments, the field in a particular region can simply be evaluated by summing all the field vectors present in that region.

In order to implement the above-described procedure, let us consider Figure 5.21a. In this the fields in regions R_1, R_2, and R_3 can be written as E_{R1}, E_{R2}, and E_{R3} respectively, where $E_{R1} = -E_1 - E_2$, $E_{R2} = E_1 - E_2$, and $E_{R3} = E_1 + E_2$. The values of E_1 and E_2 obtained from Equation 5.36 are:

$$\left.\boldsymbol{E}_1 = \frac{\rho_1}{2\varepsilon_o}\boldsymbol{a}_N \quad (\text{a}) \qquad \boldsymbol{E}_2 = \frac{\rho_2}{2\varepsilon_o}\boldsymbol{a}_N \quad (\text{b})\right\} \tag{5.37}$$

In Equations 5.37a and 5.37b the unit vector $\boldsymbol{a}_N$ will be perpendicular to the axis or direction along with which the sheet of charge is aligned.

5.6 VOLUME CHARGE DISTRIBUTION

If the discrete charges are located densely (or continuously) in a volume so as to form a (piecewise) continuous mass of charges, the distribution is referred to as volume charge distribution.

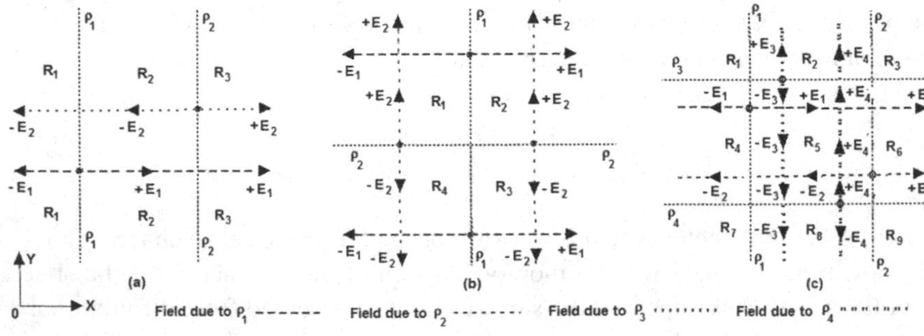

FIGURE 5.21 Field due to (a) two parallel surface charges (b) two perpendicular surface charges (c) two parallel and two perpendicular surface charges.

5.6.1 Forms of Volume Charge Distribution

Like line and surface charge distributions, the volume charge distribution may also be (piecewise) uniform or non-uniform throughout the volume. The volume containing such charges could be of finite or infinite extent. Figure 5.22 shows a charge distribution in a finite volume of arbitrary shape. As can be seen from the figure the charges are so densely located that the distribution can be considered as continuous throughout the volume. Depending on the location of field point the volume charge too may appear to be a point, or of infinite size as in the case of line and surface charge distributions.

It is to be noted that an infinite line charge completely occupies one of the dimensions, thus a field point can only be located along (any or in between) of the other two dimensions. The finite line charge partially occupies one of the dimensions, thus a field point can be located along or in between the other two dimensions and the partially unoccupied dimension. Similarly an infinite surface charge completely occupies two dimensions, thus a field point can be located along the remaining third dimension. The finite surface charge partially occupies two dimensions, thus a field point can be located along the remaining third dimension and the partially unoccupied dimension(s). The unit vectors for $\boldsymbol{E}$ field in all these cases will be in accordance with the location of the field point vis-à-vis the line or surface charges under consideration. The case of volume charge, however, differs from the above two cases. An infinite volume charge completely occupies all three dimensions and there is no place to locate the field point. A finite volume charge, however, only partially occupies the three dimensions, thus a field point can be located in any of the unoccupied locations. The unit vectors for $\boldsymbol{E}$ field in this case will be in the direction of location of the field point vis-à-vis the volume charge.

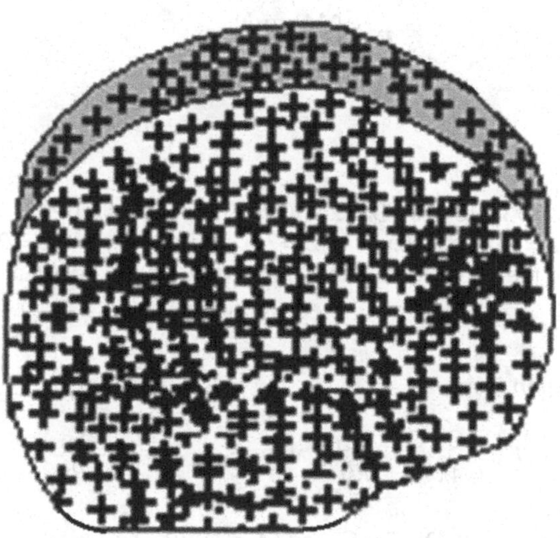

FIGURE 5.22 Volume charge distribution.

5.6.2 Volume Charge Density and the Total Charge

The volume charge density ρ_v in C/m³ is defined as:

$$\rho_v \overset{\text{def}}{=} \lim_{\Delta v \to 0} \frac{\Delta Q}{\Delta v} = \frac{dQ}{dv} \text{ C/m}^3 \tag{5.38a}$$

Due to a distributed volume charge density ρ_v, the total charge Q in a given volume v can be found by integrating the charge density over the volume as shown below:

$$Q = \iiint_v \rho_v \cdot dv \tag{5.38b}$$

It is in fact a special case of discrete distribution of charges (discussed earlier) provided they are densely located in a volume so as to form a continuous mass of charges.

5.6.3 Field Due to a Finite Volume Charge Distribution

The $\boldsymbol{E}$ field due to the volume charge distribution shown in Figure 5.22 can be obtained by replacing Q_m by $\rho_v dv$, R_m by r and the summation by the volume integral in Equation 5.11. The resulting field at some arbitrarily located field point can be written as:

$$\boldsymbol{E} = \iiint_v \frac{\rho_v dv}{4\pi\varepsilon_o r^2} a_N \tag{5.39}$$

Where a_N is the unit vector oriented towards the field point.

The symbol ρ_v for the volume charge density is often replaced by ρ. Thus

$$\boldsymbol{E} = \iiint_v \frac{\rho dv}{4\pi\varepsilon_o r^2} a_N \tag{5.40}$$

Example 5.1

Point charges $Q_1 = 3 \times 10^{-3}$ C, $Q_2 = 2 \times 10^{-3}$ C and $Q_3 = 4 \times 10^{-3}$ C are located at $(x_1 = 30, y_1 = -10, z_1 = 15)$, $(x_2 = 20, y_2 = 10, z_2 = 25)$ and $(x_3 = 0, y_3 = 0, z_3 = 0)$ respectively in vacuum. Find the force exerted on Q_1 by: (a) Q_2, (b) Q_3, (c) Q_2 and Q_3.

SOLUTION

In view of Equation 5.6b:

$$\boldsymbol{F}_{21} = \frac{Q_1 \cdot Q_2}{4\pi \cdot \varepsilon_0} \cdot \frac{\boldsymbol{a}_{R21}}{R_{21}^2} \quad \text{and} \quad \boldsymbol{F}_{31} = \frac{Q_1 \cdot Q_3}{4\pi \cdot \varepsilon_0} \cdot \frac{\boldsymbol{a}_{R31}}{R_{31}^2}$$

Since $Q_1 = 3 \times 10^{-3}$ C, $Q_2 = 2 \times 10^{-3}$ C, and $Q_3 = 4 \times 10^{-3}$ C:

$$Q_1 \cdot Q_2 = 6 \times 10^{-6} \quad \text{and} \quad Q_1 \cdot Q_3 = 12 \times 10^{-6}$$

For vacuum or free space:

$$\varepsilon_0 = 8.854\times10^{-12} = \frac{10^{-9}}{36\pi},\quad 4\pi\varepsilon_0 = 4\times\frac{10^{-9}}{36\pi},\quad \frac{1}{4\pi\varepsilon_0} = 9\times10^9.$$

As a thumb rule, to get a distance vector, subtract the coordinates of the charge which exerts the force from the coordintes of the charge on which the force is exerted. In this case the force is exerted by Q_2 and Q_3 on Q_1 and the distance vectors are $\boldsymbol{R}_{21}$ and $\boldsymbol{R}_{31}$. Which in view of Equation 3.13 we can write as:

(a) $\boldsymbol{R}_{21} = 10\boldsymbol{a}_x - 20\boldsymbol{a}_y - 10\boldsymbol{a}_z,\ |\boldsymbol{R}_{21}|^2 = 600,\ |\boldsymbol{R}_{21}| = 24.495$

$$\boldsymbol{a}_{R21} = 0.408\boldsymbol{a}_x - 0.816\boldsymbol{a}_y - 0.408\boldsymbol{a}_z$$

$$\frac{Q_1\cdot Q_2}{4\pi\varepsilon_0} = 6\times10^{-6}\cdot 9\times10^9 = 54000$$

$$\frac{Q_1\cdot Q_2}{4\pi\varepsilon_0}\cdot\frac{\boldsymbol{a}_{R21}}{R_{21}^2} = \frac{54000}{600}\boldsymbol{a}_{R21} = 90\boldsymbol{a}_{R21}$$

$$\boldsymbol{F}_{21} = 90\left(0.408\boldsymbol{a}_x - 0.816\boldsymbol{a}_y - 0.408\boldsymbol{a}_z\right)$$

(b) $\boldsymbol{R}_{31} = 30\boldsymbol{a}_x - 10\boldsymbol{a}_y + 15\boldsymbol{a}_z,\ |\boldsymbol{R}_{31}|^2 = 1225,\ |\boldsymbol{R}_{31}| = 35$

$$\boldsymbol{a}_{R31} = 0.857\boldsymbol{a}_x - 0.2857\boldsymbol{a}_y - 0.4286\boldsymbol{a}_z$$

$$\frac{Q_1\cdot Q_3}{4\pi\varepsilon_0} = 12\times10^{-6}\cdot 9\times10^9 = 108000$$

$$\frac{Q_1\cdot Q_3}{4\pi\varepsilon_0}\cdot\frac{\boldsymbol{a}_{R31}}{R_{31}^2} = \frac{108000}{1225}\boldsymbol{a}_{R31} = 88.16\boldsymbol{a}_{R31}$$

$$\boldsymbol{F}_{31} = 88.16\left(0.857\boldsymbol{a}_x - 0.2857\boldsymbol{a}_y - 0.4286\boldsymbol{a}_z\right)$$

(c) The combined force on $Q_1 = \boldsymbol{F}_{21} + \boldsymbol{F}_{31}$

Example 5.2

A point charge $Q = 3\cdot10^{-9}$ C is located in air at (–1,–2,–3). Find E and $|E|$ at (1, 2, 3).

SOLUTION

In this case the distance vector $\boldsymbol{R}_{12}$ is obtained by subtracting the coordinates of the source (or charge) which causes$\boldsymbol{E}$, from that of the field point (i.e. the point on which E is to be evaluated). Thus:

$$\boldsymbol{R}_{12} = 2\boldsymbol{a}_x + 4\boldsymbol{a}_y + 6\boldsymbol{a}_z \quad |\boldsymbol{R}_{12}| = \sqrt{56} \quad \boldsymbol{a}_{R12} = \frac{2\boldsymbol{a}_x + 4\boldsymbol{a}_y + 6\boldsymbol{a}_z}{\sqrt{56}}$$

In view of Equation 5.10b:

$$\boldsymbol{E} = \frac{1}{4\pi\varepsilon_0}\cdot\frac{Q}{(R_{12})^2}\cdot\boldsymbol{a}_{R12} = \left(9\times10^9\right)\cdot\frac{3\cdot10^{-9}}{56}\cdot\frac{2\boldsymbol{a}_x + 4\boldsymbol{a}_y + 6\boldsymbol{a}_z}{\sqrt{56}}$$

$$= 0.0644\left(2\boldsymbol{a}_x + 4\boldsymbol{a}_y + 6\boldsymbol{a}_z\right)$$

$$E = 0.1288\mathbf{a}_x + 0.2577\mathbf{a}_y + 0.3864\mathbf{a}_z$$

$$|E| = \sqrt{(0.1288)^2 + (0.2577)^2 + (0.3864)^2} \approx 0.482 \text{ V/m}$$

Example 5.3

Two point charges $Q_1 = 3\times10^{-9}$ C and $Q_2 = -3\times10^{-9}$ C are located in a vacuum at $x_1 = 2$, $x_2 = 4$ on the x-axis. Express E in terms of magnitude and unit vector at $x = 2$, $y = -2$ and $z = 2$.

SOLUTION

Let $x_1 = 2$ be point 1, $x_2 = 4$ be point 2, and $x = 2$, $y = -2$, $z = 2$ be point 3.

$$\mathbf{R}_{13} = -2\mathbf{a}_y + 2\mathbf{a}_z \quad |\mathbf{R}_{13}| = \sqrt{8} = 2.828 \quad \mathbf{a}_{R13} = \frac{-2\mathbf{a}_y + 2\mathbf{a}_z}{2.828}$$

$$\mathbf{R}_{23} = -2\mathbf{a}_x - 2\mathbf{a}_y + 2\mathbf{a}_z \quad |\mathbf{R}_{23}| = \sqrt{12} = 3.464 \quad \mathbf{a}_{R23} = \frac{-2\mathbf{a}_x - 2\mathbf{a}_y + 2\mathbf{a}_z}{3.464}$$

Since $Q_1 = 3\times10^{-9}$, $Q_2 = -3\times10^{-9}$ and $\frac{1}{4\pi\varepsilon_0} = 9\times10^9$

$$\frac{Q_1}{4\pi\varepsilon_0} = (3\times10^{-9})\times(9\times10^9) = 27 \text{ and } \frac{Q_2}{4\pi\varepsilon_0} = (-3\times10^{-9})\times(9\times10^9) = -27$$

In view of Equation 5.10b:

$$\begin{aligned} E &= \frac{Q_1}{4\pi\varepsilon_0}\cdot\frac{1}{R_{13}^2}\cdot\mathbf{a}_{R13} + \frac{Q_2}{4\pi\varepsilon_0}\cdot\frac{1}{R_{23}^2}\cdot\mathbf{a}_{R23} \\ &= \frac{27}{8}\cdot\frac{-2\mathbf{a}_y + 2\mathbf{a}_z}{2.828} - \frac{27}{12}\cdot\frac{-2\mathbf{a}_x - 2\mathbf{a}_y + 2\mathbf{a}_z}{3.464} \end{aligned}$$

Thus:

$$\begin{aligned} E &= \frac{54}{22.624}(-\mathbf{a}_y + \mathbf{a}_z) - \frac{54}{41.568}(-\mathbf{a}_x - \mathbf{a}_y + \mathbf{a}_z) \\ &= 2.387(-\mathbf{a}_y + \mathbf{a}_z) - 1.3(-\mathbf{a}_x - \mathbf{a}_y + \mathbf{a}_z) \\ &= 1.3\mathbf{a}_x - 1.087\mathbf{a}_y + 1.087\mathbf{a}_z \end{aligned}$$

$$|E| = \sqrt{(1.3)^2 + (-1.087)^2 + (1.087)^2}$$

$$|E| = \sqrt{1.69 + 1.1816 + 1.1816} = \sqrt{4.053} = 2.0132$$

$$\mathbf{a}_E = \frac{E}{|E|} = \frac{1.3\mathbf{a}_x - 1.087\mathbf{a}_y + 1.087\mathbf{a}_z}{2.0132} = 0.6457\mathbf{a}_x - 0.5399\mathbf{a}_y + 0.5399\mathbf{a}_z$$

Example 5.4

Arrange the three charges of –1, –2, and 3 nC at points (0,1,0), (0,2,0) and (0,3,0) to get the largest amplitude of E at the origin in vacuum. Also obtain the largest amplitude of E.

SOLUTION

In view of the coordinates of given points all the charges are to be located along the positive y-axis. The maximum E at the origin will result only if the charge of the maximum value is located near to it. Thus in view of Equation 5.10b, charges for maximum E are to be arranged as:

$$3\text{ nC at }0,1,0,\quad -2\text{ nC at }0,2,0\ \text{ and }\quad -1\text{ nC at }0,3,0$$

Here $\boldsymbol{R}_1 = \boldsymbol{a}_y$, $|R_1| = 1$, $\boldsymbol{R}_2 = 2\boldsymbol{a}_y$, $|R_2| = 2$, $\boldsymbol{R}_3 = 3\boldsymbol{a}_y$, $|R_3| = 3$ thus $\boldsymbol{a}_{R1} = \boldsymbol{a}_{R2} = \boldsymbol{a}_{R3} = \boldsymbol{a}_y$

Since $\dfrac{1}{4\pi\varepsilon_0} = 9\times10^9$

$$\boldsymbol{E} = \frac{Q_1}{4\pi\varepsilon_0}\cdot\frac{1}{\boldsymbol{R}_1^2}\cdot\boldsymbol{a}_{R1} + \frac{Q_2}{4\pi\varepsilon_0}\cdot\frac{1}{\boldsymbol{R}_2^2}\cdot\boldsymbol{a}_{R2} + \frac{Q_3}{4\pi\varepsilon_0}\cdot\frac{1}{\boldsymbol{R}_3^2}\cdot\boldsymbol{a}_{R3} = 9\boldsymbol{a}_y\left[\frac{3}{1} - \frac{2}{4} - \frac{1}{9}\right]$$

$$= (27 - 4.5 - 1)\,\boldsymbol{a}_y = 21.5\boldsymbol{a}_y \quad\text{or}\quad |E| = 21.5\text{ V/m}$$

Example 5.5

Evaluate the sum (a) $\sum_{m=0}^{4}\dfrac{(-2)^m}{m^2+1}$ (b) $\sum_{k=1}^{4}\left[2(k+1)\,a_x + k^2 a_y + \dfrac{4}{k}a_z\right]$.

SOLUTION

(a) $$\sum_{m=0}^{4}\frac{(-2)^m}{m^2+1} = \frac{(-2)^0}{1} + \frac{(-2)^1}{2} + \frac{(-2)^2}{5} + \frac{(-2)^3}{10} + \frac{(-2)^4}{17}$$

$$= \frac{1}{1} - \frac{2}{2} + \frac{4}{5} - \frac{8}{10} + \frac{16}{17} = 0.9412$$

(b) $$\sum_{k=1}^{4}\left[2(k+1)\,a_x + k^2 a_y + \frac{4}{k}a_z\right]$$

$$= \left[\{4a_x + a_y + 4a_z\} + \{6a_x + 4a_y + 2a_z\} + \left\{8a_x + 9a_y + \frac{4}{3}a_z\right\} + \{10a_x + 16a_y + a_z\}\right]$$

$$= 28a_x + 30a_y + 8.333a_z$$

Example 5.6

Find the magnitude and direction of $\boldsymbol{E}$ at (1,2,–3) in free space if a uniform line charge ($\rho_l = 3\times10^{-9}$ C/m) lies along the z-axis.

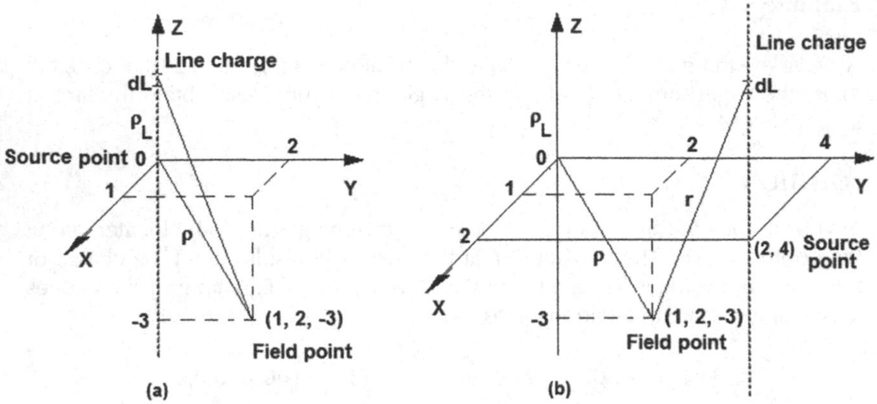

FIGURE 5.23 Uniform line charge along z-axis.

SOLUTION

In view of Equation 5.29: $\boldsymbol{E} = \boldsymbol{E}_\rho \boldsymbol{a}_\rho = \dfrac{\rho_l}{2\cdot\pi\cdot\varepsilon_0\cdot\rho}\boldsymbol{a}_\rho = \dfrac{\rho_l}{2\cdot\pi\cdot\varepsilon_0}\cdot\dfrac{1}{\rho}\boldsymbol{a}_\rho$

Since $\dfrac{1}{2\pi\varepsilon_0} = 18\times10^9$, $\rho_l = 3\times10^{-9}$,

$$\frac{\rho_l}{2\cdot\pi\cdot\epsilon_0} = \left(18\times10^9\right)\left(3\times10^{-9}\right) = 54, \quad \boldsymbol{E} = \frac{54}{\rho}\boldsymbol{a}_\rho$$

In view of Figure 5.23, $\boldsymbol{\rho} = \boldsymbol{a}_x + 2\boldsymbol{a}_y$, $|\rho| = \sqrt{5}$, $\boldsymbol{a}_\rho = \dfrac{\boldsymbol{a}_x + 2\boldsymbol{a}_y}{\sqrt{5}}$

$$\boldsymbol{E} = \frac{54}{\sqrt{5}}\boldsymbol{a}_\rho = \frac{54}{\sqrt{5}}\cdot\frac{\boldsymbol{a}_x + 2\boldsymbol{a}_y}{\sqrt{5}} = 10.8\left(\boldsymbol{a}_x + 2\boldsymbol{a}_y\right) = 10.8\boldsymbol{a}_x + 21.6\boldsymbol{a}_y$$

Example 5.7

Two infinite line charges with charge density $\rho_L = 10$ nC/m lie along the lines $y = \pm x$ in the $z = 0$ plane. Find $\boldsymbol{E}$ at (0,0,5).

SOLUTION

Equation of a straight line is written as $y = mx + c$, where m is the slope of the line and c the location of its crossing on the y axis. In the present case, $c = 0$ and $m = \pm1$, thus both the lines make $\pm45°$ angles with the x-axis and pass through the origin. The $\boldsymbol{E}$ lines emerging out from these will be perpendicular and uniform all along as these lines are taken to be infinitely long.

Figure 5.24 illustrates the two lines of charges, marked as 1 and 2, both located in the x–y plane. A pair of points (a and b) is shown on line 1 and another pair (c and d) on line 2. The points of each pair are symmetrically located on the opposite side of the origin marked as 0. Each of these points is referred to as a source point. Each such point represents an elemental length dL, which is a distance L away from

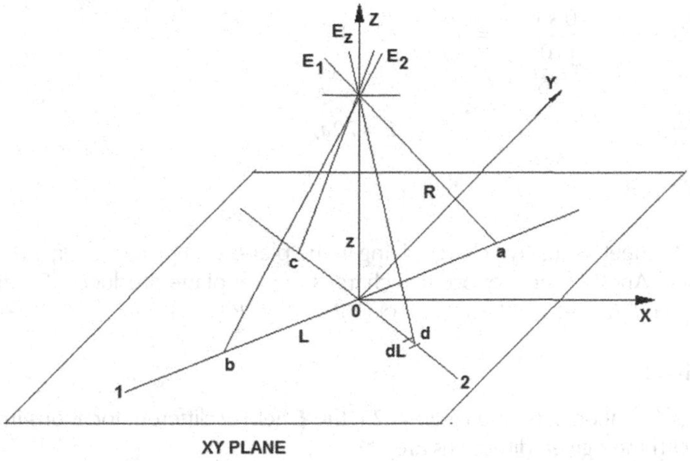

FIGURE 5.24 Two infinite line charges.

the origin. This figure also shows a point on the z-axis referred to as a field point which is at a distance R away from the source point dL. This field point is at distance $z=5$ from the origin. As illustrated each source point, in accordance with its location, will contribute towards E at the field point. In view of the symmetry only the E_z component will survive whereas the other components will get canceled.

The field intensity ***dE*** due to an elemental length *dL* on any of these line charges is given as:

$$\boldsymbol{dE} = \frac{\rho_L dL}{4\pi\varepsilon_0 R^2}\boldsymbol{a_R}$$

In view of the symmetry: (i) *dL* can be taken on any line and on any side of the line, (ii) *x* and *y* components of *E* will vanish, only the *z* component will survive, and (iii) both the infinite line charges will equally contribute towards *E*. Thus on any point on the *z*-axis the field will be twice that of the single line and can be written as:

$$\boldsymbol{dE} = \frac{2\rho_L dL}{4\pi\varepsilon_0 R^2}\boldsymbol{a_R} \quad dE_z = dE\cos\theta \quad \cos\theta = \frac{z}{R} = \frac{z}{\sqrt{L^2+z^2}} \text{ and } R = \sqrt{L^2+z^2}$$

where *dL, R,* and *z* are shown in the figure.

Thus $dE_z = \frac{2\rho_L dL}{4\pi\varepsilon_0 R^2}\cos\theta = \frac{2\rho_L dL}{4\pi\varepsilon_0}\frac{z}{\left(L^2+z^2\right)^{3/2}}$ or $E_z = \frac{\rho_L}{2\pi\varepsilon_0}\int_{-\infty}^{\infty}\frac{z}{\left(L^2+z^2\right)^{3/2}}dL$

Put $L = z\cot\theta$, $dL = -z\operatorname{cosec}^2\theta\, d\theta$, $\left(L^2+z^2\right)^{3/2} = z^3\operatorname{cosec}^3\theta$

Also the limits become: when $L=-\infty$, $\theta=\pi$ and for $L=\infty$, $\theta=0$.

$$E_z = \frac{\rho_L}{2\pi\varepsilon_0}\int_{\pi}^{0}\frac{-z^2\operatorname{cosec}^2\theta}{z^3\operatorname{cosec}^3\theta}d\theta$$

$$= \frac{\rho_L}{2\pi\varepsilon_0 z}\int_{\pi}^{0} -\sin\theta\, d\theta = \frac{\rho_L}{2\pi\varepsilon_0 z}\cos\theta\Big|_{\pi}^{0} = \frac{\rho_L}{\pi\varepsilon_0 z}$$

$$\text{At } z=0,0,5\text{: } E_z = \frac{10\times 10^{-9}}{\pi \frac{10^{-9}}{36\pi} 5} = \frac{360}{5} = 72$$

$$\boldsymbol{E} = 72\boldsymbol{a}_z$$

Example 5.8

An infinite sheet of positive charges lying in the plane $x=3$ produces $|\boldsymbol{E}|=10$ V/m in free space. Another sheet of positive charges in $y=4$ plane produces $|\boldsymbol{E}|=20$ V/m. Find $\boldsymbol{E}$ at (a) $P_a(5,5,-2)$ (b) $P_b(6,2,-2)$ (c) $P_c(2,3,3)$ (d) $P_d(1,5,4)$

SOLUTION

In view of Equation 5.36 and Figure 5.25, the $\boldsymbol{E}$ field at different locations in accordance with their given directions are:

(a) At $P_a\,(5,5,-2)$ $\boldsymbol{E} = 10\boldsymbol{a}_x + 20\boldsymbol{a}_y$

(b) At $P_b\,(6,2,-2)$ $\boldsymbol{E} = 10\boldsymbol{a}_x - 20\boldsymbol{a}_y$

(c) At $P_c\,(2,3,3)$ $\boldsymbol{E} = -10\boldsymbol{a}_x - 20\boldsymbol{a}_y$

(d) At $P_d\,(1,5,4)$ $\boldsymbol{E} = -10\boldsymbol{a}_x + 20\boldsymbol{a}_y$

Example 5.9

For the volumetric charge density ρ_v find the total charge within the specified volume.

(a) $\rho_v = 15xy^2z^3,\ 1\geq x\geq 2,\ 0\geq y\geq 1,\ 1\geq z\geq 3$

(b) $\rho_v = 20\rho z\sin\varphi,\ 1\geq \rho\geq 2,\ 0\geq \varphi\geq \frac{\pi}{4},\ 1\geq z\geq 3$

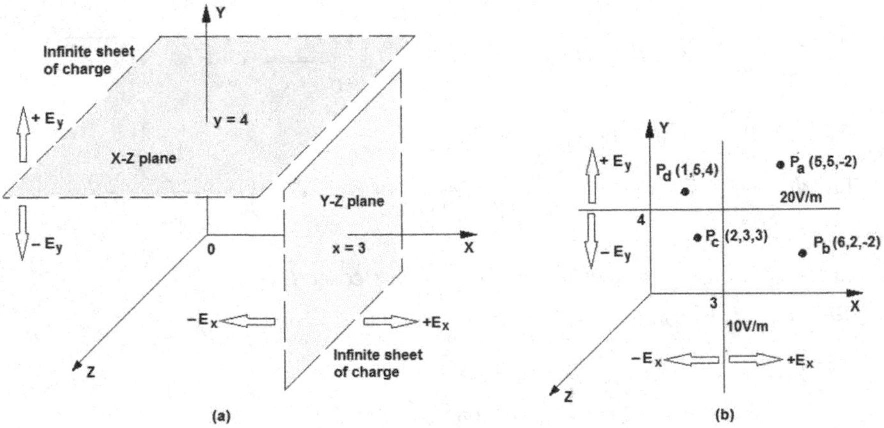

FIGURE 5.25 (a) Two infinite sheets of charges; (b) E field at different locations.

(c) $\rho_v = \frac{10}{r^2}\sin\theta\cos\varphi,\ 1 \geq r \geq 3,\ 0 \geq \theta \geq \frac{\pi}{4},\ \frac{\pi}{6} \geq \varphi \geq \frac{\pi}{3}$

SOLUTION

In view of Equation 5.38b:

(a) $$Q = \int_1^3\int_0^1\int_1^2 \left(15xy^2z^3\right)dx \cdot dy \cdot dz = 15 \cdot \frac{x^2}{2}\bigg|_1^2 \cdot \frac{y^3}{3}\bigg|_0^1 \cdot \frac{z^4}{4}\bigg|_1^3 = 150\text{ C}$$

(b) $$Q = \int_1^3\int_0^{\pi/4}\int_1^2 (20\rho \cdot z \cdot \sin\varphi)\rho d\rho \cdot d\varphi \cdot dz$$

$$= 20 \cdot \frac{\rho^3}{3}\bigg|_1^2 \cdot (-\cos\varphi)\Big|_0^{\pi/4} \cdot \frac{z^2}{2}\bigg|_1^3 = -364.16\text{ C}$$

(c) $$Q = \int_{\pi/6}^{\pi/3}\int_0^{3\pi/4}\int_1^3 \left(\frac{10}{r^2} \cdot \sin\theta \cdot \cos\varphi\right) r^2 \sin\theta dr \cdot d\theta \cdot d\varphi$$

$$= 10 \cdot r\big|_1^3 \cdot \left(\frac{\theta}{2} - \frac{\sin 2\theta}{4}\right)\bigg|_0^{\pi/4} \cdot \sin\varphi\Big|_{\pi/6}^{\pi/3}$$

$$= 10 \cdot 2 \cdot \frac{\pi}{8} \cdot 0.366 = 2.5 \cdot \pi \cdot 0.366 = 2.874\text{ C}$$

PROBLEMS

P5.1 Let the two point charges Q_1 (=3 mC) and Q_2 (=−4 mC) be located in free space at (−3,7,−4) and (2,4,−1) respectively. Find the vector force $\boldsymbol{F}$ (a) on Q_1 due to Q_2 and (b) on Q_2 due to Q_1.

P5.2 Three point charges $Q_1 = 10^{-3}$ C, $Q_2 = 3 \times 10^{-3}$ C, $Q_3 = 5 \times 10^{-3}$ C are located at ($x_1 = 10$, $y_1 = -10$, $z_1 = 10$), ($x_2 = 20$, $y_2 = 15$, $z_2 = 30$), and ($x_3 = 15$, $y_3 = 10$, $z_3 = 5$) respectively in a vacuum. Calculate the force exerted on Q_2 by: (a) Q_1 (b) Q_3 (c) Q_1 and Q_3.

P5.3 A point charge of $Q = 7 \cdot 10^{-9}$ C is located at $x_1 = -0.2$, $y_1 = -0.3$, $z_1 = 0.6$, in air. Find the magnitude of electric field intensity at $x_2 = 0.5$, $y_2 = 0.5$, $z_2 = -0.5$.

P5.4 Two point charges $Q_1 = 2 \times 10^{-9}$ C and $Q_2 = -4 \times 10^{-9}$ C are located in a vacuum at $x_1 = 1$, $x_2 = 2$ on the x-axis. Express E in terms of magnitude and unit vector at $x = 4$, $y = 2$, $z = -2$.

P5.5 A charge of $Q = 6 \cdot 10^{-9}$ C is located at $(x_1 = -0.2, y_1 = -0.3, z_1 = 0.5)$ in air. Find the magnitude of electric field intensity $|\boldsymbol{E}|$ at: (a) 3 m from the charge (b) $(x_2 = -0.4, y_2 = 0.4, z_2 = -0.3)$.

P5.6 Two point charges Q_1 (=3 μC) and Q_2 (=−5 μC) are located in free space at (−3,5,−4) and (−2,4,−1) respectively. Find (a) $\boldsymbol{E}$ (b) $|\boldsymbol{E}|$ (c) $\boldsymbol{a}_E$ at a point located at (10,15,20).

P5.7 Find the sum of the series (a) $\sum_{m=0}^{5} \frac{m}{m^2+1}$ (b) $\sum_{m=1}^{6} \frac{(-1)^m}{m\sqrt{m+1}}$.

P5.8 Evaluate the sum (a) $\sum_{m=0}^{4} \frac{(-1)^m}{m^3+1}$, (b) $\sum_{k=1}^{4} 5(k+1)a_x + k^2 a_y + \frac{6}{k} a_z$.

P5.9 Find the magnitude and direction of $\boldsymbol{E}$ at (1,2,–3) in free space if a uniform line charge ($\rho_l = 5\times10^{-9}$ C/m) lies along the line $x=2$, $y=4$.

P5.10 Two infinite line charges with charge density $\rho_L = 5$ nC/m lie along the lines $y=\pm x$ in $z=0$ plane. Find $\boldsymbol{E}$ at (0,3,0).

P5.11 Find the Cartesian components of $\boldsymbol{E}$ at (a) (0,0,0) (b) $x=2$, $y=5$, $z=3$, (c) $\rho=2$, $\varphi=\pi$, $z=2$, due to a uniform line charge $\rho_L = 25$ nC/m parallel to the y-axis and lying on the line $x=-3$, $z=4$ in free space.

P5.12 Three uniform sheets of charges of 2, –5, and 4 μC/m² are located in free space at $x=-3$, $x=1$, and $x=5$ respectively. Find E at (a) (–4,–2,–5), (b) (–2.5,–1.5,4.5), (c) (0,0,0), (d) (4,0,3), and (e) (6,3,2).

P5.13 An infinite sheet of positive charges lying in the plane x=–5 produces $|\boldsymbol{E}|=5$ V/m in free space. Another sheet of negative charges in the $y=2$ plane produces $|\boldsymbol{E}|=-10$ V/m. Find $|\boldsymbol{E}|$ at (a) P_a(2,4,–2) (b) P_b(–6,4,–2) (c) P_c(–6,–2,3) (d) P_d(2,–2,3) (e) P_e(–2.5,1,5).

P5.14 Find the total charge contents within the volume specified by:

(a) $1 \le x \le 3, 0 \le y \le 2, 1 \le z \le 2$ with $\rho_v = 30\cdot x\cdot y^2\cdot z^3$.

(b) $0 \le \rho \le 2$, $0 \le \varphi \le \pi/6$, $1 \le z \le 2$ with $\rho_v = 10\cdot\rho\cdot z\cdot\cos\varphi$

(c) $0 \le r \le 2$, $\pi/4 \le \theta \le \pi/2$, $\pi/2 \le \varphi \le 2\pi$ with $\rho_v = 5r^3\cdot\sin\theta\cdot\cos\varphi$

P5.15 Find the total charge enclosed if the volume charge density (in C/m³) and volume are specified as: (a) $\rho_v = 10ze^{-0.5x}\sin\pi y$, $-1 \le x \le 2, 0 \le y \le 1, 3 \le z \le 3.6$; (b) $\rho_v = 10xyz$, $0 \le \rho \le 2, 0 \le \varphi \le \pi/2, 0 \le z \le 3$; (c) $\rho_v = \frac{3\pi\sin\theta\cos^2\varphi}{2r^2(r^2+1)}$, $0 \le r \le \infty, 0 \le \theta \le \frac{\pi}{2}, 0 \le \varphi \le 2\pi$.

DESCRIPTIVE QUESTIONS

Q5.1 State Coulomb's law and explain the meaning of each term involved in the expression of force. Also discuss the conditions for its validity.

Q5.2 Compare electrical and gravitational fields in terms of their nature and expressions.

Q5.3 Prove that a uniform line charge results only in radial field all along the line.

FURTHER READING

Given at the end of Chapter 15.

6 Electric Flux Density

6.1 INTRODUCTION

In Chapter 5, a relation for electrostatic force between charges was obtained in *Coulomb's law*. The manipulation of this expression led to a new quantity which was referred to as *electric field intensity* $\boldsymbol{E}$. Latter expressions for $\boldsymbol{E}$ field for some simple cases of the charge distribution were obtained. In all of these cases, the charges were assumed to be uniformly distributed along a straight line, over a flat surface, or in a regular shaped volume. Any deviation from such distributions is bound to result in mathematical complications. For example, if the line or surface is curved or are of finite dimensions and the point at which the electric field is to be evaluated is asymmetrically located vis-à-vis the origin of the chosen coordinate system, it cannot be easily found. Similarly, if the volume is of an arbitrary shape, the evaluation of $\boldsymbol{E}$ may become a cumbersome task. Thus, such problems require an uncomplicated way out. This way out is provided by a law formulated by Carl Friedrich Gauss in 1835, called *Gauss' law*, which greatly simplifies the task.

6.2 GAUSS' LAW

Gauss' law states:

> The total electric flux emerging from any closed surface is equal to the total charge enclosed in it.

This simple statement involves a number of terms including the *electric flux*, *passing through*, *close surface*, and *charge enclosed*. These terms appear to be quite simple but in essence each of these has a specific meaning and needs proper explanation. In the next subsection, an attempt has been made to address the same.

6.2.1 Flux and Its Dependence

The term *flux* is an important entity. It encompasses the fields of mathematics, electricity, and magnetism. In mathematics its concept helps us to understand the divergence discussed in Section 4.3. In general, it may be conceived as the amount of "something" crossing a surface. The word "something" may include water, gas, wind, electric flux (Ψ), magnetic flux (Φ), electromagnetic energy (ϵ), or anything else that has the capability to cross. The total of this something depends on the strength of its source and the way it is crossing out. It also depends on the shape, size, and the orientation of the surface it crosses. For the rest of this chapter, the word something will be replaced by the electric flux referred in Gauss' law.

One should keep in mind that the source of flux is a vector field and the flux in itself need not necessarily be a physical object or quantity. The surface referred to in Gauss' law is the closed boundary surface through which the flux is crossing. A part of this closed boundary surface may be a part of a sphere, a flat or curved plane, or the top of a bucket. Furthermore, this boundary need not be a physical entity and may comprise an imaginary sphere, cylinder, or a plane. The timing is another parameter to be reckoned with. It is the single time instant at which the total flux is measured. This is not to be taken in terms of per unit area. This total flux is closely related to the source, as the flux passing through a surface will be doubled if the source gets doubled.

The meaning of "surface" was explained in Section 4.6.2. A closed surface may be composed of a number of open surfaces. An open surface is another important parameter which plays a significant role. The total flux through this open surface depends on the relative orientations of the field and the area of surface. Whenever a surface completely faces the field, it passes the maximum flux. An analogy can be drawn to a sail facing the wind. The wind will exert maximum pressure on the sail if the sail and wind are perpendicular to each other or if the sail completely faces the wind. When the surface of the sail tilts away from the wind, the pressure will be reduced and will be zero when the sail and wind become parallel to each other. Similarly, when the surface to be crossed by the flux tilts away from the field, the crossing flux decreases until it becomes zero when the field and boundary are parallel, i.e. the flux is along the boundary and not crossing through the boundary surface. The magnitude of total flux will also depend on the size of surface. In the same field, a larger surface will capture more flux than a smaller one.

While estimating the total flux across an open surface, the quantum of source and the direction of the flux indicated by an arrow head must be known. The flux leaving a closed surface conventionally is taken as positive and the flux entering this surface is taken as negative. For leaving flux there must be some source from which the flux emanates (i.e. positive charge) and for entering flux there must be some sink (i.e. negative charge) into which it merges. The terms "source" and "sink" often describe the nature of the field. We will now move on to Faraday's experiment.

6.2.2 Faraday's Experiment

In 1837 Michael Faraday, then the director of the Royal Society of London, performed an experiment that led to the concept of electric flux. In this experiment Faraday took one small metallic sphere and charged it to $+\,Q$ C. He then placed two dielectric hemispheres of 3/4" thickness to cover the inner sphere. Furthermore, he covered these dielectric hemispheres with two metallic hemispheres. The outer metallic spherical surface was momentarily connected to the ground. The outer metallic hemispheres were then dismantled by using tools made of dielectric material. On observation he found that both the outer metallic hemispheres were charged. The charges on the two hemispheres were separately measured and added. The sum of these charges was found to be exactly equal to the charge given to the inner sphere. The charge on the two outer hemispheres, however, was found to be of opposite polarity to that of the inner sphere. The components of Faraday's experimental setup are illustrated in Figure 6.1.

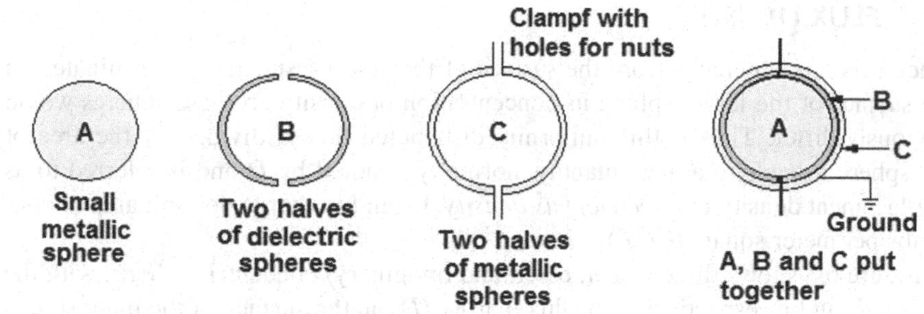

FIGURE 6.1 Components of Faraday's experiment.

Faraday concluded that some phenomenon has occurred between the two spheres which resulted in charging of outer hemispheres without any electrical contact between the inner and outer spheres. He termed this phenomenon as displacement. It later was referred to as displacement flux, electric flux, or simply the flux. Electric flux is denoted by Ψ throughout this text.

The conceived occurrence of this phenomenon between two metallic spherical surfaces shown in Figure 6.2a is illustrated in Figure 6.2b. The figure shows the outward emergence of electric flux from the positive charges on the inner sphere. After crossing the two dielectric hemispheres, located in between, this flux terminates on the negatively charged outer conducting sphere.

It is this displacement or electric flux that results in charging of the outer sphere. This flux Ψ obviously equals the source-charge Q and thus the two can be equated to yield:

$$\Psi = Q \tag{6.1}$$

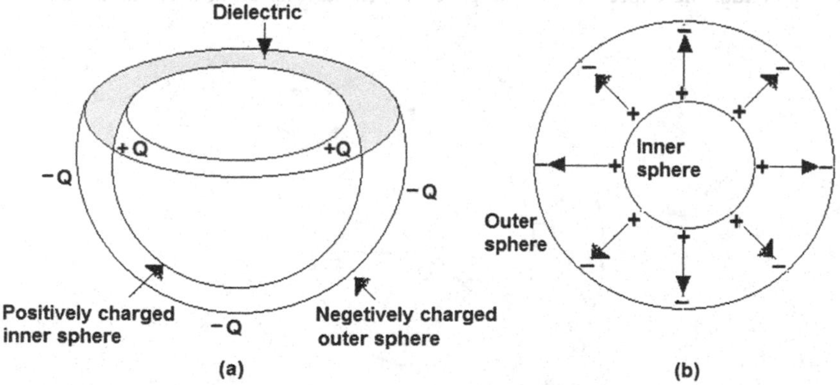

FIGURE 6.2 (a) Two charged metallic spherical surfaces separated by a dielectric. (b) Electric flux emerging from positive charges and terminating in negative charges.

6.3 FLUX DENSITY

Since this flux emanates from the surface of the smaller sphere and terminates on the surface of the larger sphere its concentration or density on these spheres would obviously differ. Thus if this uniformly distributed flux is divided by the area of the sphere it leads to a new quantity, normally denoted by ***D*** and is referred to as displacement density or (*electric*) *flux density*. Its unit is charge per unit area or coulombs per meter square (C/m^2).

Figure 6.3 shows three (inner, outer, and imaginary) concentric spheres with the radii *a*, *b*, and *r*, respectively. The flux density (*D*) on the surface of the inner sphere of radius *a* is:

$$D = \frac{\Psi}{4\pi \cdot a^2} = \frac{Q}{4\pi \cdot a^2} \tag{6.2a}$$

The density of flux on the surface of the outer sphere of radius *b* is:

$$D = \frac{\Psi}{4\pi \cdot b^2} = \frac{Q}{4\pi \cdot b^2} \tag{6.2b}$$

Similarly, the density of flux on the surface of an imaginary sphere of radius r, presumed to be located between inner and outer spheres, can be written as:

$$D = \frac{\Psi}{4\pi \cdot r^2} = \frac{Q}{4\pi \cdot r^2} \quad \text{for } a \le r \le b \tag{6.2c}$$

Equation 6.2c is a general form of expression that gives Equation 6.2a or 6.2b just by replacing "*r*" by "*a*" or "*b*", respectively. In Equation 6.1, the flux Ψ is equated to a scalar quantity Q and thus it is also taken as scalar quantity. But as is shown in Figure 6.2, while emanating from (or merging into) a charge it appears to possess some direction. This aspect is accounted by taking the flux density ***D*** as a vector quantity. Thus, the expressions of ***D*** given by Equations 6.2a, 6.2b, and 6.2c are to

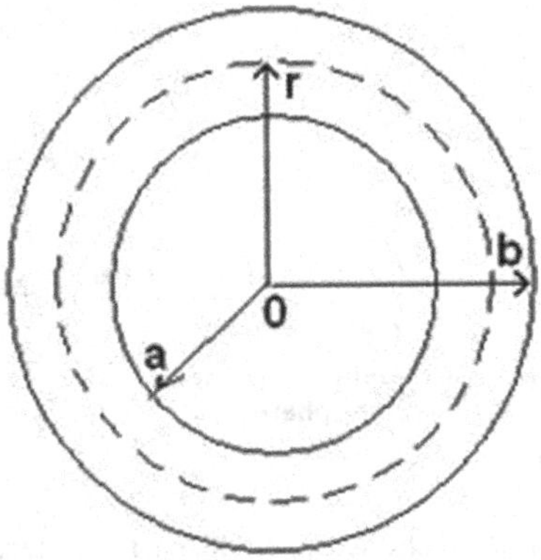

FIGURE 6.3 Radii of inner, outer, and imaginary spheres.

be represented accordingly. These expressions in vector form, at different locations can be written as:

At inner sphere ($r = a$):

$$\boldsymbol{D}\Big|_{r=a} = \frac{Q}{4\pi \cdot a^2} \cdot \boldsymbol{a}_r \tag{6.3a}$$

At outer sphere ($r = b$):

$$\boldsymbol{D}\Big|_{r=b} = \frac{Q}{4\pi \cdot b^2} \cdot \boldsymbol{a}_r \tag{6.3b}$$

And at a sphere of an arbitrary radius r ($a \le r \le b$):

$$\boldsymbol{D} = \frac{Q}{4\pi \cdot r^2} \cdot \boldsymbol{a}_r \quad \text{for } a \le r \le b \tag{6.3c}$$

Equations 6.3a, 6.3b, and 6.3c gives expressions for flux density due to the symmetrically distributed charge Q. The general definition of flux density at a point P is given as:

$$\boldsymbol{D} \stackrel{\text{def}}{=} \lim_{\Delta s \to \infty} \frac{\Delta \psi}{\Delta s} \cdot \boldsymbol{a}_n \tag{6.4}$$

where $\boldsymbol{a}_n$ is the unit vector indicating the direction of the flux at P and Δs is a small area around the point P located on the plane.

6.3.1 Relation between Flux Density and Field Intensity

Comparison of Equation 5.10a $\left(\boldsymbol{E} = \frac{1}{4\pi\varepsilon_o} \cdot \frac{Q}{r^2} \boldsymbol{a}_R\right)$ after replacing R by r with Equation 6.3c $\left(\boldsymbol{D} = \frac{Q}{4\pi \cdot r^2} \cdot \boldsymbol{a}_r\right)$ leads to a constitutive relation:

$$\boldsymbol{D} = \varepsilon_o \boldsymbol{E} \tag{6.5}$$

Equation 6.5 shows that $\boldsymbol{D}$ and $\boldsymbol{E}$ vectors have the same directions as long as ε_o is a real positive scalar quantity. Such a medium is termed as isotropic. In this chapter, the media is assumed to be isotropic unless it is specified otherwise.

Other than Equation 5.10a, which gives $\boldsymbol{E}$ due to point charge, Equations 5.16, 5.32, and 5.39 define $\boldsymbol{E}$ fields due to the line, surface, and volume charge distributions, respectively. Table 6.1 gives these equations bearing the same numbers and those obtained for $\boldsymbol{D}$ from Equation 6.5 for point, line, surface, and volume charge distribution with due modification.

6.3.2 Estimation of Electric Flux

The strength of electric field can be visualized in terms of the concentration or density of field lines or that of electric flux. Thus the number of lines passing through

TABLE 6.1
Expressions for E and D for Different Field Distributions

Point charge	$\boldsymbol{E} = \dfrac{Q}{4\pi\varepsilon_o r^2}\boldsymbol{a}_r$	(5.10a)	$\boldsymbol{D} = \dfrac{Q}{4\pi r^2}\boldsymbol{a}_r$	(6.5a)
Line charge	$\boldsymbol{E} = \int_L \dfrac{\rho_L\, d\ell}{4\pi\varepsilon_o r^2}\boldsymbol{a}_\rho$	(5.16)	$\boldsymbol{D} = \int_L \dfrac{\rho_L\, d\ell}{4\pi r^2}\boldsymbol{a}_\rho$	(6.6a)
Surface charge	$\boldsymbol{E} = \iint_s \dfrac{\rho_s ds}{4\pi\varepsilon_o r^2}\boldsymbol{a}_r$	(5.32)	$\boldsymbol{D} = \iint_s \dfrac{\rho_s ds}{4\pi r^2}\boldsymbol{a}_r$	(6.6b)
Volume charge	$E = \iiint_v \dfrac{\rho_v dv}{4\pi\varepsilon_o r^2}a_N$	(5.39)	$D = \iiint_v \dfrac{\rho_v dv}{4\pi r^2}a_N$	(6.6c)

a unit area perpendicular to the direction of field indicated the strength of the field. As noted earlier, the number of field lines passing through a geometrical surface of a given area depends on (i) the strength of field (ii) the surface area and (iii) the relative orientation of surface vis-à-vis the field lines impinging upon. The procedure to estimate the electric flux is described below.

6.3.2.1 Electric Flux and Oriented Surfaces

Figure 6.4 shows a number of surfaces each with area "S". These surfaces are oriented in different directions. All these surfaces are located in $\boldsymbol{E}$ (or $\boldsymbol{D}$) field. The normal (n) drawn on each of these surfaces makes an angle (θ) with $\boldsymbol{E}$ (or $\boldsymbol{D}$) field. In view of this figure the value of electric flux crossing a particular surface will vary in accordance with the orientation of that surface vis-à-vis the direction of $\boldsymbol{E}$ (or $\boldsymbol{D}$). In the entire region occupied by $\boldsymbol{E}$ (or $\boldsymbol{D}$), the field is assumed to be uniform and oriented in the same direction and thus the lines drawn to represent the field are parallel. This figure shows that the maximum lines are intercepted by a surface when n is parallel to $\boldsymbol{E}$ and no line pass through S when n is perpendicular to $\boldsymbol{E}$. Thus the number of $\boldsymbol{E}$ lines crossing the surface is proportional to its area (S) and cosine of the angle (θ) the normal n makes with the direction of $\boldsymbol{E}$ lines. Since $\boldsymbol{E}$ and $\boldsymbol{D}$ differ only by a factor ε_o the electric flux density vector $\boldsymbol{D}$ with magnitude, $|\boldsymbol{D}| = D$, is more convenient to proceed with.

Let us divide the entire surface S, located in the field, into small segments each with area ΔS. The flux crossing this elemental surface can be termed as incremental flux ($\Delta\Psi$). Thus, the incremental electric flux ($\Delta\Psi$) passing through the elemental surface (ΔS) is the product of electric flux density $|\boldsymbol{D}|$, the surface ΔS and cosine of the angle of orientation of ΔS with reference to the $\boldsymbol{E}$ or $\boldsymbol{D}$ field. In mathematical terms:

$$\Delta\Psi = |\boldsymbol{D}| \cdot \Delta S \cdot \cos\theta \tag{6.7a}$$

Since $\boldsymbol{D}$ is a vector quantity, the differential surface must also be represented as a vector quantity $\Delta\boldsymbol{S}$. The magnitude of $\Delta\boldsymbol{S}$ is the surface area ΔS and its orientation

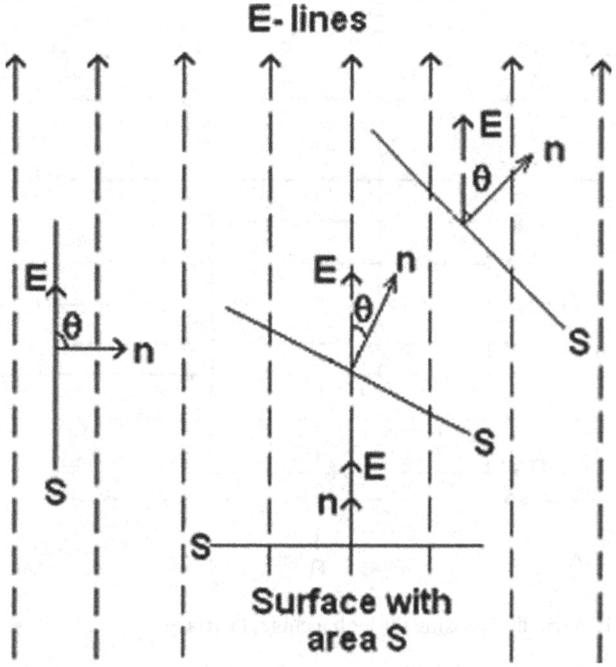

FIGURE 6.4 Surfaces with different orientations in a uniform E or D field.

is expressed in terms of normal vector $\boldsymbol{n}$. This $\boldsymbol{n}$ vector makes an angle θ with vector $\boldsymbol{E}$ or $\boldsymbol{D}$. Thus, $\Delta\Psi$ given by Equation 6.7a can be written in terms of the dot product:

$$\Delta\Psi = \boldsymbol{D} \bullet \Delta\boldsymbol{S} \tag{6.7b}$$

A unit vector normal to ΔS can point into two different directions. For a closed surface, by convention, normal points outward. Since $\Delta\Psi$ is the flux through a small area ΔS and this area may be part of a larger area S, the total flux Ψ through S can be evaluated by integrating the differential flux over the entire (open or closed) surface S:

$$\Psi = \int \Psi = \iint_S \boldsymbol{D} \bullet d\boldsymbol{S} \tag{6.8}$$

The flux passing through a closed surface can be written as:

$$\Psi = \oiint \boldsymbol{D} \bullet d\boldsymbol{S} \tag{6.9}$$

6.3.2.2 Electric Flux and a Cubical Configuration

Figure 6.5 shows a cubical configuration that is located deep inside the electric field lines. As shown in the figure, two of the faces of this cube are perpendicular to

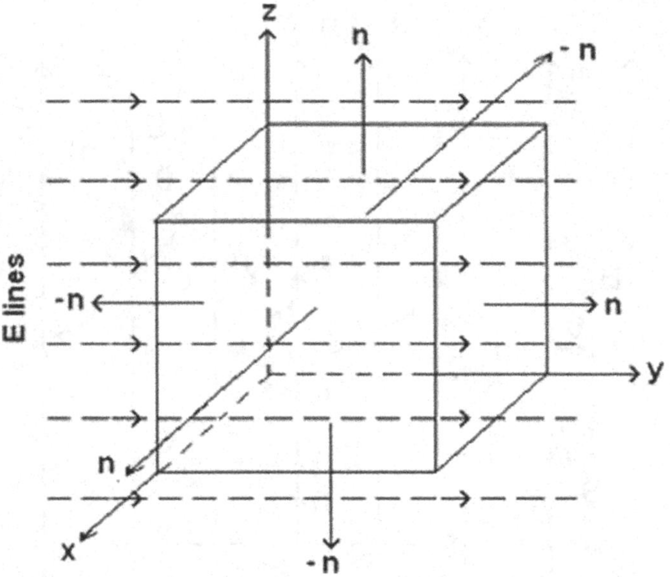

FIGURE 6.5 Electric flux passing through a cubical surface.

the electrical field which is oriented along the y-axis and assumed to be uniform throughout the space. Flux through a given surface can be positive or negative, since the cosine term involved in Equation 6.7 can be positive or negative.

Since the top, bottom, front, and back sides of the cube are parallel to $\boldsymbol{E}$ lines, these sides do not intercept any flux and thus the flux passing through these surfaces is zero. $\boldsymbol{E}$ lines are parallel to the normal vector $\boldsymbol{n}$ for the remaining two (left and right) sides. The flux through the right side is $\Psi = E{\cdot}S$. Also the field lines are anti-parallel to the normal vector $-\boldsymbol{n}$ for the left side, so the flux through this side is $\Psi = -E{\cdot}S$. The total flux through the left and right surfaces of the cube is the algebraic sum of fluxes through all sides, and is zero.

6.4 THE GAUSSIAN SURFACE

In Section 6.3.2.2, the distribution of flux was taken to be along one of the axes of the cube. Thus, flux became parallel to four of its surfaces and perpendicular to the remaining two surfaces. In this case no thought was given to the source of flux which resulted in such a distribution. With little imagination one may conclude that such flux distribution will result due to an infinite sheet containing uniformly distributed positive charges and this sheet must be located parallel to the x–z plane on the negative of the y-axis. The source is assumed to be located outside the cube. If the source is located inside a closed surface; a thought needs to be given to the nature of this surface. Obviously, the surface to be selected has to depend on the nature of the distribution of source. In the descriptions given earlier it can be concluded that if flux is perpendicular or parallel to the surface/surfaces the problem gets much simplified. This conclusion leads to the concept of a *Gaussian surface*.

The Gaussian surface is an imaginary (closed) surface that encloses a symmetrical charge distribution. This surface has to be such that $\boldsymbol{D}$ is either normal or tangential to the segments of this surface. When it is normal $\boldsymbol{D} \bullet \boldsymbol{ds} = D{\cdot}ds$ and when tangential, $\boldsymbol{D} \bullet \boldsymbol{ds} = 0$. Thus this surface is to be so chosen that some symmetry is exhibited vis-à-vis the charge distribution. Equation 6.9 is the key relation for evaluating the flux caused due to different charge configurations and passing through surfaces of different shapes. Some of these are discussed in the following subsections.

6.4.1 Gaussian Surface for a Point Charge

Let there be a positive point charge "Q" that emanates flux equally in all the possible directions. The most appropriate Gaussian surface to enclose such a configuration will be a spherical surface shown in Figure 6.6a provided this charge is located at its center. The density of these field lines gives the magnitude of $\boldsymbol{D}$ at any given point. The decrease in magnitude of density of this field will be proportional to the inverse square of distance "r" from the charge as the increase in the area of spherical surface is proportional to r^2. The total number of field (or flux) lines passing through the spherical surface, which is perpendicular to the direction of field at any point, is the product of their density and the area of the spherical surface. It is therefore independent of the size of sphere.

In addition, the number of field lines will depend only on magnitude of the charge at the center of the sphere. Because the flux is proportional to the number of field lines passing through the area, the flux is expected to be proportional to the charge inside the surface. The field of a positive point charge at the center is radial, pointing outwards or away from the point charge. It is everywhere normal to the surface of the sphere and its magnitude is given by:

$$D = \frac{Q}{4\pi \cdot r^2} \tag{6.10}$$

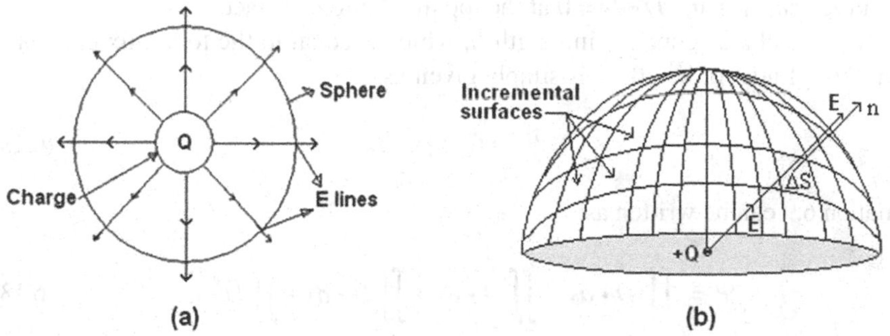

FIGURE 6.6 (a) Spherical surface enclosing a point charge. (b) Spherical surface divided into small areas and E and n vectors crossing an elemental surface.

Also, since $\boldsymbol{D} = D\boldsymbol{a}_r$ and $\boldsymbol{ds} = ds\boldsymbol{a}_r$, the total electric flux emanating from a closed spherical surface due to a point charge Q placed at its center can now be calculated by using Equation 6.9, as shown below:

$$\Psi = \oint\!\!\oint \boldsymbol{D} \bullet \boldsymbol{ds} = \oint\!\!\oint \left(\frac{Q}{4\pi \cdot r^2} \boldsymbol{a}_r \right) \bullet \left(ds\boldsymbol{a}_r \right) = \oint\!\!\oint \frac{Q}{4\pi \cdot r^2} \cdot ds \tag{6.11}$$

Since the point charge Q and the radius of the sphere r are constants, these can be taken out of the integral sign. Thus:

$$\Psi = \left(\frac{Q}{4\pi \cdot r^2} \right) \cdot \oint\!\!\oint ds = \left(\frac{Q}{4\pi \cdot r^2} \right) \cdot \left(4\pi r^2 \right) = Q \tag{6.12}$$

Thus the total flux emerging-out from the spherical surface is equal to the charge located at the center of the sphere. This is a special case of Gauss' law, which states that if electric charges are located anywhere inside a closed surface of any shape and size, the total flux emerging-out from the surface is equal to the total charge within the closed surface.

6.4.2 Gaussian Surface for a Line Charge

Consider an infinite line with uniformly distributed positive charge density ρ_L. As shown in Figure 6.7, for this line charge a cylindrical surface is the appropriate Gaussian surface which will meet the requirement of normal and tangential $\boldsymbol{D}$, provided that the axis of the cylinder coincides with the line charge. As shown in Figure 6.7, this line charge is chosen to be along the z-axis. We do not aim to find the total flux due to the infinite line charge, it is infinite any way. Our objective is to find the electric flux density $\boldsymbol{D}$ (and the electric field intensity $\boldsymbol{E}$, at any point due to this line charge. Our purpose is amicably served simply by taking a cylinder of finite length (L) and a radius ρ which encloses only a finite length (i.e. L) of the infinitely long line charge. All the flux lines emerging from the line charge are perpendicular to the curved surface and parallel to the top and bottom surfaces of the cylinder, wherever these are located. Thus, if $\boldsymbol{D}$ is taken to be the flux density $\boldsymbol{D} \bullet \boldsymbol{ds} = D \cdot ds$ on the curved surface and $\boldsymbol{D} \bullet \boldsymbol{ds} = 0$ at the top and bottom surfaces.

Thus, the charge contents in length L, which is equal to the total flux emanating from the cylinder of length L, is simply given as:

$$\Psi = Q = \rho_L \cdot L \tag{6.13a}$$

Equation 6.9 can be written as:

$$\Psi = \oint\!\!\oint_S \boldsymbol{D} \bullet \boldsymbol{ds} = \iint_{S_t} \boldsymbol{D} \bullet \boldsymbol{ds} + \iint_{S_b} \boldsymbol{D} \bullet \boldsymbol{ds} + \iint_{S_c} \boldsymbol{D} \bullet \boldsymbol{ds} \tag{6.13b}$$

Now since no flux is coming out of the top surface (S_t) and bottom surface (S_b), the first two terms on the RHS of this equation become zero. On the curved surface (S_c)

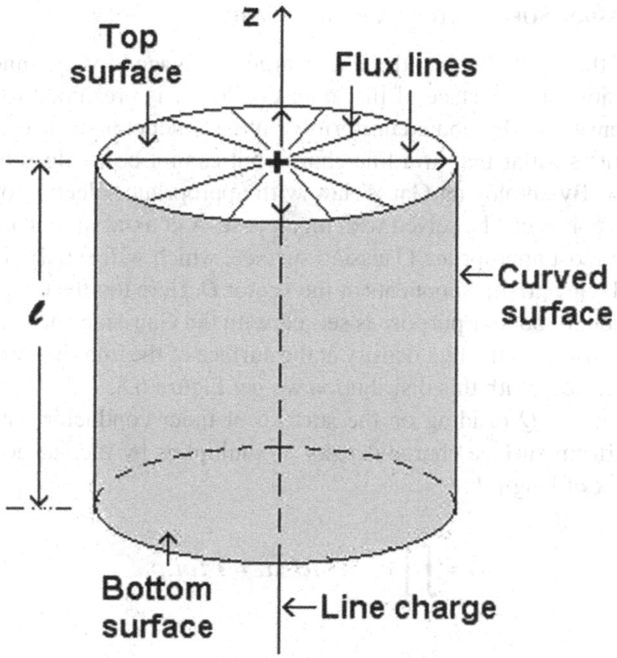

FIGURE 6.7 Cylindrical surface enclosing a line charge.

the magnitude of the flux density D is the same at every point and it is directed radially outward from the line, i.e. $\boldsymbol{D} = D_\rho \boldsymbol{a}_\rho$. Since $\boldsymbol{ds} = ds\,\boldsymbol{a}_\rho$, Equation 6.13b becomes:

$$\Psi = \iint_{S_c} \boldsymbol{D} \cdot \boldsymbol{ds} = \iint_{S_c} D_\rho \cdot ds = D_\rho \cdot \iint_{S_c} ds = D_\rho \cdot \int_0^L \int_0^{2\pi} \rho \cdot d\varphi \cdot dz = 2\pi \cdot \rho \cdot L \cdot D_\rho \tag{6.13c}$$

Therefore, from Equations 6.13a and 6.13c, we have:

$$\Psi = \rho_L \cdot L = 2\pi \cdot \rho \cdot L \cdot D_\rho \quad \text{or} \quad D_\rho = \frac{\rho_L}{2\pi \cdot \rho}$$

Thus:

$$\boldsymbol{D} = D_\rho \boldsymbol{a}_\rho = \left(\frac{\rho_L}{2\pi \cdot \rho} \right) \boldsymbol{a}_\rho \tag{6.14a}$$

Hence, the electric field intensity due to the long uniform line charge in free space is:

$$\boldsymbol{E} = E_\rho \boldsymbol{a}_\rho = \frac{D_\rho}{\varepsilon_0} \boldsymbol{a}_\rho = \left(\frac{\rho_L}{2\pi \cdot \rho \varepsilon_0} \right) \boldsymbol{a}_\rho \tag{6.14b}$$

6.4.3 Gaussian Surface for a Coaxial Cable

In the case of the coaxial cable, the line charge is replaced by the inner conductor of a coaxial cable. The surface of the inner conductor is presumed to carry a surface charge density ρ_S, distributed uniformly on the conductor surface, although this problem is quite similar that of a line charge but cannot be easily solved by using Coulomb's law. By employing Gauss' law with appropriate selection of a Gaussian surface, this problem can be solved with much ease. A coaxial imaginary cylinder of radius ρ is the most appropriate Gaussian surface, which will satisfy the conditions of normal and tangential components of the vector $\boldsymbol{D}$. Here too the coaxial cable can be of infinite length but our purpose is served with the Gaussian surface of length L. Let us assume that D_ρ is the flux density at the surface of the imaginary cylinder used as Gaussian surface. With this distribution we get Figure 6.8.

The total charge Q residing on the surface of inner conductor with radius "a" equals the uniform surface charge density ρ_S multiplies by the surface area of the inner conductor of length L:

$$Q = \int_0^{2\pi}\int_0^{L} \rho_s \cdot a \cdot d\varphi \cdot dz = 2\pi a L \rho_s \tag{6.15a}$$

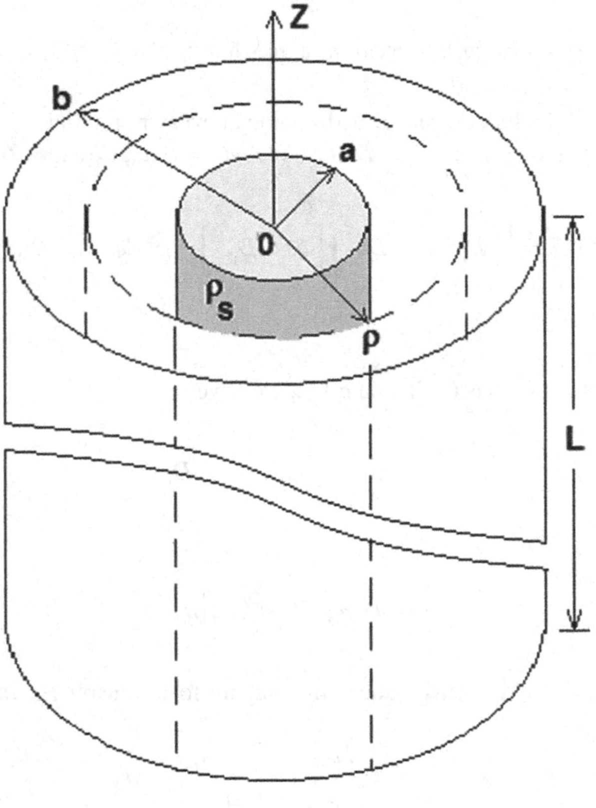

FIGURE 6.8 Imaginary coaxial cylinder of radius ρ.

In view of Equation 6.13b:

$$Q = \int_0^L \int_0^{2\pi} D_\rho \cdot \rho \cdot d\varphi \cdot dz = D_\rho \cdot \rho \int_0^L \int_0^{2\pi} d\varphi \cdot dz = D_\rho \cdot (2\pi\rho L) \tag{6.15b}$$

On equating Equations 6.15a and 6.15b we get:

$$D_\rho = \frac{a\rho_s}{\rho} \quad \text{or} \quad \boldsymbol{D} = \frac{a\rho_s}{\rho}\boldsymbol{a}_\rho \quad \text{over } a \le \rho \le b \tag{6.16}$$

6.5 ASYMMETRICAL FIELD DISTRIBUTIONS

So far, our discussion was confined to the symmetrical field configurations. In non-symmetrical cases the field will neither be tangential nor normal and nor constant at the surfaces. Without meeting these requirements the integrals arising out of mathematical formulation cannot be easily evaluated. The situation appears to be a bit hopeless unless an alternative is found. This alternative is provided by the divergence operation described in Section 4.3. The mathematical expression for divergence in a Cartesian coordinate system is given by Equation 4.22a, whereas its definition is given by Equation 4.21b. These relations and the physical interpretation thereof, given in Section 4.3.2.7, are valid for any vector including that of electric flux density *D*. Thus, on replacing vector *F* by vector *D* Equations 4.21a and 4.21b take the following form:

$$\left(\frac{\partial D_x}{\partial x} + \frac{\partial D_y}{\partial y} + \frac{\partial D_z}{\partial z}\right) = \nabla \bullet \boldsymbol{D} \stackrel{\text{def}}{=} \text{Divergence of } \boldsymbol{D} = \text{div}\,\boldsymbol{D} \tag{6.17a}$$

$$\nabla \bullet \boldsymbol{D} \stackrel{\text{def}}{=} \lim_{\Delta v \to 0}\left(\frac{\Delta\psi}{\Delta v}\right) = \lim_{\Delta v \to 0}\left(\frac{\oiint_s \boldsymbol{D} \bullet \boldsymbol{ds}}{\Delta v}\right) \tag{6.17b}$$

The terms involved in these relations can now be assigned physical meanings in conformity with the electromagnetics. Thus in these equations if *D* represents the electric flux density vector the LHS expressions of Equation 6.17b will represent the volumetric charge density ρ.

6.5.1 FIRST MAXWELL'S EQUATION

By Gauss' law for a closed surface *S* bounding a small volume Δv we have:

$$\oiint_S \boldsymbol{D} \bullet \boldsymbol{ds} \approx \Delta Q \approx \rho \cdot dv \tag{6.18a}$$

where Q is the small charge in the small volume v and ρ indicates the volume charge density at the point P. (Note that the symbol ρ is also used to indicate a coordinate in the cylindrical system of space coordinates.) With reference to the contents there should be no confusion about its local meaning. The charge density ρ is defined as:

$$\rho \overset{\text{def}}{=} \lim_{\Delta v \to 0} \left(\frac{\Delta Q}{\Delta v} \right) \tag{6.18b}$$

We have from Equations 6.18a and 6.18b:

$$\rho \approx \frac{\oiint_S \boldsymbol{D} \bullet \boldsymbol{ds}}{\Delta v} = \lim_{\Delta v \to 0} \left(\frac{\oiint_S \boldsymbol{D} \bullet \boldsymbol{ds}}{\Delta v} \right) \tag{6.18c}$$

Therefore, using Equations 6.17a and 6.18c:

$$\nabla \bullet \boldsymbol{D} = \rho \tag{6.19a}$$

In a charge-free region (i.e. when $\rho = 0$) it reduces to:

$$\nabla \bullet \boldsymbol{D} = 0 \tag{6.19b}$$

Equation 6.19a is referred to as the first Maxwell's equation. It relates the divergence of a vector field (on left hand side) to a scalar quantity on the right hand side. This equation is based on Faraday's experiments with electrostatic fields; Maxwell, however, retained this equation without any change for time varying fields as well.

6.5.2 Expressions of Divergence in Different Coordinate Systems

On replacing vector $\boldsymbol{F}$ by vector $\boldsymbol{D}$, in Equations 4.21a to 4.21c, the expressions of divergence in different coordinate systems can be given as:

$$\nabla \bullet \boldsymbol{D} \equiv \frac{\partial D_x}{\partial x} + \frac{\partial D_y}{\partial y} + \frac{\partial D_z}{\partial z} \quad \text{(Cartesian)} \tag{6.20a}$$

$$\nabla \bullet \boldsymbol{D} \equiv \frac{1}{\rho} \cdot \frac{\partial (\rho . D_\rho)}{\partial \rho} + \frac{1}{\rho} \cdot \frac{\partial D_\varphi}{\partial \varphi} + \frac{\partial D_z}{\partial z} \quad \text{(Cylindrical)} \tag{6.20b}$$

$$\begin{aligned} \nabla \bullet \boldsymbol{D} \equiv & \frac{1}{r^2} \cdot \frac{\partial (r^2 \cdot D_r)}{\partial r} + \frac{1}{r \cdot \sin\theta} \cdot \frac{\partial (\sin\theta \cdot D_\theta)}{\partial \theta} \\ & + \frac{1}{r \cdot \sin\theta} \cdot \frac{\partial (D_\varphi)}{\partial \varphi} \quad \text{(Spherical)} \end{aligned} \tag{6.20c}$$

6.5.3 Divergence Theorem

In Equations 6.1 and 6.9:

$$Q = \psi = \oiint_s \boldsymbol{D} \bullet d\boldsymbol{s} \tag{6.21a}$$

Also, in Equations 6.18b and 6.19a, total charge in a volume v bounded by the surface s is given as:

$$Q = \iiint_v \rho dv = \iiint_v (\nabla \bullet D) dv \tag{6.21b}$$

On comparing Equations 6.21a and 6.21b, we get:

$$\oiint_S \boldsymbol{D} \bullet d\boldsymbol{S} = \iiint_v (\nabla \bullet \boldsymbol{D}) \cdot dv \tag{6.22}$$

This is the same relation that was given by Equation 4.50 (see Section 4.7.1) in terms of vector $\boldsymbol{A}$.

Example 6.1

A $5C$ point charge is located at the origin in a vacuum. Find (a) $|\boldsymbol{D}|$ and $\boldsymbol{D}$ at (0.3, 0.5, −0.4) and (b) the total electric flux passing through the first octant.

SOLUTION

Given $Q = 5C$.
The vector distance between the given point and the origin is:

$$\boldsymbol{R} = 0.3\boldsymbol{a_x} + 0.5\boldsymbol{a_y} - 0.4\boldsymbol{a_z}$$

Thus $|R| = \sqrt{(0.3)^2 + (0.5)^2 + (0.4)^2} = \sqrt{0.5}$ $|R|^2 = 0.5$ and $\boldsymbol{a_R} = \dfrac{\boldsymbol{R}}{|R|} = \dfrac{0.3\boldsymbol{a_x} + 0.5\boldsymbol{a_y} - 0.4\boldsymbol{a_z}}{\sqrt{0.5}}$

(a) From Equation 6.5a $|D| = \dfrac{Q}{4\pi \cdot |R|^2} = \dfrac{5}{4 \cdot \pi \cdot 0.5} = 0.796$ c/m^2

$$\boldsymbol{D} = |D|\boldsymbol{a_R} = 0.796\frac{0.3\boldsymbol{a_x} + 0.5\boldsymbol{a_y} - 0.4\boldsymbol{a_z}}{\sqrt{0.5}} = 0.338\boldsymbol{a_x} + 0.563\boldsymbol{a_y} - 0.45\boldsymbol{a_z}$$

(b) Total flux (Ψ) coming out from the sphere = the total charge (Q) enclosed or $\Psi = Q$

The flux coming out from the first octant is $\dfrac{\Psi}{8} = \dfrac{Q}{8} = \dfrac{5}{8} = 0.625$ C

Example 6.2

Find the total charge within a cylinder of radius $\rho = 2$ and length $-1 \le z \le 1$ if $\boldsymbol{D}$=:
(a) $\frac{10}{\rho}\frac{z}{|z|}\boldsymbol{a}_z$ (b) $10z^2\sin(\varphi/2)\boldsymbol{a}_\rho$ C/m²

SOLUTION

From Equation 6.1 and 6.9 $Q = \Psi = \oiint_s \boldsymbol{D}_s \bullet \boldsymbol{ds}$.

(a) Here $\boldsymbol{D}_s = \frac{10}{\rho}\frac{\boldsymbol{z}}{|\boldsymbol{z}|}\boldsymbol{a}_z$ and $\boldsymbol{ds} = \rho d\rho d\varphi \boldsymbol{a}_z$

Thus:

$$Q = \oiint_s \boldsymbol{D}_s \bullet \boldsymbol{ds} = \oiint_s \frac{10}{\rho}\frac{z}{|z|}\boldsymbol{a}_z \bullet \rho d\rho d\varphi \boldsymbol{a}_z$$

$$= \int_0^2\int_0^{2\pi} 10\frac{\boldsymbol{z}}{|z|} d\rho d\varphi = 10\rho\Big|_0^2 \frac{\boldsymbol{z}}{|z|}\int_0^{2\pi} d\varphi = 20\frac{\boldsymbol{z}}{z}\int_0^{2\pi} d\varphi = 40\pi\frac{\boldsymbol{z}}{|z|}$$

Total charge Q = total flux = flux emerging from $z = 1$ surface + flux emerging from $z = -1$ surface. Thus $Q = 40\pi + 40\pi = 80\pi = 251.327$ C.

(b) Here $\boldsymbol{D}_s = 10z^2\sin(\varphi/2)\boldsymbol{a}_\rho$ and $\boldsymbol{ds} = \rho d\varphi dz \boldsymbol{a}_\rho$

Thus:

$$Q = \oiint_s \boldsymbol{D}_s \bullet \boldsymbol{ds} = \oiint_s 10z^2\sin(\varphi/2)\rho d\varphi dz$$

$$= \int_{-1}^{1}\int_0^{2\pi} 10z^2\sin(\varphi/2)\rho\big|_2 d\varphi dz = \int_0^{2\pi} 20\frac{z^3}{3}\Big|_{-1}^{1}\sin(\varphi/2)\rho d\varphi$$

Substitution of $\varphi/2 = \theta$ gives $\varphi = 2\theta$, $d\varphi = 2d\theta$ and limits of θ become 0 to π, thus:

$$Q = \frac{40}{3}\int_0^{2\pi}\sin(\varphi/2)d\varphi = \frac{40}{3}\int_0^{\pi}\sin(\theta)2d\theta = -\frac{80}{3}\cos\theta\Big|_0^{\pi} = \frac{160}{3} = 53.333\text{ C}$$

Example 6.3

Three cylindrical sheets of charges are located in free space with $\rho_s = 5$ C/m² at $\rho = 2$, $\rho_s = -2$ C/m² at $\rho = 4$ and $\rho_s = -3$ C/m² at $\rho = 5$. Find $\boldsymbol{D}$ at (a) $\rho = 1$ (b) $\rho = 3$ (c) $\rho = 4.5$ and (d) $\rho = 6$.

SOLUTION

In view of Figure 6.9 and Equation $\boldsymbol{D} = \frac{a\rho_s}{\rho}\boldsymbol{a}_\rho$

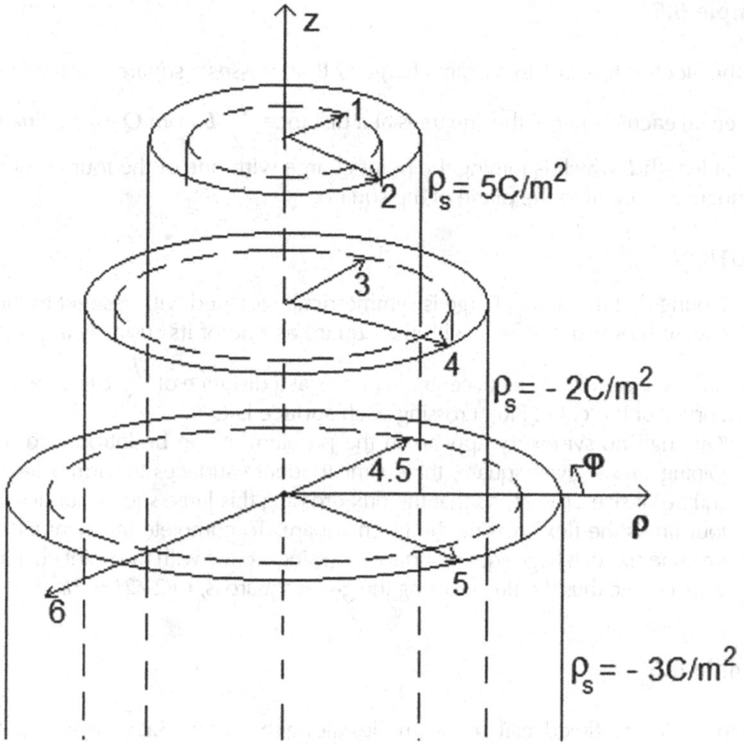

FIGURE 6.9 Three cylindrical sheets of charges.

Flux emerges radially outward from each cylindrical sheet.

(a) At $\rho = 1$, $\boldsymbol{D} = 0$ (There is no $\boldsymbol{D}$ inside the inner cylinder)

(b) At $\rho = 3$, $\boldsymbol{D} = \dfrac{a\rho_s}{\rho}\boldsymbol{a}_\rho = \dfrac{2\times 5}{3}\boldsymbol{a}_\rho = \dfrac{10}{3}\boldsymbol{a}_\rho$

(c) At $\rho = 4.5$, $\boldsymbol{D} = \left[\dfrac{2\times 5}{4.5} + \dfrac{-2\times 4}{4.5}\right]\boldsymbol{a}_\rho = \left[\dfrac{10}{4.5} + \dfrac{-8}{4.5}\right]\boldsymbol{a}_\rho = \dfrac{4}{9}\boldsymbol{a}_\rho$

(d) At $\rho = 6$, $\boldsymbol{D} = \left[\dfrac{2\times 5}{6} + \dfrac{-2\times 4}{6} + \dfrac{-3\times 5}{6}\right]\boldsymbol{a}_\rho = \left[\dfrac{10}{6} + \dfrac{-8}{6} + \dfrac{-15}{6}\right]\boldsymbol{a}_\rho = \dfrac{-13}{6}\boldsymbol{a}_\rho$

Example 6.4

A region is bounded by four identical plane surfaces. Where should a point charge Q be placed so that the flux crossing from each of its any three surfaces is $Q/6$?

SOLUTION

The region is bounded by four identical equilateral triangles. The point charge should be placed on any one triangle (inside or outside the region) where the bisectors of its angles (or right bisectors of its sides) meet.

Example 6.5

Find the electric flux due to a point charge $\mathbf{Q}$ that crosses a square surface of side $\mathbf{L}$, when (a) each corner of the square is at a distance $\frac{\sqrt{3}}{2}\mathbf{L}$ from $\mathbf{Q}$ (b) the line segment of length $\mathbf{L}$ which is joining the point charge with one of the four corners of the square is normal to the plane of the square.

SOLUTION

(a) Noting that the point charge is symmetrically located with respect to the square, construct a cube with this square as one of its surfaces and the point charge placed at its center, which is at a distance of $\frac{\sqrt{3}}{2}L$ from each corner of the cube. Flux crossing each surface is $Q/6$.

(b) Although no symmetry appears in the problem, it can be introduced by joining to the given square, three similar square surfaces to form a large square of size $2L \times 2L$, so that the flux crossing this large square surface is four times the flux crossing the given square. To complete the symmetry, imagine six such large square surfaces forming a cube with the point charge at its center, thus the flux crossing the given square is, $Q/24$ $2L \times 2L$.

Example 6.6

A uniform electric flux density of 5C/m^2 lies along the vector $5\mathbf{a}_x - 3\mathbf{a}_y + 4\mathbf{a}_z$. After expressing D as a vector, determine the magnitude of flux crossing (a) the *y–z*-plane bounded by y = 0, y = 1, z = 0, z = 1; (b) the *z–x*-plane bounded by z = 0, z = 1, x = 0, x = 1; (c) *x–y*-plane bounded by x = 0, x = 1, y = 0, y = 1.

SOLUTION

The three planes referred to above are shown in Figure 6.10. The given vector D can be expressed as a vector as follows.

$$\mathbf{R} = 5\mathbf{a}_x - 3\mathbf{a}_y + 4\mathbf{a}_z \quad |R| = \sqrt{50} \quad \mathbf{a}_R = \frac{5\mathbf{a}_x - 3\mathbf{a}_y + 4\mathbf{a}_z}{\sqrt{50}}$$

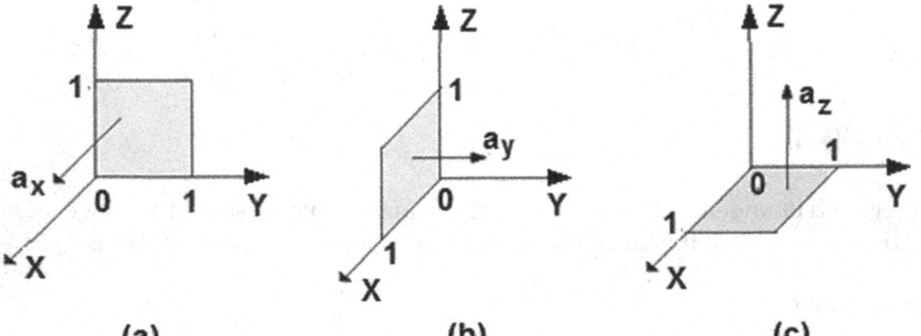

FIGURE 6.10 (a) *y–z* plane (b) *x–z* plane (c) *x–y* plane.

$$\boldsymbol{D} = |D| \cdot \boldsymbol{a_R} = 5 \cdot \frac{5\boldsymbol{a_x} - 3\boldsymbol{a_y} + 4\boldsymbol{a_z}}{\sqrt{50}} = \frac{25\boldsymbol{a_x}}{\sqrt{50}} - \frac{15\boldsymbol{a_y}}{\sqrt{50}} + \frac{20\boldsymbol{a_z}}{\sqrt{50}} = D_x\boldsymbol{a_x} + D_y\boldsymbol{a_y} + D_z\boldsymbol{a_z}$$

$$D_x = \frac{25}{\sqrt{50}} \quad D_y = \frac{-15}{\sqrt{50}} \quad D_z = \frac{20}{\sqrt{50}}$$

The magnitude of flux crossing a surface = the corresponding component multiplied by the area of the surface. Thus,

(a) The magnitude of flux crossing the y–z-plane $= \frac{25}{\sqrt{50}} \times 1 = 3.5353$ C

(b) The magnitude of flux crossing the z–x-plane $= \frac{-15}{\sqrt{50}} \times 1 = 2.1213$ C

(c) The magnitude of flux crossing the x–y-plane $= \frac{20}{\sqrt{50}} \times 1 = 2.8284$ C

Example 6.7

A uniform line charge $\rho_L = 10\pi$ C/m lies along the z-axis in free space. Find the components of D in the direction of vector $3a_x - 5a_y - 4a_z$ at the point (a) (1,0,1); (b) (1,2,0); (c) (2,3,–4).

SOLUTION

$$\boldsymbol{R} = 3\boldsymbol{a_x} - 5\boldsymbol{a_y} - 4\boldsymbol{a_z} \quad |R| = \sqrt{50} \quad \boldsymbol{a_R} = \frac{3\boldsymbol{a_x} - 5\boldsymbol{a_y} - 4\boldsymbol{a_z}}{\sqrt{50}}$$

In view of Equation 6.6b:

$$\boldsymbol{D} = \frac{\rho_L}{2\pi r}\boldsymbol{a_r} = \frac{10\pi}{2\pi r}\boldsymbol{a_r} = \frac{5}{r}\boldsymbol{a_r}$$

(a) At (1,0,1): $\boldsymbol{r} = \boldsymbol{a_x} + \boldsymbol{a_z} \quad |r| = \sqrt{2} \quad \boldsymbol{a_r} = \frac{\boldsymbol{a_x} + \boldsymbol{a_z}}{\sqrt{2}}$

$$\boldsymbol{D} = \frac{5}{r}\boldsymbol{a_r} = \frac{5}{\sqrt{2}}\boldsymbol{a_r} = \frac{5}{\sqrt{2}}\frac{\boldsymbol{a_x} + \boldsymbol{a_z}}{\sqrt{2}} = 2.5(\boldsymbol{a_x} + \boldsymbol{a_z})$$

$$\boldsymbol{D} \bullet \boldsymbol{a_R} = (2.5\boldsymbol{a_x} + 2.5\boldsymbol{a_z}) \bullet \frac{3\boldsymbol{a_x} - 5\boldsymbol{a_y} - 4\boldsymbol{a_z}}{\sqrt{50}} = \frac{7.5 - 10}{\sqrt{50}} = \frac{-2.5}{\sqrt{50}} = -0.35355$$

(b) At (1,2,0): $\boldsymbol{r} = \boldsymbol{a_x} + 2\boldsymbol{a_y} \quad |r| = \sqrt{5} \quad \boldsymbol{a_r} = \frac{\boldsymbol{a_x} + 2\boldsymbol{a_y}}{\sqrt{5}}$

$$\boldsymbol{D} = \frac{5}{r}\boldsymbol{a_r} = \frac{5}{\sqrt{5}}\boldsymbol{a_r} = \frac{5}{\sqrt{5}}\frac{\boldsymbol{a_x} + 2\boldsymbol{a_y}}{\sqrt{5}} = \boldsymbol{a_x} + 2\boldsymbol{a_y}$$

$$\mathbf{D} \cdot \mathbf{a}_R = (\mathbf{a}_x + 2\mathbf{a}_y) \cdot \frac{3\mathbf{a}_x - 5\mathbf{a}_y - 4\mathbf{a}_z}{\sqrt{50}} = \frac{3-10}{\sqrt{50}} = \frac{-7}{\sqrt{50}} = -0.9899$$

(c) At (2,3,−4): $\mathbf{r} = 2\mathbf{a}_x + 3\mathbf{a}_y - 4\mathbf{a}_z$
Since the line charge lies along the z-axis, there will be no z-component of ***E*** or ***D***.

Thus $\mathbf{r} = 2\mathbf{a}_x + 3\mathbf{a}_y \quad |r| = \sqrt{13} \quad \mathbf{a}_r = \dfrac{2\mathbf{a}_x + 3\mathbf{a}_y}{\sqrt{13}}$

$$\mathbf{D} = \frac{5}{r}\mathbf{a}_r = \frac{5}{\sqrt{13}}\mathbf{a}_r = \frac{5}{\sqrt{13}}\frac{2\mathbf{a}_x + 3\mathbf{a}_y}{\sqrt{13}} = 0.27735(2\mathbf{a}_x + 3\mathbf{a}_y)$$

$$\mathbf{D} \cdot \mathbf{a}_R = 0.27735(2\mathbf{a}_x + 3\mathbf{a}_y) \cdot \frac{3\mathbf{a}_x - 5\mathbf{a}_y - 4\mathbf{a}_z}{\sqrt{50}}$$

$$= 0.27735\frac{6-15}{\sqrt{50}} = \frac{-2.496}{\sqrt{50}} = 0.353$$

Example 6.8

A segment of infinite surface with non-uniform charge density $\rho_s = 10x^2$ C/m² is located on the z = 0 plane. Find the total electric flux leaving the following specified closed surfaces of a cube of 2 m sides with edges parallel to coordinate axes and centered at the origin.

SOLUTION

The specified closed surface is shown in Figure 6.11. The total electric flux can be obtained in view of Equation 6.22a.

Here $ds = dxdy$ and the charged surface lies in z = 0 plane. Since ***D*** and ***ds*** has the same direction the total flux is given as:

$$\Psi = \oiint \mathbf{D} \cdot d\mathbf{S} = \int_{y=-1}^{1}\int_{x=-1}^{1} 10x^2 dxdy = \int_{-1}^{1} 10\frac{x^3}{3}\Big|_{-1}^{1} dy = \frac{20}{3}\int_{-1}^{1} dy = \frac{40}{3}\text{ C}$$

Example 6.9

If $\mathbf{D} = 2xyz^3\mathbf{a}_x + x^2z^3\mathbf{a}_y + 3x^2yz^2\mathbf{a}_z$ C/m² find an approximate value of the total charge contained within (a) a small spherical volume of 10^{-12} m³ located at (1,2,3) (b) a small regular isoclines shaped volume of 10^{-10} m³ located at (2,2,2).

SOLUTION

$$\mathbf{D} = 2xyz^3\mathbf{a}_x + x^2z^3\mathbf{a}_y + 3x^2yz^2\mathbf{a}_z = D_x\mathbf{a}_x + D_y\mathbf{a}_y + D_z\mathbf{a}_z$$

$$D_x = 2xyz^3 \quad D_y = x^2z^3 \quad D_z = 3x^2yz^2$$

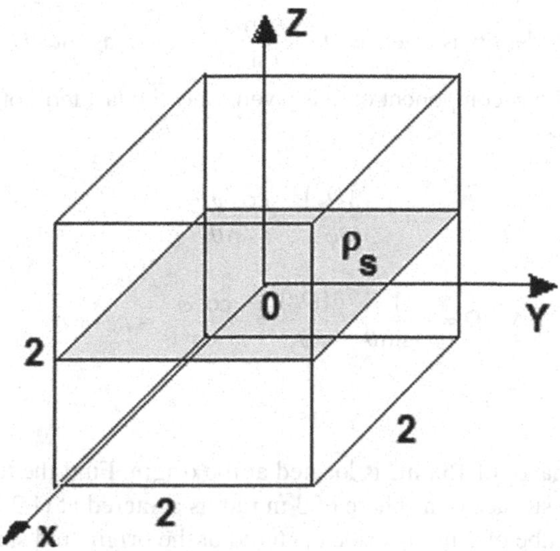

FIGURE 6.11 A charged flat surface located in a cube.

In view of Equations 6.18a, 6.18a, and 6.20a:

$$\Delta Q \approx \left(\frac{\partial D_x}{\partial x} + \frac{\partial D_y}{\partial y} + \frac{\partial D_z}{\partial z}\right) \cdot \Delta v$$

$$\frac{\partial D_x}{\partial x} = 2yz^3 \qquad \frac{\partial D_y}{\partial y} = 0 \qquad \frac{\partial D_z}{\partial z} = 6x^2yz$$

$$\Delta Q = \left(2yz^3 + 6x^2yz\right) \cdot \Delta v$$

(a) $\Delta Q\big|_{x=1,y=2,z=3} \approx (108+36)\cdot 10^{-12} = 144\cdot 10^{-12}$ F

(b) $\Delta Q\big|_{x=2,y=2,z=2} \approx (32+96)\cdot 10^{-10} = 128\cdot 10^{-10}$ F

Example 6.10

Find the volume charge distribution ρ_v that produces electric flux density $\boldsymbol{D} = \dfrac{\sin\varphi}{r\sin\theta} a_\varphi$.

SOLUTION

The problem relates to a spherical coordinate system wherein the divergence expression is given by Equation 4.20c reproduced here:

$$\nabla \cdot \boldsymbol{D} \equiv \frac{1}{r^2} \cdot \frac{\partial\left(r^2 . D_r\right)}{\partial r} + \frac{1}{r \cdot \sin\theta} \cdot \frac{\partial\left(\sin\theta \cdot D_\theta\right)}{\partial \theta} + \frac{1}{r \cdot \sin\theta} \cdot \frac{\partial\left(D_\varphi\right)}{\partial \varphi}$$

The electric flux density is given as $\boldsymbol{D} = \left(\frac{\sin\varphi}{r\sin\theta}\right)\boldsymbol{a}_\varphi = D_\varphi \boldsymbol{a}_\varphi$ thus $D_\varphi = \frac{\sin\varphi}{r\sin\theta}$

Since only the φ-component of D is given, only the last term of divergence is to be evaluated:

$$\frac{\partial(D_\varphi)}{\partial\varphi} = \frac{\cos\varphi}{r\sin\theta}$$

$$\nabla \bullet \boldsymbol{D} = \frac{1}{r\cdot\sin\theta}\cdot\frac{\partial(D_\varphi)}{\partial\varphi} = \frac{\cos\varphi}{r^2\sin^2\theta} = \rho_v \text{ or } \rho$$

PROBLEMS

P6.1 A point charge of 15π nC is located at the origin. Find the total flux leaving (a) the surface of a sphere of 3 m radius centered at (1,2,3); (b) the top face of a cube of 2 m on a side centered at the origin and sides parallel to the axes of coordinates; (c) that portion of a right circular cylinder of 3 m radius for which $z \geq 0$.

P6.2 A $3C$ point charge is located at (0.5, 0.7, −0.2) in a vacuum. Find (a) $|\boldsymbol{D}|$ and $\boldsymbol{D}$ at (0.1, 0.3, 0.4) and (b) total electric flux passing through the first octant.

P6.3 Find the total charge within a cylinder of radius $\rho = 5$ and length $0 \leq z \leq 2$ if $\boldsymbol{D}$=: (a) $\frac{10z}{\rho}\boldsymbol{a}_z$ (b) $10z^2 \sin\varphi \boldsymbol{a}_\rho$ C/m².

P6.4 Find the total charge lying in a sphere of $r = 3$ m if $\boldsymbol{D}$=: (a) $\frac{\boldsymbol{a}_r}{r^2}$ (b) $\frac{\boldsymbol{a}_r}{r}$ (c) $\frac{\sin\theta}{r}\boldsymbol{a}_r + \frac{\cos\theta\ln(r)}{r}\boldsymbol{a}_\theta$.

P6.5 Find $|D|$ at 2, −3, 4 in the field of a: (a) point charge of 0.5 μC at the origin (b) uniform line charge of 20 nC/m on the z-axis (c) uniform surface charge of 0.1π nC/m² at $x = 3$ plane.

P6.6 Find the total electric flux leaving a cylindrical surface having $\rho = 5.5$ and $z = \pm 3.5$ due to (a) point charges of 2 C each is located at $x = 0, \pm1, \pm2, \ldots$; (b) a line charge with charge density $\rho_L = 5\cos(0.1x)$ C/m located on the x-axis; (c) a surface charge with the charge density $\rho_S = 0.5\rho^2$ C/m² located on the $z = 0$ plane.

P6.7 The surfaces of three concentric spheres (Figure 6.1) of radius $r = 2$, 4 and 6 m carry surface charge densities (ρ_s) of 50, −20, and 10 μC/m², respectively. Find $|D|$ at r =: 1, 3, 5, and 7 m.

P6.8 Two infinite sheets of charges are located in free space with $\rho_s = 3$ C/m² at $x = 3$, $\rho_s = -2$ C/m² at $x = 6$ another set of two sheets are located in free space with $\rho_s = 5$ C/m² at $y = 3$, $\rho_s = -5$ C/m² at $y = 6$. Find $\boldsymbol{D}$ at: (a) $x = 2$, $y = 2$; (b) $x = 7$, $y = 2$; (c) $x = 7$, $y = 8$; (d) $x = 7$, $y = 1$; (e) $x = 4$, $y = 4$.

P6.9 A uniform electric flux density of 10C/m² lies along the vector $3\mathbf{a}_x - 5\mathbf{a}_y + 6\mathbf{a}_z$. Express D as a vector and determine the magnitude of flux crossing

(a) the y–z plane bounded by $y = 1$, $y = 3$, $z = 0$, $z = 2$; (b) the z–x plane bounded by $z = 1$, $z = 3$, $x = 0$, $x = 2$; (c) the x–y plane bounded by $x = 2$, $x = 3$, $y = 2$, $y = 3$.

P6.10 A uniform line charge $\rho_L = 3\pi$ C/m lies along the z-axis in free space. Find the components of $\boldsymbol{D}$ in the direction of vector $5a_x - 3a_y + 4a_z$ at the point (a) (3, 2, 0); (b) (2, −3, −4).

P6.11 A segment of infinite surface with non-uniform charge density $\rho_s = 10x^2$ C/m^2 is located on the $x = 0$ plane. Find the total electric flux leaving the closed surface of: (a) a 2 m radius cylinder with an axis extending from 0 to 3 m lies on the y-axis; (b) a sphere of 2 m radius centered at the origin.

P6.12 If $\boldsymbol{D} = 2xy\boldsymbol{a}_x + xy^2\boldsymbol{a}_y + 3x^2y^2z^2\boldsymbol{a}_z$ C/m^2 find an approximate value of the total charge contained in (a) a small spherical volume of 10^{-10} m^3 located at (3,2,3); (b) a small regular isoclines shaped volume of 10^{-12} m^3 located at (3,3,3).

P6.13 Find the total charge enclosed in a tiny sphere of 1 μm radius centered at (a) (3,5,1); (b) (1,2,–1) if $\boldsymbol{D} = \dfrac{100xy}{z^2+1}\boldsymbol{a}_x + \dfrac{50x^2}{z^2+1}\boldsymbol{a}_y - \dfrac{25x^2yz}{\left(z^2+1\right)^2}\boldsymbol{a}_z$.

P6.14 Find the volume charge distribution ρ_v that produces electric flux density (a) $D = \dfrac{\cos\theta}{r^2}a_\theta$ (b) $\boldsymbol{D} = r^2\boldsymbol{a}_r$.

P6.15 Find the volume charge density that results in D=: (a) $e^{2x}e^{-3y}e^{-z}$ $\left(2\boldsymbol{a}_x - 3\boldsymbol{a}_y - \boldsymbol{a}_z\right)$; (b) $e^{-3z}\left(2\rho\varphi\,\boldsymbol{a}_\rho + 3\rho\,\boldsymbol{a}_\varphi - \rho^2\varphi\boldsymbol{a}_z\right)$; (c) $5r\sin\theta\sin\varphi\,\boldsymbol{a}_r$ + $\left(\dfrac{1}{r}\right)\cos\theta\sin\varphi\,\boldsymbol{a}_\theta + r\cos\varphi\,\boldsymbol{a}_\varphi$.

DESCRIPTIVE QUESTIONS

Q6.1 Explain the terms *electric flux, passing through, close surface,* and the *charge enclosed* involved in Gauss' law.

Q6.2 Discuss the parameters that influence the flux.

Q6.3 Explain the meaning of electric flux density and its relation with the field intensity.

Q6.4 Discuss the meaning of Gaussian surface. Also discuss the criterion of selecting the Gaussian surfaces for different types of charge distributions.

FURTHER READING

Given at the end of Chapter 15.

7 Potential and Potential Energy

7.1 INTRODUCTION

In Chapter 5 a quantity referred to as electric field intensity E was introduced. Another quantity, electrical flux density D, was introduced in Chapter 6. Both of these bear simple relations with electric charge. In spite of the fact that these quantities are real physical entities, they cannot be easily visualized and appear to be merely a mathematical concept. Another quantity, related to the electrical charge, called potential or voltage, known even to commoners, cannot only be visualized but can be measured with ease. This is related to the electrical charge. This chapter is devoted to understand the meaning of electrostatic potential and its relationship to charge distributions. Although we will be discussing time-invariant electric fields produced exclusively by stationary charges, in our treatment we will make them move. This results in a simplified, albeit approximate, theory.

7.2 ELECTROSTATIC FIELDS

The meaning of electrostatic fields was briefly described in Section 2.1.1. There it was noted that it may result due to the presence of stationary (or quasi-stationary) point charges, clusters of discrete charges, line charges, surface charges, or volume charge distributions. The following subsections look at this in more detail.

7.2.1 GENESIS OF THE ELECTROSTATIC FIELDS

Coulomb discovered the interaction between stationary point charges. Sources for dc currents appeared in the form of Leclanche cell, a galvanic cell, or a voltaic cell. Oersted, from Denmark, noticed that dc current flowing in a closed conducting path deflected magnetic needles placed near it. Since permanent magnet also deflected magnetic needles placed near it, Oersted showed in experiments that the electric current behaves like a magnet. When Faraday learned about this experiment, he stated that "a permanent magnet, likewise, should cause dc current to flow in a closed conducting path placed near the magnet". Although his experiments did not support this, he discovered the well-known law of electromagnetic induction that states that *electromotive force* (*EMF*) along a closed path is equal to the rate of decrease of the magnetic flux linking the closed path. This law, for time-invariant field results:

$$emf \stackrel{\text{def}}{=} \oint_C \boldsymbol{E} \cdot d\boldsymbol{\ell} = 0 \tag{7.1}$$

where C indicates an arbitrary closed path (or contour) and $\boldsymbol{E}$ stands for the electrostatic field intensity vector on this path. This law, for time-varying fields, is discussed in detail in Chapter 15.

If a point charge Q is located at a point with electrostatic field $\boldsymbol{E}$, it will experience a force $\boldsymbol{F}$ given by Coulomb's law. The electric field intensity at a *stationary* point in a large homogeneous non-conducting region is *defined* as the force acting on a *stationary* unit point charge located at this point. The term *electric field intensity* was introduced earlier in Section 5.2.10, and the force exerted on charge Q due to electric field intensity $\boldsymbol{E}$ was given as:

$$\boldsymbol{F} \stackrel{\text{def}}{=} Q \cdot \boldsymbol{E} \tag{5.9d}$$

A law of classical mechanics states that "work is done by a force that acting on a body displaces it in the direction of this force. The work done is the product of this force and the displacement of the body in the direction of the force". Thus, the incremental work done dW by a force F displacing a body by distance dL is simply the product of the force F and the incremental distance $d\ell$. Since both the force ($\boldsymbol{F}$) and the displacement ($\boldsymbol{d\ell}$) have the directions these are to be taken as vector quantities. Furthermore, the force that must be applied has to be equal and opposite to the force due to the field. The work done can thus be written as:

$$dW = -\boldsymbol{F} \bullet \boldsymbol{d\ell} = -Q \cdot \boldsymbol{E} \bullet \boldsymbol{d\ell} \tag{7.1a}$$

On integrating both sides of Equation 7.1a, the total work done becomes:

$$W = -\int \boldsymbol{F} \bullet \boldsymbol{d\ell} = -Q \int \boldsymbol{E} \bullet \boldsymbol{d\ell} \tag{7.1b}$$

This equation can further be modified by identifying the limits of the displacement. Thus:

$$W = -\int_{\text{initial}}^{\text{final}} \boldsymbol{F} \bullet \boldsymbol{d\ell} = -Q \int_{\text{initial}}^{\text{final}} \boldsymbol{E} \bullet \boldsymbol{d\ell} \tag{7.1c}$$

If the initial and final limits belong to the same location it amounts to a closed path, thus:

$$W = -\oint_C \boldsymbol{F} \bullet \boldsymbol{d\ell} = -Q \oint_C \boldsymbol{E} \bullet \boldsymbol{d\ell} = 0 \tag{7.1d}$$

Equation 7.1d shows that no work is done to take a charge around an arbitrary closed path in a time-invariant electric field of an arbitrary space distribution. This, however, is not true and there could be many reasons, for instance:

(1) If the electric field is not uniform along the path, the moving charge finds itself in a time-varying electric field.

(2) Equation 5.9d or 7.1a is based on Coulomb's law, thus is valid only for *stationary time-invariant charges.*
(3) An alternative interpretation for Equation 7.1d may be possible.

A charge moving in a closed path must be accelerated and retarded. In 1897, *Joseph Larmor* derived a relationfor the power radiated by an accelerated charged particle. *Larmor's formula* for the radiated power $\mathbb{P}$, is as follows:

$$\boxed{\mathbb{P} = \mu_o \cdot \frac{Q^2 \cdot a^2}{6\pi \cdot c} = \eta_0 \cdot \frac{Q^2 \cdot a^2}{6\pi \cdot c^2}} \tag{7.2}$$

where Q is the point charge, c stands for the velocity of light in free space while η_0 indicates the characteristic impedance of free space. The acceleration "a" of the charged particle is defined as:

$$a \stackrel{\text{def}}{=} \frac{du}{dt} \tag{7.2a}$$

where the instantaneous velocity $u \ll c$.

Thus, work must be done if a point charge starts from a point with zero initial velocity, moves around an arbitrary open or closed path with or without encountering any electric field in the way, and then stops thereafter. A part of this work accounts for the energy radiated by the accelerated or retarded electric charge. Therefore, Equation 7.1d *must not* be interpreted that zero work will be done if the charge Q is taken around a closed path in a time-invariant electric field of intensity. In our subsequent treatment of electrostatic fields, we shall ignore this energy radiation, although it will be at the back of our minds all the time to guide us toward the correct interpretation of field equations that we will come across.

Using Stokes' theorem, Equation 7.1d reduces to:

$$\iint_S (\nabla \times \boldsymbol{E}) \bullet d\boldsymbol{s} = 0 \tag{7.3}$$

where S indicates an open surface whose boundary coincides with the closed contour C. Now, since the closed contour C and the associated surface S are arbitrary, we must have:

$$\nabla \times \boldsymbol{E} = 0 \tag{7.4}$$

for electrostatic field (also known as time-invariant electric field).

This is Maxwell's equation for time-invariant electromagnetic field. A detailed discussion of Stokes' theorem is presented in Section 4.6.2.

7.2.2 Scalar Electric Potential

For any scalar space-function V, consider the identity given by Equation 4.41e:

$$\nabla \times (\nabla V) \equiv \mathbf{0} \tag{7.5}$$

Now, since curl of the time-invariant electric field intensity is zero it is possible to express the time-invariant electric field intensity as the negative gradient of a scalar space-function V. This function is called the scalar electric potential. The choice of the minus sign is rather arbitrary; it implies that the field vector $\boldsymbol{E}$ is directed from high potential point to the low potential point. It is usually convenient to determine the expression for the distribution of scalar electric potential V, rather than that for a vector field with three scalar components. Once the distribution of the scalar potential is found(say, by solving a boundary value problem involving Laplace or Poisson's equation), the vector field $\boldsymbol{E}$ can be readily obtained from its gradient (Equation 4.10):

$$\boldsymbol{E} \stackrel{\text{def}}{=} -\nabla V \tag{7.6}$$

In this approach, no physical meaning needs to be attached with the scalar function V. Even though it has been called *the scalar electric potential*; at this stage it is simply an *intermediate by-product of mathematical operation* that leads to the determination of the vector field distribution $\boldsymbol{E}$.

As an alternative approach, one attempts to find the scalar electric potential V using a given expression of the electric field intensity vector $\boldsymbol{E}$. A vector theorem, known as the *Fundamental Theorem for Gradient*, states that for an arbitrary scalar space-function f we have:

$$\int_a^b (\nabla f) \cdot d\boldsymbol{\ell} \equiv f(b) - f(a) \tag{7.7}$$

This is purely a mathematical statement without any physical restrain imposed upon it. There is no constraint regarding the path to be taken for the line integration from point "a" to point "b". We may choose the scalar electric potential V for the arbitrary scalar function f; thus, from Equations 7.6 and 7.7 we get:

$$V_{ab} \stackrel{\text{def}}{=} V(b) - V(a) = -\int_a^b \boldsymbol{E} \cdot d\boldsymbol{\ell} \tag{7.8}$$

This gives the potential difference between two points a and b. If the point "a" is at infinity and the point "b" is any arbitrary point P with a potential, say V_P, Equation 7.8 can be rewritten as:

$$V_P - V_\infty \stackrel{\text{def}}{=} V(P) - V(\infty) = -\int_\infty^P \boldsymbol{E} \cdot d\boldsymbol{\ell} \tag{7.8a}$$

Now, the potential at infinity is arbitrarily chosen as the reference potential, i.e.:

$$V_\infty = V(\infty) \stackrel{\text{def}}{=} 0 \tag{7.9}$$

Thus, the scalar electric potential at any arbitrary point P is given as:

$$V \stackrel{\text{def}}{=} V_P = -\int_{\infty}^{P} \boldsymbol{E} \bullet d\boldsymbol{\ell} \tag{7.10}$$

Now, if we multiply both sides of Equation 7.10 by Q:

$$Q \cdot V = -\int_{\infty}^{P} (Q\boldsymbol{E}) \bullet d\boldsymbol{\ell} = -\int_{\infty}^{P} \boldsymbol{F} \bullet d\boldsymbol{\ell} \stackrel{\text{def}}{=} W \tag{7.11}$$

where V indicates the scalar electric potential at the point P.

This equation is often *interpreted*, albeit wrongly, as the *total work done* W by an external agency forcing a point charge Q against the electric field $\boldsymbol{E}$, to move from the zero potential at infinity to a point P where the potential is V.

The above interpretation of the scalar electric potential is questionable as indicated below:

(1) The scalar electric potential V has, so far, only mathematical significance (and identity).
(2) To shift a charge from one fixed point to another, it must be accelerated and also decelerated, causing power radiation. Therefore, the law of conservation of energy dictates that work must be done to move a point charge over two distinct points, even if these points were *originally* at the same potential.
(3) The interpretation of Equation 7.11 suffers from a built-in contradiction as it is based on Coulomb's law for stationary charges defining the time-invariant vector field $\boldsymbol{E}$, and in the same breath we are talking about the charges perceiving time-varying field while moving through *heterogeneous distribution* of electric field.

7.2.3 Potential Energy of Charged Particles

For a charged particle placed in electrostatic field its potential energy (**PE**), is proportional to the product of its charge and the scalar potential at the point where the charged particle is located. Therefore, Equation 7.11 does not give us the *total work done*, but only a fraction of this work in the form of the *potential energy* gained (1) by the charge Q, and (2) by the electrostatic field where this charge is placed. The rest of the work done is dissipated as the power radiated by the accelerated charge Q, during its journey from infinity to the point P. Likewise, Equation 7.1b must be interpreted as *no change* in the potential energy of a charge occurs if it is taken around a closed path in the presence of time-invariant field. The entire work done in moving the charge is dissipated as the power radiated by the accelerated charge.

Let the potential energy $\mathcal{E}$ of a point charge Q located at a point with electrostatic potential V be given as:

$$\mathcal{E} \stackrel{\text{def}}{=} k \cdot Q \cdot V \tag{7.12}$$

where the symbol k, indicates the constant of proportionality.

Consider a large non-conducting charge free homogeneous region with every point at zero electrostatic potential. If a point charge $Q_1$is brought from infinity and placed at a stationary point P_1 in this region its PE remains zero and if the power radiation is *ignored* no work is done. This charge, however, sets up a potential field on its arrival. Let the potential at a stationary point P_2 due to the point charge Q_1 be indicated as $V_{2,1}$ such that:

$$V_{2,1} = \frac{Q_1}{4\pi\varepsilon \cdot R_{1,2}} \tag{7.13}$$

where $R_{1,2}$ indicates the distance between the points P_1 and P_2.

This equation gives the value of the electrostatic potential due to the point charge Q_1 at every point in the space except at the point P_1 where the point charge Q_1 is located; thus the two points P_1 and P_2 are distinct, i.e. not same. This constraint eliminates the possibility of infinite potential if $R_{1,2}$ is zero. The potential at the point P_1 is the sum of potentials at this point due to all the charges present other, then the point charge Q_1. If Q_1 is the only charge present, the potential at the point P_1 is zero.

Now, if a second point charge Q_2 is brought from infinity and placed at the point P_2, it gains a PE as shown below:

$$\mathcal{E}_2 = k \cdot Q_2 \cdot V_{2,1} \tag{7.14}$$

The charge Q_2 sets up a potential field on its arrival. Let the potential at the point P_1 due to the point charge Q_2 be indicated as $V_{1,2}$ such that:

$$V_{1,2} = \frac{Q_2}{4\pi\varepsilon \cdot R_{1,2}} \tag{7.15}$$

Before the advent of the charge Q_2 the potential at the point P_1 was zero and so was the PEof the charge Q_1. With the arrival of the charge Q_2 the PEof the charge Q_1 is raised to $\mathcal{E}_1$ such that:

$$\mathcal{E}_1 = k \cdot Q_1 \cdot V_{1,2} \tag{7.16}$$

Thus, the total PEof the system of two point charges $\mathcal{E}$ is as follows:

$$\mathcal{E} = \mathcal{E}_1 + \mathcal{E}_2 = k \cdot Q_1 \cdot V_{1,2} + k \cdot Q_2 \cdot V_{2,1} \tag{7.17}$$

If the energy radiation is *ignored* this must be equal to the work done W in bringing the charge Q_2 from infinity and positioning it at the point P_2. Thus, using Equation 7.11:

$$W \stackrel{\text{def}}{=} -\int_{\infty}^{P_2} \boldsymbol{F} \bullet d\boldsymbol{\ell} = Q_2 \cdot V_{2,1} = \mathcal{E} \tag{7.18}$$

Therefore, in view of Equation 7.17:

$$Q_2 \cdot V_{2,1} = k \cdot Q_1 \cdot V_{1,2} + k \cdot Q_2 \cdot V_{2,1} \tag{7.19a}$$

or,

$$k \cdot Q_1 \cdot V_{1,2} = (1-k) \cdot Q_2 \cdot V_{2,1} \tag{7.19b}$$

Now, in view of Equations 7.13 and 7.15:

$$Q_1 \cdot V_{1,2} = Q_1 \cdot \frac{Q_2}{4\pi\varepsilon \cdot R_{1,2}} = Q_2 \cdot \frac{Q_1}{4\pi\varepsilon \cdot R_{1,2}} = Q_2 \cdot V_{2,1} \tag{7.20a}$$

Therefore, Equation 7.19b gives:

$$k = (1-k) \tag{7.20b}$$

or,

$$k = 1/2 \tag{7.20c}$$

Thus, half of the work done to bring a charge from infinity is stored as the PE of this charge and the remaining half accounts for the rise in PE of the charge present already. Note that when the first charge was brought no work was done and the PE of this charge remained zero. The electrostatic field of the first charge is unique as no energy is stored in this field. The energy stored in electrostatic field is *synonymous* with the PE of all charges jointly establishing the field.

Now, if we multiply both sides of Equation 7.8 by $\left(\frac{1}{2} \cdot Q\right)$, we get:

$$\frac{1}{2} \cdot Q \cdot V(b) - \frac{1}{2} \cdot Q \cdot V(a) = -\frac{1}{2} \cdot \int_a^b (Q,\mathbf{E}) \bullet d\boldsymbol{\ell} = -\frac{1}{2} \cdot \int_a^b \mathbf{F} \bullet d\boldsymbol{\ell} \overset{\text{def}}{=} \ddot{\mathcal{E}} \tag{7.21}$$

Equation 7.21 gives the gain in the potential energy $\mathcal{E}$ of point charge Q when it is shifted from point "*a*" to point "*b*". Now, we can interpret Equation 7.1b. It states that there will be no change in the potential energy of a point charge if it is taken around any arbitrary closed path encountering electrostatic field; though work must have been done to account for its power radiation. The amount of this work done is sensitive to the path followed by the point charge as well as the description of its motion on this path.

The potential energy of a point charge Q is proportional to the value of the scalar function V at the point the charge is located. This space-function V is aptly called the scalar electric potential. The time-varying electric field intensity is expressed jointly using the scalar electric potential and the vector magnetic potential.

7.3 SCALAR ELECTRIC POTENTIAL DUE TO DIFFERENT SOURCES

Having introduced the concepts of work done and potential energy, the relations for scalar electric potential due to different types of charge distributions can now be obtained.

7.3.1 Potential Due to Point Charges

The expression for the electric field intensity $\boldsymbol{E}$ at a point P_f (known as the field point), due to a stationary point charge Q located at a point P_s (known as the source point) is given by Equation 5.10a. The same is reproduced below:

$$\boldsymbol{E} = \frac{1}{4\pi\varepsilon_o} \cdot \frac{Q}{R^2} \boldsymbol{a}_R \quad \text{for } R > 0 \tag{5.10a}$$

where R indicates the distance between the source point P_s and the field point P_f, while the unit vector is directed from the source point P_s to the field point P_f.

Therefore, using Equation 7.10,we get the scalar electrostatic potential V at the field point due to a point charge Q as:

$$V = \frac{1}{4\pi\varepsilon_o} \cdot \frac{Q}{R} \quad \text{for } R > 0 \tag{7.22}$$

This gives the value of potential due to the point charge Q at every point, except at the source point ($R = 0$)where the point charge is located. The potential at this point will be the combined potential contributed by all sources other than the point charge Q located at this point.

7.3.2 Potential Due to *n* Point Charges

Consider a charge free field point P_f, the potential at this point due to n number of point charges being the algebraic sum of potentials due to all charges, is given as:

$$V = \frac{1}{4\pi\varepsilon_o} \cdot \sum_{m=1}^{n} \frac{Q_m}{R_m} \tag{7.23}$$

where R_m (> 0, for all m) indicates the distance between the source point P_m where the charge Q_m is located and the field point P_f where the potential V is being estimated.

7.3.3 Potential Due to Line Charge Distribution

Consider a charge distribution along a line of any shape and length, laying on an even or uneven surface. Around a point P on this line take a line segment of incremental length $\Delta\ell$ carrying a small charge ΔQ. The line charge density $\rho_L$at the point P is defined as follows:

$$\rho_L \stackrel{\text{def}}{=} \lim_{\Delta\ell \to 0} \frac{\Delta Q}{\Delta\ell} \text{ C/m} \tag{7.24}$$

Now, for this continuous (or piecewise continuous) charge distribution Equation 7.23 is transformed into a line integral, as given below:

$$V = \frac{1}{4\pi\varepsilon_o} \cdot \int_A^B \frac{\rho_L}{R} \cdot d\ell \tag{7.25}$$

where V is the scalar electrostatic potential at a given field point due to all charges between any two points A and B on the line charge; ρ_L indicates the line charge density at a source point P_s on the line between these two points and R indicates the distance of the point P_s from the field point P_f where the potential V is being evaluated. During integration, the location of the point P_s varies from the point A to the point B, while the location of the field point Pf remains fixed. Therefore, the distance R, in general being a variable quantity remains inside the sign of integration. A circular ring with line charge distributed over it will be an exception if the field point is anywhere on its axis; a line perpendicular to the plane of the circle passing through its center. For uniform distribution of the line charge between point A and point B, the line charge density ρ_L, being a constant, can be taken out of the integration sign.

In a closed line or contour C of any shape and size laying on an even or uneven surface, Equation 7.25 is transformed into a contour integration as shown below:

$$V = \frac{1}{4\pi\varepsilon_o} \cdot \oint_C \frac{\rho_L}{R} \cdot d\ell \tag{7.25a}$$

If the line charge density ρ_L is a constant, it can be taken out of the integration sign.

7.3.4 Potential Due to Surface Charge Distribution

Consider a charge distribution on an open surface S of any shape and size, even or uneven. Take a source point P_s on this surface. Let an incremental surface area ΔS around this point is carrying a small charge ΔQ. The surface charge density ρ_s at the point P can be defined as:

$$\rho_s \stackrel{\text{def}}{=} \lim_{\Delta S \to 0} \frac{\Delta Q}{\Delta S} \text{ C/m}^2 \tag{7.26}$$

Now, for this continuous (or piecewise continuous) charge distribution Equation 7.23 is transformed into a surface integral, as given below:

$$V = \frac{1}{4\pi\varepsilon_o} \cdot \iint_S \frac{\rho_s}{R} \cdot ds \tag{7.27}$$

where V is the scalar electrostatic potential at the given field point P_f, whereas ρ_s indicates the surface charge density at a source point P_s on the surface and R indicates the distance of the point P_s from the field point P_f where the potential is being evaluated. During integration the location of the point P_s varies over the entire surface S, while the location of the field point P_f remains fixed. Therefore, the distance R in general varies as the integration progresses. For uniform distribution of the surface charge, the surface charge density ρ_s being a constant can be taken out of the integration sign.

In a closed surface S, Equation 7.27 is transformed into a closed surface integration as shown below:

$$V = \frac{1}{4\pi\varepsilon_o} \cdot \oint\!\!\oint_S \frac{\rho_s}{R} \cdot dS \qquad (7.27a)$$

For a spherical surface of radius R, the potential at its center, therefore, is:

$$V = \frac{1}{4\pi R \cdot \varepsilon_o} \cdot \oint\!\!\oint_S \rho_s \cdot dS = \frac{Q}{4\pi R \cdot \varepsilon_o} \qquad (7.27b)$$

where Q indicates the total charge on the spherical surface.

The potential at every point inside a sphere of radius R can be given by Equation 7.27b, provided that the charge is uniformly distributed over the spherical surface.

7.3.5 Potential Due to Volume Charge Distribution

Consider a charge distribution in a volume v of any shape and size. Take a source point P_s in this volume. Let an incremental volume Δv around this point is carrying a small charge ΔQ. The volume charge density ρ_v or simply ρ, at the point P_s can be defined as:

$$\rho_v \stackrel{\text{def}}{=} \rho \stackrel{\text{def}}{=} \lim_{\Delta v \to 0} \frac{\Delta Q}{\Delta v} \text{ C/m}^3 \qquad (7.28)$$

Now for this continuous (or piecewise continuous) charge distribution Equation 7.23 is transformed into a volume integral, as given below:

$$V = \frac{1}{4\pi\varepsilon_o} \cdot \iiint_v \frac{\rho}{R} \cdot dv \qquad (7.29)$$

where V is the scalar electrostatic potential at the given field point P_f, whereas ρ indicates the volume charge density at a source point Ps in the volume and R indicates the distance of the point P_s from the field point P_f where the potential V is being evaluated. During integration the location of the point P_s varies over the entire volume v, while the location of the field point P_f remains fixed. Therefore, the distance R, varies as the integration progresses. For uniform distribution of the volume charge, the volume charge density ρ being a constant can be taken out of the integration sign.

If the field point is far from a volume of finite size, the variation of the position of the source point in the volume as the integration progresses will be small compared to the distance R between the field point and the source point. In that case the potential at a distant point can be approximately given as:

$$V \cong \frac{1}{4\pi R \cdot \varepsilon_o} \cdot \iiint_v \rho \cdot dv \stackrel{\text{def}}{=} \frac{Q}{4\pi R \cdot \varepsilon_o} \qquad (7.29a)$$

where Q indicates the total charge in the volume. Therefore, for distant fields the total charge in the volume can be treated as a point charge of the same value located somewhere near the middle of the volume.

As in the chapter on Gauss' law, we shall find that the potential at every point outside a sphere of radius R is as given by this equation provided that the charge is symmetrically distributed (i.e. the volume charge density ρ is independent of θ and φ coordinates) in the volume of the sphere. Therefore, for a point outside the sphere ($r \geq R$), electric fields (V and $\boldsymbol{E}$) will be identical to those of a point charge located at the center of the sphere ($r = 0$)provided that the value of this point charge is equal to the total charge Q in the spherical volume, i.e.:

$$Q = 4\pi \int_0^R \rho(r) \cdot r^2 \cdot dr \tag{7.29b}$$

Potentials due to various source distributions are shown in Table 7.1.

7.3.6 Conservative Field

If a charge Q moves around a close path in the presence of *electrostatic* field intensity $\boldsymbol{E}$, it amounts to a closed contour integration of the field intensity $\boldsymbol{E}$ wherein the initial and final limits are the same. The value of such an integral is zero and thus the net change in its potential energy is zero thereby the *potential energy* expanded by a charge in moving around a close path in the presence of electrostatic field is conserved, i.e. the potential energy remains unaltered; in this sense the electrostatic field is said to be conservative. If this statement is altered by substituting "work done" for "potential energy", the resulting concept of the conservation of field can at best be considered as an approximation to the reality. A charge moving around a closed path must accelerate and decelerate resulting radiation of energy. The amount of the

TABLE 7.1
Expressions for Electric Field Intensity and Electric Potential

Source	Electric Field Intensity (E)		Electric Potential (V)	
Single point charge	$\boldsymbol{E} = \frac{1}{4\pi\varepsilon_o} \cdot \frac{Q}{R^2} \boldsymbol{a}_R$	(5.10)	$V = \frac{1}{4\pi\varepsilon_o} \cdot \frac{Q}{R}$	(7.22)
n point charges	$\boldsymbol{E} = \frac{1}{4\pi\varepsilon_o} \cdot \sum_{m=1}^{n} \frac{Q_m}{R_m^2} \boldsymbol{a}_{Rm}$	(5.12)	$V = \frac{1}{4\pi\varepsilon_o} \cdot \sum_{m=1}^{n} \frac{Q_m}{R_m}$	(7.23)
Line charge distribution	$\boldsymbol{E} = \int_L \frac{\rho_L d\ell}{4\pi\varepsilon_o R^2} \boldsymbol{a}_R$	(5.24)	$V = \int_L \frac{\rho_L d\ell}{4\pi\varepsilon_o R}$	(7.25)
Surface charge distribution	$\boldsymbol{E} = \iint_s \frac{\rho_s ds}{4\pi\varepsilon_o R^2} \boldsymbol{a}_R$	(5.32)	$V = \iint_s \frac{\rho_s ds}{4\pi\varepsilon_o R}$	(7.27)
Volume charge distribution	$\boldsymbol{E} = \iiint_v \frac{\rho_v \boldsymbol{a}_r}{4\pi\varepsilon_o r^2} dv\, \boldsymbol{a}_N$	(5.39)	$V = \iiint_v \frac{\rho_v}{4\pi\varepsilon_o r} dv$	(7.29)

radiated energy depends on the motion of the charged particle as well as the shape and size of the closed path the charged particle is moved along.

7.3.7 Kirchhoff's Voltage Law

In circuit theory the Kirchhoff's voltage law is one of the most important laws which deals with the conventional circuits composed of wires, resistances and batteries. A more general form of Kirchhoff's circuital law for voltages is given by Equation 7.1, which can be applied to any region containing electrostatic field. In general, this law is not applicable to time-varying fields but it is applicable to the circuits wherein the currents and voltages vary slowly with time (say, at power frequencies), causing negligible power radiation.

7.3.8 Concept of Infinity or Ground

Earlier two terms the potential difference and the potential were given by Equations 7.8 and 7.10, respectively. The potential difference (V_{ab}) was given as the difference of two potentials $V(b)$ and $V(a)$ at points b and a, respectively. The potential V at point p was defined with reference to the potential of a point located at infinity. The potential V is also called the absolute potential of point p. Thus, it is still the potential difference between two points where one of the points is presumed to be with zero potential there. The terms "infinity" and "zero potential" are to be correlated to conform to the practical applications. In general the surface of earth is represented by an infinite plane at zero potential. Thus the surface of earth or ground is probably the most universal zero reference point used in experimental or physical potential measurements. Some of the aspects of ground vis-à-vis electrical and electronic sciences are described below.

- In an electrical circuit ground refers to the reference point (i) from which voltages are measured, (ii) which provides a common return path for electrical current and (iii) which provides a direct physical connection to the earth. In electrical circuits *mains*, powered equipment and exposed metal parts are connected to the ground to prevent user's contact with dangerous voltage if electrical insulation fails.
- While handling flammable products or electrostatic sensitive devices the connections to ground limit the build-up of static electricity.
- In some telegraph, electric traction, and transmission lines transmitting electric power, the earth itself can be used as one conductor of the circuit, saving the cost of installing a separate return conductor.
- In electronic circuit theory, the ground is idealized as an infinite source (or sink) for charge. It can absorb almost an unlimited amount of current (or charge) without changing its potential.
- Earth provides a reasonably constant potential reference against which other potentials can be measured.
- An electrical ground system should have an appropriate current-carrying capability to serve as an adequate zero-voltage reference level. When a real

ground connection has a significant resistance, the approximation of zero potential is no longer valid. Stray voltages or earth potential rise effects may occur, which may create noise in signals or if large enough may produce shock hazard.

- In portable electronic devices such as cell phones, media players, and circuits in vehicles a large conductor is attached to one side of the power supply (such as the "ground plane" on a printed circuit board). It serves as the common return path for current from many different components in the circuit. In these devices there is no actual connection to earth but the term ground (or earth) is still used. The term *common* or virtual ground is more appropriate for such connections.

7.3.9 Summary of Potential Fields

In Chapter 5, the electric field intensity was evaluated due to different charge distributions. Similarly, in Chapter 6 the expressions for flux densities for these charge distributions were obtained. Likewise the potential fields for different charge distributions can also be given. For comparison, the expressions for electric field intensity (***E***) and electric potential (V) are given in the Table 7.1. In the expressions of V, r is replaced by R in the expressions of potential distributions for the line, surface and volume charge distributions. This table also indicates the equation number of each expression for ready reference.

7.4 EQUIPOTENTIAL SURFACES FOR DIFFERENT CHARGE DISTRIBUTIONS

The equipotentialsare the surface configurations whereupon each point has the same potential. Some of these are shown in Figures 7.1 to 7.6 in the form of two-dimensional (2D) line diagrams and three-dimensional (3D) surface configurations. In these color-coded figures, the surfaces with higher potentials are shown darker whereas the ones with lower potentials by lighter colors. In all the 3D configurations, only half of the surfaces are shown.

It is worth mentioning that all these approximate sketches are drawn on the basis of imagination. Actual and accurate sketches can be obtained by assigning

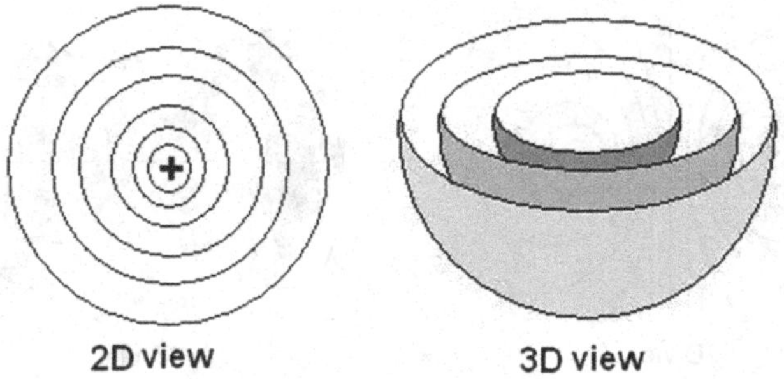

FIGURE 7.1 Configuration of equipotentials for single positive point charge.

different numerical values to the involved parameters in the potential relations given in Table 7.1.

7.4.1 Due to Single Positive Point Charge

Figure 7.1 shows the configuration of equipotentials for a single positive point charge. As illustrated, the 2D view contains a number of concentric circles, whereas that depicted as 3D view contains (only) lower half of the concentric spheres. This figure can be obtained by assigning suitable values to the parameters in Equation 7.22.

7.4.2 Due to a Pair of Opposite Point Charges

Figure 7.2 shows the configurations of equipotentials for a pair of two opposite point charges. This figure also illustrates the 2D and 3D views. In 2D view the centers of circles are gradually shifting away from positive and negative charges in the left and right directions respectively. The 3D equipotential surfaces are roughly depicted through the spheres of increasing radii, with shifting centers. This figure can be obtained by assigning suitable values to involved parameters in Equation 7.23 and summing the series for suitable *n*.

7.4.3 Due to an Infinite Uniform Line Charge

Figure 7.3 shows the configurations of equipotentials for an infinite uniform line charge. It shows only the 3D view, which contains a number of cylindrical surfaces. All these cylinders are infinite in extent along the line charge. This figure can be obtained by assigning suitable values to involved parameters in Equation 7.25 by taking the length of line charge to be infinite.

7.4.4 Due to a Finite Uniform Line Charge

Figure 7.4 shows the 2D and 3D configurations of equipotentials for a finite uniform line charge. The 3D configuration illustrates a number of paraboloidal surfaces. This figure can be obtained by assigning suitable values to involved parameters in Equation 7.25 by taking a finite length of line charge.

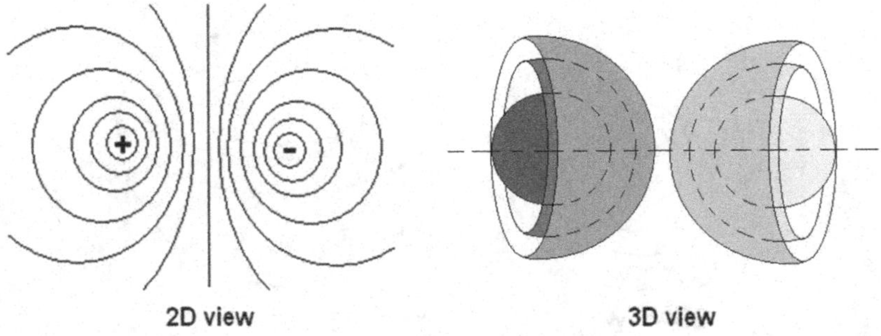

FIGURE 7.2 Configurations of equipotentials for a pair of opposite point charges.

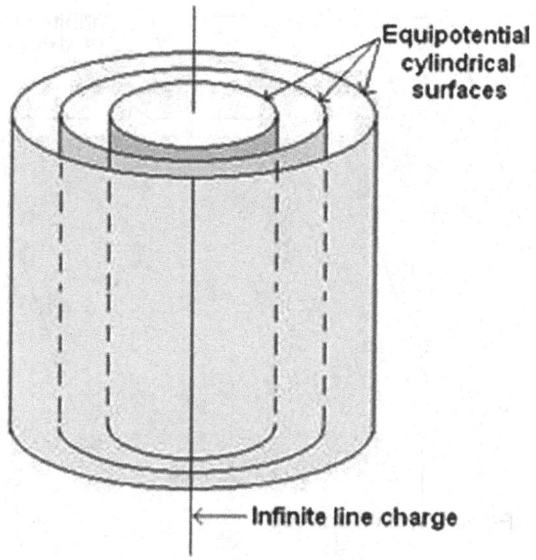

FIGURE 7.3 Configuration of equipotentials for an infinite uniform line charge.

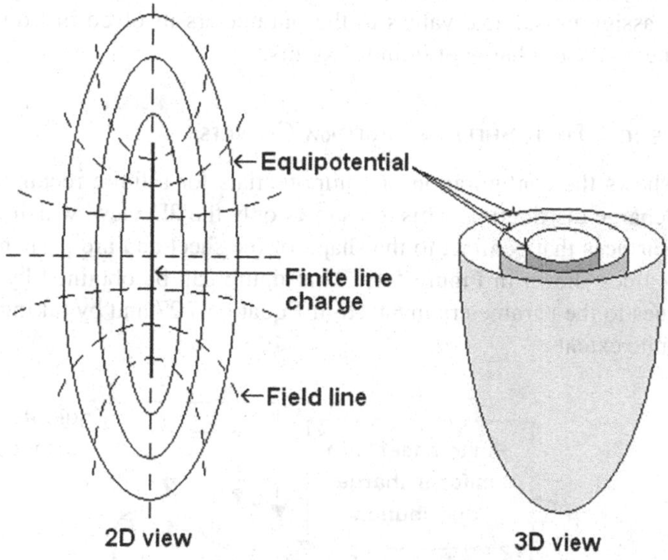

FIGURE 7.4 Configurations of equipotentials for a finite uniform line charge.

7.4.5 Due to an Infinite Sheet of Uniform Charges

Figure 7.5 shows the configuration of equipotentials for an infinite sheet of uniform charges. This shows only the 3D view, which contains a number of plane surfaces that are parallel to this sheet of charges. Such plane surfaces will appear on both the sides of sheet of charges. These surfaces in fact form the closed surfaces with

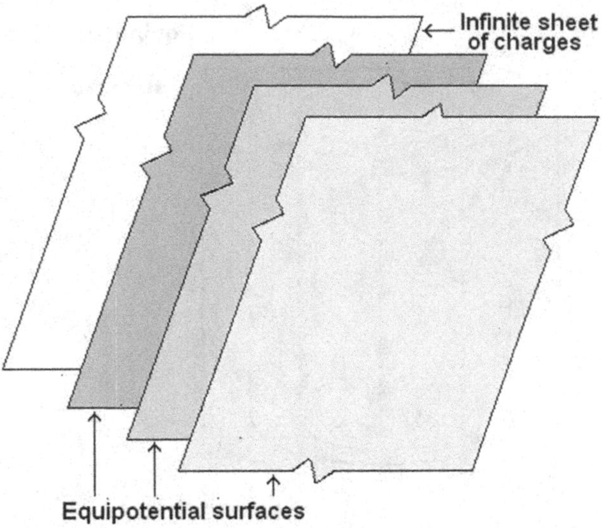

FIGURE 7.5 Configuration of equipotentials for an infinite sheet of uniform charge.

their upper, lower, left, and right end surface located at infinity. This figure can be obtained by assigning suitable values to the parameters involved in Equation 7.27 and by taking a surface charge of infinite extents.

7.4.6 Due to a Finite Sheet of Uniform Charges

Figure 7.6 shows the configuration of equipotentials for a finite rectangular sheet of uniform charge distribution. This too shows only the 3D view, which contains a number of surfaces that confirm to the shape of the sheet and are perpendicular to electric flux lines shown in Figure 5.19. This figure can be obtained by assigning suitable values to the parameters involved in Equation 7.27 and by taking a surface charge of finite extents.

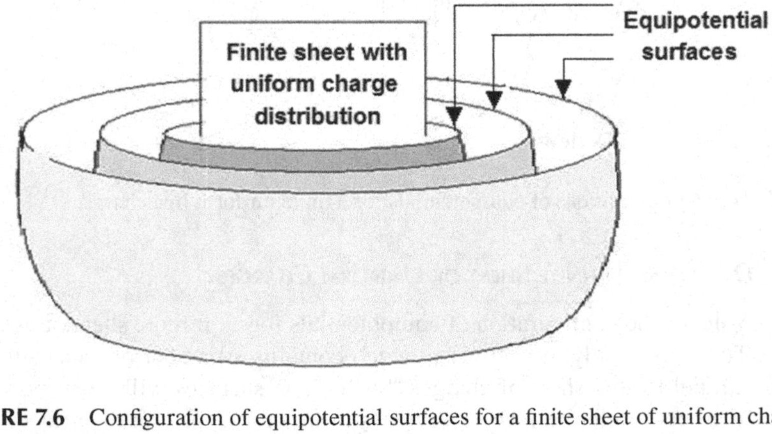

FIGURE 7.6 Configuration of equipotential surfaces for a finite sheet of uniform charge.

7.5 FIELD DUE TO AN ELECTRIC-DIPOLE

If two charges of opposite polarity are located very close to each other as compared to the distance at which the field due to the two charges is to be evaluated, such a system of charges is termed as a dipole. The resultant field due to a dipole shows a atypical behavior since the potential field decreases as the inverse of the square of the distance, and the electric field intensity decreases as the inverse of the cube of the distance from the dipole. Similar symmetrical arrangements of large numbers of point charges may be referred to as multipoles. These produce fields decreasing as the inverse of higher and higher powers of distance.

7.5.1 Potential Field

Figure 7.7 shows two point charges with equal charge contents but of opposite polarity. These charges are separated by a distance d and are symmetrically located along the z-axis. This separation (d) is much smaller than the distance (r) from the origin to the point (P) at which the field is to be evaluated, referred to as the field point. A configuration where $d \ll r$ is called a *dipole*. As shown in the figure, the distance between $+Q$ and P is r_1 and between $-Q$ and P is r_2.

In this system of charges described above the electrostatic potential at point P can be obtained just by summing the potentials due to each of the two charges.

$$V = \frac{+Q}{4\pi\varepsilon_o r_1} + \frac{-Q}{4\pi\varepsilon_o r_2} = \frac{Q}{4\pi\varepsilon_o}\cdot\left(\frac{1}{r_1} - \frac{1}{r_2}\right) = \frac{Q}{4\pi\varepsilon_o}\cdot\frac{(r_2 - r_1)}{r_1 . r_2} \tag{7.30}$$

If point P is assumed to be far away from the dipole (i.e. $r >> d$), the distances r_1 and r_2 can be taken to be almost parallel and approximately equal (i.e. $r_1 \cong r_2$). Thus in the denominator of Equation 7.30 the product $r_1 \cdot r_2$ can be equated to r^2, whereas in its numerator the relation ($r_1 \cong r_2$) is not justified. A more accurate value, in the

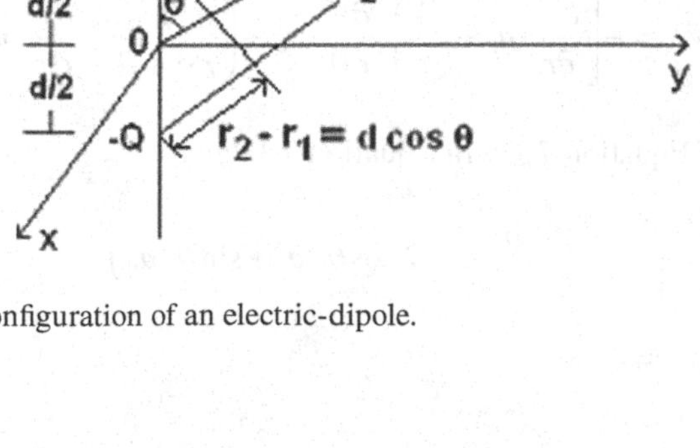

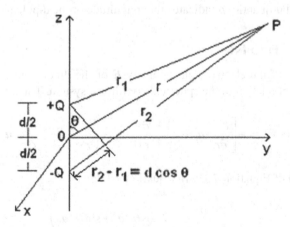

FIGURE 7.7 Configuration of an electric-dipole.

RHS of Figure 6.9 is taken in terms of separation d and the angle θ. This value is as follows:

$$r_2 - r_1 \cong d \cdot \cos\theta \tag{7.31a}$$

The electrostatic potential at point P (Equation 7.30) can now be rewritten as:

$$V \cong \frac{Q}{4\pi\varepsilon_o} \cdot \frac{d \cdot \cos\theta}{r^2} \tag{7.31b}$$

Note that the above expression for V (Equation 7.31b) is only approximate. For an exact formulation we define dipole moment vector $\mathcal{P}$ in Equation 7.22a as:

$$\boldsymbol{\mathcal{P}} \stackrel{\text{def}}{=} \lim_{\substack{Q \to \infty \\ d \to 0}} \left(Q \cdot \boldsymbol{d} \right) \tag{7.31c}$$

In Equation 7.31c, Q is the positive charge, $\boldsymbol{d}$ is the distance vector oriented from negative to the positive charge and $\boldsymbol{\mathcal{P}}$ vector is the dipole moment of this system of charge-pair.

In the involvement of cosine term this electric potential can be written as:

$$V = \frac{\boldsymbol{\mathcal{P}} \bullet \boldsymbol{a}_r}{4\pi\varepsilon_o \cdot r^2} \tag{7.32}$$

or,

$$V = \frac{\mathcal{P}}{4\pi\varepsilon_o} \cdot \frac{\cos\theta}{r^2} \tag{7.32a}$$

where $\boldsymbol{a}_r$ is the unit vector oriented along the z-axis and pointing from $-Q$ to $+Q$ and θ is the angle between the dipole moment vector $\boldsymbol{\mathcal{P}}$ and the position vector for the point P. The scalar quantity p indicates the magnitude of the dipole moment.

7.5.2 Electric Field Intensity

The distribution of the electric field intensity $\boldsymbol{E}$ of the dipole can be obtained by using the gradient relation in the spherical coordinate system. Thus:

$$\boldsymbol{E} = -\nabla V = -\left[\frac{\partial V}{\partial r} \cdot \boldsymbol{a}_r + \left(\frac{1}{r} \right) \cdot \frac{\partial V}{\partial \theta} \cdot \boldsymbol{a}_\theta + \left(\frac{1}{r \cdot \sin\theta} \right) \cdot \frac{\partial V}{\partial \varphi} \cdot \boldsymbol{a}_\varphi \right] \tag{7.33}$$

The substitution of Equation 7.32a in Equation 7.33 gives

$$\boldsymbol{E} = \frac{p}{4\pi\varepsilon_o \cdot r^3} \cdot \left(2\cos\theta \cdot \boldsymbol{a}_r + \sin\theta \cdot \boldsymbol{a}_\theta \right) \tag{7.33a}$$

Equation 7.22 shows that the potential field due to a point charge is inversely proportional to the distance, whereas for Equation 7.32a it is inversely proportional to the distance square. Similarly, the electric field intensity due to point charge (Equation 5.10) is inversely proportional to the distance square. In a dipole (Equation 7.33a), it is inversely proportional to the cube of the distance.

7.6 POTENTIAL ENERGY OF POINT CHARGES

We perform a thought experiment; bring point charges from infinity, one by one, and place them at distinct fixed points in free space. If a point charge Q is brought from infinity and placed at a point P the work done w is given as:[1, 2]

$$\boxed{w = -Q \cdot \int_{\infty}^{P} \boldsymbol{E} \bullet d\ell = Q \cdot V} \tag{7.34}$$

where $\boldsymbol{E}$ indicates the distribution of electric field intensity and V indicates the scalar electric potential at the point P; the zero reference potential is taken to be at infinity. The fields, $\boldsymbol{E}$and$V$, are due to all charges brought from infinity prior to the arrival of the charge Q. Furthermore, our objective being the "stored energy", any energy radiated[3] during the process of bringing these charges is ignored. In bringing an arbitrary n number of point charges the total work done W_n is attributed to as the potential energy of the group of n number of point charges $\mathbb{E}_n$.

7.6.1 Potential Energy of Stationary Point Charges

Consider the infinite region of free space. We bring a point charge Q_1 from infinity and position it at a fixed point P_1 inside an arbitrary finite volume v in this region. The potential energy ⓔ being proportional to the product of the point charge and the potential at the point where the charge is placed, when the first charge Q_1 was brought and placed at the point P_1, the work done w_1 was zero and its potential energy ⓔ$_1$ remained at zero value, i.e. ⓔ$_1 = w_1 = 0$.

When the second charge Q_2 was brought and placed at another fixed point P_2 inside this volume v_o, the work done w_2 is the product of this charge and the potential $V_{1,2}$ at the point P_2, caused by the charge Q_1, already present at the point P_1; thus:

$$w_2 = Q_2 \cdot V_{1,2} \tag{7.35}$$

Now, the potential at P_1, where the charge Q_1 is located, is no more zero as a potential $V_{2,1}$ has developed at this point due to the field of the newly arrived charge Q_2. Therefore, the new charge Q_2 and the existing charge Q_1 *simultaneously gained potential energy* ⓔ$_2$. This energy gain must be equal to the work done in bringing the second charge Q_2. Thus, we have:

$$ⓔ_2 = w_2 = k \cdot Q_2 \cdot V_{1,2} + k \cdot Q_1 \cdot V_{2,1} \tag{7.36}$$

Therefore, from Equations 7.35 and 7.36:

$$Q_2 \cdot V_{1,2} = k \cdot Q_1 \cdot V_{2,1} + k \cdot Q_2 \cdot V_{1,2} \tag{7.37}$$

Now, since:

$$Q_1 \cdot V_{2,1} = Q_2 \cdot V_{1,2} = \frac{Q_1 \cdot Q_2}{4\pi\varepsilon_o \cdot R_{12}} \tag{7.38}$$

where R_{12} indicates the distance between the points P_1 and P_2.

Therefore, on solving Equation 7.37 for the constant of proportionality k, we get:

$$k = 1/2 \tag{7.39}$$

Equation 7.36 giving the work done w_2, for bringing the charge Q_2 is equal to the gain in the potential energy $\textcircled{e}_2$ of the $\underbrace{\text{new charge}}$ as well as the one $\overbrace{\text{existing already}}$, can thus be rewritten as follows:

$$\textcircled{e}_2 = w_2 = \left[\underbrace{\frac{1}{2} \cdot Q_2 \cdot V_{1,2}} + \overbrace{\frac{1}{2} \cdot Q_1 \cdot V_{2,1}} \right] \tag{7.40}$$

Now, as the third point charge Q_3 is brought from infinity and placed at still another fixed point P_3 within the volume v_o; the work done w_3 is the product of this charge and the sum of the potentials $V_{1,3}$ and $V_{2,3}$ at the point P_3 caused by the charges Q_1 and Q_2, respectively. Therefore, we have:

$$w_3 = Q_3 \cdot \left(V_{1,3} + V_{2,3}\right) \tag{7.41}$$

It can be seen that the potential energy gained by this new charge Q_3 is:

$$\frac{w_3}{2} = \frac{1}{2} \cdot Q_3 \cdot \left(V_{1,3} + V_{2,3}\right) \tag{7.42}$$

while the other half of w_3 is stored as the combined *increment in the potential energy* of the previously brought charges, viz. Q_1 and Q_2, due to the potential field of the recently brought charge Q_3. This combined increment is as follows:

$$\frac{w_3}{2} = \left\{ \frac{1}{2} \cdot Q_1 \cdot V_{3,1} + \frac{1}{2} \cdot Q_2 \cdot V_{3,2} \right\} \tag{7.43}$$

Thus, from Equations 7.42 and 7.43 the work done w_3, for bringing the charge Q_3 is equal to the gain in the potential energy $\textcircled{e}_3$ of the $\underbrace{\text{new charge}}$ Q_3 as well as those $\overbrace{\text{existing already}}$, i.e. Q_1 and Q_2:

$$\textcircled{e}_3 = w_3 = \left[\underbrace{\frac{1}{2}\cdot Q_3 \cdot \left(V_{1,3} + V_{2,3}\right)} + \overbrace{\frac{1}{2}\cdot Q_1 \cdot V_{3,1} + \frac{1}{2}\cdot Q_2 \cdot V_{3,2}} \right] \tag{7.44}$$

Now, the total work done W_3 to bring and position the three point charges and thus the total potential energy $\mathbb{E}_3$ of these charges is:

$$W_3 = w_1 + w_2 + w_3 = \textcircled{e}_1 + \textcircled{e}_2 + \textcircled{e}_3 \stackrel{\text{def}}{=} \mathbb{E}_3 \tag{7.45}$$

or,

$$\begin{aligned} W_3 = \mathbb{E}_3 = &\left[\frac{1}{2}\cdot Q_1 \cdot 0\right] + \left[\underbrace{\frac{1}{2}\cdot Q_2 \cdot V_{1,2}} + \overbrace{\frac{1}{2}\cdot Q_1 \cdot V_{2,1}} \right] \\ &+ \left[\underbrace{\frac{1}{2}\cdot Q_3 \cdot \left(V_{1,3} + V_{2,3}\right)} + \overbrace{\frac{1}{2}\cdot Q_1 \cdot V_{3,1} + \frac{1}{2}\cdot Q_2 \cdot V_{3,2}} \right] \end{aligned} \tag{7.46}$$

On rearranging terms on its RHS, we get:

$$W_3 = \mathbb{E}_3 = \frac{1}{2}\cdot\left[Q_1 \cdot \left(V_{2,1} + V_{3,1}\right) + Q_2 \cdot \left(V_{1,2} + V_{3,2}\right) + Q_3 \cdot \left(V_{1,3} + V_{2,3}\right)\right]$$

or,

$$W_3 = \mathbb{E}_3 \stackrel{\text{def}}{=} \frac{1}{2}\cdot\left[V_1 \cdot Q_1 + V_2 \cdot Q_2 + V_3 \cdot Q_3\right] \tag{7.47}$$

where the symbols V_1, V_2, and V_3, indicate the potentials at points P_1, P_2, and P_3 respectively, due to all charges except those located at respective points.

Now Equation 7.47 can also be written as:

$$W_3 = \mathbb{E}_3 = \frac{1}{2}\cdot\sum_{m=1}^{3} V_m^{(3)} \cdot Q_m \tag{7.48}$$

where the symbol Q_m indicates the mth point charge brought from infinity and placed at the distinct fixed point P_m in the region of an arbitrary finite volume v_o, while $V_m^{(3)}$ indicates the potential at the point P_m due to all charges except the point charge Q_m.

For n number of point charges, Equation 7.48 is modified as follows:

$$W_n = \boxed{\mathbb{E}_n = \frac{1}{2}\cdot\sum_{m=1}^{n} V_m^{(n)} \cdot Q_m} \tag{7.49}$$

the symbol $V_m^{(n)}$ indicates the potential at the point P_m due to all n- charges except the charge Q_m, located at the point P_m.

Equation 7.49 gives the total work done W_n in terms of the point charges and their potential fields. The same is also the total potential energy $\mathbb{E}_n$ attained by all point charges brought from infinity. *Since a charge and its potential could be either a positive or a negative quantity, the combined potential energy* $\mathbb{E}_n$ *could also be either a positive or a negative quantity.*

Since V_1 is zero if n is one, Equation 7.49 shows that the potential energy of *an isolated single point charge* $\mathbb{E}_1$ is zero.

7.6.2 Energy Expression in Terms of Field Quantities

Consider the delta-dirac function (see Appendix A4 online) defined as follows:

$$\begin{cases} \delta\left(x-x_0\right) \overset{\text{def}}{=} 0 \ \text{ for } \ x \neq x_0 \\ \text{and} \quad \displaystyle\int_{-\infty}^{+\infty} \delta\left(x-x_0\right)\cdot dx \overset{\text{def}}{=} 1 \end{cases} \tag{7.50a}$$

Thus, for an arbitrary function $F(x)$, which is continuous at $x = x_0$:

$$\int_{x_1}^{x_2} F\left(x\right)\delta\left(x-x_0\right)dx = \begin{cases} 0, & \text{for } x_1 < x_2 < x_0 \\ F\left(x_0\right), & \text{for } x_1 < x_0 < x_2 \\ 0, & \text{for } x_0 < x_1 < x_2 \end{cases} \tag{7.50b}$$

Using delta-dirac functions, point charges can be expressed as a distribution of volume charge density ρ; as given below:

$$\rho = \sum_{m=1}^{n} Q_m \cdot \delta\left(x-x_m\right)\cdot\delta\left(y-y_m\right)\cdot\delta\left(z-z_m\right) \tag{7.50c}$$

where x_m, y_m, and z_m indicate the coordinates of the point P_m.

Thus, Equation 7.49 can be transformed as follows:

$$W_n = \mathbb{E}_n = \boxed{\frac{1}{2}\cdot\sum_{m=1}^{n} V_m \cdot Q_m \doteq \frac{1}{2}\cdot\iiint_{v\to\infty}\left(V\cdot\rho\right)\cdot dv} \tag{7.51}$$

This equation is identically satisfied if we substitute the expression for the charge density ρ from Equation 7.50c on the RHS of Equation 7.51 and then the integration is performed over infinite volume, as shown below:

$$\frac{1}{2}\cdot\iiint_{v\to\infty}(V\cdot\rho)\cdot dv=\frac{1}{2}\cdot\int_{x=-\infty}^{\infty}\int_{y=-\infty}^{\infty}\int_{z=-\infty}^{\infty}\sum_{m=1}^{n}f_m\cdot dx\cdot dy\cdot dz$$

$$=\frac{1}{2}\cdot\sum_{m=1}^{n}\int_{x=-\infty}^{\infty}\int_{y=-\infty}^{\infty}\int_{z=-\infty}^{\infty}f_m\cdot dx\cdot dy\cdot dz \tag{7.52}$$

$$=\frac{1}{2}\cdot\sum_{m=1}^{n}V\left(x_m,y_m,z_m\right)\cdot Q_m\overset{\text{def}}{=}\frac{1}{2}\cdot\sum_{m=1}^{n}V_m\cdot Q_m$$

where

$$f_m\overset{\text{def}}{=}V\left(x,y,z\right)\cdot Q_m\cdot\delta\left(x-x_m\right)\cdot\delta\left(y-y_m\right)\cdot\delta\left(z-z_m\right) \tag{7.53}$$

The integration over infinite volume ensures that all n point charges brought from infinity and placed at distinct stationary points are included in the integration. On the other hand, if the integration is performed over an arbitrary finite volume v_ℓ, containing only ℓ number of point charges while some charges, viz. $(n-\ell)$ are left outside this volume; the resulting volume integration will give the total potential energy $\mathbb{E}_\ell$ of only those *ℓ number of point charges located within the volume v_ℓ*. With point charges labeled afresh, Equation 7.51 is modified for the arbitrary volume v_ℓ as follows:

$$\boxed{\mathbb{E}_\ell=\frac{1}{2}\cdot\sum_{k=1}^{\ell}V_k^{(n)}\cdot Q_k=\frac{1}{2}\cdot\iiint_{v_\ell}(V\cdot\rho)\cdot dv} \tag{7.54}$$

where

$$0\le\ell\le n \tag{7.54a}$$

and the symbol $V_k^{(n)}$ indicates the potential at the point P_k in the volume v_ℓ due to all n- charges, internal as well as external, except the charge Q_k, located at the point P_k in the volume v_ℓ.

Note that if $n>1$, $\mathbb{E}_\ell\big|_{\ell=1}\neq 0$. Furthermore, *if no point charge is located inside the finite arbitrary volume v_ℓ, i.e. for charge free regions, the volume integration in* Equation 7.54 *being zero*:

$$\mathbb{E}_\ell\big|_{\ell=0}=\frac{1}{2}\cdot\iiint_{v_\ell}(V\cdot\rho)\cdot dv\Bigg|_{\ell=0}=0. \tag{7.54b}$$

Using Maxwell's equation $\rho=\underbrace{\nabla\bullet\boldsymbol{D}}$, Equation 7.54 can be rewritten as:

$$\mathbb{E}_\ell=\frac{1}{2}\cdot\sum_{k=1}^{\ell}V_k^{(n)}\cdot Q_k=\frac{1}{2}\cdot\iiint_{v_\ell}\left(V\cdot\underbrace{\nabla\bullet\boldsymbol{D}}\right)\cdot dv \tag{7.55}$$

Next, consider the identity given by Equation 4.40b:

$$(V \cdot \nabla \bullet \boldsymbol{D}) \equiv (-\nabla V) \bullet \boldsymbol{D} + \nabla \bullet (V\boldsymbol{D}) \tag{7.56}$$

Since for electrostatic fields $-\nabla V = \boldsymbol{E}$ and $\nabla \bullet \boldsymbol{D} = \rho$
we get,

$$(V \cdot \rho) = (V \cdot \nabla \bullet \boldsymbol{D}) = \boldsymbol{E} \bullet \boldsymbol{D} + \nabla \bullet (V\boldsymbol{D}) \tag{7.57}$$

Therefore, Equation 7.55 can be rewritten as:

$$\mathbb{E}_\ell = \frac{1}{2} \cdot \sum_{k=1}^{\ell} V_k^{(n)} \cdot Q_k = \iiint_{v_\ell} \left(\frac{1}{2} \cdot \boldsymbol{E} \bullet \boldsymbol{D} \right) \cdot dv + \iiint_{v_\ell} \nabla \bullet \left(\frac{1}{2} \cdot V\boldsymbol{D} \right) \cdot dv \tag{7.58}$$

Equation 7.58 gives the potential energy of point charges exclusively in terms of field quantities. *This prompts one to conclude* that “the potential energy of point charges may also be considered as stored in the electrostatic fields”.

The combined potential energy of all the point charges brought from infinity, in Equations 7.51 and 7.58 is given as follows:

$$\boxed{\mathbb{E}_n = \iiint_{v \to \infty} \left(\frac{1}{2} \cdot \boldsymbol{E} \bullet \boldsymbol{D} \right) \cdot dv + \iiint_{v \to \infty} \nabla \cdot \left(\frac{1}{2} \bullet V\boldsymbol{D} \right) \cdot dv} \tag{7.59}$$

Using Gauss’ theorem, the second term on the RHS of this equation is transformed into a surface integration, i.e.:

$$\iiint_{v \to \infty} \nabla \bullet \left(\frac{1}{2} \cdot V\boldsymbol{D} \right) \cdot dv \equiv \oiint_{s \to \infty} \left(\frac{1}{2} \cdot V\boldsymbol{D} \right) \bullet \boldsymbol{ds} \stackrel{\text{def}}{=} \oiint_{s \to \infty} \boldsymbol{\mathcal{E}}_s \bullet \boldsymbol{ds} \tag{7.60}$$

where the symbol s stands for the surface of the volume v and the surface density of the stored energy is defined as:

$$\boldsymbol{\mathcal{E}}_s \stackrel{\text{def}}{=} \left(\frac{1}{2} \cdot V\boldsymbol{D} \right) \tag{7.60a}$$

As the volume v increases, its surface s also increases; the average distance r of this surface from point charges also increases. Point charges being the source for the electric field; on the surface s, V varies as $1/r$ and $\boldsymbol{D}$ as $1/r^2$. Hence $(V\boldsymbol{D})$ in the expression of $\boldsymbol{\mathcal{E}}_s$ as given in Equation 7.60a, must vary as $1/r^3$. Since s varies as r^2, with the increase in r the integrand decreases at a faster rate than the rate of increase in the surface area. Therefore, as the surface tends to infinity, the integration over infinite surface must tend to zero,[1, 2] i.e.:

$$\boxed{\oiint_{s \to \infty} \boldsymbol{\mathcal{E}}_s \bullet \boldsymbol{ds} \equiv 0} \tag{7.60b}$$

Therefore, Equation 7.60 reduces to:

$$\boxed{\mathbb{E}_n = \iiint_{v\to\infty} \left(\frac{1}{2} \cdot \boldsymbol{E} \bullet \boldsymbol{D} \right) \cdot dv} \tag{7.61}$$

In this equation, the classical expression for the density of the energy stored in the electrostatic fields $\mathcal{E}$, is *defined* as indicated below:

$$\mathcal{E} \stackrel{\text{def}}{=} \frac{1}{2} \cdot \boldsymbol{E} \bullet \boldsymbol{D} \tag{7.61a}$$

Thus:

$$\boxed{\mathbb{E}_n = \iiint_{v\to\infty} \mathcal{E} \cdot dv} \tag{7.61b}$$

7.6.3 Energy Stored in Electrostatic Field

The statement that "the potential energy of point charges may also be considered as stored in the electrostatic fields" is speculative in nature rather than a fact. Consider the following observations:

(i) If energy is stored in the electrostatic field spreading over infinite space, we must be able to estimate the energy stored in any arbitrary finite volume for a given field distribution therein. The expression for the energy density defined by Equations 7.61a and 7.61b is valid only for the region of infinite volume.

(ii) When the first charge Q_1 was brought from infinity no energy was stored, even though electrostatic field is established in the infinite space surrounding the point charge.

(iii) In Equation 7.49, it has been argued that since a charge and its potential could be either a positive or a negative quantity, the combined potential energy $\mathbb{E}_n$ could also be either a positive or a negative quantity. The energy density defined by Equation 7.61a is nowhere negative; therefore, Equation 7.61b gives erroneous results if the combined potential energy $\mathbb{E}_n$ is negative.

The dilemma presented leads to the conclusion that no energy is stored in electrostatic field even though the potential energy of point charges, if it is positive, is expressible in terms of field quantities. If this is true, what about the time-varying electromagnetic fields? This point is discussed in Section 15.7 of Chapter 15.

7.6.4 Energy for Displacement of Point Charges

In this subsection, we shall consider a simple case involving only three point charges Q_1, Q_2, and Q_3 placed, respectively, at points P_1, P_2, and P_3. The potential energy for these charges is as follows:

$$\left.\begin{aligned} w_1 &= \frac{1}{2}\cdot Q_1\cdot\left(V_{1,2}+V_{1,3}\right) \quad \text{(a)}, \quad w_2 = \frac{1}{2}\cdot Q_2\cdot\left(V_{2,1}+V_{2,3}\right) \quad \text{(b)},\\ w_3 &= \frac{1}{2}\cdot Q_3\cdot\left(V_{3,1}+V_{3,2}\right) \quad \text{(c)} \end{aligned}\right\} \tag{7.62}$$

The potential at the point P_4 due to the point charges Q_1 and Q_2 is given as:

$$V_4 \overset{\text{def}}{=} V_{4,1}+V_{4,2} = \frac{Q_1}{4\pi\cdot\varepsilon_o\cdot R_{4,1}} + \frac{Q_2}{4\pi\cdot\varepsilon_o\cdot R_{4,2}} \tag{7.63}$$

Now, if the point charge Q_3 is shifted to the point P_4, it attains a potential energy:

$$\overline{w}_3 = \frac{1}{2}\cdot Q_3\cdot\left(V_{4,1}+V_{4,2}\right) = \frac{1}{2}\cdot Q_3\cdot\left[\frac{Q_1}{4\pi\cdot\varepsilon_o\cdot R_{4,1}} + \frac{Q_2}{4\pi\cdot\varepsilon_o\cdot R_{4,2}}\right] \tag{7.64a}$$

Thus, the gain in the potential energy for the point charge Q_3 is:

$$\Delta w_3 = \overline{w}_3 - w_3 = \frac{1}{2}.Q_3\cdot\left[\left(V_{4,1}-V_{3,1}\right)+\left(V_{4,2}-V_{3,2}\right)\right]$$

$$\Delta w_3 = \frac{1}{2}\cdot Q_3\cdot\left[\left(V_{4,1}+V_{4,2}\right)-\left(V_{3,1}+V_{3,2}\right)\right]$$

or,

$$\Delta w_3 = \frac{1}{2}\cdot Q_3\cdot\left[\left\{\frac{Q_1}{4\pi\cdot\varepsilon_o\cdot R_{4,1}} + \frac{Q_2}{4\pi\cdot\varepsilon_o\cdot R_{4,2}}\right\} - \left\{\frac{Q_1}{4\pi\cdot\varepsilon_o\cdot R_{3,1}} + \frac{Q_2}{4\pi\cdot\varepsilon_o\cdot R_{3,2}}\right\}\right] \tag{7.64b}$$

This shift in the position for the charge Q_3 alters the values of the potential energy for the charges Q_1 and Q_2. The revised values for the potential energy of the point charges Q_1 and Q_2 are, respectively, given as:

$$\left.\overline{w}_1 = \frac{1}{2}\cdot Q_1\cdot\left(V_{1,2}+V_{1,4}\right) \ \text{(a)}, \quad \overline{w}_2 = \frac{1}{2}\cdot Q_2\cdot\left(V_{2,1}+V_{2,4}\right) \ \text{(b)}\right\} \tag{7.65}$$

Therefore, the gains in the potential energy for the two point charges are, respectively, as follows:

$$\left.\begin{aligned} \Delta w_1 &= \overline{w}_1 - w_1 = \frac{1}{2}\cdot Q_1\cdot\left(V_{1,4}-V_{1,3}\right) \quad \text{(a)}\\ \Delta w_2 &= \overline{w}_2 - w_2 = \frac{1}{2}\cdot Q_2\cdot\left(V_{2,4}-V_{2,3}\right) \quad \text{(b)} \end{aligned}\right\} \tag{7.66a}$$

Thus, the gain in the potential energy for the electrostatic field of two point charges is as follows:

$$\Delta w \stackrel{\text{def}}{=} \Delta w_1 + \Delta w_2 = \frac{1}{2} \cdot Q_1 \cdot \left(V_{1,4} - V_{1,3}\right) + \frac{1}{2} \cdot Q_2 \cdot \left(V_{2,4} - V_{2,3}\right) \tag{7.67a}$$

or,

$$\Delta w = \frac{1}{2} \cdot \left[\left\{ Q_1 \cdot V_{1,4} + Q_2 \cdot V_{2,4} \right\} - \left\{ Q_1 \cdot V_{1,3} + Q_2 \cdot V_{2,3} \right\} \right]$$

or,

$$\Delta w = \frac{1}{2} \cdot \left[\left\{ Q_1 \cdot \frac{Q_3}{4\pi \cdot \varepsilon_o \cdot R_{1,4}} + Q_2 \cdot \frac{Q_3}{4\pi \cdot \varepsilon_o \cdot R_{2,4}} \right\} - \left\{ Q_1 \cdot \frac{Q_3}{4\pi \cdot \varepsilon_o \cdot R_{1,3}} + Q_2 \cdot \frac{Q_3}{4\pi \cdot \varepsilon_o \cdot R_{2,3}} \right\} \right]$$

or,

$$\Delta w = \frac{1}{2} \cdot Q_3 \cdot \left[\left\{ \frac{Q_1}{4\pi \cdot \varepsilon_o \cdot R_{1,4}} + \frac{Q_2}{4\pi \cdot \varepsilon_o \cdot R_{2,4}} \right\} - \left\{ \frac{Q_1}{4\pi \cdot \varepsilon_o \cdot R_{1,3}} + \frac{Q_2}{4\pi \cdot \varepsilon_o \cdot R_{2,3}} \right\} \right] \tag{7.67b}$$

Now, since distances: $R_{1,4} = R_{4,1}$, $R_{2,4} = R_{4,2}$, $R_{1,3} = R_{3,1}$, and $R_{2,3} = R_{3,2}$, we have:

$$\Delta w_3 = \Delta w \tag{7.68}$$

Thus, in general, if a point charge is shifted to a new position, it gains a potential energy which is equal to the combined gain in the potential energy of all remaining point charges. Therefore, energy needed to shift a point charge is double the energy gained by the shifted point charge. This energy is added to the energy already stored in the electrostatic field. Therefore, if some energy is to be extracted from or implanted into the electrostatic field of a set of point charges, the potential energy of one or more point charges may be changed by suitable displacements. The agency shifting the point charges needs to supply an extra energy, which is radiated during the shift.

7.6.5 Analogy of Energy Stored

A somewhat crude analogy for the energy stored in an electrostatic field can be drawn in the system of springs shown in Figure 7.8. While Figure 7.8a shows two wooden (or iron) rods (or bars) marked as A and B. On each of these bars, n number of rings are fixed. If one spring is attached to each ring, n springs can be attached to

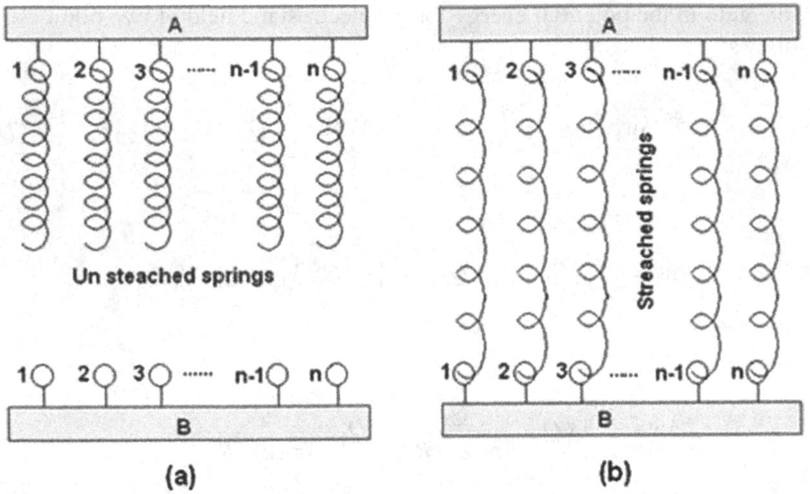

FIGURE 7.8 (a) Freely hanging springs; (b) stretched springs.

bar A. If the bar A is fixed, this bar will experience no force except that due to the weights of these hanging springs. If these springs are pulled onebyone and attached to the rings of the fixed bar B, both the bars will experience a force that will try to pull these bars toward each other. More is the value of n more will be the force. Figure 7.8b shows these stretched springs connected to bar B.

Let us now explore the process of stretching and the amount of forces involved if a spring is connected between the two bars. When this spring is stretched from its original length (say L_1) to another length (say L_1') a force (say F_1) is to be applied. This force depends on the amount "x" the spring is stretched. This elongation process requires work W to be done which can be given by the following equation:

$$W = \int_{x=0}^{(L_1'-L_1)} F_1(x)\cdot dx \tag{7.69}$$

The energy expended in doing this work will remain stored in the stretched spring as long as it remains attached to the two fixed bars. The moment the spring is detached from the lower or the upper bar it will jump back to its original length and release the stored energy. Here if we assume the shape of the bar B to be such that for attaching to bar B every spring requires different lengths (say $L_1, L_2, L_3, \ldots, L_n$) and that the elasticity of the different spring is different, the force exerted on each spring will be different (say $F_1, F_2, \ldots, F_n$). Also the energy expended (or the energy stored) in each spring will be different.

Let us now look back at our description of energy stored in electrostatic field. The bringing of charges from infinity to different locations in the empty space is analogous to the stretching of springs by different lengths. For each charge, the work done or energy stored will be in general different. If any of these charges is set free, it will go back to its earlier location at infinity. Thus, roughly speaking, the process of the energy stored in electrostatic field is qualitatively analogous to the energy stored in stretched springs. This analogy however, breaks down on closer examination. It assumes that the power radiated by electric charges is negligible, or springs are ideal and do not oscillate.

For certain types of forces, the "potential" of a field can be defined such that the potential energy of an object depends only on its position with respect to the field. The effect on objects due to these forces must depend only on the intrinsic properties of the objects and their locations in the field. Two such well-known forces are the gravitational force and the electrostatic force due to stationary electric charges. The potential of an electric field in the present context is called the scalar electrostatic potential. In the study of time-varying electric fields we shall come across time-varying scalar electric potential. In the study of magnetic fields in current-carrying regions, we shall come across vector magnetic potential. Likewise, we can define vector electric potential for charge free regions, although it is rarely done.

To consolidate what we have learned, let us consider the following examples.

Example 7.1

A 10 μC charge is moving along a 10^{-4} m incremental path directed along $a_x - 0.5a_y - 0.2a_z$ in the field $E = x^2a_x + 2y^2a_y + 5z^2a_z$ V/m. Find the work done if this path is located at (a) (1,1,0); (b) (2,0,–2).

SOLUTION

Given $Q = 20\times10^{-6}$, $\boldsymbol{E} = x^2\boldsymbol{a}_x + 2y^2\boldsymbol{a}_y + 5z^2\boldsymbol{a}_z$,

$$\boldsymbol{dL} = 10^{-4}\left(\boldsymbol{a}_x - 0.5\boldsymbol{a}_y \quad 0.2\boldsymbol{a}_z\right)$$

In view of Equation 7.1a, the incremental work done can be written as:

$$dW = -Q\boldsymbol{E}\bullet\boldsymbol{dL} = -10\times10^{-6}\left(x^2a_x + 2y^2a_y + z^2a_z\right)\cdot10^{-4}\left(a_x - 0.5a_y - 0.2a_z\right)$$

$$= -10^{-9}\left(x^2 - y^2 - z^2\right)$$

(a) $dW\big|_{1,1,1} = -10^{-9}\left(x^2 - y^2 - z^2\right)_{1,1,1} = 10^{-9}$ Joules

(b) $dW\big|_{2,1,-2} = -10^{-9}\left(x^2 - y^2 - z^2\right)_{2,1,-2} = 10^{\;9}$ Joules

Example 7.2

Find the work done in carrying 5 μC charge from $B(1,1,0)$ to $A(0.3,0.4,0.5)$ along the path identified by $x^2 + y^2 = 1$, $z = 1$ in a field $\boldsymbol{E} = y\boldsymbol{a}_x - x\boldsymbol{a}_y + \boldsymbol{a}_z$.

SOLUTION

Given $Q = 5\times10^{-6}$, $\boldsymbol{E} = y\boldsymbol{a}_x - x\boldsymbol{a}_y + \boldsymbol{a}_z$, $\boldsymbol{dL} = dx\boldsymbol{a}_x + dy\boldsymbol{a}_y + dz\boldsymbol{a}_z$

In view of Equation 7.1c, the work done can be written as:

$$W = -Q\int_B^A \mathbf{E}\cdot d\mathbf{L} = -5\times10^{-6}\int_B^A (y\mathbf{a}_x - x\mathbf{a}_y + \mathbf{a}_z)\cdot(dx\mathbf{a}_x + dy\mathbf{a}_y + dz\mathbf{a}_z)$$

$$= -5\times10^{-6}\int_B^A (ydx - xdy + dz) = -5\times10^{-6}\left[\left(\int_1^{0.3} ydx - \int_1^{0.4} xdy + \int_0^{0.5} dz\right)\right]$$

Since $x^2 + y^2 = 1$, $x = \sqrt{1-y^2}$ and $y = \sqrt{1-x^2}$

$$W = -5\times10^{-6}\left[\left(\int_1^{0.3}\sqrt{1-x^2}\,dx - \int_1^{0.34}\sqrt{1-y^2}\,dy + \int_0^{0.5} dz\right)\right]$$

$$= -5\times10^{-6}\left[\left(\left\{x\sqrt{1-x^2} + \sin^{-1}x\right\}_1^{0.3} - \left\{y\sqrt{1-y^2} + \sin^{-1}y\right\}_1^{0.4} + z\right\}_0^{0.5}\right)\right]$$

$$= -5\times10^{-6}\left[(0.3\times0.954 + 0.3047 - 1.571)\right.$$
$$\left. -(0.4\times0.9165 + 0.4115 - 1.571) + (0.5 - 0)\right]$$

$$= -1.564\ \mu\text{Joules}$$

Example 7.3

Find the work done in carrying 3mC charge from $B(0,0,0)$ to $A(0.4,0.6,0.8)$ along the straight line path, $z = 1$ in a field $\mathbf{E} = 3yx^2\mathbf{a}_x + 2y^2x\mathbf{a}_y + 5z^2\mathbf{a}_z$.

SOLUTION

Given $Q = 3\times10^{-3}$, $\mathbf{E} = 3yx^2\mathbf{a}_x + 2y^2x\mathbf{a}_y + 5z^2\mathbf{a}_z$,

$$d\mathbf{L} = (dx\mathbf{a}_x + dy\mathbf{a}_y + dz\mathbf{a}_z)$$

In view of Equation 7.1c, the work done can be written as:

$$W = -Q\int_B^A \mathbf{E}\cdot d\mathbf{L} = -3\times10^{-3}\int_B^A (3yx^2\mathbf{a}_x + 2y^2x\mathbf{a}_y + 5z^2\mathbf{a}_z)\cdot(dx\mathbf{a}_x + dy\mathbf{a}_y + dz\mathbf{a}_z)$$

$$= -3\times10^{-3}\int_B^A (3yx^2dx + 2y^2xdy + 5z^2dz)$$

$$= -3\times10^{-3}\left[\left(\int_0^{0.4} 3yx^2dx + \int_0^{0.6} 2y^2xdy + \int_0^{0.8} 5z^2dz\right)\right]$$

In view of Equation 4.48:

$$y = \frac{y_A - y_B}{x_A - x_B}(x - x_B) = \frac{0.6 - 0}{0.4 - 0}(x - 0) = 1.5x \text{ thus } y = 1.5x \text{ and } x = y/1.5.$$

$$W = -3\times10^{-3}\left[\left(\int_1^{0.4} 4.5x^3 dx + \int_0^{0.6} \frac{2y^3}{1.5} dy + \int_0^{0.8} 5z^2 dz\right)\right]$$

$$= -3\times10^{-3}\left[\left(4.5\frac{x^4}{4}\right)_0^{0.4} + \left(\frac{2}{1.5}\cdot\frac{y^4}{4}\right)_0^{0.6} + \left(\frac{5z^3}{3}\right)_0^{0.8}\right]$$

$$= -3\times10^{-3}\left[0.0288 + 0.0432 + 0.8533\right] = -2.776 \text{ milli-Joules}$$

Example 7.4

The field due to a point charge at the origin results in the potential difference V_{AB} = 12 V between point A and B located at (2,0,0) and (4,0,0), respectively. Find c for point $C(c,0,0)$ such that V_{BC} = 8 V.

SOLUTION

In view of Equation 7.22, $V = \frac{1}{4\pi\varepsilon_o}\cdot\frac{Q}{R}$

Thus $V_{A \text{ or } B} = \frac{1}{4\pi\varepsilon_o}\cdot\frac{Q}{A \text{ or } B}$ and $V_{AB} = \frac{Q}{4\pi\varepsilon_o}\left(\frac{1}{A} - \frac{1}{B}\right)$

$$V_{AB} = \frac{Q}{4\pi\varepsilon_o}\cdot\left(\frac{1}{2} - \frac{1}{4}\right) = \frac{1}{4}\cdot\frac{Q}{4\pi\varepsilon_o} = 12 \quad\text{or}\quad Q = 192\pi\varepsilon_o$$

$$V_{BC} = \frac{Q}{4\pi\varepsilon_o}\cdot\left(\frac{1}{4} - \frac{1}{c}\right) = \frac{192\pi\varepsilon_o}{4\pi\varepsilon_o}\cdot\left(\frac{1}{4} - \frac{1}{c}\right)$$

$$= 48\cdot\left(\frac{1}{4} - \frac{1}{c}\right) = 12 - \frac{48}{c} = 8 \quad\text{or}\quad \frac{48}{c} = 4 \quad\text{or}\quad c = 12$$

Example 7.5

Find k in the relation $V = \frac{Q}{4\pi\varepsilon_0 r} + k$ if $Q = 10^{-6}$C and (a) V = 5 volt at r = 40; (b) V = 10 volt at r = 20; and (c) V = 50 volt at r = 15.

SOLUTION

Since $4\pi\varepsilon_0 = 4\pi \times \frac{10^{-9}}{36\pi} = \frac{10^{-9}}{9}$ $\quad \frac{Q}{4\pi\varepsilon_0} = \frac{10^{-6}}{\frac{10^{-9}}{9}} = 9000$

(a) $V\big|_{r=40} = \left.\dfrac{Q}{4\pi\varepsilon_0 r}\right|_{r=40} + k = \left.\dfrac{9000}{r}\right|_{r=40} + k = 225 + k = 5 \text{ or } k = -220$

(b) $V\big|_{r=20} = \left.\dfrac{Q}{4\pi\varepsilon_0 r}\right|_{r=20} + k = \left.\dfrac{9000}{r}\right|_{r=20} + k = 450 + k = 10 \text{ or } k = -440$

(c) $V\big|_{r=10} = \left.\dfrac{Q}{4\pi\varepsilon_0 r}\right|_{r=15} + k = \left.\dfrac{9000}{r}\right|_{r=15} + k = 600 + k = 50 \text{ or } k = -550$

Example 7.6

The potential fields of $+Q$ and $-Q$, located at (0,0,1) and (0,0,–1), respectively, have zero reference at infinity. Find the ratio of the total potential at (0,0,50) to that at (a) (0,0,100); (b) (0,0,150); and (c) (0,0,500).

SOLUTION

$$V\big|_{r=50} = \left.\frac{Q}{4\pi\varepsilon_0 r}\right|_{r=49} + \left.\frac{-Q}{4\pi\varepsilon_0 r}\right|_{r=51} = \frac{Q}{4\pi\varepsilon_0}\left(\frac{51-49}{51\times 49}\right) = \frac{Q}{4\pi\varepsilon_0}\cdot\frac{2}{51\times 49}$$

(a) $V\big|_{r=100} = \left.\dfrac{Q}{4\pi\varepsilon_0 r}\right|_{r=99} + \left.\dfrac{-Q}{4\pi\varepsilon_0 r}\right|_{r=101} = \dfrac{Q}{4\pi\varepsilon_0}\cdot\dfrac{2}{99\times 101} \quad \dfrac{V\big|_{r=50}}{V\big|_{r=100}} = \dfrac{99\times 101}{51\times 49} \approx 4$

(b) $V\big|_{r=150} = \left.\dfrac{Q}{4\pi\varepsilon_0 r}\right|_{r=149} + \left.\dfrac{-Q}{4\pi\varepsilon_0 r}\right|_{r=151} = \dfrac{Q}{4\pi\varepsilon_0}\cdot\dfrac{2}{149\times 151} \quad \dfrac{V\big|_{r=50}}{V\big|_{r=150}} = \dfrac{149\times 151}{51\times 49} \approx 9$

(c) $V\big|_{r=500} = \left.\dfrac{Q}{4\pi\varepsilon_0 r}\right|_{r=499} + \left.\dfrac{-Q}{4\pi\varepsilon_0 r}\right|_{r=501} = \dfrac{Q}{4\pi\varepsilon_0}\cdot\dfrac{2}{499\times 501} \quad \dfrac{V\big|_{r=50}}{V\big|_{r=500}} = \dfrac{499\times 501}{51\times 49} \approx 100$

Example 7.7

Find the potential at (0,0,10) due to a uniformly distributed charge of 10^{-8} C along a configuration lying on the z = 0 plane and centered on the origin in the form of a circular ring of line of 3 m radius.

SOLUTION

The specified charged configuration is shown in Figure 7.9.

As shown in Figure 7.9 the total charge of 10^{-8} C is distributed over a ring of 3 m radius in the form of a line charge.

The circumference of the ring is $2\pi r = 6\pi$, thus $\rho_L = \dfrac{10^{-8}}{6\pi}$

Since $\dfrac{1}{4\pi\varepsilon_o} = 9\times 10^9 \quad \dfrac{\rho_L}{4\pi\varepsilon_o} = 9\times 10^9 \times \dfrac{10^{-8}}{6\pi} = \dfrac{90}{6\pi}$

$$dL = r d\varphi = 3 d\varphi \quad R = \sqrt{r^2 + z^2} = \sqrt{3^2 + 10^2} = \sqrt{9 + 100} = \sqrt{109}$$

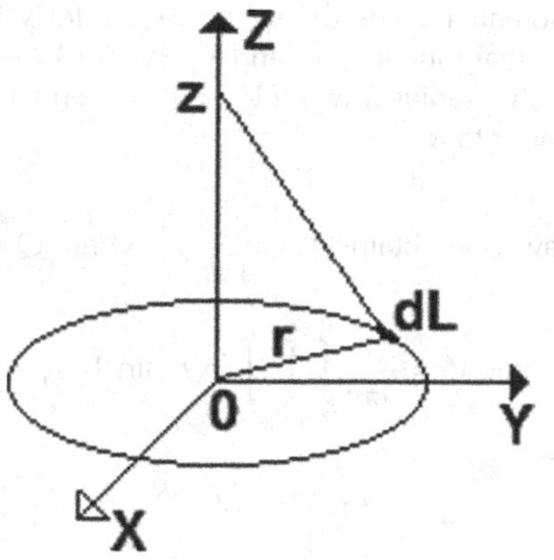

FIGURE 7.9 The charged ring configuration.

In view of Equation 7.25:

$$V = \frac{1}{4\pi\varepsilon_o} \cdot \int_A^B \frac{\rho_L}{R} . dL = \frac{\rho_L}{4\pi\varepsilon_o} \cdot \int_A^B \frac{1}{R} \cdot dL = \frac{90}{6\pi} \cdot \frac{1}{\sqrt{109}} \int_0^{2\pi} 3d\varphi$$

$$= \frac{90}{6\pi} \cdot \frac{6\pi}{\sqrt{109}} = \frac{90}{10.44} = 8.62 \text{ V}$$

Example 7.8

Find scalar electrical potential V, everywhere due to: (a) a uniformly distributed surface charge density ρ_s, over a spherical surface of radius R; (b) a uniformly distributed volume charge density ρ_v, in a spherical region of radius R.

SOLUTION

Let the origin of a spherical coordinate system be at the center of the sphere of radius R.

(a) Total charge due to the surface charge distribution $Q = \rho_s 4\pi R^2 C$.

From Gauss's law, potential at r is: $V = \frac{Q}{4\pi\varepsilon} \cdot \frac{1}{r}$ for $r \geq R$ and $V =$ constant for $r \leq R$.

Thus, $V = \frac{Q}{4\pi\varepsilon} \cdot \frac{1}{R}$ for $r \leq R$ (since potential is continuous at r = R).

(b) Total charge due to the volume charge distribution $Q = \rho_v \frac{4}{3}\pi R^3$

Using Gauss' Law, $V = \frac{Q}{4\pi\varepsilon} \cdot \frac{1}{r} = \frac{R^3 / r}{3\varepsilon} \rho_v$, for $r \geq R$.

To obtain the potential inside the sphere, $r \leq R$, let V be expressed as a sum of two potential function, V_1 and V_2; where V_1 is the potential due to all charges within radius r, while V_2 is the potential due to all charges over a radius from r to R.

Using Gauss' Law, one obtains: $V = \frac{Q}{4\pi\varepsilon} \cdot \frac{1}{r}$ where $Q = \frac{4}{3}\pi r^3 \rho_v$

Thus, $V_1 = \frac{r^2}{3\varepsilon}\rho_v$ and $V_2 = \frac{1}{4\pi\varepsilon}\int_r^R \int_{\theta=0}^{\pi} \int_{\varphi=0}^{2\pi} \rho_v r \cdot \sin\theta \cdot d\varphi \cdot d\theta \cdot dr = \frac{(R^2 - r^2)}{2\varepsilon} \cdot \rho_v$

Thus $V = V_1 + V_2 = \frac{3R^2 - r^2}{6\varepsilon} \cdot \rho_v$ for $0 \leq r \leq R$

Example 7.9

A dipole moment at (0,0,0) in free space is given as $\boldsymbol{p} = 50\pi\varepsilon_0 \left(0.5\mathbf{a}_x - 0.5\mathbf{a}_y + \mathbf{a}_z\right)$ $C - m$. Find potential V at (a) (3,0,0); (b) (0,3,0); (c) (0,0,3); (d) (1,2,3).

SOLUTION

Given $\boldsymbol{p} = 50\pi\varepsilon_0 \left(0.5\mathbf{a}_x - 0.5\mathbf{a}_y + \mathbf{a}_z\right)$

In view of Equation 7.32, $V = \frac{\boldsymbol{p} \bullet \mathbf{a}_r}{4\pi\varepsilon_o \cdot r^2}$

(a) At (3,0,0): $\boldsymbol{r} = 3\mathbf{a}_x \quad |\boldsymbol{r}| = 3 \quad \mathbf{a}_r = \mathbf{a}_x \quad \boldsymbol{p} \bullet \mathbf{a}_r = 25\pi\varepsilon_0$

$$V = \frac{\boldsymbol{p} \bullet \mathbf{a}_r}{4\pi\varepsilon_o \cdot r^2} = \frac{25\pi\varepsilon_0}{4\pi\varepsilon_{o.}\left(3^2\right)} = 0.6944 \text{ Volt}$$

(b) At (0,3,0): $\boldsymbol{r} = 3\mathbf{a}_y \quad |\boldsymbol{r}| = 3 \quad \mathbf{a}_r = \mathbf{a}_y \quad \boldsymbol{p} \bullet \mathbf{a}_r = -25\pi\varepsilon_0$

$$V = \frac{\boldsymbol{p} \bullet \mathbf{a}_r}{4\pi\varepsilon_{o.}r^2} = \frac{-25\pi\varepsilon_0}{4\pi\varepsilon_{o.}\left(3^2\right)} = -0.6944 \text{ Volt}$$

(c) At (0,0,3): $\boldsymbol{r} = 3\mathbf{a}_z \quad |\boldsymbol{r}| = 3 \quad \mathbf{a}_r = \mathbf{a}_z \quad \boldsymbol{p} \bullet \mathbf{a}_r = 50\pi\varepsilon_0$

$$V = \frac{\boldsymbol{p} \bullet \mathbf{a}_r}{4\pi\varepsilon_o \cdot r^2} = \frac{50\pi\varepsilon_0}{4\pi\varepsilon_o \cdot \left(3^2\right)} = 1.3889 \text{ Volt}$$

(d) At (1,2,3): $\boldsymbol{r} = (\mathbf{a}_x + 2\mathbf{a}_y + 3\mathbf{a}_z) \quad |\boldsymbol{r}| = \sqrt{14} \quad \mathbf{a}_r = \frac{(\mathbf{a}_x + 2\mathbf{a}_y + 3\mathbf{a}_z)}{\sqrt{14}}$

$$\boldsymbol{p} \bullet \mathbf{a}_r = 50\pi\varepsilon_0 \left(0.5\mathbf{a}_x - 0.5\mathbf{a}_y + \mathbf{a}_z\right) \bullet \frac{(\mathbf{a}_x + 2\mathbf{a}_y + 3\mathbf{a}_z)}{\sqrt{14}}$$

$$= \frac{50\pi\varepsilon_0}{\sqrt{14}}(0.5 - 1 + 3) = \frac{125\pi\varepsilon_0}{\sqrt{14}}$$

$$V = \frac{\boldsymbol{p} \bullet \boldsymbol{a}_r}{4\pi\varepsilon_o \cdot r^2} = \frac{125\pi\varepsilon_0}{\sqrt{14}} \frac{1}{4\pi\varepsilon_o} \cdot \frac{1}{14} = 0.5966 \text{ Volts}$$

Example 7.10

A dipole in free space is formed at $r = 0.5$, $\theta = 60°$, $\varphi = 0°$ by two charges of 2nC and –2nC located at (0,0,0.01) and (0,0,–0.01), respectively. Find (a) $\boldsymbol{E}$ and (b) $|E|$ due to this dipole.

SOLUTION

$$Q = \pm 2 \times 10^{-9},\ d = 0.02,\ Qd = 0.04 \times 10^{-9},\ 4\pi\varepsilon_o = \frac{10^{-9}}{9}$$

$$2\cos\theta = 2\cos 60° = 2 \times 0.5 = 1, \quad \sin\theta = \sin 60° = 0.866,$$

In view of Equation 7.33b:

$$\boldsymbol{E} = \frac{Qd}{4\pi\varepsilon_o \cdot r^3} \cdot \left(2\cos\theta \cdot \boldsymbol{a}_r + \sin\theta \cdot \boldsymbol{a}_\theta\right) = \frac{0.04 \times 10^{-9}}{\frac{10^{-9}}{9}} \cdot \frac{1}{r^3}\left(\boldsymbol{a}_r + 0.866\boldsymbol{a}_\theta\right)$$

$$= \frac{1}{r^3}\left(0.36\boldsymbol{a}_r + 0.312\boldsymbol{a}_\theta\right)$$

(a) $r = 0.5 \quad \boldsymbol{E} = \frac{1}{(0.5)^3}\left(0.36\boldsymbol{a}_r + 0.311\boldsymbol{a}_\theta\right) = 2.88\boldsymbol{a}_r + 2.488\boldsymbol{a}_\theta$

(b) $|\boldsymbol{E}| = \left\{(2.88)^2 + (2.488)^2\right\}^{1/2} = \sqrt{8.2944 + 6.1901} = 3.8 \text{ V/m}$

Example 7.11

A potential field in free space is given by $V = 10xy^2 + 20yz^2 + 30zx^2$. Find (a) V; (b) E; (c) ρ; (d) dV/dN; and (e) $\boldsymbol{a}_N$, at (1,1,1).

SOLUTION

Given $V = 10xy^2 + 20yz^2 + 30zx^2$

(a) $V\big|_{x=1,y=1,z=1} = \left(10xy^2 + 20yz^2 + 30zx^2\right)_{x=1,y=1,z=1} = 10 + 20 + 30 = 60 \text{ V}$

(b) In view of Equation 7.6:

$$\boldsymbol{E} = -\nabla V = -\left[\frac{\partial V}{\partial x}\boldsymbol{a}_x + \frac{\partial V}{\partial y}\boldsymbol{a}_y + \frac{\partial V}{\partial z}\boldsymbol{a}_z\right]$$

$$\boldsymbol{E} = -\left[\left(10y^2 + 60zx\right)\boldsymbol{a}_x + \left(20xy + 20z^2\right)\boldsymbol{a}_y + \left(40yz + 30x^2\right)\boldsymbol{a}_z\right]$$

$$\boldsymbol{E}\Big|_{x=1,y=1,z=1} = -\left(70\boldsymbol{a}_x + 40\boldsymbol{a}_y + 70\boldsymbol{a}_z\right) \text{ V/m}$$

(c) In view of Equations 6.19a and 6.20a:

$$\boldsymbol{D} = \varepsilon_0 \boldsymbol{E} = -\varepsilon_0 \left[\left(10y^2 + 60zx\right)\boldsymbol{a}_x + \left(20xy + 20z^2\right)\boldsymbol{a}_y + \left(40yz + 30x^2\right)\boldsymbol{a}_z\right]$$

$$\rho = \nabla \bullet \boldsymbol{D} = -\varepsilon_0 \left\{\frac{\partial D_x}{\partial x} + \frac{\partial D_y}{\partial y} + \frac{\partial D_z}{\partial z}\right\}$$

$$= -\varepsilon_0 \left\{\frac{\partial\left(10y^2 + 60zx\right)}{\partial x} + \frac{\partial\left(20xy + 20z^2\right)}{\partial y} + \frac{\partial\left(40yz + 30x^2\right)}{\partial z}\right\}$$

$$= -\varepsilon_0 \left[(60z) + (20x) + (40y)\right]_{x=1,y=1,z=1}$$

$$= -120\varepsilon_0 = -120 \times 8.854 \times 10^{-12} = -1.0628 \text{ nC/m}^2$$

(d) In view of Equation 4.8e, $\boldsymbol{E} = -\dfrac{dV}{dN}\boldsymbol{a}_N$

$$\boldsymbol{E}\Big|_{x=1,y=1,z=1} = -\left(70\boldsymbol{a}_x + 40\boldsymbol{a}_y + 70\boldsymbol{a}_z\right)$$

$$\left|\boldsymbol{E}\right|_{x=1,y=1,z=1} = -\frac{dV}{dN}$$

$$\frac{dV}{dN} = \left|\boldsymbol{E}\right| = \sqrt{4900 + 1600 + 4900} = \sqrt{10400} = 106.77 \; V/m$$

(e) $$\boldsymbol{a}_N = \frac{\boldsymbol{E}\Big|_{x=1,y=1,z=1}}{\left|\boldsymbol{E}\right|} = \frac{-\left(70\boldsymbol{a}_x + 40\boldsymbol{a}_y + 70\boldsymbol{a}_z\right)}{106.77}$$

$$= -\left(0.6556\boldsymbol{a}_x + 0.3746\boldsymbol{a}_y + 0.6556\boldsymbol{a}_z\right)$$

Example 7.12

Figure 7.10 shows two point charges Q_1 and Q_2 located at points $P_1(0,0,c)$ and $P_2(0,0,-c)$, respectively. If $Q_1 = -Q$ and $Q_2 = +Q$, show that the energy stored in the electrostatic field remains unaltered if a third point charge $Q_3 = \mathbb{Q}$, is placed at a point $P_3(a,b,0)$. Find the potential distribution on the $z = 0$ plane. Show that it is independent of the value of Q, and also the z-coordinate.

SOLUTION

Since the $z = 0$ plane is a zero potential surface thus the potential energy of the charge Q_3 is zero. Therefore, the energy stored in the electrostatic field remains unaltered, even though the $z = 0$ plane with the arrival of the charge Q_3 no longer remains equipotential. However, because of the potential field due to the charge Q_3 the potential energy of the first two charges are modified, while their sum remains unchanged.

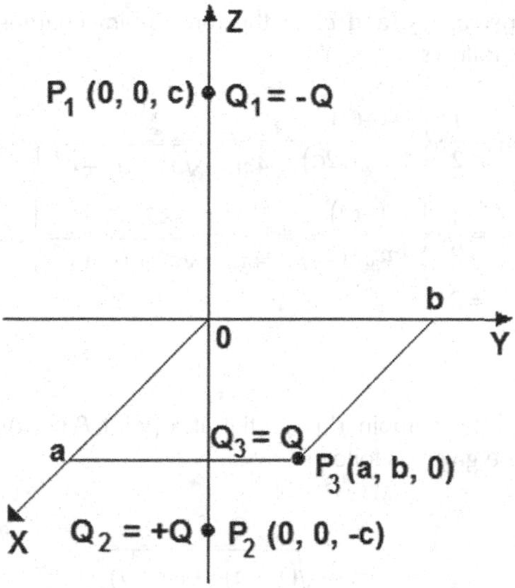

FIGURE 7.10 Location of charges and the field point.

On substituting expressions for electrostatic potentials in Equation 7.51, we get:

$$W = \frac{1}{8\pi\varepsilon_o}\cdot\left[\left\{\frac{Q_2}{(2c)}+\frac{Q_3}{\sqrt{a^2+b^2+c^2}}\right\}\cdot Q_1+\left\{\frac{Q_3}{\sqrt{a^2+b^2+c^2}}+\frac{Q_1}{(2c)}\right\}\cdot Q_2\right.$$

$$\left.+\left\{\frac{Q_1}{\sqrt{a^2+b^2+c^2}}+\frac{Q_2}{\sqrt{a^2+b^2+c^2}}\right\}\cdot Q_3\right]$$

On inserting the given values for the point charges, we get:

$$W = \frac{1}{8\pi\varepsilon_o}\cdot\left[\left\{\frac{Q}{(2c)}+\underbrace{\frac{Q}{\sqrt{\boldsymbol{a}^2+\boldsymbol{b}^2+\boldsymbol{c}^2}}}\right\}\cdot(-Q)+\left\{\underbrace{\frac{Q}{\sqrt{\boldsymbol{a}^2+\boldsymbol{b}^2+\boldsymbol{c}^2}}}+\frac{(-Q)}{(2c)}\right\}\cdot Q\right.$$

$$\left.+\left\{\overbrace{\frac{\boldsymbol{(-Q)}}{\sqrt{\boldsymbol{a}^2+\boldsymbol{b}^2+\boldsymbol{c}^2}}}+\overbrace{\frac{\boldsymbol{Q}}{\sqrt{\boldsymbol{a}^2+\boldsymbol{b}^2+\boldsymbol{c}^2}}}\right\}\cdot Q\right]$$

Each pair of terms shown in $\underbrace{\textbf{\textit{bold}}}$ / $\overbrace{\textbf{\textit{bold}}}$ are canceled out, thus resulting in the total energy stored, W, as independent of the third charge q:

$$W = -\frac{Q^2}{4\pi\varepsilon_o\cdot(2c)}$$

The potential energy e_1, e_2, and e_3 of the three point charges Q_1, Q_2, and Q_3 respectively, are as follows:

$$\begin{cases} e_1 = \dfrac{1}{2} \cdot \left\{ \dfrac{(-Q)}{4\pi\varepsilon_o \cdot (2c)} - \dfrac{Q}{4\pi\varepsilon_o \cdot \sqrt{a^2 + b^2 + c^2}} \right\} \cdot Q \\ e_2 = \dfrac{1}{2} \cdot \left\{ \dfrac{(-Q)}{4\pi\varepsilon_o \cdot (2c)} + \dfrac{Q}{4\pi\varepsilon_o \cdot \sqrt{a^2 + b^2 + c^2}} \right\} \cdot Q \\ e_3 = 0 \end{cases}$$

It may be seen that the combined potential at a point $P_0(x,y,0)$, due to all three point charges can be given as follows:

$$V_0 = \frac{Q}{4\pi\varepsilon_o \cdot \sqrt{(x-a)^2 + (y-b)^2}}$$

Due to the presence of the point charge Q, the x–y-plane ceases to be an equipotential surface. This potential distribution on the x–y-plane is independent of the charges $\mp Q$ and the z-coordinates $\pm c$ of their locations.

In this example, we find that the total potential energy W stored in the point charges, as given by Equation 7.4, is a negative quantity. Therefore, the *energy stored in the electrostatic field spread over the infinite space is also negative.*

PROBLEMS

P7.1 A 1 mm long differential linear path is directed along a vector $2\boldsymbol{a_x} - 3\boldsymbol{a_y} - 4\boldsymbol{a_z}$. Find the work done in moving a –5 μC charge along this path in the field of $\boldsymbol{E} = 2\boldsymbol{a_x} - 3y^2\boldsymbol{a_y} + x\boldsymbol{a_z}$ V/m if the path is located at (a) $P_a = (1,1,3)$; (b) $P_b = (2,0,-2)$; (c) $P_c = (3,1,-4)$.

P7.2 Find the work done in moving a 20 μC charge in the field $\boldsymbol{E} = 3x^2\boldsymbol{a_x} + 5y^2\boldsymbol{a_y} - 2z^2\boldsymbol{a_z}$ V/m, along a 10^{-3} m incremental path directed along $-0.5\boldsymbol{a_x} + 0.5\boldsymbol{a_y}$ $0.5\boldsymbol{a_z}$ and located at (a) (1,1,2) and (b) (2,–1,–1).

P7.3 Find the work done in carrying 5C charge from $B(1,1,1)$ to $A(0.5,0.5,0.5)$ along the path identified by $x^2 + y^2 = 1$, $z = 1$ in a field $\boldsymbol{E} = y\boldsymbol{a_x} + x\boldsymbol{a_y} + z\boldsymbol{a_z}$.

P7.4 Find the work done in carrying 10 μC charge from (0,0,0) to (1,2,3) in the vector field $\boldsymbol{E} = 2x^2y\boldsymbol{a_x} - 3x^2\boldsymbol{a_y} + 4z\boldsymbol{a_z}$ V/m along (a) straight line segments (0,0,0) to (1,0,0) to (1,2,0) to (1,2,3); (b) a straight line specified by $y = 2x$, $z = 3x$; (c) a curve characterized by $y = 2x$, $z = 3x^4$.

P7.5 Find the work done in carrying 5C charge from $B(0.6,0.4,0.8)$ to $A(1,1,1)$ along the straight line path, $z = 1$ in a non-uniform field $\boldsymbol{E} = y^2\boldsymbol{a_x} + x^2\boldsymbol{a_y} + 5z\boldsymbol{a_z}$.

P7.6 A point charge of 5nC is located at the origin in free space. Find the potential at $r = 0.5$ m if the: (a) zero reference is at infinity; (b) zero reference is at $r = 0.8$ m; (c) $V = 5$ V at $r = 1$.

P7.7 If the potential difference between point A and B located at (1,0,0) and (2,0,0) respectively is 10 V determine c for point $C(c,0,0)$ such that $V_{BC} = 3$ V. Assume the source of field to be a point charge located at the origin.

P7.8 In a relation $V = \dfrac{Q}{4\pi\varepsilon_0 r} + k^2$ find k if $Q = 0.5 \times 10^{-6}$C and (a) $V = 0$ volt at $r = 5$ (b) $V = 100$ volt at $r = 10$.

P7.9 The potential fields of $+Q$ and $-Q$ located at (0,0,5) and (0,0,–5), respectively, has zero reference at infinity. Find the ratio of the total potential at (0,0,10) to that at (a) (0,0,20); (b) (0,0,30); and (c) (0,0,40).

P7.10 Find the potential at (0,0,20) due to the total charge 5×10^{-9}C which is uniformly distributed along a configuration lying on the z = 0 plane and centered the origin in the form of a disc of 4 m radius.

P7.11 Find the potential in free space at 0,0,4 due to: (a) a ring of radius $\rho = 3$, at $z = 0$ with $\rho_l = 1$nC/m; (b) a disc $0 \leq \rho \leq 3$, at $z = 0$ with $\rho_s = 2$nC/m²; (c) a disc $2 \leq \rho \leq 3$, at $z = 0$ with $\rho_s = 3$nC/m². Configurations for parts a, b, and c are shown in Figure 7.3

P7.12 A dipole moment at (0,0,0) in free space is given as $\boldsymbol{p} = 100\pi\varepsilon_0 \left(0.3\boldsymbol{a}_x + 0.5\boldsymbol{a}_y - 0.4\boldsymbol{a}_z\right)$ C-m. Find the potential V at (a) (5,0,0); (b) (0,4,0); (c) (0,0,3); (d) (5,4,3).

P7.13 A dipole moment is given free space as $p = 20\pi\varepsilon_0 \left(0.5a_x - 0.7a_y + 0.9a_z\right)$ Cm. Find the potential at: (a) $P_A(0,0,5)$; (b) $P_B(0,5,0)$; (c) $P_C(5,0,0)$; (d) $P_D(1,2,3)$.

P7.14 A dipole is formed in free space by two charges of 1nC and –1nC located at (0,0,0.01) and (0,0,–0.01), respectively. Find $\boldsymbol{E}$ and $|\boldsymbol{E}|$ at $P\left(r = 0.5, \theta = 60°, \varphi = 0°\right)$.

P7.15 Find $\boldsymbol{E}$ and $|\boldsymbol{E}|$ due to a dipole in free space formed at $r = 0.8, \theta = 30°$, $\varphi = 30°$ by a charge of 5 nC at (0,0,0.02) and –5 nC at (0,0,–0.02).

P7.16 The potential field in free space is given as $V = 5x^2y^2 + 10y^2z^2 - 5z^2x^2$. Find at (1,2,–1) (a) V; (b) $\boldsymbol{E}$; (c) ρ; (d) dV/dN; and (e) $\boldsymbol{a}_N$.

P7.17 A potential field $V = 10x^2yz + 5y_2$V is given in free space. Find: (a) V at P(2,1,3); (b) E_P; (c) ρ_P; (d) dV/dN at P; (e) a_N at P.

P7.18 Find the energy stored in free space in the spherical region $r \leq 10$ for the potential field given as (a) $V = 5r^2$ (b) $V = 5r^2\sin\theta$.

P7.19 Three point charges Q_1, Q_2, and Q_3 located at corners of an equilateral triangle marked as P_1, P_2, and P_3 respectively. The length of each side is ℓ. If $Q_1 = Q_2 = +Q$, and $Q_3 = -Q$, find the potential energy of each charge.

P7.20 Figure 7.5 shows three point charges. Charges Q_1 and Q_2, each of value Q, are placed at $P_1(-a,0,0)$ and $P_2(+a,0,0)$ respectively, and a third point charge $Q_3 = -Q$ is located at $P_3(0,y,0)$. Find the potential energy $ⓔ_1$, $ⓔ_2$ and $ⓔ_3$ of these charges if (a) $y = 0$; (b) $y = \infty$; (c) Find the value of y for zero net potential energy.

DESCRIPTIVE QUESTIONS

Q7.1 Differentiate between force, work done, energy expended and the potential difference.

Q7.2 Explain the concept of infinity vis-à-vis the definition of potential.

Q7.3 Obtain the relation between the electric potential and electric field intensity.

Q7.4 Draw the equipotential surfaces due to two equal point charges of the same polarity.

Q7.5 Draw the equipotential surfaces due to two parallel uniform line charges with opposite polarity and the same charge density.

Q7.6 Draw the equipotential surfaces due to two parallel uniform surface charges with same polarity and the same charge density.

Q7.7 Derive the relations for electric potential and electric field intensity due to a dipole.

REFERENCES

1. William H. Hayt, Jr., *Engineering Electromagnetics*, 5th Ed, Tata McGraw-Hill Publishing Company Limited, New Delhi, pp. 106–108, 1997.
2. Matthew N. O. Sadiku, *Principles of Electromagnetics*, 4th Ed, Oxford University Press, New Delhi & Oxford, pp. 114–115, 2009.
3. David J. Griffiths, *Introduction to Electrodynamics*, 3rd Ed, PHI Learning Private Limited, New Delhi, p. 458, 2012.
4. ibid, p. 29.

FURTHER READING

More references are given at the end of Chapter 15.

8 Electrical Field in Materials

8.1 INTRODUCTION

In Chapter 7, a new quantity referred to as potential was introduced and its relations to different types of charge distributions and the electric field intensity were discussed. This quantity is frequently used even though the user may not be aware of its technical aspects and real meaning. Similarly, there is another quantity that is also known to the masses is the current. This chapter describes the meaning, types, and relations associated with the current and the current density. This chapter also includes a brief description of the nature of conducting, semiconducting, dielectric, and superconducting materials. This chapter also describes the conditions satisfied by different field quantities at the boundaries between different types of materials.

8.2 ELECTRICAL CURRENT AND CURRENT DENSITY

This section describes two commonly used terms, *current* and *current density.*

8.2.1 Electrical Current

The flow of electric charges is referred to as an *electric current.* This charge is carried by moving electrons in wires, by ions in electrolytes and by both (ions and electrons) in plasma. In electronic devices, the electric current may be caused due to the flow of electrons through resistors or vacuum in vacuum tubes, flow of ions in a battery or a neuron, and the flow of holes within a semiconductor. The current is usually denoted by the symbol I and is measured in *ampères* (abbreviated by A) through a device called *ammeter.* The one ampere current is the flow of electric charge across a surface at the rate of one coulomb per second. The electric currents can have many effects including generation of heat and creation of magnetic field used in inductors, motors, and generators. In wires and conductors made of metals, positive charges are immobile and charge carriers are electrons. As the electrons carry negative charge, their motion in a metal conductor is in the direction opposite to that of conventional current, or the direction of current is taken in the opposite direction to that of the movement of electrons. The current can be classified in many ways. Some of these are discussed in the following subsections.

8.2.1.1 Direct and Alternating Currents

Current can be classified as direct current (dc) or alternating current (ac). In *direct current*, the flow of electric charges is unidirectional. Such currents are produced by batteries, thermocouples, solar cells, and by commutator-type electrical machines.

Direct current may not only flow in conductors but also through semiconductors or even through a vacuum as in electron or ion beams. This type of current is also called the *galvanic current*. The current flowing in a conductor is called *conduction current*. Electron or ion beams constitute what is called *convection current*. In *alternating current*, the direction of movement of charge carriers and hence of the current reverses periodically. The flow of an oscillating electric current in an antenna transmits power as an electromagnetic wave. Oscillating electric current indicates acceleration and retardation of free electrons in the current carrying conductor. In 1897, Joseph Larmor derived the following relation from Maxwell's equations:

$$\boxed{P = \frac{\mu_o \cdot q^2 \cdot a^2}{6\pi \cdot c}} \tag{8.1}$$

This gives the power P radiated by accelerated charged particle. In this relation "a" is the *acceleration*, "q" is the *charge*, "c" is the velocity of light in free space, and "μ_o" is the permeability of free space (discussed in the next chapter).

8.2.1.2 Conduction, Convection, and Displacement Currents

Current can also be classified as conduction, convection, and displacement current. The *conduction current* is the movement or flow of free electrons in a conductor due to an application of electric field. It is the ordered movement of electrons in the conductor. The *convection current* is the flow of charges (electrons and/or ions) in a fluid or in vacuum subjected to an electric field. *Convection currents* may also flow on conductor surfaces carrying distributions of free surface charges. Lastly, the *displacement current;* a quantity which first appeared in Maxwell's equations, is defined in terms of the rate of change of electric displacement field (in a limited sense, also known as the field of electric flux density). It has an associated magnetic field just as actual currents do. It is not an electric current of moving charges but instead a time-varying electric field. In materials, there is also contribution toward the *displacement* current from the slight motion of bound charges or due to dielectric polarization caused by time-varying electric field.

8.2.1.3 Eddy Current and Surface Current

A current may be caused due to a natural phenomenon such as lightning, solar winds, etc. A current may also be caused by manmade systems. The flow of electrons in overhead power lines and within electrical and electronic equipment are some examples of manmade occurrences. latter case it is called *conduction current*. When a conductor is exposed to a changing magnetic field, it gives rise to another form of current called *eddy current*. *Time-invariant magnetic fields* produce eddy currents on the surfaces of *superconductors*, blocking the entry of magnetic flux in the volume. *Surface currents* may result due to the exposure of a conductor to an electromagnetic wave.

8.2.2 Current Density

In the preceding section, current was classified in many ways. This section deals with *current density*. A detailed discussion on displacement current density is included in Chapter 15, and that of eddy current density in Chapter 21.

Current density is normally denoted by the letter $\boldsymbol{J}$. It is a measure of the density of an electric current. It is defined as a vector whose magnitude is the electric current (I) per cross-sectional area of a conductor, and its direction is the same as the direction of the current flow. It is measured in amperes per square meter (A/m^2). The current density and the current flowing in a conductor with a cross-sectional area "s" are related by an integral equation:

$$I = \iint_s \boldsymbol{J} \bullet d\boldsymbol{s} \tag{8.2}$$

The volume and surface current density, $\boldsymbol{J}$ and $\boldsymbol{k}$, are defined respectively as:

$$\boldsymbol{J} \overset{\text{def}}{=} \lim_{\Delta s \to 0} \frac{\Delta I}{\Delta s} \boldsymbol{a}_n \text{ A/m}^2 \tag{8.2a}$$

and

$$\boldsymbol{K} \overset{\text{def}}{=} \lim_{\Delta \ell \to 0} \frac{\Delta I}{\Delta \ell} \boldsymbol{a}_n \text{ A/m} \tag{8.2b}$$

In Equation 8.2a, ΔI is the small current flowing through a small area Δs, the unit vector $\boldsymbol{a}_n$ is normal to the area Δs and in the direction of the flow of the current ΔI. In Equation 8.2b, ΔI is the small current flowing across a small line segment $\Delta \ell$, the unit vector $\boldsymbol{a}_n$ is normal to this line segment and in the direction of the flow of the current ΔI.

8.2.2.1 Convection Current Density

A convection current occurs in insulators or dielectrics such as liquid, vacuum, and (rarified) gas. It results from the motion of electrons or ions in an insulating fluid. This current doesn't satisfy hm's law, as it does not involve conductors.

Figure 8.1 shows an elemental volume Δv with elemental length ΔL and elemental cross-sectional area ΔS. This volume contains a continuous charge distribution with charge density ρ_v(or simply ρ). In Equation 5.38a and this figure the incremental charge ΔQ in the volume can be given as:

$$\Delta Q = \rho_v \cdot \Delta v = \rho \cdot \Delta s \cdot \Delta L \tag{8.3a}$$

Assume that these charges are moving in the x-direction with a velocity $\boldsymbol{U} = U_x \boldsymbol{a}_x$. In time Δt the charge contents $\Delta Q'$ in a volume $\Delta s \cdot \Delta x$ move along the x-axis. This shifting is indicated by the new location of ΔS. The charge contents of this shifted volume leave the original volume Δv and can be given by:

$$\Delta Q' = \rho \cdot \Delta s \cdot \Delta x \tag{8.3b}$$

This shifting of charges in time Δt is regarded as current ΔI, which can be written as:

$$\Delta I = \frac{\Delta Q'}{\Delta t} = \rho \cdot \Delta s \cdot \frac{\Delta x}{\Delta t} \tag{8.4a}$$

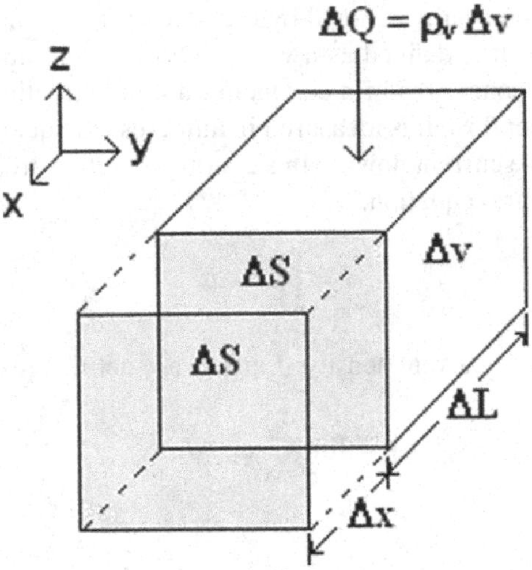

FIGURE 8.1 Movement of charges in an elemental volume.

Thus, from Equation 8.4a:

$$\Delta I = \rho \cdot \Delta s \cdot U_x \tag{8.4b}$$

Although current has a direction of flow, it is always treated as a scalar quantity. A new quantity, therefore, needs to be defined to spell the direction of current. As noted above, such a quantity is denoted by $\boldsymbol{J}$ and is referred to as current density with the unit of ampere per unit area. In this particular case, the movement of charges is assumed to be only in the x-direction and thus current-density $\boldsymbol{J}$ will only have an x-component (denoted by J_x). From Equation 8.4b, this can be written as:

$$J_x = \lim_{\Delta s \to 0} \frac{\Delta I}{\Delta s} = \rho \cdot U_x \tag{8.5}$$

In this particular case, we have assumed the movement of charges along the x-axis. If a similar movement is assumed in y or z directions, the resulting current density will have J_y or J_z components. To generalize the result let us assume that charges are moving in all the three positive directions of the space coordinates (viz. x, y, and z). This situation leads to the relation which spells the convection current density.

$$\boldsymbol{J} = \rho \cdot \boldsymbol{U} \tag{8.6}$$

In this relation,

$$\boldsymbol{U} = U_x \cdot \boldsymbol{a}_x + U_y \cdot \boldsymbol{a}_y + U_z \cdot \boldsymbol{a}_z \tag{8.7}$$

8.2.2.2 Conduction Current Density

It occurs in conductors where there are a large number of free electrons, which act as charge carriers. The conduction current obeys Ohm's law. When an external electric field is applied to a metallic conductor, conduction current occurs due to the drift motion of electrons. The moment an electric field is applied, the charges inside the conductor experience force. This force compels the charges to move and tries to accelerate their motion,but due to continuous collision with atomic lattice, their velocity is reduced. As a result, the electrons move or drift with an average velocity called the drift velocity (v_d). This velocity is proportional to the applied electric field ($\boldsymbol{E}$). Thus, in accordance with Newton's law, if an electron with a mass m is moving in an electric field $\boldsymbol{E}$ with an average drift velocity v_d, the average change in momentum of the free electron must be equal to the applied force ($\boldsymbol{F} = -e\boldsymbol{E}$); the symbol ($-e$) stands for the charge on each electron. If τ is the average time interval between successive collisions the two forces can be equated to yield:

$$\frac{m\boldsymbol{v}_d}{\tau} = -e\boldsymbol{E} \tag{8.8a}$$

From Equation 8.8a:

$$v_d = -\left(\frac{e\tau}{m}\right)\boldsymbol{E} \stackrel{\text{def}}{=} -\mu_e\boldsymbol{E} \tag{8.8b}$$

The RHS of Equation 8.7b involves a new quantity (μ_e), which is called the *mobility of electrons*. It can be defined as the drift velocity per unit applied electric field. From Equation 8.8b, the mobility of electrons is:

$$\mu_e = \frac{e\tau}{m} \tag{8.9}$$

In a conducting wire, the charges subjected to an electric field move with the drift velocity v_d. If there are n_e number of free electrons per cubic meter of conductor, then the free volume charge density(ρ_v or ρ) within the wire can be given as:

$$\rho = -e \cdot n_e \tag{8.10}$$

The charge $\Delta Q'$ in the shifted incremental volume $\Delta v'$ shown in Figure 8.1 can be given as:

$$\Delta Q' = \rho \cdot \Delta v' = -e \cdot n_e \cdot \Delta s \cdot \Delta x = -e \cdot n_e \cdot \Delta s \cdot v_d \cdot \Delta t \tag{8.11}$$

In this equation since charges are drifting with velocity v_d and will shift by distance Δx in time Δt thus Δx is replaced by $v_d{\cdot}\Delta t$ as $v_d = \Delta x/\Delta t$.

The value of incremental current can thus be written as:

$$\Delta I = \frac{\Delta Q'}{\Delta t} = -e \cdot n_e \cdot \Delta s \cdot v_d \tag{8.12a}$$

In Equations 8.8b and 8.12a we get:

$$\Delta I = e \cdot n_e \cdot \Delta s \cdot \mu_e E \tag{8.12b}$$

Finally, the magnitude of the *conduction current*-density (J_c) can be defined as:

$$J_c = \frac{\Delta I}{\Delta s} = e \cdot n_e \cdot \mu_e \cdot E \tag{8.13a}$$

The current density (on replacing J_c by J) can be written in the following commonly used form:

$$\boldsymbol{J} = \sigma \cdot \boldsymbol{E} \tag{8.13b}$$

In Equation 8.13b, σ is referred to as the *conductivity* of the material. It is measured in mhos per meter. Its value from Equations 8.13a, 8.13b, and 8.10 is:

$$\sigma = e \cdot n_e \cdot \mu_e = -\rho \cdot \mu_e \tag{8.14a}$$

In semiconductors, the current flow is due to the movement of both electrons and holes, thus the conductivity is given as:

$$\sigma = -\rho_e \cdot \mu_e + \rho_h \cdot \mu_h \tag{8.14b}$$

Where ρ_e and ρ_h indicate charge densities for electrons and holes, while μ_e and μ_h indicatethe mobility of free electrons and holes respectively. The values of μ_e for copper and aluminum, the two main conducting materials, are 0.0052 and 0.0014, respectively.

8.2.2.3 Drift Velocity

This term drift velocity given by Equation 8.8b needs some clarification. It was earlier noted that there is a large number of free electrons in conducting material. These electrons act as charge carriers in metals. Like the particles of a gas, these mobile (or free) charged particles keep on continuously moving in random directions within conductors. For a net flow of charge, these particles must also move together with an average drift rate. These electrons follow an erratic path, bouncing from atom to atom, but generally drifting in the opposite direction to that of the electric field. In order to estimate the speed at which they drift let us assume that there are n number of charged particlesper unit volume (or charge carrier density), e is the charge on each particle, s is the cross-sectional area of the conductor, and $I$is the electric current then the drift velocity v_d can be given by the relation:

$$v_d = \frac{I}{n \cdot e \cdot s} \tag{8.15}$$

The substitution of suitable values of the parameters involved in Equation 8.15 reveals that the movement of electric charges in solids is quite slow. As described in Section 5.2.8,1 cm^3 of copper contains about 8.5×10^{22} free electrons (each with charge of -1.6×10^{-19}C). The multiplication $n{\cdot}e$ gives a figure of 13600C of potentially mobile charge in 1 cm^3 or 13.6C in 1 mm^3. In a copper wire of cross-sectional area of 0.5 mm^2, carrying 5A current the electrons drift with a velocity of $5/(13.6 \times 0.5) =$ 0.735 mm/s. In a cathode ray tube having near-vacuum inside, the electrons adopt near-straight path. In these tubes the electrons travel at about a tenth of the speed of light.

An accelerating charge or the changing electric current, gives rise to an electromagnetic wave. Such wave propagates at very high speed outside the surface of the conductor. This speed is a significant fraction of the speed of light and many times faster than the drift velocity of the electrons. In ac power lines the energy is carried by the electromagnetic wave from the source to the destination. This wave propagates through the space between the wires at a significant speed but the electrons in the wires only move (or drift) back and forth over a tiny distance.

8.3 CURRENT FLOW IN METALLIC CONDUCTORS

A conductor is an object or type of material that permits the flow of electrical current in one or more directions. In metallic conductors such as those made of copper or aluminum, the movable charged particles are electrons. In these conductors the current distributes itself in different forms and follows the ohms law as per the description given below.

8.3.1 Forms of Current Distributions

Before we switch over to the magnetic field laws which spell the effect of the electric currents it seems to be appropriate to have a look at the type of current distributions we may come across in the practical applications. These are briefly described below.

8.3.1.1 Line Current

The line current is the flow of a current in a uniform cylindrical conductor of circular cross-section. The filamentary current is the limiting case of line current when the radius of this conductor approaches zero. Both the line and filamentary current denoted by word I is taken to be a scalar quantity which is measured in ampères (A). Unless there is some outlet (in between) for the flow or leakage of the current it can be assumed to be uniform. Figure 8.2a illustrates a line (or filamentary) current configuration.

8.3.1.2 Linear (or Surface) Current Density

Figure 8.2b illustrates a current in the distributed form over a surface of a conductor of finite length and breadth. This current is confined to a vanishingly small thickness. This distributed form of the current is referred to as surface current density. It is denoted by letter ***K***, which is taken as a vector quantity. This current density may be uniform or non-uniform and is measured in ampère per meter (A/m). At a given point P on the surface, the surface current density is defined as follows:

$$\boldsymbol{K} \stackrel{\text{def}}{=} \lim_{\Delta L \to 0} \frac{\Delta I}{\Delta L} \boldsymbol{a} \tag{8.16a}$$

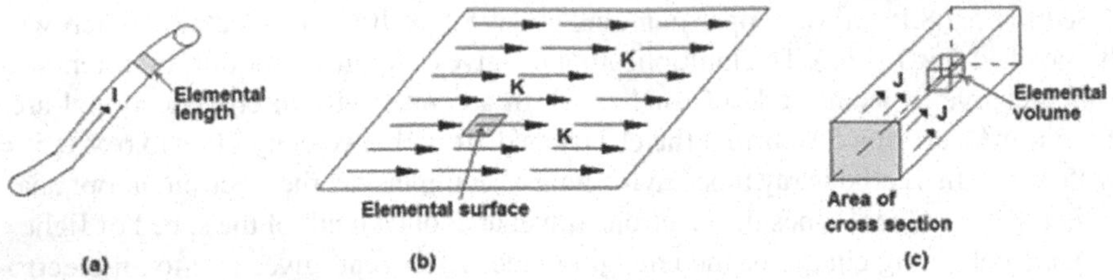

FIGURE 8.2 (a) Line current I; (b) surface current-density K; (c) current-density **J**.

where ΔL indicates an elemental length, at the point P, on the surface oriented normal to the direction of the small current ΔI flowing across this elemental length. The vector $\boldsymbol{a}$ is the unit vector showing the direction of the current flow at the point P.

8.3.1.3 (Volumetric) Current Density

The current density (also referred to as volumetric current density) is obtained by dividing a current by the area of cross-section across which it is flowing at a particular location. This may be uniform or non-uniform across the cross-section. This current density is denoted by a vector quantity $\boldsymbol{J}$ and is measured in ampere per meter square (A/m^2). If one of the dimensions of cross-section becomes zero the current density becomes infinite. Figure 8.2c illustrates the current-density $\boldsymbol{J}$ across the cross-section of a rectangular conductor.

At a given point P in the volume, the current-density $\boldsymbol{J}$ is defined as follows:

$$\boldsymbol{J} \overset{\text{def}}{=} \lim_{\Delta S \to 0} \frac{\Delta I}{\Delta S} \boldsymbol{a} \tag{8.16b}$$

where ΔS indicates an elemental surface area, at the point P; the small current ΔI is flowing in a perpendicular direction across this elemental surface area. The vector $\boldsymbol{a}$ is the unit vector showing the direction of the current flow at the point P. The reverse relation is:

$$\iint_S \boldsymbol{J} \bullet \boldsymbol{ds} = I \tag{8.16c}$$

where the current I is flowing across the surface S, with a current-density $\boldsymbol{J}$ on the surface.

8.3.2 Ohm's Law

Equation 8.13b is known as *Ohm's law in point form* and is valid at every point in space. To understand the essence of this law, consider Figure 8.3.

This figure shows a metallic conductor with cross-section S, length L, and conductivity σ. It carries a current I under the influence of potential difference V. Equation 8.1 with the uniform current-density J can be written as:

$$I = \iint_S \boldsymbol{J} \bullet \boldsymbol{ds} = \boldsymbol{J} \bullet \iint_S \boldsymbol{ds} = J \cdot S \tag{8.17a}$$

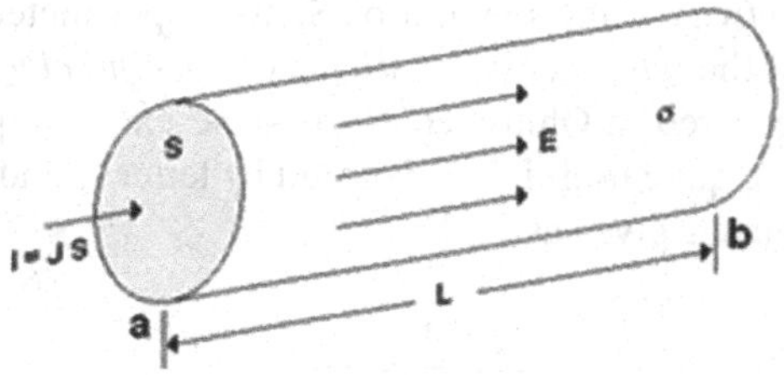

FIGURE 8.3 Current in a metallic conductor.

or

$$J = \frac{I}{S} \tag{8.17b}$$

In Equation 7.8 the potential difference between points a and b of the conductor can be written as:

$$V_{ab} = -\int_b^a \mathbf{E} \bullet d\mathbf{L} = -\mathbf{E} \bullet \int_b^a d\mathbf{L} = -\mathbf{E} \bullet \mathbf{L}_{ba} = \mathbf{E} \bullet \mathbf{L}_{ab} \tag{8.18a}$$

On replacing V_{ab} by V and $\mathbf{L}_{ab}$ by L we get:

$$V = E \cdot L \tag{8.18b}$$

or

$$E = \frac{V}{L} \tag{8.18c}$$

In Equations 8.17b and 8.18c:

$$J = \frac{I}{S} = \sigma E = \sigma \frac{V}{L} \tag{8.19a}$$

or

$$V = I \cdot \frac{L}{\sigma S} \tag{8.19b}$$

From Equation 7.18b, we get:

$$V = I \cdot R \tag{8.20a}$$

8.3.2.1 Resistance and Conductance

In Equation 8.20a, the letter R indicates a quantity called the resistance. It is the first of the three (passive) circuit parameters used in circuit theory. From Equations 8.20a and 8.19b, the resistance can be given as:

$$R = \frac{L}{\sigma S} \tag{8.20b}$$

The *electrical conductivity*, σ has the unit of Siemens per meter. The reciprocal of *conductivity* is called the (*electrical*) *resistivity* or *specific electrical resistance* of the material. It is measured in Ohm-meters. In some cases, a quantity called *conductance* is found to be quite useful. It is denoted by letter G and is the reciprocal of resistance R. It can thus be given as:

$$G = \frac{1}{R} = \frac{\sigma S}{L} \tag{8.20c}$$

Equation 8.20a is another form of Ohm's law. This law involves terms V (the potential difference in volts), I (the current in ampères), and R (the resistance in ohms). These terms are commonly used in circuit theory. This law states that the current is directly proportional to the potential difference between two ends (across) of a material, called a resistor (or any other ohmic device). Non-linear materials with conductivity varying with the current through or voltage across it do not obey Ohm's law.

Equation 8.20b is also a well-known relation which gives the value of resistance in terms of the area of cross-section "S" (in square meters) and length "L" (in meters) of the conductor and the *electrical conductivity* "σ" or the *electrical resistivity* "($1/\sigma$)" of the material. Resistivity and conductivity are reciprocal. Resistivity is defined as a measure of the material's ability to oppose the flow of electric current. From Equation 8.20b, it can be concluded that the resistance of a conductor depends on the material and its dimensions. For a given material, the resistance of a conductor is inversely proportional to the cross-sectional area and directly proportional to the length.

While deriving the above relations, the current density is assumed to be totally uniform in the conductor which can be taken as true in most of the practical situations. This condition prevails in linear materials such as metals, and under low frequencies. In the case of alternating current, these relations do not yield the true value of resistance due to *skin effect* and *proximity effect* which are briefly described below. Further description of these effects is included after the introduction of inductance in Chapter 14.

8.3.2.2 Skin Effect

The *skin effect* inhibits the flow of alternating current near the center of the conductor, resulting in an uneven distribution of the current (density) in the conductor section. Thus the *effective* cross-section in which current actually flows reduces and the resistance becomes higher than that in case of direct current.

8.3.2.3 Proximity Effect

The *proximity effect* is observed when two conductors carrying alternating currents are near to each other. In this case the current confines to one side of the conductor, the current carrying area gets modified and the distribution of current across the conductor cross-section becomes non-uniform. As a result the resistance and inductance of the conductor increases.

8.3.2.4 Ampacity

The amount of current a conductor can carry is referred to as its *ampacity*, a brief form for *ampère–capacity*, i.e. current carrying capacity. It bears an inverse relation to the electrical resistance of a conductor. The resistance of a conductor is a function of material property and its size. For a given material, conductors with a larger cross-sectional area have less resistance than conductors with a smaller cross-sectional area. Low resistance conductors can carry more current.

8.3.2.5 Isotropic and Anisotropic Conductors

If an electric field is applied to a material, and the resulting electric current is in the same direction, the material is said to be an *isotropic electrical conductor.* If this current is in a different direction from the electric field, the material is said to be an *anisotropic conductor.* For such materials the conductivity is not a scalar quantity.

8.3.3 Continuity Equation

In electromagnetics the continuity equation is an empirical law expressing (local) charge conservation. Mathematically, it is a consequence of Maxwell's equations, although charge conservation is more fundamental than these equations. The continuity equation can be expressed in both integral and differential forms.

As described earlier the flow of electric charge in a conductor constitutes conduction current, or simply a current. Imagine that there is a closed surface bounding a volume v, and a current I is emerging out of this closed surface. This outward flow of current (or charges) implies that the total electric charge Q, stored inside this closed surface will decrease with time. This decrease will be justified only if there is neither creation nor destruction of electric charge inside the closed surface. In mathematical term,

$$I = -\frac{dQ}{dt} \tag{8.21}$$

8.3.3.1 Integral Form

In Equations 8.2 and 5.38b, Equation 8.21 can be written as:

$$\oiint_s J \bullet ds = -\frac{d}{dt}\iiint_v \rho \cdot dv \tag{8.22}$$

Equation 8.22 is the *integral form of the continuity equation.* In this equation $\boldsymbol{J}$ is the current density and ρ is the charge density in the volume v bounded by the closed surface s.

8.3.3.2 Point Form

If divergence theorem is applied on the LHS of Equation 8.22 and the time derivative is taken inside the integral sign on the RHS, this equation gives:

$$\iiint_{v} (\nabla \cdot \boldsymbol{J}) \cdot dv = \iiint_{v} \left(-\frac{\partial \rho}{\partial t} \right) \cdot dv \tag{8.23}$$

Since the volume v is arbitrary, the integrands on the two sides can be equated, this gives:

$$\nabla \cdot \boldsymbol{J} = -\frac{\partial \rho}{\partial t} \tag{8.24}$$

Equation 8.24 is called the *point form of continuity equation*. It states that the divergence of the current-density $\boldsymbol{J}$ (in ampères per square meter) is equal to the negative rate of change, i.e. the rate of decrease of the charge density ρ (in coulombs per cubic meter). As the movement of charge constitutes the current thus if charge is moving out of a differential volume (i.e. divergence of current density is positive) then the amount of charge within that volume is bound to decrease, so the rate of change of charge density is negative. Therefore the continuity equation amounts to spell the conservation of charge.

8.3.4 Relaxation Time

One of the unique properties of a conductor is that there is no *electrostatic field* inside it due to any external source. Any external *electrostatic field* will set the free charges in the conductor to move and reorient till the net field inside the conductor reduces to zero. If a charge is somehow injected into the interior of a conductor, this charge will soon appear on the surface. An analogy can be drawn with the transplant of an organ of some donor into the body of the acceptor. After transplant, unless the body of acceptor is immunized through medication, the implanted organ will be rejected. Similarly, the conductor will reject any injected charge to maintain net zero field inside it. The process of rejection is quite simple. The moment a charge is injected into the interior of a conductor, it will setup an electrical field in its own vicinity and will disturb the electrical neutrality of that region. This field will exert force on its neighboring charges. Under the influence of this force, another charge of equal magnitude must get displaced to maintain the electrical neutrality at that localized region. The movement of this new charge to another location will again disturb the neutrality in the vicinity of its new location, which will force another charge to move. This process will continue unless the same or an equivalent charge appears on the surface and *the conductor gets relaxed*. The time taken by a charge to appear on the surface is called the *relaxation time*. This time depends on the properties of the material.

8.3.4.1 Mathematical Formulation

From Equation 8.21 and LHS of Equation 8.22, we get:

$$\frac{dQ}{dt} = -I = -\oiint_{s} \boldsymbol{J} \cdot d\boldsymbol{s} \tag{8.25a}$$

In the relation: $\boldsymbol{J} = \sigma\boldsymbol{E} = \sigma(\boldsymbol{D}/\varepsilon)$, RHS of Equation 8.25a can be written as:

$$\frac{dQ}{dt} = -\frac{\sigma}{\varepsilon} \oiint_{s} \boldsymbol{D} \bullet d\boldsymbol{s} \tag{8.25b}$$

In the divergence theorem:

$$\frac{dQ}{dt} = -\frac{\sigma}{\varepsilon} \iiint_{v} (\nabla \bullet \boldsymbol{D}) \cdot dv \tag{8.25c}$$

Using Maxwell's second equation $(\nabla \bullet \boldsymbol{D} = \rho)$, to get:

$$\frac{dQ}{dt} = -\frac{\sigma}{\varepsilon} \iiint_{v} \rho \cdot dv \tag{8.25d}$$

On invoking Equation 5.38b, we get:

$$\frac{dQ}{dt} = -\frac{\sigma}{\varepsilon} \cdot Q \tag{8.25e}$$

The solution of Equation 8.25e can be written as:

$$Q = Q_o \cdot e^{-t/\tau} \tag{8.26a}$$

where Q_o is the initial charge in the volume v and τ is the *relaxation time* defined as:

$$\tau \overset{\text{def}}{=} \varepsilon / \sigma \tag{8.26b}$$

According to Equation 8.26a, the *relaxation time* is defined as the time taken by the charge $\left\{\left(1 - \frac{1}{e}\right) \cdot Q_o\right\}$, to appear on the conductor surface, where e is the base of the natural logarithm.

For any point in v, Equation 8.26a can also be written in terms of the charge density as:

$$\rho = \rho_o \cdot e^{-t/\tau} \tag{8.27}$$

where the initial charge density at that point is indicated by ρ_o which in general may be a function of space coordinates.

The exponential term involved in Equation 8.27 will take infinite time to become zero thus for all practical purposes the relaxation time is taken as that time in which the charge (or charge density) reduces to $1/e$ times of its original value.

8.3.4.2 Relaxation Time for Good Conductors

In Equation 8.26b, the estimated time for some good conductors are:

- Silver: $\sigma = 6.17\times 10^7$, $\varepsilon = \varepsilon_0 = 8.854 \times 10^{-12}$ thus $\tau \approx 14.35$ μμμ sec.
- Copper: $\sigma = 5.8\times 10^7$, $\varepsilon = \varepsilon_0 = 8.854 \times 10^{-12}$ thus $\tau \approx 15.2675$ μμμ sec.
- Aluminum: $\sigma = 3.72\times 10^7$, $\varepsilon = \varepsilon_0 = 8.854 \times 10^{-12}$ thus $\tau \approx 23.80$ μμμ sec.

As can be seen from these values of relaxation times, the charge reduces in all the cases to $1/e$ times of its original value in micro-micro-micro seconds.

8.4 CLASSIFICATION OF ELECTRICAL MATERIALS

The electrical materials can crudely be classified as conductors, semiconductors, and insulators. This classification is based on the energy levels which are labeled as the valance, conduction, and forbidden energy bands (or energy-gaps). These bands play a key role vis-à-vis the properties of electrical materials. The genesis of these bands is related to the atomic structure and the behavior of electrons therein. A brief description about their formation is given below.

All matters are made of small particles called atoms. An atom in itself contains a nucleus and electrons. These electrons spin on their own axis and simultaneously revolve around the nucleus in different orbits. The number of orbits in an isolated atom is defined by the quantum mechanics. Each of these orbits corresponds to some level of potential energy. These energy levels are determined by accounting the energy required to move an electron from an inner orbit to those which are farther from the nucleus. The electron of an atom prefers the innermost orbit. However, in accordance with Pauli's Exclusion Principle, which spells a fundamental law of nature, not more than two electrons in an atom may occupy the same energy level and that too when these electrons have the opposite spin. Thus in a normal atom, the electron occupies lower energy levels, while the higher energy levels remain unoccupied.

An electron in an isolated atom is in quite definite state. The energy of this state can be represented by a straight line marked as E_1. Figure 8.4a illustrates a single atom and the line representing energy level E_1. If the two similar atoms, with the same energy levels, are brought closer together their electron orbits will interact. In

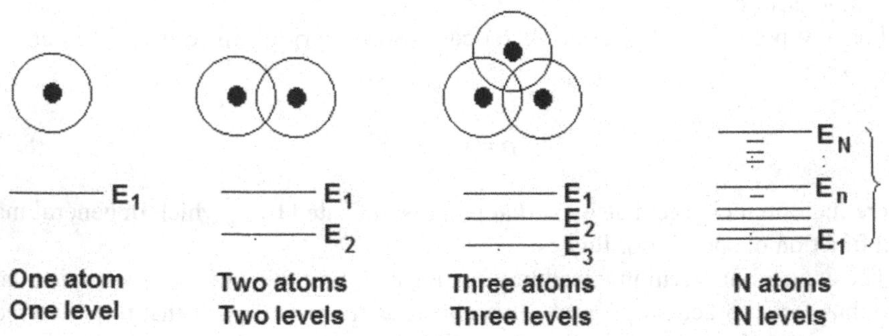

FIGURE 8.4 Energy levels for (a) one atom, (b) two atoms, (c) three atoms, (d) N atoms.

accordance with the quantum mechanics, the lines representing the energy levels of the two atoms will move slightly apart due to this interaction and will result in two different energy levels E_1 and E_2. This situation is depicted in Figure 8.4b. Similarly, when three identical atoms are brought together the interaction of the electron orbits will result in three different energy levels. Figure 8.4c shows the three different energy levels E_1, E_2, and E_3, etc. Figure 8.4d shows N energy levels due to N closely located similar atoms. As evident from this last figure, the number of energy levels becomes so large and so close that these may be regarded as a continuous energy band. Furthermore, the process of such addition of similar atoms may result in the formation of a solid body.

In the above process, bands are formed. Furthermore, the details of quantum mechanics reveal that a band may be fully filled, partially filled or empty. The first two types of these are referred to as valence bands, whereas the last one is called the conduction band. Furthermore, the two categories of these bands may overlap, may be adjacent, or may be separated by distance on the energy scale. In the following paragraphs the description relating to the type and separation of bands, referred to as forbidden energy bands, is used to classify the electrical materials.

In conductors, the conduction band is just above the valance band. With the application of an electrical excitation the electrons can easily crossover from the valence band to the conduction band. Figure 8.5a shows a half-filled (valance) band, which relates to the materials consisting of atoms with only one valence electron per atom. Most highly conducting metals including silver, copper, and gold satisfy this condition. Figure 8.5b shows a situation wherein, the filled (valance) band overlaps with an empty band. Such materials contain atoms with two valence electrons. These too may exhibit high conductivity.

Figures 8.5c illustrates a band-gap diagram for the materials referred to as semiconductors. In these, there is a small forbidden energy region between conduction

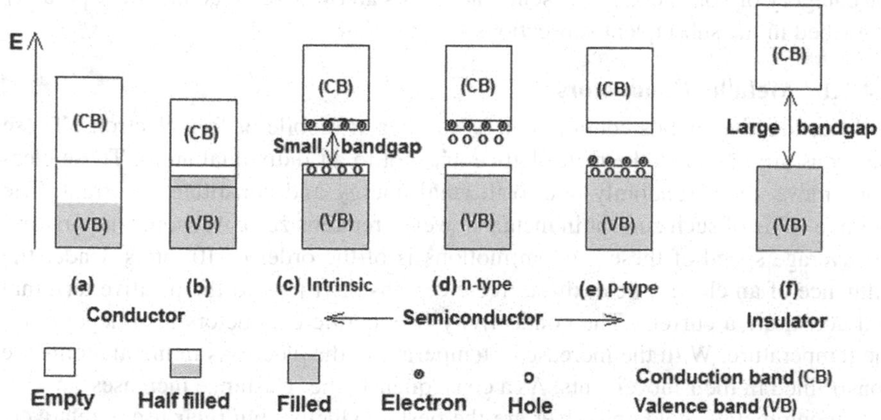

FIGURE 8.5 Possible energy band diagrams of a crystal (a) a half-filled band, (b) two overlapping bands, (c) an almost full band separated by a small band-gap from an almost empty band, (d) band-gap structure for p-type semiconductor (e) band-gap structure for n-type semiconductor, and (f) a full band and an empty band separated by a large gap.

and valance bands. This is the difference between the energy level from the top of the valence to the bottom of conduction band. In these materials, the charge carriers may include both, electrons and holes. These charge carriers under the influence of electrical excitation, may travel from valence band to conduction band only on acquiring sufficient energy to crossover the forbidden energy gap. The semiconductors, in essence, differ from metals and insulators since they contain "almost empty" conduction band and "almost full" valence band. This unique situation allows the transport of charge carriers in both bands. The holes referred to above are in fact the missing electrons. They behave as particles with the same properties as the electrons would have behaved, when occupying the same states except that they carry a positive charge. Figure 8.5c illustrates a pure (or intrinsic) semiconductor. To enhance the conductivity an intrinsic semiconducting material is added with some impurity. Such mixing results in two types of semiconductors. Figure 8.5d shows n (or negative)-type semiconductor wherein holes act as donors and electrons as acceptors. Similarly, Figure 8.5e illustrates a p (or positive)-type semiconductor wherein holes act as acceptors and electrons as donors. These are further explained while dealing with the properties of semiconductors.

Figure 8.5f shows a completely filled band separated from the next higher empty band by a large energy gap. This large band-gap does not permit electrons to crossover from valence band to conduction band. Thus in these materials, referred to as insulators, no conduction is possible unless there is insulation breakdown. At this stage, the dielectric loses its property and starts conducting.

8.4.1 Conducting Materials

The conducting materials may include metals, non-metals, electrolytes, superconductors, semiconductors, and plasmas. Silver, copper, gold, and aluminum are some of the metallic conductors, whereas graphite and conductive polymers fall under the category of non-metals. The semiconductors and superconductors are separately described in the subsequent subsections.

8.4.1.1 Metallic Conductors

Solid conductive metals contain a large number of mobile or free electrons. These electrons are bound to the metal lattice and not to an individual atom. These electrons move about randomly due to thermal energy and constitute a current. The average value of such current in metals however remains zero. At room temperature, the average speed of these random motions is of the order of 10^6 cm/s. Under the influence of an electric field, these free electrons drift toward the positive terminal and constitute a current. The conductivity of metallic conductors also depends on the temperature. With the increase of temperature, the electrons in metal atoms are constrained in their movements. As a consequence, the resistance increases.

Among metals, gold and silver are the best conductors but their use is relatively rare because of high costs. Their use is limited to some specialized equipment and applications. Copper is another metallic conductor with high conductivity. It is a commonly used material but its conductivity is lower than that of gold and silver.

Aluminum is another metallic conductorwith 64% conductivity of copper. It is relatively cheaper than copper and used where cost is one of the deciding factors.

8.4.1.2 Non-Metallic Conductors

Besides metals, salts can also conduct electricity. In these, there are no free electrons and their conductivity depends on ions. When a salt melts or dissolves its ions become free to move.

8.4.2 Semiconducting Materials

The electrical conductivities of semiconducting materials fall in between those of conductors and insulators and can be widely varied by changing their temperature, excitation, and impurity contents. These lie roughly in the range of 10^{-2} to 10^{4} Siemens per centimeter ($S{\cdot}cm^{-1}$). The transport of charges through a semiconductor is not only the function of properties of electrons but strongly linked to the arrangement of atoms in solids. The pure semiconductor is referred to as an intrinsic semiconductor. Semiconductors mixed with impurities can be classified in terms of type of impurity. These impurities, also called extrinsic materials, are referred to as n- or p-type. The n (or negative) type materials are those wherein electrons are in excess whereas in p (or positive) type holes are in excess. Thus in n-type, electrons are the charge carriers and in p-type, holes are the charge carriers. Holes, the mobile positive charge carriers are identified as the places where a valence electron is missing in the semiconducting crystal.

In classic crystalline semiconductors, electrons can have energies only within certain bands. These bands exist between the energy of the ground state and the free electron energy. In the ground state the electrons are tightly bound to the atomic nuclei of the material and the free electron energy describes the energy required for an electron to escape entirely from the material.

The energy bands correspond to a number of discrete quantum states of the electrons. Most of the states with low energy are occupied, up to a particular band called the valence band. The state of the low energy of electrons refers to those electrons orbiting closer to the nucleus.

The valence band in any given metal is nearly filled with electrons under usual operating conditions. In semiconductors, very few valence bands have such a filling. In insulators, virtually none of the valence band is so filled. The ease with which the electrons in a semiconductor can be excited from the valence band to the conduction band depends on the band-gap energy.

When the energy band-gap is large, for example 5 or 10 volts, corresponding to very strong valence and other bonds, it is very unlikely that an electron will acquire sufficient energy from thermal effects at room temperature to jump to the unfilled band. The solid material can thus be regarded as a good insulator. However, if the energy gap is of the order of about 1 volt or less, corresponding to bonds of only moderate strength, enough electrons may acquire the energy necessary to jump the gap to result in appreciable conductivity. This situation can be noticed in semiconductors. The Germanium and Silicon have energy-gaps of 0.75 and 1.12 volts, respectively.

Thus, nearly 4 electron volts (eV) can be taken as the dividing line between the semiconductors and insulators.

This is often stated that the filled bands do not contribute to the electrical conductivity. However, as temperature is raised above absolute zero, a semiconductor acquires more energy to spend on lattice vibration and on exciting electrons into the conduction band.

8.4.3 Dielectric Materials

These materials are electrical *insulators* or non-conducting materials. Unlike metals these have no loosely bound or free electrons that may drift through the material. Under the influence of electric field there is no flow of electric charges through these materials and thus these support insignificant electric current. A *perfect dielectric* has zero electrical conductivity and infinite permittivity, thus exhibits only a *displacement current*. It can store and return electrical energy as if it were an ideal capacitor. "Free space" is a perfect insulator but a poor dielectric.

8.4.3.1 Classical Approach to the Dielectric Model

This approach assumes that the material is made up of atoms wherein each atom consists of a cloud of negative charge (electrons) bound to and surrounding a positive point charge (proton) at its center. This cloud gets distorted with the application of an electric field and reduces to a simple dipole in the superposition principle. A dipole is characterized by its dipole moment, a vector quantity defined by Equation 7.31b. The relationship between the electric field $\boldsymbol{E}$ and the dipole moment $\boldsymbol{p}$ determines the behavior of a dielectric. In general with the removal of field the atom returns to its original state. This time of return is referred to as *dielectric relaxation time.*

The behavior of the dielectric depends on many factors which include the direction of applied field, the nature of field vis-à-vis time variation, the homogeneity of the material, and the linearity of response with respect to the electric field.

Dielectrics can be solids, liquids, or gases. Solid dielectrics are the most commonly used and many solids are very good insulators as well. Solid dielectrics include porcelain, glass, and most of the plastics. The air, nitrogen, and sulfur hexafluoride are the most commonly used gaseous dielectrics. A high vacuum with unity relative dielectric constant is a nearly lossless dielectric.

It needs to clarify that in general the terms insulator and dielectric are considered to be synonyms. However, the term insulator is used to indicate electrical obstruction while the term dielectric indicates the energy storing capability of the material. A common example of a dielectric is the electrically insulating material between the metallic plates of a capacitor. The polarization of the dielectric by the applied electric field increases the capacitor's surface charge for the given electric field strength.

8.4.3.2 Polar and Non-Polar Molecules

In Chapter 7, the field due to a dipole was evaluated and the electric dipole moment was represented by vector $\boldsymbol{p}$ ($Q\boldsymbol{d}$). In this relation Q is the positive charge and $\boldsymbol{d}$ is a distance vector oriented from negative to positive charge. In general these dipole moments are arbitrarily oriented throughout a volume in dielectrics.

Since matter is composed of atoms and molecules the study of dielectrics requires understanding of the nature of these constituents. In all matters every atom contains positive and negative charges. Unlike conducting materials there are no free charges in dielectrics. The charges in dielectrics can thus be referred to as bound charges. These positive and negative charges have a minimum natural spacing. The molecules with almost zero minimum natural spacing are called *non-polar molecules.* The molecules wherein this minimum natural spacing is greater than zero are called *polar molecules.* Figures 8.6a and b show these two classes of molecules with displacements d_1 and d $(d \to 0)$ respectively. These displacements are the average equilibrium positions of polar and non-polar molecules. These configurations of closely located opposite charges in fact represent electric dipoles.

8.4.3.3 Polarization

Under the influence of an electric field, the positive and negative charges in a dielectric shift from their average equilibrium positions. Figures 8.6c and d shows this shifting in polar and non-polar molecules as d_1' $(d_1' > d_1)$ and $d'(d' > 0)$. As can be seen, this shifting is such that the positive charges are displaced toward the field E and the negative in the opposite direction. The internal electric field created by these dipoles reduces the overall field within the dielectric.

The shifting between positive and negative charges is referred to as *dielectric polarization* and is denoted by a vector $\boldsymbol{P}$. If there are n identical dipoles per unit volume then the polarization can be written as:

$$\boldsymbol{P} = n \cdot Q\boldsymbol{d} = n \cdot \boldsymbol{p} \quad \text{or} \quad \boldsymbol{P} = n \cdot Q\boldsymbol{d}' = n \cdot \boldsymbol{p}' \tag{8.28a}$$

The polarization $\boldsymbol{P}$ is identified as dipole moment per unit volume. In the case of non-identical dipoles it can be written as an average over a sample volume v:

$$\boldsymbol{P} = \frac{1}{v}\left(Q_1\boldsymbol{d}_1 + Q_2\boldsymbol{d}_2 + \cdots + Q_n\boldsymbol{d}_n\right) = \frac{1}{v}\sum_{m=1}^{n} Q_m\boldsymbol{d}_m = \frac{1}{v}\sum_{m=1}^{n} \boldsymbol{p}_m \tag{8.28b}$$

If a dielectric is composed of weakly bonded molecules, the application of an electric field not only polarizes the molecules but also tries to reorient the dipole moments to align them along the field. Figures 8.7a and b show the orientation of dipoles before and after the application of electric field.

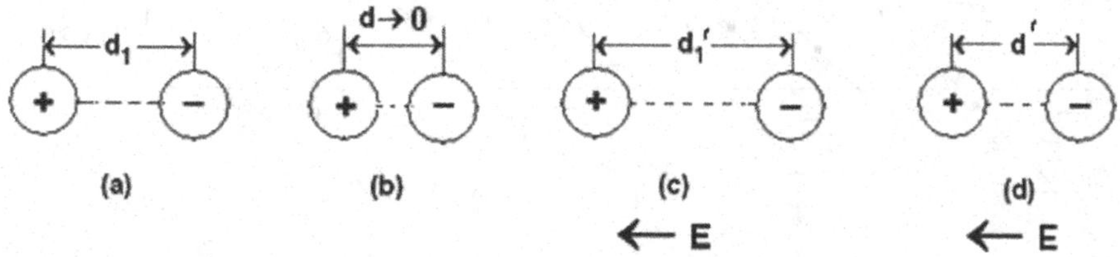

FIGURE 8.6 Displacement between charges: without electric field in (a) polar and (b) non-polar molecules, under the influence of electric field in (c) polar (d) non-polar molecules.

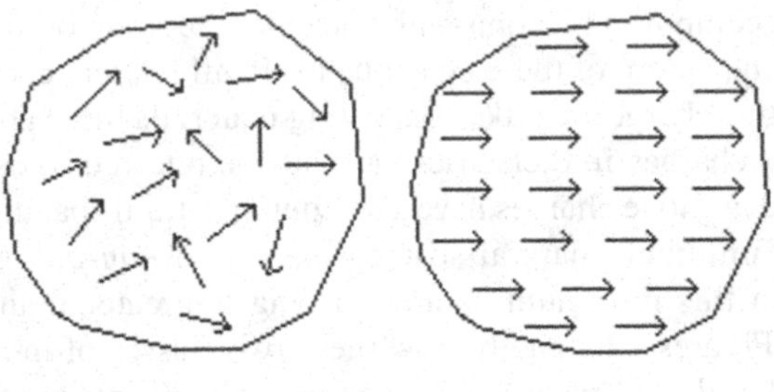

FIGURE 8.7 Orientation of dipoles (a) before and (b) after application of E field.

8.4.3.4 Bound Charge Density

Figure 8.8 shows an incremental volume with surface area ΔS in the interior of a dielectric. This volume is assumed to contain n number of non-polar molecules. Since no molecule has a dipole moment the polarization $\boldsymbol{P}$ is zero throughout the material. The application of an electric field to this sample will elongate the distances between positive and negative charges and thus will produce dipole moment $\boldsymbol{p} = Q\boldsymbol{d}$ in each molecule. Assume that these reoriented dipoles, well within the volumes, make an angle θ with surface vector $\boldsymbol{\Delta S}$. In this angle, the height (Δh), shown in the figure, and the volume Δv of this sample can be written as:

$$\Delta h = d \cdot \cos\theta \quad (\text{a}) \text{ and } \Delta v = \Delta S \cdot \Delta h = \Delta S \cdot d \cdot \cos\theta \quad (\text{b})\} \tag{8.29}$$

Assuming that every molecule was initially centered in the volume Δv at a distance Δh below the surface ΔS. Due to the elongation of the distances between positive and negative charges $+Q$ charges will cross the surface ΔS and move upward and $-Q$ charges will cross the surface ΔS and move downward. Since there are n molecules per m^3, the net total charge, which can now be referred to as bound charge ΔQ_b, crossing the elemental surface in upward direction are:

$$\Delta Q_b = n \cdot Q \cdot d \cdot \cos\theta \cdot \Delta S = n \cdot Q \cdot \boldsymbol{d} \bullet \boldsymbol{\Delta S} \tag{8.30a}$$

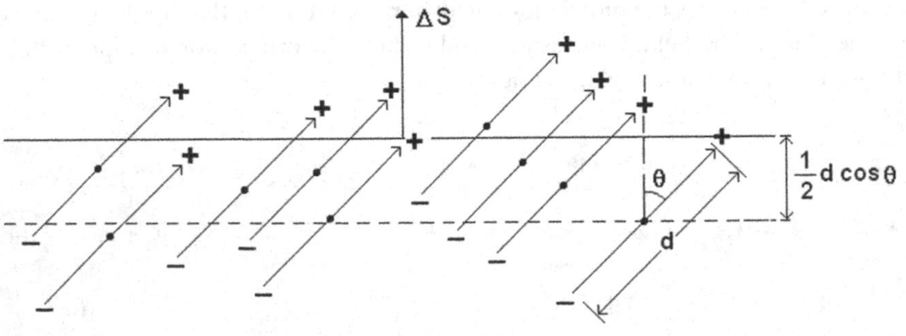

FIGURE 8.8 Transfer of bound charges across incremental surface ΔS in the interior of dielectric (arrows indicate direction of the applied field and +/– signs indicate polarized molecules).

In Equation 8.28a:

$$\Delta Q_b = \boldsymbol{P} \bullet \Delta \boldsymbol{S} \tag{8.30b}$$

If this surface is assumed to be the closed surface bounding the volume, the surface vector will be directed outward and the net increase of bound charges within the volume can be written as:

$$Q_b = -\oint\!\!\oint_s \boldsymbol{P} \bullet \Delta \boldsymbol{S} \tag{8.31a}$$

If this charge is assumed to be distributed as a bound charge density ρ_b, Q_b can be written as:

$$Q_b = \iiint_v \rho_b \cdot dv \tag{8.31b}$$

In Equations 8.31a and 8.31b:

$$\iiint_v \rho_b \cdot dv = -\oint\!\!\oint_s \boldsymbol{P} \bullet \Delta \boldsymbol{S} \tag{8.32a}$$

The application of the divergence theorem on the RHS of Equation 8.32a gives:

$$\iiint_v \rho_b \cdot dv = -\iiint_v (\nabla \bullet \boldsymbol{P}) \cdot dv \tag{8.32b}$$

From Equation 8.32b:

$$\nabla \bullet \boldsymbol{P} = -\rho_b \tag{8.33}$$

8.4.3.5 Impact of Polarization and Bound Charges

This relation is similar to that given by Equation 6.20a ($\nabla \bullet \boldsymbol{D} = \rho$). Let us write this equation in terms of E, for free space:

$$\nabla \bullet (\varepsilon_o \boldsymbol{E}) = \rho \tag{8.34}$$

If we assume that the volume charge density comprises both the charge density due to free electrons ρ_f and that due to bound electrons ρ_b, the RHS of Equation 8.34 gets modified to:

$$\nabla \bullet (\varepsilon_o \boldsymbol{E}) = \rho_f + \rho_b \tag{8.35a}$$

In Equation 8.33:

$$\nabla \bullet (\varepsilon_o \boldsymbol{E}) = \rho_f - \nabla \bullet \boldsymbol{P} \tag{8.35b}$$

This equation can be manipulated to yield:

$$\nabla \bullet (\varepsilon_o \boldsymbol{E} + \boldsymbol{P}) = \rho_f \tag{8.35c}$$

Thus $\boldsymbol{D}$ can be written as:

$$\boldsymbol{D} = \varepsilon_o \boldsymbol{E} + \boldsymbol{P} \tag{8.36}$$

As can be seen from Equation 8.36, there is an added term in the relation of $\boldsymbol{D}$ given by Equation 6.5 ($\boldsymbol{D} = \boldsymbol{\varepsilon}_o \boldsymbol{E}$). This added term is due to the presence of polarizable material. In Equation 8.35c and 8.36, a new relation emerges as:

$$\nabla \bullet \boldsymbol{D} = \rho_f \tag{8.37}$$

8.4.3.6 Electric Susceptibility

The electric susceptibility of a dielectric material is a measure of its sensitivity to the polarization in the presence of an electric field E. It is denoted by a Greek letter χ_e (chi) and can be considered as the constant of proportionality in the relation of electric field $\boldsymbol{E}$ and the induced dielectric polarization density $\boldsymbol{P}$. These are related as:

$$\boldsymbol{P} = \varepsilon_o \cdot \chi_e \cdot \boldsymbol{E} \tag{8.38}$$

Thus, in Equation 8.36 and 8.38, we get:

$$\boldsymbol{D} = \varepsilon_o \cdot (1 + \chi_e) \cdot \boldsymbol{E} \tag{8.39}$$

The susceptibility of a medium is also related to its relative permittivity $\boldsymbol{\varepsilon}_r$ as:

$$\chi_e = \varepsilon_r - 1 \tag{8.10}$$

Its value for free space or vacuum is zero. This influences many phenomena in the medium, from the capacitance of capacitors to the speed of light.

The substitution of Equation 8.40 into Equation 8.39 gives:

$$\boldsymbol{D} = \varepsilon_o \cdot \varepsilon_r \cdot \boldsymbol{E} \overset{\text{def}}{=} \varepsilon \cdot \boldsymbol{E} \tag{8.41}$$

From Equation 8.41:

$$\varepsilon = \varepsilon_o \cdot \varepsilon_r \tag{8.42}$$

8.4.3.7 Dielectric Constant

The polarizability of a dielectric material is expressed by a term called the *relative permittivity* or *relative dielectric constant*. This parameter is normally denoted by ε_r. The value of permittivity for the free space or the vacuum is denoted by ε_o. The values of dielectric constants for all other materials are given in terms of the relative dielectric constant. Thus the absolute value of permittivity ε for any dielectric material is taken to be the product of ε_o and ε_r as given by Equation 8.42.

8.4.4 Superconductors

The phenomenon of superconductivity was discovered by the Dutch physicist Heike Kamerlingh Onnes in 1911. This quantum mechanical phenomenon of exactly zero electrical resistance and expulsion of magnetic fields is exhibited by certain materials below a characteristic critical temperature. It is characterized by the Meissner effect, the complete ejection of magnetic field lines from the interior of the superconductor as it transitions into the superconducting state. The occurrence of this effect indicates that the superconductivity cannot be understood simply as the idealization of *perfect conductivity* in classical physics. The electrical resistivity of a metallic conductor decreases gradually as temperature is lowered. This decrease is limited by impurities and other defects and a normal conductor shows some resistance even near absolute zero. In a superconductor, the resistance abruptly drops to zero when the material is cooled below its critical temperature. The flow of a time-invariant electric current through a loop of superconducting wire can persist indefinitely without any power source (since there is no I^2R loss and no power radiation with dc currents).

8.4.4.1 Applications of Superconductors

- Superconducting magnets find applications in MRI/NMR machines, mass spectrometers, and the beam-steering magnets used in particle accelerators.
- These can also be used for magnetic separation, where weakly magnetic particles are extracted from a background of less or non-magnetic particles, as in pigment industries.
- These are used to make digital circuits based on rapid single flux quantum technology and RF and microwave filters for mobile phone base stations.
- These are used to build Josephson junctions the building blocks of superconducting quantum interference devices (SQUIDs), the most sensitive magnetometers known. SQUIDs are used in scanning SQUID microscopes and magnetoencephalography.
- Depending on the particular mode of operation, a superconductor-insulator-superconductor Josephson junction can be used as a photon detector or as a mixer.
- The large resistance change at the transition from the normal to the superconducting state is used to build thermometers in cryogenic micro-calorimeter photon detectors and in ultrasensitive bolometers.
- Its promising future applications may include high-performance smart grid, electric power transmission, transformers, power storage devices, electric motors (e.g. for vehicle propulsion, as in maglev trains), magnetic levitation devices, fault current limiters, and superconducting magnetic refrigeration. Since superconductivity is sensitive to moving magnetic fields some

of these applications will be more difficult to develop than those rely on conventional methods.
- The relative efficiency, size, and weight advantages of devices based on high-temperature superconductivity outweigh the additional costs involved.

8.5 BOUNDARY CONDITIONS

In many applications, a field quantity (viz. *D*, *E*, *B*, or *H*) may come across materials with different characterizing parameters. An electrical machine with static or rotating metallic parts separated by an airgap, an electromagnetic wave propagating through a media containing trees, building, hills, and other obstacles, and the devices and system wherein conductors and dielectrics, different dielectrics, or different magnetic materials are simultaneously present may be some such examples. All these situations demand to know the behavior of field quantities when they cross from one medium to another. Obviously the behavior of electric field quantities (*D* and *E*) will depend on the conductivity and the dielectric constants, whereas that of magnetic field quantities (*B* and *H*) will depend on the relative permeabilities of the media. In addition, the behavior in case of electrostatic, magnetostatic, and time-varying fields will also differ. This behavior of field quantities due to the presence of materials of differing nature is accounted by the so-called *boundary conditions*. The study of boundary conditions revolves around the continuity conditions of electric or magnetic fields at the interface of two differing media.

8.5.1 COMMON CONFIGURATIONS

Figure 8.9 shows a simple configuration which can be used for evaluating boundary conditions to be satisfied by a particular vector field quantity. In this configuration the entire space is divided into two regions which are marked as Region 1 and Region 2. Both of these regions are in the form of volumes which are filled with

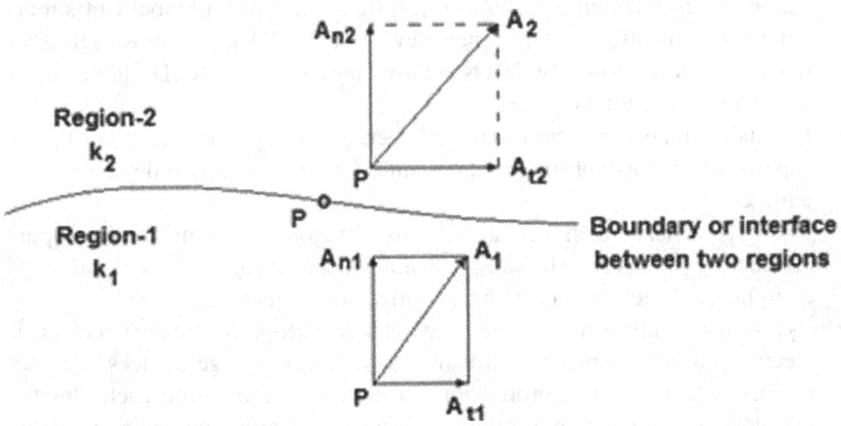

FIGURE 8.9 Common configuration for boundary conditions.

two different materials. The upper surface of Region 1 meets the lower surface of Region 2. The meeting location of these surface layers shown by a line is referred to as a boundary or an interface.

This figure also shows characteristic parameters of Regions 1 and 2 marked as k_1 and k_2. These may represent two different conductivities or permittivities in the case of electrical fields and two different permeabilities in the case of magnetic fields. The vector quantity shown by $\boldsymbol{A}$ may indicate $\boldsymbol{D}$ and $\boldsymbol{E}$ in electrical (or $\boldsymbol{B}$ and $\boldsymbol{H}$ in magnetic) field, both at point P. The suffixes 1 and 2 with $\boldsymbol{A}$ represent the number of region (say $\boldsymbol{A}_1$ for Region 1 and $\boldsymbol{A}_2$ for Region 2). Since the values of characterizing parameters of the two regions differ, the orientations of $\boldsymbol{A}_1$ and $\boldsymbol{A}_2$ will also differ. Both $\boldsymbol{A}_1$ and $\boldsymbol{A}_2$ exists on the two sides of the interface at the same point P. Furthermore, in any of the regions the arbitrarily oriented vector can be resolved into normal and tangential components which can be indicated by putting corresponding suffixes along with the number of the respective region. Thus, the tangential components of vector quantity in these regions are written as A_{t1} and A_{t2} and the normal components as A_{n1} and A_{n1}. Alternatively, these can also be written as, A_{tan1}, A_{tan2}, A_{norm1}, and A_{norm2}.

8.5.2 Common Forms of Equations

In general, there are two common forms of Maxwell's equations, which are used to evaluate the boundary conditions in electrostatic and magnetostatic fields. One of them involves a closed line integral (also called a contour integral) and the other a closed surface integral. The closed line integral can be evaluated by taking a small rectangular loop with width Δw and height Δh. This loop may be marked "a-b-c-d-a" and may be assumed to be embedded in both the regions in equal halves. In the description of Section 8.5.1, the line integrals involving $\boldsymbol{E}$ (or $\boldsymbol{H}$) can be evaluated from the configuration shown in Figure 8.10a.

For evaluating the closed surface integral normally a small circular cylinder, called pillbox, with height Δh and having top and bottom surface areas ΔS. This pillbox is also assumed to be embedded in both the regions in equal halves. In the description given in Section 8.5.1, the surface integral, involving $\boldsymbol{D}$ (or $\boldsymbol{B}$), can be evaluated from the configuration shown in Figure 8.10b.

As noted earlier the boundary conditions spell the continuity of electric (or magnetic) fields at the interface of two different media. To arrive at this interface, in both,

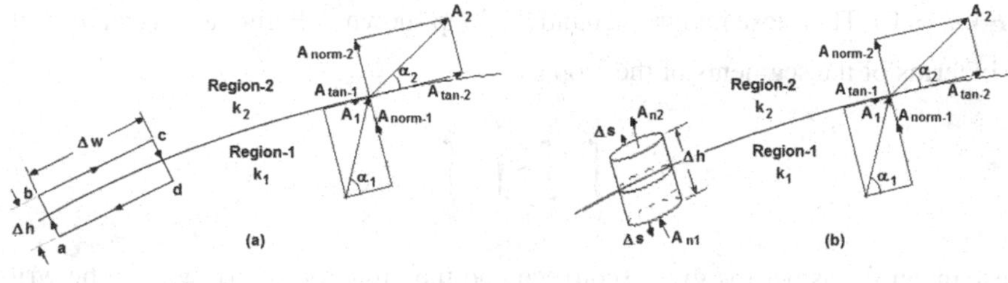

FIGURE 8.10 Configuration for evaluating (a) line integral and (b) surface integral.

Figures 8.10a and b, height Δh must shrink to zero. This condition greatly simplifies the evaluation of both, the line integral and the surface integral as all the integrals involving Δh will vanish.

To use these figures one has to simply replace A by the relevant field quantity as noted above. In the following subsections the boundary conditions for electric field intensity (***E***) and electric flux density (***D***) for electrostatic fields are given for two different cases. These include (i) the boundary conditions at the conductor-dielectric interface and (ii) at the dielectric-dielectric interface. The boundary conditions for magnetostatic and time-varying fields are discussed in Chapters 12 and 15, respectively, after introducing ***B*** and ***H*** vectors.

Since the field vectors may be obliquely oriented at the interface between the two regions these need to be resolved into normal and tangential components. The values of these components will depend on the striking angle. When a field vector crosses from one region to another, due to the differing characterizing parameters of the two regions, it may make a different angle with the boundary. This is called the refractive phenomenon and is shown in Figures 8.10a and b.

8.5.3 Conductor-Dielectric Interface

As discussed earlier, no charge and thus no electric field can exist at any point within a conducting material. Charges may however appear on the surface as surface charge and an electric field related to these may exist on the surface. If the external field intensity on the surface of a conductor is decomposed into tangential and normal components, the tangential component comes out to be zero. If it were not the tangential force would be applied to the surface charges, resulting in the motion of charges which in turn would lead to a non-static condition. Since static conditions are assumed, the tangential electric field intensity and electric flux density are zero.

In relation to the normal component electric flux density let us consider Gauss's law. The electric flux leaving a small incremental surface must be equal to the charge residing on that surface. As the electric field inside the conductor is zero the flux leaves the conductor surface perpendicularly in outward/inward directions depending on the polarity of surface charges. The above description related to the normal and tangential fields can further be verified in the following mathematical equations.

8.5.3.1 Boundary Conditions for E Field

The boundary conditions to be satisfied by the electric field intensity between a conductor (Region 1) and a dielectric (Region 2) can be derived from Equation 7.1 ($\oint \boldsymbol{E} \bullet d\ell = 0$). This close integral around the loop shown in Figure 8.10a can be written in terms of the segments of the loop as:

$$\int_a^b + \int_b^c + \int_c^d + \int_d^a$$

These integrals, as per the given sequence and the direction of arrows, can be written as:

$$E_{n1}\Delta h + E_{t2}\Delta w - E_{n2}\Delta h - E_{t1}\Delta w = 0 \tag{8.43a}$$

In this equation, as Δh shrinks to zero, the first and third terms vanish. The fourth term vanishes as there is no field inside the conductor (i.e. $E_{t1} = 0$). Thus, we get:

$$E_{t2} = D_{t2} = 0 \tag{8.43b}$$

Equation 8.42a indicates that on the surface of the conductor $E_t = D_t = 0$.

8.5.3.2 Boundary Conditions for D Field

The boundary conditions to be satisfied by the electric flux density (**D**) between the two regions can be derived in Equation 6.9 ($\oint\!\!\oint \boldsymbol{D} \bullet \boldsymbol{ds} = \psi = Q$). As before Region 1 is conducting and Region 2 contains dielectric. In Figure 8.10b the closed surface integral can be written in terms of its components as below.

$$\iint_{\text{top}} + \iint_{\text{bottom}} + \iint_{\text{curved-surface}}$$

Maintaining the sequence of the segments, the integrals for a tiny pillbox, can be written as:

$$D_{n2} \cdot \Delta S - D_{n1} \cdot \Delta S + D_{t1} \cdot \frac{\Delta h}{2} + D_{t2} \cdot \frac{\Delta h}{2} = Q = \rho_s \cdot \Delta S \tag{8.44a}$$

In Equation 8.44a, the last two segments vanish when Δh shrinks to zero and the integral over bottom surface vanishes as there no field inside the conductor ($D_{n1} = 0$). The integral results

$$D_{n2} \cdot \Delta S = \Delta Q = \rho_s \cdot \Delta S \tag{8.44b}$$

Equation 8.44b indicates that on the conductor surface:

$$D_n = \varepsilon_o E_n = \rho_s \tag{8.45}$$

8.5.4 Dielectric-Dielectric Interface

In this case, the boundary conditions get modified since electric field may exist inside the dielectric material. To obtain boundary conditions between two dielectrics (1 and 2) having permittivities ε_1 and ε_2, simply replace k_1 and k_2 by ε_1 and ε_2 in the configurations shown in Figure 8.9 and Figure 8.10. Also replace A_1, A_2 by E_1, E_2 in Figure 8.10a and by D_1, D_2 in Figure 8.10b.

8.5.4.1 Boundary Conditions for E Field

The boundary conditions to be satisfied by the electric field intensity between these two dielectrics can again be derived from relation ($\oint \boldsymbol{E} \bullet d\ell = 0$) and Figure 8.10a (after replacing A_1, A_2 by E_1, E_2). Thus in the description given earlier with $\Delta \boldsymbol{h}$ shrinking to zero, we get:

$$E_{t2} \cdot \Delta W - E_{t1} \cdot \Delta W = 0 \quad \text{(a)} \quad \text{or} \quad E_{t1} = E_{t2} \quad \text{(b)}\} \tag{8.46}$$

In this relation, the tangential component of $\boldsymbol{D}$ in the two regions satisfy:

$$\frac{D_{t1}}{D_{t2}} = \frac{\varepsilon_1}{\varepsilon_2} \tag{8.47}$$

In Figure 8.10a if $\boldsymbol{A}$ is replaced by $\boldsymbol{E}$ it will illustrate the refraction of $\boldsymbol{E}$. As shown in the figure, E_1 is making an angle α_1 and E_2 angle α_2 with their respective tangential components. In the continuity of tangential components of $\boldsymbol{E}$:

$$E_1 \cdot \cos\alpha_1 = E_2 \cdot \cos\alpha_2 \tag{8.48}$$

8.5.4.2 Boundary Conditions for D Field

As before, the boundary conditions for $\boldsymbol{D}$ field between the two dielectrics can be derived from the relation ($\oiint_s \boldsymbol{D} \bullet ds = Q$) and Figure 8.10b (after replacing A_1, A_2 by D_1, D_2). Thus, in the earlier description with Δh shrinking to zero gives:

$$D_{n2} \cdot \Delta S - D_{n1} \cdot \Delta S = \Delta Q = \rho_s \cdot \Delta S \tag{8.49}$$

Depending on the presence and absence of surface charge density (ρ_s), Equation 8.49 yields the following two results:

$$D_{n2} - D_{n1} = \rho_s \quad \text{(a)} \qquad \text{and} \qquad D_{n2} - D_{n1} = 0 \quad \text{(b)}\} \tag{8.50}$$

If $\rho_s = 0$, we get the following relation from Equation 8.50:

$$\frac{E_{n1}}{E_{n2}} = \frac{\varepsilon_2}{\varepsilon_1} \tag{8.51}$$

In Figure 8.10b, if vector **A** is replaced by vector $\boldsymbol{D}$ it will illustrate the refraction of $\boldsymbol{D}$. As shown in the figure, D_1 is making an angle α_1 and D_2 is making angle α_2 with their respective tangential components. In the continuity of normal components of $\boldsymbol{D}$:

$$\boldsymbol{D}_1 \cdot \sin\alpha_1 = \boldsymbol{D}_2 \cdot \sin\alpha_2 \tag{8.52}$$

On multiplying both sides of Equation 4.47 ($E_1 \cdot \cos\alpha_1 = E_2 \cdot \cos\alpha_2$) by ε_1, ε_2, we get:

$$\varepsilon_2 \cdot \varepsilon_1 \cdot \boldsymbol{E}_1 \cdot \cos\alpha_1 = \varepsilon_1 \cdot \varepsilon_2 \cdot \boldsymbol{E}_2 \cdot \cos\alpha_2$$

On substituting $\varepsilon_1 \cdot \boldsymbol{E}_1 = \boldsymbol{D}_1$ and with $\varepsilon_2 \cdot \boldsymbol{E}_2 = \boldsymbol{D}_2$, we have:

$$\varepsilon_2 \cdot D_1 \cdot \cos\alpha_1 = \varepsilon_1 \cdot D_2 \cdot \cos\alpha_2$$

or, from Equation 8.46b:

$$\frac{D_{t1}}{D_{t2}} = \frac{\varepsilon_1 \cdot E_{t1}}{\varepsilon_2 \cdot E_{t2}} = \frac{\varepsilon_1}{\varepsilon_2} = \frac{D_1 \cdot \cos\alpha_1}{D_2 \cdot \cos\alpha_2} \tag{8.53}$$

Thus, in Equations 8.53 and 8.52, at any point on the boundary we have:

$$\frac{\varepsilon_1}{\varepsilon_2} = \frac{D_1}{D_2} \cdot \frac{\cos\alpha_1}{\cos\alpha_2} = \frac{\sin\alpha_2}{\sin\alpha_1} \cdot \frac{\cos\alpha_1}{\cos\alpha_2} = \frac{\tan\alpha_2}{\tan\alpha_1} \tag{8.54}$$

8.5.5 An Alternative Approach

Figure 8.11 shows a regular point P on the boundary surface (extending along the x- and y-axes) between two regions with different values of permittivity, say, ε_1 and ε_2 for Regions 1 and 2, respectively. On this boundary, let there be a surface charge density ρ_s. The z-direction is normal to the boundary, directed from Region 2 to Region 1. At this point, the following boundary conditions are well known:

$$\left.\begin{matrix} E_{1x}\big|_P - E_{2x}\big|_P = 0 \quad (a) & E_{1y}\big|_P - E_{2y}\big|_P = 0 \quad (b)\end{matrix}\right\} \tag{8.55}$$

where E_{1x}, E_{2x} are the x-components and E_{1y}, E_{2y} are the y components in Regions 1 and 2, respectively. We therefore have, for the tangential derivatives on the boundary:

CASE I:

$$\left.\begin{matrix} \left.\frac{\partial E_{1x}}{\partial x}\right|_P - \left.\frac{\partial E_{2x}}{\partial x}\right|_P = 0 \quad (a) & \left.\frac{\partial E_{1y}}{\partial y}\right|_P - \left.\frac{\partial E_{2y}}{\partial y}\right|_P = 0 \quad (b)\end{matrix}\right\} \tag{8.56}$$

Thus, on adding, we get:

$$\left[\frac{\partial E_{1x}}{\partial x} + \frac{\partial E_{1y}}{\partial y}\right]_P - \left[\frac{\partial E_{2x}}{\partial x} + \frac{\partial E_{2y}}{\partial y}\right]_P = 0 \tag{8.57}$$

These boundary conditions are valid for linear, non-linear, homogeneous, or inhomogeneous media, with or without any charge density distribution.

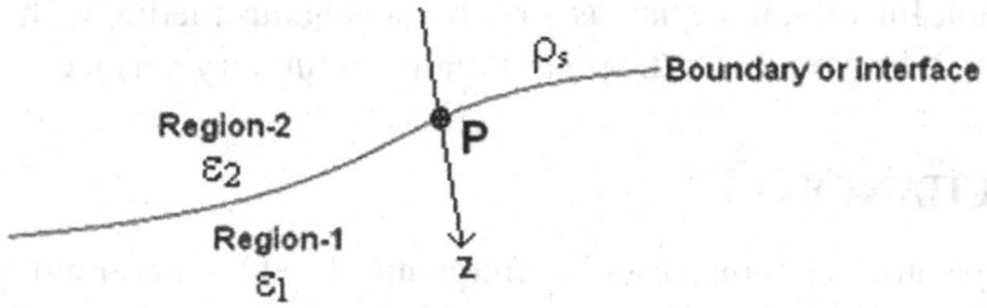

FIGURE 8.11 Modified version of Figure 8.8.

Let us assume that at least near the boundary surface both regions are linear, homogeneous and free from any volume charge density distribution. Thus, near the boundary surface:

$$\nabla \bullet \boldsymbol{E}_1 = \nabla \bullet \boldsymbol{E}_2 = 0 \tag{8.58}$$

Therefore, Equation 7.56 may be written as:

$$\left.\frac{\partial E_{1z}}{\partial z}\right|_P - \left.\frac{\partial E_{2z}}{\partial z}\right|_P = 0 \tag{8.59}$$

Thus, the normal derivative of the normal component (E_{1z} and E_{2z}) of electric field intensity will be continuous across the boundary between the two regions, even though the boundary surface may or may not carry any surface charge distribution. This boundary condition may find application in solving boundary value problems involving electrostatic field.

CASE II:

$$\left.\begin{array}{l}\left.\dfrac{\partial E_{1x}}{\partial y}\right|_P - \left.\dfrac{\partial E_{2x}}{\partial y}\right|_P = 0 \quad (\mathrm{a}) \qquad \left.\dfrac{\partial E_{1y}}{\partial x}\right|_P - \left.\dfrac{\partial E_{2y}}{\partial x}\right|_P = 0 \quad (\mathrm{b})\end{array}\right\} \tag{8.60}$$

Therefore,

$$\left.\frac{\partial E_{1y}}{\partial x}\right|_P - \left.\frac{\partial E_{1x}}{\partial y}\right|_P = \left.\frac{\partial E_{2y}}{\partial x}\right|_P - \left.\frac{\partial E_{2x}}{\partial y}\right|_P$$

or,

$$\left.\left[\nabla \times \boldsymbol{E}_1\right]_z\right|_P = \left.\left[\nabla \times \boldsymbol{E}_2\right]_z\right|_P \tag{8.61a}$$

Thus, the normal component of the curl of electric field intensity is continuous across the boundary between the two regions. For *time-invariant fields*, the curl of electric field intensity is zero. This implies that:

$$\left.\left[\nabla \times \boldsymbol{E}_1\right]_z\right|_P = \left.\left[\nabla \times \boldsymbol{E}_2\right]_z\right|_P = 0 \tag{8.61b}$$

The boundary conditions given by Equations 8.61a and 8.61b are valid for linear, non-linear, homogeneous, or inhomogeneous media, with or without any charge distribution at the point P on the boundary surface.

8.6 CAPACITANCE

The word "capacitor" or "condenser" is frequently used by a commoner in connection with the domestic electrical appliances. A capacitor is a device that introduces capacitance in an electrical circuit. A capacitor opposes the instantaneous change of voltage. It comprises two parallel surfaces whereupon the charges of opposite polarity accumulate. It stores energy in electric field during charging by a battery which is analogous to filling of the air (or water) tank. The charging (or discharging on removal of source) of the capacitor depends on the time constant of the circuit, ($\tau = R{\cdot}C$). The surfaces of the capacitor may be in the form of two parallel flat plates, two concentric cylinders or spheres, etc. These surfaces are separated by some dielectric material or simply by air.

The capacitance is that property of a system of conductors and dielectrics that permits storage of electrical energy when potential differences exist between the conductors. Its value, which is always positive, is expressed as the ratio of electrical charge Q to a potential difference V (or V_o). In mathematical terms:

$$C = \frac{Q}{V_o} \tag{8.62}$$

8.6.1 Electric Field between Two Parallel Plates

Figure 8.12 shows various aspects of a parallel plate capacitor. Figure 8.12a shows two flat parallel plates, marked as A and B, located parallel to the y–z-plane and separated by a distance d. The gap between these plates, each of surface area S, is filled with a dielectric of permittivity ε. These plates contain charges $+Q$ and $-Q$, which are uniformly distributed over the inner surfaces of these plates in the form of surface charge densities $+\rho_s$ and $-\rho_s$ respectively. This charge distribution results in $\boldsymbol{E}$ field between the plates, which can be obtained by using Equation 5.36. It is to be noted that the field outside the plates is zero in the Figures 8.12b as per the description given in Section 5.5. The field obtained from Equation 5.36 is:

$$\left.\boldsymbol{E} = \frac{\rho_s}{\varepsilon}\boldsymbol{a}_x \quad \text{(a)} \quad \text{and} \quad \boldsymbol{D} = \rho_s \boldsymbol{a}_x \quad \text{(b)}\right\} \tag{8.63}$$

Figure 8.12c illustrates $\boldsymbol{E}$ between the two plates.

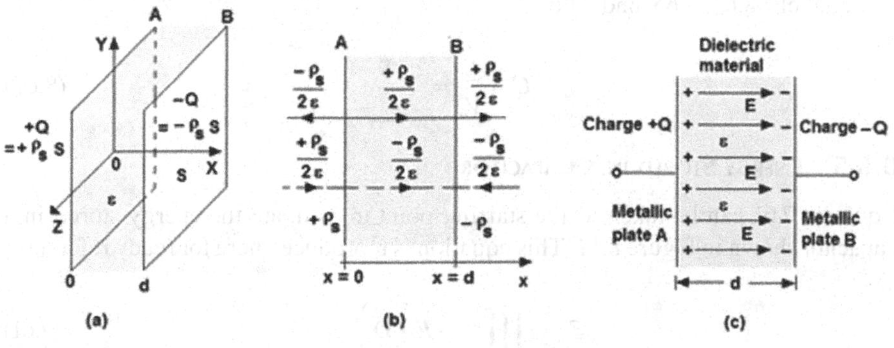

FIGURE 8.12 (a) A parallel plate capacitor (b) and (c) fields between two parallel plates.

8.6.2 Charge Densities from Given Electric Field

The charge densities at the two plates can be evaluated from **E** and **D** given by Equations 8.63a and 8.63b. In Equation 8.63b, the charge on plate A must be positive since **D** is directed rightward. The normal value D_n of **D** can be written as:

$$\rho_s\Big|_{\text{plate-}A} = D_n = D_x \tag{8.64a}$$

The surface charge density at plate A is equal to the *x*-component of the displacement vector. At plate B, this can be written as the negative of the *x*-component of the displacement vector, since for this plate the normal direction on the inner surface of the plate is reversed:

$$\rho_s\Big|_{\text{plate-}B} = D_n = -D_x \tag{8.64b}$$

8.6.3 Potential Difference and Total Charge

Once **E** is known the potential difference V_o between the plates A and B can be evaluated. Thus, in Equations 8.64a and 7.8:

$$V_o = -\int_B^A \boldsymbol{E} \bullet d\boldsymbol{\ell} = -\int_d^0 \frac{\rho_s}{\varepsilon} \boldsymbol{a}_x \bullet dx\boldsymbol{a}_x = -\int_d^0 \frac{\rho_s}{\varepsilon} dx = \frac{\rho_s}{\varepsilon} \cdot d \tag{8.65}$$

If the plates are taken to be of infinite dimensions the charge accumulation on such a plate will also be infinite as it is simply the product of ρ_s and the area of the plate. A more practical approach is to consider a plate of finite area "S". Thus the total charge on a plate is:

$$Q = \rho_s \cdot S \tag{8.66}$$

8.6.4 Evaluation of Capacitance

In Equations 8.62, 8.65, and 8.66:

$$C = \frac{Q}{V_o} = \frac{\varepsilon \cdot S}{d} \tag{8.67}$$

8.6.5 Energy Stored in a Capacitor

Equation 7.61 can be taken as the starting point to estimate the energy stored in a capacitor shown in Figure 8.12. This equation is reproduced here for ready reference:

$$\mathbb{E}_n = \iiint_{v \to \infty} \left(\frac{1}{2} \cdot \boldsymbol{E} \bullet \boldsymbol{D} \right) \cdot dv \tag{7.61}$$

Since energy stored is equivalent to the work done, the above equation can be written as:

$$W_E = \iiint_{v\to\infty} \left(\frac{1}{2}\cdot \boldsymbol{E} \bullet \boldsymbol{D}\right)\cdot dv = \frac{1}{2}\iint_s \left[\int_l \varepsilon E^2 d\ell\right] ds$$

Substitution of $E = \dfrac{\rho_s}{\varepsilon}$ from Equation 8.63a gives:

$$W_E = \frac{1}{2}\iint_S \left[\int_0^d \varepsilon \cdot \left(\frac{\rho_s}{\varepsilon}\right)^2 \cdot d\ell\right]\cdot dS = \frac{1}{2}\cdot\frac{\rho_s^2}{\varepsilon}\cdot S\cdot d = \frac{1}{2}\cdot\frac{\rho_s^2}{\varepsilon}\cdot v \tag{8.68a}$$

Equation 8.68a can be manipulated using Equations 8.67 and 8.65 to the following form:

$$W_E = \left\{\frac{\varepsilon\cdot d}{\varepsilon\cdot d}\right\}\cdot\frac{1}{2}\cdot\frac{\rho_s^2}{\varepsilon}\cdot S\cdot d = \frac{1}{2}\cdot\left(\frac{\varepsilon\cdot S}{d}\right)\cdot\left[\frac{\rho_s^2\cdot d^2}{\varepsilon^2}\right] = \frac{1}{2}\cdot C\cdot V_o^2 \tag{8.68b}$$

In Equation 8.68b, if C is replaced by Q/V_o, the following relation results:

$$W_E = \frac{1}{2}\cdot Q\cdot V_o \tag{8.68c}$$

In Equation 8.68c, if V_o is replaced by Q/C, it gives:

$$W_E = \frac{1}{2}\cdot\frac{Q^2}{C} \tag{8.68d}$$

8.6.6 Capacitance of a Coaxial Cable

Figure 8.13 shows a configuration of two concentric cylindrical conductors of radii *a* and *b*. Radius *a* is the outer radius of inner conductor and *b* the inner radius of outer conductor. This configuration is referred to as coaxial cable. The capacitance of coaxial cable of length L can be obtained as under.

Let us assume that this capacitance configuration has the following features:

- The space between the two cylindrical conductors is filled by a material of dielectric constant ε_r
- The inner and outer conductors carry surface charge distributions with density $+\rho_s$ and $-\rho_s$, respectively
- This charge distribution results in a uniform electric field throughout the space
- This field has only radial component, i.e. $E = E_\rho a_\rho$

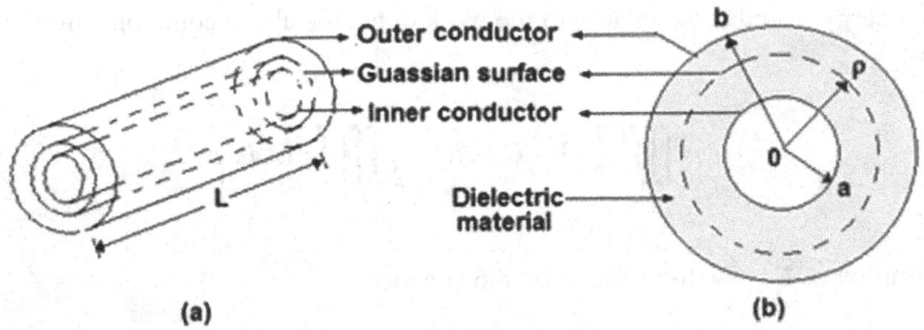

FIGURE 8.13 (a) Coaxial cable configuration and (b) cross-sectional coaxial cable.

The total charge Q residing over the surface of inner conductor of radius "a" can be obtained by multiplying the uniform surface charge density ρ_s by the surface area of this conductor of length L. Thus:

$$Q = 2\pi \cdot a \cdot L \cdot \rho_s \tag{8.69}$$

The potential difference between the conductors of radius a and b ($b > a$), is:

$$V_{ab} = -\int_a^b \boldsymbol{E} \bullet d\boldsymbol{\ell} = -\int_a^b E_\rho \boldsymbol{a}_\rho \bullet d\rho \boldsymbol{a}_\rho - \int_a^b E_\rho d\rho \tag{8.70a}$$

Also, in Equation 6.16:

$$E_\rho = \frac{a \cdot \rho_s}{\rho \varepsilon} \quad \text{over} \quad a \le \rho \le b \tag{8.70b}$$

Thus:

$$V_{ab} = -\int_a^b \frac{a \cdot \rho_s}{\rho \varepsilon} d\rho = -\frac{a \cdot \rho_s}{\varepsilon} \log \rho \Big|_a^b = -\frac{a \cdot \rho_s}{\varepsilon_0 \varepsilon_r} \log\left(\frac{b}{a}\right) \tag{8.70c}$$

$$C = \frac{Q}{V_{ab}} = \frac{2\pi \cdot a \cdot L \cdot \rho_s}{\frac{a \cdot \rho_s}{\varepsilon_0 \varepsilon_r} \log\left(\frac{b}{a}\right)} = \frac{2\pi \cdot \varepsilon_0 \cdot \varepsilon_r \cdot L}{\log\left(\frac{b}{a}\right)} \text{ Farad} \quad \text{or} \quad C = \frac{2\pi . \varepsilon_0 \cdot \varepsilon_r}{log\left(\frac{b}{a}\right)} \text{ F/m} \tag{8.71}$$

8.6.7 Capacitance between Two Spherical Surfaces

Figure 8.14 shows a configuration formed by two concentric spherical shells of radius a and b ($b > a$). The capacitance of this configuration can be obtained as under:

The electric field between the two spheres is:

$$\boldsymbol{E} = \frac{Q}{4\pi \varepsilon r^2} \boldsymbol{a}_r = E_r \boldsymbol{a}_r \quad \text{over} \quad a \le r \le b \tag{8.72}$$

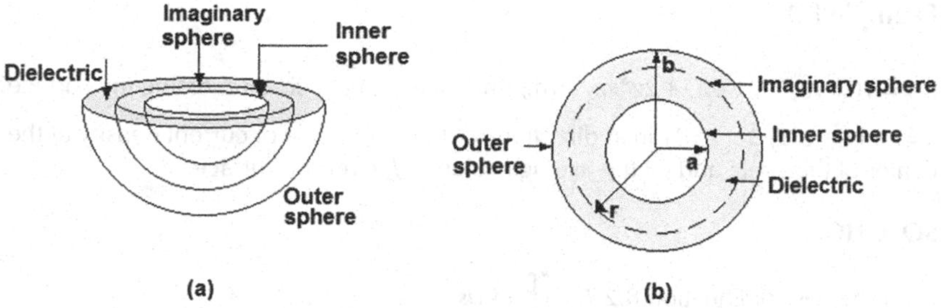

FIGURE 8.14 Configuration of spherical capacitor.

The potential difference between these spheres is:

$$V_{ab} = -\int_b^a \mathbf{E} \cdot \mathbf{dl} = -\int_b^a E_r \mathbf{a}_r \cdot dr\mathbf{a}_r = -\int_b^a E_r dr = -\int_b^a \frac{Q}{4\pi\varepsilon r^2} dr$$
$$= \frac{Q}{4\pi\varepsilon_o\varepsilon_r} \cdot \left(\frac{1}{a} - \frac{1}{b}\right) \tag{8.73}$$

where Q is the total charge on the outer surface of inner sphere. Thus, the capacitance is:

$$C = \frac{Q}{V_{ab}} = \frac{4\pi\varepsilon_o\varepsilon_r}{\left(\frac{1}{a} - \frac{1}{b}\right)} \tag{8.74}$$

If the outer sphere is made infinitely large the capacitance of an isolated sphere becomes:

$$C = 4\pi\varepsilon_o\varepsilon_r a \tag{8.75}$$

8.6.8 Classification of Capacitors

The capacitors can be classified in accordance with the dielectrics involved in their construction. These may be an air capacitor, a mica capacitor, a paper capacitor, etc. These can also be classified in accordance with the shapes of electrodes, e.g. a parallel plate capacitor, a cylindrical capacitor, etc.

8.6.9 Applications of Capacitors

The importance of capacitors is realized in ac circuits, wherein its frequency-dependant impedance contributes to the resonant circuits. Its other important applications include filters circuits wherein the time constant plays a pivotal role. In power, distribution, and other circuits these are used to improve the power factor. Bioelectricity is another important sphere of capacitor applications.

Example 8.1

If $J = 10x^3ya_x - 2y^3za_y + 2z^3xa_z$ A/m, find (a) the total current crossing the surface $x = 2, 1 \le y \le 2, 3 \le z \le 4$ in a_x direction; (b) the magnitude of current density at the center of this area; and (c) the average value of J_x over the surface.

SOLUTION

(a) In view of Equation 8.2 $I = \iint_s \boldsymbol{J} \cdot \boldsymbol{ds}$

Where $\boldsymbol{J} = 10x^3y\boldsymbol{a_x} - 2y^3z\boldsymbol{a_y} + 2z^3x\boldsymbol{a_z}$ and $\boldsymbol{ds} = dydz\boldsymbol{a_x}$

$$\boldsymbol{J} \cdot \boldsymbol{ds} = \left(10x^3y\boldsymbol{a_x} - 2y^3z\boldsymbol{a_y} + 2z^3x\boldsymbol{a_z}\right) \cdot dxdy\boldsymbol{a_x} = \left(10x^3y\right)dydz$$

$$I = \int_{z=3}^{4}\int_{y=1}^{2}\left(10x^3y\right)dydz = 5x^3y^2z\Big|_{y=1,z=3}^{y=2,z=4} = 15x^3\Big|_{x=2} = 120 \text{ A}$$

(b) $$\boldsymbol{J}\Big|_{\text{at center}} = 10x^3y\boldsymbol{a_x} - 2y^3z\boldsymbol{a_y} + 2z^3x\boldsymbol{a_z}\Big|_{x=2,y=1.5,z=3.5}$$

$$= 120\boldsymbol{a_x} - 23.625\boldsymbol{a_y} + 171.5\boldsymbol{a_z}$$

$$|J| = \sqrt{(120)^2 + (23.625)^2 + (171.5)^2} = 271.6296 \text{ A/m}^2$$

(c) Surface area $S = \int_{z=3}^{4}\int_{y=1}^{2} dydz = 1$

Average value of $J_x = \frac{I}{S} = \frac{120}{1} = 120$ A/m^2

Example 8.2

The current density in the (outward) radial direction, in the region near the origin, is given as $\boldsymbol{J} = 15r^{-1.5}\boldsymbol{a_r}$ A/m^2. Calculate the current crossing the spherical surface (a) $r = 1.6$ mm; (b) $r = 3.2$ mm.

SOLUTION

In view of Equation 8.2, $I = \iint_s \boldsymbol{J} \cdot \boldsymbol{ds}$,

Also, in view of Equation 3.33a, $\boldsymbol{ds_r} = r^2 \sin\theta \cdot d\theta \cdot d\varphi \cdot \boldsymbol{a_r}$
Thus

$$I = \iint_s 15r^{-1.5}\boldsymbol{a_r} \bullet r^2 \sin\theta \cdot d\theta \cdot d\varphi \cdot \boldsymbol{a_r} = \int_{\theta=0}^{\pi}\int_{\varphi=0}^{2\pi} 15 \cdot r^{1/2} \cdot \sin\theta \cdot d\theta \cdot d\varphi$$

$$= 30\pi r^{1/2}\int_{\theta=0}^{\pi} \sin\theta \cdot d\theta = -30\pi r^{1/2}\left(\cos\theta\right)\Big|_{\theta=0}^{\pi} = 60\pi r^{1/2}$$

(a) $I\Big|_{r=1.6\text{ mm}} = 60\pi r^{1/2}\Big|_{r=1.6\times10^{-3}} = 60\pi\times0.04 = 7.54\text{ Amp}$

(b) $I\Big|_{r=3.2\text{ mm}} = 60\pi r^{1/2}\Big|_{r=3.2\times10^{-3}} = 60\pi\times0.05657 = 10.663\text{ Amp}$

Example 8.3

At a specified temperature, for intrinsic germanium, the magnitude of both electron and hole volume charge densities are 2.35 C/m³, mobility $\mu_e = 0.36$, $\mu_h = 0.17$ in $\dfrac{\text{m}^2}{\text{V-sec}}$. Calculate its conductivity.

SOLUTION

Given $\rho_e = \rho_h = 2.35\text{ C/m}^3$, $\mu_e = 0.36$ and $\mu_h = 0.17\dfrac{\text{m}^2}{\text{V-sec}}$

In view of Equation 8.14b, $\sigma = -\rho_e\cdot\mu_e + \rho_h\cdot\mu_h$

$$\sigma = -2.35\times0.36 + 2.35\times0.17 = -0.4465\ \mho/\text{m}$$

Example 8.4

Find the magnitude of the current density within an aluminum sample if: (a) the electric field intensity E = 50 mV/m; (b) the drift velocity $v_d = 10^{-5}$ m/s; (c) the sample is in the form of a cube 1 mm on a side carrying a total current of 5A; (d) sample is in the form of a cube 1 mm on a side with a potential difference of 50 μV between two opposite faces.

SOLUTION

For Aluminum $\sigma = 3.82\times10^7\ \mho/\text{m}$, $\mu_e = 0.0012$

(a) In view of Equation 8.19a:

$$|J| = \sigma E = 3.82\times10^7\times50\times10^{-3} = 191\times10^4\text{ A/m}^2$$

(b) In view of Equation 8.8b: $|E| = \dfrac{v_d}{\mu_e} = \dfrac{10^{-5}}{0.0012} = 8.333\times10^{-3}$

$$|J| = \sigma E = 3.82\times10^7\times8.333\times10^{-3} = 31.83\times10^4\text{ A/m}^2$$

(c) In view of Equation 8.17b: $|J| = \dfrac{I}{S} = \dfrac{5}{10^{-6}} = 5\times10^6\text{ A/m}^2$

(d) In view of Equation 8.19a: $|J| = \sigma\dfrac{V}{L} = 3.82\times10^7\cdot\dfrac{50\times10^{-6}}{10^{-3}}$
$= 19.1\times10^5\text{ A/m}^2$

Example 8.5

A 50 m long circular copper conductor carries 10 mA current. Find the potential difference between its two ends if it is a: (a) solid conductor of 0.5 cm diameter; (b) hollow conductor of inner and outer diameter of 2 and 5 cm, respectively. For copper, $\sigma = 5.8\times10^{7}$ ℧/m.

SOLUTION

(a) For solid conductor:

$$S=\frac{\pi}{4}d^2=\frac{\pi}{4}\left(5\times10^{-3}\right)^2=19.635\times10^{-6}\ \text{m}^2$$

In view of Equation 8.20b: $R=\dfrac{L}{\sigma S}=\dfrac{10}{5.8\times10^{7}\times19.635\times10^{-6}}=8.78\ \text{m}\Omega$

In view of Equation 8.20a: $V=I\cdot R=10^{-2}\times8.78\times10^{-3}=8.78\times10^{-5}$ Volts

(b) For hollow conductor:

$$S=\frac{\pi}{4}\left\{\left(d_{\text{outer}}\right)^2-\left(d_{\text{inner}}\right)^2\right\}$$

$$=\frac{\pi}{4}\left\{\left(5\times10^{-3}\right)^2-\left(2\times10^{-3}\right)^2\right\}=\frac{\pi}{4}\left\{21\times10^{-6}\right\}$$

$$=16.493\times10^{-6}\ \text{m}^2$$

$$R=\frac{L}{\sigma S}=\frac{10}{5.8\times10^{7}\times16.493\times10^{-6}}=0.01045\ \Omega$$

$$V=I\cdot R=10^{-2}\times0.01045=0.1045\times10^{-3}\ \text{Volts}$$

Example 8.6

The current density in a region near the origin is given as $\boldsymbol{J}=10r^{-1}\mathbf{a}_r$ A/m². Find (a) the rate of increase of charge density ρ at the point where r = 1 mm; (b) the rate of increase of total charge within the sphere of 2 mm radius.

SOLUTION

(a) In view of Equation 8.24: $\dfrac{\partial\rho}{\partial t}=-\nabla\bullet\boldsymbol{J}$

Also, in view of Equation 4.21c: $\nabla\bullet\boldsymbol{J}=-\dfrac{1}{r^2}\cdot\dfrac{\partial\left(r^2\cdot J_r\right)}{\partial r}$

Thus, $\dfrac{\partial\rho}{\partial t}=-\dfrac{1}{r^2}\cdot\dfrac{\partial\left(r^2\cdot10r^{-1}\right)}{\partial r}=-\dfrac{1}{r^2}\cdot\dfrac{\partial(10r)}{\partial r}=-\dfrac{10}{r^2}$

$$\left.\frac{\partial\rho}{\partial t}\right|_{r=10^{-3}} = -\left.\frac{10}{r^2}\right|_{r=10^{-3}} = -10^7\ \text{C/m}^2\text{-sec}$$

(b) In view of Equation 8.25a, $\frac{dQ}{dt} = -I = -\oiint_s \boldsymbol{J} \bullet ds$

$$\frac{dQ}{dt} = -\int_0^{\pi}\int_0^{2\pi} 10r^{-1}\mathbf{a}_r \bullet r^2 \sin\theta \cdot d\theta \cdot d\varphi \cdot \mathbf{a}_r = -\int_0^{\pi}\int_0^{2\pi} 10r\sin\theta \cdot d\theta \cdot d\varphi$$

$$= -20\pi r\int_0^{\pi} \sin\theta \cdot d\theta = 40\pi r\Big|_{r=2\times10^{-3}} \approx 0.25\ \text{C/sec}$$

Example 8.7

Find the relaxation time for (a) bakelite ($\varepsilon_r = 4.9$, $\sigma = 10^{-9}$); (b) porcelain ($\varepsilon_r = 6$, $\sigma = 2 \times 10^{-13}$).

SOLUTION

In view of Equation 8.26b, $\tau = \frac{\varepsilon}{\sigma} = \frac{\varepsilon_0\varepsilon_r}{\sigma} = \frac{8.854\times10^{-12}\times\varepsilon_r}{\sigma}$

(a) For Bakelite:

$$\tau = \frac{8.854\times10^{-12}\times4.9}{10^{-9}} = 4.3385\times10^{-2}\ \text{second}$$

(b) For Porcelain:

$$\tau = \frac{8.854\times10^{-12}\times6}{2\times10^{-13}} = 2.656\times10^{2}\ \text{second}$$

Example 8.8

Find polarization (***P***) in a material that has: (a) an electric flux density of 2.5 μC/m² and an electric field intensity of 5kV/m; (b) an electric flux density of 3 μC/m² and electric susceptibility of 1.7.

SOLUTION

(a) In view of Equations 8.40, 8.41, 8.39, and 8.37 respectively:

$$\varepsilon = \frac{D}{E} = \frac{2.5\times10^{-6}}{5\times10^{3}} = 0.5\times10^{-9}$$

$$\varepsilon_r = \frac{\varepsilon}{\varepsilon_0} = \frac{0.5\times10^{-9}}{8.854\times10^{-12}} = 56.47$$

$$\chi_e = \varepsilon_r - 1 = 56.47 - 1 = 55.47$$

$$P = \varepsilon_o \cdot \chi_e \cdot E = 8.854 \times 10^{-12} \times 55.47 \times 5 \times 10^3 = 2.455\ \mu\text{C/m}$$

(b) In view of Equations 8.39, 8.40, and 8.37 respectively:

$$\chi_e = \varepsilon_r - 1 = 1.7 \quad \varepsilon_r = \chi_e + 1 = 2.7$$

$$E = \frac{D}{\varepsilon} = \frac{3 \times 10^{-6}}{8.854 \times 10^{-12} \times 2.7} = 12.55 \times 10^4\ \text{V/m}$$

$$P = \varepsilon_o \cdot \chi_e \cdot E = 8.854 \times 10^{-12} \times 1.7 \times 12.55 \times 10^4 = 1.889\ \mu\text{C/m}$$

Example 8.9

A point $P(-2,0,1)$ lies on the surface of a conductor where $E = 100a_x - 200a_y + 300a_z$ V/m. The conductor lies in free space. Find at P the magnitude of (a) E_n; (b) E_t; (c) ρ_s.

SOLUTION

(a) $E_n = |E| = \sqrt{(100)^2 + (200)^2 + (300)^2} = 100\sqrt{14} = 374.166$ V/m

(b) In view of Equation 8.43c:

$$E_t = 0\ \text{V/m}$$

(c) In view of Equation 8.45:

$$\rho_s = D_n = \varepsilon_0 E_n = \frac{10^{-9}}{36\pi} \cdot 374.166 = 3.308\ \text{nC/m}^2$$

Example 8.10

Two materials with the dielectric constants $\varepsilon_{R1} = 3$ and $\varepsilon_{R2} = 5$ are located in the regions $z \geq 0$ and $z \leq 0$, respectively. If $\mathbf{E}_2 = 10\mathbf{a}_x - 20\mathbf{a}_y + 30\mathbf{a}_z$ V/m, find (a) $\mathbf{D}_2$; (b) $\mathbf{D}_1$; (c) $\mathbf{E}1$; and (d) $\mathbf{P}_1$.

SOLUTION

In the linear relation $\mathbf{D} = \varepsilon\mathbf{E}$, $\mathbf{D}$ and $\mathbf{E}$ are in the same direction or parallel to each other. Furthermore, since z-axis is perpendicular to both the dielectric regions the normal components of all the field vectors will be their z-components. Also, for all the field vectors x- and y-components will be the tangential components. Thus, for a given $\mathbf{E}_2$, the normal component $\mathbf{E}_{n2} = 30\mathbf{a}_z$ and the tangential component $\mathbf{E}_{\tan 2} = 10\mathbf{a}_x - 20\mathbf{a}_y$.

(a) $\mathbf{D}_2 = \varepsilon_0 \varepsilon_{R2} \mathbf{E}_2 = 5\varepsilon_0 (10\mathbf{a}_x - 20\mathbf{a}_y + 30\mathbf{a}_z) = \varepsilon_0 (50\mathbf{a}_x - 100\mathbf{a}_y + 150\mathbf{a}_z)$

Thus $\boldsymbol{D}_{n2} = 150\varepsilon_0 \boldsymbol{a}_z$ and $\boldsymbol{D}_{t2} = 50\varepsilon_0 \boldsymbol{a}_x - 100\varepsilon_0 \boldsymbol{a}_y$

(b) In view of Equation 8.50b, $\boldsymbol{D}_{n1} = \boldsymbol{D}_{n2} = 150\varepsilon_0 \boldsymbol{a}_z$

In view of Equation 8.47, $\boldsymbol{D}_{t1} = \dfrac{\varepsilon_1}{\varepsilon_2}\boldsymbol{D}_{t2} = \dfrac{3\varepsilon_0}{5\varepsilon_0}\left(50\varepsilon_0 \boldsymbol{a}_x - 100\varepsilon_0 \boldsymbol{a}_y\right) = 30\varepsilon_0 \boldsymbol{a}_x - 60\varepsilon_0 \boldsymbol{a}_y$

Thus, $\boldsymbol{D}_1 = 30\varepsilon_0 \boldsymbol{a}_x - 60\varepsilon_0 \boldsymbol{a}_y + 150\varepsilon_0 \boldsymbol{a}_z$

(c) $\boldsymbol{E}_1 = \dfrac{\boldsymbol{D}_1}{\varepsilon_0 \varepsilon_{R1}} = \dfrac{\boldsymbol{D}_1}{3\varepsilon_0} = 10\boldsymbol{a}_x - 20\boldsymbol{a}_y + 50\boldsymbol{a}_z$

(d) In view of Equations 8.38 and 8.40:

$$\boldsymbol{P}_1 = \varepsilon_o \cdot \chi_e \cdot \boldsymbol{E} = \varepsilon_o \cdot \left(\varepsilon_{R1} - 1\right) \cdot \boldsymbol{E}_1 = 2\varepsilon_o \boldsymbol{E}_1 = \varepsilon_o \left(20\boldsymbol{a}_x - 40\boldsymbol{a}_y + 100\boldsymbol{a}_z\right)$$

Example 8.11

Find the capacitance of a parallel plate capacitor if (a) $\varepsilon_r = 500$, $S = 20$ cm^2, and $d = 1$ mm; (b) $V_0 = 200$ V and $W_E = 0.002$ Joules; (c) S and d given in (a) are doubled; (d) every linear dimension given in (a) is doubled; (e) every linear dimension given in (a) is halved.

SOLUTION

(a) In view of Equation 8.67:

$$C = \frac{\varepsilon \cdot S}{d} = \frac{\varepsilon_r \varepsilon_0 \cdot S}{d} = \frac{500 \times \dfrac{10^{-9}}{36\pi} \times 20 \times 10^{-4}}{1 \times 10^{-3}} = 0.00884\ \mu\text{F}$$

(b) In view of Equation 8.68b:

$$C = \frac{2W_E}{V_o^2} = \frac{2 \times 0.002}{(200)^2} = 0.1\ \mu\text{F}$$

(c) In the case that S and d given in (a) are doubled, there will be no change in the value of capacitor.

(d) When every linear dimension is doubled, the new area will become $4S$ and the new separation of plates will become 2d. Thus, the numerator in Equation 8.67 will get multiplied by 4 and the denominator by 2. As a result, the new capacitance will be twice that of case (a) or $C = 0.01768$ μF.

(e) When every linear dimension is halved, the new area will become $S/4$ and the new separation of plates will become $d/2$. Thus, the numerator in Equation 8.67 will get multiplied by 1/4 and the denominator by 1/2. As a result, the new capacitance will be half that of case (a) or $C = 0.00442$ μF.

PROBLEMS

P8.1 Find: (a) the total current crossing the surface located at $x = 5$ and bounded by $3 \le y \le 4, 4 \le z \le 5$ in a_x direction; (b) the magnitude of current density at the center of this area; and (c) the average value of J_x over the surface, if the current density $J = 8x^2ya_x - 4y^2za_y + 6z^2xa_z$ A/m.

P8.2 The current density in the (outward) radial direction, in the region near the origin, is given as $\boldsymbol{J} = 2.5r^{-2.5}\boldsymbol{a_r}$ A/m^2. Calculate the current crossing the spherical surface (a) $r = 2$ mm; (b) $r = 3$ mm.

P8.3 Calculate the conductivity for silicon if at a specified temperature the magnitude of volume charge density for both electrons and holes is 4.85mC/m^3 mobility $\mu_e = 0.12$, $\mu_h = 0.025$ both in $\dfrac{\text{m}^2}{\text{V-sec}}$.

P8.4 Find the magnitude of the current density within a copper sample with $\sigma = 5.8 \times 10^7$ ℧/m, $\mu_e = 0.0052$ if: (a) the electric field intensity $E = 100$ mV/m; (b) the drift velocity $v_d = 5 \times 10^{-5}$ m/s; (c) the sample is in the form of a cube 1.5 mm on a side carrying a total current of 3A; (d) sample is in the form of a cube 2 mm on a side with a potential difference of 40 μV between two opposite faces.

P8.5 Find the potential difference between the two ends of a 100 mlong, circular aluminum conductor carrying 10 mA current, if it is: (a) solid with 0.8 cm diameter; (b) hollow with inner and outer radii of 2 and 4 cm respectively. For aluminum, $\sigma = 3.72 \times 10^7$ ℧/m.

P8.6 The current density in a region near the origin is $\boldsymbol{J} = 5r^{-1.5}\boldsymbol{a_r}$ A/m^2. Find (a) the rate of increase of charge density ρ at the point where $r = 1.5$ mm; (b) the rate of increase of total charge within the sphere of 3 mm radius.

P8.7 Find the relaxation time for (a) Quartz ($\varepsilon_r = 3.8$, $\sigma = 10^{-17}$); (b) Soil (sandy) ($\varepsilon_r = 3$, $\sigma = 10^{-5}$); (c) Distilled water ($\varepsilon_r = 80$, $\sigma = 2 \times 10^{-4}$).

P8.8 Find polarization ($\boldsymbol{P}$) in a material that has (a) $n = 10^{20}$ molecules/m^2, each with a dipole moment $p = 3.5 \times 10^{-26}$ C·m; (b) $E = 7.5$kV/m and a relative permittivity of 2.6.

P8.9 A point P(−1,2,1) lies on the surface of a conductor whereat $E = 50a_x + 100a_y - 200a_z$ V/m. The conductor lies in free space. Find at P the magnitude of (a) E_n; (b) E_t; (c) ρ_s.

P8.10 Two materials with the dielectric constants $\varepsilon_{R1} = 2$ and $\varepsilon_{R2} = 4$ are located in the regions $z \le 0$ and $z \ge 0$, respectively. If $\boldsymbol{E}_2 = 5\boldsymbol{a_x} + 30\boldsymbol{a_y} - 10\boldsymbol{a_z}$ V/m find (a) $\boldsymbol{D}_2$; (b) $\boldsymbol{D}_1$; (c) $\boldsymbol{E}_1$; and (d) $\boldsymbol{P}_1$.

P8.11 Find the capacitance (a) between two spheres of 4 cm and 5 cm radii separated by a dielectric having $\varepsilon_r = 3.6$; (b) of an isolated sphere of 5 cm diameter in free space.

P8.12 Find the capacitance of a 1 m long coaxial cable having outer radius of inner conductor of 0.1 cm and inner radius of outer conductor of 0.5 cm. The space between these conductors is filled by (a) air; (b) polyethylene ($\varepsilon_r = 2.26$).

DESCRIPTIVE QUESTIONS

Q8.1 List different types of currents and describe the essential features of each.

Q8.2 Prove that the convection current density can be given by the relation $\boldsymbol{J} = \rho \cdot \boldsymbol{U}$.

Q8.3 Prove that the conduction current density bears the relation $\boldsymbol{J} = \sigma \cdot \boldsymbol{E}$.

Q8.4 Explain the meaning of drift velocity. How it is related to current?

Q8.5 Discuss the current flow in metallic conductors and obtain the expression for *Ohm's law* in point form.

Q8.6 Explain the meaning of *skin effect*, *proximity effect*, and *ampacity*.

Q8.7 Derive the *continuity equation* in integral and point forms.

Q8.8 Obtain the boundary conditions for E and D fields at the conductor-dielectric interface. Why these conditions differ from those at dielectric-dielectric interface?

Q8.9 Derive the following relations and explain the meaning of terms involved therein.

(a) $E_2 = E_1\sqrt{\cos^2 \alpha_1 + \left(\frac{\varepsilon_1}{\varepsilon_2}\right)^2 \sin^2 \alpha_1}$

(b) $D_2 = D_1\sqrt{\sin^2 \alpha_1 + \left(\frac{\varepsilon_2}{\varepsilon_1}\right)^2 \cos^2 \alpha_1}$

Q8.10 Give the meaning, types, and applications of capacitor. Derive the relation for capacitance between two parallel plates separated by a distance d.

Q8.11 Derive the relation for the energy stored in a capacitor.

FURTHER READING

Given at the end of Chapter 15.

9 Electrostatic Boundary Value Problems Involving Laplacian Fields

9.1 INTRODUCTION

In mathematical terms, a *boundary value problem* is a differential equation with a set of additional constraints. These constraints are called the *boundary conditions*. Thus, a solution to a boundary value problem is a solution to the differential equation that satisfies some specified boundary conditions. Boundary value problems arise in several disciplines, including that of electromagnetics. It is necessary to address a boundary value problem in such a way that for a given input it yields a unique solution. Much theoretical work in the field of partial differential equations has been done to address such problems.

In Chapters 5 to 8, many field quantities (viz. charge, electric field intensity, electric flux, electric flux density, potential, current and current density, etc.) were introduced. In most of these problems, neither the charge distribution nor the potential distribution are known. The problems wherein only electrostatic conditions (viz. the distribution of electric charges and potentials at some boundaries) are known can be solved by using Poisson and Laplace's equations to obtain E and V fields. The method of images, another method of solving boundary problems, is also sometimes helpful in tackling such cases.

This chapter begins with the definition of a Laplacian field, its relations in different coordinate systems, and their solutions. It also introduces the concept of uniqueness of the solution and describes the related theorem, called the Uniqueness Theorem. The methods to obtain solutions to one-dimensional (1D), two-dimensional (2D), and three-dimensional (3D) Laplacian fields in a Cartesian coordinate system are discussed at length. The solutions to these problems in cylindrical and spherical space coordinate systems are also briefly discussed.

This chapter is divided into two parts. The first, Sections 9.2 and 9.3, deals with general descriptions and field relations. The second, Sections 9.4 to 9.12, includes examples of Laplacian fields in one, two, and three dimensions.

9.2 NATURE OF FIELD RELATIONS AND UNIQUENESS OF SOLUTIONS

9.2.1 LAPLACIAN AND POISSON'S FIELDS

Consider the following field relations derived earlier in Chapters 6 and 7:

$$\boldsymbol{D} = \varepsilon \cdot \boldsymbol{E} \tag{6.5}$$

$$\nabla \bullet \boldsymbol{D} = \rho \tag{6.19a}$$

$$\boldsymbol{E} = -\nabla V \tag{7.6}$$

In view of Equations 6.5, 6.19a, and 7.6, one gets:

$$\nabla \bullet \boldsymbol{D} = \nabla \bullet (\varepsilon \boldsymbol{E}) = \varepsilon \cdot (\nabla \bullet \boldsymbol{E}) = \varepsilon \cdot \nabla \bullet (-\nabla V) = -\varepsilon \cdot \nabla^2 V = \rho \tag{9.1}$$

or

$$\nabla^2 V = -\rho / \varepsilon \tag{9.1a}$$

The RHS of Equation 9.1a is a known function of space coordinates.

In a charge-free region ($\rho = 0$) Equation 9.1a reduces to:

$$\nabla^2 V = 0 \tag{9.1b}$$

The relation given by Equation 9.1a is called *Poisson's equation*, whereas that given by Equation 9.1b is referred to as *Laplace's equation*.

The *Laplacian field* is defined as a field that gives zero value if it is operated upon by the Laplacian operator, ∇^2. Thus, the field expressed by Equation 9.1b can be called the Laplacian field.

Poisson's field is defined as a field that does not give zero value if it is operated upon by the Laplacian operator, ∇^2. Thus, the field expressed by Equation 9.1a can be referred to as Poisson's field.

9.2.2 Laplacian Field as Scalar or Vector Function

In view of Equation 9.1a, it is evident that when the operand is a scalar function of space coordinates, the Laplacian operator, ∇^2, is defined as the divergence of gradient (Equation 4.40d). Thus, if V is an arbitrary scalar function of space coordinates:

$$\nabla^2 V \stackrel{\text{def}}{=} \nabla \bullet (\nabla V) \tag{9.2}$$

In the Laplace equation, Equation 9.1b, the function V is called the *scalar Laplacian field.*

On the other hand, if the operand (say, $\boldsymbol{A}$) is a vector function of space coordinates, then in view of the following vector identity, Equation 4.41d is defined as:

$$\nabla^2 \boldsymbol{A} \stackrel{\text{def}}{=} \nabla (\nabla \bullet \boldsymbol{A}) - \nabla \times \nabla \times \boldsymbol{A} \tag{9.3}$$

Now, if vector $\boldsymbol{A}$ is a Laplacian field, it must satisfy the Laplace equation:

$$\nabla^2 \boldsymbol{A} = 0 \tag{9.4a}$$

If vector $\boldsymbol{A}$ is not a Laplacian field, it satisfies Poisson's equation:

$$\nabla^2 \boldsymbol{A} = -\mu \boldsymbol{J} \tag{9.4b}$$

where the RHS quantity is a known function of space coordinates.

Similar to scalar Laplacian field, function **A** in Equation 9.4a can be referred to as a *vector Laplacian field.*

9.2.3 Laplacian Operator

The Laplacian operator is a scalar differential operator. In the Cartesian system of space coordinates, its value was given by Equation 4.34. The Laplacian of the scalar electric potential V, in a Cartesian, cylindrical, and spherical system of space coordinates was given by Equations 4.35a through 4.35c, respectively. These are reproduced here:

$$\nabla^2 V \equiv \frac{\partial^2 V}{\partial x^2} + \frac{\partial^2 V}{\partial y^2} + \frac{\partial^2 V}{\partial z^2} \quad \text{(Cartesian)} \tag{9.5}$$

$$\nabla^2 V = \frac{1}{\rho} \cdot \frac{\partial}{\partial \rho}\left(\rho \frac{\partial V}{\partial \rho}\right) + \frac{1}{\rho^2} \cdot \frac{\partial^2 V}{\partial \varphi^2} + \frac{\partial^2 V}{\partial z^2} \quad \text{(Cylindrical)} \tag{9.6}$$

$$\nabla^2 V = \frac{1}{r^2} \cdot \frac{\partial}{\partial r}\left(r^2 \frac{\partial V}{\partial r}\right) + \frac{1}{r^2 \sin\theta} \cdot \frac{\partial}{\partial \theta}\left(\sin\theta \frac{\partial V}{\partial \theta}\right) + \frac{1}{r^2 \sin^2\theta} \cdot \frac{\partial^2 V}{\partial \varphi^2} \quad \text{(Spherical)} \tag{9.7}$$

The scalar electric potential V and the vector magnetic potential $\boldsymbol{A}$ in source free homogeneous regions are examples of Laplacian field.

9.2.4 Scalar Electric Potential

It is usually convenient to determine the distribution of scalar electric potential rather than a vector field with three scalar components. In view of the identity:

$$\nabla \times (\nabla V) \equiv 0 \tag{9.8}$$

it may be concluded that since for the time invariant electromagnetic fields, curl of the electric field intensity is zero; it is possible to express the electric field intensity as the negative gradient of the scalar electric potential. The choice of the minus sign is rather arbitrary. This, however, shows that the field vector is directed from high potential point to the low potential point. Once the distribution of the scalar potential is found, the vector field can be readily obtained from its gradient relation given as:

$$\boldsymbol{E} = -\nabla V \tag{7.6}$$

As an accelerated charge radiates power in the form of an electromagnetic wave, the following points need to be noted.

1. The scalar electric potential V is only a mathematical tool for the calculation of Electric field intensity vector $\boldsymbol{E}$, and it may not necessarily have any physical meaning as often associated with the work done or energy stored in electric field.
2. The equipotential surfaces are the family of those surfaces where for each such surface the scalar electric potential has a constant value at each and every point on the surface. This value, however, shall be different for the different surfaces.
3. The conjecture that no work will be done if a point charge is placed on an equipotential surface and moved around without leaving this surface is patently false. The field ceases to be electrostatic if any charge, including a test charge, is moving. Thus, if a charge is moved in a closed path, work will be done even if no other charge exists within a finite distance.
4. The equipotential surfaces are the family of those surfaces that are orthogonal to the flux lines. Indeed, it is purely a mathematical notion.

One of the important applications of the distribution of scalar electric potential is the determination of capacitance of a system of conductors.

The attributes cited above, however, are restricted to static electric fields. For high-frequency fields, both the scalar electric potential as well as the vector magnetic potential are needed to express the electric field intensity vector $\boldsymbol{E}$. Let us consider again Maxwell's equation given by Equation 6.19a:

$$\nabla \cdot \boldsymbol{D} = \rho \tag{6.19a}$$

In this equation, ρ is the volume charge density in coulombs per meter cube and the vector $\boldsymbol{D}$ is the displacement or electric flux density vector. Equation 6.5 relating vector $\boldsymbol{E}$ to these quantities is again reproduced here:

$$\boldsymbol{D} = \varepsilon \boldsymbol{E} \tag{6.5}$$

In Equation 6.5, the symbol ε indicates the permittivity of the medium at the point the electric field is specified. For a homogeneous region, ε is constant at every point in the region.

For a charge-free region, Equation 6.19a becomes:

$$\nabla \cdot \boldsymbol{D} = 0 \tag{9.9}$$

Therefore, for charge-free regions:

$$\nabla \cdot (\varepsilon \boldsymbol{E}) \equiv \boldsymbol{E} \cdot \nabla \varepsilon + \varepsilon (\nabla \cdot \boldsymbol{E}) = -\nabla V \cdot \nabla \varepsilon - \varepsilon (\nabla^2 V) = 0 \tag{9.10}$$

This equation shows that the scalar electric potential V will satisfy Laplace equation provided that either, (i) the region is homogenous, thus the gradient of the permittivity ε is zero, or (ii) the gradients of the scalar potential V and that of the permittivity ε, are mutually perpendicular, resulting their zero scalar product. It may be noted that the scalar electric potential V, in charge-free regions satisfies homogenous differential equation. For regions with charge density distributions, the zero on the RHS of Equation 9.10 needs to be replaced by the volume charge density ρ. Consequently, for homogeneous regions the scalar potential will satisfy Poisson's equation instead of the Laplace equation. Poisson's equation is an inhomogeneous differential equation.

9.2.5 Uniqueness Theorem

In general, there can be infinite different solutions for Laplace and Poisson equations. However, under certain boundary conditions, unique solutions for these equations can be obtained. Thus, in order to obtain the unique solution for the Laplace equation, it is necessary that certain boundary conditions must be specified in the problem. A solution of the Laplace equation that satisfies these boundary conditions is referred to as the *unique solution.* This point can be better elaborated in view of the *Uniqueness Theorem.*

The Uniqueness Theorem states:

> If a solution exists for a given equation under certain specified boundary conditions, it is the only possible solution and may be referred to as unique.

The objective of the Uniqueness Theorem is to identify those boundary conditions a solution must satisfy in order to render the solution unique.

To verify the above statement, let V_1 and V_2 be any two distinct solutions for the Laplace equation for potential distribution in a given volume v. The difference potential V_o is defined as:

$$V_o \stackrel{\text{def}}{=} V_1 - V_2 \tag{9.11}$$

Since V_1 and V_2 individually satisfy the Laplace equation, their difference V_o also satisfies the Laplace equation:

$$\nabla^2 V_o = 0 \tag{9.12}$$

Consider the identity:

$$\nabla \bullet (V_o \nabla V_o) \equiv |\nabla V_o|^2 + V_o \cdot \nabla^2 V_o \tag{9.13}$$

In view of Equation 9.12, Equation 9.13 reduces to:

$$\nabla \bullet (V_o \nabla V_o) = |\nabla V_o|^2 \tag{9.14}$$

Furthermore, the integration of both sides of Equation 9.14 over the volume v, and the application of the divergence theorem gives:

$$\oiint_s (V_o \nabla V_o) \bullet ds = \iiint_v |\nabla V_o|^2 \, dv \tag{9.15}$$

Where s indicates the surface which internally as well as externally bounds the volume v.

Under certain boundary conditions satisfied by the two solutions (V_1 and V_2) the LHS of Equation 9.15 reduces to zero. The zero value for the volume integration on the RHS of this equation implies that since the integrand $|\nabla V_o|^2$ is not negative anywhere, ∇V_o must be zero at every point in the volume v. Therefore, the zero value for the LHS implies that the difference potential must be a constant, i.e. the two solutions, at every point in the volume as well as *at every point on the bounding surface* of this volume, must only differ by a constant.

The conditions for the LHS of Equation 9.15 to be zero are: (1) V_o is zero on the entire boundary surface, or (2) V_o is zero on parts of the bounding surface s, and the normal derivative of V_o is zero on its remaining parts. This results in zero value for the constant difference potential. Therefore, to obtain the unique solution, potentialdistribution must be defined over parts of the boundary surface and its normal derivative over the remaining parts. Once the unique expression for potential distribution in a region is found, using this expression its normal derivative over the entire boundary can be uniquely obtained. Therefore, at no part on the boundary surface, both potential and its normal derivative can be specified independently. However, there is an exception. If the normal derivative of the potential is defined over the entire boundary surface, the value of the potential *at one, and only one, point* on the boundary surface, or anywhere in the volume, needs to be defined in order to obtain the unique solution. This is because solutions found using exclusively the normal derivative of the potential defined over the entire boundary surface differ from one another by an arbitrary constant. This constant can be evaluated from the known potential at any given point in the region. It is important to note that one or the other (but not both) boundary condition must be specified over the entire bounding surface, i.e. no part of the boundary surface should be left with undefined boundary condition.

9.2.6 Classification of Field Problems

The problems involving scalar electric potential can be classified in accordance with the coordinate systems and the dimensions. Depending on the shape of physical configuration wherein the field distribution is to be obtained a problem may be solved either by using Cartesian, cylindrical, or spherical coordinates. There are other useful coordinate systems; these are, however, not considered here. The field problems, in general, involve three-dimensional (3D) physical configurations. In cases where there is no field variation in one of the directions the problem becomes a two-dimensional (2D) field problem. Finally, if the field varies only along one coordinate the problem becomes a one-dimensional (1D) field problem. The possible variations of

V with one or more coordinates, in the three systems of space coordinates are given in Table 9.1.

With reference to this table the following points need to be noted:

- V = $f(\ldots)$ indicates that V is a function of the parameter(s) within the small bracket.
- In the 1D case the partial derivatives are to be replaced by complete (or total) derivatives as and where they appear.
- The solution for a 1D Cartesian coordinate problem {viz. V = $f(x)$, V = $f(y)$ V = $f(z)$} and 1D cylindrical coordinate problem {viz. V = $f(z)$} is similar. Thus, only one of these cases is considered.
- The solution for a 2D Cartesian coordinate problem {viz. V = $f(x, y)$, V = $f(y, z)$, and V = $f(z, x)$} is similar. Thus, only one of these cases is considered in the subsequent sections.
- All other cases of 2D and 3D problems are discussed separately.

The following sections describe the methods for obtaining solutions to the problems as per the above classification.

9.3 ONE-DIMENSIONAL FIELD PROBLEMS

This section deals with the one-dimensional field problems in different coordinate systems. Solutions are obtained by taking specific examples.

9.3.1 Field Problems in Cartesian Coordinates

The Laplacian for the scalar function V in Cartesian coordinates given by Equation 9.5 is reproduced here:

$$\nabla^2 V \equiv \frac{\partial^2 V}{\partial x^2} + \frac{\partial^2 V}{\partial y^2} + \frac{\partial^2 V}{\partial z^2} \tag{9.5}$$

This equation contains three terms on the RHS. In a one-dimensional case, the potential V is a function of only one variable. If it is a function of, say, x, the terms involving y and z derivatives will disappear. Likewise, if V is a function of either y

TABLE 9.1
Variation of V in Accordance with the Coordinate Systems and Dimensions

	Cartesian Coordinates	Cylindrical Coordinates	Spherical Coordinates
1D	V = f\{(x) or (y) or (z)\} alone	V = f\{(ρ) or (ϕ) or (z)\} alone	V = f\{(r) or (θ) or (ϕ)\} alone
2D	V = f\{(x and y) or (y and z), or (x and z)\}	V = f\{(ρ and ϕ) or (ϕ and z) or (ρ and z)\}	V = f\{(r and θ) or (θ and ϕ) or (r and ϕ)\}
3D	V = $f(x, y, \text{and } z)$	V = $f(\rho, \phi, \text{and } z)$	V = $f(r, \theta, \text{and } \phi)$

or z, the terms related with x and z or x and y derivatives will disappear. In view of Equation 9.1b, the appropriate term is to be equated to zero. Thus, on replacing partial derivative with total derivative we get:

$$\left.\frac{d^2V}{dx^2}=0\quad(\mathrm{a})\quad\text{or}\quad\frac{d^2V}{dy^2}=0\quad\text{or}\quad\frac{d^2V}{dz^2}=0\quad(\mathrm{c})\right\} \tag{9.16}$$

A one-dimensional field problem will require the solution of one of these three equations. The selection will depend on the coordinate along which the field varies. Since the procedure of solution to any of these relations will remain the same only one of these needs to be considered. In the following subsections only the last of these three equations is considered through various examples. In these examples the potential field is evaluated for both homogeneous and heterogeneous media. The medium between two parallel plates may become heterogeneous in different ways. A few of such cases are discussed in the subsequent subsections.

9.3.1.1 Potential Fields in Homogeneous Medium

Figure 9.1 shows a pair of large parallel conducting plates with a separation distance d between them. The upper surface of the bottom plate is located at $z = 0$ plane, while the lower surface of the top plate is located at $z = d$ plane. The top plate is at zero potential and the bottom plate is maintained at a potential V_o. The region between the two plates extends over $0 < z < d$, for all values of x and y; is free space. This becomes the configuration of an idealized air capacitor. In this configuration the distribution of the electrostatic potential between the two plates is a function of the z-coordinate only. In this case the potential distribution is to be obtained by solving Equation 9.16c.

The solution for Equation 9.16c can be written as:

$$V = a \cdot z + b \tag{9.17}$$

where a and b indicate two arbitrary constants that are to be determined in order to complete the solution. These can be obtained using the following boundary conditions:

$$\left.V\right|_{z=d}=0\quad(\mathrm{a})\quad\text{and}\quad\left.\left.V\right|_{z=0}=V_o\quad(\mathrm{b})\right\} \tag{9.18}$$

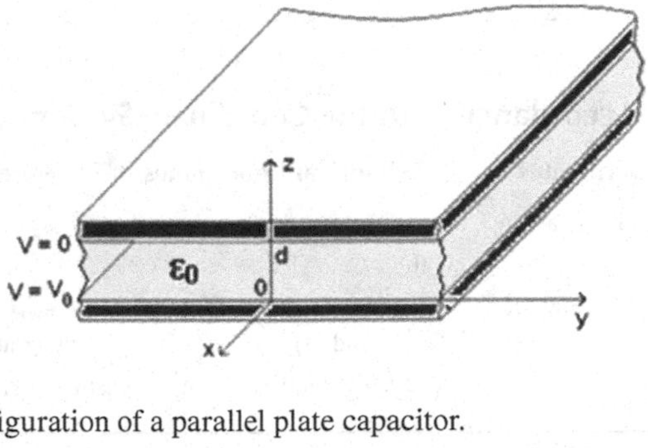

FIGURE 9.1 Configuration of a parallel plate capacitor.

In view of Equations 9.18a and 9.18b, we get:

$$\left. a = -\frac{V_o}{d} \quad (\text{a}) \quad \text{and} \quad b = V_o \quad (\text{b}) \right\} \tag{9.19}$$

Thus,

$$V = -V_o \cdot \left(\frac{z-d}{d} \right) \tag{9.20}$$

From Equation 9.20, the electric field intensity in the free space between the two plates can be obtained as:

$$\boldsymbol{E} = -\nabla V = -\frac{d}{dz}\left[-V_o \cdot \left(\frac{z-d}{d} \right) \right] \boldsymbol{a}_z = \left(V_o \frac{1}{d} \right) \boldsymbol{a}_z \tag{9.21}$$

This shows that the electric field intensity between the two plates has only the z component and is a constant, viz. V_o/d. The densities of electric charges deposited on the two plate surfaces are:

$$\left. \rho_s\big|_{z=0} = D_z\big|_{z=0} = V_o \frac{1}{d}\varepsilon_o \quad (\text{a}), \quad \text{and} \quad \rho_s\big|_{z=d} = -D_z\big|_{z=d} = -V_o \frac{1}{d}\varepsilon_o \quad (\text{b}) \right\} \tag{9.22}$$

Thus, capacitance per unit plate area is:

$$C = \frac{\left(\rho_s\big|_{z=0} \right)}{V_o} = \frac{1}{d}\varepsilon_o \tag{9.23}$$

9.3.1.2 Two Homogeneous Subregions along the *y*-Axis

This case is shown in Figure 9.2. In this case the space between the two plates is divided into two homogeneous subregions named Regions 1 and 2. Region 1 extends over $-\infty < y < 0$ and Region 2 extends over $0 < y < \infty$, for $-\infty < x < \infty$ and $0 \le z \le d$. These regions contain two dielectrics with permittivity ε_1 for Region 1 and ε_2 for Region 2. This results in piecewise homogeneous medium between the two plates. The scalar electric potential in each of these subregions satisfies the Laplace

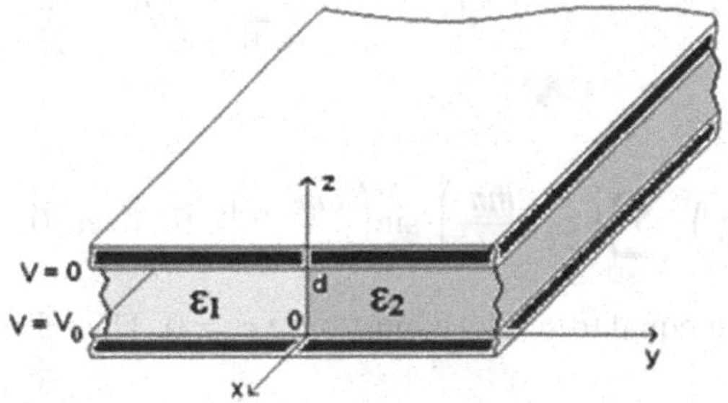

FIGURE 9.2 A parallel plate capacitor with two subregions along the *y*-axis.

equation. To determine the distributions of scalar electric potentials in these subregions it is necessary to specify the potential distribution at the boundary surface between the two subregions, i.e. at the surface $y = 0$. Let the unknown potential distribution be given as follows:

$$V\Big|_{y=0} = -V_o \cdot \left(\frac{z-d}{d}\right) + \sum_{m\text{-odd}}^{\infty} c_m \cdot \sin\left(\frac{m\pi}{d} \cdot z\right) \tag{9.24}$$

where c_m indicates the Fourier coefficient of the half range Fourier series.

This equation aims to synthesize the potential function that satisfies the given potentials at $z = 0$ and at $z = d$. This variation in the z-direction may not be linear. The half range Fourier sine series with unknown Fourier coefficients introduces an arbitrary variation without perturbing the potential values at $z = 0$ and d.

Subject to the boundary conditions given by Equations 9.18a, 9.18b, and 9.24, distributions of potentials in the two subregions obtained by solving the Laplace equation are:

$$V_1 = -V_o \cdot \left(\frac{z-d}{d}\right) + \sum_{m\text{-odd}}^{\infty} c_m \cdot \sin\left(\frac{m\pi}{d} \cdot z\right) \cdot e^{\frac{m\pi}{d} y} \tag{9.25a}$$

and

$$V_2 = -V_o \cdot \left(\frac{z-d}{d}\right) + \sum_{m\text{-odd}}^{\infty} c_m \cdot \sin\left(\frac{m\pi}{d} \cdot z\right) \cdot e^{-\frac{m\pi}{d} y} \tag{9.25b}$$

The Fourier coefficient c_m can be determined by using the boundary condition that states the continuity of the normal component of the electric flux density between two dielectric subregions, i.e.:

$$-\varepsilon_1 \frac{\partial V_1}{\partial y}\bigg|_{y=0} = -\varepsilon_2 \frac{\partial V_2}{\partial y}\bigg|_{y=0} \tag{9.26}$$

Therefore, using Equations 9.25a, 9.25b, and 9.26, we get:

$$-\varepsilon_1 \sum_{m\text{-odd}}^{\infty} c_m \cdot \frac{m\pi}{d} \cdot \sin\left(\frac{m\pi}{d} \cdot z\right) = \varepsilon_2 \sum_{m\text{-odd}}^{\infty} c_m \cdot \frac{m\pi}{d} \cdot \sin\left(\frac{m\pi}{d} \cdot z\right) \tag{9.27a}$$

or,

$$(\varepsilon_1 + \varepsilon_2) \cdot \sum_{m\text{-odd}}^{\infty} \left(c_m \cdot \frac{m\pi}{d}\right) \cdot \sin\left(\frac{m\pi}{d} \cdot z\right) = 0 \quad \text{over } 0 < z < d \tag{9.27b}$$

Since $(\varepsilon_1 + \varepsilon_2)$ is not equal to zero, we must have $c_m = 0$. Therefore, we still have:

$$\left. V_1 = V_2 = -V_o \cdot \left(\frac{z-d}{d}\right) \; (\text{a}) \quad \text{and} \quad E_{1z} = E_{2z} = \frac{V_o}{d} \; (\text{b}) \right\} \tag{9.28}$$

while

$$D_{1z} = \varepsilon_1 \cdot \frac{V_o}{d} \quad (\mathrm{a}) \quad \text{and} \quad D_{2z} = \varepsilon_2 \cdot \frac{V_o}{d} \quad (\mathrm{b})\Big\} \tag{9.29}$$

This shows that all flux lines are straight lines perpendicular to the plate surfaces.

The densities of electric charges deposited on the two plate surfaces are:

$$\rho_s\Big|_{z=0} = \begin{cases} D_{1z}\Big|_{z=0} = V_o \dfrac{1}{d}\varepsilon_1 & \text{over } -\infty < y < 0 \\ D_{2z}\Big|_{z=0} = V_o \dfrac{1}{d}\varepsilon_2 & \text{over } 0 < y < +\infty \end{cases} \tag{9.30a}$$

and

$$\rho_s\Big|_{z=d} = \begin{cases} -D_{1z}\Big|_{z=d} = -V_o \dfrac{1}{d}\varepsilon_1 & \text{over } -\infty < y < 0 \\ -D_{2z}\Big|_{z=d} = -V_o \dfrac{1}{d}\varepsilon_2 & \text{over } 0 < y < +\infty \end{cases} \tag{9.30b}$$

Note that the distribution of the surface charge density is discontinuous at $y = 0$.

9.3.1.3 Two Homogeneous Subregions along the z-Axis

In Figure 9.3, we can see that the space between the two plates is again divided into two homogeneous subregions, this time along the z-axis. Thus, Region 1 extends over $0 < z < d_1$ and Region 2 over $d_1 < z < (d_1 + d_2)$, for $-\infty < x < \infty$ and $-\infty < y < \infty$. These regions are filled with homogeneous dielectric materials such that the permittivity for Region 1 is ε_1 and that for the Region 2 is ε_2. This is another case of the piecewise homogeneous medium. The scalar electric potential in each of these subregions satisfies the Laplace equation. Furthermore, it may be seen that the surface $z = z_0$, where $0 \le z_o \le (d_1 + d_2)$; is an equipotential surface. To find the potential distribution in these subregions, let us assume that:

$$V_1\Big|_{z=d_1} = V_2\Big|_{z=d_1} \stackrel{\text{def}}{=} V_o' \tag{9.31}$$

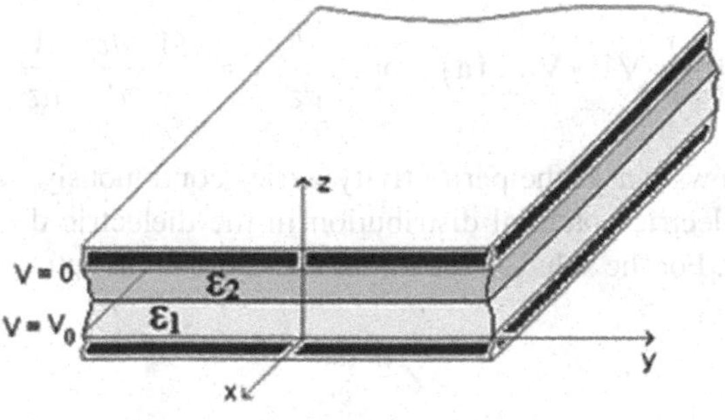

FIGURE 9.3 A parallel plate capacitor with two subregions along the z-axis.

where V_o' indicates the unknown potential at the boundary surface $z = d_1$, between the two subregions. Solutions of the Laplace equation for the two subregions can now be given as:

$$\left. V_1 = z \cdot \frac{V_o'}{d_1} \quad (a) \qquad V_2 = -(z - d_1 - d_2) \cdot \frac{V_o'}{d_2} + (z - d_1) \cdot \frac{V_o}{d_2} \quad (b) \right\} \tag{9.32}$$

To determine the unknown potential V_o', at the boundary surface $z = d_1$, we can use the boundary condition that states the continuity of the normal component of the electric flux density between two dielectric subregions, i.e.:

$$-\varepsilon_1 \left. \frac{\partial V_1}{\partial z} \right|_{z=d_1} = -\varepsilon_2 \left. \frac{\partial V_2}{\partial z} \right|_{z=d_1} \tag{9.33}$$

Therefore,

$$\left. \varepsilon_1 \cdot \frac{V_o'}{d_1} = \varepsilon_2 \cdot \left[\frac{V_o}{d_2} - \frac{V_o'}{d_2} \right] \quad (a) \quad \text{or} \quad V_o' \cdot \left[\frac{\varepsilon_1}{d_1} + \frac{\varepsilon_2}{d_2} \right] = V_o \cdot \frac{\varepsilon_2}{d_2} \quad (b) \right\} \tag{9.34}$$

From Equation 9.34b, we get:

$$V_o' = V_o \cdot \frac{d_1 \cdot \varepsilon_2}{(d_1 \cdot \varepsilon_2 + d_2 \cdot \varepsilon_1)} \tag{9.35}$$

9.3.1.4 Continuously Varying Permittivity between Plates

In the last two examples, the variation of permittivity resulted in piecewise homogeneous dielectric region between the two large parallel conducting plates. Thus, the scalar electric potential distribution in each subregion satisfies the Laplace equation. In the present example it is assumed that the permittivity of the dielectric between the two parallel plates is continuously varying, $\varepsilon(z)$ and thus it does not result in piecewise homogeneous region. This configuration is shown in Figure 9.4. Because this variation is in the z-direction, and invariant in the x- and y-directions, the potential distribution V will be a function of only the z coordinates. Thus, from Equation 9.10, we get:

$$\left. \nabla^2 V = -\frac{1}{\varepsilon} \cdot \nabla V \bullet \nabla \varepsilon \quad (a) \quad \text{or} \quad \frac{d^2}{dz^2} V = -\frac{1}{\varepsilon} \cdot \frac{d\varepsilon}{dz} \cdot \frac{dV}{dz} \quad (b) \right\} \tag{9.36a}$$

This equation shows that if the permittivity varies continuously, say in the z-direction, the scalar electric potential distribution in the dielectric does not satisfy the Laplace equation. For the solution for Equation 9.36b, let us set:

$$Z \overset{\text{def}}{=} \frac{dV}{dz} \tag{9.37a}$$

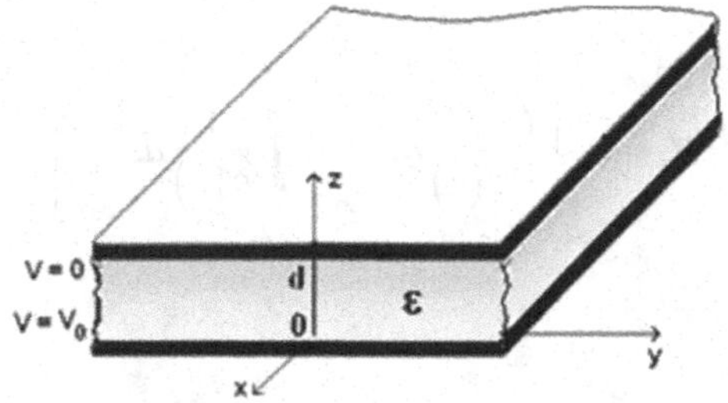

FIGURE 9.4 A parallel plate capacitor with continuously varying permittivity.

In view of Equation 9.37a, Equation 9.36b can be rewritten as:

$$\frac{d}{dz}Z = -\frac{1}{\varepsilon}\cdot\frac{d\varepsilon}{dz}\cdot Z \tag{9.37b}$$

or,

$$\frac{1}{Z}\cdot\frac{d}{dz}Z = -\frac{1}{\varepsilon}\cdot\frac{d\varepsilon}{dz} \tag{9.37c}$$

After integrating both sides of Equation 9.37c, we get:

$$\left.\ln(Z) = \ln(c_1) - \ln(\varepsilon) \quad (\text{a}) \quad \text{or} \quad Z = \frac{c_1}{\varepsilon} \quad (\text{b})\right\} \tag{9.38a}$$

where c_1 is a constant of integration.

Substituting Z from Equation 9.38b into Equation 9.37a, and integrating with respect to z, we find:

$$V(z) = c_2 + c_1\cdot\int\frac{1}{\varepsilon(z)}dz \tag{9.39}$$

where c_2 indicates yet another constant of integration.

For the evaluation of the two unknowns, c_1 and c_2, consider the following boundary condition:

$$\left.V(z)\right|_{z=0} = V_o \quad (\text{a}) \quad \text{and} \quad \left.V(z)\right|_{z=d} = 0 \quad (\text{b})\} \tag{9.40a}$$

Therefore,

$$V(0) = V_o = c_2 + c_1\cdot\left.\int\frac{1}{\varepsilon(z)}dz\right|_{z=0} \tag{9.41a}$$

$$V(d) = 0 = c_2 + c_1\cdot\left.\int\frac{1}{\varepsilon(z)}dz\right|_{z=d} \tag{9.41b}$$

Thus,

$$V_o = c_1 \cdot \left[\int \frac{1}{\varepsilon(z)} dz \bigg|_{z=0} - \int \frac{1}{\varepsilon(z)} dz \bigg|_{z=d} \right] \tag{9.41c}$$

From this we get:

$$\left. c_1 = -\frac{V_o}{\int_0^d \frac{1}{\varepsilon(z)} dz} \quad (a) \quad \text{and} \quad c_2 = \frac{V_o}{\int_0^d \frac{1}{\varepsilon(z)} dz} \cdot \int \frac{1}{\varepsilon(z)} dz \bigg|_{z=d} \quad (b) \right\} \tag{9.42}$$

Thus, the expression for the scalar electric potential as a function of the z-coordinate is as follows:

$$V(z) = \frac{V_o}{\int_0^d \frac{1}{\varepsilon(z)} dz} \cdot \int \frac{1}{\varepsilon(z)} dz \bigg|_{z=d} - \frac{V_o}{\int_0^d \frac{1}{\varepsilon(z)} dz} \cdot \int \frac{1}{\varepsilon(z)} dz$$

or

$$V(z) = \frac{V_o}{\int_0^d \frac{1}{\varepsilon(z)} dz} \cdot \left[\int \frac{1}{\varepsilon(z)} dz \bigg|_{z=d} - \int \frac{1}{\varepsilon(z)} dz \right] = V_o \cdot \left[\frac{\int_z^d \frac{1}{\varepsilon(z)} dz}{\int_0^d \frac{1}{\varepsilon(z)} dz} \right] \tag{9.43}$$

The solution assumes that for a given function $\varepsilon(z)$, the two integrals on the RHS exist.

9.3.2 Field Problems in Cylindrical Coordinates

The Laplacian for the scalar function V in cylindrical coordinates, given by Equation 9.6 is reproduced below:

$$\nabla^2 V = \frac{1}{\rho} \cdot \frac{\partial}{\partial \rho}\left(\rho \frac{\partial V}{\partial \rho} \right) + \frac{1}{\rho^2} \cdot \frac{\partial^2 V}{\partial \varphi^2} + \frac{\partial^2 V}{\partial z^2} \tag{9.6}$$

In a one-dimensional problem, the potential V is a function of only one of the three coordinates. Its derivative with respect to the other two coordinates must be zero. Let us assume that V is a function of ρ only. Therefore, in view of Equations 9.1b and 9.6, on replacing partial derivatives by complete derivatives we get:

$$\nabla^2 V = \frac{1}{\rho} \cdot \frac{d}{d\rho}\left(\rho \frac{dV}{d\rho} \right) = 0 \tag{9.44a}$$

Similarly, if V is a function of either φ or z, we have:

$$\nabla^2 V = \frac{1}{\rho^2} \cdot \frac{d^2V}{d\varphi^2} = 0 \tag{9.44b}$$

or

$$\nabla^2 V = \frac{d^2V}{dz^2} = 0 \tag{9.44c}$$

A one-dimensional field problem will require the solution of one of these components. The selection will depend on the coordinate along which the field varies. The procedure of obtaining solutions of these equations is given in the following subsections.

9.3.2.1 Scalar Potential between Two Constant ρ Surfaces

The capacitance between two coaxial cylinders or coaxial cable is such an example that requires the evaluation of scalar potential between two constant ρ surfaces. In this section, the Laplace equation given by Equation 9.44a is solved. The configuration for such a capacitor is shown in Figure 9.5. This figure shows a long coaxial capacitor with inner and outer radii of conductors as a and b, respectively. The inner cylinder is maintained at a potential V_o and the outer at zero potential. The space between the two cylinders is filled with a homogeneous dielectric of constant permittivity ε.

In this configuration, scalar electric potential satisfies:

$$\frac{d}{d\rho}\left(\rho \frac{dV}{d\rho}\right) = 0 \tag{9.45a}$$

The potential varies only in the radial direction. Equation 9.45a implies that its bracketed term $\left(\rho \frac{dV}{d\rho}\right)$ is a constant, say k. Hence:

$$\frac{dV}{d\rho} = \frac{k}{\rho} \tag{9.45b}$$

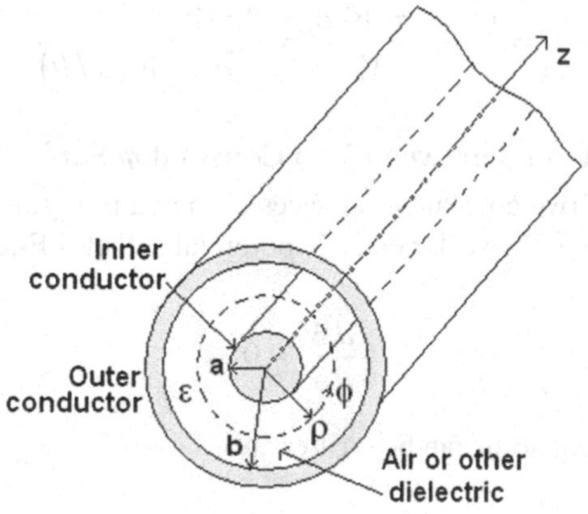

FIGURE 9.5 Configuration of a coaxial cable.

Now, on integrating both sides, one gets:

$$V = k \cdot \ln(c) + k \cdot \ln(\rho) = k \cdot \ln(c \cdot \rho) \tag{9.45c}$$

In Equation 9.45c, c is the constant of integration and k is an unknown quantity. To evaluate these quantities, the following boundary conditions can be used:

$$\left. \begin{array}{l} V\big|_{\rho=a} = 0 \quad (\mathrm{a}) \quad \text{and} \quad V\big|_{\rho=b} = V_o \quad (\mathrm{b}) \end{array} \right\} \tag{9.46}$$

In view these boundary conditions one gets,

$$\left. c = 1/a \quad (\mathrm{a}) \quad \text{and} \quad k = V_o / \ln(b/a) \quad (\mathrm{b}) \right\} \tag{9.47}$$

Therefore, from Equations 9.45c, 9.47a, and 9.47b,

$$V = V_o \cdot \frac{\ln(\rho/a)}{\ln(b/a)} \tag{9.48}$$

In view of Equation 9.48, the electric flux density on the inner surface of the outer cylinder is given by:

$$D_\rho\big|_{\rho=b} = -\varepsilon \frac{dV}{d\rho}\bigg|_{\rho=b} = -\frac{\varepsilon}{a} \cdot \frac{V_o}{\ln(b/a)} \cdot \frac{1}{\rho}\bigg|_{\rho=b} = \frac{\varepsilon}{a \cdot b} \cdot \frac{V_o}{\ln(a/b)} \tag{9.49a}$$

The surface charge density can be written as:

$$\rho_s\big|_{\rho=b} = D_\rho\big|_{\rho=b} = \frac{\varepsilon}{a \cdot b} \cdot \frac{V_o}{\ln(a/b)} \tag{9.49b}$$

Hence, the capacitance per unit cylinder length, C, is given as:

$$C = \frac{Q}{V_o} = \frac{2\pi \cdot b \cdot \rho_s}{V_o} = \frac{2\pi \cdot \varepsilon}{a} \cdot \frac{1}{\ln(a/b)} \tag{9.50}$$

9.3.2.2 Scalar Potential between Two Constant φ Surfaces

The configuration of two constant φ surfaces is shown in Figure 9.6.

In this configuration the scalar electric potential satisfies Equation 9.44b:

$$\frac{d^2V}{d\varphi^2} = 0 \tag{9.44b}$$

The solution of this equation can be written as:

$$V = A \cdot \varphi + B \tag{9.51}$$

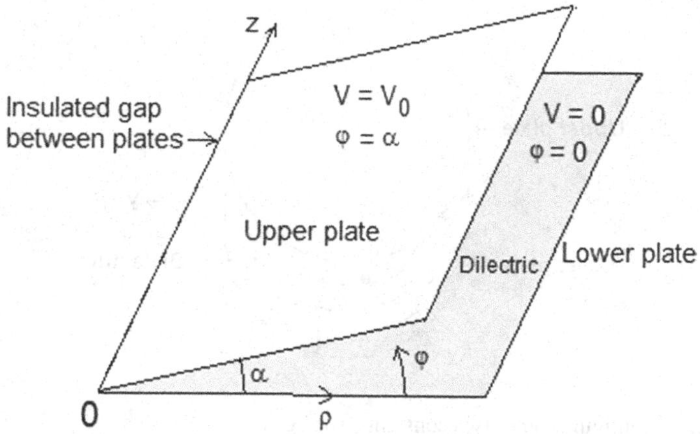

FIGURE 9.6 Configuration of two constant ϕ surfaces.

As shown in the figure, the potential field has to satisfy the following boundary conditions:

$$\left. V\right|_{\varphi=0} = 0 \quad (a) \quad \text{and} \quad \left. V\right|_{\varphi=\alpha} = V_o \quad (b) \Big\} \tag{9.52}$$

By applying these boundary conditions, we get:

$$V = V_o \cdot \frac{\varphi}{\alpha} \tag{9.53}$$

The electric field intensity can be evaluated in the gradient relation ($\boldsymbol{E} = -V$). Thus,

$$\boldsymbol{E} = -\frac{V_o}{\alpha \cdot \rho} \cdot \boldsymbol{a}_\varphi \tag{9.54}$$

A cursory look at the relation given by Equations 9.54 and 9.53 reveals that $\boldsymbol{E}$ field is a function of ρ and not that of φ, whereas V is a function of φ and not that of ρ.

9.3.2.3 Scalar Potential between Two Constant z Surfaces

The configuration of two constant z surfaces is shown in Figure 9.7.

For this configuration:

$$\frac{d^2V}{dz^2} = 0 \tag{9.44c}$$

The solution of this equation can be written as:

$$V = A \cdot z + B \tag{9.55}$$

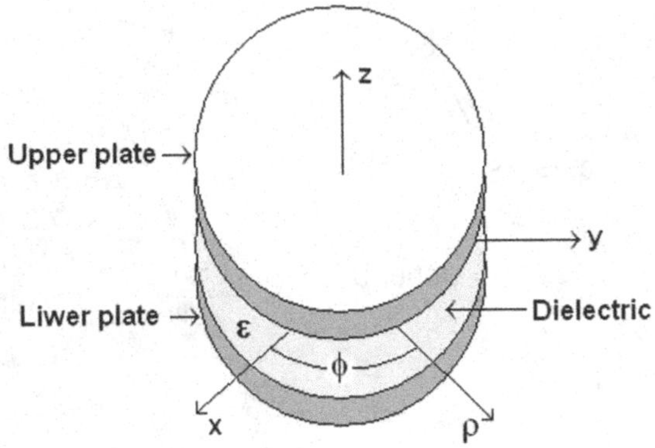

FIGURE 9.7 Configuration of two constant z surfaces.

As shown in the figure, the potential field has to satisfy the following boundary conditions:

$$V\big|_{z=0} = 0 \quad (a) \quad \text{and} \quad V\big|_{z=d} = V_o \quad (b)\} \tag{9.56}$$

By applying these boundary conditions:

$$V = \frac{V_o}{d} \cdot z \tag{9.57}$$

Here also the E field can be evaluated from the gradient relation.

9.3.3 Field Problems in Spherical Coordinates

The Laplacian equation for the scalar function V in spherical coordinates, given by Equation 9.7, is reproduced here:

$$\nabla^2 V = \frac{1}{r^2} \cdot \frac{\partial}{\partial r}\left(r^2 \frac{\partial V}{\partial r}\right) + \frac{1}{r^2 \sin\theta} \cdot \frac{\partial}{\partial \theta}\left(\sin\theta \frac{\partial V}{\partial \theta}\right) + \frac{1}{r^2 \sin^2\theta} \cdot \frac{\partial^2 V}{\partial \varphi^2} \tag{9.7}$$

This equation contains three terms on the RHS. Depending on the nature of the field variation, only one of these terms may be retained. The resulting Laplace equation is:

$$\nabla^2 V = \frac{1}{r^2} \cdot \frac{d}{dr}\left(r^2 \frac{dV}{dr}\right) = 0 \tag{9.58a}$$

or,

$$\nabla^2 V = \frac{1}{r^2 \sin\theta} \cdot \frac{\partial}{d\theta}\left(\sin\theta \frac{dV}{d\theta}\right) = 0 \tag{9.58b}$$

or,

$$\nabla^2 V = \frac{1}{r^2 \sin^2\theta} \cdot \frac{d^2 V}{d\varphi^2} = 0 \tag{9.58c}$$

A one-dimensional field problem will require the solution of one of these components. The selection will depend on the coordinate along which the field varies. The procedure of obtaining solutions of these equations is given in the following subsections.

9.3.3.1 Potential Fields in Homogeneous Medium between Two Constant *r* Surfaces

In Figure 9.8, two concentric spherical equipotential surfaces are shown. The radii for the inner and outer sphere are a and b, respectively. The inner sphere is maintained at a potential V_o and the outer sphere at zero potential. If the space between the two concentric spherical equipotential surfaces is filled with a homogeneous dielectric of constant permittivity ε, the scalar electric potential will satisfy the Laplace equation and it will vary only in the radial direction.

The Laplace equation for this case is:

$$\nabla^2 V = \frac{1}{r^2} \cdot \frac{d}{dr}\left(r^2 \frac{dV}{dr}\right) = 0 \tag{9.58a}$$

This equation implies that $\left(r^2 \dfrac{dV}{dr}\right)$ is a constant, say k. Hence:

$$\frac{dV}{dr} = \frac{k}{r^2} \tag{9.59a}$$

Now, on integrating both sides, one gets:

$$V = c - \frac{k}{r} \tag{9.59b}$$

where c is the constant of integration. To evaluate the two unknown quantities, viz. k and c, consider the following boundary conditions:

$$\left. V\right|_{r=a} = V_o \;\; (\text{a}) \quad \text{and} \quad \left. V\right|_{r=b} = 0 \;\; (\text{b})\} \tag{9.60}$$

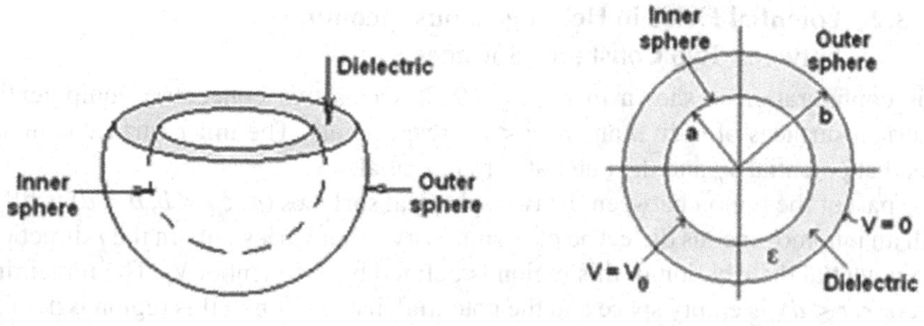

FIGURE 9.8 Configuration of two constant *r* surfaces with homogeneous medium.

Using these boundary conditions:

$$k = V_o \cdot \frac{a \cdot b}{(a-b)} \quad (\text{a}) \quad \text{and} \quad c = V_o \cdot \frac{b}{(a-b)} \quad (\text{b}) \Bigg\} \tag{9.61}$$

Therefore, from Equations 9.59b, 9.61a, and 9.61b:

$$V = V_o \cdot \frac{a}{(a-b)}\left(1 - \frac{b}{r}\right) = V_o \cdot \frac{a \cdot b}{(a-b)} \cdot \left(\frac{1}{b} - \frac{1}{r}\right) \tag{9.62}$$

Therefore, the electric flux density on the inner sphere is given by:

$$D_\rho\Big|_{r=a} = -\varepsilon \frac{dV}{dr}\Big|_{r=a} = -\varepsilon \cdot V_o \cdot \frac{a.b}{(a-b)} \cdot \frac{1}{r^2}\Big|_{r=a} = \varepsilon \cdot V_o \cdot \frac{b/a}{(b-a)} \tag{9.63}$$

This being equal to the charge density on the inner sphere, the capacitance is given as:

$$C = \varepsilon - \frac{b/a}{(b-a)} - 4\pi a^2 = \varepsilon - \frac{a-b}{(b-a)} - 4\pi \tag{9.64}$$

If the outer radius b, tends to infinity, the capacitance of a sphere of radius a, is:

$$C\Big|_{b\to\infty} = \varepsilon \cdot 4\pi a \tag{9.65}$$

For the Earth, with $\varepsilon = \varepsilon_o = 8.854 \times 10^{-12} \cong \frac{10^{-9}}{36\pi}$, and its radius as say, 6000 km:

$$C_{\text{Earth}} \cong \frac{10^{-9}}{36\pi} \cdot 4\pi \times 6.10^6 \cong \frac{2}{3} 10^{-3} \cong \frac{2}{3}\ \text{mF} \tag{9.66}$$

Thus, we can have a one Farad capacitor by connecting 1,500 Earth-size–conducting spheres in parallel! Clearly, one Farad is too big a unit for capacitance.

9.3.3.2 Potential Fields in Heterogeneous Medium between Two Constant *r* Surfaces

This configuration is shown in Figure 9.9. It shows two concentric equipotential spherical surfaces of radii a and d ($a < d$), respectively. The inner surface is maintained at potential V_o and the outer at zero potential.

A part of the region between the two spherical surfaces ($a < r < b$, $b < d$) is filled with an inhomogeneous dielectric of permittivity ε that varies only in the r direction. The potential distribution in this region is defined by the symbol V_1. The remaining part ($b < r < d$), is empty space and the potential distribution in this region is defined as V_2. Clearly, all equipotential surfaces are concentric spheres.

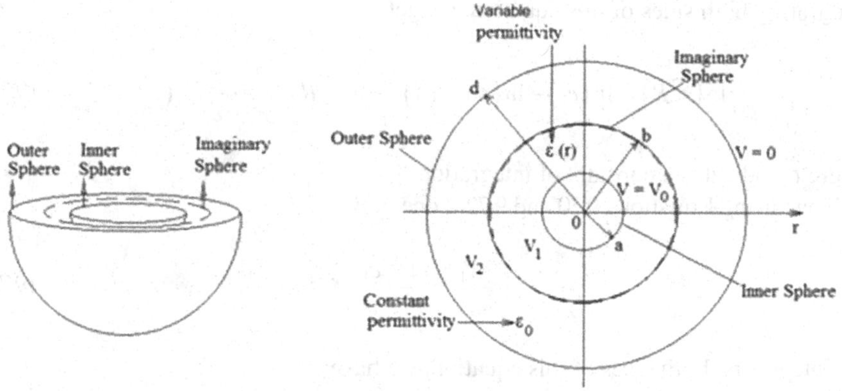

FIGURE 9.9 Configuration of two constant *r* surfaces with heterogeneous medium.

Since the potential is a function of only *r* coordinates, i.e. independent of θ and φ, the Laplacian equation of potential is given in view of Equation 8.58a, as follows:

$$\nabla^2 V = \frac{1}{r^2} \cdot \frac{d}{dr}\left(r^2 \frac{dV}{dr}\right) \tag{9.67a}$$

Also, as the permittivity ε is a function exclusively of the *r* coordinates, then:

$$\nabla \varepsilon = \frac{d\varepsilon}{dr} \boldsymbol{a}_r \tag{9.67b}$$

Therefore, from Equation 9.10, we have, for Region 1:

$$\nabla^2 V_1 = -\frac{1}{\varepsilon} \nabla V_1 \bullet \nabla \varepsilon \tag{9.68}$$

thus,

$$\frac{1}{r^2} \cdot \frac{d}{dr}\left(r^2 \frac{dV_1}{dr}\right) = -\frac{1}{\varepsilon} \cdot \frac{d\varepsilon}{dr} \cdot \frac{dV_1}{dr} \tag{9.69}$$

Note that none of these equations is a Laplace equation.

Setting:

$$R \stackrel{\text{def}}{=} \frac{dV_1}{dr} \tag{9.70}$$

one may rewrite Equation 9.69 as follows:

$$\frac{1}{\left(r^2 R\right)} \cdot \frac{d}{dr}\left(r^2 R\right) = -\frac{1}{\varepsilon} \cdot \frac{d\varepsilon}{dr} \tag{9.71}$$

Integrating both sides of this equation, we get:

$$\ln\left(r^2R\right)=\ln\left(c_1\right)-\ln\left(\varepsilon\right)\quad(\mathrm{a})\quad\text{or}\quad R=\frac{1}{r^2}\cdot\frac{c_1}{\varepsilon}\quad(\mathrm{b})\Big\}\tag{9.72}$$

where c_1 indicates a constant of integration.

Now, using Equations 9.70 and 9.72a, one gets:

$$\frac{dV_1}{dr}=\frac{1}{r^2}\cdot\frac{c_1}{\varepsilon}\tag{9.73}$$

On integrating both sides of this equation, we have:

$$V_1=c_2+c_1\cdot\int\left\{\frac{1}{r^2}\cdot\frac{1}{\varepsilon}\right\}dr,\quad\text{over }a<r<b\tag{9.74}$$

where c_2 indicates another constant of integration. For the determination of these constants, the following boundary conditions may be used:

$$V_1\big|_{r=a}=V_o\quad(\mathrm{a})\quad\text{and}\quad V_1\big|_{r=b}=V_o'\quad(\mathrm{b})\Big\}\tag{9.75}$$

where V_o' in an unknown potential on the surface $r=b$.

Using these boundary conditions:

$$V_o=c_2+c_1\cdot\int\left\{\frac{1}{r^2}\cdot\frac{1}{\varepsilon}\right\}dr\bigg|_{r=a}\tag{9.76a}$$

$$V_o'=c_2+c_1\cdot\int\left\{\frac{1}{r^2}\cdot\frac{1}{\varepsilon}\right\}dr\bigg|_{r=b}\tag{9.76b}$$

From Equations 9.76a and 9.76b:

$$V_o-V_o'=c_1\cdot\left[\int\left\{\frac{1}{r^2}\cdot\frac{1}{\varepsilon}\right\}dr\bigg|_{r=a}-\int\left\{\frac{1}{r^2}\cdot\frac{1}{\varepsilon}\right\}dr\bigg|_{r=b}\right]=-c_1\cdot\int_a^b\left\{\frac{1}{r^2}\cdot\frac{1}{\varepsilon}\right\}dr\tag{9.76c}$$

From Equation 9.76c:

$$c_1=-\left(V_o-V_o\right)\bigg/\int_a^b\left\{\frac{1}{r^2}\cdot\frac{1}{\varepsilon}\right\}dr\tag{9.77a}$$

From Equations 9.76b and 9.77a:

$$c_2 = V_o' + \left(V_o - V_o'\right) \cdot \frac{\left.\int \left\{\frac{1}{r^2} \cdot \frac{1}{\varepsilon}\right\} dr\right|_{r=b}}{\int_a^b \left\{\frac{1}{r^2} \cdot \frac{1}{\varepsilon}\right\} dr}$$

$$= V_o \cdot \frac{\left.\int \left\{\frac{1}{r^2} \cdot \frac{1}{\varepsilon}\right\} dr\right|_{r=b}}{\int_a^b \left\{\frac{1}{r^2} \cdot \frac{1}{\varepsilon}\right\} dr} + V_o' \cdot \frac{\int_a^b \left\{\frac{1}{r^2} \cdot \frac{1}{\varepsilon}\right\} dr - \left.\int \left\{\frac{1}{r^2} \cdot \frac{1}{\varepsilon}\right\} dr\right|_{r=b}}{\int_a^b \left\{\frac{1}{r^2} \cdot \frac{1}{\varepsilon}\right\} dr} \tag{9.77b}$$

$$= \left[V_o \cdot \frac{\left.\int \left\{\frac{1}{r^2} \cdot \frac{1}{\varepsilon}\right\} dr\right|_{r=b}}{\int_a^b \left\{\frac{1}{r^2} \cdot \frac{1}{\varepsilon}\right\} dr} - V_o' \cdot \frac{\left.\int \left\{\frac{1}{r^2} \cdot \frac{1}{\varepsilon}\right\} dr\right|_{r=a}}{\int_a^d \left\{\frac{1}{r^2} \cdot \frac{1}{\varepsilon}\right\} dr} \right]$$

Hence, using Equation 9.74:

$$V_1 = \left[V_o \cdot \frac{\left.\int \left\{\frac{1}{r^2} \cdot \frac{1}{\varepsilon}\right\} dr\right|_{r=b}}{\int_a^b \left\{\frac{1}{r^2} \cdot \frac{1}{\varepsilon}\right\} dr} - V_o' \cdot \frac{\left.\int \left\{\frac{1}{r^2} \cdot \frac{1}{\varepsilon}\right\} dr\right|_{r=a}}{\int_a^d \left\{\frac{1}{r^2} \cdot \frac{1}{\varepsilon}\right\} dr} \right]$$

$$-\left(V_o - V_o'\right) \cdot \frac{\int \left\{\frac{1}{r^2} \cdot \frac{1}{\varepsilon}\right\} dr}{\int_a^b \left\{\frac{1}{r^2} \cdot \frac{1}{\varepsilon}\right\} dr} \tag{9.78}$$

$$= V_o \cdot \frac{\int_r^b \left\{\frac{1}{r^2} \cdot \frac{1}{\varepsilon}\right\} dr}{\int_a^b \left\{\frac{1}{r^2} \cdot \frac{1}{\varepsilon}\right\} dr} + V_o' \cdot \frac{\int_a^r \left\{\frac{1}{r^2} \cdot \frac{1}{\varepsilon}\right\} dr}{\int_a^b \left\{\frac{1}{r^2} \cdot \frac{1}{\varepsilon}\right\} dr} \qquad (\text{over } a < r < b)$$

For Region 2, being a homogenous region, the distribution of the scalar electric potential can be derived by using the procedure of the preceding section. The resulting expression is:

$$V_2 = V_o' \cdot \frac{\left(\frac{1}{r} - \frac{1}{d}\right)}{(d-b)} \cdot d \cdot b \tag{9.79}$$

Now to find the unknown, V_o', we shall use the following boundary condition:

$$-\varepsilon \left.\frac{dV_1}{dr}\right|_{r=b} = -\varepsilon_o \left.\frac{dV_2}{dr}\right|_{r=b} \tag{9.80}$$

Thus, using Equations 9.78 and 9.79, we get:

$$\varepsilon\left[V_o \cdot \frac{\left\{\frac{1}{r^2}\cdot\frac{1}{\varepsilon}\right\}}{\int_a^b \left\{\frac{1}{r^2}\cdot\frac{1}{\varepsilon}\right\}dr} - V_o' \cdot \frac{\left\{\frac{1}{r^2}\cdot\frac{1}{\varepsilon}\right\}}{\int_a^b \left\{\frac{1}{r^2}\cdot\frac{1}{\varepsilon}\right\}dr}\right]_{r=b} = \left.\varepsilon_o V_o' \cdot \frac{\left(\frac{1}{r^2}\right)}{(d-b)} \cdot b \cdot d\right|_{r=b}$$

or,

$$V_o' = V_o \cdot \frac{\left[\frac{\left\{\frac{1}{b^2}\right\}}{\int_a^b \left\{\frac{1}{r^2}\cdot\frac{1}{\varepsilon}\right\}dr}\right]}{\left[\frac{\left\{\frac{1}{b^2}\right\}}{\int_a^b \left\{\frac{1}{r^2}\cdot\frac{1}{\varepsilon}\right\}dr} + \frac{\left(\frac{d}{b}\right)\cdot\varepsilon_o}{(d-b)}\right]} = V_o \cdot \frac{1}{\left[1 + \frac{d \cdot b}{(d-b)} \cdot \varepsilon_o \cdot \int_a^b \left\{\frac{1}{r^2}\cdot\frac{1}{\varepsilon}\right\}dr\right]} \tag{9.81}$$

Therefore, the potential distributions in the two regions found from Equations 9.78 and 9.79, respectively, are:

$$V_1 = V_o \cdot \left[\frac{\int_r^b \left\{\frac{1}{r^2}\cdot\frac{1}{\varepsilon}\right\}dr}{\int_a^b \left\{\frac{1}{r^2}\cdot\frac{1}{\varepsilon}\right\}dr} + \frac{\int_a^r \left\{\frac{1}{r^2}\cdot\frac{1}{\varepsilon}\right\}dr}{\int_a^b \left\{\frac{1}{r^2}\cdot\frac{1}{\varepsilon}\right\}dr} \cdot \frac{1}{\left[1 + \frac{d \cdot b}{(d-b)} \cdot \varepsilon_o \cdot \int_a^b \left\{\frac{1}{r^2}\cdot\frac{1}{\varepsilon}\right\}dr\right]}\right]$$

$$= V_o \cdot \frac{1}{\int_a^b \left\{\frac{1}{r^2}\cdot\frac{1}{\varepsilon}\right\}dr} \cdot \left[\int_r^b \left\{\frac{1}{r^2}\cdot\frac{1}{\varepsilon}\right\}dr + \frac{\int_a^r \left\{\frac{1}{r^2}\cdot\frac{1}{\varepsilon}\right\}dr}{\left[1 + \frac{d \cdot b}{(d-b)} \cdot \varepsilon_o \cdot \int_a^b \left\{\frac{1}{r^2}\cdot\frac{1}{\varepsilon}\right\}dr\right]}\right] \tag{9.82a}$$

and

$$V_2 = V_o' \cdot \frac{d \cdot b}{(d-b)} \cdot \left(\frac{1}{r} - \frac{1}{d}\right) = V_o \cdot \frac{\dfrac{d \cdot b}{(d-b)}}{\left[1 + \dfrac{d \cdot b}{(d-b)} \cdot \varepsilon_o \cdot \displaystyle\int_a^b \left\{\frac{1}{r^2} \cdot \frac{1}{\varepsilon}\right\} dr\right]} \cdot \left(\frac{1}{r} - \frac{1}{d}\right) \tag{9.82b}$$

As a special case, as d tends to infinity:

$$V_o'\Big|_{d \to \infty} = \frac{V_o}{\left[1 + b \cdot \varepsilon_o \cdot \displaystyle\int_a^b \left\{\frac{1}{r^2} \cdot \frac{1}{\varepsilon}\right\} dr\right]} \tag{9.83a}$$

$$V_1 = V_o \cdot \frac{1}{\displaystyle\int_a^b \left\{\frac{1}{r^2} \cdot \frac{1}{\varepsilon}\right\} dr} \cdot \left[\int_r^b \left\{\frac{1}{r^2} \cdot \frac{1}{\varepsilon}\right\} dr + \frac{\displaystyle\int_a^r \left\{\frac{1}{r^2} \cdot \frac{1}{\varepsilon}\right\} dr}{\left[1 + b \cdot \varepsilon_o \cdot \displaystyle\int_a^b \left\{\frac{1}{r^2} \cdot \frac{1}{\varepsilon}\right\} dr\right]}\right] \tag{9.83b}$$

and

$$V_2 = V_o' \cdot b \cdot \frac{1}{r} = V_o \cdot \frac{b}{\left[1 + b \cdot \varepsilon_o \cdot \displaystyle\int_a^b \left\{\frac{1}{r^2} \cdot \frac{1}{\varepsilon}\right\} dr\right]} \cdot \frac{1}{r} \tag{9.83c}$$

The surface charge density on the inner sphere of radius a is given by:

$$\rho_s = D_{1r}\Big|_{r=a} = -\varepsilon \cdot \frac{d}{dr} V_1 \Bigg|_{r=a} \tag{9.84}$$

or,

$$\rho_s = V_o \cdot \frac{\dfrac{1}{r^2}}{\displaystyle\int_a^b \left\{\frac{1}{r^2} \cdot \frac{1}{\varepsilon}\right\} dr} \cdot \left[1 - \frac{1}{\left[1 + b \cdot \varepsilon_o \cdot \displaystyle\int_a^b \left\{\frac{1}{r^2} \cdot \frac{1}{\varepsilon}\right\} dr\right]}\right]\Bigg|_{r=a} \tag{9.85}$$

$$= \frac{V_o \cdot b \cdot \varepsilon_o \cdot \dfrac{1}{a^2}}{\left[1 + b \cdot \varepsilon_o \cdot \displaystyle\int_a^b \left\{\frac{1}{r^2} \cdot \frac{1}{\varepsilon}\right\} dr\right]}$$

Hence, the capacitance is given by:

$$C = \frac{4\pi a^2}{V_o} \cdot \rho_s = \frac{4\pi a^2}{V_o} \cdot \frac{V_o \cdot b \cdot \varepsilon_o \cdot \dfrac{1}{a^2}}{\left[1 + b \cdot \varepsilon_o \cdot \displaystyle\int_a^b \left\{\frac{1}{r^2} \cdot \frac{1}{\varepsilon}\right\} dr\right]} \tag{9.86}$$

Therefore, as $d \rightarrow \infty$:

$$C = \left[\frac{4\pi b \cdot \varepsilon_o}{1 + b \cdot \varepsilon_o \cdot \displaystyle\int_a^b \left\{\frac{1}{r^2} \cdot \frac{1}{\varepsilon}\right\} dr}\right] \tag{9.87}$$

9.3.3.3 Scalar Potential between Two Constant θ Surfaces

This configuration is shown in Figure 9.10. This shows two conical surfaces with inner and outer cones $\theta = \alpha$ ($\alpha < \pi/2$) and $\theta = \pi/2$ angles, respectively. The inner cone is maintained at a potential V_o and the outer cone at zero potential. If the space between the two conical equipotential surfaces is filled with a homogeneous dielectric of constant permittivity ε, the scalar electric potential will satisfy Laplace's equation and the field will vary only in the θ direction. The Laplace equation for this case is:

$$\nabla^2 V = \frac{1}{r^2 \sin\theta} \cdot \frac{d}{d\theta}\left(\sin\theta \frac{dV}{d\theta}\right) = 0 \tag{9.58b}$$

From Equation 9.8b, we have:

$$\frac{d}{d\theta}\left(\sin\theta \frac{\partial V}{\partial \theta}\right) = 0 \quad \text{or} \quad \sin\theta \frac{dV}{d\theta} = A \tag{9.88a}$$

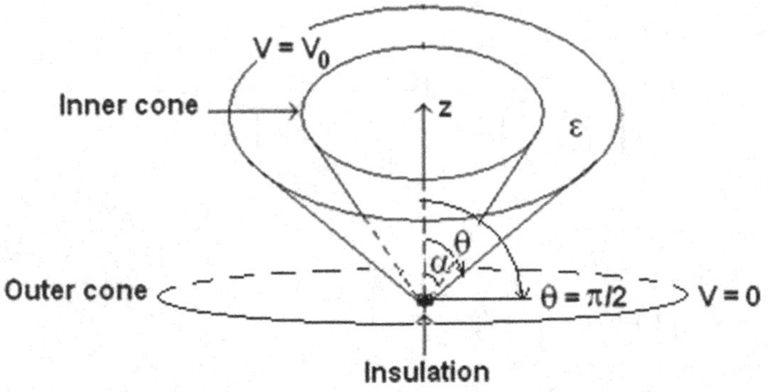

FIGURE 9.10 Configuration of two constant θ surfaces.

$$\frac{dV}{d\theta} = \frac{A}{\sin\theta} \tag{9.88b}$$

From this equation we get:

$$V = \int \frac{A}{\sin\theta} d\theta + B \tag{9.89a}$$

Equation 9.79a gives:

$$V = A \cdot \ln\left\{\tan\left(\frac{\theta}{2}\right)\right\} + B \tag{9.89b}$$

In Figure 9.10, the boundary conditions are:

$$\left.\begin{matrix} V\big|_{\theta=\alpha} = V_o \quad (a) \quad \text{and} \quad V\big|_{\theta=\pi/2} = 0 \quad (b) \end{matrix}\right\} \tag{9.90}$$

Application of these boundary conditions leads to the following relation:

$$V = V_o \cdot \frac{\ln\left\{\tan\left(\frac{\theta}{2}\right)\right\}}{\ln\left\{\tan\left(\frac{\alpha}{2}\right)\right\}} \tag{9.91}$$

9.3.3.4 Scalar Potential between Two Constant φ Surfaces

This configuration is shown in Figure 9.11. This shows two semicircular surfaces with radius tending to infinity. Both are semi-infinite surfaces, one covering half of the x–z-plane, over the entire positive x-axis ($\varphi = 0$), the other semi-infinite surface meets the first surface at the z-axis subtending an angle ($\varphi = \alpha$) with this surface.

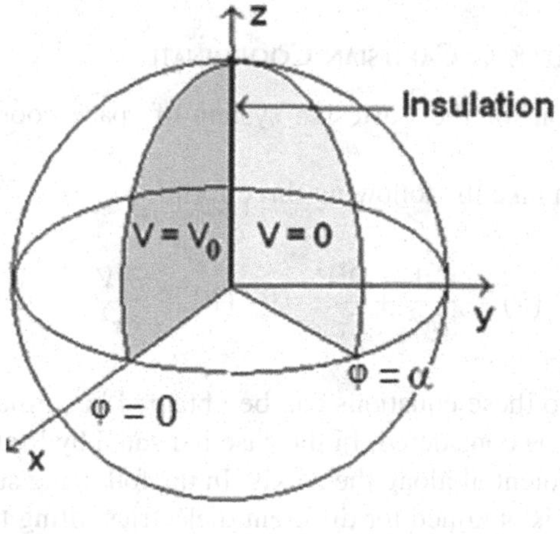

FIGURE 9.11 Configuration of two constant φ surfaces.

The former surface is maintained at a potential V_o and the later one is at zero potential. If the space between the two equipotential surfaces is filled with a homogeneous dielectric of constant permittivity ε, the scalar electric potential will satisfy the Laplace equation and the field will vary only in the φ direction. The Laplace equation for this case is:

$$\nabla^2 V = \frac{1}{r^2 \sin^2\theta} \cdot \frac{d^2V}{d\varphi^2} = 0 \tag{9.58c}$$

From Equation 9.58c, we have:

$$\frac{d^2V}{d\varphi^2} = 0 \tag{9.92}$$

Its solution can simply be given as:

$$V = A \cdot \varphi + B \tag{9.93}$$

In view of Figure 9.11, the boundary conditions are:

$$\left. V \right|_{\varphi=0} = V_o \quad (\text{a}) \quad \text{and} \quad \left. V \right|_{\varphi=\alpha} = 0 \quad (\text{b}) \Big\} \tag{9.94}$$

Application of these boundary conditions gives to the following relation:

$$V = V_o \cdot \left(1 - \frac{\varphi}{\alpha}\right) \tag{9.95}$$

9.4 TWO-DIMENSIONAL FIELD PROBLEMS

This section deals with two-dimensional field problems. The method of solutions in different coordinates is illustrated by taking specific examples.

9.4.1 Field Problems in Cartesian Coordinates

The Laplace equation for the Cartesian system of space coordinates is given by Equation 9.5.

This equation can take the following three forms:

$$\frac{\partial^2 V}{\partial x^2} + \frac{\partial^2 V}{\partial y^2} = 0 \ (\text{a}), \quad \frac{\partial^2 V}{\partial y^2} + \frac{\partial^2 V}{\partial z^2} = 0 \ (\text{b}), \quad \frac{\partial^2 V}{\partial x^2} + \frac{\partial^2 V}{\partial z^2} = 0 \ (\text{c}) \Big\} \tag{9.96a}$$

Since the solutions to these equations can be obtained in similar ways, only one of these cases needs to be considered. In the case indicated by Equation 9.96a, there is no variation of the potential along the z-axis. In the following subsections, the solution of this equation is obtained for different dielectrics filling the two-dimensional space.

9.4.1.1 Potential Fields in Homogeneous Medium

Figure 9.12 shows a two-dimensional rectangular region with corners located at $(-w/2,h)$, $(w/2,h)$, $(-w/2,0)$, and $(-w/2,0)$. Each of the three sides of this rectangle, viz. at $x = \pm w/2$ and at $y = 0$, is at zero potential. The potential at the fourth side, i.e. at $y = h$, is an arbitrarily prescribed even function of x, such that the potential is continuous along the boundaries of the rectangular region. It is required to find the distribution of scalar electric potential inside this rectangular region.

The given potential distribution along the top side of the rectangle, i.e. at $y = h$, can be expressed as a half range Fourier cosine series. Thus, giving:

$$V\Big|_{y=h} = \sum_{m\text{-odd}}^{\infty} a_m \cdot \cos\left(m\pi \cdot \frac{x}{w}\right) \tag{9.97}$$

where the Fourier coefficient a_m is known from the prescribed even function.

Let us *assume* that the solution of the Laplace equation giving the potential function is expressed as a product two unknown functions X and Y, where X is a function of x alone and Y is a function of y alone. Thus, provisionally we take:

$$V(x,y) \overset{\text{def}}{=} X(x) \cdot Y(y) \tag{9.98}$$

Substitute Equation 9.98 into Equation 9.96a and then divide both sides of the resulting equation by V, to get:

$$\frac{1}{X \cdot Y} \cdot \frac{\partial^2 (X \cdot Y)}{\partial x^2} + \frac{1}{X \cdot Y} \cdot \frac{\partial^2 (X \cdot Y)}{\partial y^2} = 0 \tag{9.99}$$

From the first term on the LHS of this equation, the function Y can be canceled out as the term involves partial derivation with respect to x. Likewise, from the second term, the function X can be canceled out, thus:

$$\frac{1}{X} \cdot \frac{\partial^2 X}{\partial x^2} + \frac{1}{Y} \cdot \frac{\partial^2 Y}{\partial y^2} = 0 \tag{9.100a}$$

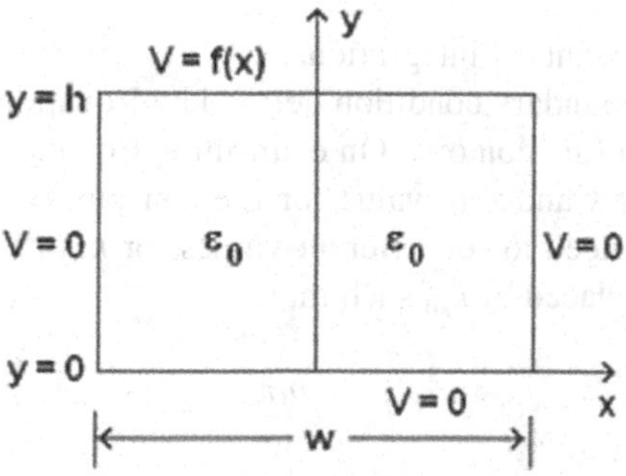

FIGURE 9.12 Rectangular region filled with homogeneous dielectric.

Now, since the partial derivations are redundant, we replace these by ordinary derivations. The resulting equation is as follows:

$$\frac{1}{X}\cdot\frac{d^2X}{dx^2}+\frac{1}{Y}\cdot\frac{d^2Y}{dy^2}=0 \tag{9.100b}$$

It is an interesting equation. The first term on the LHS of this equation is a function of x, while the second term is a function of y. Since x and y are independent of each other, how can these two functions, one of x and the other of y, sum up to zero? There exists one and only one possibility, i.e. if:

$$\left.\frac{1}{Y}\cdot\frac{d^2Y}{dy^2}=k^2.y^0 \quad (\text{a}) \quad \text{and} \quad \frac{1}{X}\cdot\frac{d^2X}{dy^2}=-k^2\cdot x^0 \quad (\text{b})\right\} \tag{9.101}$$

where k indicates an unknown constant. Thus, we have succeeded in replacing a partial differential equation by two ordinary differential equations by separating the variables x and y. The unknown constant k is therefore called the constant of separation. The constant k can be in general a complex constant, depending upon the nature of the boundary conditions required to be satisfied. The general solutions for the above equations are:

$$Y=c_1\cdot e^{ky}+c_2\cdot e^{-ky} \tag{9.102a}$$

or

$$Y=c_1\cdot\cosh(ky)+c_2\cdot\sinh(ky) \tag{9.102b}$$

and

$$X=c_1\cdot e^{jkx}+c_2\cdot e^{-jkx} \tag{9.103a}$$

or

$$X=c_1\cdot\cos(kx)+c_2\cdot\sin(kx) \tag{9.103b}$$

where c_1 and c_2 are constants of integrations.

Now, consider the boundary condition defined by Equation 9.97. This shows that the potential is an even function of x. On examining Equations 9.102a to 9.103b, we choose a real value for k and zero value for the constant c_2. We further note from Equation 9.97 that we need to set different values for k, one for each harmonic m. Therefore, k is to be replaced by k_m, such that:

$$k_m=\frac{m\pi}{w} \tag{9.104a}$$

Furthermore, replace c_1 and X, respectively by c_{1m} and X_m, such that:

$$X_m = c_{1m} \cdot \cos\left(k_m \cdot x\right) \tag{9.104b}$$

Now, since the potential is zero at $y = 0$, for the function Y, we tentatively choose:

$$Y_m = c_{2m} \cdot \sinh\left(k_m \cdot y\right) \tag{9.104c}$$

Therefore, Equation 9.98 is to be modified as:

$$V_m\left(x, y\right) \stackrel{\text{def}}{=} X_m\left(x\right) \cdot Y_m\left(y\right) \tag{9.104d}$$

Since, for each value of m, the above expression satisfies the Laplace equation; the complete solution can be taken as the sum of all these solutions. Thus:

$$V = \sum_m V_m\left(x, y\right) = \sum_m X_m\left(x\right) \cdot Y_m\left(y\right) \tag{9.105a}$$

Therefore, from Equations 9.104a, 9.104b, and 9.104c, since the potential is an even function x with zero value at $x = \pm w/2$, we get:

$$V = \sum_{m\text{-odd}}^{\infty} c_m \cdot \cos\left(\frac{m\pi}{w} \cdot x\right) \cdot \sinh\left(\frac{m\pi}{w} \cdot y\right) \tag{9.105b}$$

where

$$c_m \stackrel{\text{def}}{=} c_{1m} \cdot c_{2m} \tag{9.105c}$$

To determine the arbitrary constant c_m, we consider the boundary condition given by Equation 9.97, thus:

$$\sum_{m\text{-odd}}^{\infty} a_m \cdot \cos\left(m\pi \cdot \frac{x}{w}\right) = \sum_{m\text{-odd}}^{\infty} c_m \cdot \sinh\left(\frac{m\pi}{w} \cdot h\right) \cdot \cos\left(m\pi \cdot \frac{x}{w}\right) \tag{9.106a}$$

Thus,

$$c_m = a_m \cdot \text{cosech}\left(\frac{m\pi}{w} \cdot h\right) \tag{9.106b}$$

Note that it is neither possible nor required to determine c_{1m} and c_{2m} separately, it is only their product that matters. The final solution can therefore be given as follows:

$$V = \sum_{m\text{-odd}}^{\infty} a_m \cdot \cos\left(\frac{m\pi}{w} \cdot x\right) \cdot \frac{\sinh\left(\frac{m\pi}{w} \cdot y\right)}{\sinh\left(\frac{m\pi}{w} \cdot h\right)} \tag{9.107}$$

It may also be noted that the expression for the potential distribution V is independent of the constant permittivity of the homogeneous region.

9.4.1.2 Potential Field in Heterogeneous Medium (Case I)

Figure 9.13 shows a rectangular region which is filled up with two different dielectric materials, say with permittivity ϵ_1 (for $0 < y < g$ and $-w/2 < x < w/2$) and with permittivity ϵ_2 (for $g < y < h$ and $-w/2 < x < w/2$). This configuration has the same boundary conditions as given in Figure 9.12. The expression for the potential distribution in the piecewise homogeneous dielectric region will however differ from that given by Equation 9.107 found for a homogeneous dielectric region. For obtaining the solution this rectangular region is divided into two homogeneous subregions with the potential distributions $V_1$and V_2, each obeying Laplace's equation. Therefore, from Figure 9.12:

$$\nabla^2 V_1 = 0, \quad \text{for} \quad 0 < y < g \quad \text{and} \quad -w/2 < x < w/2 \tag{9.108a}$$

and

$$\nabla^2 V_2 = 0, \quad \text{for} \quad g < y < h \quad \text{and} \quad -w/2 < x < w/2 \tag{9.108b}$$

Consider the common boundary between the two subregions (at $y = g$, over $-w/2 < x < w/2$). Using the boundary conditions stated in the preceding example, the potential distribution along the common boundary can be expressed in Fourier cosine series with unknown coefficients as:

$$V\Big|_{y=g} \overset{\text{def}}{=} \sum_{m\text{-odd}}^{\infty} b_m \cdot \cos\left(\frac{m\pi}{w} \cdot x\right) \tag{9.109}$$

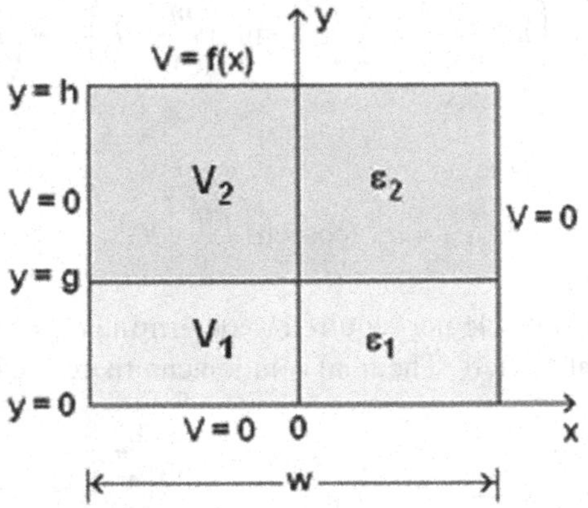

FIGURE 9.13 A rectangular region with heterogeneous (or piecewise homogeneous) dielectric.

The solutions of Equations 9.96a, 9.108a, and 9.108b, consistent with prescribed boundary conditions, respectively, are:

$$V_1 = \sum_{m\text{-odd}}^{\infty} b_m \cdot \frac{\sinh\left(\frac{m\pi}{w}\cdot y\right)}{\sinh\left(\frac{m\pi}{w}\cdot g\right)} \cdot \cos\left(\frac{m\pi}{w}\cdot x\right) \tag{9.110a}$$

$$V_2 = \sum_{m\text{-odd}}^{\infty} \left[a_m \cdot \frac{\sinh\left\{\frac{m\pi}{w}\cdot(y-g)\right\}}{\sinh\left\{\frac{m\pi}{w}\cdot(h-g)\right\}} - b_m \cdot \frac{\sinh\left\{\frac{m\pi}{w}\cdot(y-h)\right\}}{\sinh\left\{\frac{m\pi}{w}\cdot(h-g)\right\}} \right] \cdot \cos\left(\frac{m\pi}{w}\cdot x\right) \tag{9.110b}$$

To find the Fourier coefficient, b_m, the following boundary condition needs to be used:

$$\left. D_{1y}\right|_{y=g} = \left. D_{2y}\right|_{y=g} \;(\text{a}) \quad \text{thus} \quad \left. \epsilon_1 \cdot \frac{\partial V_1}{\partial y}\right|_{y=g} = \left. \epsilon_2 \cdot \frac{\partial V_2}{\partial y}\right|_{y=g} \;(\text{b}) \Bigg\} \tag{9.111}$$

Now, using Equations 9.110a and 9.110b, we get:

$$\begin{aligned} &\varepsilon_1 \cdot \left[a_m \cdot \operatorname{cosech}\frac{m\pi}{w}\cdot(h-g) - b_m \cdot \coth\frac{m\pi}{w}\cdot(h-g)\right] \\ &= \varepsilon_2 \cdot b_m \cdot \coth\left(\frac{m\pi}{w}\cdot g\right) \end{aligned} \tag{9.112a}$$

Thus,

$$b_m = a_m \cdot \frac{\epsilon_1 \cdot \operatorname{cosech}\left\{\frac{m\pi}{w}\cdot(h-g)\right\}}{\epsilon_1 \cdot \coth\left\{\frac{m\pi}{w}\cdot(h-g)\right\} + \epsilon_2 \cdot \coth\left(\frac{m\pi}{w}\cdot g\right)} \quad (\text{for } m = 1,3,5,\ldots) \tag{9.112b}$$

It may be noted that to find the potential distribution in the rectangular region we assumed a suitable distribution along the common boundary between the two subregions, in terms of a set of unknown Fourier coefficients. These coefficients were subsequently evaluated using a hitherto unused boundary condition at the same common boundary. It is further to be noted that now the potential distribution in each subregion is a function of both regional permittivities, viz. ϵ_1 and ϵ_2. It may also be noted that a different symbol is used to represent the permittivity, i.e. ϵ instead of ε, in order that the reader may become familiar with both symbols.

9.4.1.3 Potential Field in Heterogeneous Medium (Case II)

This case relates to the rectangle shown in Figure 9.12, with potential distributions along three of its four sides as given below:

$$V\Big|_{y=0} = 0, \quad \text{over } -w/2 < x < w/2 \tag{9.113a}$$

$$V\Big|_{x=-w/2} = 0, \quad \text{over } 0 < y < h \tag{9.113b}$$

$$V\Big|_{x=w/2} = 0, \quad \text{over } 0 < y < h \tag{9.113c}$$

Along the fourth side, namely at $y = h$, neither the potential nor its normal derivative is given. However, as an additional boundary condition the normal derivative of the potential at the boundary $y = 0$ is given in terms of Fourier series, as shown below:

$$\frac{\partial V}{\partial y}\Big|_{y=0} = \sum_{m\text{-odd}}^{\infty} a_m \cdot \cos\left(\frac{m\pi}{w}\cdot x\right) + \sum_{n=1}^{\infty} b_n \cdot \sin\left(\frac{n2\pi}{w}\cdot x\right) \text{ over } -w/2 < x < w/2 \tag{9.113d}$$

where a_m and b_n indicate two sets of *known* Fourier coefficient.

It is required to find the unique solution, if exist, for the Laplace's equation giving the potential distribution in the homogeneous rectangular region shown in Figure 9.12.

Let a function describing the unknown distribution of potential along the side $y = h$, of the rectangular region be defined in terms of the following Fourier series:

$$V\Big|_{y=h} \stackrel{\text{def}}{=} \sum_{m\text{-odd}}^{\infty} c_m \cdot \cos\left(\frac{m\pi}{w}\cdot x\right) + \sum_{n=1}^{\infty} d_n \cdot \sin\left(\frac{n2\pi}{w}\cdot x\right) \tag{9.114}$$

where c_m and d_n indicate two sets of unknown Fourier coefficients. Now, the potential distribution in the rectangular region, satisfying Equations 9.113a, 9.113c, and 9.114 can be obtained in terms of c_m and d_n as:

$$V = \sum_{m\text{-odd}}^{\infty} c_m \cdot \cos\left(\frac{m\pi}{w}\cdot x\right) \cdot \frac{\sinh\left(\frac{m\pi}{w}\cdot y\right)}{\sinh\left(\frac{m\pi}{w}\cdot h\right)} + \sum_{n=1}^{\infty} d_n \cdot \sin\left(\frac{n2\pi}{w}\cdot x\right) \cdot \frac{\sinh\left(\frac{n2\pi}{w}\cdot y\right)}{\sinh\left(\frac{n2\pi}{w}\cdot h\right)} \tag{9.115}$$

Therefore,

$$\left.\frac{\partial V}{\partial y}\right|_{y=0} = \sum_{m\text{-odd}}^{\infty}\left[c_m \cdot \frac{m\pi}{w} \cdot \operatorname{cosech}\left(\frac{m\pi}{w}\cdot h\right)\right]\cdot \cos\left(\frac{m\pi}{w}\cdot x\right) + \sum_{n=1}^{\infty}\left[d_n \cdot \frac{n2\pi}{w} \cdot \operatorname{cosech}\left(\frac{n2\pi}{w}\cdot h\right)\right]\cdot \sin\left(\frac{n2\pi}{w}\cdot x\right) \tag{9.116}$$

Equating the RHS of Equations 9.113d and 9.116, the two sets of the unknown Fourier coefficients found are as follows:

$$c_m = a_m \cdot \left[\frac{w}{m\pi} \cdot \sinh\left(\frac{m\pi}{w}\cdot h\right)\right] \tag{9.117a}$$

and

$$d_n = b_n \cdot \left[\frac{w}{n2\pi} \cdot \sinh\left(\frac{n2\pi}{w}\cdot h\right)\right] \tag{9.117b}$$

Therefore, the potential distribution in the rectangular region of Figure 9.12, consistent with all boundary conditions can be given as follows:

$$V = \sum_{m\text{-odd}}^{\infty} a_m \cdot \frac{w}{m\pi} \cdot \cos\left(\frac{m\pi}{w}\cdot x\right)\cdot \sinh\left(\frac{m\pi}{w}\cdot y\right) + \sum_{n=1}^{\infty} b_n \cdot \frac{w}{n2\pi} \cdot \sin\left(\frac{n2\pi}{w}\cdot x\right)\cdot \sinh\left(\frac{n2\pi}{w}\cdot y\right) \tag{9.118}$$

Two points may be noted. First, the potential distribution is assumed along a part of the boundary, while it is evaluated using a boundary condition elsewhere. Second, the Fourier coefficients describing the potential distribution, in this problem, could be computed in a non-iterative fashion. As shown in the next example, this may not always be the case.

9.4.1.4 Potential Field in Heterogeneous Medium (Case III)

This case also relates to the rectangle shown in Figure 9.12 with the same potential distributions along three of its four sides, as in the preceding example; (see Equations 9.113a–9.113c). Again, along the fourth side, namely at $y = h$, neither the potential nor its normal derivative is given. However, as additional boundary conditions, it is given that:

$$\left.\frac{\partial V}{\partial x}\right|_{x=w/2} = \sum_{p\text{-odd}}^{\infty} a_p \cdot \sin\left(\frac{p\pi}{h}\cdot y\right) \quad \text{over } 0 < y < h \tag{9.119a}$$

and

$$\left.\frac{\partial V}{\partial x}\right|_{x=-w/2} = \sum_{p\text{-odd}}^{\infty} b_p \cdot \sin\left(\frac{p\pi}{h}\cdot y\right) \quad \text{over } 0 < y < h \tag{9.119b}$$

where a_p and b_p indicate two sets of *known* Fourier coefficients.

It is required to find, if it exists, the unique solution for Laplace's equation giving the potential distribution in the rectangular region.

Since Fourier coefficients a_p and b_p are unrelated (i.e. $a_p \pm b_p \neq 0$), the potential function distributed on the top side of the rectangle is neither an odd nor an even function of x. Let the potential distribution at $y = h$ be defined as:

$$\left.V\right|_{y=h} \overset{\text{def}}{=} \sum_{m\text{-odd}}^{\infty} c_m \cdot \cos\left(\frac{m\pi}{w}\cdot x\right) + \sum_{n=1}^{\infty} d_n \cdot \sin\left(\frac{n2\pi}{w}\cdot x\right) \tag{9.120}$$

where c_m and d_n indicate two sets of unknown Fourier coefficients.

Note that this function vanishes at $x = \pm w/2$. Furthermore, it is a combination of even and odd functions of x. In view of Equations 9.113a– 9.113c, the distribution of the potential V can therefore be expressed as:

$$V = \sum_{m\text{-odd}}^{\infty} c_m \cdot \cos\left(\frac{m\pi}{w}\cdot x\right)\cdot\frac{\sinh\left(\frac{m\pi}{w}\cdot y\right)}{\sinh\left(\frac{m\pi}{w}\cdot h\right)} + \sum_{n=1}^{\infty} d_n \cdot \sin\left(\frac{n2\pi}{w}\cdot x\right)\cdot\frac{\sinh\left(\frac{n2\pi}{w}\cdot y\right)}{\sinh\left(\frac{n2\pi}{w}\cdot h\right)} \tag{9.121}$$

Therefore:

$$\left.\frac{\partial V}{\partial x}\right|_{x=w/2} = -\sum_{m\text{-odd}}^{\infty} c_m \cdot \frac{m\pi}{w}\cdot \sin\left(\frac{m\pi}{2}\right)\cdot\frac{\sinh\left(\frac{m\pi}{w}\cdot y\right)}{\sinh\left(\frac{m\pi}{w}\cdot h\right)} + \sum_{n=1}^{\infty} d_n \cdot \frac{n2\pi}{w}\cdot \cos(n\pi)\cdot\frac{\sinh\left(\frac{n2\pi}{w}\cdot y\right)}{\sinh\left(\frac{n2\pi}{w}\cdot h\right)} \tag{9.122a}$$

and

$$\left.\frac{\partial V}{\partial x}\right|_{x=-w/2} = \sum_{m\text{-odd}}^{\infty} c_m \cdot \frac{m\pi}{w} \cdot \sin\left(\frac{m\pi}{2}\right) \cdot \frac{\sinh\left(\frac{m\pi}{w} \cdot y\right)}{\sinh\left(\frac{m\pi}{w} \cdot h\right)} + \sum_{n=1}^{\infty} d_n \cdot \frac{n2\pi}{w} \cdot \cos(n\pi) \cdot \frac{\sinh\left(\frac{n2\pi}{w} \cdot y\right)}{\sinh\left(\frac{n2\pi}{w} \cdot h\right)} \tag{9.122b}$$

Now, in view of Equations 9.119a and 9.119b, we get:

$$\sum_{p\text{-odd}}^{\infty} a_p \cdot \sin\left(\frac{p\pi}{h} \cdot y\right) = -\sum_{m\text{-odd}}^{\infty} c_m \cdot \frac{m\pi}{w} \cdot \sin\left(\frac{m\pi}{2}\right) \cdot \frac{\sinh\left(\frac{m\pi}{w} \cdot y\right)}{\sinh\left(\frac{m\pi}{w} \cdot h\right)} + \sum_{n=1}^{\infty} d_n \cdot \frac{n2\pi}{w} \cdot \cos(n\pi) \cdot \frac{\sinh\left(\frac{n2\pi}{w} \cdot y\right)}{\sinh\left(\frac{n2\pi}{w} \cdot h\right)} \quad \text{over } 0 < x < h \tag{9.123a}$$

and

$$\sum_{p\text{-odd}}^{\infty} b_p \cdot \sin\left(\frac{p\pi}{h} \cdot y\right) = \sum_{m\text{-odd}}^{\infty} c_m \cdot \frac{m\pi}{w} \cdot \sin\left(\frac{m\pi}{2}\right) \cdot \frac{\sinh\left(\frac{m\pi}{w} \cdot y\right)}{\sinh\left(\frac{m\pi}{w} \cdot h\right)} + \sum_{n=1}^{\infty} d_n \cdot \frac{n2\pi}{w} \cdot \cos(n\pi) \cdot \frac{\sinh\left(\frac{n2\pi}{w} \cdot y\right)}{\sinh\left(\frac{n2\pi}{w} \cdot h\right)} \quad \text{over } 0 < x < h \tag{9.123b}$$

Therefore, using the orthogonal property of Fourier series, we get for p = 1, 3, 5, …:

$$a_p = -\sum_{m\text{-odd}}^{\infty} c_m \cdot \frac{2}{w} \cdot \sin\left(\frac{m\pi}{2}\right) \cdot \left[\frac{p \cdot m}{(p)^2 + \left(\frac{h}{w} \cdot m\right)^2}\right]$$
$$+\sum_{n=1}^{\infty} d_n \cdot \frac{h}{w} \cdot \cos(n\pi) \cdot \left[\frac{p \cdot n2}{(p)^2 + \left(\frac{h}{w} \cdot n2\right)^2}\right] \quad (9.124a)$$

and

$$b_p = \sum_{m\text{-odd}}^{\infty} c_m \cdot \frac{2}{w} \cdot \sin\left(\frac{m\pi}{2}\right) \cdot \left[\frac{p \cdot m}{(p)^2 + \left(\frac{h}{w} \cdot m\right)^2}\right]$$
$$+\sum_{n=1}^{\infty} d_n \cdot \frac{h}{w} \cdot \cos(n\pi) \cdot \left[\frac{p \cdot n2}{(p)^2 + \left(\frac{h}{w} \cdot n2\right)^2}\right] \quad (9.124b)$$

From Equations 9.124a and 9.124b, we get for p = 1, 3, 5, …,:

$$b_p + a_p = \sum_{n=1}^{\infty} d_n \cdot \frac{2h}{w} \cdot \cos(n\pi) \cdot \left[\frac{p \cdot n2}{(p)^2 + \left(\frac{h}{w} \cdot n2\right)^2}\right] \quad (9.125a)$$

and

$$b_p - a_p = \sum_{m\text{-odd}}^{\infty} c_m \cdot \frac{4}{w} \cdot \sin\left(\frac{m\pi}{2}\right) \cdot \left[\frac{p \cdot m}{(p)^2 + \left(\frac{h}{w} \cdot m\right)^2}\right] \quad (9.125b)$$

for $p = 1, 3, 5,\ldots$

The solution of the above two sets of simultaneous equations determines the Fourier coefficients c_m and d_n involved in the distribution of the potential V

(see Equation 9.121). For the numerical evaluation of the Fourier coefficients c_m and d_n, we must work with truncated Fourier series. Let the p-series be truncated after P number of terms, thus the upper limit is $(2P - 1)$. Therefore, for the m-series, the upper limit will also be $(2P - 1)$. While, for the n-series, its upper limit will be P. Two sets of simultaneous equations need to be solved, each for P number of unknowns. Since the two sets of simultaneous equations are linearly independent to each other, we get unique values for c_m and d_n. Hence, the solution giving the potential distribution V, is unique, although it is approximate.

If the coefficients a_p and b_p are identical, the coefficient c_m will be zero. On the other hand, if a_p is equal to the negative of b_p, the coefficient d_n will be zero. In the former case the potential function will be an odd function of x, while in the latter case it will be an even function of x. A unique solution cannot be found if either a_p or b_p is not given.

9.4.1.5 Potential Field in Triangular Region

Figure 9.14 illustrates a square abcd in dotted lines. A part of this square is a right-angled isosceles triangular region with one of its sides, located along the x-axis over $0 \le x \le \ell$, and another side located along the y-axis over $0 \le y \le \ell$. Clearly, these two line segments meet at the origin. The hypotenuse of the triangular is formed by joining the free ends of the two sides. Its length is $\ell\sqrt{2}$ and it lies along the line $y = \ell - x$. In this homogeneous triangular region the scalar electric potential V obeys the Laplace equation:

$$\frac{\partial^2 V}{\partial x^2} + \frac{\partial^2 V}{\partial y^2} = 0 \qquad (9.126)$$

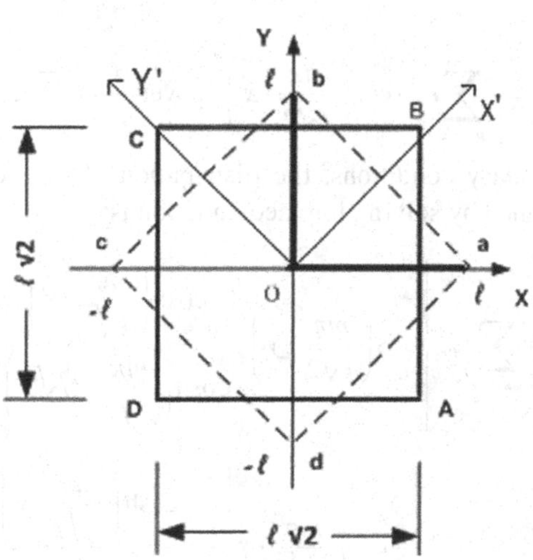

FIGURE 9.14 A square region.

It is required to find the solution of the Laplace equation giving the distribution of the potential in the triangular region (i.e. aboa) subjected to the following boundary conditions:

$$V\Big|_{x=0} = \sum_{n=1}^{\infty} a_n \cdot \sin\left(\frac{n\pi}{\ell} \cdot y\right) \quad \text{over } 0 \le y \le \ell \tag{9.127a}$$

$$V\Big|_{y=0} = \sum_{n=1}^{\infty} b_n \cdot \sin\left(\frac{n\pi}{\ell} \cdot x\right) \quad \text{over } 0 \le y \le \ell \tag{9.127b}$$

$$V\Big|_{x'=\ell/\sqrt{2}} = \sum_{m\text{-odd}}^{\infty} c_m \cdot \cos\left(\frac{m\pi}{\ell\sqrt{2}} \cdot y'\right) \tag{9.127c}$$

where the y'-axis is parallel to the hypotenuse ab of the right-angled isosceles triangle aboa. The middle of the hypotenuse corresponds to $y' = 0$.

Let us split up the potential field V in the triangular region into two partial field components, each is a solution of the Laplace equation and subjected to different sets of boundary conditions;

$$V = V_1 + V_2 \tag{9.128}$$

Consider the square region ABCD in Figure 9.14. Let the boundary conditions for the scalar electric potential V_1 in this region be as indicated below:

$$V_1\Big|_{x=\pm\ell/\sqrt{2}} = \sum_{m\text{-odd}}^{\infty} c_m \cdot \cos\left(\frac{m\pi}{\ell\sqrt{2}} \cdot y\right) \quad \text{over } -\ell/\sqrt{2} < y < l/\sqrt{2} \tag{9.129a}$$

and

$$V_1\Big|_{y=\pm\ell/\sqrt{2}} = -\sum_{m\text{-odd}}^{\infty} c_m \cdot \cos\left(\frac{m\pi}{\ell\sqrt{2}} \cdot x\right) \quad \text{over } -\ell/\sqrt{2} < x < l/\sqrt{2} \tag{9.129b}$$

For this set of boundary conditions, the distribution of the potential field in the square ABCD, obtained by solving Laplace equation is:

$$V_1 = \sum_{m\text{-odd}}^{\infty} c_m \cdot \left[\cos\left(\frac{m\pi}{\ell\sqrt{2}} \cdot y\right) \cdot \frac{\cosh\left(\frac{m\pi}{\ell\sqrt{2}} \cdot x\right)}{\cosh\left(\frac{m\pi}{\ell\sqrt{2}} \cdot \ell/\sqrt{2}\right)} - \cos\left(\frac{m\pi}{\ell\sqrt{2}} \cdot x\right) \cdot \frac{\cosh\left(\frac{m\pi}{\ell\sqrt{2}} \cdot y\right)}{\cosh\left(\frac{m\pi}{\ell\sqrt{2}} \cdot \ell/\sqrt{2}\right)}\right] \tag{9.130}$$

It may be noted that the potential field V_1, along the two diagonals $y = \pm x$, is zero. If the square ABCD in Figure 9.14, is subjected to an angular displacement about its center in the counterclockwise direction by $\pi/4$ (the resulting square abcd is shown in this figure by dotted lines), the distribution of potential field V_1, in the square abcd found from Equation 9.130, will be:

$$V_1 = \sum_{m\text{-odd}}^{\infty} c_m \cdot \left[\cos\left(\frac{m\pi}{\ell\sqrt{2}} \cdot y'\right) \cdot \frac{\cosh\left(\frac{m\pi}{\ell\sqrt{2}} \cdot x'\right)}{\cosh\left(\frac{m\pi}{\ell\sqrt{2}} \cdot \ell/\sqrt{2}\right)} - \cos\left(\frac{m\pi}{\ell\sqrt{2}} \cdot x'\right) \cdot \frac{\cosh\left(\frac{m\pi}{\ell\sqrt{2}} \cdot y'\right)}{\cosh\left(\frac{m\pi}{\ell\sqrt{2}} \cdot \ell/\sqrt{2}\right)} \right] \tag{9.131a}$$

This expression satisfies the boundary condition given by Equation 9.127c.

where

$$\left. x' = \frac{1}{\sqrt{2}} \cdot (x + y) \quad (b) \quad \text{and} \quad y' = -\frac{1}{\sqrt{2}} \cdot (x - y) \quad (c) \right\} \tag{9.131}$$

Thus,

$$V_1 = \sum_{m\text{-odd}}^{\infty} c_m \cdot \left[\cos\left\{\frac{m\pi}{2\ell}(x - y)\right\} \cdot \frac{\cosh\left\{\frac{m\pi}{2\ell}(x + y)\right\}}{\cosh\left(\frac{m\pi}{2\ell} \cdot \ell\right)} - \cos\left\{\frac{m\pi}{2\ell}(x + y)\right\} \cdot \frac{\cosh\left\{\frac{m\pi}{2\ell}(x - y)\right\}}{\cosh\left(\frac{m\pi}{2\ell} \cdot \ell\right)} \right] \tag{9.132a}$$

Note that:

$$V_1\Big|_{x \text{ or } y=0} = 0 \tag{9.132b}$$

and

$$V_1\Big|_{y=\ell-x} = \sum_{m\text{-odd}}^{\infty} c_m \cdot \sin\left(\frac{m\pi}{2}\right) \cdot \sin\left(\frac{m\pi}{\ell} \cdot x\right) \tag{9.132c}$$

Consider the square ABCD shown in Figure 9.14, with the origin shifted to the point D and the length of each side set to ℓ, instead of $\ell\sqrt{2}$. Further, let the potential field V_2 in the square region satisfy the following boundary conditions:

$$V_2\big|_{x=0} = \sum_{n=1}^{\infty} a_n \cdot \sin\left(\frac{n\pi}{\ell}\cdot y\right) \quad \text{over } 0<y<l \tag{9.133a}$$

$$V_2\big|_{y=0} = \sum_{n=1}^{\infty} b_n \cdot \sin\left(\frac{n\pi}{\ell}\cdot x\right) \quad \text{over } 0<x<l \tag{9.133b}$$

$$V_2\big|_{x=\ell} = \sum_{n=1}^{\infty} b_n \cdot \cos(n\pi)\cdot \sin\left(\frac{n\pi}{\ell}\cdot y\right) \quad \text{over } 0<y<l \tag{9.133c}$$

$$V_2\big|_{y=\ell} = \sum_{n=1}^{\infty} a_n \cdot \cos(n\pi)\cdot \sin\left(\frac{n\pi}{\ell}\cdot x\right) \quad \text{over } 0<y<l \tag{9.133d}$$

The solution of Equation 9.126, subjected to the above boundary conditions is given as:

$$\begin{aligned} V_2 = -\sum_{n=1}^{\infty} \Bigg[\!\Bigg[a_n \cdot &\left[\sin\left(\frac{n\pi}{\ell}\cdot y\right)\cdot \frac{\sinh\left\{\frac{n\pi}{\ell}\cdot(x-\ell)\right\}}{\sinh\left(\frac{n\pi}{\ell}\cdot \ell\right)} - \cos(n\pi)\cdot \sin\left(\frac{n\pi}{\ell}\cdot x\right)\cdot \frac{\sinh\left\{\frac{n\pi}{\ell}\cdot y\right\}}{\sinh(n\pi)} \right] \\ + b_n \cdot &\left[\sin\left(\frac{n\pi}{\ell}\cdot x\right)\cdot \frac{\sinh\left\{\frac{n\pi}{\ell}\cdot(y-\ell)\right\}}{\sinh\left(\frac{n\pi}{\ell}\cdot \ell\right)} - \cos(n\pi)\cdot \sin\left(\frac{n\pi}{\ell}\cdot y\right)\cdot \frac{\sinh\left\{\frac{n\pi}{\ell}\cdot x\right\}}{\sinh(n\pi)} \right] \Bigg]\!\Bigg] \end{aligned} \tag{9.134a}$$

The potential field along the diagonal, $y = \ell - x$, found from Equation 9.134a, is:

$$V_2\Big|_{y=\ell-x} = 0 \tag{9.134b}$$

In view of Equations 9.128, 9.132, and 9.134a the distribution of the potential field in the triangular region can be written as:

$$V = \sum_{m\text{-odd}}^{\infty} c_m \cdot \left[\cos\frac{m\pi}{2\ell}(x-y) \cdot \frac{\cosh\left\{\frac{m\pi}{2\ell}(x+y)\right\}}{\cosh\left(\frac{m\pi}{2\ell}\cdot\ell\right)} - \cos\frac{m\pi}{2\ell}(x+y) \cdot \frac{\cosh\left\{\frac{m\pi}{2\ell}(x-y)\right\}}{\cosh\left(\frac{m\pi}{2\ell}\cdot\ell\right)} \right]$$

$$-\sum_{n=1}^{\infty} \left[\!\left[a_n \cdot \left[\sin\left(\frac{n\pi}{\ell}\cdot y\right) \cdot \frac{\sinh\left\{\frac{n\pi}{\ell}\cdot(x-\ell)\right\}}{\sinh\left(\frac{n\pi}{\ell}\cdot\ell\right)} - \cos(n\pi)\cdot\sin\left(\frac{n\pi}{\ell}\cdot x\right) \cdot \frac{\sinh\left\{\frac{n\pi}{\ell}\cdot y\right\}}{\sinh(n\pi)} \right] \right.\right. \tag{9.135}$$

$$\left.\left. + b_n \cdot \left[\sin\left(\frac{n\pi}{\ell}\cdot x\right) \cdot \frac{\sinh\left\{\frac{n\pi}{\ell}\cdot(y-\ell)\right\}}{\sinh\left(\frac{n\pi}{\ell}.\ell\right)} - \cos(n\pi)\cdot\sin\left(\frac{n\pi}{\ell}\cdot y\right) \cdot \frac{\sinh\left\{\frac{n\pi}{\ell}\cdot x\right\}}{\sinh(n\pi)} \right] \right]\!\right]$$

9.4.2 Field Problems in Cylindrical Coordinates

The Laplacian for the scalar function V in cylindrical coordinates is given by Equation 9.6. For two-dimensional fields, it yields the following three forms of Laplacian relations:

$$\nabla^2 V = \frac{1}{\rho^2} \cdot \frac{\partial^2 V}{\partial \varphi^2} + \frac{\partial^2 V}{\partial z^2} \tag{9.136a}$$

$$\nabla^2 V = \frac{1}{\rho}\cdot\frac{\partial}{\partial\rho}\left(\rho\frac{\partial V}{\partial\rho}\right)+\frac{\partial^2 V}{\partial z^2} \tag{9.136b}$$

$$\nabla^2 V = \frac{1}{\rho}\cdot\frac{\partial}{\partial\rho}\left(\rho\frac{\partial V}{\partial\rho}\right)+\frac{1}{\rho^2}\cdot\frac{\partial^2 V}{\partial\varphi^2} \tag{9.136c}$$

9.4.2.1 Potential Field Independent of ρ

This case corresponds to Equation 9.136a, and the Laplace equation can be written as:

$$\nabla^2 V = \frac{1}{\rho^2}\cdot\frac{\partial^2 V}{\partial\varphi^2}+\frac{\partial^2 V}{\partial z^2}=0 \tag{9.137}$$

Therefore:

$$\frac{\partial^2 V}{\partial\varphi^2}=\frac{\partial^2 V}{\partial z^2}=0 \tag{9.138}$$

The solutions are obtained by setting:

$$V(\varphi,z)\stackrel{\text{def}}{=}\Phi(\varphi)\cdot Z(z) \tag{9.139a}$$

where

$$\Phi(\varphi)=c_1\varphi+c_2 \tag{9.139b}$$

and

$$Z(z)=d_1 z+d_2 \tag{9.139c}$$

where c_1, c_2, d_1, and d_2 indicate arbitrary constants.

Note that for a single-valued solution, the range for the variable φ must be less than 2π.

9.4.2.2 Potential Field Independent of φ

This case corresponds to Equation 9.136b, and the Laplace equation can be written as:

$$\nabla^2 V = \frac{1}{\rho}\cdot\frac{\partial}{\partial\rho}\left(\rho\frac{\partial V}{\partial\rho}\right)+\frac{\partial^2 V}{\partial z^2}=0 \tag{9.140}$$

Consider Equation 9.140. Let the potential V be expressed as a product of $R(\rho)$, a function of ρ, and of $Z(z)$, a function of z. Thus,

$$V(\rho,z) \stackrel{\text{def}}{=} R(\rho)\cdot Z(z) \tag{9.141}$$

On substituting Equation 9.141 in Equation 9.140 and then dividing the resulting equation by V, we get:

$$\frac{1}{\rho\cdot R}\cdot\frac{d}{d\rho}\left(\rho\frac{dR}{d\rho}\right)+\frac{1}{Z}\cdot\frac{d^2Z}{dz^2}=0 \tag{9.142a}$$

Let:

$$\frac{1}{\rho R}\cdot\frac{d}{d\rho}\left(\rho\frac{dR}{d\rho}\right)=k^2 \tag{9.142b}$$

Thus:

$$\frac{1}{Z}\cdot\frac{d^2Z}{dz^2}=-k^2 \tag{9.142c}$$

Now, Equation 9.142b can be rewritten as:

$$\frac{\partial^2 R}{\partial(k\rho)^2}+\frac{1}{(k\rho)}\cdot\frac{\partial R}{\partial(k\rho)}-R=0 \tag{9.143a}$$

This equation is identified as the modified Bessel equation of zero order. Its solution is given in terms of modified Bessel functions: $I_o(k\rho)$ and $K_o(k\rho)$, as shown below:

$$R=c_1\cdot I_o(k\rho)+c_2\cdot K_o(k\rho) \tag{9.143b}$$

The solution for Equation 9.142c is readily found as:

$$Z=d_1\cdot\cos(kz)+d_2\cdot\sin(kz) \tag{9.143c}$$

where c_1, c_2, d_1, and d_2 indicate arbitrary constants. These can be evaluated in accordance with the boundary conditions specified in the problem. Thereafter the substitution of Equation 9.143b and 9.143c in Equation 9.141 gives the value of V.

9.4.2.3 Potential Field Independent of z

This case corresponds to Equation 9.136c and the Laplace equation can be written as:

$$\nabla^2 V=\frac{1}{\rho}\cdot\frac{\partial}{\partial\rho}\left(\rho\frac{\partial V}{\partial\rho}\right)+\frac{1}{\rho^2}\cdot\frac{\partial^2 V}{\partial\varphi^2}=0 \tag{9.144a}$$

Therefore:

$$\rho \cdot \frac{\partial}{\partial \rho}\left(\rho \frac{\partial V}{\partial \rho}\right) + \frac{\partial^2 V}{\partial \varphi^2} = 0 \tag{9.144b}$$

Now, considering Equation 9.144a, let us assume:

$$V(\rho,\varphi) \overset{\text{def}}{=} R(\rho)\cdot \Phi(\varphi) \tag{9.145}$$

On substituting Equation 9.145 into Equation 9.144a and then dividing the resulting equation by V, we get:

$$\frac{\rho}{R}\cdot\frac{d}{d\rho}\left(\rho\frac{dR}{d\rho}\right) + \frac{1}{\Phi}\cdot\frac{d^2\Phi}{d\varphi^2} = 0 \tag{9.146a}$$

Let:

$$\frac{\rho}{R}\cdot\frac{d}{d\rho}\left(\rho\frac{dR}{d\rho}\right) = k^2 \tag{9.146b}$$

Thus:

$$\frac{1}{\Phi}\cdot\frac{d^2\Phi}{d\varphi^2} = -k^2 \tag{9.146c}$$

The solution for Equation 9.146b found by inspection is given as:

$$R = c_1 \cdot \rho^k + c_2 \cdot \rho^{-k} \quad \text{for } k \neq 0 \tag{9.147a}$$

$$= c_1 \cdot \ln(\rho) + c_2 \quad \text{for } k = 0 \tag{9.147b}$$

And that for Equation 9.146c is:

$$\Phi = d_1 \cdot \cos(k\varphi) + d_2 \cdot \sin(k\varphi) \quad \text{for } k \neq 0 \tag{9.148a}$$

$$= d_1 \cdot \varphi + d_2 \quad \text{for } k = 0 \tag{9.148b}$$

In Equations 9.147a, 9.147b, 9.148a and 9.148b, c_1, c_2, d_1, and d_2 indicate arbitrary constants.

It is worth mentioning that if $k = 0$ for a single-valued solution, the range for the variable φ has to be less than 2π.

9.4.2.4 Periodic Field in a Long Coaxial Cylinder

Figure 9.15 shows a long coaxial cylinder with R_1 and R_2 as internal and external radii, respectively. The potential on the inner cylinder is periodically varying with the φ-coordinate as given here:

$$V\Big|_{\rho=R_1} = \sum_{m\text{-odd}}^{\infty} F_m \cdot \cos(m\varphi) \tag{9.149}$$

where F_m is the Fourier coefficient for the known waveform.

The outer cylinder is kept at zero potential. It is required to determine the distribution of the potential in the dielectric region between the two cylinders.

In view of Equation 9.147a, and the stated boundary conditions, let:

$$R\Big|_{\rho=R_1} = c_{1m} \cdot R_1^{\ m} + c_{2m} \cdot R_1^{\ -m} = 1 \tag{9.150a}$$

and

$$R\Big|_{\rho=R_2} = c_{1m} \cdot R_2^{m} + c_{2m} \cdot R_2^{-m} = 0 \tag{9.150b}$$

Therefore:

$$c_{1m} = \frac{1}{R_1^{\ m} \cdot \left[1-(R_2/R_1)^{2m}\right]} \ \text{(a)} \ \text{and} \ c_{2m} = -\frac{R_2^{2m}}{R_1^{m} \cdot \left[1-(R_2/R_1)^{2m}\right]} \ \text{(b)} \tag{9.151}$$

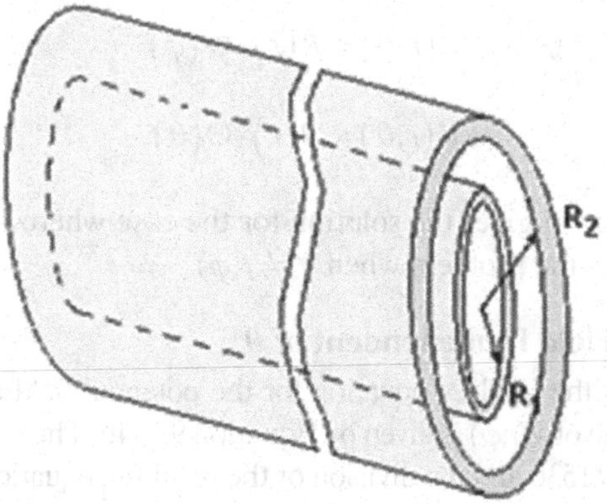

FIGURE 9.15 A long cylindrical capacitor.

Hence, the potential distribution in the dielectric region is as follows:

$$V=\sum_{m\text{-odd}}^{\infty}F_m\cdot\frac{\rho^m-\rho^{-m}\cdot R_2^{2m}}{R_1^m\cdot\left[1-\left(R_2/R_1\right)^{2m}\right]}\cdot\cos\left(m\varphi\right) \tag{9.152}$$

9.4.3 Field Problems in Spherical Coordinates

The Laplacian equation for the scalar function V in spherical coordinates is given by Equation 9.7, which is reproduced here:

$$\nabla^2V=\frac{1}{r^2}\cdot\frac{\partial}{\partial r}\left(r^2\frac{\partial V}{\partial r}\right)+\frac{1}{r^2\sin\theta}\cdot\frac{\partial}{\partial\theta}\left(\sin\theta\frac{\partial V}{\partial\theta}\right)+\frac{1}{r^2\sin^2\theta}\cdot\frac{\partial^2V}{\partial\varphi^2} \tag{9.7}$$

This equation leads to the following three equations on eliminating variation along one of the coordinates:

$$\nabla^2V=\frac{1}{r^2\sin\theta}\cdot\frac{\partial}{\partial\theta}\left(\sin\theta\frac{\partial V}{\partial\theta}\right)+\frac{1}{r^2\sin^2\theta}\cdot\frac{\partial^2V}{\partial\varphi^2}=0\quad\left\{\text{for }V\neq f\left(r\right)\right\} \tag{9.153a}$$

$$\nabla^2V=\frac{1}{r^2}\cdot\frac{\partial}{\partial r}\left(r^2\frac{\partial V}{\partial r}\right)+\frac{1}{r^2\sin^2\theta}\cdot\frac{\partial^2V}{\partial\varphi^2}=0\quad\left\{\text{for }V\neq f(\theta)\right\} \tag{9.153b}$$

$$\nabla^2V=\frac{1}{r^2}\cdot\frac{\partial}{\partial r}\left(r^2\frac{\partial V}{\partial r}\right)+\frac{1}{r^2\sin\theta}\cdot\frac{\partial}{\partial\theta}\left(\sin\theta\frac{\partial V}{\partial\theta}\right)=0\quad\left\{\text{for }V\neq f(\varphi)\right\} \tag{9.153c}$$

All of these equations can be solved by the method of separation of variables. Set the following in Equations 9.153a, 9.153b, and 9.153c, respectively:

$$V\left(\theta,\varphi\right)=\Theta\left(\theta\right)\cdot\Phi\left(\varphi\right) \tag{9.154a}$$

$$V\left(r,\varphi\right)=R\left(r\right)\cdot\Phi\left(\varphi\right) \tag{9.154b}$$

$$V\left(r,\theta\right)=R\left(r\right)\cdot\Theta\left(\theta\right) \tag{9.154c}$$

The following subsection gives the solution for the case where $V\neq f(\theta)$, whereas the next subsection solves the problem when $V\neq f(\varphi)$.

9.4.3.1 Potential Field I Independent of θ

Equation 9.153b gives the Laplace equation for the potential field that is independent of θ coordinate. The value of $V(r,\varphi)$ is given by Equation 9.154b. The substitution of Equation 9.154b into Equation 9.153b and the division of the resulting equation by V gives:

$$\frac{1}{R}\cdot\frac{d}{dr}\left(r^2\frac{dR}{dr}\right)+\frac{1}{\sin^2\theta}\cdot\frac{1}{\Phi}\cdot\frac{d^2\Phi}{d\varphi^2}=0 \tag{9.155}$$

Since the potential V is independent of θ, therefore it follows that:

$$\frac{1}{R}\cdot\frac{d}{dr}\left(r^2\frac{dR}{dr}\right)=\frac{1}{\Phi}\cdot\frac{d^2\Phi}{d\varphi^2}=0 \tag{9.156}$$

giving:

$$\left.r^2\frac{dR}{dr}=c_1 \quad (a) \quad \text{and} \quad \frac{d\Phi}{d\varphi}=c_2 \quad (b)\right\} \tag{9.157}$$

where c_1 and c_2 are constants of integrations.

On further integrating, we have:

$$\left.R=c_1'-\frac{c_1}{r} \quad (a) \quad \text{and} \quad \Phi=c_2'+c_2\cdot\varphi \quad (b)\right\} \tag{9.158}$$

where c_1' and c_2' are constants of integration.

9.4.3.2 Concentric Spheres with Inhomogeneous Dielectric

Consider a concentric spherical surface with internal and external radius, respectively, as R_1 and R_2. The potential on the inner spherical surface is given as V_o and the outer spherical surface is at zero potential. It is required to find the potential distribution in the inhomogeneous dielectric region between the two spherical surfaces. The permittivity ε for the dielectric is a given function of θ:

$$\varepsilon=\varepsilon_o\cdot\operatorname{cosec}(\theta) \tag{9.159}$$

Equation 9.10 is reproduced here:

$$\nabla^2V+\frac{1}{\varepsilon}\cdot\nabla V\cdot\nabla\varepsilon=0 \tag{9.160}$$

Since the potential field is independent of the φ coordinate, the corresponding Laplacian equation is:

$$\nabla^2V=\frac{1}{r^2}\cdot\frac{\partial}{\partial r}\left(r^2\frac{\partial V}{\partial r}\right)+\frac{1}{r^2\sin\theta}\cdot\frac{\partial}{\partial\theta}\left(\sin\theta\frac{\partial V}{\partial\theta}\right) \tag{9.161a}$$

and

$$\nabla V\cdot\nabla\varepsilon=\left(\frac{1}{r}\cdot\frac{\partial V}{\partial\theta}\right)\cdot\left(\frac{1}{r}\cdot\frac{d\varepsilon}{d\theta}\right)=\frac{1}{r^2}\cdot\frac{\partial V}{\partial\theta}\cdot\frac{d\varepsilon}{d\theta} \tag{9.161b}$$

Thus, Equation 9.160 can be written as:

$$\left[\frac{1}{r^2}\cdot\frac{\partial}{\partial r}\left(r^2\frac{\partial V}{\partial r}\right)+\frac{1}{r^2\sin\theta}\cdot\frac{\partial}{\partial\theta}\left(\sin\theta\frac{\partial V}{\partial\theta}\right)\right]+\frac{1}{\varepsilon}\cdot\left[\frac{1}{r^2}\cdot\frac{\partial V}{\partial\theta}\cdot\frac{d\varepsilon}{d\theta}\right]=0$$

In view of Equation 9.159,

$$\frac{\partial}{\partial r}\left(r^2\frac{\partial V}{\partial r}\right)+\left(\frac{\partial^2 V}{\partial \theta^2}+\cot\theta\cdot\frac{\partial V}{\partial \theta}\right)+\left(-\cot\theta\right)\cdot\frac{\partial V}{\partial \theta}=0$$

or

$$\frac{\partial}{\partial r}\left(r^2\frac{\partial V}{\partial r}\right)+\frac{\partial^2 V}{\partial \theta^2}=0 \tag{9.162}$$

The solution to Equation 9.162 can be written as:

$$V\left(r,\theta\right)\overset{\text{def}}{=}R\left(r\right)\cdot\Theta\left(\theta\right) \tag{9.163}$$

Putting Equation 9.163 into Equation 9.162 and then dividing both sides of the resulting equation by V, we get:

$$\frac{1}{R}\cdot\frac{d}{dr}\left(r^2\frac{dR}{dr}\right)=-\frac{1}{\Theta}\cdot\frac{d^2\Theta}{d\theta^2}\overset{\text{def}}{=}k^2 \tag{9.164}$$

where k is the *constant of separation.*

From Equation 9.164, we have:

$$\left.\frac{d}{dr}\left(r^2\frac{dR}{dr}\right)-k^2\cdot R=0\quad(\text{a})\quad\text{and}\quad\frac{d^2\Theta}{d\theta^2}+k^2\cdot\Theta=0\quad(\text{b})\right\} \tag{9.165a}$$

Now, let us substitute $n\cdot(n+1)$ for the *constant* k^2, i.e.:

$$k^2\overset{\text{def}}{=}n\cdot\left(n+1\right)=n^2+n \tag{9.166}$$

Then, for any real value of k taking the positive value for n:

$$n=\frac{\sqrt{1+4k^2}-1}{2} \tag{9.167}$$

The solution for Equation 9.165a can be shown to be:

$$R=c_1\cdot r^n+c_2\cdot r^{-n-1} \tag{9.168}$$

where c_1 and c_2 indicate arbitrary constants.

Now, the solution of Equation 9.165b can be given as:

$$\Theta=d_1\cdot\cos\left(k\cdot\theta\right)+d_2\cdot\sin\left(k\cdot\theta\right) \tag{9.169}$$

where d_1 and d_2 are two arbitrary constants.

It may be noted that there can be more than one value for the constant k therefore, in view of Equations 9.165 to 9.169, the solution for one such value for k, i.e. k_m, is given as:

$$V_m \stackrel{\text{def}}{=} \left[c_{1m} \cdot r^{n_m} + c_{2m} \cdot r^{-n_m-1}\right] \cdot \left[d_{1m} \cdot \cos\left(k_m \cdot \theta\right) + d_{2m} \cdot \sin\left(k_m \cdot \theta\right)\right] \quad (9.170a)$$

where in view of Equation 9.167:

$$n_m = \frac{1}{2} \cdot \left[\sqrt{1 + 4k_m^2} - 1\right] \quad (9.170b)$$

The complete solution is found in view of the partial solutions giving by Equations 9.168 and 9.169. Thus in view of Equation 9.170a, the complete solution can be written as:

$$V = \sum_{m}^{\infty} \left[c_{1m} \cdot r^{n_m} + c_{2m} \cdot r^{-n_m-1}\right] \cdot \left[d_{1m} \cdot \cos\left(k_m \cdot \theta\right) + d_{2m} v \sin\left(k_m \cdot \theta\right)\right] \quad (9.171)$$

Now consider the boundary conditions:

$$\left. V(r,\theta)\right|_{r=R_1} = V_o \quad (a) \quad \text{and} \quad \left. V(r,\theta)\right|_{r=R_2} = 0 \quad (b) \Big\} \quad (9.172)$$

Therefore:

$$c_{2m} = -\frac{V_o \cdot R_2^{(2n_m+1)}}{\left[R_1^{(n_m)} - R_2^{(2n_m+1)} \cdot R_1^{-(n_m+1)}\right]} \quad (9.173a)$$

and

$$c_{1m} = \frac{V_o}{\left[R_1^{(n_m)} - R_2^{(2n_m+1)} \cdot R_1^{-(n_m+1)}\right]} \quad (9.173b)$$

Thus:

$$\left[c_{1m} \cdot r^{n_m} + c_{2m} \cdot r^{-n_m-1}\right] = V_o \cdot \frac{\left[r^{(n_m)} - R_2^{(2n_m+1)} \cdot r^{-(n_m+1)}\right]}{\left[R_1^{(n_m)} - R_2^{(2n_m+1)} \cdot R_1^{-(n_m+1)}\right]} \quad (9.174)$$

It may be noted that for any value of n_m, boundary conditions specified in the problem will be satisfied provided that the factor $\left[d_{1m} \cdot \cos\left(k_m \cdot \theta\right) + d_{2m} \cdot \sin\left(k_m \cdot \theta\right)\right]$ in Equation 9.169 has unity value over $0 < \theta < \pi$. This can be readily achieved say, by setting:

$$\left. d_{1m} \stackrel{\text{def}}{=} 0 \quad (\text{a}) \quad \text{and} \quad d_{2m} \stackrel{\text{def}}{=} d_m \quad (\text{b}) \right\} \tag{9.175}$$

for $m = 1, 3, 5, 7, \ldots$

Therefore:

$$V = \sum_m^{\infty} V_o \cdot \frac{\left[r^{(n_m)} - R_2^{(2n_m+1)} \cdot r^{-(n_m+1)} \right]}{\left[R_1^{(n_m)} - R_2^{(2n_m+1)} \cdot R_1^{-(n_m+1)} \right]} \cdot d_m \cdot \sin\left(k_m \cdot \theta\right) \tag{9.176}$$

so that:

$$V\Big|_{r=R_2} = 0 \tag{9.177a}$$

and:

$$V\Big|_{r=R_1} = V_o \cdot \sum_m^{\infty} d_m \cdot \sin\left(k_m \cdot \theta\right) = V_o \tag{9.177b}$$

and therefore:

$$\sum_m^{\infty} d_m . \sin\left(k_m . \theta\right) = 1 \quad \text{over} \quad 0 < \theta < \pi \tag{9.178}$$

This equation can be identified as the Fourier series expansion for unity, over $0 < \theta < \pi$.

Therefore, we choose:

$$k_m = m \tag{9.179a}$$

And thus, in view of Equation 9.167:

$$n_m = \frac{1}{2} \cdot \left[\sqrt{1 + 4m^2} - 1 \right] \tag{9.179b}$$

where m is an integer.

Thus, Equation 9.178 can be rewritten as:

$$\sum_m^{\infty} d_m \cdot \sin\left(m \cdot \theta\right) = 1 \quad \text{over} \quad 0 < \theta < \pi \tag{9.180}$$

Now, to evaluate the Fourier coefficient d_m, multiply both sides of this equation by $\sin(p \cdot \theta)$, where p is an integer, then integrate over $0 < \theta < \pi$. Thus:

$$\sum_m^{\infty} \frac{d_m}{2} \cdot \int_0^{\pi} 2 \sin\left(m \cdot \theta\right) \cdot \sin\left(p \cdot \theta\right) d\theta = \int_0^{\pi} \sin\left(p \cdot \theta\right) d\theta$$

or

$$\sum_{m}^{\infty}\frac{d_m}{2}\cdot\left[\frac{\sin\{(m-p)\theta\}}{(m-p)}-\frac{\sin\{(m+p)\theta\}}{(m+p)}\right]_0^{\pi}=\left[-\frac{\cos(p\cdot\theta)}{p}\right]_0^{\pi}$$

or

$$\sum_{m}^{\infty}\frac{d_m}{2}\cdot\left[\frac{\sin\{(m-p)\pi\}}{(m-p)}\right]=\frac{1-\cos(p\cdot\pi)}{p} \tag{9.181a}$$

Thus:

$$\frac{\pi}{2}\cdot d_p=\frac{1-\cos(p\cdot\pi)}{p} \tag{9.181b}$$

or

$$d_p=\frac{2}{p\cdot\pi}\cdot\left[1-\cos(p\cdot\pi)\right] \tag{9.182a}$$

Thus:

$$d_p=\begin{cases}0 & \text{for } p\text{-even}\\ \dfrac{4}{p\cdot\pi} & \text{for } p\text{-odd}\end{cases} \tag{9.182b}$$

Therefore, Equation 9.176 can finally be given as:

$$V=\sum_{m\text{-odd}}^{\infty}V_o\cdot\frac{\left[r^{(n_m)}-R_2^{(2n_m+1)}\cdot r^{-(n_m+1)}\right]}{\left[R_1^{(n_m)}-R_2^{(2n_m+1)}\cdot R_1^{-(n_m+1)}\right]}\cdot\frac{4}{m\cdot\pi}\cdot\sin(m\cdot\theta) \tag{9.183}$$

where the expression for n_m is given in Equation 9.179b.

9.5 THREE-DIMENSIONAL FIELD PROBLEMS

The Laplacian equation of the scalar electric potential V, in a Cartesian, cylindrical, and spherical system of space coordinates were given by Equations 9.5, 9.6, and 9.7, respectively. This section deals with the methods used to obtain solutions of these equations. The solution in Cartesian coordinates is illustrated by taking an example of a cuboid, whereas in cylindrical and spherical systems, only the general solutions are given.

9.5.1 Field Problem in Cartesian Coordinates

9.5.1.1 Field Inside a Cuboid

The evaluation of a field inside a cuboid bounded by six rectangular surfaces is a three-dimensional problem. Figure 9.16 illustrates a cuboid wherein the front and back surfaces are located at $x = \pm a/2$; the right and the left surfaces are at $y = \pm b/2$, and the top and bottom surfaces are, respectively, at $z = c$, and $z = 0$. On the top surface of this cuboid, the potential distribution is given by the following double Fourier series:

$$V\Big|_{z=c} = \sum_{m\text{-odd}}^{\infty}\sum_{n=1}^{\infty} d_{m,n} \cdot \cos\left(\frac{m\pi}{a}\cdot x\right)\cdot \sin\left(\frac{n2\pi}{b}\cdot y\right) \tag{9.184}$$

Each of the remaining five surfaces of this cuboid is at zero potential. It is required to find the distribution of scalar electric potential inside the homogeneous region of this cuboid.

The potential distribution inside the homogeneous region of the cuboid satisfies a three-dimensional Laplace equation in a Cartesian system of space coordinates, thus:

$$\nabla^2 V \equiv \frac{\partial^2 V}{\partial x^2} + \frac{\partial^2 V}{\partial y^2} + \frac{\partial^2 V}{\partial z^2} = 0 \tag{9.185}$$

Let us try the method of separation of variables, so we assume that:

$$V(x,y,z) \overset{\text{def}}{=} X(x)\cdot Y(y)\cdot Z(z) \tag{9.186}$$

Now, divide Equation 9.185 by V, on substituting from Equation 9.186, we get:

$$\frac{1}{X}\cdot\frac{d^2X}{dx^2} + \frac{1}{Y}\cdot\frac{d^2Y}{dy^2} + \frac{1}{Z}\cdot\frac{d^2Z}{dz^2} = 0 \tag{9.187}$$

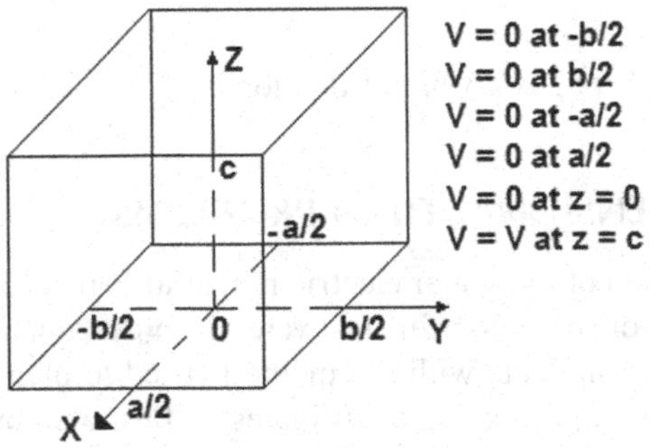

FIGURE 9.16 Cuboid with given dimensions and potentials.

Now, since each term on the LHS must be a constant, and they sum up to zero, thus, using Equation 9.184, let us assume that:

$$\frac{1}{X} \cdot \frac{d^2X}{dx^2} = -\left(\frac{m\pi}{a}\right)^2 \tag{9.188a}$$

$$\frac{1}{Y} \cdot \frac{d^2Y}{dy^2} = -\left(\frac{n2\pi}{b}\right)^2 \tag{9.188b}$$

and

$$\frac{1}{Z} \cdot \frac{d^2Z}{dz^2} = k_{m,n}^2 \tag{9.188c}$$

where

$$k_{m,n}^2 = \left(\frac{m\pi}{a}\right)^2 + \left(\frac{n2\pi}{b}\right)^2 \tag{9.189}$$

or

$$k_{m,n} = \sqrt{\left(\frac{m\pi}{a}\right)^2 + \left(\frac{n2\pi}{b}\right)^2} \tag{9.189a}$$

Therefore, a provisional solution for the Equation 9.185 can be given as follows:

$$V = \sum_{m\text{-odd}}^{\infty} \sum_{n=1}^{\infty} f_{m,n} \cdot \cos\left(\frac{m\pi}{a} \cdot x\right) \cdot \sin\left(\frac{n2\pi}{b} \cdot y\right) \cdot \sinh\left(k_{m,n} \cdot z\right) \tag{9.190}$$

where the arbitrary constant $f_{m,n}$ found from Equation 9.184, is:

$$f_{m,n} = \frac{d_{m,n}}{\sinh\left(k_{m,n} \cdot c\right)} \tag{9.191}$$

The complete solution satisfying all given boundary conditions is therefore found as:

$$V = \sum_{m\text{-odd}}^{\infty} \sum_{n=1}^{\infty} d_{m,n} \cdot \cos\left(\frac{m\pi}{a} \cdot x\right) \cdot \sin\left(\frac{n2\pi}{b} \cdot y\right) \cdot \frac{\sinh\left(k_{m,n}^2 \cdot z\right)}{\sinh\left(k_{m,n}^2 \cdot c\right)} \tag{9.192}$$

This method is used to obtain field components in a rectangular waveguide and a cavity resonator in Chapter 19.

9.5.2 Field Problem in Cylindrical Coordinates

In cylindrical coordinates, the three-dimensional Laplace equation is given as follows:

$$\nabla^2 V \equiv \frac{1}{\rho}\cdot\frac{\partial}{\partial\rho}\left(\rho\frac{\partial V}{\partial\rho}\right)+\frac{1}{\rho^2}\cdot\frac{\partial^2 V}{\partial\varphi^2}+\frac{\partial^2 V}{\partial z^2}=0 \tag{9.193}$$

To solve this equation, let us try the method of separation of variables. Thus we assume that:

$$V(\rho,\varphi,z)\overset{\text{def}}{=}F_1(\rho)\cdot F_2(\varphi)\cdot F_3(Z) \tag{9.194}$$

We divide Equation 9.193 by V; then in view of Equation 9.194, we get:

$$\left[\frac{1}{F_1}\cdot\frac{1}{\rho}\cdot\frac{d}{d\rho}\left(\rho\frac{dF_1}{d\rho}\right)+\frac{1}{F_2}\cdot\frac{1}{\rho^2}\cdot\frac{d^2F_2}{d\varphi^2}\right]+\left[\frac{1}{F_3}\cdot\frac{d^2F_3}{dz^2}\right]=0 \tag{9.195}$$

The LHS of this equation has been divided into two terms, each enclosed into a rectangular bracket. The first term is independent of the z-coordinate, while the second term is a function of the z-coordinate only. Since these two terms sum up to zero, each must be the same constant quantity with opposite sign. Therefore, let:

$$\frac{1}{F_1}\cdot\frac{1}{\rho}\cdot\frac{d}{d\rho}\left(\rho\frac{dF_1}{d\rho}\right)+\frac{1}{F_2}\cdot\frac{1}{\rho^2}\cdot\frac{d^2F_2}{d\varphi^2}=-k^2 \tag{9.196a}$$

and

$$\frac{1}{F_3}\cdot\frac{d^2F_3}{dz^2}=k^2 \tag{9.196b}$$

the symbol k is the constant of separation.

The solution of Equation 9.196b is:

$$F_3=c_1\cdot\cosh(k\cdot z)+c_2\cdot\sinh(k\cdot z) \tag{9.197}$$

where c_1 and c_2 are arbitrary constants.

Now, consider Equation 9.196a. On multiplying both sides by ρ^2, this equation can be rewritten as:

$$\left[\frac{1}{F_1}\cdot\rho\cdot\frac{d}{d\rho}\left(\rho\frac{dF_1}{d\rho}\right)+\rho^2\cdot k^2\right]+\left[\frac{1}{F_2}\cdot\frac{d^2F_2}{d\varphi^2}\right]=0 \tag{9.198}$$

There are two terms on the LHS of this equation. The first term is a function of only ρ coordinate, while the second term is a function of φ coordinate only. Although ρ and φ coordinates are independent of each other, still these two terms sum up to zero. This implies that each of these terms equal to the same constant but with opposite sign. Let us assume:

$$\frac{1}{F_1}\cdot\rho\cdot\frac{d}{d\rho}\left(\rho\frac{dF_1}{d\rho}\right)+\rho^2\cdot k^2=\ell^2 \tag{9.199a}$$

and

$$\frac{1}{F_2} \cdot \frac{d^2 F_2}{d\varphi^2} = -\ell^2 \tag{9.199b}$$

the symbol ℓ is the second separation constant.

The solution of Equation 9.199b is:

$$F_2 = c_3 \cdot \cos(\ell \cdot \varphi) + c_4 \cdot \sin(\ell \cdot \varphi) \tag{9.200}$$

where c_3 and c_4 are arbitrary constants.

Consider Equation 9.199a. Let us substitute:

$$\xi \stackrel{\text{def}}{=} \rho \cdot k \tag{9.201}$$

in this equation; the resulting equation is:

$$\frac{1}{F_1} \cdot \xi \cdot \frac{d}{d\xi}\left(\xi \cdot \frac{dF_1}{d\xi}\right) + \xi^2 = \ell^2 \tag{9.202}$$

or

$$\xi^2 \cdot \frac{d^2 F_1}{d\xi^2} + \xi \cdot \frac{dF_1}{d\xi} + \left(\xi^2 - \ell^2\right) \cdot F_1 = 0$$

or

$$\frac{d^2 F_1}{d\xi^2} + \frac{1}{\xi} \cdot \frac{dF_1}{d\xi} + \left(1 - \frac{\ell^2}{\xi^2}\right) \cdot F_1 = 0 \tag{9.203}$$

This is known as *Bessel's equation*. Its solution for non-integer values of ℓ is given as:

$$F_1 = c_5 \cdot J_\ell(\xi) + c_6 \cdot J_{-\ell}(\xi) \tag{9.204a}$$

or

$$F_1 = c_5 \cdot J_\ell(\rho \cdot k) + c_6 \cdot J_{-\ell}(\rho \cdot k) \tag{9.204b}$$

while for integer values of ℓ:

$$F_1 = c_5 \cdot J_\ell(\xi) + c_6 \cdot N_\ell(\xi) \tag{9.205a}$$

or

$$F_1 = c_5 \cdot J_\ell(\rho \cdot k) + c_6 \cdot N_\ell(\rho \cdot k) \tag{9.205b}$$

The symbols c_5 and c_6 indicate arbitrary constants, values of which may vary with the values of k and ℓ. The function, $J_\ell(\xi)$ is known as the *Bessel function of the first kind of order* ℓ, while $N_\ell(\xi)$ is known as the *Bessel function of the second kind of order* ℓ. This latter function is also called the *Neumann function.*

Since there can be many values for k and ℓ, the complete solution for the Laplace equation can be written as:

$$V(\rho,\varphi,z)=\sum_{k,\ell} F_{1k,\ell}(\rho\cdot k)\cdot F_{2\ell}(\ell\cdot\varphi)\cdot F_{3k}(k\cdot z) \tag{9.206}$$

This method is used to obtain field components in the circular waveguide in Chapter 19.

9.5.3 Field Problems in Spherical Coordinates

In the spherical system of space coordinates, the three-dimensional Laplace equation is given as follows:

$$\nabla^2 V \equiv \frac{1}{r^2}\cdot\frac{\partial}{\partial r}\left(r^2\frac{\partial V}{\partial r}\right)+\frac{1}{r^2\sin\theta}\cdot\frac{\partial}{\partial\theta}\left(\sin\theta\frac{\partial V}{\partial\theta}\right)+\frac{1}{r^2\sin^2\theta}\cdot\frac{\partial^2 V}{\partial\varphi^2}=0 \tag{9.207}$$

Solutions of this equation can be found using the method of separation of variables. Let us assume:

$$V(r,\theta,\varphi)\overset{\text{def}}{=}R(r)\cdot\Theta(\theta)\cdot\Phi(\varphi) \tag{9.208}$$

Multiply both sides of Equation 9.207 by $r^2\sin^2\theta$, divide the resulting equation by V, then using Equation 9.208, we get:

$$\left[\frac{\sin^2\theta}{R}\cdot\frac{d}{dr}\left(r^2\frac{dR}{dr}\right)+\frac{\sin\theta}{\Theta}\cdot\frac{d}{d\theta}\left(\sin\theta\frac{d\Theta}{d\theta}\right)\right]+\left[\frac{1}{\Phi}\cdot\frac{d^2\Phi}{d\varphi^2}\right]=0 \tag{9.209}$$

On the LHS of this equation there are two terms, each inside a rectangular bracket. The first term is independent of φ coordinate, while the second term is a function of φ coordinate only. Since these two terms sum up to zero, each must be equal to the same constant quantity but with opposite sign. Therefore, let:

$$\frac{\sin^2\theta}{R}\cdot\frac{d}{dr}\left(r^2\frac{dR}{dr}\right)+\frac{\sin\theta}{\Theta}\cdot\frac{d}{d\theta}\left(\sin\theta\frac{d\Theta}{d\theta}\right)=k^2 \tag{9.210a}$$

and

$$\frac{1}{\Phi}\cdot\frac{d^2\Phi}{d\varphi^2}=-k^2 \tag{9.210b}$$

The solution for the Equation 9.210b is as follows:

$$\Phi = c_1 \cdot \cos(k \cdot \varphi) + c_2 \cdot \sin(k \cdot \varphi) \tag{9.211}$$

where c_1 and c_2 indicate arbitrary constants.

Now consider Equation 9.210a. Divide both sides of this equation by $\sin^2\theta$. The resulting equation is:

$$\left[\frac{1}{R} \cdot \frac{d}{dr}\left(r^2 \frac{dR}{dr}\right)\right] + \left[\frac{1}{\Theta} \cdot \frac{1}{\sin\theta} \cdot \frac{d}{d\theta}\left(\sin\theta \frac{d\Theta}{d\theta}\right) - \frac{k^2}{\sin^2\theta}\right] = 0 \tag{9.212}$$

On the LHS of this equation there are two terms, each inside a rectangular bracket. The first term is a function of r coordinate, while the second term is a function of θ coordinate. Noting that these two coordinates are independent of each other and since these two terms sum up to zero, each term must be equal to the same constant quantity but with opposite sign. Therefore, let:

$$\frac{1}{R} \cdot \frac{d}{dr}\left(r^2 \frac{dR}{dr}\right) = \ell \cdot (\ell + 1) \tag{9.213a}$$

and

$$\frac{1}{\Theta} \cdot \frac{1}{\sin\theta} \cdot \frac{d}{d\theta}\left(\sin\theta \frac{d\Theta}{d\theta}\right) - \frac{k^2}{\sin^2\theta} = -\ell \cdot (\ell + 1) \tag{9.213b}$$

The RHS of these equations indicates the constant of separation.

It can be seen that the solution for Equation 9.213a can be given as:

$$R = c_3 \cdot r^{\ell} + c_4 \cdot r^{-(\ell+1)} \tag{9.214}$$

where c_3 and c_4 indicate arbitrary constants.

Now, consider Equation 9.213b. This equation can be rewritten as:

$$\frac{1}{\sin\theta} \cdot \frac{d}{d\theta}\left(\sin\theta \frac{d\Theta}{d\theta}\right) + \left[\ell \cdot (\ell + 1) - \frac{k^2}{\sin^2\theta}\right] \cdot \Theta = 0 \tag{9.215}$$

This equation is transformed by setting: $\eta = \cos\theta$.

Thus, $\dfrac{d}{d\theta} = \dfrac{d\eta}{d\theta} \cdot \dfrac{d}{d\eta} = -\sin\theta \cdot \dfrac{d}{d\eta}$

Therefore, from Equation 9.215, we get:

$$\frac{d}{d\eta}\left\{\left(1 - \eta^2\right) \cdot \frac{d\Theta}{d\eta}\right\} + \left\{\ell \cdot (\ell + 1) - \frac{k^2}{\left(1 - \eta^2\right)}\right\} \cdot \Theta = 0 \tag{9.216}$$

This equation with k equal to zero is called the *Legendre equation*, while for any non-zero value for k, this equation is known as the *generalized Legendre equation*. The solution of the Legendre equation is given in terms of *Legendre functions*. If ℓ is a positive integer, Legendre functions degenerate into Legendre polynomials, $P_\ell(\eta)$. The solution of the generalized Legendre equation is given in terms of the associated Legendre polynomials, $P_\ell^k(\eta)$.

In Sections 9.2 to 9.4, our study remained confined to the source-free regions. The problems of potential fields in the source filled regions are discussed in the next chapter.

Example 9.1

A potential field in cylindrical coordinate system is given as $V = 5\left(\rho^k + \rho^{-p}\right)\cos(6\varphi)$. Find k and p so that V satisfies Laplace's equation. Also find V and E at $\rho = 1$, $\varphi = \dfrac{\pi}{4}$.

SOLUTION

$$\nabla^2 V = \frac{1}{\rho}\cdot\frac{\partial}{\partial\rho}\left(\rho\frac{\partial V}{\partial\rho}\right) + \frac{1}{\rho^2}\cdot\frac{\partial^2 V}{\partial\varphi^2} + \frac{\partial^2 V}{\partial z^2} = \frac{1}{\rho}\cdot\frac{\partial}{\partial\rho}\left(\rho\frac{\partial V}{\partial\rho}\right) + \frac{1}{\rho^2}\cdot\frac{\partial^2 V}{\partial\varphi^2}$$

since $V \neq f(z)$, $\dfrac{\partial^2 V}{\partial z^2} = 0$

Given: $V = 5\left(\rho^k + \rho^{-p}\right)\cos(6\varphi)$.

Term 1: $\dfrac{\partial V}{\partial\rho} = 5\left(k\rho^{k-1} - p\rho^{-p-1}\right)\cos(6\varphi)$

$$\rho\frac{\partial V}{\partial\rho} = 5\left(k\rho^k - p\rho^{-p}\right)\cos(6\varphi)$$

$$\frac{\partial}{\partial\rho}\left(\rho\frac{\partial V}{\partial\rho}\right) = 5\left(k^2\rho^{k-1} + p^2\rho^{-p-1}\right)\cos(6\varphi)$$

$$\frac{1}{\rho}\cdot\frac{\partial}{\partial\rho}\left(\rho\frac{\partial V}{\partial\rho}\right) = 5\left(k^2\rho^{k-2} + p^2\rho^{-p-2}\right)\cos(6\varphi)$$

Term 2: $\dfrac{\partial V}{\partial\varphi} = -30\left(\rho^k + \rho^{-p}\right)\sin(6\varphi)$

$$\frac{\partial^2 V}{\partial\varphi^2} = -180\left(\rho^k + \rho^{-p}\right)\cos(6\varphi) \quad \frac{1}{\rho^2}\cdot\frac{\partial^2 V}{\partial\varphi^2} = -180\left(\rho^{k-2} + \rho^{-p-2}\right)\cos(6\varphi)$$

$$\begin{aligned}\frac{1}{\rho}\cdot\frac{\partial}{\partial\rho}\left(\rho\frac{\partial V}{\partial\rho}\right) + \frac{1}{\rho^2}\cdot\frac{\partial^2 V}{\partial\varphi^2} &= 5\left(k^2\rho^{k-2} + p^2\rho^{-p-2}\right)\cos(6\varphi)\\ &\quad -180\left(\rho^{k-2} + \rho^{-p-2}\right)\cos(6\varphi)\\ &= 0\end{aligned}$$

On comparing the like terms $k^2 = 36$ or $k = \pm 6$ also $p^2 = 36$ or $p = \pm 6$
Selecting only positive values for k and p:

$$V\Big|_{\rho=1,\varphi=\frac{\pi}{4}} = 5\left(1^k + 1^{-p}\right)\cos\left(6\frac{\pi}{4}\right) = 5\left(1^6 + 1^{-6}\right)\cos\left(\frac{3\pi}{2}\right) = 10\text{ V}$$

$$E = -\nabla V = -\frac{\partial V}{\partial \rho}a_\rho - \left(\frac{1}{\rho}\right)\frac{\partial V}{\partial \varphi}a_\varphi$$

$$= 30\left(\rho^k + \rho^{-p}\right)\sin(6\varphi)\,a_\rho - 30\left(\rho^{k-1} + \rho^{-p-1}\right)\sin(6\varphi)\,a_\varphi$$

$$E\Big|_{\rho=1,\,\varphi=\frac{\pi}{4}} = 30\left(1^6 + 1^{-6}\right)\sin\left(\frac{3\pi}{2}\right)a_\rho - 30\left(1^5 + 1^{-5}\right)\sin\left(\frac{3\pi}{2}\right)a_\varphi = 60\mathbf{a}_\rho - 60\mathbf{a}_\varphi$$

Example 9.2

Identify the potential function that does not satisfy Laplace's equation $V =$: (a) $C{\cdot}x{\cdot}y{\cdot}z$; (b) $C{\cdot}\rho{\cdot}\varphi{\cdot}z$; (c) $C{\cdot}r{\cdot}\theta{\cdot}\varphi$.

SOLUTION

(a) In view of Equation 9.5, Laplace's equation is:

$$\nabla^2 V \equiv \frac{\partial^2 V}{\partial x^2} + \frac{\partial^2 V}{\partial y^2} + \frac{\partial^2 V}{\partial z^2} = 0 \quad \text{where } V = C\cdot x\cdot y\cdot z$$

Term 1: $\frac{\partial^2 V}{\partial x^2} = 0$ **Term 2:** $\frac{\partial^2 V}{\partial y^2} = 0$ **Term 3:** $\frac{\partial^2 V}{\partial z^2} = 0$

The sum of the terms of Equation 9.5 is zero, thus the function satisfies Laplace's equation.

(b) In view of Equation 9.6, Laplace's equation is:

$$\nabla^2 V = \frac{1}{\rho}\cdot\frac{\partial}{\partial \rho}\left(\rho\frac{\partial V}{\partial \rho}\right) + \frac{1}{\rho^2}\cdot\frac{\partial^2 V}{\partial \varphi^2} + \frac{\partial^2 V}{\partial z^2} = 0 \quad \text{where } V = C\cdot\rho\cdot\varphi\cdot z$$

Term 1: $\frac{\partial V}{\partial \rho} = C\cdot\varphi\cdot z \quad \rho\frac{\partial V}{\partial \rho} = C\cdot\rho\cdot\varphi\cdot z \quad \frac{\partial}{\partial \rho}\left(\rho\frac{\partial V}{\partial \rho}\right) = C\cdot\varphi\cdot z$

$$\frac{1}{\rho}\cdot\frac{\partial}{\partial \rho}\left(\rho\frac{\partial V}{\partial \rho}\right) = \frac{C\cdot\varphi\cdot z}{\rho} \neq 0,$$

Term 2: $\frac{\partial^2 V}{\partial \varphi^2} = 0 \quad \frac{1}{\rho^2}\cdot\frac{\partial^2 V}{\partial \varphi^2} = 0,$

Term 3: $\frac{\partial^2 V}{\partial z^2} = 0$

The sum of three terms is not zero, thus the function does not satisfy Laplace's equation.

(c) In view of Equation 9.7, Laplace's equation is:

$$\nabla^2 V = \frac{1}{r^2}\cdot\frac{\partial}{\partial r}\left(r^2\frac{\partial V}{\partial r}\right)+\frac{1}{r^2\sin\theta}\cdot\frac{\partial}{\partial\theta}\left(\sin\theta\frac{\partial V}{\partial\theta}\right)+\frac{1}{r^2\sin^2\theta}\cdot\frac{\partial^2 V}{\partial\varphi^2}=0$$

where $V = C{\cdot}r{\cdot}\theta{\cdot}\varphi$

Term 1: $\frac{\partial V}{\partial r} = C\theta\varphi \quad r^2\frac{\partial V}{\partial r} = Cr^2\theta\varphi \quad \frac{\partial}{\partial r}\left(r^2\frac{\partial V}{\partial r}\right) = 2Cr\theta\varphi$

$$\frac{1}{r^2}\cdot\frac{\partial}{\partial r}\left(r^2\frac{\partial V}{\partial r}\right) = \frac{2C\theta\varphi}{r} \neq 0$$

Term 2: $\frac{\partial V}{\partial\theta} = Cr\varphi \quad \sin\theta\frac{\partial V}{\partial\theta} = Cr\varphi\sin\theta \quad \frac{\partial}{\partial\theta}\left(\sin\theta\frac{\partial V}{\partial\theta}\right) = Cr\varphi\cos\theta$

$$\frac{1}{r^2\sin\theta}\cdot\frac{\partial}{\partial\theta}\left(\sin\theta\frac{\partial V}{\partial\theta}\right) = \frac{C\varphi}{r}\cot\theta \neq 0,$$

Term 3: $\frac{\partial^2 V}{\partial\varphi^2} = 0 \quad \frac{1}{r^2\sin^2\theta}\cdot\frac{\partial^2 V}{\partial\varphi^2} = 0$

The sum of the three terms is not zero, thus the function does not satisfy Laplace's equation.

Example 9.3

Show that both the potential functions $V_1 = \ln\rho$ and $V_2 = \cosh^{-1}\left[\frac{\rho}{2}+\frac{1}{2\rho}\right]$ (a) satisfy the boundary conditions $V = 0$ at $\rho = 1$ and $V = 0.693$ at $\rho = 2$; (b) satisfy Laplace's equation in cylindrical coordinates; and (c) are the two identical solutions.

SOLUTION

(a) $V_1\big|_{\rho=1} = \ln 1 = 0 \qquad V_1\big|_{\rho=2} = \ln 2 = 0.693$

$$V_2\big|_{\rho=1} = \cosh^{-1}\left[\frac{1}{2}+\frac{1}{2}\right] = \cosh^{-1}[1] = 0$$

$$V_2\big|_{\rho=2} = \cosh^{-1}\left[1+\frac{1}{4}\right] = \cosh^{-1}[1.25] = 0.693$$

(b) In cylindrical coordinates

$$\nabla^2 V = \frac{1}{\rho}\cdot\frac{\partial}{\partial\rho}\left(\rho\frac{\partial V}{\partial\rho}\right)+\frac{1}{\rho^2}\cdot\frac{\partial^2 V}{\partial\varphi^2}+\frac{\partial^2 V}{\partial z^2} = \frac{1}{\rho}\cdot\frac{\partial}{\partial\rho}\left(\rho\frac{\partial V}{\partial\rho}\right)$$

As V_1 and V_2 are functions of ρ only, $\nabla^2 V = \frac{1}{\rho}\cdot\frac{\partial}{\partial\rho}\left(\rho\frac{\partial V}{\partial\rho}\right)$

$$V_1 = \ln\rho \quad \frac{\partial V_1}{\partial\rho} = \frac{1}{\rho} \quad \rho\frac{\partial V_1}{\partial\rho} = 1 \quad \frac{\partial}{\partial\rho}\left(\rho\frac{\partial V_1}{\partial\rho}\right) = 0 \quad \frac{1}{\rho}\cdot\frac{\partial}{\partial\rho}\left(\rho\frac{\partial V}{\partial\rho}\right) = 0$$

$$V_2 = \cosh^{-1}\left[\frac{\rho}{2} + \frac{1}{2\rho}\right]. \text{ Put } \cosh^{-1}\left[\frac{\rho}{2} + \frac{1}{2\rho}\right] = \theta$$

$$\cosh\theta = \frac{\rho}{2} + \frac{1}{2\rho} = \frac{1}{2}\left(\frac{\rho}{1} + \frac{1}{\rho}\right) = \frac{e^{\theta} + e^{-\theta}}{2}$$

By inspection $e^{\theta} = \rho$ and $e^{-\theta} = 1/\rho$, $\log(e^{\theta} = \rho)$ gives $\theta = \ln\rho$.

$$\text{Thus, } \frac{\partial V_2}{\partial\rho} = \frac{\partial}{\partial\rho}\cosh^{-1}\left[\frac{\rho}{2} + \frac{1}{2\rho}\right] = \frac{\partial\theta}{\partial\rho} = \frac{\partial}{\partial\rho}\ln\rho = \frac{1}{\rho} \quad \rho\frac{\partial V_2}{\partial\rho} = 1$$

$$\frac{\partial}{\partial\rho}\left(\rho\frac{\partial V}{\partial\rho}\right) = 0 \quad \frac{1}{\rho}\cdot\frac{\partial}{\partial\rho}\left(\rho\frac{\partial V}{\partial\rho}\right) = 0$$

V_1 and V_2 both satisfy Laplace's equation.

(c) In view of the above results, both potential functions are identical solutions.

Example 9.4

Find V at $P(2,1,3)$ in the field of two conducting (a) coaxial cylinders, $V = 50$ V at $\rho = 2$ m and $V = 20$ V at $\rho = 3$ m; (b) radial planes, $V = 50$ V at $\varphi = 10°$ and $V = 20$ V at $\varphi = 30°$; (c) concentric spheres, $V = 50$ V at $r = 3$ m and $V = 20$ V at $r = 5$ m; (d) coaxial cones, $V = 50$ V at $\theta = 30°$ and $V = 20$ V at $\theta = 50°$.

SOLUTION

(a) In this case V is a function of ρ only, thus Equation 9.44a is the relevant relation. This can be solved as here:

$$\nabla^2 V = \frac{1}{\rho}\cdot\frac{d}{d\rho}\left(\rho\frac{dV}{d\rho}\right) = 0 \text{ or } \frac{d}{d\rho}\left(\rho\frac{dV}{d\rho}\right) = 0. \text{ It leads to the solution} \rightarrow$$

$V = A\ln\rho + B$

Arbitrary constants A and B can be obtained by substituting values of V at given ρs.
Thus $50 = A\ln 2 + B$ and $20 = A\ln 3 + B$.
Manipulation of these gives $A = -73.96$ and $B = 101.265$.

Since $\rho = \sqrt{x^2 + y^2} = \sqrt{2^2 + 1^2} = \sqrt{5}$

$$V = A\ln\rho + B = -73.96\ln\left(\sqrt{5}\right) + 101.265 = 41.75\ \text{V}$$

(b) In this case V is a function of φ only, thus Equation 9.44b is the relevant relation. This can be solved as here:

$\nabla^2 V = \dfrac{1}{\rho^2}\cdot\dfrac{d^2V}{d\varphi^2} = 0$ or $\dfrac{d^2V}{d\varphi^2} = 0$. It leads to the solution → $V = A\varphi + B$

A and *B* can be obtained by substituting values of *V* at given φs.

Thus $50 = 10A + B$ and $20 = 30A + B$.

These give $A = -1.5$ and $B = 65$, also $\varphi = \tan^{-1}\left(\dfrac{y}{x}\right) = \tan^{-1}\left(\dfrac{1}{2}\right) = 26.565°$.

Thus, $V = A\varphi + B = -1.5 \times 26.565 + 65 = 25.2\ \text{V}$.

(c) In this case V is a function of *r* only, thus Equation 9.58a is the relevant relation. This can be solved as below.

$\nabla^2 V = \dfrac{1}{r^2}\cdot\dfrac{d}{dr}\left(r^2\dfrac{dV}{dr}\right) = 0$ or $\dfrac{d}{dr}\left(r^2\dfrac{dV}{dr}\right) = 0$. It leads to the solution →

$$V = -\frac{A}{r} + B$$

A and *B* can be obtained by substituting values of *V* at given *r*s.

Thus $50 = -\dfrac{A}{3} + B$ and $20 = -\dfrac{A}{5} + B$.

These give $A = -225$ and $B = -25$, also $r = \sqrt{x^2 + y^2 + z^2} = \sqrt{2^2 + 1^2 + 3^2}$ $= \sqrt{14}$.

$$\text{Thus } V = -\frac{A}{r} + B = \frac{225}{\sqrt{14}} - 25 = 35.134\ \text{V}.$$

(d) In this case, V is function of θ only, thus Equation 9.58b is the relevant relation. This can be solved as here:

$\nabla^2 V = \dfrac{1}{r^2\sin\theta}\cdot\dfrac{\partial}{d\theta}\left(\sin\theta\dfrac{dV}{d\theta}\right) = 0$ or $\dfrac{\partial}{d\theta}\left(\sin\theta\dfrac{dV}{d\theta}\right) = 0$ It gives

$$V = A\cdot\ln\left(\frac{\tan\theta}{2}\right) + B$$

A and *B* are obtained by substituting values of *V* at given θs.

$$V = A\cdot\ln(\tan 15) + B \qquad V = A\cdot\ln(\tan 25) + B$$

These give $A = -54.7$ and $B = 21.34$, also

$$\theta = \cos^{-1}\frac{Z}{\sqrt{x^2 + y^2 + z^2}} = \cos^{-1}\frac{3}{\sqrt{14}} = 36.7°$$

$$V = A \cdot \ln\left(\tan\frac{\theta}{2}\right) + B = -54.7 \cdot \ln\left(\tan\frac{36.7}{2}\right) + 21.34 = 38.44 \text{ V}$$

PROBLEMS

P9.1 Find V at $P(1,2,3)$ for the field of two coaxial conducting cylinders with radii of $\rho = 2$ m and $\rho = 3$ m maintained at 40 and 20 volts, respectively.

P9.2 Find V at $P(1,2,3)$ for the field of (a) two coaxial conducting cylinders with radii of $\rho = 2$ m and $\rho = 3$ m maintained at 40 and 20 volts, respectively; (b) two radial conducting planes at $\varphi = 10°$ and $\varphi = 30°$ maintained at 40 and 20 volts, respectively.

P9.3 Find V at $P(2,1,3)$ for the field of two concentric conducting spheres of $r = 3$ m and $r = 5$ m maintained at 40 and 30 volts, respectively.

P9.4 Find V at $P(2,1,3)$ for the field of two coaxial conducting cones at $\theta = 30°$ and $\theta = 45°$ maintained at 40 and 30 volts, respectively.

P9.5 Solve Laplace's equation for the potential field in the homogeneous region between two concentric conducting spheres with radius a and ($b > a$), if $V = 0$ at $r = b$, and $V = V_0$ at $r = a$. Also find the capacitance between these spheres.

P9.6 Two spherical regions are specified as $1 \le r \le 2$ with $\varepsilon_{r1} = 3$ and $3 \le r \le 4$ with $\varepsilon_{r2} = 1$. In these regions the potentials are specified as $V = 100$ volts at $r = 1$, and $V = 0$ volts at $r = 4$. Solve Laplace's equation in both the regions so as to yield an identical potential at $r = 3$ and which appropriately accounts for the dielectric boundary conditions. Find V at (a) $r = 2$; (b) $r = 3$; and (c) $r = 3.5$.

P9.7 The region between two concentric conducting cylinders with radii of 2 cm and 5 cm contains a volume charge distribution of $\rho_v = -10^{-8}(1+10r)$ C/m^3. If E_r and V both are zero at the inner cylinder find V at the outer cylinder. Assume $\varepsilon = \varepsilon_0$.

DESCRIPTIVE QUESTIONS

Q9.1 Identify the situations that lead to Laplace's or Poisson's fields.

Q9.2 State and explain the Uniqueness Theorem for Laplace's or Poisson's equations.

Q9.3 Classify the field problems in terms of shapes of configurations and variation of field parameters with the coordinates.

FURTHER READING

References are given at the end of Chapter 15.

10 Electrostatic Boundary Value Problems Involving Poisson's Fields

10.1 INTRODUCTION

In Chapter 9, the electrostatic boundary problems related with Laplacian fields (i.e. for the source free regions) were discussed at length. This chapter is entirely devoted to the problems of the potential fields in source filled regions. These problems involve Poisson's equation defined Equation 9.1a.

In Chapter 9, the Laplacian field was defined as a field that gives zero value if it is operated upon by the Laplacian operator, ∇^2. As evident from Equation 9.1a, for Poisson's field the operation of Laplacian operator, ∇^2 on a field does not yield a zero value. For Laplacian fields, 1D, 2D, and 3D problems belonging to different coordinate systems were solved. This chapter deals only with 1D and 2D problems.

10.2 SOLUTION OF POISSON'S EQUATION FOR SCALAR ELECTRIC POTENTIAL

For homogeneous regions, the scalar potential V satisfies Poisson's equation, instead of the Laplace equation. Poisson's equation is an inhomogeneous differential equation. Its solution is in two parts, viz. particular integral V_p and the complementary function V_c and thus can be written as:

$$V = V_p + V_c \tag{10.1}$$

The determination of the complementary function has been discussed through various examples presented in Chapter 9. In this and the following section, solutions for the particular integrals are presented through examples. It may be noted that the expressions for the particular integral V_p are free from any arbitrary constants. All arbitrary constants found in the expression for the complementary function V_c are evaluated by imposing boundary conditions on the complete solution V. Particular integrals V_p are defined as functions that satisfy the following equation identically:

$$\nabla^2 V_p \equiv -\frac{1}{\varepsilon} \cdot \rho \tag{10.2}$$

It may also be noted that particular integrals V_p need not be unique. However, the complete solution V must be unique.

The field problems can again be categorized as one-, two-, and three-dimensional field problems.

The following text includes only one- and two-dimensional problems.

10.3 ONE-DIMENSIONAL FIELD PROBLEMS

This section deals with 1D problems in Cartesian, cylindrical, and spherical coordinates.

10.3.1 Field Problems in Cartesian Coordinates

The expression for the Laplacian of potential in Cartesian coordinates is reproduced here:

$$\nabla^2 V = \frac{\partial^2 V}{\partial x^2} + \frac{\partial^2 V}{\partial y^2} + \frac{\partial^2 V}{\partial z^2} \tag{9.5}$$

Since the method of solution for the problem involving any of the three coordinates is identical we need to consider only one of the coordinates. To proceed with let us assume that the charge density and the potential field vary only along the x-axis. The following section deals with the same 1D problem for three different types of boundary conditions. These include (i) the Dirichlet boundary conditions, (ii) the particular integral vanishing on the boundary, and (iii) the Neumann boundary conditions. In *Dirichlet boundary conditions*, the potentials are known at the boundary whereas in *Neumann boundary conditions*, instead of the value of the potential at the boundary, the derivative of the potential normal to the boundary is specified.

10.3.1.1 Problem with Dirichlet Boundary Conditions

Since the charge density and the potential field vary only with x-coordinates, we can write:

$$\nabla^2 V_p(x) \equiv F(x) \quad \text{over } x_2 \le x \le x_1 \tag{10.3}$$

where $F(x)$ is a given function of the variable $-x$.

Equation 10.3 can be written as:

$$\frac{d^2}{dx^2} V_p(x) = F(x) \tag{10.4}$$

Integrate both sides of this equation twice with respect to x, on ignoring all constants of integrations, we get the required solution for the particular integral as:

$$V_p(x) = \int \left[\int F(x)\, dx \right] dx \tag{10.5a}$$

The complementary function for Equation 10.4 is:

$$V_c = a \cdot x + b \tag{10.5b}$$

where a and b indicate arbitrary constants. Let the boundary conditions be given as:

$$V = \begin{cases} V_1 & \text{at } x = x_1 \\ V_2 & \text{at } x = x_2 \end{cases} \tag{10.6}$$

Now, the complete solution is as follows:

$$V = V_p + V_c = \int \left[\int F(x)\, dx \right] dx + a \cdot x + b \tag{10.7}$$

On imposing the boundary conditions, we get:

$$\left. \int \left[\int F(x)\, dx \right] dx \right|_{x=x_1} + a \cdot x_1 + b = V_1 \tag{10.7a}$$

and

$$\left. \int \left[\int F(x)\, dx \right] dx \right|_{x=x_2} + a \cdot x_2 + b = V_2 \tag{10.7b}$$

The expressions for the arbitrary constants found on solving these two equations are:

$$a = \frac{(V_1 - V_2) - \left(\left. \int \left[\int F(x)\, dx \right] dx \right|_{x=x_1} - \left. \int \left[\int F(x)\, dx \right] dx \right|_{x=x_2} \right)}{(x_1 - x_2)} \tag{10.8a}$$

and

$$b = \frac{\left\{ V_2 - \left. \int \left[\int F(x)\, dx \right] dx \right|_{x=x_2} \right\} \cdot x_1 - \left\{ V_1 - \left. \int \left[\int F(x)\, dx \right] dx \right|_{x=x_1} \right\} \cdot x_2}{(x_1 - x_2)} \tag{10.8b}$$

10.3.1.2 Problem with Particular Integral Vanishing on the Boundary

An alternative method to solve the above problem involves choosing the particular integral that vanishes on the boundary. This approach is often convenient for the determination of arbitrary constants found in the complementary function.

Let the Fourier series expansion for the function $F(x)$ over $x_2 < x < x_1$, be given as:

$$F(x) = \sum_{m=1}^{\infty} A_m \cdot \sin\left\{ (x - x_2) \cdot \frac{m\pi}{(x_1 - x_2)} \right\} \tag{10.9}$$

To find the Fourier coefficient A_m, let us multiply both sides of this equation by $\sin\left\{(x-x_2)\cdot\frac{n\pi}{(x_1-x_2)}\right\}$ than integrate over $x_2 < x < x_1$. These steps result in:

$$\int_{x_2}^{x_1} F(x)\cdot\sin\left\{(x-x_2)\cdot\frac{n\pi}{(x_1-x_2)}\right\}\cdot dx$$

$$=\frac{1}{2}\sum_{m=1}^{\infty}A_m\cdot\left[\frac{\sin\left\{(x-x_2)\cdot\frac{(m-n)\pi}{(x_1-x_2)}\right\}}{\frac{(m-n)\pi}{(x_1-x_2)}}-\frac{\sin\left\{(x-x_2)\cdot\frac{(m+n)\pi}{(x_1-x_2)}\right\}}{\frac{(m+n)\pi}{(x_1-x_2)}}\right]_{x_2}^{x_1}$$

For any integer $-n$, we get:

$$\int_{x_2}^{x_1} F(x)\cdot\sin\left\{(x-x_2)\cdot\frac{n\pi}{(x_1-x_2)}\right\}\cdot dx = A_n\cdot\frac{(x_1-x_2)}{2} \tag{10.9a}$$

Therefore, setting $n = m$, we have:

$$A_m=\frac{2}{(x_1-x_2)}\cdot\left[\int_{x_2}^{x_1} F(x)\cdot\sin\left\{(x-x_2)\cdot\frac{m\pi}{(x_1-x_2)}\right\}\cdot dx\right] \tag{10.9b}$$

Thus, the particular integral found, in view of Equations 10.4 and 10.9 is:

$$V_p(x)=-\sum_{m=1}^{\infty}A_m\cdot\left[\frac{x_1-x_2}{m\pi}\right]^2\cdot\sin\left\{(x-x_2)\cdot\frac{m\pi}{(x_1-x_2)}\right\} \tag{10.10}$$

Using Equations 10.5b and 10.10, the complete solution is as follows:

$$V = V_c + V_p$$

or,

$$V=a\cdot x+b-\sum_{m=1}^{\infty}A_m\cdot\left[\frac{x_1-x_2}{m\pi}\right]^2\cdot\sin\left\{(x-x_2)\cdot\frac{m\pi}{(x_1-x_2)}\right\} \tag{10.11}$$

Using boundary conditions given by Equation 10.6, one gets:

$$V_1 = a\cdot x_1 + b \quad \text{and} \quad V_2 = a\cdot x_2 + b.$$

On solving these equations, the arbitrary constants found are:

$$a = \frac{V_1 - V_2}{x_1 - x_2} \;(\text{a}) \quad \text{and} \quad b = \frac{V_2 \cdot x_1 - V_1 \cdot x_2}{x_1 - x_2} \;(\text{b}) \Bigg\} \tag{10.11}$$

10.3.1.3 Problem with Neumann Boundary Conditions

With *Neumann boundary conditions*, we have:

$$\frac{\partial V}{\partial x} = \begin{cases} f_1 & \text{at } x = x_1 \\ f_2 & \text{at } x = x_2 \end{cases} \tag{10.12}$$

Subjected to these boundary conditions, we aim to solve the 1D equation as here:

$$\nabla^2 V_p(x) \equiv F(x) \quad \text{over} \quad x_2 \le x \le x_1 \tag{10.13}$$

Using Equation 10.5b, the complete solution is:

$$V = V_c + V_p$$

or

$$V = a \cdot x + b + V_p \tag{10.14a}$$

Thus,

$$\frac{\partial V}{\partial x} = \frac{dV}{dx} = a + \frac{dV_p}{dx} \tag{10.14b}$$

This shows, using Equations 10.13 and 10.12, that we cannot determine the arbitrary constant b, and the arbitrary constant a has two different values. Clearly, the Neumann boundary conditions cannot be applied to the one-dimensional problem.

The third possibility is the application of the mixed boundary conditions, as indicated here:

$$\begin{cases} V = \not{f} & \text{at } x = x_1 \\ \dfrac{dV}{dx} = f & \text{at } x = x_2 \end{cases} \tag{10.15}$$

where $\not{f}$ and f are given constants.

Using this equation, we may express $F(x)$ as the following Fourier series expansion over $x_2 < x < x_1$:

$$F(x) = \sum_{m\text{-odd}}^{\infty} A_m \cdot \cos\left\{ (x - x_2) \cdot \frac{m\pi/2}{(x_1 - x_2)} \right\} \tag{10.16a}$$

After multiplying both sides of this equation by $\left[\frac{1}{2}\cdot\cos\left\{(x-x_2)\cdot\frac{n\pi/2}{(x_1-x_2)}\right\}\right]$, and then integrating over $x_2 < x < x_1$, we get:

$$\frac{1}{2}\int_{x_2}^{x_1} F(x)\cdot\cos\left\{(x-x_2)\cdot\frac{n\pi/2}{(x_1-x_2)}\right\}\cdot dx = A_n\cdot(x_1-x_2)$$

with odd values for n.

Thus, substituting m for n we get:

$$A_m = \frac{1}{2(x_1-x_2)}\cdot\int_{x_2}^{x_1} F(x)\cdot\cos\left\{(x-x_2)\cdot\frac{m\pi/2}{(x_1-x_2)}\right\}\cdot dx \tag{10.16b}$$

Thus, the particular integral V_p can be given as:

$$V_p = -\sum_{m\text{-odd}}^{\infty} A_m\cdot\left[\frac{x_1-x_2}{m\pi/2}\right]^2 \cos\left\{(x-x_2)\cdot\frac{m\pi/2}{(x_1-x_2)}\right\} \tag{10.17}$$

The complete solution is therefore given as:

$$V = a\cdot x + b + V_p \tag{10.18a}$$

or,

$$V = a\cdot x + b - \sum_{m\text{-odd}}^{\infty} A_m\cdot\left[\frac{x_1-x_2}{m\pi/2}\right]^2 \cos\left\{(x-x_2)\cdot\frac{m\pi/2}{(x_1-x_2)}\right\} \tag{10.18b}$$

Now, to find the two arbitrary constants use Equation 10.15. Thus:

$$V\Big|_{x=x_1} = a\cdot x_1 + b = \not{f} \tag{10.19a}$$

and

$$\frac{dV}{dx}\Big|_{x=x_2} = a = f \tag{10.19b}$$

Therefore, Equation 10.19a gives:

$$b = \left(\not{f} - f\cdot x_1\right) \tag{10.19c}$$

This completes the solution.

10.3.2 Field Problems in Cylindrical Coordinates

The expression for the Laplacian equation of potential in cylindrical coordinates is reproduced below.

$$\nabla^2 V = \frac{1}{\rho}\cdot\frac{\partial}{\partial \rho}\left(\rho\frac{\partial V}{\partial \rho}\right) + \frac{1}{\rho^2}\cdot\frac{\partial^2 V}{\partial \varphi^2} + \frac{\partial^2 V}{\partial z^2} \tag{9.6}$$

In cylindrical systems, if the potential distribution is a function of the z-coordinate only, the three methods discussed in Section 10.3.1 are equally applicable. The methods of solution for the problems involving ρ- and φ-coordinates are discussed in the following subsections.

10.3.2.1 Variation of V Only with ρ (Dirichlet Boundary Conditions)

Let the scalar electric potential V be a function of the ρ-coordinate only. Thus, Poisson's equation reduces to the following form:

$$\frac{1}{\rho}\cdot\frac{d}{d\rho}\left(\rho\frac{dV}{d\rho}\right) = F(\rho) \quad \text{over} \quad \rho_2 < \rho < \rho_1 \tag{10.20}$$

where $F(\rho)$ is a given function of the ρ-coordinate.

On integrating the above equation we get, ignoring the constant of integration:

$$\rho\frac{dV}{d\rho} = \int\{\rho\cdot F(\rho)\}\cdot d\rho$$

The second integration leads to the particular integral:

$$V_p = \int\left[\frac{1}{\rho}\cdot\int\{\rho\cdot F(\rho)\}\cdot d\rho\right]\cdot d\rho \tag{10.21}$$

The complementary function for Equation 10.20 is readily found as:

$$V_c = k\cdot\ln(\rho/\rho_o)$$

where k and ρ_o indicate two arbitrary constants.

Thus, the complete solution is:

$$V = V_c + V_p$$

or

$$V = k\cdot\ln(\rho/\rho_o) + \int\left[\frac{1}{\rho}\cdot\int\{\rho\cdot F(\rho)\}\cdot d\rho\right]\cdot d\rho \tag{10.22}$$

Let the Dirichlet boundary condition be given as:

$$V = \begin{cases} V_1 & \text{at } \rho = \rho_1 \\ V_2 & \text{at } \rho = \rho_2 \end{cases} \tag{10.23}$$

Thus,

$$V_1 = k \cdot \ln\left(\rho_1/\rho_o\right) + \int \left[\frac{1}{\rho} \cdot \int \{\rho \cdot F(\rho)\} \cdot d\rho \right] \cdot d\rho \Bigg|_{\rho=\rho_1} \tag{10.23a}$$

and

$$V_2 = k \cdot \ln\left(\rho_2/\rho_o\right) + \int \left[\frac{1}{\rho} \cdot \int \{\rho \cdot F(\rho)\} \cdot d\rho \right] \cdot d\rho \Bigg|_{\rho=\rho_2} \tag{10.23b}$$

Thus,

$$k = \frac{\left(V_1 - V_2\right) - \int_{\rho_2}^{\rho_1} \left[\frac{1}{\rho} \cdot \int \{\rho \cdot F(\rho)\} \cdot d\rho \right] \cdot d\rho}{\ln\left(\rho_1/\rho_2\right)} \tag{10.24a}$$

and

$$\rho_o = \sqrt{\rho_1 \cdot \rho_2} \cdot \exp\left[\!\left[\frac{1}{2k} \cdot \int \left[\frac{1}{\rho} \cdot \int \{\rho \cdot F(\rho)\} \cdot d\rho \right] \cdot d\rho \Bigg|_{\rho=\rho_1} + \frac{1}{2k} \cdot \int \left[\frac{1}{\rho} \cdot \int \{\rho \cdot F(\rho)\} \cdot d\rho \right] \cdot d\rho \Bigg|_{\rho=\rho_2} - \frac{\left(V_1 + V_2\right)}{2k} \right]\!\right] \tag{10.24b}$$

10.3.2.2 Variation of V Only with ρ (Mixed Boundary Conditions)

Let the scalar electric potential V be again a function of the ρ-coordinate only. Thus, Poisson's equation reduces to the following form:

$$\frac{1}{\rho} \cdot \frac{d}{d\rho}\left(\rho \frac{dV}{d\rho} \right) = F(\rho) \quad \text{over} \quad \rho_2 < \rho < \rho_1 \tag{10.20}$$

where $F(\rho)$ is a given function of the ρ-coordinate. The *mixed boundary conditions* given are as indicated below:

$$\begin{cases} V = \not{f} & \text{at } \rho = \rho_1 \\ \dfrac{\partial V}{\partial \rho} = f & \text{at } \rho = \rho_2 \end{cases} \tag{10.25}$$

where $\not{f}$ and f are given constants.

The particular integral V_p is as follows:

$$V_p = \int \left[\frac{1}{\rho} \cdot \int \{\rho.F(\rho)\} \cdot d\rho \right] \cdot d\rho \tag{10.21}$$

The complementary function for Equation 10.20 is readily found as:

$$V_c = k \cdot \ln(\rho/\rho_o)$$

where k and ρ_o indicate two arbitrary constants.

Thus, the complete solution is:

$$V = V_c + V_p$$

or

$$V = k \cdot \ln(\rho/\rho_o) + \int \left[\frac{1}{\rho} \cdot \int \{\rho \cdot F(\rho)\} \cdot d\rho \right] \cdot d\rho \tag{10.22}$$

Therefore:

$$\frac{\partial V}{\partial \rho} = \frac{dy}{d\rho} = \frac{k}{\rho} + \frac{1}{\rho} \cdot \int \{\rho \cdot F(\rho)\} \cdot d\rho \tag{10.26}$$

Now, using the boundary conditions as given by Equation 10.25:

$$\not{f} = V\big|_{\rho=\rho_1} = k \cdot \ln(\rho_1/\rho_o) + \int \left[\frac{1}{\rho} \cdot \int \{\rho \cdot F(\rho)\} \cdot d\rho \right] \cdot d\rho \bigg|_{\rho=\rho_1} \tag{10.27a}$$

and

$$f = \frac{dy}{d\rho}\bigg|_{\rho=\rho_2} = \frac{k}{\rho_2} + \left[\frac{1}{\rho} \cdot \int \{\rho \cdot F(\rho)\} \cdot d\rho \right]_{\rho=\rho_2} \tag{10.27b}$$

Therefore:

$$k = f \cdot \rho_2 - \int \{\rho \cdot F(\rho)\} \cdot d\rho \bigg|_{\rho=\rho_2} \tag{10.28a}$$

and

$$\rho_o = \rho_1 \cdot \exp\left[\left[\frac{1}{k} \cdot \int \left[\frac{1}{\rho} \cdot \int \{\rho \cdot F(\rho)\} \cdot d\rho \right] \cdot d\rho \bigg|_{\rho=\rho_1} - \not{f} \cdot \frac{1}{k} \right]\right] \tag{10.28b}$$

10.3.2.3 Variation of *V* Only with φ (Dirichlet Boundary Conditions)

Let the scalar electric potential *V* be a function of the φ-coordinate only. Thus Poisson's equation in view of Equation 9.6 reduces to the following form:

$$\frac{1}{\rho^2}\cdot\frac{\partial^2 V}{\partial\varphi^2} = F(\varphi,\rho) \quad \text{over} \quad \varphi_2 < \varphi < \varphi_1 \tag{10.29}$$

where $F(\varphi,\rho)$ is a given function of the ρ- and φ-coordinates. It may be noted that $V = V(\varphi)$, implies that the function at the RHS of Equation 10.29 must be a given function of φ, $\mathbb{F}(\varphi)$ divided by ρ squared, i.e.:

$$F(\varphi,\rho) \equiv \mathbb{F}(\varphi)/\rho^2 \tag{10.30}$$

Therefore Equation 10.29 reduces to:

$$\frac{d^2V}{d\varphi^2} = \mathbb{F}(\varphi) \quad \text{over} \quad \varphi_2 < \varphi < \varphi_1 \tag{10.31}$$

Let the Dirichlet boundary conditions be given as:

$$V = \begin{cases} V_1 & \text{at } \varphi = \varphi_1 \\ V_2 & \text{at } \varphi = \varphi_2 \end{cases} \tag{10.32}$$

The Fourier series expansion for the known function $\mathbb{F}(\varphi)$ can be given as:

$$\mathbb{F}(\varphi) = \sum_{m=1}^{\infty} A_m \sin\left\{m\pi\cdot\frac{(\varphi-\varphi_2)}{(\varphi_1-\varphi_2)}\right\} \tag{10.33a}$$

where

$$A_m = \frac{1}{2\cdot(\varphi_1-\varphi_2)}\cdot\int_{\varphi_2}^{\varphi_1}\mathbb{F}(\varphi)\cdot\frac{1}{2}\cdot\sin\left\{n\pi\cdot\frac{(\varphi-\varphi_2)}{(\varphi_1-\varphi_2)}\right\}\cdot d\varphi \tag{10.33b}$$

Thus, the particular integral V_p is found as:

$$V_p = -\sum_{m=1}^{\infty} A_m\cdot\left[\frac{\varphi_1-\varphi_2}{m\pi}\right]^2\cdot\sin\left\{m\pi\cdot\frac{(\varphi-\varphi_2)}{(\varphi_1-\varphi_2)}\right\} \tag{10.34a}$$

The complementary function for Equation 10.29 is readily found as:

$$V_c = a\cdot\varphi + b \tag{10.34b}$$

where *a* and *b* indicate arbitrary constants.

Thus, the complete solution is:

$$V = V_c + V_p$$

or

$$V = a \cdot \varphi + b - \sum_{m=1}^{\infty} A_m \cdot \left[\frac{\varphi_1 - \varphi_2}{m\pi}\right]^2 \cdot \sin\left\{m\pi \cdot \frac{(\varphi - \varphi_2)}{(\varphi_1 - \varphi_2)}\right\} \tag{10.35}$$

Now, using the boundary conditions given by Equation 10.32, we get:

$$\left. V_1 = a \cdot \varphi_1 + b \quad (\text{a}) \quad \text{and} \quad V_2 = a \cdot \varphi_2 + b \quad (\text{b}) \right\} \tag{10.36}$$

Therefore, on solving we get:

$$\left. a = \frac{(V_1 - V_2)}{(\varphi_1 - \varphi_2)} \quad (\text{a}) \quad \text{and} \quad b = \frac{(V_2 \cdot \varphi_1 - V_1 \cdot \varphi_2)}{(\varphi_1 - \varphi_2)} \quad (\text{b}) \right\} \tag{10.37}$$

10.3.2.4 Variation of *V* Only with φ (Mixed Boundary Conditions)

Let the scalar electric potential V be a function of the φ-coordinate only. Thus, as we have seen in the preceding example, Poisson's equation reduces to the following form:

$$\frac{d^2V}{d\varphi^2} = \mathbb{F}(\varphi) \quad \text{over } \varphi_2 < \varphi < \varphi_1 \tag{10.31}$$

where $\mathbb{F}(\varphi)$ is a known function of φ-coordinates.

Let the *mixed boundary conditions* given be as indicated below:

$$\begin{cases} V = \not{f} & \text{at } \varphi = \varphi_1 \\ \dfrac{dV}{d\varphi} = f & \text{at } \varphi = \varphi_2 \end{cases} \tag{10.38}$$

where $\not{f}$ and f are given constants.

Using this equation, we may express $\mathbb{F}(\varphi)$ as the following Fourier series expansion over $\varphi_2 < \varphi < \varphi_1$:

$$\mathbb{F}(\varphi) = \sum_{m\text{-odd}}^{\infty} A_m \cdot \cos\left\{(\varphi - \varphi_2) \cdot \frac{m\pi/2}{(\varphi_1 - \varphi_2)}\right\} \tag{10.39a}$$

After multiplying both sides of this equation by $\left[\frac{1}{2} \cdot \cos\left\{(\varphi - \varphi_2) \cdot \frac{n\pi/2}{(\varphi_1 - \varphi_2)}\right\}\right]$, and then integrating over $\varphi_2 < \varphi < \varphi_1$, we get:

$$\frac{1}{2}\int_{\varphi_2}^{\varphi_1}\mathbb{F}(\varphi)\cdot\cos\left\{(\varphi-\varphi_2)\cdot\frac{n\pi/2}{(\varphi_1-\varphi_2)}\right\}\cdot d\varphi = A_n\cdot(\varphi_1-\varphi_2)$$

with odd values for n.

Thus, substituting m for n we get:

$$A_m = \frac{1}{2(\varphi_1-\varphi_2)}\cdot\int_{\varphi_2}^{\varphi_1}\mathbb{F}(\varphi)\cdot\cos\left\{(\varphi-\varphi_2)\cdot\frac{m\pi/2}{(\varphi_1-\varphi_2)}\right\}\cdot d\varphi \tag{10.39b}$$

Thus, the particular integral V_p can be given as:

$$V_p = -\sum_{m\text{-odd}}^{\infty} A_m\cdot\left[\frac{\varphi_1-\varphi_2}{m\pi/2}\right]^2 \cos\left\{(\varphi-\varphi_2)\cdot\frac{m\pi/2}{(\varphi_1-\varphi_2)}\right\} \tag{10.40}$$

The complete solution is therefore given as:

$$V = a\cdot\varphi + b + V_p \tag{10.41a}$$

or

$$V = a\cdot\varphi + b - \sum_{m\text{-odd}}^{\infty} A_m\cdot\left[\frac{\varphi_1-\varphi_2}{m\pi/2}\right]^2 \cos\left\{(\varphi-\varphi_2)\cdot\frac{m\pi/2}{(\varphi_1-\varphi_2)}\right\} \tag{10.41b}$$

Now, to find the two arbitrary constants use Equation 10.38. Thus:

$$V\Big|_{\varphi=\varphi_1} = a\cdot\varphi_1 + b = \not{f} \tag{10.42}$$

and

$$\frac{dV}{d\varphi}\bigg|_{\varphi=\varphi_2} = a = f \tag{10.42a}$$

Thus:

$$b = \not{f} - f\cdot\varphi_1 \tag{10.42b}$$

10.3.3 Field Problems in Spherical Coordinates

The expression for the Laplacian equation of potential in spherical coordinates is reproduced here:

$$\nabla^2 V = \frac{1}{r^2}\cdot\frac{\partial}{\partial r}\left(r^2\frac{\partial V}{\partial r}\right) + \frac{1}{r^2\sin\theta}\cdot\frac{\partial}{\partial\theta}\left(\sin\theta\frac{\partial V}{\partial\theta}\right) + \frac{1}{r^2\sin^2\theta}\cdot\frac{\partial^2 V}{\partial\varphi^2} \tag{9.7}$$

In this case, the methods of solution for the problems involving different coordinates are not identical. Thus, all of the cases are to be separately described.

10.3.3.1 Potential Is a Function of *r* Only (Dirichlet Boundary Conditions)

Let the scalar electric potential V be a function of the r-coordinate only. Thus Poisson's equation in view of Equation 9.7 can be reduced to the following form:

$$\frac{1}{r^2}\cdot\frac{d}{dr}\left(r^2\frac{dV}{dr}\right)=F(r) \text{ over } r_2<r<r_1 \tag{10.43}$$

where $F(r)$ is a given function of the r-coordinate.

Therefore, from Equation 10.43, it follows that the particular integral V_p can be given as:

$$V_p=\int\left[\frac{1}{r^2}\cdot\int\left\{r^2\cdot F(r)\right\}\cdot dr\right]\cdot dr \tag{10.44a}$$

and the complementary function V_c as:

$$V_c=a\cdot\frac{1}{r}+b \tag{10.44b}$$

where a and b indicate arbitrary constants.

The complete solution is therefore given as:

$$V=a\cdot\frac{1}{r}+b+V_p \tag{10.45a}$$

or

$$V=a\cdot\frac{1}{r}+b+\int\left[\frac{1}{r^2}\cdot\int\left\{r^2\cdot F(r)\right\}\cdot dr\right]\cdot dr \tag{10.45b}$$

Let the *Dirichlet boundary conditions* be given as:

$$V=\begin{cases}V_1 & \text{at } r=r_1\\ V_2 & \text{at } r=r_2\end{cases} \tag{10.46}$$

Thus:

$$V_1=a\cdot\frac{1}{r_1}+b+\mathbb{F}(r_1) \tag{10.47a}$$

and

$$V_2=a\cdot\frac{1}{r_2}+b+\mathbb{F}(r_2) \tag{10.47b}$$

where

$$\mathbb{F}(r_1) \stackrel{\text{def}}{=} \int \left[\frac{1}{r^2} \cdot \int \left\{ r^2 \cdot F(r) \right\} \cdot dr \right] \cdot dr \Bigg|_{r=r_1} \tag{10.48a}$$

and

$$\mathbb{F}(r_2) \stackrel{\text{def}}{=} \int \left[\frac{1}{r^2} \cdot \int \left\{ r^2 \cdot F(r) \right\} \cdot dr \right] \cdot dr \Bigg|_{r=r_2} \tag{10.48b}$$

On solving Equations 10.47a and 10.47b, we get:

$$a = \left[\left\{ \mathbb{F}(r_1) - \mathbb{F}(r_2) \right\} - \left(V_1 - V_2 \right) \right] \cdot \left(\frac{r_1 \cdot r_2}{r_1 - r_2} \right) \tag{10.49a}$$

and

$$b = \frac{\left[\left(r_1 V_1 - r_2 V_2 \right) - \left\{ r_1 \mathbb{F}(r_1) - r_2 \mathbb{F}(r_2) \right\} \right]}{\left(r_1 - r_2 \right)} \tag{10.49b}$$

10.3.3.2 Potential Is a Function of *r* Only (Mixed Boundary Conditions)

The scalar electric potential V be a function of the r-coordinate only. Poisson's equation in Equation 9.7 therefore reduces to the following form:

$$\frac{1}{r^2} \cdot \frac{d}{dr} \left(r^2 \frac{dV}{dr} \right) = F(r) \quad \text{over} \quad r_2 < r < r_1 \tag{10.43}$$

where $F(r)$ is a given function of the r-coordinate.

The solution found in the preceding example is reproduced here:

$$V = a \cdot \frac{1}{r} + b + \int \left[\frac{1}{r^2} \cdot \int \left\{ r^2 \cdot F(r) \right\} \cdot dr \right] \cdot dr \tag{10.45b}$$

Let the *mixed boundary conditions* given are as indicated here:

$$\begin{cases} V = \not{f} & \text{at } r = r_1 \\ \dfrac{dV}{dr} = f & \text{at } r = r_2 \end{cases} \tag{10.50}$$

where $\not{f}$ and f are given constants.

Thus,

$$\not{f} = a \cdot \frac{1}{r_1} + b + \mathbb{F}_1(r_1) \tag{10.51a}$$

where

$$\mathbb{F}_1(r_1) \stackrel{\text{def}}{=} \int \left[\frac{1}{r^2} \cdot \int \left\{ r^2 \cdot F(r) \right\} \cdot dr \right] \cdot dr \Bigg|_{r=r_1} \tag{10.51b}$$

and

$$f = -a \frac{1}{r_2^2} + \mathbb{F}_2(r_2) \tag{10.52a}$$

where

$$\mathbb{F}_2(r_2) \stackrel{\text{def}}{=} \frac{1}{r^2} \cdot \int \left\{ r^2 \cdot F(r) \right\} \cdot dr \Bigg|_{r=r_2} \tag{10.52b}$$

On solving Equations 10.51a and 10.52a, we get:

$$a = \left[\mathbb{F}_2(r_2) - f \right] \cdot r_2^2 \tag{10.53a}$$

and

$$b = \not{f} - \left[\mathbb{F}_2(r_2) - f \right] \cdot \frac{r_2^2}{r_1} - \mathbb{F}_1(r_1) \tag{10.53b}$$

10.3.3.3 Potential a Function of θ Only

Let the scalar electric potential V be a function of the θ-coordinate only. Thus, Poisson's equation in view of Equation 9.7 reduces to the following form:

$$\frac{1}{r^2 \sin\theta} \cdot \frac{\partial}{\partial\theta} \left(\sin\theta \frac{\partial V}{\partial\theta} \right) = F(r,\theta) \text{ over } \theta_2 < \theta < \theta_1 \tag{10.54a}$$

where $F(r,\theta)$ is a given function of the r- and θ-coordinates.

Since the potential distribution is a function of only θ-coordinate, we must have:

$$F(r,\theta) \equiv \frac{1}{r^2} \cdot \mathcal{F}(\theta) \tag{10.54b}$$

where $\mathcal{F}(\theta)$ indicates a given arbitrary function of θ. Equation 10.54a can therefore be rewritten as follows:

$$\frac{1}{\sin\theta} \cdot \frac{d}{d\theta} \left(\sin\theta \frac{dV}{d\theta} \right) = \mathcal{F}(\theta) \text{ over } \theta_2 < \theta < \theta_1 \tag{10.55}$$

The particular integral can be shown to be:

$$V_p = \int \left[\frac{1}{\sin\theta} \cdot \int \left\{ \sin\theta \cdot \mathcal{F}(\theta) \right\} d\theta \right] d\theta \tag{10.56a}$$

while the complementary function found is:

$$V_c = a \cdot \ln\{\cot(\theta/2)\} + b \tag{10.56b}$$

where a and b indicate two arbitrary constants.

The complete solution is therefore given as:

$$V = V_c + V_p \tag{10.57a}$$

or

$$V = a \cdot \ln\{\cot(\theta/2)\} + b + \int\left[\frac{1}{\sin\theta} \cdot \int\{\sin\theta \cdot \mathcal{F}(\theta)\}\, d\theta\right] d\theta \tag{10.57b}$$

Let the *Dirichlet boundary conditions* be given as:

$$V = \begin{cases} V_1 & \text{at } \theta = \theta_1 \\ V_2 & \text{at } \theta = \theta_2 \end{cases} \tag{10.58}$$

Thus:

$$V_1 = a \cdot \ln\{\cot(\theta_1/2)\} + b + \mathbb{F}(\theta_1) \tag{10.59a}$$

and

$$V_2 = a \cdot \ln\{\cot(\theta_2/2)\} + b + \mathbb{F}(\theta_2) \tag{10.59b}$$

where

$$\mathbb{F}(\theta_1) \stackrel{\text{def}}{=} \int\left[\frac{1}{\sin\theta} \cdot \int\{\sin\theta \cdot \mathcal{F}(\theta)\}\, d\theta\right] d\theta\Bigg|_{\theta=\theta_1} \tag{10.60a}$$

and

$$\mathbb{F}(\theta_2) \stackrel{\text{def}}{=} \int\left[\frac{1}{\sin\theta} \cdot \int\{\sin\theta . \mathcal{F}(\theta)\}\, d\theta\right] d\theta\Bigg|_{\theta=\theta_2} \tag{10.60b}$$

Solving Equations 10.59a and 10.59b, we get:

$$a = \frac{\left[\{\mathbb{F}(\theta_1) - \mathbb{F}(\theta_2)\} - (V_1 - V_2)\right]}{\left[\ln\left\{\dfrac{\tan(\theta_1/2)}{\tan(\theta_2/2)}\right\}\right]} \tag{10.61a}$$

and

$$b = \frac{\left[\mathbb{F}(\theta_1)\cdot\ln\{\tan(\theta_2/2)\} - \mathbb{F}(\theta_2)\cdot\ln\{\tan(\theta_1/2)\}\right] - \left[V_1\cdot\ln\{\tan(\theta_2/2)\} - V_2\cdot\ln\{\tan(\theta_1/2)\}\right]}{\left[\ln\left\{\dfrac{\tan(\theta_1/2)}{\tan(\theta_2/2)}\right\}\right]} \tag{10.61b}$$

10.3.3.4 Potential Is a Function of φ Only

Let the scalar electric potential V be a function of the φ-coordinate only. Thus, Poisson's equation in view of Equation 9.7 reduces to the following form:

$$\frac{1}{r^2\sin^2\theta}\cdot\frac{\partial^2 V}{\partial\varphi^2} = F(r,\theta,\varphi) \quad \text{over } \varphi_2 < \varphi < \varphi_1 \tag{10.62a}$$

where $F(r,\theta,\varphi)$ is a given function of the r-, θ-, and φ-coordinates.

Since the potential distribution is a function of only the φ-coordinate, we must have:

$$F(r,\theta,\varphi) \equiv \frac{1}{r^2\sin^2\theta}\cdot\mathcal{F}(\varphi) \tag{10.62b}$$

where $\mathcal{F}(\varphi)$ indicates a given arbitrary function of φ. Equation 10.62a can therefore be rewritten as follows:

$$\frac{\partial^2 V}{\partial\varphi^2} = \frac{d^2 V}{d\varphi^2} = \mathcal{F}(\varphi) \quad \text{over } \varphi_2 < \varphi < \varphi_1 \tag{10.63}$$

This equation is similar to the following equation:

$$\frac{d^2 V}{d\varphi^2} = \mathbb{F}(\varphi) \quad \text{over } \varphi_2 < \varphi < \varphi_1 \tag{10.31}$$

Therefore, the solution of Equation 10.63 is similar to that of Equation 10.31, discussed earlier in this section.

10.4 TWO-DIMENSIONAL FIELD PROBLEMS

This section demonstrates methods for solving two-dimensional field problems. It includes five problems involving Cartesian, three cylindrical, and three spherical coordinates.

10.4.1 Field Problems in Cartesian Coordinates

The expression for the Laplacian equation of V in Cartesian coordinates is reproduced here:

$$\nabla^2 V = \frac{\partial^2 V}{\partial x^2} + \frac{\partial^2 V}{\partial y^2} + \frac{\partial^2 V}{\partial z^2} \tag{9.5}$$

Using Equation 9.5, a two-dimensional Poisson's equation can be given as:

$$\frac{\partial^2 V}{\partial x^2} + \frac{\partial^2 V}{\partial y^2} = F(x, y) \tag{10.64}$$

where $F(x,y)$ is a given function of x- and y-coordinates.

The complementary function for this equation is the solution of the Laplace equation with unresolved arbitrary constants. An exhaustive discussion for the solution of the Laplace equation with examples has been presented earlier. The method of separation of variables finds extensive applications in these solutions.

10.4.1.1 Potential Distribution in Finite Rectangular Region (Case I)

Figure 10.1 shows a rectangular region extending over $0 \le x \le X_o$ and $-Y_o/2 \le y \le Y_o/2$. Three of its sides, namely $x = 0$ and $y = \pm Y_o/2$, are at zero potential. On the fourth side, i.e. at $x = X_0$, an arbitrarily distributed potential $f(y)$ is specified. The potential $V(x,y)$ in this region satisfies the following Poisson's equation:

$$\frac{\partial^2 V}{\partial x^2} + \frac{\partial^2 V}{\partial y^2} = F(x, y) \quad \text{over} \quad 0 \le x \le X_o \text{ and } -Y_o/2 \le y \le Y_o/2 \tag{10.64}$$

where $F(x,y)$ is a given function of x- and y-coordinates.

Noting that the potential is zero at $y = \pm Y_o/2$, the potential function $f(y)$ can be resolved into its even and odd components; each of these components can be expressed as a Fourier series. The resulting expression for the potential function $f(y)$ is:

$$V\big|_{x=X_o} = f(y) = \sum_{m\text{-odd}}^{\infty} A'_m \cdot \cos\left(m\pi \cdot \frac{y}{Y_o}\right) + \sum_{n=1}^{\infty} A''_n \cdot \sin\left(2n\pi \cdot \frac{y}{Y_o}\right) \tag{10.65}$$

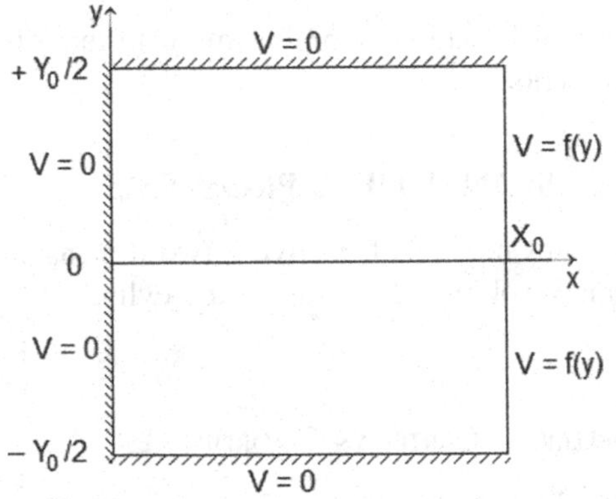

FIGURE 10.1 2D problem of Poisson's equation (Case I).

To find the Fourier coefficient A'_m, let us multiply both sides of this equation by $\cos\left(M\pi\cdot\frac{y}{Y_o}\right)$; on integrating over $-Y_o/2 \le y \le Y_o/2$ we get, for the LHS:

$$\text{L.H.S.} = \int_{-Y_o/2}^{Y_o/2} f(y)\cdot\cos\left(M\pi\cdot\frac{y}{Y_o}\right)\cdot dy \tag{10.66}$$

while for the first term on the RHS:

$$\begin{aligned}
(\text{R.H.S.})_1 &= \sum_{m\text{-odd}}^{\infty} A'_m \cdot \int_{-Y_o/2}^{Y_o/2} \cos\left(m\pi\cdot\frac{y}{Y_o}\right)\cdot\cos\left(M\pi\cdot\frac{y}{Y_o}\right)dy \\
&= \sum_{m\text{-odd}}^{\infty} \frac{1}{2}\cdot A'_m \cdot \int_{-Y_o/2}^{Y_o/2} \left[\cos(m-M)\pi\cdot\frac{y}{Y_o} + \cos(m-M)\pi\cdot\frac{y}{Y_o}\right]dy \\
&= \sum_{m\text{-odd}}^{\infty} \frac{Y_o}{2\pi}\cdot A'_m \cdot \left[\frac{\sin(m-M)\pi\cdot\frac{y}{Y_o}}{(m-M)} + \frac{\sin(m+M)\pi\cdot\frac{y}{Y_o}}{(m+M)}\right]_{-Y_o/2}^{Y_o/2} \\
&= \frac{Y_o}{2}\cdot A'_m \cdot \quad \text{for } m = M \text{ (odd values)}
\end{aligned} \tag{10.66a}$$

$$\begin{aligned}
(\text{R.H.S.})_2 &= \sum_{n=1}^{\infty} A''_n \cdot \int_{-Y_o/2}^{Y_o/2} \sin\left(2n\pi\cdot\frac{y}{Y_o}\right)\cdot\cos\left(M\pi\cdot\frac{y}{Y_o}\right)dy \\
&= \sum_{n=1}^{\infty} \frac{1}{2}\cdot A''_n \cdot \int_{-Y_o/2}^{Y_o/2} \left[\sin(2n-M)\pi\cdot\frac{y}{Y_o} + \sin(2n+M)\pi\cdot\frac{y}{Y_o}\right]dy \\
&= -\sum_{n=1}^{\infty} \frac{Y_o}{2\pi}\cdot A''_n \cdot \left[\frac{\cos(2n-M)\pi\cdot\frac{y}{Y_o}}{(2n-M)} + \frac{\cos(2n+M)\pi\cdot\frac{y}{Y_o}}{(2n+M)}\right]_{-Y_o/2}^{Y_o/2} \\
&= 0 \quad \text{for } M \text{ (odd values)}
\end{aligned} \tag{10.66b}$$

Thus,

$$A'_m = \frac{2}{Y_o}\cdot \int_{-Y_o/2}^{Y_o/2} f(y)\cdot\cos\left(m\pi\cdot\frac{y}{Y_o}\right)\cdot dy \tag{10.67}$$

To find the Fourier coefficient A_n'', let us multiply both sides of this equation by $\sin\left(2N\pi\cdot\frac{y}{Y_o}\right)$; on integrating over $-Y_o/2 \le y \le Y_o/2$ we get, for the LHS:

$$\text{L.H.S.} = \int_{-Y_0/2}^{Y_0/2} f(y)\cdot\sin\left(2N\pi\cdot\frac{y}{Y_o}\right)\cdot dy \tag{10.68}$$

while for the first term on the RHS:

$$\begin{aligned}(\text{R.H.S.})_1 &= \sum_{m\text{-odd}}^{\infty} A_m'\cdot\int_{-Y_0/2}^{Y_0/2}\sin\left(2N\pi\cdot\frac{y}{Y_o}\right)\cdot\cos\left(m\pi\cdot\frac{y}{Y_o}\right)dy\\ &= \sum_{m\text{-odd}}^{\infty}\frac{1}{2}\cdot A_m'\cdot\int_{-Y_0/2}^{Y_0/2}\left[\sin(2N-m)\pi\cdot\frac{y}{Y_o}+\sin(2N+m)\pi\cdot\frac{y}{Y_o}\right]dy\\ &= -\sum_{m\text{-odd}}^{\infty}\frac{Y_o}{2\pi}\cdot A_m'\cdot\left[\frac{\cos(2N-m)\pi\cdot\frac{y}{Y_o}}{(2n-M)}+\frac{\cos(2N+m)\pi\cdot\frac{y}{Y_o}}{(2n+M)}\right]_{-Y_0/2}^{Y_0/2}\\ &= 0 \quad \text{for } N \text{ (integer values)}\end{aligned} \tag{10.68a}$$

$$\begin{aligned}(\text{R.H.S.})_2 &= \sum_{n=1}^{\infty} A_n''\cdot\int_{-Y_0/2}^{Y_0/2}\sin\left(2N\pi\cdot\frac{y}{Y_o}\right)\cdot\sin\left(2n\pi\cdot\frac{y}{Y_o}\right)dy\\ &= \sum_{n=1}^{\infty}\frac{1}{2}\cdot A_n''\cdot\int_{-Y_0/2}^{Y_0/2}\left[\cos(N-n)2\pi\cdot\frac{y}{Y_o}-\cos(N+n)2\pi\cdot\frac{y}{Y_o}\right]dy\\ &= \sum_{n=1}^{\infty}\frac{Y_o}{4\pi}\cdot A_n''\cdot\left[\frac{\sin(N-n)2\pi\cdot\frac{y}{Y_o}}{(N-n)}-\frac{\sin(N+n)2\pi\cdot\frac{y}{Y_o}}{(N+n)}\right]_{-Y_0/2}^{Y_0/2}\\ &= \frac{Y_o}{2}\cdot A_n'' \quad \text{for } n=N \text{ (integer values)}\end{aligned} \tag{10.68b}$$

Thus,

$$A_n'' = \frac{2}{Y_o}\cdot\int_{-Y_0/2}^{Y_0/2} f(y)\cdot\sin\left(2n\pi\cdot\frac{y}{Y_o}\right)\cdot dy \tag{10.69}$$

The two Fourier coefficients involved in Equation 10.30 are thus given by Equations 10.67 and 10.69.

Next, consider the function $F(x,y)$. Let function be expressed as a double Fourier series over $0 < x < X_0$ and $-Y_o/2 \le y \le Y_o/2$:

$$F(x,y) = \sum_{p=1}^{\infty}\sum_{m\text{-odd}}^{\infty} A_{p,m}^{(1)} \cdot \sin\left(p\pi \cdot \frac{x}{X_o}\right) \cdot \cos\left(m\pi \cdot \frac{y}{Y_o}\right) + \sum_{p=1}^{\infty}\sum_{n=1}^{\infty} A_{p,n}^{(2)} \cdot \sin\left(p\pi \cdot \frac{x}{X_o}\right) \cdot \sin\left(2n\pi \cdot \frac{y}{Y_o}\right) \tag{10.70}$$

To determine the Fourier coefficients $A_{p,m}^{(1)}$ and $A_{p,n}^{(2)}$, multiply both sides of this equation by and then integrate over $0 < x < X_o$. This results in:

$$\text{L.H.S.} = \int_0^{X_o} F(x,y) \cdot \sin\left(p\pi \cdot \frac{x}{X_o}\right) \cdot dx \tag{10.71a}$$

$$\left(\text{R.H.S.}\right)_1 = \sum_{m\text{-odd}}^{\infty} A_{p,m}^{(1)} \cdot \frac{X_o}{2} \cdot \cos\left(m\pi \cdot \frac{y}{Y_o}\right) \tag{10.71b}$$

$$\left(\text{R.H.S.}\right)_2 = \sum_{n=1d}^{\infty} A_{p,n}^{(2)} \cdot \frac{X_o}{2} \cdot \sin\left(2n\pi \cdot \frac{y}{Y_o}\right) \tag{10.71c}$$

Thus,

$$\int_0^{X_o} F(x,y) \cdot \sin\left(p\pi \cdot \frac{x}{X_o}\right) .dx = \sum_{m\text{-odd}}^{\infty} A_{p,m}^{(1)} \cdot \frac{X_o}{2} \cdot \cos\left(m\pi \cdot \frac{y}{Y_o}\right) + \sum_{n=1d}^{\infty} A_{p,n}^{(2)} \cdot \frac{X_o}{2} \cdot \sin\left(2n\pi \cdot \frac{y}{Y_o}\right) \tag{10.72}$$

Furthermore, multiply the resulting equation by $\cos\left(M\pi \cdot \frac{y}{Y_o}\right)$ and then integrate over $-Y_o/2 < y < Y_o/2$. Substituting m for M, these result in:

$$\text{L.H.S.} \overset{\text{def}}{=} \mathcal{F}_L'(p,m) = \int_{-Y_o/2}^{Y_o/2} \left\{ \int_0^{X_o} F(x,y) \cdot \sin\left(p\pi \cdot \frac{x}{X_o}\right) \cdot dx \right\} \cdot \cos\left(m\pi \cdot \frac{y}{Y_o}\right) \cdot dy \tag{10.73a}$$

$$\left(\text{R.H.S.}\right)_1 \overset{\text{def}}{=} \mathcal{F}_{R1}'(p,m) = A_{p,m}^{(1)} \cdot \frac{X_o}{2} \cdot \frac{Y_o}{2} \tag{10.73b}$$

$$\left(\text{R.H.S.}\right)_2 \stackrel{\text{def}}{=} \mathcal{F}'_{R2}\left(p,m\right) = 0 \tag{10.73c}$$

Thus, $\mathcal{F}'_L\left(p,m\right) = A^{(1)}_{p,m} \cdot \frac{X_o}{2} \cdot \frac{Y_o}{2} + 0$

Therefore,

$$A^{(1)}_{p,m} = \frac{4}{X_o \cdot Y_o} \mathcal{F}'_L\left(p,m\right) \tag{10.74}$$

Let us multiply Equation 10.70 by $\sin\left(2N\pi \cdot \frac{y}{Y_o}\right)$ and then integrate over $-Y_o/2 < y < Y_o/2$. Substituting n for N, these result:

$$\text{L.H.S.} \stackrel{\text{def}}{=} \mathcal{F}''_L\left(p,n\right) = \int_{-Y_o/2}^{Y_o/2} \left\{ \int_0^{X_o} F(x,y) \cdot \sin\left(p\pi \cdot \frac{x}{X_o}\right) \cdot dx \right\} \cdot \sin\left(2n\pi \cdot \frac{y}{Y_o}\right) \cdot dy \tag{10.75a}$$

$$\left(\text{R.H.S.}\right)_1 \stackrel{\text{def}}{=} \mathcal{F}''_{R1}\left(p,n\right) = 0 \tag{10.75b}$$

$$\left(\text{R.H.S.}\right)_2 \stackrel{\text{def}}{=} \mathcal{F}''_{R1}\left(p,n\right) = A^{(2)}_{p,n} \cdot \frac{X_o}{2} \cdot \frac{Y_o}{2} \tag{10.75c}$$

Thus,

$$\mathcal{F}''_L\left(p,n\right) = 0 + A^{(2)}_{p,n} \cdot \frac{X_o}{2} \cdot \frac{Y_o}{2}$$

Hence,

$$A^{(2)}_{p,n} = \frac{4}{X_o \cdot Y_o} \mathcal{F}''_L\left(p,m\right) \tag{10.76}$$

The two coefficients in the double Fourier series of Equation 10.70 are thus given by Equations 10.74 and 10.76.

Now, in view of Equation 10.70, the particular integral for Poisson's equation, i.e. Equation 10.64, obtained is:

$$V_p = -\sum_{p=1}^{\infty} \sin\left(p\pi \cdot \frac{x}{X_o}\right) \cdot \left[\sum_{m-odd}^{\infty} A^{(1)}_{p,m} \cdot \frac{1}{\left(\frac{p\pi}{X_o}\right)^2 + \left(\frac{m\pi}{Y_o}\right)^2} \cdot \cos\left(m\pi \cdot \frac{y}{Y_o}\right) \right. \\ \left. + \sum_{n=1}^{\infty} A^{(2)}_{p,n} \cdot \frac{1}{\left(\frac{p\pi}{X_o}\right)^2 + \left(\frac{2n\pi}{Y_o}\right)^2} \cdot \sin\left(2n\pi \cdot \frac{y}{Y_o}\right) \right] \tag{10.77a}$$

The complementary function for Poisson's equation obtained by solving the two-dimensional Laplace equation is as follows:

$$V_c = \sum_{m\text{-odd}}^{\infty} A'_m \cdot \cos\left(\frac{m\pi}{Y_o}\cdot y\right)\cdot\frac{\sinh\left(\dfrac{m\pi}{Y_o}\cdot x\right)}{\sinh\left(\dfrac{m\pi}{Y_o}\cdot X_o\right)} + \sum_{n=1}^{\infty} A''_n \cdot \sin\left(\frac{2n\pi}{Y_o}\cdot y\right)\cdot\frac{\sinh\left(\dfrac{2n\pi}{Y_o}\cdot x\right)}{\sinh\left(\dfrac{2n\pi}{Y_o}\cdot X_o\right)} \tag{10.77b}$$

Therefore, the complete solution for Poisson's equation is given as:

$$V(x,y) = V_c + V_p \tag{10.78}$$

Note that we have taken cognizance of the boundary conditions from the very beginning.

10.4.1.2 Potential Distribution in Finite Rectangular Region (Case II)

Figure 10.2 shows a rectangular region extending over $0 \le x \le X_o$ and $-Y_o/2 \le y \le Y_o/2$. Three of its sides, namely $x = 0$, X_o, and $y = -Y_o/2$, are at zero potential. On the fourth side, i.e. at $y = Y_o/2$, an arbitrarily distributed potential $f(x)$ is specified. The potential $V(x,y)$ in this region satisfies the following Poisson's equation:

$$\frac{\partial^2 V}{\partial x^2} + \frac{\partial^2 V}{\partial y^2} = F(x,y) \quad \text{over } 0 \le x \le X_o \text{ and } -Y_o/2 \le y \le Y_o/2 \tag{10.64}$$

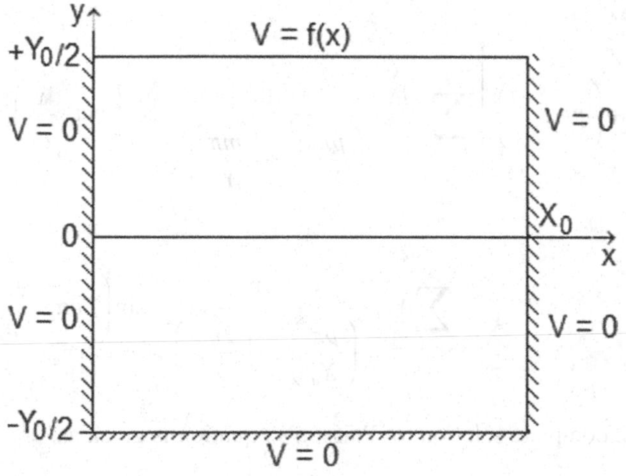

FIGURE 10.2 2D problem of Poisson's equation (Case II).

where $F(x,y)$ is a given function of x- and y-coordinates

This problem is a variation of the example discussed in Section 10.6.1. Since the potential is zero at $x = 0$ as well as at $x = X_o$ the specified potential function $f(x)$ can be expressed by the following Fourier series:

$$V\Big|_{y=Y_o/2} = f(x) = \sum_{p=1}^{\infty} A_p \cdot \sin\left(\frac{p\pi}{X_o} \cdot x\right) \tag{10.78a}$$

Since

$$\int_0^{X_o} f(x) \cdot \sin\left(\frac{P\pi}{X_o} \cdot x\right) \cdot dx = \frac{X_o}{2} \cdot A_P$$

Therefore, the Fourier coefficient A_p is:

$$A_p = \frac{2}{X_o} \cdot \int_0^{X_o} f(x) \cdot \sin\left(\frac{p\pi}{X_o} \cdot x\right) \cdot dx \tag{10.78b}$$

In view of Equation 10.77a, the complementary function subjected to all boundary conditions is given as follows:

$$V_c = \sum_{p=1}^{\infty} A_p \cdot \sin\left(\frac{p\pi}{X_o} \cdot x\right) \cdot \frac{\sinh\left\{\frac{p\pi}{X_o} \cdot (y + Y_o/2)\right\}}{\sinh\left(p\pi \cdot \frac{Y_o}{X_o}\right)} \tag{10.79a}$$

The particular integral is the same as that found in the preceding example. It is reproduced as: below:

$$V_p = -\sum_{p=1}^{\infty} \sin\left(p\pi . \frac{x}{X_o}\right) \cdot \left[\sum_{m-odd}^{\infty} A_{p,m}^{(1)} \cdot \frac{1}{\left(\frac{p\pi}{X_o}\right)^2 + \left(\frac{m\pi}{Y_o}\right)^2} \cdot \cos\left(m\pi \cdot \frac{y}{Y_o}\right) + \sum_{n=1}^{\infty} A_{p,n}^{(2)} \cdot \frac{1}{\left(\frac{p\pi}{X_o}\right)^2 + \left(\frac{2n\pi}{Y_o}\right)^2} \cdot \sin\left(2n\pi \cdot \frac{y}{Y_o}\right)\right] \tag{10.80}$$

Therefore, the complete solution for Poisson's equation is given as:

$$V(x, y) = V_c + V_p \tag{10.81}$$

Note that we have taken cognizance of the boundary conditions from the very beginning.

10.4.1.3 Potential Distribution in Semi-Finite Rectangular Region

Consider Figure 10.3, which shows a semi-infinite rectangular region extending over $0 \le x < \infty$ and $-Y_o/2 \le y \le Y_o/2$. Two of its sides, namely $y = \pm Y_o/2$, are at zero potential. On the side at $x = 0$, an arbitrarily distributed potential $f(y)$ is specified. The potential $V(x,y)$ in this region satisfies the following Poisson's equation:

$$\frac{\partial^2 V}{\partial x^2} + \frac{\partial^2 V}{\partial y^2} = F(x,y) \quad \text{over } 0 \le x < \infty \text{ and } -Y_o/2 \le y \le Y_o/2 \tag{10.82}$$

where $F(x,y)$ is a given function of x- and y-coordinates, as shown here:

$$F(x,y) = \sum_{m\text{-odd}}^{\infty} A_m^{(1)} \cdot \cos\left(m\pi \cdot \frac{y}{Y_o}\right) \cdot e^{-\left(k_{1m} \cdot \frac{x}{Y_o}\right)} + \sum_{n=1}^{\infty} A_n^{(2)} \cdot \sin\left(2n\pi \cdot \frac{y}{Y_o}\right) \cdot e^{-\left(k_{2n} \cdot \frac{x}{Y_o}\right)} \tag{10.83}$$

where $A_m^{(1)}$ and $A_n^{(2)}$ are given real quantities; k_{1m} and k_{2n} are given positive real quantities.

Noting that the potential is zero at $y = \pm Y_o/2$, the potential function $f(y)$ can be resolved into its even and odd components; each of these components can be expressed as Fourier series. The resulting expression for the potential function $f(y)$ is:

$$V\Big|_{x=0} = f(y) = \sum_{m\text{-odd}}^{\infty} A'_m \cdot \cos\left(m\pi \cdot \frac{y}{Y_o}\right) + \sum_{n=1}^{\infty} A''_n \cdot \sin\left(2n\pi \cdot \frac{y}{Y_o}\right) \tag{10.84}$$

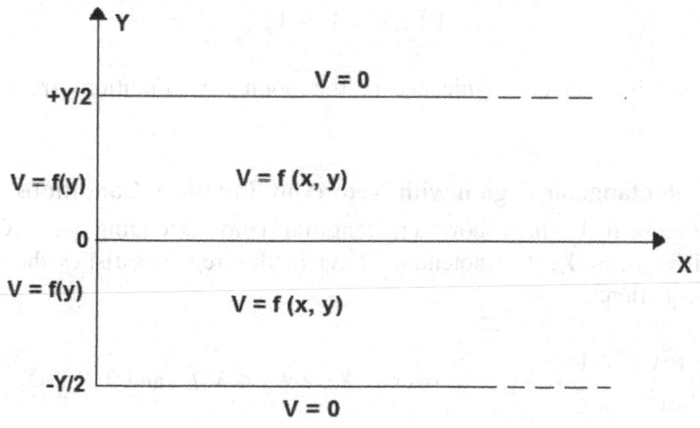

FIGURE 10.3 Configuration of semi-infinite rectangular region with specified potentials.

where

$$A'_m = \frac{2}{Y_o} \cdot \int_{-Y_o/2}^{Y_o/2} f(y) \cdot \cos\left(m\pi \cdot \frac{y}{Y_o}\right) \cdot dy \qquad (10.84a)$$

and

$$A''_n = \frac{2}{Y_o} \cdot \int_{-Y_o/2}^{Y_o/2} f(y) \cdot \sin\left(2n\pi \cdot \frac{y}{Y_o}\right) \cdot dy \qquad (10.84b)$$

The complementary function is therefore given as:

$$V_c = \sum_{m\text{-odd}}^{\infty} A'_m \cdot \cos\left(m\pi \cdot \frac{y}{Y_o}\right) \cdot e^{-\left(m\pi \cdot \frac{x}{Y_o}\right)} + \sum_{n=1}^{\infty} A''_n \cdot \sin\left(2n\pi \cdot \frac{y}{Y_o}\right) \cdot e^{-\left(2n\pi \cdot \frac{x}{Y_o}\right)} \qquad (10.85a)$$

In view of Equation 10.83, the particular integral is as follows:

$$V_p = \sum_{m\text{-odd}}^{\infty} A_m^{(1)} \cdot \frac{1}{k_{1m}^2 - \left(\frac{m\pi}{Y_o}\right)^2} \cdot \cos\left(m\pi \cdot \frac{y}{Y_o}\right) \cdot e^{-\left(k_{1m} \cdot \frac{x}{Y_o}\right)}$$
$$+ \sum_{n=1}^{\infty} A_n^{(2)} \cdot \frac{1}{k_{2m}^2 - \left(\frac{2n\pi}{Y_o}\right)^2} \cdot \sin\left(2n\pi \cdot \frac{y}{Y_o}\right) \cdot e^{-\left(k_{2n} \cdot \frac{x}{Y_o}\right)} \qquad (10.85b)$$

Therefore, the complete solution for Poisson's equation is given as:

$$V(x, y) = V_c + V_p \qquad (10.86)$$

Note that we have taken cognizance of the boundary conditions from the very beginning.

10.4.1.4 Rectangular Region with Neumann Boundary Conditions

Consider Figure 10.4, which shows a rectangular region extending over $-X_o/2 \le x \le X_o/2$ and $0 \le y \le Y_o$. The potential $V(x,y)$ in this region satisfies the following Poisson's equation:

$$\frac{\partial^2 V}{\partial x^2} + \frac{\partial^2 V}{\partial y^2} = F(x, y) \quad \text{over} \ -X_o/2 \le x \le X_o/2, \text{and } 0 \le y \le Y_o \qquad (10.87)$$

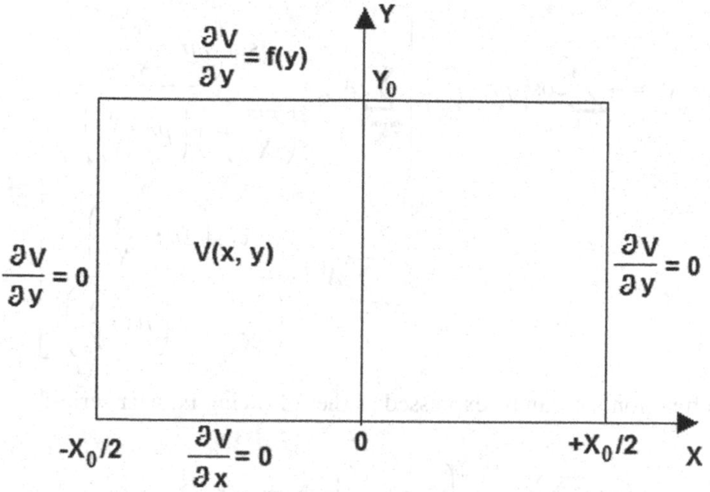

FIGURE 10.4 Rectangular region with Neumann boundary conditions.

where $F(x,y)$ is a given function of x- and y-coordinates.

The Neumann boundary conditions are given as:

$$\frac{\partial V}{\partial x} = 0 \quad \text{at } x = \pm X_o/2 \tag{10.88a}$$

$$\frac{\partial V}{\partial y} = \begin{cases} 0 & \text{at } y = 0 \\ f(x) & \text{at } y = Y_o \end{cases} \tag{10.88b}$$

where $f(x)$ is an arbitrarily specified function of the x-coordinate.

Let us express the given function $F(x,y)$ as the following double Fourier series, so that its normal derivatives at the four boundary lines are identically zero:

$$F(x,y) = \sum_{p=1}^{\infty}\sum_{m\text{-odd}}^{\infty} A_{p,m}^{(1)} \cdot \sin\left(m\pi \cdot \frac{x}{X_o}\right) \cdot \cos\left(p\pi \cdot \frac{y}{Y_o}\right)$$
$$+ \sum_{p=1}^{\infty}\sum_{n=1}^{\infty} A_{p,n}^{(2)} \cdot \cos\left(2n\pi \cdot \frac{x}{X_o}\right) \cdot \cos\left(p\pi \cdot \frac{y}{Y_o}\right) \tag{10.89}$$

over $-X_o/2 < x < X_o/2$ and $0 \le y \le Y_o$.

The coefficients of this double Fourier series, $A_{p,m}^{(1)}$ and $A_{p,n}^{(2)}$, can be readily found for the given function $F(x,y)$, as discussed earlier.

Now, in view of Equations 10.87 and 10.89, the particular integral V_p found is given as follows:

$$V_p = -\sum_{p=1}^{\infty} \cos\left(p\pi \cdot \frac{y}{Y_o}\right) \cdot \left[\sum_{m\text{-odd}}^{\infty} A_{p,m}^{(1)} \cdot \frac{\sin\left(m\pi \cdot \frac{x}{X_o}\right)}{\left(\frac{m\pi}{X_o}\right)^2 + \left(p\pi \cdot \frac{y}{Y_o}\right)^2} - \sum_{n=1}^{\infty} A_{p,n}^{(2)} \cdot \frac{\cos\left(2n\pi \cdot \frac{x}{X_o}\right)}{\left(\frac{2n\pi}{X_o}\right)^2 + \left(p\pi \cdot \frac{y}{Y_o}\right)^2} \right] \tag{10.90}$$

Now, the function $f(x)$ can be expressed as the following Fourier series:

$$f(x) = \sum_{n=1}^{\infty} C_n \cdot \cos\left(2n\pi \cdot \frac{x}{X_o}\right) \quad \text{over} \ -X_o/2 \le x \le X_o/2 \tag{10.91}$$

where the Fourier coefficient C_n can be easily found for the given function $f(x)$.

Therefore, the complementary function V_c for Equation 10.87, subjected to the specified boundary conditions found using Equations 10.88a, 10.88b, and 10.91, is as follows:

$$V_c = \sum_{n=1}^{\infty} \frac{C_n}{2n\pi} \cdot \cos\left(2n\pi \cdot \frac{x}{X_o}\right) \cdot \frac{\cosh\left(2n\pi \cdot \frac{y}{X_o}\right)}{\sinh\left(2n\pi \cdot \frac{Y_0}{X_o}\right)} \tag{10.92}$$

Thus, the complete solution for Equation 10.87 is:

$$V = V_c + V_p \tag{10.93}$$

It may be noted that the solution obtained is not unique since if we add an arbitrary constant to the RHS of this equation, the resulting expression for the potential distribution shall still satisfy Poisson's equation and all given boundary conditions.

10.4.1.5 Case of Mixed Boundary Value Problem

This is an example of a mixed boundary value problem. Consider Figure 10.5, which shows a rectangular region extending over $0 \le x \le X_o$ and $-Y_o/2 \le y \le Y_o/2$. Three of its sides, namely $x = 0$ and $y = \pm Y_o/2$, are at zero potential. On the fourth side, i.e. at $x = X_o$, an arbitrarily distributed normal derivative of the potential, $f(y)$ is specified. The potential $V(x,y)$ in this region satisfies the following Poisson's equation:

$$\frac{\partial^2 V}{\partial x^2} + \frac{\partial^2 V}{\partial y^2} = F(x, y) \quad \text{over } 0 \le x \le X_o \text{ and } -Y_o/2 \le y \le Y_o/2 \tag{10.94}$$

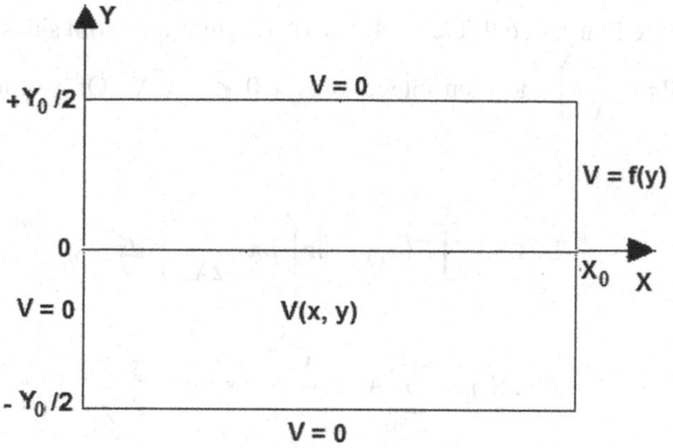

FIGURE 10.5 A rectangular region with mixed boundary conditions.

where $F(x,y)$ is a given function of x- and y-coordinates.

Noting that the potential is zero at $y = \pm Y_o/2$, the given normal derivative of the potential function $f(y)$, can be resolved into its even and odd components; each of these components can be expressed as Fourier series. The resulting expression for the normal derivative of the potential function $f(y)$, therefore is:

$$\left.\frac{\partial V}{\partial x}\right|_{x=X_o} = f(y) = \sum_{m\text{-odd}}^{\infty} A'_m \cdot \cos\left(m\pi \cdot \frac{y}{Y_o}\right) + \sum_{n=1}^{\infty} A''_n \cdot \sin\left(2n\pi \cdot \frac{y}{Y_o}\right) \quad (10.95)$$

The two Fourier coefficients involved in Equation 10.95, given by Equations 10.67 and 10.69, are reproduced here:

$$A'_m = \frac{2}{Y_o} \cdot \int_{-Y_o/2}^{Y_o/2} f(y) \cdot \cos\left(m\pi \cdot \frac{y}{Y_o}\right) \cdot dy \quad (10.67)$$

and

$$A''_n = \frac{2}{Y_o} \cdot \int_{-Y_o/2}^{Y_o/2} f(y) \cdot \sin\left(2n\pi \cdot \frac{y}{Y_o}\right) \cdot dy \quad (10.69)$$

Next, consider the function $F(x,y)$. Let function be expressed as a double Fourier series over $0 < x < X_o$ and $-Y_o/2 \le y \le Y_o/2$, such that the function vanishes at $x = 0$ and $y = \pm Y_o/2$, while its normal derivative is zero at $x = X_o$:

$$F(x,y) = \sum_{p\text{-odd}}^{\infty} \sum_{m\text{-odd}}^{\infty} A^{(1)}_{p,m} \cdot \sin\left(p\pi \cdot \frac{x}{2X_o}\right) \cdot \cos\left(m\pi \cdot \frac{y}{Y_o}\right) + \sum_{p\text{-odd}}^{\infty} \sum_{n=1}^{\infty} A^{(2)}_{p,n} \cdot \sin\left(p\pi \cdot \frac{x}{2X_o}\right) \cdot \sin\left(2n\pi \cdot \frac{y}{Y_o}\right) \quad (10.96)$$

To determine the Fourier coefficients $A^{(1)}_{p,m}$ and $A^{(2)}_{p,n}$, multiply both sides of this equation by $\sin\left(P\pi\cdot\dfrac{x}{2X_o}\right)$ and then integrate over $0 < x < X_o$. On setting $P = p$, this results in:

$$\text{L.H.S.} = \int_0^{X_o} F(x,y)\cdot\sin\left(p\pi\cdot\frac{x}{2X_o}\right)\cdot dx \tag{10.97a}$$

$$(\text{R.H.S.})_1 = \sum_{m\text{-odd}}^{\infty} A^{(1)}_{p,m}\cdot\frac{X_o}{2}\cdot\cos\left(m\pi\cdot\frac{y}{Y_o}\right) \tag{10.97b}$$

$$(\text{R.H.S.})_2 = \sum_{n=1d}^{\infty} A^{(2)}_{p,n}\cdot\frac{X_o}{2}\cdot\sin\left(2n\pi\cdot\frac{y}{Y_o}\right) \tag{10.97c}$$

Thus,

$$\int_0^{X_o} F(x,y)\cdot\sin\left(p\pi\cdot\frac{x}{2X_o}\right)\cdot dx$$
$$= \sum_{m\text{-odd}}^{\infty} A^{(1)}_{p,m}\cdot\frac{X_o}{2}\cdot\cos\left(m\pi\cdot\frac{y}{Y_o}\right) + \sum_{n=1d}^{\infty} A^{(2)}_{p,n}\cdot\frac{X_o}{2}\cdot\sin\left(2n\pi\cdot\frac{y}{Y_o}\right) \tag{10.98}$$

Furthermore, multiply the resulting equation by $\cos\left(M\pi\cdot\dfrac{y}{Y_o}\right)$ and then integrate over $-Y_o/2 < y < Y_o/2$. On setting $M = m$, these results:

$$\text{L.H.S.} \stackrel{\text{def}}{=} \mathcal{F}'_L(p,m) = \int_{-Y_o/2}^{Y_o/2}\left\{\int_0^{X_o} F(x,y)\cdot\sin\left(p\pi\cdot\frac{x}{2X_o}\right)\cdot dx\right\}\cdot\cos\left(m\pi\cdot\frac{y}{Y_o}\right)\cdot dy \tag{10.99a}$$

$$(\text{R.H.S.})_1 \stackrel{\text{def}}{=} \mathcal{F}'_{R1}(p,m) = A^{(1)}_{p,m}\cdot\frac{X_o}{2}\cdot\frac{Y_o}{2} \tag{10.99b}$$

$$(\text{R.H.S.})_2 \stackrel{\text{def}}{=} \mathcal{F}'_{R2}(p,m) = 0 \tag{10.99c}$$

Thus, $\mathcal{F}'_L(p,m) = A^{(1)}_{p,m}\cdot\dfrac{X_o}{2}\cdot\dfrac{Y_o}{2} + 0$

Therefore,

$$A_{p,m}^{(1)} = \frac{4}{X_o \cdot Y_o} \mathcal{F}_L'(p,m) \tag{10.100}$$

where $\mathcal{F}_L'(p,m)$ is given by Equation 10.99a.

Let us multiply Equation 10.98 by $\sin\left(2N\pi \cdot \frac{y}{Y_o}\right)$ and then integrate over $-Y_o/2 < y < Y_o/2$. On setting $N = n$, this results in:

$$\text{L.H.S.} \stackrel{\text{def}}{=} \mathcal{F}_L''(p,n) = \int_{-Y_o/2}^{Y_o/2} \left\{ \int_0^{X_o} F(x,y) \cdot \sin\left(p\pi \cdot \frac{x}{2X_o}\right) \cdot dx \right\} \cdot \sin\left(2n\pi \cdot \frac{y}{Y_o}\right) \cdot dy \tag{10.101a}$$

$$(\text{R.H.S.})_1 \stackrel{\text{def}}{=} \mathcal{F}_{R1}''(p,n) = 0 \tag{10.101b}$$

$$(\text{R.H.S.})_2 \stackrel{\text{def}}{=} \mathcal{F}_{R1}''(p,n) = A_{p,n}^{(2)} \cdot \frac{X_o}{2} \cdot \frac{Y_o}{2} \tag{10.101c}$$

Thus,

$$\mathcal{F}_L''(p,n) = 0 + A_{p,n}^{(2)} \cdot \frac{X_o}{2} \cdot \frac{Y_o}{2}$$

Hence,

$$A_{p,n}^{(2)} = \frac{4}{X_o \cdot Y_o} \mathcal{F}_L''(p,m) \tag{10.102}$$

where $\mathcal{F}_L''(p,m)$ is given by Equation 10.101a.

The two coefficients in the double Fourier series of Equation 10.96 are thus given by Equations 10.100 and 10.102.

Now, in view of Equation 10.96, the particular integral for Poisson's equation, i.e. Equation 10.94, obtained is:

$$V_p = -\sum_{p\text{-odd}}^{\infty} \sin\left(p\pi \cdot \frac{x}{2X_o}\right) \cdot \left[\sum_{m\text{-odd}}^{\infty} A_{p,m}^{(1)} \cdot \frac{1}{\left(\frac{p\pi}{2X_o}\right)^2 + \left(\frac{m\pi}{Y_o}\right)^2} \cdot \cos\left(m\pi \cdot \frac{y}{Y_o}\right) \right. \tag{10.103a}$$

$$\left. + \sum_{n=1}^{\infty} A_{p,n}^{(2)} \cdot \frac{1}{\left(\frac{p\pi}{2X_o}\right)^2 + \left(\frac{2n\pi}{Y_o}\right)^2} \cdot \sin\left(2n\pi \cdot \frac{y}{Y_o}\right) \right]$$

The complementary function for Poisson's equation obtained by solving the two-dimensional Laplace equation is as follows:

$$V_c = \sum_{m\text{-odd}}^{\infty} A'_m \cdot \left(\frac{Y_o}{m\pi}\right) \cdot \cos\left(\frac{m\pi}{Y_o}\cdot y\right) \cdot \frac{\sinh\left(\frac{m\pi}{Y_o}\cdot x\right)}{\cosh\left(\frac{m\pi}{Y_o}\cdot X_o\right)}$$

$$+ \sum_{n=1}^{\infty} A''_n \cdot \left(\frac{Y_o}{2n\pi}\right) \cdot \sin\left(\frac{2n\pi}{Y_o}\cdot y\right) \cdot \frac{\sinh\left(\frac{2n\pi}{Y_o}\cdot x\right)}{\cosh\left(\frac{2n\pi}{Y_o}\cdot X_o\right)} \tag{10.103b}$$

Therefore, the complete solution for Poisson's equation is given as:

$$V(x, y) = V_c + V_p \tag{10.104}$$

Note that we have taken cognizance of the boundary conditions from the very beginning.

10.4.2 Field Problems in Cylindrical Coordinates

The expression for the Laplacian equation of V in cylindrical coordinates is reproduced here:

$$\nabla^2 V = \frac{1}{\rho}\cdot\frac{\partial}{\partial \rho}\left(\rho\frac{\partial V}{\partial \rho}\right) + \frac{1}{\rho^2}\cdot\frac{\partial^2 V}{\partial \varphi^2} + \frac{\partial^2 V}{\partial z^2} \tag{9.6}$$

This equation contains three segments which relate to three different coordinates. The following subsections contain three cases shown in Figure 10.6. Each of these cases corresponds to the combination of two of the segments. The shaded parts of

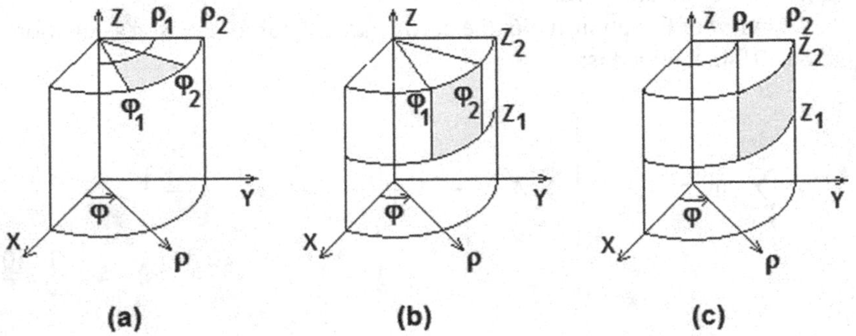

FIGURE 10.6 Regions identified by (a) ρ–φ coordinates; (b) φ–z coordinates; and (c) ρ–z coordinates.

Figure 10.6a, b, and c show regions wherein the potential is a function of ρ and φ, φ and z, and ρ and z, respectively.

10.4.2.1 Potential Function of ρ and φ

Poisson's equation:

$$\frac{1}{\rho}\cdot\frac{\partial}{\partial\rho}\left(\rho\frac{\partial V}{\partial\rho}\right)+\frac{1}{\rho^2}\cdot\frac{\partial^2 V}{\partial\varphi^2}=F(\rho,\varphi) \tag{10.105a}$$

over $\rho_2 < \rho < \rho_1$ and $\varphi_2 < \varphi < \varphi_1$.

The function $F(\rho,\varphi)$ is given in the form:

$$F(\rho,\varphi)=\left[F_1(\rho)+\frac{1}{\rho^2}\cdot F_2(\varphi)\right] \tag{10.105b}$$

where F_1 and F_2 are specified functions, each of single variable.

The complementary function for Equation 10.105a was discussed in Section 9.7.2.3, given in the initial part of this chapter. In these examples, only particular integral will be considered.

We can rewrite Equation 10.105a as:

$$\left[\frac{1}{\rho}\cdot\frac{\partial}{\partial\rho}\left(\rho\cdot\frac{\partial V}{\partial\rho}\right)-F_1(\rho)\right]+\frac{1}{\rho^2}\cdot\left[\frac{\partial^2 V}{\partial\varphi^2}-F_2(\varphi)\right]=0$$

or

$$\left[\rho\cdot\frac{\partial}{\partial\rho}\left(\rho\cdot\frac{\partial V}{\partial\rho}\right)-\rho^2\cdot F_1(\rho)\right]+\left[\frac{\partial^2 V}{\partial\varphi^2}-F_2(\varphi)\right]=0 \tag{10.106}$$

To determine the particular integral, let:

$$V_p=V(\rho,\varphi)\stackrel{\text{def}}{=}V_1(\rho)+V_2(\varphi) \tag{10.107}$$

Therefore, Equation 10.106 can be rewritten as:

$$\left[\rho\cdot\frac{d}{d\rho}\left(\rho\cdot\frac{dV_1(\rho)}{d\rho}\right)-\rho^2\cdot F_1(\rho)\right]+\left[\frac{d^2V_2(\varphi)}{d\varphi^2}-F_2(\varphi)\right]=0 \tag{10.108}$$

The first term on the LHS is a function of ρ while the second term is a function of φ. Since ρ and φ are independent variables, while the two LHS terms sum up to zero; each of these terms must be a constant. Therefore, let:

$$\rho\cdot\frac{d}{d\rho}\left(\rho\cdot\frac{dV_1(\rho)}{d\rho}\right)-\rho^2\cdot F_1(\rho)=k^2 \tag{10.109a}$$

and

$$\frac{d^2V_2(\varphi)}{d\varphi^2} - F_2(\varphi) = -k^2 \tag{10.109b}$$

where k is the constant of separation.

The above equations can be rewritten as:

$$\frac{d}{d\rho}\left(\rho \cdot \frac{dV_1(\rho)}{d\rho}\right) = \left[\rho \cdot F_1(\rho) + \frac{1}{\rho} \cdot K^2\right] \tag{10.110a}$$

And

$$\frac{d^2V_2(\varphi)}{d\varphi^2} = \left[F_2(\varphi) - K^2\right] \tag{10.110b}$$

These equations can be readily solved by repetitive integration ignoring the constants of integrations. Having thus obtained the functions, $V_1(\rho)$ and $V_2(\varphi)$, the particular integral can be found from Equation 10.107. The complete solution is obtained if the complementary function is added to the particular integral. The value for the separation constant K must be found while arbitrary constants in the complementary function are being evaluated using boundary conditions imposed on the complete solution.

10.4.2.2 Potential Function of ρ and z

The potential distribution in cylindrical coordinates is a function of Poisson's equation:

$$\frac{1}{\rho} \cdot \frac{\partial}{\partial \rho}\left(\rho \frac{\partial V}{\partial \rho}\right) + \frac{\partial^2 V}{\partial z^2} = F(\rho, z) \tag{10.111}$$

over $\rho_2 < \rho < \rho_1$ and $z_2 < z < z_1$. It is given that:

$$F(\rho, z) \stackrel{\text{def}}{=} F_1(\rho) + F_2(z) \tag{10.112}$$

Let us assume that the particular integral is given as:

$$V_p = V(\rho, z) = V_1(\rho) + V_2(z) \tag{10.113}$$

Therefore, Equation 10.111 can be rewritten as follows:

$$\frac{1}{\rho} \cdot \frac{d}{d\rho}\left(\rho \frac{dV_1(\rho)}{d\rho}\right) + \frac{d^2V_2(z)}{dz^2} = F_1(\rho) + F_2(z)$$

Thus:

$$\left[\frac{1}{\rho}\cdot\frac{d}{d\rho}\left(\rho\frac{dV_1(\rho)}{d\rho}\right)-F_1(\rho)\right]+\left[\frac{d^2V_2(z)}{dz^2}-F_2(z)\right]=0 \tag{10.114}$$

Using the constant of separation K we can split up this equation into two:

$$\left[\frac{1}{\rho}\cdot\frac{d}{d\rho}\left(\rho\frac{dV_1(\rho)}{d\rho}\right)-F_1(\rho)\right]=K^2 \tag{10.114a}$$

and

$$\left[\frac{d^2V_2(z)}{dz^2}-F_2(z)\right]=-K^2 \tag{10.114b}$$

Therefore:

$$\frac{d}{d\rho}\left(\rho\frac{dV_1(\rho)}{d\rho}\right)=\rho\cdot\left[F_1(\rho)+K^2\right] \tag{10.115a}$$

and

$$\frac{d^2V_2(z)}{dz^2}=\left[F_2(z)-K^2\right] \tag{10.115b}$$

These equations can be readily solved by repetitive integration ignoring the constants of integrations. Having thus obtained the functions, $V_1(\rho)$ and $V_2(z)$, the particular integral can be found from Equation 10.113. The complete solution is obtained if the complementary function is added to the particular integral. The value for the separation constant K must be found while arbitrary constants are being evaluated using boundary conditions imposed on the complete solution.

10.4.2.3 Potential Function of φ and z

The potential distribution in cylindrical coordinates is a function of φ and z. It satisfies Poisson's equation:

$$\frac{1}{\rho^2}\cdot\frac{\partial^2V}{\partial\varphi^2}+\frac{\partial^2V}{\partial z^2}=F(\rho,\varphi,z) \tag{10.116}$$

over $\varphi_2<\varphi<\varphi_1$ and $z_2<z<z_1$. It is given that:

$$F(\rho,\varphi,z)\stackrel{\text{def}}{=}\frac{1}{\rho^2}\cdot F_1(\varphi)+F_2(z) \tag{10.117}$$

Let us assume that the particular integral is as follows:

$$V_p = V(\varphi, z) = V_1(\varphi) + V_2(z) \tag{10.118}$$

Therefore, Equation 10.116 can be rewritten as:

$$\frac{1}{\rho^2} \cdot \frac{d^2V_1(\varphi)}{d\varphi^2} + \frac{d^2V_2(z)}{dz^2} = \frac{1}{\rho^2} \cdot F_1(\varphi) + F_2(z)$$

or

$$\frac{1}{\rho^2} \cdot \left[\frac{d^2V_1(\varphi)}{d\varphi^2} - F_1(\varphi)\right] + \left[\frac{d^2V_2(z)}{dz^2} - F_2(z)\right] = 0 \tag{10.119a}$$

The first term is a function of ρ and φ, while the second term is a function of z only, and these two terms sum to zero. This is only possible if:

$$\left[\frac{d^2V_1(\varphi)}{d\varphi^2} - F_1(\varphi)\right] = \left[\frac{d^2V_2(z)}{dz^2} - F_2(z)\right] = 0 \tag{10.119b}$$

Therefore,

$$\frac{d^2V_1(\varphi)}{d\varphi^2} = F_1(\varphi) \tag{10.120a}$$

and

$$\frac{d^2V_2(z)}{dz^2} = F_2(z) \tag{10.120b}$$

These equations can be readily solved by repetitive integration ignoring the constants of integrations. Having thus obtained the functions, $V_1(\varphi)$ and $V_2(z)$, the particular integral can be found from Equation 10.118.

$$V_p = \int\left\{\int F_1(\varphi) \cdot d\varphi\right\} \cdot d\varphi + \int\left\{\int F_2(z) \cdot dz\right\} \cdot dz \tag{10.121}$$

The value for the separation constant K is obviously zero.

10.4.3 Field Problems in Spherical Coordinates

The expression for the Laplacian of V in spherical coordinates is reproduced here:

$$\nabla^2 V = \frac{1}{r^2} \cdot \frac{\partial}{\partial r}\left(r^2 \frac{\partial V}{\partial r}\right) + \frac{1}{r^2 \sin\theta} \cdot \frac{\partial}{\partial \theta}\left(\sin\theta \frac{\partial V}{\partial \theta}\right) + \frac{1}{r^2 \sin^2\theta} \cdot \frac{\partial^2 V}{\partial \varphi^2} \tag{9.7}$$

This equation contains three segments relating to three different coordinates. The following subsections include three cases. Each of these corresponds to the combination of two of the segments. The shaded parts of Figure 10.7a, b, and c show regions wherein the potential is a function of r and θ, r and φ, and θ and φ, respectively.

10.4.3.1 Potential Function of r and θ

The potential distribution in spherical coordinates is a function of r and θ. It satisfies Poisson's equation:

$$\frac{1}{r^2}\cdot\frac{\partial}{\partial r}\left(r^2\frac{\partial V}{\partial r}\right)+\frac{1}{r^2\sin\theta}\cdot\frac{\partial}{\partial\theta}\left(\sin\theta\frac{\partial V}{\partial\theta}\right)=F(r,\theta) \tag{10.122}$$

over $r_2 < r < r_1$ and $\theta_2 < \theta < \theta_1$. It is given that:

$$F(r,\theta)\overset{\text{def}}{=}F_1(r)+\frac{1}{r^2}\cdot F_2(\theta) \tag{10.123}$$

Let the particular integral V_p be expressed as:

$$V_p = V(r,\theta)=V_1(r)+V_2(\theta) \tag{10.124}$$

Therefore, Equation 10.122 may be rewritten as:

$$\frac{1}{r^2}\cdot\frac{d}{dr}\left(r^2\frac{dV_1(r)}{dr}\right)+\frac{1}{r^2\sin\theta}\cdot\frac{d}{d\theta}\left(\sin\theta\frac{dV_2(\theta)}{d\theta}\right)=F_1(r)+\frac{1}{r^2}\cdot F_2(\theta)$$

or

$$\left[\frac{d}{dr}\left(r^2\frac{dV_1(r)}{dr}\right)-r^2\cdot F_1(r)\right]+\left[\frac{1}{\sin\theta}\cdot\frac{d}{d\theta}\left(\sin\theta\frac{dV_2(\theta)}{d\theta}\right)-F_2(\theta)\right]=0 \tag{10.125}$$

Therefore, we can split this equation into two:

$$\frac{d}{dr}\left(r^2\frac{dV_1(r)}{dr}\right)=\left\{r^2\cdot F_1(r)+K^2\right\} \tag{10.126a}$$

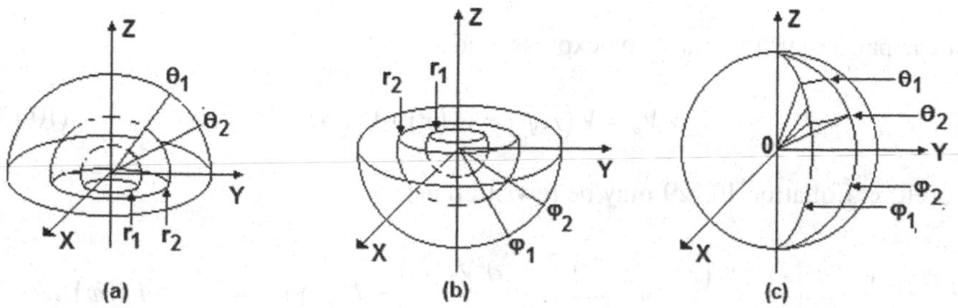

FIGURE 10.7 Regions identified by (a) r and θ; (b) r and φ; and (c) θ and φ coordinates.

and

$$\frac{d}{d\theta}\left(\sin\theta\frac{dV_2(\theta)}{d\theta}\right)=\left\{F_2(\theta)-K^2\right\}\cdot\sin\theta \tag{10.126b}$$

where K indicates the constant of separation. From Equations 10.102a and 10.102b, we get:

$$V_1(r)=\int\left[\frac{1}{r^2}\cdot\int\left\{r^2\cdot F_1(r)+K^2\right\}\cdot dr\right]\cdot dr \tag{10.127a}$$

and

$$V_2(r)=\int\left[\frac{1}{\sin\theta}\cdot\int\left\{F_2(\theta)-K^2\right\}\cdot\sin\theta\cdot d\theta\right]\cdot d\theta \tag{10.127b}$$

Therefore, in view of Equation 10.124, we have:

$$\begin{aligned}V_p&=\int\left[\frac{1}{r^2}\cdot\int\left\{r^2\cdot F_1(r)+K^2\right\}\cdot dr\right]\cdot dr\\&\quad+\int\left[\frac{1}{\sin\theta}\cdot\int\left\{F_2(\theta)-K^2\right\}\cdot\sin\theta\cdot d\theta\right]\cdot d\theta\end{aligned} \tag{10.128}$$

10.4.3.2 Potential Function of r and φ

The potential distribution in spherical coordinates is a function of r and φ. It satisfies Poisson's equation:

$$\frac{1}{r^2}\cdot\frac{\partial}{\partial r}\left(r^2\frac{\partial V}{\partial r}\right)+\frac{1}{r^2\sin^2\theta}\cdot\frac{\partial^2 V}{\partial\varphi^2}=F(r,\varphi) \tag{10.129}$$

over $r_2<r<r_1$ and $\varphi_2<\varphi<\varphi_1$. It is given that:

$$F(r,\varphi)\overset{\text{def}}{=}F_1(r)+\frac{1}{r^2\sin^2\theta}\cdot F_2(\varphi) \tag{10.130}$$

Let the particular integral V_p be expressed as:

$$V_p=V(r,\varphi)=V_1(r)+V_2(\varphi) \tag{10.131}$$

Therefore, Equation 10.129 may be rewritten as:

$$\frac{1}{r^2}\cdot\frac{d}{dr}\left(r^2\frac{dV_1(r)}{dr}\right)+\frac{1}{r^2\sin^2\theta}\cdot\frac{\partial^2 V_2(\varphi)}{\partial\varphi^2}=F_1(r)+\frac{1}{r^2\sin^2\theta}\cdot F_2(\varphi)$$

or

$$\left[\frac{d}{dr}\left(r^2\frac{dV_1(r)}{dr}\right)-r^2\cdot F_1(r)\right]+\frac{1}{r^2\sin^2\theta}\cdot\left[\frac{\partial^2 V_2(\varphi)}{\partial\varphi^2}-F_2(\varphi)\right]=0 \quad (10.132)$$

Therefore, we can split this equation into two:

$$\frac{d}{dr}\left(r^2\frac{dV_1(r)}{dr}\right)=r^2\cdot F_1(r) \quad (10.132a)$$

and

$$\frac{\partial^2 V_2(\varphi)}{\partial\varphi^2}=F_2(\varphi) \quad (10.132b)$$

The constant of separation is obviously zero. Solving Equations 10.132a and 10.132b, we get:

$$V_1(r)=\int\left[\frac{1}{r^2}\int\left\{r^2\cdot F_1(r)\right\}\cdot dr\right]\cdot dr \quad (10.133a)$$

and

$$V_2(\varphi)=\int\left[\int F_2(\varphi)\cdot d\varphi\right]\cdot d\varphi \quad (10.133b)$$

Therefore, the particular integral found using Equation 10.131 is given as:

$$V_p=\int\left[\frac{1}{r^2}\int\left\{r^2\cdot F_1(r)\right\}\cdot dr\right]\cdot dr+\int\left[\int F_2(\varphi)\cdot d\varphi\right]\cdot d\varphi \quad (10.134)$$

The complete solution is obtained if the complementary function is added to the particular integral. Arbitrary constants in the complementary function are to be evaluated using boundary conditions imposed on the complete solution.

10.4.3.3 Potential Function of θ and φ

The potential distribution in spherical coordinates is a function of θ and φ. It satisfies Poisson's equation:

$$\frac{1}{r^2\sin\theta}\cdot\frac{\partial}{\partial\theta}\left(\sin\theta\frac{\partial V}{\partial\theta}\right)+\frac{1}{r^2\sin^2\theta}\cdot\frac{\partial^2 V}{\partial\varphi^2}=F(r,\theta,\varphi) \quad (10.135)$$

over $\theta_2<\theta<\theta_1$ and $\varphi_2<\varphi<\varphi_1$. It is given that:

$$F(r,\varphi)\stackrel{\text{def}}{=}\frac{1}{r^2}\cdot F_1(\theta)+\frac{1}{r^2\sin^2\theta}\cdot F_2(\varphi) \quad (10.136)$$

Let the particular integral V_p be expressed as:

$$V_p = V(\theta,\varphi) = V_1(\theta) + V_2(\varphi) \tag{10.137}$$

Therefore, Equation 10.135 may be rewritten as:

$$\frac{1}{r^2 \sin\theta} \cdot \frac{d}{d\theta}\left(\sin\theta \frac{dV_1(\theta)}{d\theta}\right) + \frac{1}{r^2 \sin^2\theta} \cdot \frac{d^2V_2(\varphi)}{d\varphi^2}$$

$$= \frac{1}{r^2} \cdot F_1(\theta) + \frac{1}{r^2 \sin^2\theta} \cdot F_2(\varphi)$$

or

$$\frac{1}{r^2} \cdot \left[\frac{1}{\sin\theta} \cdot \frac{d}{d\theta}\left(\sin\theta \frac{dV_1(\theta)}{d\theta}\right) - F_1(\theta)\right] + \frac{1}{r^2 \sin^2\theta} \cdot \left[\frac{d^2V_2(\varphi)}{d\varphi^2} - F_2(\varphi)\right] = 0 \tag{10.138}$$

We can split this equation into two:

$$\frac{d}{d\theta}\left(\sin\theta \frac{dV_1(\theta)}{d\theta}\right) = \sin\theta \cdot F_1(\theta) \tag{10.139a}$$

$$\frac{d^2V_2(\varphi)}{d\varphi^2} = F_2(\varphi) \tag{10.139b}$$

Thus,

$$V_1(\theta) = \int\left[\frac{1}{\sin\theta} \cdot \int\{\sin\theta \cdot F_1(\theta)\} \cdot d\theta\right] \cdot d\theta \tag{10.140a}$$

and

$$V_2(\varphi) = \int\left[\int F_2(\varphi) \cdot d\varphi\right] \cdot d\varphi \tag{10.140b}$$

The constant of separation is obviously zero.

Thus, the particular integral V_p in view of Equation 10.137 is given as:

$$V_p = \int\left[\frac{1}{\sin\theta} \cdot \int\{\sin\theta \cdot F_1(\theta)\} \cdot d\theta\right] \cdot d\theta + \int\left[\int F_2(\varphi) \cdot d\varphi\right] \cdot d\varphi \tag{10.141}$$

The complete solution is obtained if the complementary function is added to the particular integral. Arbitrary constants in the complementary function are to be evaluated using boundary conditions imposed on the complete solution.

FURTHER READING

S. K. Mukerji, A. S. Khan and Y. P. Singh, *Electromagnetics for Electrical Machines*, CRC Press, New York, Mar. 2015.

David J. Griffiths, *Introduction to Electrodynamics*, PHI Learning Private Limited, New Delhi, 2012, pp. iii.

D. Polyanin, *Handbook of Linear Partial Differential Equation for Engineers and Scientists*, Chapman & Hall/CRC Press, Boca Raton, FL, 2002. ISBN 1-58488-299-9.

M. Hazewinkel (ed.), "Poisson equation", *Encyclopedia of Mathematics*, Springer, Verlag Berlin, 2001. ISBN 978-1-55608-010-4.

L. C. Evans, *Partial Differential Equations*, American Mathematical Society, Providence, RI, 1998. ISBN 0-8218-0772-2.

Naum. S. Landkof, *Foundations of Modern Potential Theory*, Springer-Verlag, Berlin, 1972 (Translated from Russian by Doohovskoy, A.P.).

Corneliu. Constantinescu, Aurel Cornea, *Potential Theory on Harmonic Spaces*, Springer, Verlag Berlin, 1972.

N. M. Günter, *Potential Theory and Its Applications to Basic Problems of Mathematical Physics*, F. Ungar, New York, 1967 (Translated from the French).

I. G. Petrovsky, *Partial Differential Equations*, W. B. Saunders, Philadelphia, PA, 1967.

A. Sommerfeld, *Partial Differential Equation in Physics*, Academic Press, New York, 1949.

Also see references given at the end of Chapter 15.

11 Steady Magnetic Fields

11.1 INTRODUCTION

So far, our study of the fields was based on the assumption that electric charges are stationary in dielectric media. Any conducting body if placed in electrostatic field, the field sets the free electrons present in the conductor to move towards the (outer) surface of the conductor. Eventually, the flow of free electrons ceases and a distribution of charge density on the conductor surface results that totally cancel the external electric field inside the conductor rendering it as an equipotential body. Alternatively, if a conductor is located in a field free region, any charge given to it from outside must entirely reside on the (outer) surface of the conductor in order to ensure that the electric field inside a conductor remains zero.

This chapter describes the magnetic phenomena associated with the steady electric currents. In contrast to the stationary charges which result in a (time-invariant) electric field, the steady flow of charges resulting in dc currents gives rise to a new form of field called the (time-invariant) magnetic field. It is assumed that all media parameters are independent of field and time.

This chapter deals with the different aspects of this field including the laws (viz. Biot-Savart's law and Ampère's law) governing its behavior, the terminology (viz. magnetic field intensity, magnetic flux density, and permeability) used to specify its various facets and the applications of governing laws referred to above.

11.2 MAGNETIC FIELD

Before dealing with the phenomenon associated with the steady electric current it is necessary to understand the meaning of magnetic field. "*Magnetic field*" is made up of two words. The first represents a cause (the magnet) and the second its effect (the field).

In Chapter 1, the field was described to be the region of influence. The phenomenon of magnetism has been known to mankind for ages. A magnet swiftly catches (or exerts a force of attraction to) the pieces of *iron* (and some other materials, known as magnetic materials) from a distance even though it possesses no tentacles. Another phenomenon of magnetism, rather lately discovered, is the force experienced by current-carrying conductors (even of non-magnetic materials), placed near a magnet. These unique phenomena motivate one to think about the possible mean through which the magnet transfers its force to the iron pieces or the current-carrying conductors. One may conceive that there is something emerging out of a magnet, which does the job. This leads to the imagination of magnetic lines of force. The magnetic phenomenon can further be visualized if we dwell in our own surroundings including our own planet. The common knowledge that Earth possesses a magnetic field indicates that its cause must be present in one form or the other, somewhere within the earth. It is also well known that a magnet contains two poles referred to as north and south poles and there are flux lines between them.

Figure 11.1a shows a rough sketch of Earth's magnetic field, representing the source as a hypothetical giant bar magnet, with its axis along the earth's spin axis. The (geographic) North Pole of Earth is near the top, whereas the South Pole is near the bottom of the diagram. The magnetic South Pole is deep in Earth's interior below Earth's geographic North Pole. The magnetic field lines come out of the earth near Antarctica and enter near Canada. The two magnetic poles are known to flip and interchange their positions. This happens once in many thousands of years. Earth's magnetic field is presumed to be produced in the outer liquid part of its core due to a self-excited dynamo, a phenomenon that produces electrical currents there. Earth's magnetic field protects Earth's surface and its inhabitants from the harmful cosmic radiations (energetic protons) from the Sun. In accordance with the right-hand rule, the positively charged particles coming from the Sun are deflected in an easterly direction.

11.2.1 Features of Magnets

The statement that Earth possesses a magnetic field further calls for an explanation of the term magnet. A magnet is a body which produces a field or region of influence outside of itself. As illustrated in Figure 11.1b, magnets always contain poles in pairs called north (N) and south (S) poles. There are no isolated magnetic poles (or magnetic charges or magnetic monopoles). Electric charges can be separated, but in the case of magnets such a separation is not possible. Like electric charges, opposite poles (N and S) attract and the like poles (N and N, or S and S) repel each other. Figure 11.1b also illustrates the flux lines between the two unlike poles, outside as well as inside the bar magnet.

Figure 11.2a illustrates a field due to two unlike aligned poles and Figure 11.2b shows a field due to two displaced unlike and non-aligned poles. Here the bar magnets are assumed to be placed perpendicular to the paper.

In physics labs the effect of a magnet is commonly demonstrated by putting it in the region sprinkled with iron filings. Figure 11.2c illustrates this effect wherein the magnetic field lines align the iron fillings in a particular pattern referred to as field lines. The pattern formed by these aligned iron filling is similar to that shown in

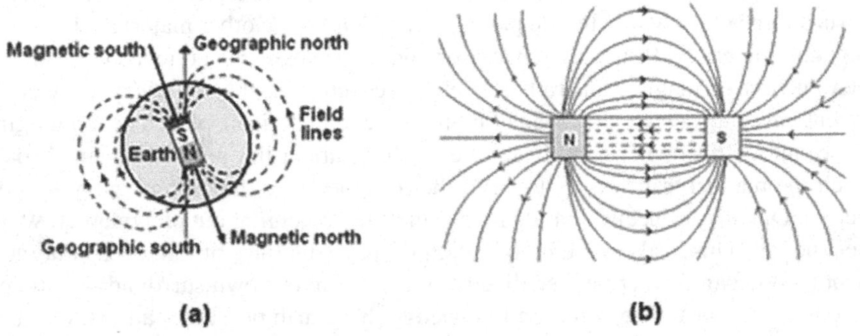

FIGURE 11.1 (a) Earth's magnetic field; (b) flux lines between the north and south poles.

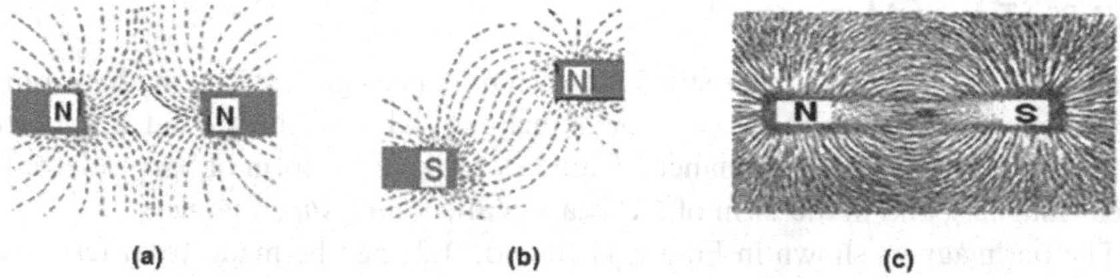

FIGURE 11.2 Flux lines between two (a) aligned, like poles; (b) non-aligned, unlike poles; (c) field lines through alignment of iron fillings.

Figure 11.1b. In these figures the bar magnets are assumed to be placed perpendicular to the paper.

The high magnetic permeability of the individual filings causes their magnetic fields to concentrate at their ends. The mutual attraction of opposite poles results in the formation of elongated clusters of filings along the field lines. These lines do not precisely represent the field lines of the magnet, as the presence of iron filings somewhat alters its magnetic field. The term permeability referred to above is the property of material which is discussed in the subsequent text.

11.2.2 Comparison of Electric and Magnetic Fields

An electric charge is an entity that emanates electric lines of force or electric flux lines. Similarly, a magnet is a body that emanates magnetic flux lines. In the case of electric charge, these flux lines emerge from positive electric charge and terminate on negative electric charge. In a *permanent* magnet, the flux lines outside the magnet emerge from north pole and terminate on the south pole. Inside the *permanent* magnet, these lines emanate from the south pole and terminate on the north pole. Thus, magnetic flux lines are in the form of closed curves. From these aspects it can be concluded that the behavior of electric and magnetic flux lines is altogether different. The basic cause of this difference can be attributed to their existence. The positive and negative electric charges can exist independently. This makes it possible that only one type of electric charge may reside inside a closed surface. On the other hand, the north and south magnetic poles cannot independently exist, and no close surface can enclose only one type of magnetic pole. This description of dissimilarity between charge and magnet leads to the following conclusions.

- The total electric flux emanating from a closed surface is equal to the total charge inside the surface. If charges of both the polarities are present inside a closed surface; the net charge will be the algebraic sum of discrete charges. In the case of distributed charges, the net charge will be the integrated value of the charge density.
- In the case of a magnet, the total flux emanating from any closed surface is always zero.

These conclusions culminate in Gauss' law, or the integral form of Maxwell's equation for and electric and magnetic fields.

11.2.3 Type of Magnets

In general, the magnets are classified into two broad categories. The magnets of the first category are referred to as permanent magnets, whereas the second are referred to as electromagnets. The permanent magnets can be in the form of bars, referred to as bar magnets, and in the form of a U shape, called *horseshoe* magnets.

The bar magnets shown in Figure 11.1b and 11.2c can be made from ferromagnetic materials (e.g. iron, nickel, or cobalt). As noted earlier there is an attractive force between unlike and repulsive force between like poles. It was also noted that the lines of the magnetic field from a bar magnet form closed lines. The configuration of field lines forming the closed paths between two magnetic poles is further shown in Figure 11.3a. By convention, the field direction is taken to be outward from North Pole and inward in South Pole. These bars remain "permanently" magnetized until the alignment of poles at micro level gets disturbed.

The horseshoe (or U-shaped) magnet is shown in Figure 11.3b. Its north and south poles are closer and lie in the same plane. The magnetic lines of force or flux lines are directed from pole to pole just like in the bar magnet. However, due to the closeness of poles, a direct path exists for the lines of flux to travel and the field is more concentrated between the poles.

Permanent magnets are not the only source of magnetic fields. In 1820, Oersted demonstrated that the current in a wire can generate a magnetic field. He observed that the magnetic field existed in coaxial circular form around the long isolated wire and that its intensity is directly proportional to the amount of current carried by the wire. As per his observation, the strength of this field was strongest in the vicinity of the wire and weakens with the distance from the wire. In most conductors, the magnetic field exists only as long as the current flows. However, in ferromagnetic materials the removal of electric current leaves behind a residual magnetic field. This phenomenon is known as "hysteresis".

Oersted noticed that the direction of the field was dependent on the direction of current in the wire. A three-dimensional (3D) magnetic field is shown in Figure 11.3c.

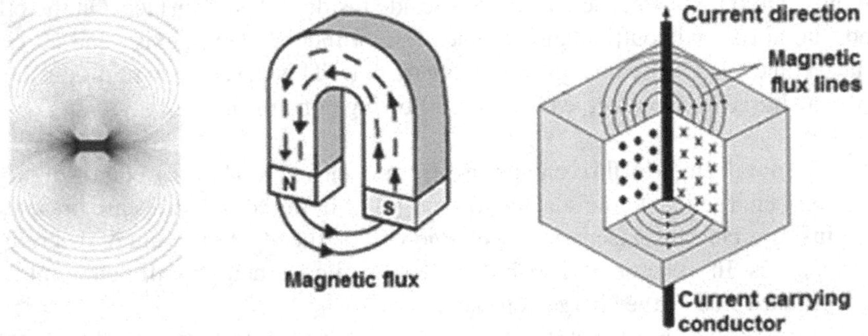

(a) (b) (c)

FIGURE 11.3 (a) Configuration of magnetic flux lines due to a bar magnet; (b) horseshoe magnet with two flux lines; (c) 3D magnetic field lines.

It also gives the idea for remembering the direction of the magnetic field around a conductor. It is called the *right-hand clasp rule*. If a person grasps a conductor in one's right hand with the thumb pointing in the direction of the current, the fingers will circle the conductor in the direction of the magnetic field. It is also known as the *right-handed corkscrew rule*, the screw moves forward if it is rotated in the clockwise direction. Thus, the flux lines are in the clockwise direction if the current is flowing in the forward direction.

11.2.4 Properties of Magnetic Lines of Force

The magnetic flux lines shown in Figure 11.1b to 11.3a possess the following properties.

- These always adopt the path of least reluctance between opposite poles. They form closed loops.
- The magnetic flux lines neither cross nor even touch one another.
- They all have the same strength.
- Their density decreases as they spread out, i.e. when they move from a region of higher permeability to that of lower permeability.
- Their density reduces with the distance from the poles.
- They are considered to have direction, the same as the direction of the magnetic flux density vector at various points on the line of force.
- These lines are directed from the south pole to the north pole within the magnet and from the north pole to the south pole outside the magnet.

11.2.5 Terms Related to Magnetic Field

After knowing the magnets, the nature of their fields and the properties of magnetic lines of force it seems to be proper to define some important quantities related to field phenomenon. These quantities, which frequently appear in the subsequent text, are described below.

11.2.5.1 Magnetic Flux

As shown in Figure 11.3a, like electric flux, the magnetic flux is also a bunch of imaginary lines. In *permanent magnets*, these are assumed to originate from the north pole (or positive polarity) and sink into the south pole (or negative polarity). An *electromagnet* is obtained by passing an electric current through a conducting coil. Its polarity depends on the direction of current flow. In electrostatics, the charge is the source of the lines of electric flux, and these lines begin and terminate on positive and negative charges respectively, no such source has ever been discovered for the lines of magnetic flux. The magnetic flux is generally denoted by a Greek letter Φ and is measured in webers. Earlier, the magnetic flux was used to be measured in Maxwell in CGS unit. Our discussion will be mostly limited to the magnetic field caused by electric currents. The mathematical aspects of magnetic flux are described in a subsequent section.

11.2.5.2 Magnetic Field Intensity

The magnetic field intensity as the term itself spell is the strength or intensity of magnetic field at a point, in a region of surface or that of a volume caused by a current distribution. The magnetic field intensity in a particular location or at a point can be obtained by applying Biot-Savart's law or Ampère's circuital law (both discussed in the subsequent sections). It is generally denoted by symbol ***H*** and has the unit of ampères per meter (A/m). Earlier the magnetic field was given in Oersted in CGS units.

11.2.5.3 Magnetic Flux Density

The magnetic flux density is the measure of concentration of flux lines around a point where the magnetic field is present. The magnetic flux density also referred to as magnetic induction is denoted by the letter B. In electrostatics, Equation 6.5 ($D = \varepsilon E$) spelled a relation between electric flux density D and electric field intensity E. A similar relation exists between magnetic flux density B and the magnetic field intensity H. It is written as:

$$B = \mu H \tag{11.1}$$

The magnetic flux density is measured in webers per square meter (Wb/m^2) or in Tesla (T) where Tesla is the SI unit. In non-SI (or CGS) units it is measured in Gauss (G), where 1 Tesla = 10^4 Gauss.

In terms of SI unit the approximate values of B for different cases are: (1) Earth's magnetic field ~12 Gauss or ~12×10^{-4} T, (2) small bar magnet ~10^{-2} T (3) MRI body scanner magnet ~2 T, (4) hair dryer ~(10^{-7}–10^{-3}) T, (4) colour TV ~10^{-6} T, (5) magnets in Physics labs up to ~50 T, (6) at a Sunspot ~0.3 T, (7) Sunlight ~3×10^{-6} T, and (8) the field at the surface of a neutron star is thought to be ~10^8 T. As reported in *Scientific American* (p. 36, February 2003), the magnetic field of some astronomical objects such as magnetars ranges from 10^8 to 10^{11} T.

11.2.5.4 Permeability

In Equation 11.1 the magnetic flux density and the magnetic field intensity, are related to each other through a media parameter called the permeability. The permeability is (generally) taken as a constant and scalar quantity. It is the property of a material that describes the ease with which a magnetic flux is established in it. It is commonly denoted by the Greek letter μ. For free space it is represented by the symbol μ_0 and its value in SI units is taken to be

$$\mu_0 = 4\pi \times 10^{-7} \text{ Tesla-meter/ampere or Henry/meter (H/m)} \tag{11.2a}$$

The permeabilities of all magnetic materials are generally expressed in terms of relative permeability (μ_r) which is defined as:

$$\mu_r = \frac{\mu}{\mu_o} \tag{11.2b}$$

For any linear, homogenous, isotropic medium, the relative permeability μ_r is a positive real quantity, which is always greater than one.

11.3 STEADY MAGNETIC FIELD LAWS

This section deals with two laws which govern the behavior of steady magnetic fields.

11.3.1 Biot-Savart's Law

Coulomb's law relates point charges approximated by small electric sources like elemental line charge ($\rho_L.dL$), elemental surface charge ($\rho_S.ds$), and elemental volume charge (($\rho.dv$), at a *source point* to the elemental electrostatic field intensity $\boldsymbol{dE}$ at a *field point*. The elemental field is due entirely to the elemental charge. Likewise, Biot-Savart's law relates elemental magnetic sources like line current element ($\boldsymbol{IdL}$), surface current element ($\boldsymbol{Kds}$), and volume current element ($\boldsymbol{J}$dv) at a *source point* to the elemental magnetostatic field intensity $\boldsymbol{dH}$ at a *field point*. The elemental field is due entirely to the elemental magnetic source. In both cases the resulting field at a given *field point* is obtained by performing certain integration involving the source and its location. This law can therefore be treated as parallel to Coulomb's law. It may, however, be noted that though the electric field intensity $\boldsymbol{E}$ is caused by *scalar sources*, the magnetic field intensity $\boldsymbol{H}$ is produced by *vector sources*.

With the knowledge of the above aspects of similarity and the properties of the magnetic field the phenomenon associated with the steady electric currents can now be explored through a law formulated by Jean-Baptiste Biot, a French physicist (1774–1862), and Félix Savart, a French professor (1791–1841), in collaboration. This law, which is stated below, is used to deal with the magnetostatic problems wherein steady currents produce time-invariant magnetic fields. This law leads to the evaluation of magnetic field intensity.

11.3.1.1 Statement of the Law

> At any point P the magnitude of the magnetic field intensity produced by the differential current element is proportional to the product of the current, the magnitude of the differential length and the sine of the angle between the filament and the line connecting the filament to the point P where the field is desired. The magnitude of the magnetic field intensity is inversely proportional to the square of the distance from the differential element to the point P. The constant of proportionality (in MKS units) is $1/4\pi$.

11.3.1.2 Mathematical Form of Biot-Savart's Law

To translate the statement of Biot-Savart's law into the mathematical language let us consider a current-carrying filamentary wire of an arbitrary geometry shown in Figure 11.4a. Since steady current can flow only in a close conducting circuit, this filament is in fact a small segment of the close circuit shown in Figure 11.4b, which also contains a source. Let I be the current in the wire and dL be an incremental length. The field due to this current is to be evaluated at point P which is located at a distance R from dL.

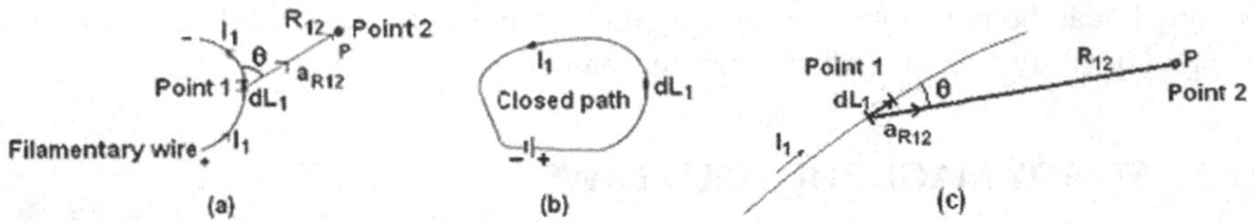

FIGURE 11.4 (a) Filametary current carying wire; (b) closed loop; (c) close-up view.

In Figure 11.4a and also in its close-up view Figure 11.4c the location of current I marked as 1 and that of point P is marked as 2. In view of these locations the current I can be replaced by I_1, dL by dL_1, R by R_{12}, and the incremental value of magnetic field intensity at P can be written as dH_2. Thus in view of Biot-Savart's law:

$$dH_2 = \frac{I_1 \cdot \boldsymbol{dL}_1 \times \boldsymbol{a}_{R12}}{4\pi \cdot R_{12}^{\ 2}} \tag{11.3}$$

According to the statement of Biot-Savart's law, the magnetic field intensity is proportional to the product of current I_1 and the differential length $\boldsymbol{dL}_1$. This law also spells about the sine of an angle θ between the elemental filament $\boldsymbol{dL}_1$ and the line R_{12} drawn from this filament (source point) to the point P (field point) at which the field is desired. The involvement of $sin\theta$ amounts to the magnitude of a vector product as $|\boldsymbol{A} \times \boldsymbol{B}| = |A||B| \sin\theta$. In view of the vector's nature, $\boldsymbol{dL}_1$ is to be assigned a direction which can be taken as the direction of the current I_1 which itself is a scalar quantity. As R_{12} represents the distance between two points, 1 and 2, this too is to be assigned a direction to make it a vector quantity. As an alternative a unit vector $\boldsymbol{a}_{R12}$ in association with the magnitude $|\boldsymbol{R}_{12}|$ can serve the purpose. This unit vector is directed from point 1 to point 2. If no suffixes are used with the locations of current, elemental length, the distance between the source point and the field point Equation 11.3 can simply be written as:

$$\boldsymbol{dH} = \frac{I \cdot \boldsymbol{dL} \times \boldsymbol{a}_R}{4\pi \cdot R^2} \tag{11.4}$$

Biot-Savart's law is sometimes referred to as Ampère's law for the current element, but this nomenclature is no longer in use so as to avoid confusion, as there is another law in the name of Ampère. Here it is pertinent to note that it is not possible to experimentally verify the expressions given by Equation 11.3 or Equation 11.4. The basic stumbling block in such verification is that no current-carrying elemental length can ever be isolated.

Since presently our study is confined to the dc current wherein the charge density is not a function of time. For such a case the continuity equation given by Equation 8.23 reduces to:

$$\nabla \bullet J = -\frac{\partial \rho}{\partial t} = 0 \tag{8.23}$$

Integration of Equation 8.23 over an arbitrary volume and the application of divergence theorem results:

$$\oiint_s \boldsymbol{J} \bullet \boldsymbol{ds} = 0 \tag{11.5}$$

where $\boldsymbol{s}$ indicates the surface of the arbitrary volume.

This equation states that the total current crossing any close surface is zero. This condition can only be met if the current is flowing around a closed path. It is this current which is flowing in a closed path shown in Figure 11.4b and the elemental length given in Figure 11.4a is simply a small segment of this closed path. Thus, if Equations 11.3 and 11.4 are integrated the resulting quantity can be experimentally verified. Such a quantity is the total magnetic field intensity caused at point P due to the entire closed path of current-carrying conductor. The integration on both sides of Equation 11.3 gives:

$$H_2 = \oint_c \frac{I_1 \cdot \boldsymbol{dL}_1 \times \boldsymbol{a}_{R12}}{4\pi \cdot R_{12}^{\ 2}} \tag{11.6}$$

where the symbol c indicates the closed path taken by the current.

In view of Equation 11.4, it can also be written as:

$$H = \oint_c \frac{I \cdot \boldsymbol{dL} \times \boldsymbol{a}_R}{4\pi \cdot R^2} \tag{11.7}$$

As noted earlier, the unit of magnetic field intensity H in view of these relations emerges as Ampères per meter (A/m).

Equations 11.6 and 11.7 are derived for current filament. In actual problems the current could be distributed over a surface or in a volume. These forms were discussed in Section 8.3.1. In view of the involvement of such currents, the above relations are modified in the following subsection.

11.3.1.3 Magnetic Field Intensities Due to Different Current Distributions

The terms related to the current distributions shown in Figure 8.2 and described in Section 8.3.1 can be related as:

$$I \cdot \boldsymbol{dL} = \boldsymbol{K} \cdot ds = \boldsymbol{J} \cdot dv \tag{11.8}$$

In Equation 11.8, it is to be noted that since I is a scalar quantity, $\boldsymbol{dL}$ is taken as a vector quantity. In the second and third segments, $\boldsymbol{K}$ and $\boldsymbol{J}$ are taken as the vector quantities, thus ds and dv have to be scalars. As required the units of all the products involved in this equation yield the same dimension i.e. A/m for the magnetic field intensity H. Equation 11.7 gives the magnetic field intensity due to the first segment of Equation 11.8. In view of the second and third segments of Equations 11.7 and 11.8 gets modified to the following forms:

For surface current density $\boldsymbol{K}$:

$$\boldsymbol{H} = \iint_{s} \frac{\boldsymbol{K} \times \boldsymbol{a_R}}{4\pi \cdot R^2} \cdot ds \tag{11.9}$$

For volumetric current density $\boldsymbol{J}$:

$$\boldsymbol{H} = \iiint_{v} \frac{\boldsymbol{J} \times \boldsymbol{a_R}}{4\pi \cdot R^2} \cdot dv \tag{11.10}$$

It may be noted that in Equation 11.9 the surface s is not *necessarily* a closed one, while in Equation 11.10 the volume v *cannot* be a closed one.

11.3.1.4 Field Due to an Infinitely Long Current Filament

To assess the usefulness of Biot-Savart's law, let us consider an infinitely long current filament shown in Figure 11.5a. This filament carries a current I ampère and is aligned along the z-axis.

Figure 11.5a shows the vector distance $\boldsymbol{R}_{12}$ between the elemental length vector $\boldsymbol{dL}$ (or source point) and the field point P. Its magnitude $|\boldsymbol{R}_{12}|$, the unit vector a_{R12}, and the elemental source vector $\boldsymbol{I.dL}$ can be written as:

$$\left.\boldsymbol{R}_{12} = \rho \cdot \boldsymbol{a}_\rho - z \cdot \boldsymbol{a}_z \quad \text{(a)} \quad |\boldsymbol{R}_{12}| = \sqrt{\rho^2 + z^2} \quad \text{(b)} \quad \boldsymbol{a}_{R12} \overset{\text{def}}{=} \frac{\boldsymbol{R}_{12}}{|\boldsymbol{R}_{12}|} = \frac{\rho \cdot \boldsymbol{a}_\rho - z \cdot \boldsymbol{a}_z}{\sqrt{\rho^2 + z^2}} \quad \text{(c)}\right\} \tag{11.11}$$

$$I \cdot \boldsymbol{dL} = I_1 \cdot \boldsymbol{dL}_1 = I \cdot dz \cdot \boldsymbol{a}_z \tag{11.12}$$

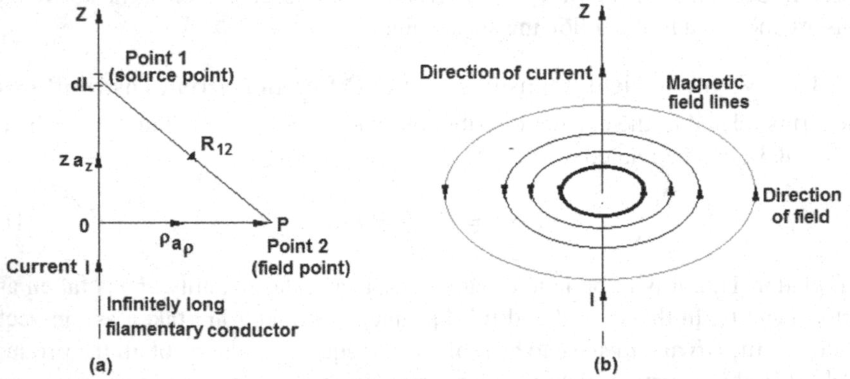

FIGURE 11.5 An infinitely long current-carrying filamentary conductor with (a) locations of source and field points (b) resulting in field lines.

The substitution of these values in Equation 11.3 gives

$$dH_2 \stackrel{\text{def}}{=} dH = \frac{I \cdot dz \cdot \boldsymbol{a}_z \times \left(\rho \cdot \boldsymbol{a}_\rho - z \cdot \boldsymbol{a}_z\right)}{4\pi \cdot \left(\rho^2 + z^2\right)^{3/2}} \tag{11.13}$$

Its integral encompassing the entire length of the wire becomes:

$$\boldsymbol{H} = \int_{-\infty}^{+\infty} \frac{I \cdot dz \cdot \boldsymbol{a}_z \times \left(\rho \cdot \boldsymbol{a}_\rho - z \cdot \boldsymbol{a}_z\right)}{4\pi \cdot \left(\rho^2 + z^2\right)^{3/2}} = \frac{I}{4\pi} \cdot \left[\int_{-\infty}^{+\infty} \frac{\rho \cdot dz}{\left(\rho^2 + z^2\right)^{3/2}}\right] \boldsymbol{a}_\varphi \tag{11.14}$$

Equation 11.16 involves a unit vector $\boldsymbol{a}_\varphi$, which changes for each value of φ. Since it is independent of ρ and z and the integration in the equation is with respect to z it has been taken out of the integral sign. Setting:

$$z \stackrel{\text{def}}{=} \rho \cdot \tan\theta \tag{11.15}$$

we have:

$$\theta = \tan^{-1}\left(z/\rho\right) = \begin{cases} \dfrac{\pi}{2} & \text{for } z = +\infty \\ -\dfrac{\pi}{2} & \text{for } z = -\infty \end{cases} \tag{11.15a}$$

$$\left(\rho^2 + z^2\right)^{3/2} = \rho^3 \cdot \left(1 + \tan^2\theta\right)^{3/2} = \left(\rho \cdot \sec\theta\right)^3 \tag{11.15b}$$

$$\text{and } dz = \rho \cdot \sec^2 d\theta \tag{11.15c}$$

Equation 11.16 can be written as:

$$\boldsymbol{H} = \frac{I}{4\pi} \cdot \left[\int_{-\infty}^{+\infty} \frac{\rho \cdot dz}{\left(\rho^2 + z^2\right)^{3/2}}\right] \boldsymbol{a}_\varphi = \frac{I}{4\pi} \cdot \left[\int_{-\pi/2}^{+\pi/2} \frac{\cos\theta}{\rho} \cdot d\theta\right] \boldsymbol{a}_\varphi = \frac{I}{2\pi \cdot \rho} \cdot \boldsymbol{a}_\varphi \tag{11.16}$$

11.3.1.5 Field Due to a Finite Current Filament

Consider Figure 11.6 wherein a straight current-carrying filamentary conductor AB of finite length lying along the z-axis is shown. This also shows angles α_1 and α_2 subtended by the lines drawn from the lower and upper ends of the conductor located at A (z_1) and B (z_2), respectively, toward the field point P, the point at which $\boldsymbol{H}$ field is to be obtained. Let $\boldsymbol{dH}$ be the contribution of an element $\boldsymbol{dL}$ located at (0, 0, z). The line joining $\boldsymbol{dL}$ and point P subtend an angle α.

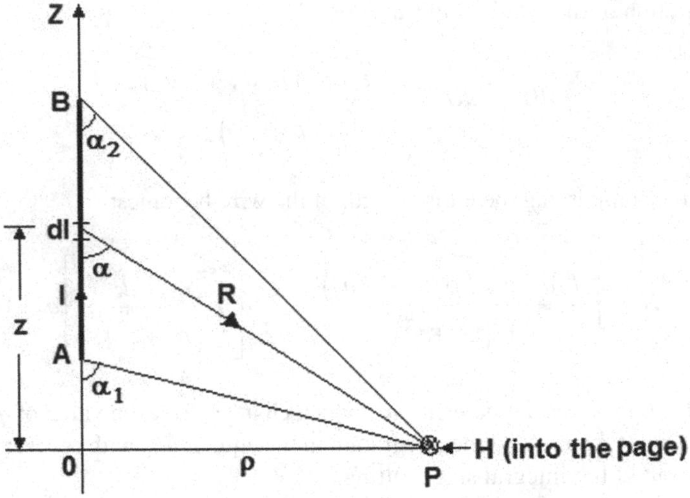

FIGURE 11.6 A finite current-carrying filamentary conductor with source and field points.

In view of Equation 11.4:

$$\boldsymbol{dH} = \frac{I \cdot \boldsymbol{dL} \times \boldsymbol{a_R}}{4\pi \cdot R^2}$$

From Figure 11.6 (similar to Equation 11.11):

$$\boldsymbol{dL} = dz\boldsymbol{a}_z,\ \boldsymbol{R} = \rho\boldsymbol{a}_\rho - z\boldsymbol{a}_z,\ |\boldsymbol{R}| = \sqrt{\rho^2 + z^2},\ \boldsymbol{a_R} = \frac{\rho\boldsymbol{a}_\rho - z\boldsymbol{a}_z}{\sqrt{\rho^2 + z^2}} \text{ and}$$

$$\boldsymbol{dL} \times \boldsymbol{a_R} = \frac{\rho dz}{\sqrt{\rho^2 + z^2}} \boldsymbol{a}_\varphi$$

$$\text{Thus } \boldsymbol{dH} = \frac{I \cdot \boldsymbol{dL} \times \boldsymbol{a_R}}{4\pi \cdot R^2} = \frac{I \cdot \dfrac{\rho dz}{\sqrt{\rho^2 + z^2}} \boldsymbol{a}_\varphi}{4\pi \cdot \left(\rho^2 + z^2\right)} = \frac{I \cdot \rho dz \boldsymbol{a}_\varphi}{4\pi \cdot \left(\rho^2 + z^2\right)^{3/2}}$$

or

$$\boldsymbol{H} = \int_{z_1}^{z_2} \frac{I \cdot \rho dz}{4\pi \cdot \left(\rho^2 + z^2\right)^{3/2}} \boldsymbol{a}_\varphi \tag{11.17a}$$

As in Equation 11.15, set $z = \rho \cot \alpha$, $dz = -\rho\, \text{cosec}^2\alpha\, d\alpha \left(\rho^2 + z^2\right)^{3/2} = \rho^3 \text{cosec}^3 \alpha$

$$\boldsymbol{H} = -\frac{I}{4\pi}\int_{\alpha_1}^{\alpha_2} \frac{\rho^2 \text{cosec}^2\alpha \, d\alpha}{\rho^3 \cos ec^3\alpha} \boldsymbol{a}_\varphi = -\frac{I}{4\pi\rho}\int_{\alpha_1}^{\alpha_2} \sin\alpha \, d\alpha \, \boldsymbol{a}_\varphi$$

or

$$\boldsymbol{H} = \frac{I}{4\pi\rho}\left(\cos\alpha_2 - \cos\alpha_1\right)\boldsymbol{a}_\varphi \tag{11.17b}$$

Case I: When the conductor is semi-infinite (with respect to *P*) so that point *A* be located at 0(0, 0, 0) and *B* at (0, 0, ∞), $\alpha_1 = 90°$, $\alpha_2 = 0°$, Equation 11.17b becomes:

$$\boldsymbol{H} = \frac{I}{4\pi\rho}\boldsymbol{a}_\varphi \tag{11.18a}$$

Case II: When the conductor is finite (with respect to *P*), i.e. point *A* is located at 0(0, 0, −∞) and *B* at (0, 0, +∞), $\alpha_1 = 180°$, $\alpha_2 = 0°$, Equation 11.17b becomes:

$$\boldsymbol{H} = \frac{I}{4\pi\rho}\boldsymbol{a}_\varphi \tag{11.18b}$$

In view of the above, Equation 11.17b reduces to Equations 11.18a and 11.18b or 11.16. Thus, expression of Equation 11.17b is applicable to any straight filamentary conductor. It is to be noted that in both Equations 11.16 and 11.17 $\boldsymbol{H}$ is always along concentric circular path or along the unit vector $\boldsymbol{a}_\varphi$ irrespective of length of the wire or the field point *P*.

Equation 11.16 (or Equations 11.18a and 11.18b) leads to some interesting conclusions. These are:

- The magnitude of the magnetic field intensity is independent of φ and z.
- It is directly proportional to the current *I*, thus more is the current more is the magnetic field intensity.
- It is inversely proportional to the distance ρ from the filament, thus farther is the location of point P from the filament weaker is the magnetic field intensity. This variation can be observed from the diminishing thickness of the lines in Figure 11.5b.
- It is symmetrical along the φ-axis for all values of ρ and z. Thus the lines drawn to depict the magnetic field intensity will be in the form of circles of different radii. These lines will encircle the filamentary current as shown in Figure 11.5b.

Here it may be observed that all the relations given by Equations 11.7, 11.9, 11.10, and 11.18 are independent of the permeability μ of the medium wherein both source points and field points are located. However, this may not be so for heterogeneous media.

11.3.2 Ampère's Circuital Law

This law is named after a French physicist and mathematician André Marie Ampère (1775–1836), who is remembered for the unit of electric current in his name. This law provides an elegant method for calculating the magnetic field but only in cases of symmetry. For non-symmetrical cases Biot-Savart's law is a better alternative. This law is a fundamental contribution to the electromagnetic theory. Ampère's circuital law (also called Ampère's work law) is stated below.

11.3.2.1 Statement of the Law

> The line integral of magnetic field intensity H about any closed path c is exactly equal to the current enclosed by that path.

This statement can be compared with that of Gauss' law given in Section 6.2. Ampère's circuital law for magnetostatic fields can be treated as parallel to Gauss' law for electrostatic fields. Both these laws give the fields in integrated form. Ampère's law involves line integration of magnetostatic field intensity ***H*** over closed contours and Gauss' law involves surface integration of electric flux density ***D*** over closed surfaces. The phrase "current enclosed by that path" in Ampère's Law is significant. The meaning of this phrase needs proper understanding to make use of this law.

Figure 11.7a shows a current-carrying conductor and its cross-section. Figure 11.7b shows that the currents flowing in all the three conductors are completely encircled/enclosed by the circular, rectangular, and arbitrary closed paths. Figure 11.7c illustrates that the currents in these conductors are only partially enclosed. The encircled parts of the currents across the conductor cross-sections are shown as shaded. In Figure 11.7d no current is enclosed by the arbitrary closed path.

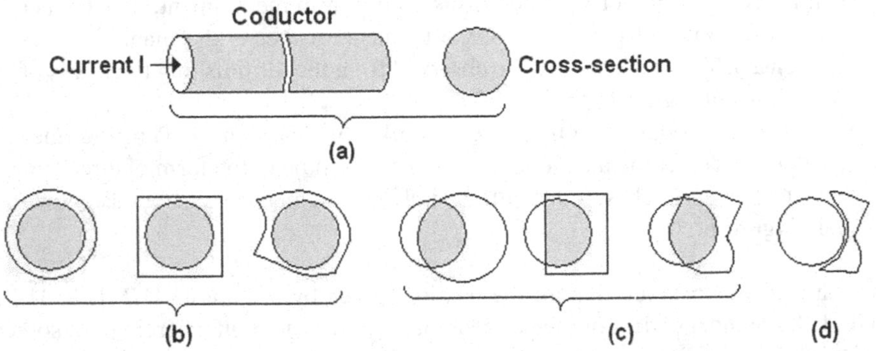

FIGURE 11.7 (a) A current-carrying conductor and its cross-section; (b) totally enclosed currents; (c) partially enclosed currents; (d) no enclosed current.

11.3.2.2 Mathematical Form of Ampère's Circuital Law

The statement of Ampère's law can be given in a simple mathematical form:

$$\oint_c \boldsymbol{H} \bullet d\ell = I \tag{11.19}$$

where the symbol c indicates the given closed path.

In view of the above description, the enclosed current does not depend on the shape or contour of the enclosing path but on the enclosed segments of the conductor cross-section and the current flowing through this cross-section. At this stage, the following two aspects pertaining to the line integral involved in Equation 11.19 need special attention.

In cases where the integrand of Equation 11.19 becomes complicated and its integration becomes difficult the close integral may be approximated by the summation. The accuracy of such approximation will depend on the number of small straight line segments used to construct the encircling contour. In such summation the dot products of the terms involved in the integrand are to be separately evaluated for each segment of the contour. Here it is to be noted that for different contours the values of the dot products at different segments are likely to differ but the overall summation of these products will yield almost the same result. Thus the total current enclosed is independent of the shape of the contour.

The selection of Gaussian surface was described in Section 6.4 for easy implementation of Equation 6.9. There it was noted that this surface has to be such that $\boldsymbol{D}$ is either normal or tangential to the Gaussian surface. As a result for normal $\boldsymbol{D}$ the dot product $\boldsymbol{D} \cdot \boldsymbol{ds}$ was simply replaced by $D \cdot ds$ and for tangential $\boldsymbol{D}$ this product was zero. Further, for D constant over the surface, it was taken out of the integral sign. Thus if the chosen surface exhibits symmetry vis-à-vis the charge distribution, the problem becomes much simplified. Similarly, in Equation 11.19 if the components ($\boldsymbol{H}$ and $\boldsymbol{dL}$) of the integrand are perpendicular to each other, the product becomes zero. In case these components are parallel $\boldsymbol{H} \bullet \boldsymbol{dL} = H \cdot dL$. In the parallel case, if $\boldsymbol{H}$ is constant at the contour it can be taken out of the integral sign. Thus with the proper selection of contour this problem also becomes much simplified.

11.3.2.3 Field Due to an Infinitely Long Conductor

With reference to the above description let us consider the filamentary conductor used in Section 11.3.1.4 and shown by Figure 11.5. As before this conductor aligned along the z-axis carries I ampère current in the positive z direction. In view of symmetry, the field due to this infinitely long filamentary conductor does not vary with φ and z. Furthermore, in Biot-Savart's Law it was noted that the field is perpendicular to the direction of the elemental length (i.e. $\boldsymbol{a}_z$) and the direction of the line drawn from this length to the field point (i.e. $\boldsymbol{a}_{R12}$). Thus, in view of Equation 11.7, the field obviously lies along the φ direction or $\boldsymbol{H}$ only has an H_φ component.

The knowledge of the component of $\boldsymbol{H}$ makes the selection of contour quite simple. Since $\boldsymbol{H}$ has only H_φ component it will be parallel to the contour of a circle for all values of φ provided the conductor passes through its center. Furthermore, the value H_φ will also be the same or constant along the perimeter of this circle. In view of this description, Equation 11.19 can be written as:

$$\oint_c \boldsymbol{H} \bullet d\boldsymbol{\ell} = \int_0^{2\pi} (H_\varphi \boldsymbol{a}_\varphi) \bullet (\rho . d\varphi \boldsymbol{a}_\varphi) = \int_0^{2\pi} H_\varphi . \rho . d\varphi$$
$$= H_\varphi \cdot \rho \cdot \int_0^{2\pi} d\varphi = H_\varphi \cdot \rho \cdot 2\pi = I \tag{11.20a}$$

or

$$H_\varphi = \frac{I}{2\pi \cdot \rho} \tag{11.20b}$$

Thus $\boldsymbol{H}$ can be written as:

$$\boldsymbol{H} = H_\varphi \boldsymbol{a}_\varphi = \frac{I}{2\pi \cdot \rho} \boldsymbol{a}_\varphi \tag{11.20c}$$

Equation 11.20c is the same as Equation 11.16 obtained from Biot-Savart's Law.

11.3.2.4 Ampère's Law in Point Form

The mathematical definition of divergence (Equation 4.20b), leads to Maxwell's equation ($\nabla \bullet \boldsymbol{D} = \rho$). This equation is also referred to as Gauss' law in point form. Similarly, the mathematical formulation of curl given in Section 4.4 leads to another Maxwell's equation, which represents Ampère's law in point form. To arrive at such a relation, some physical meaning is to be assigned to vector **F**. Thus if F is replaced by the magnetic field intensity vector H, Equations 4.25a, 4.25b, and 4.25c can be written as:

$$\oint_{1,2,3,4,1} \boldsymbol{H} \bullet d\boldsymbol{\ell} \approx \left(\frac{\partial H_z}{\partial y} - \frac{\partial H_y}{\partial z} \right) \cdot \Delta y \cdot \Delta z \tag{11.21a}$$

$$\oint_{5,6,7,8,5} \boldsymbol{H} \bullet d\boldsymbol{\ell} \approx \left(\frac{\partial H_x}{\partial z} - \frac{\partial H_z}{\partial x} \right) \cdot \Delta x \cdot \Delta z \tag{11.21b}$$

$$\oint_{9,10,11,12,9} \boldsymbol{H} \bullet d\boldsymbol{\ell} \approx \left(\frac{\partial H_y}{\partial x} - \frac{\partial H_x}{\partial y} \right) \cdot \Delta x \cdot \Delta y \tag{11.21c}$$

These expressions represent only the LHS of Ampère's law in integral form ($\oint_c \boldsymbol{H} \bullet d\boldsymbol{\ell} = I$) given by Equation 11.19 along the loops located in constant x, y, and z planes. The current I, on the RHS of Equation 11.19, can be replaced by the expression ($I = \iint_s \boldsymbol{J} \bullet d\boldsymbol{s}$) given by Equation 8.16a. In this relation, $\boldsymbol{J}$ is the vector current density, which contains three components ($\boldsymbol{J} = J_x\boldsymbol{a_x} + J_y\boldsymbol{a_y} + J_z\boldsymbol{a_z}$) along three different coordinates. Similarly the elemental surface vector $\boldsymbol{dS}$ can be resolved into three components (viz. $\Delta y \cdot \Delta z \cdot \boldsymbol{a_x}$, $\Delta z \cdot \Delta x \cdot \boldsymbol{a_y}$, $\Delta x \cdot \Delta y \cdot \boldsymbol{a_z}$) corresponding to three different planes. The product $\boldsymbol{J} \bullet \boldsymbol{ds}$ gives $J_x \cdot \Delta y \cdot \Delta z$, $J_y \cdot \Delta x \cdot \Delta z$ and $J_z \cdot \Delta x \cdot \Delta y$. Substitution of these on RHS of Equation 11.19, and equating the same with the RHS of Equations 11.21a, 11.21b, and 11.21c gives:

$$\left.\left(\frac{\partial H_z}{\partial y} - \frac{\partial H_y}{\partial z}\right) = J_x \ (a) \ \left(\frac{\partial H_x}{\partial z} - \frac{\partial H_z}{\partial x}\right) = J_y \ (b) \ \left(\frac{\partial H_y}{\partial x} - \frac{\partial H_x}{\partial y}\right) = J_z \ (c)\right\} \tag{11.22}$$

The summation of these components in view of Equation 4.29 leads to the relation:

$$\left(\frac{\partial H_z}{\partial y} - \frac{\partial H_y}{\partial z}\right)\boldsymbol{a_x} + \left(\frac{\partial H_x}{\partial z} - \frac{\partial H_z}{\partial x}\right)\boldsymbol{a_y} + \left(\frac{\partial H_y}{\partial x} - \frac{\partial H_x}{\partial y}\right)\boldsymbol{a_z} = J_x\boldsymbol{a_x} + J_y\boldsymbol{a_y} + J_z\boldsymbol{a_z} = \boldsymbol{J} \tag{11.23}$$

The LHS of Equation 11.23 represents the curl of vector $\boldsymbol{H}$ in Cartesian coordinates. The curl expressions for $\boldsymbol{H}$ in different coordinate systems can be obtained simply by replacing vector $\boldsymbol{F}$ by vector $\boldsymbol{H}$ in Equations 4.34a, 4.34b, and 4.34c. Equation 11.23 can now be written as:

$$\nabla \times \boldsymbol{H} = \boldsymbol{J} \tag{11.24}$$

Both of these laws are obtained by dealing with the divergence and curl operations in Cartesian coordinates. Equation 11.24 is also referred to as the *second Maxwell's equation* for a magnetostatic field in point form. It relates magnetic field intensity $\boldsymbol{H}$ with current density $\boldsymbol{J}$ in a steady magnetic field. This equation is obtained from Equation 11.19 ($\oint_c \boldsymbol{H} \bullet \boldsymbol{d\ell} = I$), the mathematical form of Ampère's law. Similarly, Equation 7.1 ($\oint_c \boldsymbol{E} \bullet \boldsymbol{d\ell} = 0$) leads to:

$$\nabla \times \boldsymbol{E} = \boldsymbol{0} \tag{11.25}$$

This is another Maxwell's equation (noted earlier as Equation 7.4) relates to the electrostatic field.

In the mathematical form of Stokes' theorem (Equation 4.51a) if vector **A** is replaced by the magnetic field intensity vector $\boldsymbol{H}$ it gives:

$$\iint_s (\nabla \times \boldsymbol{H}) \bullet \boldsymbol{ds} = \oint_c \boldsymbol{H} \bullet \boldsymbol{d\ell} \tag{11.26}$$

The validity of this expression is evident in view of Equations 8.15c, 11.19, and 11.24.

If the curl expression in the integrand on LHS of Equation 4.51a is replaced by another vector $\boldsymbol{B}$ it reduces to:

$$\iint_s \boldsymbol{B} \bullet d\boldsymbol{s} = \oint_c \boldsymbol{A} \bullet d\boldsymbol{\ell} \tag{11.27}$$

Where

$$\boldsymbol{B} \overset{\text{def}}{=} \nabla \times \boldsymbol{A} \tag{11.28}$$

In Equation 11.28, both the vectors $\boldsymbol{B}$ and $\boldsymbol{A}$ are of physical significance. Vector $\boldsymbol{B}$ refers to the magnetic flux density whereas $\boldsymbol{A}$ indicates a new quantity called vector magnetic potential. The magnetic flux density was discussed earlier in Section 11.2.5.3, whereas the vector magnetic potential is discussed in Section 11.5.2 in detail.

11.3.3 Another Maxwell's Equation Related to the Magnetostatic Field

The study of Ampère's law in point form, in Section 11.2.2.4, led to two of Maxwell's equations. The first one of these given by Equation 11.24 relates to the magnetostatic field while the second given by Equation 11.25 is related to electrostatic field. One more Maxwell's equation related to the magnetostatic field can be obtained through the relation of magnetic flux and magnetic flux density.

In Section 11.2.5 some terms related to the magnetic fields were described. These were the magnetic flux, magnetic field intensity, magnetic flux density, and permeability. Therefore Equation 11.1 describes the relation between magnetic field intensity and the magnetic flux density, whereas the media parameter permeability, involved in this equation was elaborated by Equation 11.2. One of these terms, magnetic flux, was only theoretically described and was not supplemented with any mathematical relation. This section first gives the mathematical relation pertaining to the flux and then extends this relation to arrive at another Maxwell's equation.

The nature of magnetic flux reveals that the magnetic flux lines or lines of force are a family of curves orthogonal to scalar magnetic equipotential surfaces. The flux Φ passing through any designated surface $\boldsymbol{s}$ is measured in webers. We can thus relate the magnetic flux with the magnetic flux density B as:

$$\Phi = \int_s \boldsymbol{B} \bullet d\boldsymbol{s} \tag{11.29a}$$

In view of this equation, as a reverse relation, the magnetic flux density $\boldsymbol{B}$ at any point P can be defined as:

$$\boldsymbol{B} \overset{\text{def}}{=} \lim_{\Delta S_n \to 0} \frac{\Delta\Phi}{\Delta S_n} \boldsymbol{a} \tag{11.29b}$$

where the small magnetic flux $\Delta\Phi$ is crossing perpendicularly to the small area ΔS_n around the point P, while the unit vector $\boldsymbol{a}$ indicates the direction of the flux line at P.

Unlike electric charges, magnetic monopoles do not exist. Therefore, we can infer that the total magnetic flux leaving a closed surface is zero. Thus:

$$\oiint_s \boldsymbol{B} \bullet \boldsymbol{ds} = 0 \tag{11.29c}$$

Using the Divergence Theorem, we get:

$$\iiint_v (\nabla \bullet \boldsymbol{B})\, dv = 0 \tag{11.29d}$$

where v is the total volume within the closed surface s.

Since the surface s is arbitrary, the integrand in Equation 11.29d must be zero, i.e.:

$$\nabla \bullet \boldsymbol{B} = 0 \tag{11.30}$$

Equation 11.30 is another Maxwell's equation related with the magnetostatic field. It needs to be stated that Maxwell retained this equation even for time-varying electromagnetic fields.

11.4 APPLICATION OF FIELD LAWS

This section deals with some applications of the two field laws discussed above. Biot-Savart's law can be used to evaluate the fields due to both symmetrical and asymmetrical configurations, whereas in the case of Ampère's law, the symmetry of the configuration is the basic requirement.

11.4.1 Field Due to a Magnetic Dipole

Figure 11.8a shows a circular loop of radius ρ with its center located at the origin and lying on the $z = 0$ plane. This loop carries a dc current I in the direction of the unit vector $\boldsymbol{a}_\varphi$. It forms a magnetic dipole. The magnitude of the dipole moment m is defined as:

$$m \overset{\text{def}}{=} \lim_{I \to \infty,\, \rho \to 0} \left(I \cdot \pi\rho^2\right) \tag{11.31}$$

The dipole moment referred to above is discussed in detail in Section 12.3.3 and is given as $m = I \cdot S$. In this relation I is the current and S is the area enclosed by the contour in which this current is flowing.

Consider a field point P_f with coordinates (r, θ, φ). Since the distribution of the magnetic field due to this dipole is independent of the φ coordinate, we may arbitrarily choose $\varphi = 0$, for the field point. The position vector $\boldsymbol{r}$, for the field point will therefore lie on the $\varphi = 0$ plane. This position vector is shown by drawing a line

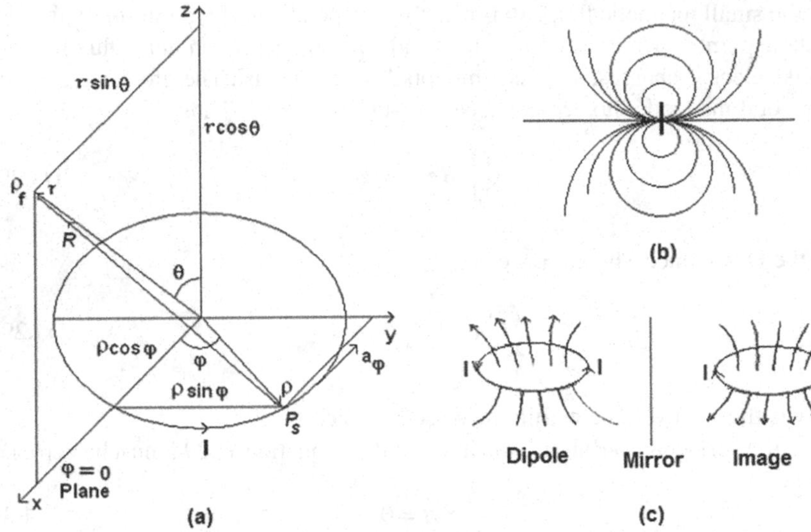

FIGURE 11.8 Magnetic dipole (a) dimensional parameters, (b) magnetic field lines due to a magnetostatic dipole pointing to the right, (c) magnetic field as a pseudovector.

segment from the origin to the point P_f, having an arrow head at P_f. The vector $\boldsymbol{r}$ subtends an angle θ with the positive z-axis. The vector $\boldsymbol{r}$ has no component in the y direction, thus it can be expressed in terms of constant unit vectors $\boldsymbol{a}_x$ and $\boldsymbol{a}_z$ as:

$$\boldsymbol{r} = r\boldsymbol{a}_r = r\left[\sin\theta\boldsymbol{a}_x + \cos\theta\boldsymbol{a}_z\right] \tag{11.32}$$

Furthermore, on the current-carrying circular loop, take another position vector $\boldsymbol{\rho}$ subtending an angle φ with the positive x-axis. This position vector indicates the position of source point P_s. Since this position vector $\boldsymbol{\rho}$ lies on the $z = 0$ plane, it has no component in the z-direction; thus, it can be expressed in terms of constant unit vectors $\boldsymbol{a}_x$ and $\boldsymbol{a}_y$ as:

$$\boldsymbol{\rho} = \rho\boldsymbol{a}_\rho = \rho\left[\cos\varphi\boldsymbol{a}_x + \sin\varphi\boldsymbol{a}_y\right] \tag{11.33}$$

Now, let the vector $\boldsymbol{R}$ indicate the distance from the source point P_s to the field point P_f, with the arrow head at the field point P_f. Therefore, we have:

$$\boldsymbol{r} = \boldsymbol{\rho} + \boldsymbol{R} \text{ or } \boldsymbol{R} = \boldsymbol{r} - \boldsymbol{\rho} \tag{11.34}$$

In view of Equations 11.32 and 11.33, Equation 11.34 becomes:

$$\boldsymbol{R} = \left[r\sin\theta - \rho\cos\varphi\right]\boldsymbol{a}_x + \left[-\rho\sin\varphi\right]\boldsymbol{a}_y + \left[r\cos\theta\right]\boldsymbol{a}_z \tag{11.35a}$$

Thus, $\left|\boldsymbol{R}\right| = \sqrt{\left[r\sin\theta - \rho\cos\varphi\right]^2 + \left[-\rho\sin\varphi\right]^2 + \left[r\cos\theta\right]^2}$

$$|\boldsymbol{R}| \stackrel{\text{def}}{=} R = \sqrt{r^2 + \rho^2 - 2r\rho \sin\theta \cos\varphi} \tag{11.35b}$$

For small loop $\rho \cong 0$ and for $r >> \rho$, Equation 11.35b becomes:

$$R \cong \sqrt{r^2 - 2r\rho \sin\theta \cos\varphi} \cong r\left[1 - 2\frac{\rho}{r}\cdot \sin\theta \cdot \cos\varphi\right] \tag{11.36}$$

In Equation 11.36, on substituting $x = 2\frac{\rho}{r}\cdot\sin\theta\cdot\cos\varphi$ we get:

$$R \cong r[1-x], \quad R^{-3} \cong r^{-3}[1-x]^{-3} \cong \frac{1}{r^3}\cdot[1+3x]$$

or

$$\frac{1}{R^3} \cong \frac{1}{r^3}\cdot\left[1 + 6\cdot\frac{\rho}{r}\cdot\sin\theta\cdot\cos\varphi\right] \tag{11.37a}$$

In view of Equation 11.4, on substituting $\boldsymbol{a}_R = \frac{\boldsymbol{R}}{|R|}$ we get:

$$d\boldsymbol{H} = \frac{I\cdot d\boldsymbol{L}\times\boldsymbol{a}_R}{4\pi\cdot R^2} = \frac{I\cdot d\boldsymbol{L}\times\boldsymbol{R}}{4\pi\cdot R^3} \tag{11.37b}$$

An elemental length ($d\boldsymbol{L}$) of the current loop can be written in vector form as:

$$d\boldsymbol{L} = (\rho\cdot d\varphi)\boldsymbol{a}_\varphi \tag{11.38}$$

Where

$$\boldsymbol{a}_\varphi = -\sin\varphi\,\boldsymbol{a}_x + \cos\varphi\,\boldsymbol{a}_y \tag{11.39}$$

Thus, in view of Equations 11.37b and 11.38, we get:

$$d\boldsymbol{H} = \frac{I\cdot d\boldsymbol{L}\times\boldsymbol{R}}{4\pi\cdot R^3} = \frac{I}{4\pi}\cdot\frac{1}{r^3}\cdot\left\{1 + 6\cdot\frac{\rho}{r}\cdot\sin\theta\cdot\cos\varphi\right\}(\rho\cdot d\varphi)(\boldsymbol{a}_\varphi\times\boldsymbol{R}) \tag{11.40}$$

while in view of Equations 11.39 and 11.35a:

$$\begin{aligned}\boldsymbol{a}_\varphi\times\boldsymbol{R} &= \left[-\sin\varphi\,\boldsymbol{a}_x + \cos\varphi\,\boldsymbol{a}_y\right]\times\left[\{r\sin\theta - \rho\cos\varphi\}\boldsymbol{a}_x - \{\rho\sin\varphi\}\boldsymbol{a}_y + \{r\cos\theta\}\boldsymbol{a}_z\right]\\ &= r\cdot\cos\theta\cdot\cos\varphi\boldsymbol{a}_x + r\cdot\cos\theta\cdot\sin\varphi\boldsymbol{a}_y + \{\rho - r\cdot\sin\theta\cdot\cos\varphi\}\,\boldsymbol{a}_z\\ &= r\cdot\{\cos\theta\cdot\cos\varphi\boldsymbol{a}_x + \cos\theta\cdot\sin\varphi\boldsymbol{a}_y - \sin\theta\cdot\cos\varphi\boldsymbol{a}_z\} + \rho\cdot\boldsymbol{a}_z\end{aligned} \tag{11.41}$$

Therefore, Equation 11.40 can be written as follows:

$$dH = \frac{I}{4\pi}\cdot\frac{\rho}{r^3}\cdot\Big[r\cdot\{\cos\theta\cdot\cos\varphi \boldsymbol{a}_x + \cos\theta\cdot\sin\varphi \boldsymbol{a}_y - \sin\theta\cdot\cos\varphi \boldsymbol{a}_z\}$$

$$+6\cdot\{(\rho^2/r)\cdot\sin\theta\cdot\cos\varphi \boldsymbol{a}_z\}\Big]d\varphi$$

$$+\frac{I}{4\pi}\cdot\frac{\rho}{r^3}\cdot\Big[6\cdot\frac{\rho}{r}\cdot\sin\theta\cdot\cos\varphi\cdot r\cdot\{\cos\theta\cdot\cos\varphi \boldsymbol{a}_x + \cos\theta\cdot\sin\varphi \boldsymbol{a}_y - \sin\theta\cdot\cos\varphi \boldsymbol{a}_z\}\Big]$$

$$d\varphi + \frac{I}{4\pi}\cdot\frac{\rho}{r^3}\cdot\rho \boldsymbol{a}_z d\varphi \tag{11.42}$$

Therefore, the magnetic field intensity $\boldsymbol{H}$at the field point P_f, due to the circular current loop is found by integrating the RHS of this equation over $\varphi = 0$ to $\varphi = 2\pi$, resulting:

$$\boldsymbol{H} = \frac{3}{2\pi}\cdot\frac{I\cdot\pi\rho^2}{r^3}\cdot\Big[\sin\theta\cdot\cos\theta \boldsymbol{a}_x - \{1-\cos^2\theta\}\boldsymbol{a}_z\Big] + \frac{1}{2\pi}\cdot\frac{I\pi\rho^2}{r^3}\boldsymbol{a}_z \tag{11.43}$$

In view of Equation 12.34b ($m = I\cdot S = \mathrm{I}\cdot\pi\rho^2$), Equation 11.43 can be written as:

$$\boldsymbol{H} = \frac{m}{4\pi r^3}\cdot\Big[3\sin\theta\cdot\cos\theta \boldsymbol{a}_x - \{1-3\cos^2\theta\}\boldsymbol{a}_z\Big] \tag{11.44}$$

Now, since we have chosen φ to be zero for the field point P_f; in the expression for the magnetic field $\boldsymbol{H}$ at this point, unit vectors of Cartesian coordinates can be replaced by unit vectors of Spherical system of space coordinates as given below:

$$\boldsymbol{a}_x = \sin\theta \boldsymbol{a}_r + \cos\theta \boldsymbol{a}_\theta \text{ (a) and } \boldsymbol{a}_z = \cos\theta \boldsymbol{a}_r - \sin\theta \boldsymbol{a}_\theta \text{ (b)}\} \tag{11.45}$$

Therefore, Equation 11.44 can be rewritten as:

$$\boldsymbol{H} = \frac{m}{4\pi r^3}\cdot\big[2\cdot\cos\theta \boldsymbol{a}_r + \sin\theta \boldsymbol{a}_\theta\big] \tag{11.46}$$

In view of Equation 11.46, the magnetic field intensity due to a *current dipole* varies as the inverse cube of the distance from the dipole. Also, the field is φ independent and has no φ-component. The resulting field due to this dipole is shown in Figure 11.8b.

Here it may be stated that Equation 11.46 can also be derived by using the vector magnetic potential (discussed in Section 11.5.2). However, this alternative method is chosen to indicate a typical application of Biot-Savart's law.

In the above example it is seen that a loop of wire carrying a current creates a magnetic field. When the wire is reflected in a mirror, this generated magnetic field is *not* reflected in the mirror as such but is reflected *and reversed*. The position of the wire and its current are (polar) vectors, so they reflect normally. Figure 11.8c shows that the magnetic field is a pseudo-vector thus changes its sign on reflection.

11.4.2 Field Due to an Infinite Current Sheet

Figure 11.9a shows a large sheet of current lying in the $z = 0$ plane. This sheet carries a uniform current density $\boldsymbol{K} = k_y \boldsymbol{a}_y$. The return path for the current may be presumed to be through two distantly located sheets each carrying half of this current. Figure 11.9b shows that one of these presumed sheets is located above and the other below this sheet.

A critical look at Figure 11.9a leads to the following conclusions:

- As the uniform current sheet is taken to be of infinite in extent along the x- and y-axes, H cannot vary with x and y coordinates.
- If this sheet is divided into a number of current filaments (along the y-axis), no filament can produce a y-component of the magnetic field (i.e. $H_y = 0$), since the current is flowing in this direction.
- In view of the Biot-Savart's law, the contribution to H_z produced by two symmetrically located pair of current filaments, at the field point, will get canceled (i.e. $H_z = 0$).
- Finally, H_x is the only component that survives.

Figure 11.9a shows a closed path 1-1′-2′-2-1, comprising straight line segments. It encloses a segment of current sheet. Using Ampère's circuital law, the current enclosed can be given as:

$$H_{x1} \cdot L + H_{x2} \cdot (-L) = k_y \cdot L \text{ (a) or } H_{x1} - H_{x2} = k_y \text{ (b)}\Big\} \tag{11.47}$$

where H_{x1} and H_{x2} indicate the x-component of the magnetic field, respectively, above and below the current sheet.

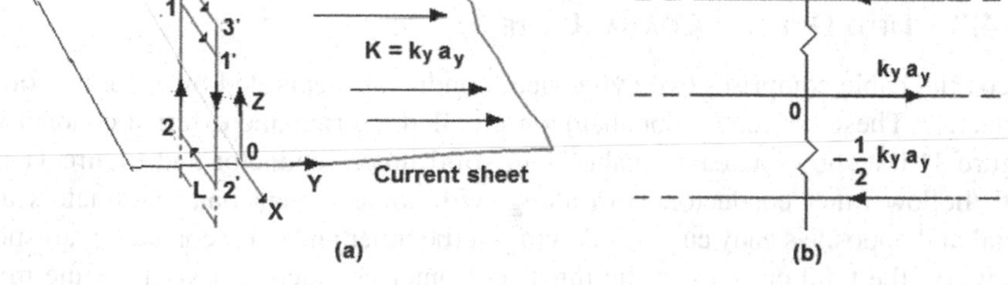

FIGURE 11.9 Infinite current sheets (a) close path comprising straight line segments; (b) current densities and the directions in the presumed current sheets.

If we select another closed path 3-3′-2′-2-3, shown in the same figure and calculate the current enclosed by this path we get:

$$H_{x3} - H_{x2} = k_y \tag{11.48a}$$

In view of Equations 11.47b and 11.48a, we have:

$$H_{x3} = H_{x1} \tag{11.48b}$$

Equation 11.48b shows that H_x above the current sheet is independent of z. Similarly, it may be seen that H_x below the sheet will remain the same for all z. However, in view of symmetry H_x on upper and lower sides of sheet will have opposite polarities. Thus, in view of Equation 11.47b:

$$H_x = \begin{cases} \frac{1}{2} \cdot k_y \ \text{ for } \ z > 0 \\ -\frac{1}{2} \cdot k_y \ \text{ for } \ z < 0 \end{cases} \tag{11.48c}$$

These components of $\boldsymbol{H}$ can be expressed in terms of a unit vector $\boldsymbol{a}_n$, which is perpendicular to the current sheet (i.e. in the positive of the z-direction). The resulting expression is:

$$\boldsymbol{H} = \frac{1}{2} \boldsymbol{K} \times \boldsymbol{a}_n \tag{11.48d}$$

where the vector surface current density is given as:

$$\boldsymbol{K} = k_y \boldsymbol{a}_y \tag{11.48e}$$

If another sheet is placed at $z = h$ with current in the opposite direction (i.e. $\boldsymbol{K} = -k_y \boldsymbol{a}_y$) Equation 11.48d leads to the following result:

$$\boldsymbol{H} = \begin{cases} \boldsymbol{K} \times \boldsymbol{a}_n = k_y \boldsymbol{a}_x & \text{for } \ 0 < z < h \\ 0 & \text{for } \ z < 0 \,\text{and}\, z > h \end{cases} \tag{11.48f}$$

11.4.3 Field Due to a Coaxial Cable

A coaxial cable comprises two cylindrical conductors separated by air or any other dielectric. These concentric conductors are called internal and external conductors. Figure 11.10a shows a coaxial cable with solid inner conductor and Figure 11.10b with hollow inner conductor, both along with some dimensional parameters and equal and opposite steady currents flowing in the inner and outer conductors, respectively. As the total currents in the inner and outer conductors have the same magnitude these will result in zero magnetic fields outside the coaxial cable. Due to

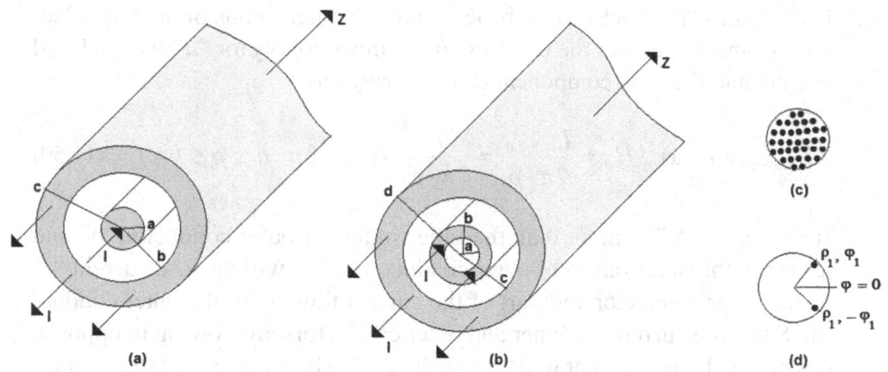

FIGURE 11.10 Coaxial cable geometry of with (a) solid inner conductor; (b) hollow inner conductor; (c) filamentary elements forming the cross-section; (d) symmetrically located filaments.

the absence of a field outside the cable, there will be no interference in the nearby electrical equipment and circuits. Anyway, the interference due to steady magnetic field is rarely of any serious matter unless the magnetic field is exceptionally strong. Cables carrying high-frequency signals are invariably coaxial with solid inner conductor. For time-varying currents the field outside the coaxial cable may not be zero, however, it could be small.

For obtaining the field due to a coaxial cable, first the components of field and their variation with the coordinates are to be ascertained. The symmetry of the problem reveals that H is not a function of φ and z. Let us assume that the solid inner conductor is composed of a large number of filamentary conductors as shown in Figure 11.10c. This assumption allows us to use Equation 11.18 to obtain the field. This also gives that no filament has a z component of $\boldsymbol{H}$.

Figure 11.10d shows two filaments symmetrically located at ρ_1, φ_1 and $\rho_1, -\varphi_1$. Any ρ component of $\boldsymbol{H}$ caused by one of these filaments will be effectively canceled by the other. If we consider similar symmetrically located filaments throughout the conductor, it will yield that $H_\rho = 0$. Thus, only the H_φ component survives, which varies with radius ρ. This component can be obtained by using Equation 11.18, in steps. These steps are required since our selected closed loop have different radii and thus will enclose different amounts of currents. In the case of a solid inner conductor, the enclosed currents and the resulting field components can be given as below:

1. If the radius (ρ) of the closed loop is selected to be less than the outer radius of inner conductor (a), the enclosed current will obviously be less than the total current I carried by the inner conductor. Thus enclosed current (I_{enclosed}) and the field component H_φ for $\rho \leq a$ can be written as:

$$\left. I_{\text{enclosed}} = I \cdot \frac{\rho^2}{a^2} \quad \text{(a)} \quad H_\varphi = \frac{I_{\text{enclosed}}}{2\pi \cdot \rho} = \frac{I \cdot \rho}{2\pi \cdot a^2} \quad \text{(b)} \right\} \tag{11.49}$$

2. If the radius "ρ" is selected to be less than the inner radius of outer conductor "b" and more than the outer radius of inner conductor "a" the enclosed current and the field component can be written as:

$$\left.I_{\text{enclosed}} = I \text{ (a)} \quad H_\varphi = \frac{I_{\text{enclosed}}}{2\pi \cdot \rho} = \frac{I}{2\pi \cdot \rho} \text{ (b)}\right\} \text{ for } a \le \rho \le b \qquad (11.50)$$

3. If the radius "ρ" is more than the inner radius of outer conductor "b" and less than the outer radius of outer conductor "c", it will enclose current "I" of the inner conductor and part of the current flowing in the outer conductor. Since the currents in inner and outer conductors are flowing in opposite directions the net current will be less than "I". The current enclosed and H_φ can be given as:

$$\left.I_{\text{enclosed}} = I \cdot \frac{c^2 - \rho^2}{c^2 - b^2} \text{ (a)} \quad H_\varphi = \frac{I}{2\pi \cdot \rho} \cdot \frac{c^2 - \rho^2}{c^2 - b^2} \text{ (b)}\right\} \text{ for } b \le \rho \le c \qquad (11.51)$$

4. If radius "ρ" of the loop is more than the outer radius of outer conductor "c" it will now enclose two equal currents flowing in opposite directions. Thus, the total enclosed current will be zero, therefore H_φ will also be zero.
5. Since H is directly related with $I_{enclosed}$, H is zero when $I_{enclosed} = 0$. If the inner conductor is hollow, $I_{enclosed} = 0$ if radius "ρ" is less than the inner radius of the inner conductor.

In view of Equations 11.49b, 11.50b, and 11.51b, the variation of H_φ can be illustrated by taking some suitable values of a, b, and c.

11.4.4 Field Due to a Solenoid

Let us consider the flat current sheet shown in Figure 11.9a to be of finite dimensions. This current sheet may be folded to form a cylindrical structure shown in Figure 11.11a. This figure also illustrates the surface current density $\boldsymbol{K}$ as it would appear after folding the sheet of Figure 11.9a after an appropriate change of coordinate axis. This configuration can be approximated by making a spring-like cylindrical structure from a wire as shown in Figure 11.11b. Both of these configurations are referred to as solenoids.

A solenoid may be termed as long (or infinite) or short (finite) in accordance with the length of its core. Figure 11.12a shows a long solenoid whereas Figure 11.12b shows short solenoid. The magnetic field in long solenoid remains almost uniformly concentrated in its center. There will be little or no divergence at its ends and almost negligible field outside. If it is infinitely long, the field calculations become easier. In a short solenoid the magnetic field lines are divergent at its ends. Thus, the shorter the lengths, the more divergent the field lines.

If the wires (of Figure 11.11b) are wrapped over a cylindrical iron core, the structure takes the form of Figure 11.12c and is referred to as an iron-cored solenoid. It is an air-cored solenoid if there is no core inside the coil. The insertion of an iron

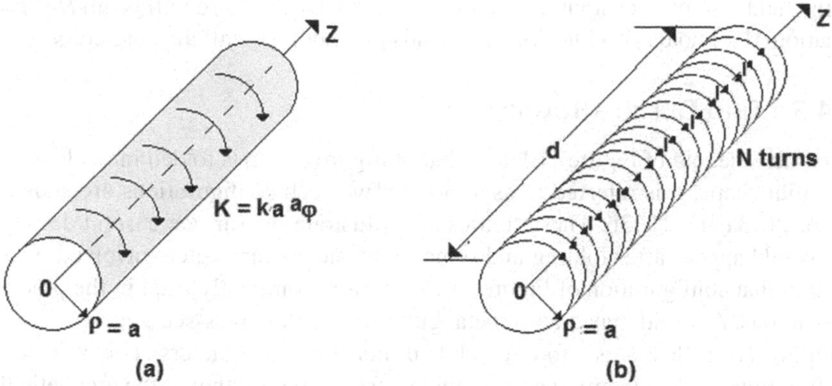

FIGURE 11.11 Solenoid made of (a) current sheet; (b) wire.

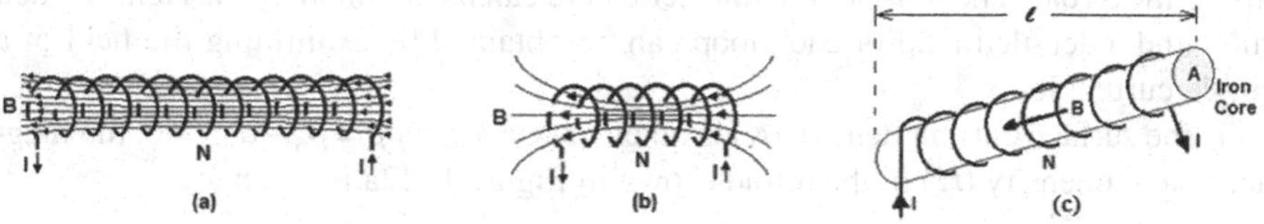

FIGURE 11.12 Field of an (a) air-cored long solenoid; (b) air-cored short solenoid; (c) iron-cored long solenoid.

core greatly enhances its magnetic field in comparison to that of the air core. The iron-cored solenoids are more frequently used as they create a strong magnetic field inside the coil. The electromagnets are usually iron-cored solenoids. It is to mention that the magnetic field produced by an electric current in a solenoid is similar to that of a bar magnet. By wrapping the same wire many times around the cylinder, the magnetic field becomes quite strong. A solenoid can also be formed of a number of rings placed closer to each other.

The magnetic field intensity (***H***) within the solenoid shown in Figure 11.11a is given as:

$$H = k_a a_z \quad (\text{for } \rho < a) \text{ (a)} \quad H = 0 \quad (\text{for } \rho > a) \text{ (b)}\Big\} \tag{11.52}$$

where a is the radius and k_a is the surface current density shown in the figure.

The magnetic field intensity (***H***) in the case shown in Figure 11.11b is given as:

$$H = \frac{NI}{d} a_z \quad (\text{for } \rho < a) \text{ (c)} \quad H = 0 \quad (\text{for } \rho > a) \text{ (d)}\Big\} \tag{11.52}$$

where I is the current and N is the number of turns in the coil.

The magnetic field inside a solenoid is proportional to both the applied current and the number of turns per unit length. There is no dependence on the diameter of the solenoid, and the field strength doesn't depend on the position inside the solenoid, i.e. the field inside is constant. In view of Equation 11.52c, there will be no H if $I = 0$. Equation 11.52 holds good for long solenoids and not for small air-core coils.

11.4.5 Field Due to a Toroid

When the solenoid of Figure 11.11a or that of Figure 11.11b is folded into a loop, like a circular shape, it is referred to as a toroid. Two such configurations are shown in Figure 11.13a and 11.13b. These figures also illustrate the surface current density $\boldsymbol{K}$ as it would appear after folding and some dimensional parameters involved. It is to mention that configuration of Figure 11.13b is more commonly used in the practical applications. A toroid may have a rectangular or circular cross-section.

Figure 11.14 illustrates a toroid and its dimensional parameters. This also shows the directions of the incoming and outgoing currents (I) and that of the magnetic flux density (B) in the circular core. The core is wound with N number turns. All loops of wire that make a toroid contribute toward the magnetic field in the same direction inside the toroid. The sense of the magnetic field can be obtained by the right-handed rule, and a detailed field of each loop can be obtained by examining the field of a single current loop.

If the surface current density (K) is taken as $K = k_a a_z$ at $\rho = \rho_0 - a$, $z = 0$, the magnetic field intensity ($\boldsymbol{H}$) in the toroid shown in Figure 11.12a is given as:

$$\left. H = k_a \frac{\rho_0 - a}{\rho} a_\varphi \;\; \left(\text{inside toroid}\right) \text{ (a)} \quad H = 0 \;\; \left(\text{outside}\right) \text{ (b)} \right\} \tag{11.53}$$

where a, ρ_0, and k_a are shown in the figure.

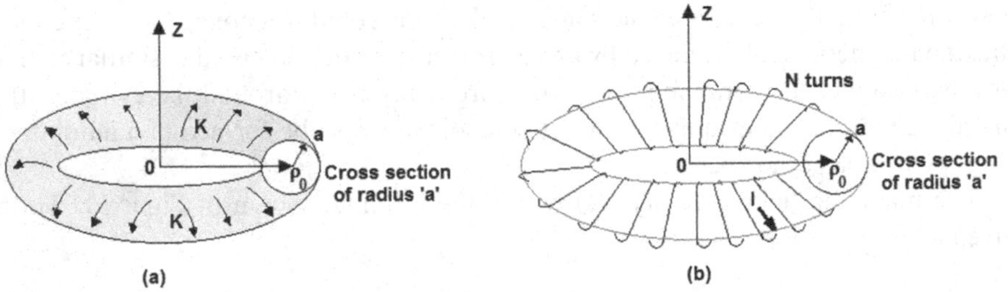

FIGURE 11.13 Toroid made of (a) current sheet; (b) wires.

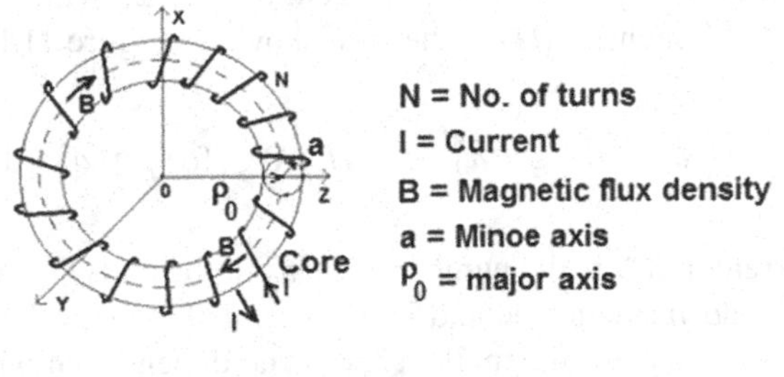

FIGURE 11.14 Toroid with dimensional and other parameters.

The magnetic field intensity (***H***) in case shown in Figure 11.13b is given as:

$$H = \frac{NI}{2\pi\rho} a_\varphi \ \left(\text{well inside}\right) \ \text{(c)} \quad H = 0 \ \left(\text{outside}\right) \ \text{(d)}\Bigg\} \tag{11.53}$$

where current *I* and the number of turns in the coil *N* are shown in the figure.

Here, too, *H* becomes zero wherever there is no current. Thus, the field outside the core of toroid is zero.

As noted earlier, the solenoid and the toroid, generally of wires, are more commonly used. Since these may be air- or iron-cored Equations 11.52c and 11.53c can be multiplied by μ or μ_0 (as the case may be) to get the flux density *B*. Thus:

$$B = \mu \frac{NI}{2\pi\rho} a_\varphi \quad \left(\text{well inside solenoid}\right) \tag{11.54a}$$

$$B = \mu \frac{NI}{d} a_z \quad (\text{for } \rho < a \text{ for Toroid}) \tag{11.54b}$$

11.5 MAGNETIC POTENTIALS

This section describes the scalar and vector magnetic potentials. The scalar magnetic potential is analogous to the scalar electric potential whereas for vector magnetic potential no such analogy exists in the electrostatic field. This is because the scalar electric potential can be defined only for source free regions (i.e. for regions free from electric charges), whereas the vector magnetic potential can be defined even for regions with current sources.

11.5.1 Scalar Magnetic Potential

On the analogy of the scalar electric potential *V* related to *E* as $\boldsymbol{E} = -\nabla V$, the scalar magnetic potential $\mathcal{V}$ is defined by the relation $\boldsymbol{H} = -\nabla \mathcal{V}$. Since this leads to the relation: $\boldsymbol{J} = \nabla \times \boldsymbol{H} = -\nabla \times \nabla \mathcal{V} \equiv 0$, the scalar magnetic potential $\mathcal{V}$ is defined only where the current density is zero (i.e. in current free regions). In those advanced magnetic problems where the current-carrying conductors occupy a relatively small fraction of the total region of interest, the concept of scalar magnetic potential with the introduction of barrier surfaces, is found to be quite useful as it satisfies the Laplace equation. The scalar magnetic potential is also applicable in the case of permanent magnets. The unit of this potential is an ampère. Unlike the scalar electric potential *V*, $\mathcal{V}$ is not always a single-valued function. Thus it does not always represent a conservative field, whereas *V* is a conservative field. The mathematical aspects of the scalar magnetic potential are discussed in the following paragraphs.

11.5.1.1 Mathematical Description

In view of Equation 11.1, for homogeneous regions Equation 11.30 can be written as:

$$\nabla \bullet \boldsymbol{H} = 0 \tag{11.55}$$

On taking curl at both sides of Equation 11.24, with $\boldsymbol{J}$ replaced by $\boldsymbol{J_o}$ and using a vector identity given by Equation 4.41d, we get for homogeneous regions:

$$\nabla \times \nabla \times \boldsymbol{H} \equiv \nabla(\nabla \bullet \boldsymbol{H}) - \nabla^2 \boldsymbol{H} = \nabla \times \boldsymbol{J_o} \tag{11.56}$$

Thus, in view of Equation 11.55, Equation 11.56 becomes:

$$\nabla^2 \boldsymbol{H} = -\nabla \times \boldsymbol{J_o} \tag{11.57}$$

11.5.1.2 Defining Relation

For a given time-invariant current density distribution $\boldsymbol{J_o}$, this Poisson's equation is to be solved for the three components of the vector $\boldsymbol{H}$. In an electrostatic field, since the curl of $\boldsymbol{E}$ is zero, we could define a scalar electric potential V. In a magnetic field (see Equation 11.24), the curl of magnetic field intensity, in general, is not zero. Therefore, scalar magnetic potential $\mathcal{V}$ can be *defined only for current free regions* as:

$$H \stackrel{\text{def}}{=} -\nabla \mathcal{V} \tag{11.58}$$

This definition identically satisfies Equation 11.24 provided the region is current-free; thus the current density $\boldsymbol{J}$ *or* $\boldsymbol{J_o}$ is zero.

11.5.1.3 Laplace's Equation

If we take divergence on both sides of Equation 11.58 and use identity given by Equation 4.40d and Equation 11.55, we get the Laplace equation:

$$\nabla^2 \mathcal{V} = 0 \tag{11.59}$$

Therefore, one may conclude that for *current-free regions* a solution of the Laplace equation gives the distribution of scalar magnetic potential, there from the magnetic field intensity in the region can be readily found by using Equation 11.59. This has simplified the problem as we have to solve for a single scalar function. However, its usefulness is limited as this approach cannot be made for current-carrying regions.

11.5.1.4 Multivalued Nature

The multivalued nature of scalar magnetic potential can be explained in view of Figure 11.15. This figure shows the cross-section of a coaxial cable with outer radii of the inner conductor "*a*", and inner and outer radii of the outer conductor "*b*" and "*c*", respectively. The inner conductor carries a current I in the positive z-direction and the outer in the negative z-direction. The field H between the two conductors obtained earlier is reproduced here in the modified form:

$$\boldsymbol{H} = \frac{I}{2\pi.\rho} \boldsymbol{a}_\varphi \quad \text{for} \quad b \geq \rho \geq a \tag{11.60}$$

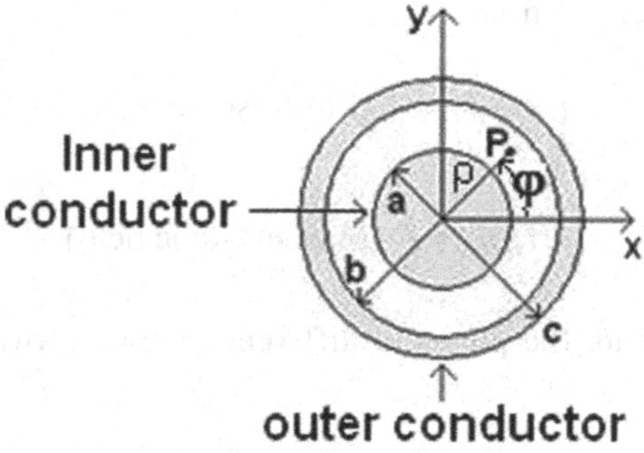

FIGURE 11.15 Cross-section of coaxial cable.

In view of Equations 11.60 and 11.59, we get:

$$\left.\frac{I}{2\pi\cdot\rho}=-\nabla\mathcal{V}\right|_{\varphi}=-\frac{1}{\rho}\cdot\frac{\partial\mathcal{V}}{\partial\varphi}\ \text{(a)}\quad\text{or}\quad\frac{\partial\mathcal{V}}{\partial\varphi}=\frac{d\mathcal{V}}{d\varphi}=-\frac{I}{2\pi}\ \text{(b)}\Bigg\}\qquad(11.61)$$

Integration of both sides of Equation 11.61b, with respect to φ gives:

$$\mathcal{V}=k-\frac{I}{2\pi}\cdot\varphi\qquad(11.62a)$$

The value of constant of integration k can be evaluated by taking $\mathcal{V}=0$ at some reference value of φ. If $\mathcal{V}$ is taken to be zero at $\varphi=0$, $k=0$ and Equation 11.62a reduces to:

$$\mathcal{V}=-\frac{I}{2\pi}\cdot\varphi\qquad(11.62b)$$

At an arbitrary point $P(\rho,\varphi)$ located between inner and outer conductors, the potential $\mathcal{V}$ can be readily obtained simply by substituting the value of φ. The constitution of Equation 11.62b, however, leads to a peculiar situation. This peculiarity can be visualized by substituting (say) $\varphi=\pi/4, 5\pi/4, 9\pi/4$, etc. All these substitutions result in different values of $\mathcal{V}$ though all these values of φ belong to the same point P. This shows that $\mathcal{V}$ is not a single-valued function.

11.5.1.5 Condition for $\mathcal{V}$ to be Single-Valued

The multivalued nature of $\mathcal{V}$ can be understood by comparing the corresponding relations belonging to electrostatic and magnetostatic fields derived earlier. These relations are:

$$\nabla\times\boldsymbol{E}=\boldsymbol{0}\ \text{(Electrostatic field)}\qquad(11.25)$$

$$\nabla\times\boldsymbol{H}=\boldsymbol{J}\ \text{(Magnetostatic field)}\qquad(11.24)$$

These relations are derived from:

$$\oint \boldsymbol{E} \bullet d\boldsymbol{\ell} = 0 \text{ (Electrostatic field)} \tag{7.1}$$

$$\oint \boldsymbol{H} \bullet d\boldsymbol{\ell} = I \text{ (Magnetostatic field)} \tag{11.19}$$

In an electrostatic field, the potential difference between two points A and B is given as:

$$V_{AB} = V_A - V_B = \int_A^B \boldsymbol{E} \bullet d\boldsymbol{\ell} \tag{7.8}$$

In view of Equation 11.19, no such relation is possible for magnetostatic field. In view of Equation 11.62b the current enclosed keeps on increasing with the value of φ. A single valued $\mathcal{V}$ can, however, be obtained by creating a barrier in the closed path at some appropriate location between A and B. The scalar magnetic potential between these points can thus be written, without crossing this barrier, as:

$$\mathcal{V}_{AB} = \int_A^B \boldsymbol{H} \bullet d\boldsymbol{\ell} \tag{11.63}$$

Table 11.1 summarizes the descriptions and the mathematical relations related with the electric and magnetic scalar potentials.

11.5.2 Vector Magnetic Potential

The vector magnetic potential $\boldsymbol{A}$ is defined by the relation $\boldsymbol{B} \stackrel{\text{def}}{=} \nabla \times \boldsymbol{A}$ (Equation 11.28). Since the divergence of the vector $\boldsymbol{D}$ is in general not zero, the vector electric

TABLE 11.1
Comparison of Electric and Magnetic Scalar Potentials

Designating Term	Scalar Electric Potential	Scalar Magnetic Potential
Symbol used	V	$\mathcal{V}$
Unit	Volt	Ampère
Defining region	With or without charge	Current free
Defining relation	$\boldsymbol{E} \stackrel{\text{def}}{=} -\nabla V$	$\boldsymbol{H} \stackrel{\text{def}}{=} -\nabla \mathcal{V}$
Equation satisfied	Poisson $(\nabla^2 V = -\rho / \varepsilon)$ Laplace $(\nabla^2 V = 0)$	Laplace $(\nabla^2 \mathcal{V} = 0)$
Nature of field	Single-valued	Multi-valued
Type of field	Conservative	Non-conservative

potential is not defined. The vector magnetic potential can be defined for both current-free and current-carrying regions. It is in general taken as the starting point in the analysis of radiation and other problems involving time variations. In current free regions it satisfies the Laplace equation whereas in regions with current density it satisfies the Poisson equation. Its unit is Ampères. The mathematical aspects of the vector magnetic potential are discussed below.

11.5.2.1 Mathematical Description

Earlier, it was noted that for *electrostatic fields* in a given region, the curl of electric field intensity is zero irrespective of the presence or absence of its source, i.e. electric charge distribution, in the region. It was this fact that prompted us to define the electrostatic potential *V*. For magnetostatic field in a given region, the curl of magnetic field intensity is zero only if the current distribution in the region is zero, i.e. the source for the field is elsewhere. This limits the utility of the magnetostatic potential $\mathcal{V}$. This calls for defining a different kind of magnetic potential that can be used regardless of the presence or absence of the current distribution in the region. We find that the divergence of magnetic flux density is zero *without any exception* (this is true even for the time-varying fields). Let us try to make use of this fact. Consider the vector identity:

$$\nabla \bullet (\nabla \times \boldsymbol{A}) \equiv 0 \tag{11.64}$$

for any arbitrary vector $\boldsymbol{A}$.

On comparing this equation with Equation 11.30 ($\nabla \bullet \boldsymbol{B} = 0$), it appears that we may define the vector magnetic potential $\boldsymbol{A}$ through Equation 11.28 ($\boldsymbol{B} \stackrel{\text{def}}{=} \nabla \times \boldsymbol{A}$), which is rewritten as:

$$\nabla \times \boldsymbol{A} \stackrel{\text{def}}{=} \boldsymbol{B} \tag{11.65a}$$

Thus, in view of Equation 11.1 ($\boldsymbol{B} = \mu \boldsymbol{H}$), we have:

$$\nabla \times \boldsymbol{A} \stackrel{\text{def}}{=} \mu \boldsymbol{H} \tag{11.65b}$$

Now, taking curl on both sides of this equation and using the identity:

$$\nabla \times \nabla \times \boldsymbol{A} \equiv \nabla(\nabla \bullet \boldsymbol{A}) - \nabla^2 \boldsymbol{A} \tag{11.66}$$

We therefore, get from Equations 11.24 and 11.65b:

$$\nabla^2 \boldsymbol{A} = \nabla(\nabla \bullet \boldsymbol{A}) - \nabla \times (\mu \boldsymbol{H}) = \nabla(\nabla \bullet \boldsymbol{A}) - \mu(\nabla \times \boldsymbol{H}) \tag{11.67a}$$

Thus

$$\nabla^2 \boldsymbol{A} = \nabla(\nabla \bullet \boldsymbol{A}) - \mu \cdot \boldsymbol{J}_o \tag{11.67b}$$

So far we have defined the vector $\boldsymbol{A}$ through Equation 11.65a. This equation, however, does not define the vector potential $\boldsymbol{A}$ uniquely. There is a vector theorem known as *Helmholtz theorem*, which states that for a unique definition of a vector quantity, it is necessary to define both its curl as well as its divergence. At this point, we reassert that the object for introducing an intermediate (vector) function $\boldsymbol{A}$, is only to make the calculation for magnetic field $\boldsymbol{B}$ or $\boldsymbol{H}$ simple. So long this vector satisfies the Equation 11.67b, we can determine the magnetic fields by using Equations 11.65a and 11.66. It is immaterial whether the solution of Equation 11.67b is unique or not. Apparently, the solution of Equation 11.67b is more involved than the original Equation 11.58. Since the unique definition of the vector $\boldsymbol{A}$ is more of a mathematical necessity, we take advantage of this by choosing the divergence of $\boldsymbol{A}$ in such a way that simplifies Equation 11.67b. At a glance, it seems to be a good idea to define:

$$\nabla \bullet \boldsymbol{A} \overset{\text{def}}{=} 0 \tag{11.68}$$

Therefore, Equation 11.67b reduces to:

$$\nabla^2 \boldsymbol{A} = -\mu \cdot \boldsymbol{J_o} \tag{11.69}$$

For current-free regions $\boldsymbol{J_o}$ is zero, thus Equation 11.69 reduces to:

$$\nabla^2 \boldsymbol{A} = 0 \tag{11.70}$$

Indeed, Equation 9.70 is slightly less involved then Equation 11.57. This intermediate vector function $\boldsymbol{A}$ defined through Equations 11.65a and 11.68, is called the *vector magnetic potential* for magnetostatic fields. The vector magnetic potential is extensively used with advantage for the determination of magnetic fields in those two-dimensional problems where only one component of this vector exists. Although this equation is developed for time-invariant fields, it is also used for slowly time-varying fields (say, at power frequencies) where reasonable approximate solutions are also acceptable. We shall see later that in the study of general time-varying electromagnetic fields (say, at high frequencies) we choose a different definition for the divergence of the vector magnetic potential.

11.6 VERIFICATION OF STEADY MAGNETIC FIELD LAWS

Earlier, it was seen that the curl of $\boldsymbol{A}$ is related to $\boldsymbol{B}$ or $\boldsymbol{H}$ through Equations 11.65a and 11.65b. Furthermore, $\boldsymbol{H}$ given by Equation 11.16 for a long current filament can be substituted in these relations. The resulting equation is:

$$\nabla \times \boldsymbol{A} = \frac{\mu_o \cdot \boldsymbol{I}}{2\pi \cdot \rho} \cdot \boldsymbol{a_\varphi} \tag{11.71}$$

Even if all the parameters on RHS of Equation 11.71 are known it is difficult to evaluate the value of $\boldsymbol{A}$. Thus, one may solve Poisson's equation 11.69 subject to specified boundary conditions. The value of $\boldsymbol{A}$ can alternatively be obtained by assuming suitable expressions which satisfy the basic magnetic field laws. Such expressions can normally be expressed in terms of line integral, surface integral, and the volume integral, all involving suitable sources. These, in view of Equation 11.8 ($\boldsymbol{I} \cdot d\boldsymbol{L} = \boldsymbol{K}ds = \boldsymbol{J}dv$), are as given below:

$$\left.\boldsymbol{A} = \int_L \frac{\mu_o \cdot \boldsymbol{I}}{4\pi \cdot R} \cdot dL \quad \text{(a)} \quad \boldsymbol{A} = \iint_s \frac{\mu_o \cdot \boldsymbol{K}}{4\pi \cdot R} \cdot ds \quad \text{(b)} \quad \boldsymbol{A} = \iiint_v \frac{\mu_o \cdot \boldsymbol{J}}{4\pi \cdot R} \cdot dv \quad \text{(c)}\right\} \tag{11.72}$$

The following two subsections are devoted to proving that the expressions given by Equations 11.72 satisfy both Biot-Savart's and Ampère's circuital laws.

11.6.1 Verification of Biot-Savart's Law

Figure 11.16 illustrates two elemental volumes ($dv_1 = dx_1.dy_1.dz_1$ and $dv_2 = dx_2.dy_2.dz_2$) both located in free space ($\mu = \mu_o$) at point $P_1(x_1, y_1, z_1)$ and $P_2(x_2, y_2, z_2)$, respectively. The presence of current density $\boldsymbol{J}_1$ at P_1 in volume dv_1 results in vector magnetic potential $d\boldsymbol{A}_2$ at P_2 in volume dv_2. In view of Equation 11.72c, the vector potential $\boldsymbol{A}_2$ due to the distributed current in the volume v_1 can be written as:

$$\boldsymbol{A}_2 = \iiint_{v_1} \frac{\mu_o \cdot \boldsymbol{J}_1}{4\pi \cdot R_{12}} \cdot dv_1 \tag{11.73}$$

In view of Equations 11.1, ($B = \mu H$) and 11.65a ($\nabla \times \boldsymbol{A} \overset{\text{def}}{=} \boldsymbol{B}$) H can be written as:

$$\boldsymbol{H} = \frac{\boldsymbol{B}}{\mu_o} = \frac{\nabla \times \boldsymbol{A}}{\mu_o} \tag{11.74}$$

Further in view of Equations 11.73 and 11.74:

$$\boldsymbol{H}_2 = \frac{\nabla_2 \times \boldsymbol{A}_2}{\mu_o} = \nabla_2 \times \iiint_{v_1} \frac{\boldsymbol{J}_1}{4\pi \cdot R_{12}} \cdot dv_1 = \frac{1}{4\pi} \iiint_{v_1} \left[\nabla_2 \times \left(\frac{\boldsymbol{J}_1}{R_{12}}\right)\right] \cdot dv_1 \tag{11.75}$$

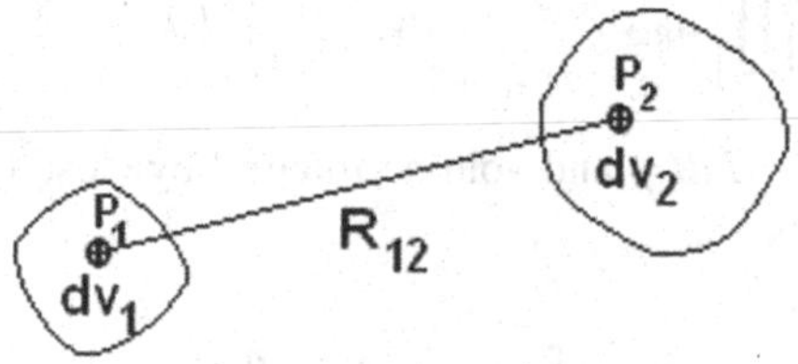

FIGURE 11.16 Two distantly located elemental volumes.

The bracketed term in Equations 11.75 contains a vector $\boldsymbol{J}_1$ and a scalar R_{12}. This equation can be modified by using the identity (Equation 4.41b), which gives the curl of the product of a scalar S and a vector $\boldsymbol{V}$:

$$\nabla\times(S\cdot\boldsymbol{V})\equiv(\nabla S)\times\boldsymbol{V}+S\cdot(\nabla\times\boldsymbol{V}) \tag{11.76}$$

The modified form of Equation 11.75 is:

$$\boldsymbol{H}_2=\frac{1}{4\pi}\iiint_{v_1}\left[\left\{\nabla_2\left(\frac{1}{R_{12}}\right)\right\}\times\boldsymbol{J}_1+\left(\frac{1}{R_{12}}\right)\cdot(\nabla_2\times\boldsymbol{J}_1)\right]\cdot dv_1 \tag{11.77}$$

In Equation 11.77, the second term becomes zero as $\boldsymbol{J}_1$ is a function of x_1, y_1, z_1, and ∇_2 of x_2, y_2, z_2. Thus, Equation 11.77 reduces to:

$$\boldsymbol{H}_2=\frac{1}{4\pi}\iiint_{v_1}\left[\left\{\nabla_2\left(\frac{1}{R_{12}}\right)\right\}\times\boldsymbol{J}_1\right]\cdot dv_1 \tag{11.78}$$

From Figure 11.15, we have:

$$\boldsymbol{R}_{12}=(x_2-x_1)\boldsymbol{a}_x+(y_2-y_1)\boldsymbol{a}_y+(z_2-z_1)\boldsymbol{a}_z \tag{11.79a}$$

$$R_{12}=|\boldsymbol{R}_{12}|=\sqrt{(x_2-x_1)^2+(y_2-y_1)^2+(z_2-z_1)^2} \tag{11.79b}$$

and

$$\boldsymbol{a}_{R12}=\frac{\boldsymbol{R}_{12}}{|\boldsymbol{R}_{12}|}=\frac{(x_2-x_1)\boldsymbol{a}_x+(y_2-y_1)\boldsymbol{a}_y+(z_2-z_1)\boldsymbol{a}_z}{\sqrt{(x_2-x_1)^2+(y_2-y_1)^2+(z_2-z_1)^2}} \tag{11.79c}$$

The gradient of $(1/R_{12})$ involved in Equation 11.78 can be written as:

$$\nabla_2\left(\frac{1}{R_{12}}\right)=-\frac{1}{R_{12}^2}\cdot(\nabla_2 R_{12})=-\frac{1}{R_{12}^2}\cdot\frac{\boldsymbol{R}_{12}}{R_{12}}=-\frac{1}{R_{12}^2}\cdot\boldsymbol{a}_{R12} \tag{11.80}$$

Substitution of Equation 11.80 in Equation 11.78 gives:

$$\boldsymbol{H}_2=-\frac{1}{4\pi}\iiint_{v_1}\left[\boldsymbol{a}_{R12}\times\frac{\boldsymbol{J}_1}{R_{12}^2}\right]\cdot dv_1=\iiint_{v_1}\left[(\boldsymbol{J}_1\cdot dv_1)\times\frac{\boldsymbol{a}_{R12}}{4\pi\cdot R_{12}^2}\right] \tag{11.81a}$$

On replacing $(\boldsymbol{J}_1\cdot dv_1)$ by $(I_1\boldsymbol{dL}_1)$ and volume integral by closed line integral over the contour c, we get:

$$\boldsymbol{H}_2=\oint_c\left[(I_1.\boldsymbol{dL}_1)\times\frac{\boldsymbol{a}_{R12}}{4\pi.R_{12}^2}\right] \tag{11.81b}$$

The mathematical relation obtained from Biot-Savart's law is reproduced here:

$$H_2 = \oint_c \left[\frac{I_1 \cdot dL_1 \times a_{R12}}{4\pi \cdot R_{12}^{\ 2}} \right] \tag{11.3}$$

As can be seen, Equation 11.81b is the same as Equation 11.3. Thus, the presumed expression for vector magnetic potential given by Equation 11.72 fully satisfies Biot-Savart's law.

11.6.2 Verification of Ampère's Circuital Law

The expression for Ampère's circuital law in point form is given as:

$$\nabla \times H = J \tag{11.24}$$

In view of Equations 11.24 and 11.65a, the LHS of Equation 11.1 can be written as:

$$\nabla \times H = \nabla \times \left(\frac{B}{\mu_o} \right) = \frac{1}{\mu_o} \cdot (\nabla \times \nabla \times A) \tag{11.82}$$

In view of the identity:

$$\nabla \times \nabla \times A \equiv \nabla(\nabla \bullet A) - \nabla^2 A \tag{11.83}$$

Equation 11.82 becomes:

$$\nabla \times H = \frac{1}{\mu_o} \cdot \left[\nabla(\nabla \bullet A) - \nabla^2 A \right] \tag{11.84}$$

The RHS of Equation 11.84 involves the divergence and the Laplacian equation of vector A. The evaluation of these carried out in view of Figure 11.15 is given below.

11.6.2.1 Evaluation of Divergence

The divergence of A_2 given by Equation 11.81 can be written as:

$$\nabla_2 \bullet A_2 = \frac{\mu_o}{4\pi} \iiint_{v1} \left[\nabla_2 \bullet \left(\frac{J_1}{R_{12}} \right) \right] dv_1 \tag{11.85}$$

In view of the identity involving a scalar function S and a vector function V:

$$\nabla \bullet (SV) \equiv V \bullet (\nabla S) + S(\nabla \bullet V) \tag{11.86}$$

Equation 11.85 can be written as:

$$\nabla_2 \bullet A_2 = \frac{\mu_o}{4\pi} \iiint_{v1} \left[J_1 \bullet \left\{ \nabla_2 \left(\frac{1}{R_{12}} \right) \right\} + \frac{1}{R_{12}} . (\nabla_2 \bullet J_1) \right] dv_1 \tag{11.87a}$$

In Equation 11.87a, the second term becomes zero as $\boldsymbol{J}_1$ is a function of (x_1, y_1, z_1) and ∇_2 of (x_2, y_2, z_2). Thus, Equation 11.87a reduces to:

$$\nabla_2 \bullet \boldsymbol{A}_2 = \frac{\mu_o}{4\pi} \iiint_{v_1} \left[\boldsymbol{J}_1 \bullet \left\{ \nabla_2 \left(\frac{1}{R_{12}} \right) \right\} \right] dv_1 \tag{11.87b}$$

The gradient term involved in Equation 11.87b is the same as given by Equation 11.80. This gradient evaluated at P_2 and is given here:

$$\nabla_2 \left(\frac{1}{R_{12}} \right) = -\frac{1}{R_{12}^{\;2}} \cdot \boldsymbol{a}_{R12} \tag{11.87c}$$

If the gradient of $(1/R_{12})$ is taken at location P_1, in view of Equation 11.79a it can be given as:

$$\nabla_1 \left(\frac{1}{R_{12}} \right) = \frac{1}{R_{12}^{\;2}} \cdot \boldsymbol{a}_{R12} \tag{11.88}$$

From Equations 11.87c and 11.88 we get:

$$\nabla_1 \left(\frac{1}{R_{12}} \right) = -\nabla_2 \left(\frac{1}{R_{12}} \right) \tag{11.89}$$

In view of Equation 11.89, Equation 11.87b becomes:

$$\nabla_2 \bullet \boldsymbol{A}_2 = -\frac{\mu_o}{4\pi} \iiint_{v_1} \left[\boldsymbol{J}_1 \bullet \left\{ \nabla_1 \left(\frac{1}{R_{12}} \right) \right\} \right] dv_1 \tag{11.90}$$

The vector identity of Equation 11.86 can be manipulated to the following form:

$$\boldsymbol{V} \bullet (\nabla S) \equiv \nabla \bullet (S\boldsymbol{V}) - S(\nabla \bullet \boldsymbol{V}) \tag{11.91}$$

In view of Equation 11.91, Equation 11.90 can be written as:

$$\nabla_2 \bullet \boldsymbol{A}_2 = \frac{\mu_o}{4\pi} \iiint_{v_1} \left[\left(\frac{1}{R_{12}} \right) \cdot (\nabla_1 \bullet \boldsymbol{J}_1) - \nabla_1 \bullet \left(\frac{\boldsymbol{J}_1}{R_{12}} \right) \right] dv_1 \tag{11.92}$$

For steady (or time-invariant) fields, the first integral term becomes zero as the divergence term is zero in view of the continuity equation (see Equation 8.23). Thus, Equation 11.92 reduces to:

$$\nabla_2 \bullet \boldsymbol{A}_2 = -\frac{\mu_o}{4\pi} \iiint_{v_1} \left[\nabla_1 \bullet \left(\frac{\boldsymbol{J}_1}{R_{12}} \right) \right] dv_1 \tag{11.93}$$

Application of the divergence theorem on the RHS of Equation 11.93 gives:

$$\nabla_2 \bullet \boldsymbol{A}_2 = -\frac{\mu_o}{4\pi} \oiint_{s_1} \left(\frac{1}{R_{12}}\right) \cdot \boldsymbol{J}_1 \bullet \boldsymbol{ds}_1 \tag{11.94a}$$

In Equation 11.94, s_1 is the surface that encloses volume v_1 involved in Equation 11.93.

The vector potential $\boldsymbol{A}_2$at the point P_2 is due to the time-invariant distributed current in the volume v_1, while all currents outside v_1 are assumed to be zero. Therefore, at every point on the boundary surface $\boldsymbol{s}_1$ the normal component of the current density $\boldsymbol{J}_1$ is zero. This results in a zero value on the RHS of Equation 11.94a. Therefore:

$$\nabla_2 \bullet \boldsymbol{A}_2 = 0 \tag{11.94b}$$

Thus, in general:

$$\nabla \bullet \boldsymbol{A} = 0 \tag{11.94c}$$

Equation 11.94b is consistent with Equation 11.68. Equation 11.84 now reduces to:

$$\nabla \times \boldsymbol{H} = -\frac{1}{\mu_o} \cdot \nabla^2 \mathbf{A} \tag{11.95}$$

Thus, for free space, in view of Equations 11.69 and 11.95, we have Ampère's circuital law in point form:

$$\nabla \times \boldsymbol{H} = \boldsymbol{J} \tag{11.24}$$

Therefore, the presumed expression for vector magnetic potential given by Equation 11.72 fully satisfies Ampère's circuital law.

11.6.2.2 Evaluation of Laplacian of Vector *A*

The Laplacian equation of a vector **A** can be expanded in Cartesian coordinates as:

$$\nabla^2 \boldsymbol{A} = \nabla^2 \left(A_x \boldsymbol{a}_x + A_y \boldsymbol{a}_y + A_z \boldsymbol{a}_z\right) = \left(\nabla^2 A_x\right) \boldsymbol{a}_x + \left(\nabla^2 A_y\right) \boldsymbol{a}_y + \left(\nabla^2 A_z\right) \boldsymbol{a}_z \tag{11.96}$$

Note that in the Cartesian system of space coordinates the unit vectors are constant quantities.

The bracketed terms indicate scalar components of the resulting vector.

Let us have a look at the similarity between Equations 7.29 and 11.72c, which are reproduced here:

$$\left.\begin{aligned} V = \iiint_v \frac{\rho}{4\pi\varepsilon_o \cdot R} \cdot dv \qquad (7.29) \qquad \boldsymbol{A} = \iiint_v \frac{\mu_o \cdot \boldsymbol{J}}{4\pi \cdot R} \cdot dv \qquad (11.72c) \end{aligned}\right\}$$

The electric potential V and other parameters involved in Equation 7.29 are all scalar quantities. In Equation 11.72a, the magnetic potential $\boldsymbol{A}$ and the current density $\boldsymbol{J}$ are the vector quantities. Since $\boldsymbol{J}$ is the cause and $\boldsymbol{A}$ is the effect J_x will result in A_x, J_y in A_y and J_z in A_z. The current density $\boldsymbol{J}$ must have all the three components to result in Equation 11.95. Since the potential V satisfies Poisson's equation ($\nabla^2 V = -\rho / \varepsilon$) the analogy demands that $\boldsymbol{J}$ must also satisfy Poisson's equation.

In view of Equations 11.83 and 11.94c, we have: $\nabla \times \boldsymbol{B} \overset{\text{def}}{=} \nabla \times \nabla \times \boldsymbol{A} \equiv \nabla(\nabla \bullet \boldsymbol{A}) - \nabla^2 \mathbf{A} = -\nabla^2 \mathbf{A}$. Also, for free space: $\nabla \times \boldsymbol{B} = \mu_o \cdot \nabla \times \boldsymbol{H} = \mu_o \cdot \boldsymbol{J}$.

Therefore:

$$\nabla^2 \boldsymbol{A} = -\mu_o \cdot \boldsymbol{J} = -\mu_o \cdot \left(J_x \boldsymbol{a}_x + J_y \boldsymbol{a}_y + J_z \boldsymbol{a}_z\right) \tag{9.97}$$

Since for time invariant fields $\boldsymbol{J} = \boldsymbol{J}_o$, Equation 11.97 is the same as Equation 11.69 derived earlier. Thus in view of Equation 11.96, Poisson's equation satisfied by different components of $\boldsymbol{A}$ can be written as:

$$\nabla^2 A_x = -\mu_o J_x \;\text{(a)} \quad \nabla^2 A_y = -\mu_o J_y \;\text{(b)} \quad \nabla^2 A_z = -\mu_o J_z \;\text{(c)}\Big\} \tag{11.98}$$

Note: A summary of some important relations is given in Table 14 of Appendix A3 in the eResources.

Example 11.1

Find the incremental contribution to the magnetic field intensity ($\Delta \boldsymbol{H}$) at the origin caused by a current element ($I\Delta \boldsymbol{L}$) in free space equal to $2\pi \boldsymbol{a}_z$ $\mu A - m$ located at (3, −4, 0).

SOLUTION

Give the source point at (3, −4, 0) and the field at (0, 0, 0).

$$\boldsymbol{R}_{12} = -3\boldsymbol{a}_x + 4\boldsymbol{a}_y \quad |\boldsymbol{R}_{12}| = 5 \quad \boldsymbol{R}_{12}^{\;2} = 25$$

$$\boldsymbol{a}_R = \boldsymbol{a}_{R12} = \frac{-3\boldsymbol{a}_x + 4\boldsymbol{a}_y}{5} = -0.6\boldsymbol{a}_x + 0.8\boldsymbol{a}_y$$

$$d\boldsymbol{H}_2 = \frac{Id\boldsymbol{L} \times \boldsymbol{a}_R}{4\pi R^2} = \frac{2\pi \boldsymbol{a}_z \times \left(-0.6\boldsymbol{a}_x + 0.8\boldsymbol{a}_y\right)}{4\pi \times 25} \times 10^{-6}$$

$$= \frac{\boldsymbol{a}_z \times \left(-0.6\boldsymbol{a}_x + 0.8\boldsymbol{a}_y\right)}{50} \times 10^{-6}$$

$$= -16a_x - 12a_y \text{ nA/m}$$

Example 11.2

A filament carrying 0.5A current passes through point (2, −4, 0). Find $\boldsymbol{H}$ at (0, 2, 0) if this filament, parallel to the z-axis, is in the interval (a) $3 \leq z \leq 3$; (b) $-\infty \leq z \leq \infty$; (c) $0 \leq z \leq \infty$; (d) $-\infty \leq z \leq 0$.

SOLUTION

The distance vector ($\rho \mathbf{a}_\rho$) between point (2, −4, 0) and the field point (0, 2, 0) is:

$$\rho \mathbf{a}_\rho = -2\mathbf{a}_x + \{2-(-4)\}\,\mathbf{a}_y = -2\mathbf{a}_x + 6\mathbf{a}_y \; |\rho| = \sqrt{40} \; \rho^2 = 40 \; IdL = 0.5\mathbf{a}_z$$

The vector ($\boldsymbol{R}_{12}$) directed from the source point $\mathbf{IdL}$ to the field point (0, 2, 0) is:

$$\boldsymbol{R}_{12} = \rho \mathbf{a}_\rho - z\mathbf{a}_z \; |R_{12}| = \sqrt{\rho^2 + z^2} \; \mathbf{a}_{R12} = \frac{\rho \mathbf{a}_\rho - z\mathbf{a}_z}{\sqrt{\rho^2 + z^2}} = \frac{-2\mathbf{a}_x + 6\mathbf{a}_y - z\mathbf{a}_z}{\sqrt{\rho^2 + z^2}}$$

$$\boldsymbol{H} = \int \frac{0.5\mathbf{a}_z \times (-2\mathbf{a}_x + 6\mathbf{a}_y - z\mathbf{a}_z)}{4\pi(\rho^2 + z^2)^{3/2}} dz = \int \frac{(-\mathbf{a}_y - 3\mathbf{a}_x)}{4\pi(\rho^2 + z^2)^{3/2}} dz$$

$$= -\int \frac{(3\mathbf{a}_x + \mathbf{a}_y)}{4\pi(\rho^2 + z^2)^{3/2}} dz$$

Let $z = \rho \cot\theta \; dz = -\rho \csc^2\theta d\theta \; \cot\theta = \dfrac{z}{\rho} \; \cos\theta = \dfrac{z}{\sqrt{\rho^2 + z^2}}$

$$\rho^2 + z^2 = \rho^2(1 + \cot^2\theta) = \rho^2 \csc^2\theta \, (\rho^2 + z^2)^{3/2} = \rho^3 \csc^3\theta$$

$$\frac{dz}{(\rho^2 + z^2)^{3/2}} = \frac{-\rho \csc^2\theta d\theta}{\rho^3 \csc^3\theta} = -\frac{\sin\theta d\theta}{\rho^2}$$

$$\boldsymbol{H} = -\int \frac{(3\mathbf{a}_x + \mathbf{a}_y)}{4\pi(\rho^2 + z^2)^{3/2}} dz = \int \frac{(3\mathbf{a}_x + \mathbf{a}_y)}{4\pi} \frac{\sin\theta d\theta}{\rho^2} = \int \frac{(3\mathbf{a}_x + \mathbf{a}_y)}{160\pi} \sin\theta d\theta$$

$$= -\frac{(3\mathbf{a}_x + \mathbf{a}_y)}{160\pi} \cos\theta \Big|_a^b$$

Limits: (for $z = -\infty \; \theta = \pi \; \cos\theta = -1$), (for $z = \infty \; \theta = 0 \; \cos\theta = 1$),

$\left(\text{for } z = 0 \; \theta = \dfrac{\pi}{2} \cos\theta = 0\right)$ and $\left(\text{for } z = \pm 3 \; \cos\theta = \dfrac{\pm 3}{\sqrt{40 + 9}} = \dfrac{\pm 3}{7}\right)$

(a) $$\boldsymbol{H} = -\frac{(\mathbf{a}_y + 3\mathbf{a}_x)}{160\pi} \cos\theta \Big|_{-3}^{3} = -\frac{(\mathbf{a}_y + 3\mathbf{a}_x)}{160\pi} \cdot \frac{6}{7} \approx -(1.7\mathbf{a}_x + 5.1\mathbf{a}_y) \text{ mA/m}$$

(b) $H = -\frac{(\mathbf{a}_x + 3\mathbf{a}_y)}{160\pi} \cos\theta\Big|_{\pi}^{0} = -2 \cdot \frac{(\mathbf{a}_x + 3\mathbf{a}_y)}{160\pi} \approx -(4\mathbf{a}_x + 12\mathbf{a}_y)\ \text{mA/m}$

(c) $H = -\frac{(\mathbf{a}_x + 3\mathbf{a}_y)}{160\pi} \cos\theta\Big|_{\pi}^{\pi/2} = \frac{(\mathbf{a}_x + 3\mathbf{a}_y)}{160\pi} \approx -(2\mathbf{a}_x + 6\mathbf{a}_y)\ \text{mA/m}$

(d) $H = -\frac{(\mathbf{a}_x + 3\mathbf{a}_y)}{160\pi} \cos\theta\Big|_{\pi/2}^{\pi} = \frac{(\mathbf{a}_x + 3\mathbf{a}_y)}{160\pi} \approx (2\mathbf{a}_x + 6\mathbf{a}_y)\ \text{mA/m}$

Example 11.3

Find the magnetic field intensity (H) everywhere caused by two uniformly distributed current sheets with the current density $K_0\mathbf{a}_y$, on the surface $z = 0$, $y > 0$, and $-K_0\mathbf{a}_z$ on the surface $y = 0$, $z > 0$, where K_0 is a constant.

SOLUTION

The given current density distributions on two specified sheets are shown in Figure 11.17.

In view of Equation 11.48c, ***H*** obtained for different regions is given here:

_H_ due to sheet (i)

On the right side of the sheet:

$$-\frac{1}{2}K_0\mathbf{a}_z \times \mathbf{a}_y = \frac{1}{2}K_0\mathbf{a}_x \quad \text{for} \quad 0 < z < \infty \quad \text{and} \quad = 0 \quad \text{for} \quad 0 > z > -\infty$$

On the left side of the sheet:

$$-\frac{1}{2}K_0\mathbf{a}_z \times -\mathbf{a}_y = -\frac{1}{2}K_0\mathbf{a}_x \quad \text{for} \quad 0 < z < \infty \quad \text{and} \quad = 0 \quad \text{for} \quad 0 > z > -\infty$$

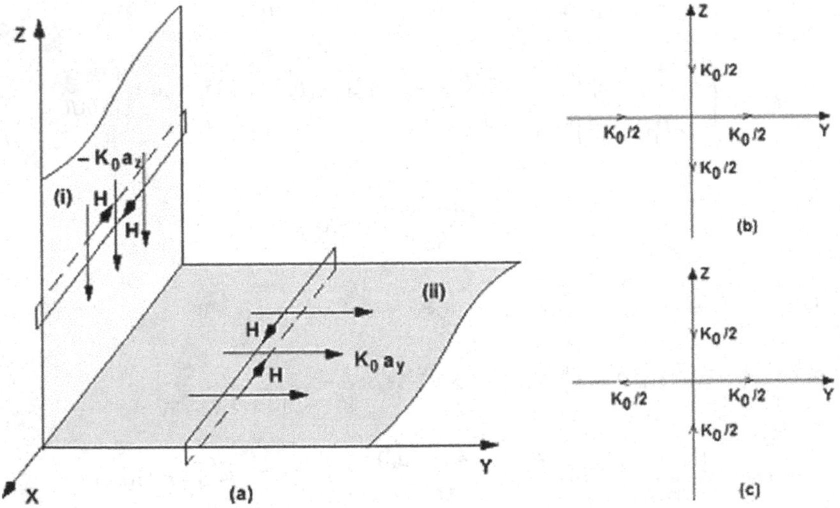

FIGURE 11.17 Two uniformly distributed current sheets.

***H* due to sheet (ii)**

Above the sheet:

$$\frac{1}{2}K_0\mathbf{a}_y \times \mathbf{a}_z = \frac{1}{2}K_0\mathbf{a}_x \text{ for } 0 < y < \infty \text{ and } = 0 \text{ for } 0 > y > -\infty$$

Below the sheet:

$$\frac{1}{2}K_0\mathbf{a}_y \times -\mathbf{a}_z = -\frac{1}{2}K_0\mathbf{a}_x \text{ for } 0 > y > \infty \text{ and } = 0 \text{ for } 0 > y > -\infty$$

Thus the field component H_x in different quadrants due to first, second, and both of the currents sheets are:

Quadrant	I Current Sheet	II Current Sheet	Both Sheets
I $0 > y > \infty, 0 < z < \infty$	$\frac{1}{2}K_0$	$\frac{1}{2}K_0$	K_0
II $0 < y < -\infty, 0 < z < \infty$	$-\frac{1}{2}K_0$	0	$-\frac{1}{2}K_0$
III $0 < y < -\infty, 0 > z > -\infty$	0	0	0
IV $0 > y > \infty, 0 > z > -\infty$	0	$-\frac{1}{2}K_0$	$-\frac{1}{2}K_0$

Example 11.4

Find $\boldsymbol{H}$ in rectangular components at $P(0, 0.008, 0.1)$ in the field of two infinite current filaments carrying currents $I_1 = 50$ mA on z-axis in the $-\mathbf{a}_z$ direction and $I_2 = 25$ mA at $x = 0$, $y = 0.01$, in the $+\mathbf{a}_z$ direction.

SOLUTION

Let $\boldsymbol{R}_1$ be the distance vector from line 1 (at 0, 0, 0) to the field point at (0, 0.008, 0.1) and $\boldsymbol{R}_2$ the distance vector from line 2 (at 0, 0.01, 0) to the same field point. In view of Figure 11.18:

$$\boldsymbol{R}_1 = 0.008\mathbf{a}_y \quad \mathbf{a}_{R1} = \mathbf{a}_y \quad \boldsymbol{R}_2 = -0.002\mathbf{a}_y \quad \mathbf{a}_{R2} = -\mathbf{a}_y \quad \mathbf{a}_\varphi = \mathbf{a}_y$$

In view of Equation 11.4, $\boldsymbol{H}$ at the field point is:

$$\boldsymbol{H} = \frac{(I_1\mathbf{a}_z)\times\mathbf{a}_y}{2\pi R_1^2} + \frac{(I_2\mathbf{a}_z)\times(-\mathbf{a}_y)}{2\pi R_2^2} = \frac{(50\times10^{-3})\mathbf{a}_x}{2\pi\times(8\times10^{-3})^2} + \frac{(-25\times10^{-3})\mathbf{a}_x}{2\pi\times(2\times10^{-3})^2}$$

$$= \frac{(50\times10^{-3})\mathbf{a}_x}{2\pi\times64\times10^{-6}} - \frac{(25\times10^{-3})\mathbf{a}_x}{2\pi\times4\times10^{-6}}$$

$$= (123.449 - 994.178)\mathbf{a}_x = 870.729\mathbf{a}_x \text{ A/m}$$

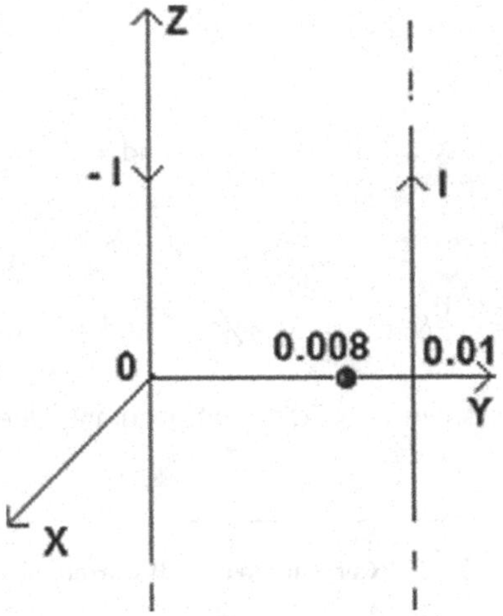

FIGURE 11.18 Two infinite current filaments.

Example 11.5

Determine ***H*** in rectangular components at $P(0, 0.007, 0.1)$ in the field of two sheets having current density $K_1 = 8\boldsymbol{a}_x$ A/m at $y = 3$ mm, and $K_2 = -2\pi\boldsymbol{a}_x$ A/m at $y = 10$ mm.

SOLUTION

Figure 11.19 illustrates two current-carrying sheets.

In view of Equation 11.48c, $\boldsymbol{H} = \frac{1}{2}\boldsymbol{K} \times \boldsymbol{a}_n$.

Since K has an x-component and the y direction is perpendicular to the sheets, thus at $y = 0.007$, which lies between 3 mm and 10 mm, the field due to sheet 1 and 2 located at $y = 3$ mm and $y = 10$ mm is:

$$\boldsymbol{H}_1 = \frac{1}{2}\boldsymbol{K}_1 \times \boldsymbol{a}_n = \left(\frac{1}{2} \cdot 8\boldsymbol{a}_x\right) \times \boldsymbol{a}_y = 4\boldsymbol{a}_z$$

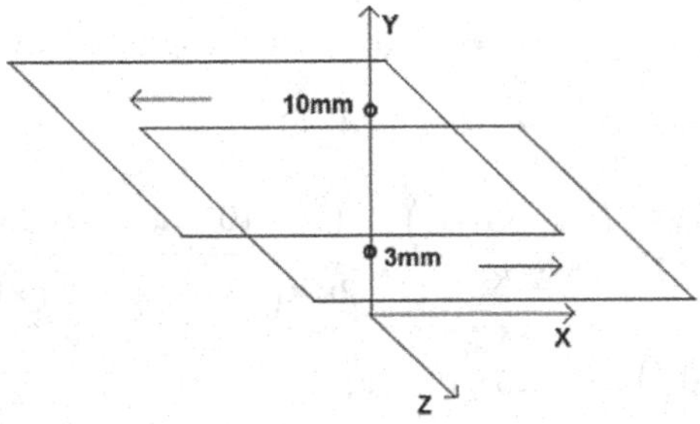

FIGURE 11.19 Two current-carrying sheets.

$$H_2 = \frac{1}{2} K_2 \times a_n = \left\{ \frac{1}{2}(-2\pi a_x) \right\} \times (-a_y) = \pi a_z$$

Total field at 0.007: $H_1 + H_2 = (4 + 3.1415) a_z = 7.1314 a_z$ A/m

Example 11.6

A coaxial cable with inner hollow conductor carrying $I = 1000\ A$ dc current in the z direction has radii of $a = 5$ mm, $b = 7$ mm, $c = 19$ mm and $d = 21$ mm. Find H and B in each conductor and between the conductors. Also determine the total flux in $L = 1$ m length of (a) inner conductor; (b) outer conductor; and (c) the space between the conductors.

SOLUTION

Figure 11.20 shows the configuration of a coaxial cable. In this figure, $a = 5$ mm, $b = 7$ mm, $c = 19$ mm, $d = 21$ mm, $L = 1$ m and $I = 1000\ A$.

Since current is in the z direction, $H = H_\varphi a_\varphi$. Also, since both the conductors are hollow, Equation 11.49b can be modified as under:

(a) For $a \le \rho \le b$ $H_\varphi = \frac{I}{2\pi \cdot \rho} \cdot \frac{\rho^2 - a^2}{b^2 - a^2} = \frac{I}{2\pi}\left(\frac{\rho}{b^2 - a^2} - \frac{1}{\rho}\frac{a^2}{b^2 - a^2}\right) =$

$$\frac{I}{2\pi(b^2 - a^2)}\left(\rho - \frac{a^2}{\rho}\right)$$

$$H = \frac{I}{2\pi(b^2 - a^2)}\left(\rho - \frac{a^2}{\rho}\right) a_\varphi \quad B = \mu H = \frac{\mu_0 I}{2\pi(b^2 - a^2)}\left(\rho - \frac{a^2}{\rho}\right) a_\varphi$$

$$\Phi = \int_s B \bullet ds = \int_0^L \int_a^b \frac{\mu_0 I}{2\pi(b^2 - a^2)}\left(\rho - \frac{a^2}{\rho}\right) a_\varphi \bullet d\rho dz a_\varphi = \frac{\mu_0 I}{2\pi(b^2 - a^2)} \int_0^1 \int_{5\times10^{-3}}^{7\times10^{-3}} \left(\rho - \frac{a^2}{\rho}\right) d\rho dz$$

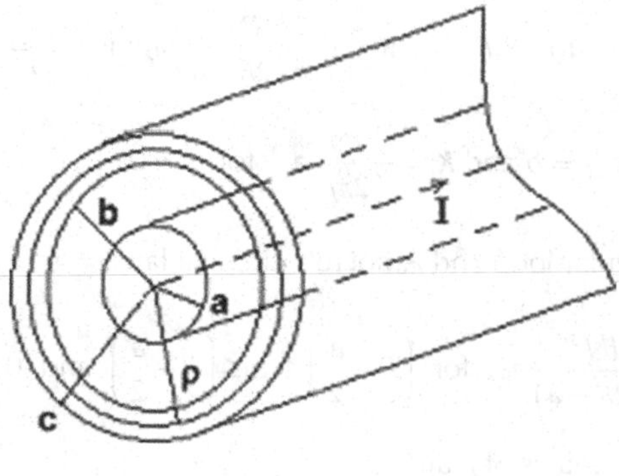

FIGURE 11.20 A coaxial cable.

$$= \frac{\mu_0 I}{2\pi\left(b^2 - a^2\right)} \int_{5\times10^{-3}}^{7\times10^{-3}} \left(\rho - \frac{a^2}{\rho}\right) d\rho = \frac{4\pi\times10^{-7}\times1000}{2\pi\left\{\left(7\times10^{-3}\right)^2 - \left(5\times10^{-3}\right)^2\right\}} \left\{\frac{\rho^2}{2} - a^2 \ln\rho\right\}_{5\times10^{-3}}^{7\times10^{-3}}$$

$$= 29.9\mu\,\text{wb}$$

(b) For $b \le \rho \le c$ $\boldsymbol{H}_\varphi = \frac{I}{2\pi\cdot\rho}$ $\boldsymbol{H} = \frac{I}{2\pi.\rho}\mathbf{a}_\varphi$ $\boldsymbol{B} = \mu\boldsymbol{H} = \frac{\mu_0 I}{2\pi\cdot\rho}\mathbf{a}_\varphi$

$$\Phi = \frac{\mu_0 I}{2\pi} \int_0^{1} \int_{7\times10^{-3}}^{19\times10^{-3}} \frac{1}{\rho} d\rho dz = \frac{4\pi\times10^{-7}\times1000}{2\pi} \ln\rho\Big|_{7\times10^{-3}}^{19\times10^{-3}} = 2\times10^{-4}\ln\left(\frac{19}{7}\right)$$

$$= 199.7\mu\,\text{wb}$$

(c) For $c \le \rho \le d$ $\boldsymbol{H}_\varphi = \frac{I}{2\pi\cdot\rho}\cdot\frac{d^2 - \rho^2}{d^2 - c^2}$ $\boldsymbol{B} = \mu\boldsymbol{H} = \frac{\mu_0 I}{2\pi\cdot\rho}\frac{d^2 - \rho^2}{d^2 - c^2}\mathbf{a}_\varphi$

$$\Phi = \frac{\mu_0 I}{2\pi\left(d^2 - c^2\right)} \int_0^1 \int_{19\times10^{-3}}^{21\times10^{-3}} \left(\frac{d^2}{\rho} - \rho\right) d\rho dz$$

$$= \frac{4\pi\times10^{-7}\times1000}{2\pi\times10^{-6}\left\{(21)^2 - (19)^2\right\}} \left\{441\ln\rho - \frac{\rho^2}{2}\right\}_{19\times10^{-3}}^{21\times10^{-3}} = 10.342\ \mu\,\text{wb}$$

Example 11.7

A uniformly wound toroid with N number of turns has the mean radius R. Its cross-section is rectangular with radial-and axial-thickness a and b, respectively. Find magnetic field $\boldsymbol{H}$ if the toroid current is I.

SOLUTION

In view of Figure 11.21 only the φ-component of H exists. The surface current density is as follows:

$$\boldsymbol{K} = \frac{IN}{\pi(2R - a)}\mathbf{a}_z \text{ for } = R - \frac{a}{2}, \boldsymbol{K} = -\frac{IN}{\pi(2R - a)}\mathbf{a}_z \text{ for } \rho = R + \frac{a}{2}$$

$$\boldsymbol{K} = \frac{IN}{2\pi\rho}\mathbf{a}_\rho \text{ for } z = b \text{ and } \boldsymbol{K} = -\frac{IN}{2\pi\rho}\mathbf{a}_\rho \text{ for } z = 0$$

Using a rectangular loop and Ampère's circuital law:

$$\boldsymbol{H} = \frac{IN}{\pi(2R - a)}\mathbf{a}_\theta, \text{ for } \left(R - \frac{a}{2}\right) \le \rho \le \left(R + \frac{a}{2}\right) \text{ and } 0 \le z \le b$$

$\boldsymbol{H} = 0$, for all other values of ρ and z.
H is independent of the φ-coordinate.

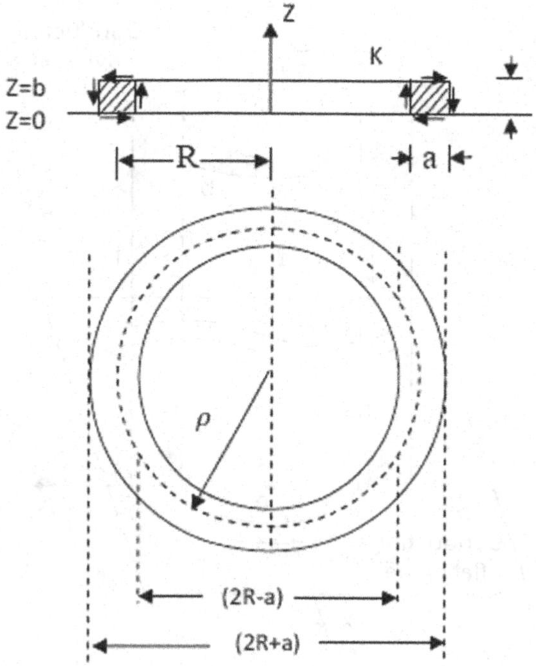

FIGURE 11.21 A uniformly wound toroid.

Example 11.8

Two large coaxial conducting cylinders, each extending over $0 \leq z \leq \infty$, are placed on the conducting x–y plane with z as their common axis. The inner cylinder carries a uniformly distributed surface current I flowing from $z = \infty$ to $z = 0$. The outer cylinder provides the path for the return current, from $z = 0$ to $z = \infty$. If the radii of these cylinders are a and b such that $a < b$, find H everywhere.

SOLUTION

The configuration for the problem is shown in Figure 11.22. Using Ampère's circuital law, we find the solution to be:

$H = 0$ for $\rho < a$, $H = 0$ for $\rho > b$, $H = 0$ for $z < 0$, and

$$H = \frac{I}{2\pi\rho} \text{ for } a \leq \rho \leq b \text{ and for } z \geq 0.$$

Example 11.9

A toroid is specified by $\rho_0 = 10$ cm, $a = 3$ cm and $K_a = 100$ A/m. Find the scalar magnetic potential at $\rho = 12$ cm, $\varphi = 0.2\pi$, $z = 1.5$ cm if (a) $\mathcal{V} = 0$ at $\varphi = 0$ and a barrier established at $\varphi = \pi$; (b) $\mathcal{V} = 0$ at $\varphi = -\frac{\pi}{2}$ and a barrier established at $\varphi = \pi$; and (c) $\mathcal{V} = 0$ at $\varphi = 0$ and a barrier established at $\varphi = 0.1\pi$.

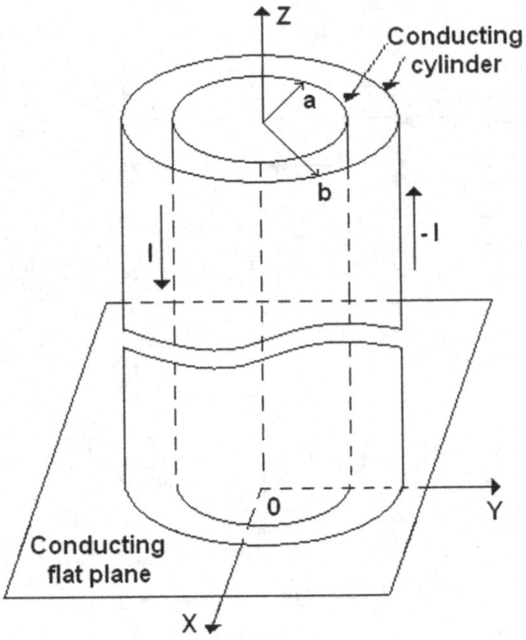

FIGURE 11.22 Two large co-axial conducting cylinders.

SOLUTION

The toroid configuration along with relevant parameters is shown in Figure 11.23 whereas Figure 11.24 illustrates further details of the problem.

In view of Equation 11.53:

$$\boldsymbol{H} = K_a \frac{\rho_0 - a}{\rho} \boldsymbol{a}_\varphi = 100 \frac{10-3}{12} \boldsymbol{a}_\varphi = \frac{700}{12} \boldsymbol{a}_\varphi$$

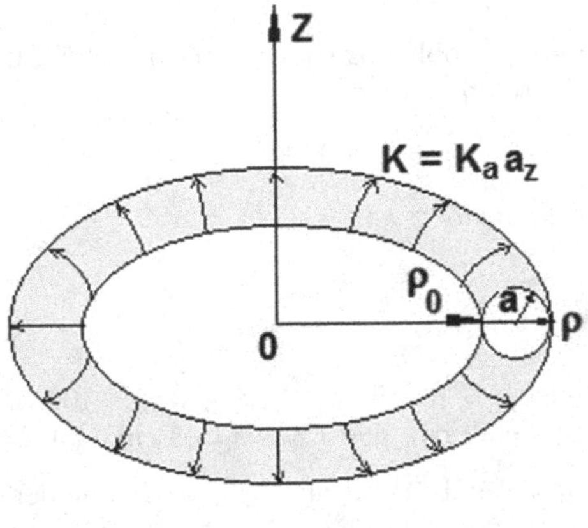

FIGURE 11.23 A toroid configuration.

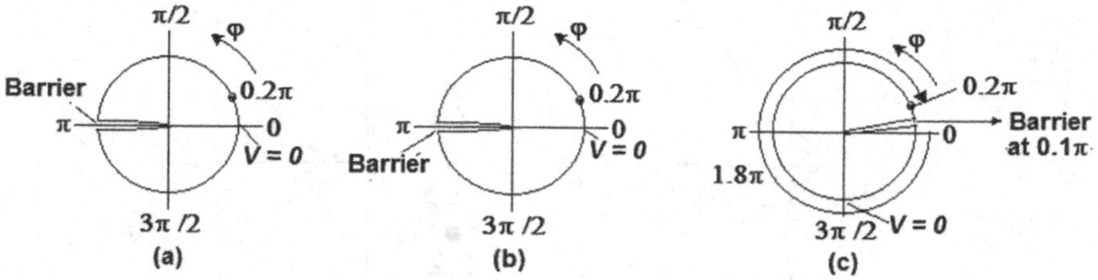

FIGURE 11.24 A toroid configuration with details.

$$H = -\nabla \mathcal{V}\Big|_{\varphi} = -\frac{1}{\rho}\frac{\delta \mathcal{V}}{\delta \varphi} \cdot \frac{\delta \mathcal{V}}{\delta \varphi} = -\rho H \mathbf{a}_{\varphi} = -\left(12\times 10^{-2}\right)\left(\frac{700}{12}\right)\mathbf{a}_{\varphi} = -7\mathbf{a}_{\varphi}$$

$$\mathcal{V} = -7\varphi + C \quad C = \mathcal{V} + 7\varphi$$

(a) $\mathcal{V} = 0$ at $\varphi = 0$, $C = 0$

$$\mathcal{V}\Big|_{\varphi=0.2\pi} = -7\varphi = -7\times 0.2\pi = -4.398A \text{ (for barrier at } \varphi = \pi)$$

(b) $\mathcal{V} = 0$ at $\varphi = -\dfrac{\pi}{2}$, $C = -7\times 0.5\pi = -3.5\pi$

$$\mathcal{V}\Big|_{\varphi=0.2\pi} = -7\times 0.2\pi - 3.5\pi = -1.4\pi - 3.5\pi = -15.39A \text{ (for barrier at } \varphi = \pi)$$

(c) $\mathcal{V} = 0$ at $\varphi = 0$, $C = 0$

$$\mathcal{V}\Big|_{\varphi=-1.8\pi} = -7\times\left(-1.8\pi\right) = 39.58A \text{ (for barrier at } \varphi = 0.1\pi)$$

Example 11.10

Four conductors each of 1 cm radius are located parallel to the z-axis. Two of these carry 10mA current in $\mathbf{a}_z$ direction are centered at (–3, –3) and (3, 3) while the other two carrying –10mA current are centered at (–3, 3) and (3, –3). Compute the vector magnetic potential $\boldsymbol{A}$ at (a) (2, 2, 0) (b) (2, 4, 0) and (c) ∞ by setting $\boldsymbol{A}$ = 0 at the origin, using the relation $A_z = \dfrac{\mu_0 I}{2\pi}\ln\left(\dfrac{b}{\rho}\right)$ and by setting zero reference at $\rho = b$.

SOLUTION

The configuration is shown in Figure 11.25.
$\boldsymbol{A}$ is set to zero at the origin of the coordinate system.
Let b be the distance from the origin to the center of each conductor.
In view of the given coordinates: $b = \sqrt{3^2 + 3^2} = \sqrt{18}$.

$$A_z = \frac{\mu_0 I}{2\pi}\ln\left(\frac{b}{\rho}\right) = \frac{\left(4\pi\times 10^{-7}\right)\times 10\times 10^{-3}}{2\pi}\ln\left(\frac{\sqrt{18}}{\rho}\right) = 2\times 10^{-9}\times \ln\left(\frac{\sqrt{18}}{\rho}\right)$$

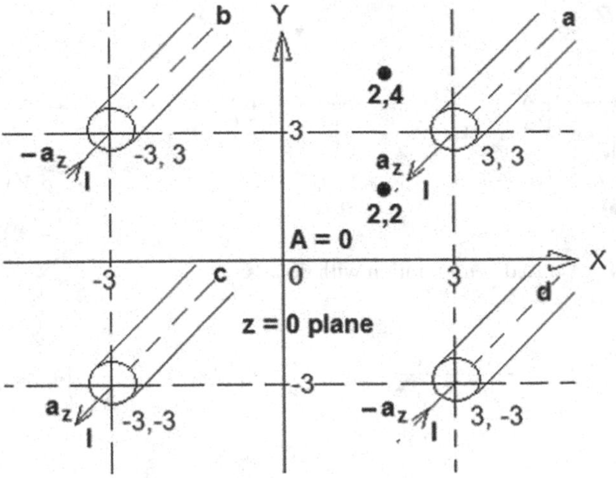

FIGURE 11.25 Four conductors located parallel to the z-axis.

(a) The values of ρ for point (2, 2) from centers of conductor a, b, c, and d are $\sqrt{2}$, $\sqrt{26}$, $\sqrt{50}$, $\sqrt{26}$,

Thus, at point (2, 2, 0):

$$A_z = 2\times 10^{-9}\times\left[\ln\left(\frac{\sqrt{18}}{\sqrt{2}}\right)-\ln\left(\frac{\sqrt{18}}{\sqrt{26}}\right)+\ln\left(\frac{\sqrt{18}}{\sqrt{50}}\right)-\ln\left(\frac{\sqrt{18}}{\sqrt{26}}\right)\right]=1.914$$

or $\boldsymbol{A} = 1.914\boldsymbol{a_z}$ nwb/m

(b) The values of ρ for point (2, 4) from centers of conductor a, b, c, and d are $\sqrt{2}$, $\sqrt{26}$, $\sqrt{74}$, $\sqrt{50}$

Thus, at point (2, 4, 0):

$$A_z = 2\times 10^{-9}\times\left[\ln\left(\frac{\sqrt{18}}{\sqrt{2}}\right)-\ln\left(\frac{\sqrt{18}}{\sqrt{26}}\right)+\ln\left(\frac{\sqrt{18}}{\sqrt{74}}\right)-\ln\left(\frac{\sqrt{18}}{\sqrt{50}}\right)\right]=2.1763$$

or $\boldsymbol{A} = 2.1763\boldsymbol{a_z}$ nwb/m

(c) The values of ρ for point at ∞ is infinity from centers of all the conductors. Thus, at infinity $A_z = 0$ or $A = 0$:

PROBLEMS

P11.1 Find the incremental magnetic field intensity ($\Delta\boldsymbol{H}$) at the origin due to a current element ($I\Delta\boldsymbol{L}$) in free space equal to: (a) $3\pi\left(\boldsymbol{a_x}-\boldsymbol{a_y}+5\boldsymbol{a_z}\right)\mu A-m$ located at (0, 2, 0) (b) $5\pi\left(\boldsymbol{a_x}-2\boldsymbol{a_y}+3\boldsymbol{a_z}\right)\mu A.m$ located at (4, 0, 0).

P11.2 A filament carrying a current of 1 A passes through a point (2, 3, 0) and is parallel to the z-axis. Find $\boldsymbol{H}$ at (1, 2, 0) if this filament is in the interval (a) $-5 \le z \le 5$; (b) $-5 \le z \le 0$; (c) $0 \le z \le 5$; (d) $0 \le z \le \infty$.

P11.3 Two surface currents with current density $\boldsymbol{K} = 100\boldsymbol{a}_y$ and $\boldsymbol{K} = -100\boldsymbol{a}_y$ A/m are located at $z = 3$ and $z = -3$ respectively. The region $|z| > 3$ is free space. Find H and B in the region $|z| < 3$ if: (a) $\mu = \mu_0$ for $|z| < 3$; (b) $\mu = 15\mu_0$ for $|z| < 3$; (c) $\mu = \mu_0$ for $|z| < 1$ and $\mu = 15\mu_0$ for $1 < |z| < 3$; (d) $\mu = \mu_0$ for $y < 0$ and $\mu = 1.5\mu_0$ for $y > 0$.

P11.4 Find the magnetic field intensity (H) everywhere due to two uniformly distributed current sheets with the current density $-K_0\boldsymbol{a}_y$, on the surface $z = 0$, $y < 0$, and $K_0 a_z$ on the surface $y = 0$, $z < 0$, where K_0 is a constant.

P11.5 Find $\boldsymbol{H}$ in rectangular components at $P(0, 0.006, 0.1)$ in the field of a coaxial cable with $a = 3$ cm, $b = 5$ cm, $c = 7$ cm, $I = 0.9A$ centered on the z-axis and in the a_z direction in the central conductor.

P11.6 Determine $\boldsymbol{H}$ in rectangular components at $P(0, 0.005, 0.1)$ in the field of a solenoid with axes at $x = 1$ cm, $y = 2$ cm, extending from $z = -5$ cm to $z = 20$ cm, 5 cm diameter, 2000 turns, $I = 1$ mA in the clockwise direction when viewed from $z = 10$ cm.

P11.7 Determine $\boldsymbol{H}$ in rectangular components at P(0, 0.006, 0.2) in the field of a toroid centered at the origin with its axis on the x-axis, $\rho_0 = 3$ cm, $a = 1$ mm, $N = 500$, $I = 3$mA in the a_x direction at the outer radius.

P11.8 A coaxial cable having two hollow conductors with radii $a = 3$ mm, $b = 5$ mm, $c = 9$ mm, *and* $d = 11$ mm. It carries a dc current $I = 500$ A in the z direction. Find H and B in each conductor and in between. Also find the total flux in $L = 2$ m length of (a) inner conductor; (b) outer conductor; and (c) the space between the conductors.

P11.9 A long solenoid of radius $\boldsymbol{R}$ carries a uniform surface current sheet of density K_φ. Find the magnetic field intensity everywhere.

P11.10 A toroid has $\rho_0 = 15$ cm, $a = 5$ cm, and $K_a = 50\dfrac{A}{m}$. Find the scalar magnetic potential $\mathcal{V}$ at $\rho = 20$ cm, $\varphi = 0.4\pi$, $z = 2$ cm if (a) $\mathcal{V} = 0$ at $\varphi = 0$ and a barrier established at $\varphi = \pi$; (b) $\mathcal{V} = 0$ at $\varphi = 0.6\pi$ and a barrier established at $\varphi = 0.5\pi$; and (c) $\mathcal{V} = 0$ at $\varphi = 0.5\pi$ and a barrier established at $\varphi = 0.3\pi$.

P11.11 Find the vector magnetic potential $\boldsymbol{A}$ within a solid non-magnetic conductor of radius a carrying a total current I in the a_z direction. Use the known value of $\boldsymbol{H}$ or $\boldsymbol{B}$ for $\rho < a$ and the relation $\boldsymbol{B} = \nabla \times \boldsymbol{A}$. If $\boldsymbol{A} = \dfrac{\mu_0 I \cdot \ln 5}{2\pi}$ at $\rho = a$, find $\boldsymbol{A}$ at (a) $\rho = 0$; (b) $\rho = 0.3a$; (c) $\rho = 0.6a$; and (d) $\rho = 0.9a$.

P11.12 A conducting triangular loop is shown in Figure 11.10. It carries a current $I = 10$ A. Find $\boldsymbol{H}$ at (0, 0, 5) due to side A of the loop.

P11.13 Find the magnetic flux density at point P located at distance h from the center of a rectangular loop of wire with sides a and b and I ampère current.

DESCRIPTIVE QUESTIONS

Q11.1 Compare the electric and magnetic fields.

Q11.2 Discuss the properties of magnetic lines of force.

Q11.3 Explain the terms magnetic field intensity, magnetic flux density, and magnetic flux.

Q11.4 State and explain Biot-Savart's law. Write the expressions for magnetic field intensity obtained from Biot-Savart's law for various forms of current distributions.

Q11.5 State Ampère's circuital law and explain the meaning of the words (a) line integral; (b) closed path; and (c) current enclosed by the path.

Q11.6 In Ampère's law, derive the relation for the field due to (a) coaxial cable; (b) current sheet.

Q11.7 Define scalar magnetic potential and discuss its properties.

Q11.8 Define vector magnetic potential and discuss its practical applications.

Q11.9 Prove that the vector magnetic potential $\boldsymbol{A} = \int_v \frac{\mu_0 \boldsymbol{J} dv}{4\pi R}$ satisfies Biot-Savart's law.

Q11.10 Prove that the vector potential $\boldsymbol{A} = \int_v \frac{\mu_0 \boldsymbol{J} dv}{4\pi R}$ satisfies Ampère's circuital law.

Q11.11 Dc current flowing in a closed conducting path produces time-invariant magnetic field. Explain why time-invariant magnetic field does not cause dc currents to flow in a closed conducting path placed in the magnetic field.

FURTHER READING

Given at the end of Chapter 15.

12 Fields in Magnetic Materials

12.1 INTRODUCTION

In Chapter 11, it was noted that irrespective of its nature of distribution, the flow of current results in a magnetic field. To assess this field, certain parameters, referred to as magnetic field quantities (viz. $\boldsymbol{H}$, $\boldsymbol{B}$, φ, $\mathcal{V}$, and $\boldsymbol{A}$), were introduced. As fire affects the environment by raising the temperature, so too does the field affect its surroundings, by exerting force on charged particles. The effect of an electric field, discussed in Chapters 5 to 7, ultimately resulted in force, work done, scalar electric potential, and the energy stored. Similarly, the magnetic field culminates in force, torque, and dipole moment. The material involved in all of the devices comprises tiny charged particles. In the case of an electric field, these particles are also affected by the presence of a magnetic field. Furthermore, their behavior is likely to differ in moving and stationary states. The effect of a magnetic field on charged particles ultimately gets transferred to the matter itself. This chapter therefore deals with different types of materials, which are susceptible to the magnetic field. This chapter also describes the boundary conditions that are to be satisfied by different magnetic field quantities.

12.2 FORCE ON CHARGES AND CURRENT

This section deals with the forces exerted on charges and on differential current elements. It also accounts for the force between the current elements.

12.2.1 Force on a Moving Charge

As noted earlier both the electric and magnetic fields exert force on charged particles. The following subsections indicate the difference between the forces exerted by these two fields.

12.2.1.1 Force Due to a Static Electric Field

In Chapter 5, it was noted that the electric field intensity ($\boldsymbol{E}$) exerts a force ($\boldsymbol{F}$) on a particle with charge (Q). The mathematical relation (Equation 5.9d) for this force is reproduced below:

$$\boldsymbol{F} = Q \cdot \boldsymbol{E} \tag{12.1}$$

Equation 12.1 implies that the static electric field exerts force on a charged particle irrespective of its stationary or slowly moving state.

12.2.1.2 Force Due to a Steady Magnetic Field

In a steady magnetic field the force acting on the point charge is given by:

$$\boldsymbol{F} = Q \cdot (\boldsymbol{U} \times \boldsymbol{B}) \tag{12.2}$$

where $\boldsymbol{B}$ is the magnetic flux density vector and $\boldsymbol{U}$ is the velocity vector of the motion of the charged particle Q. In Equation 10.2, it is evident that the steady magnetic field exerts no force on the stationary charges that is when $\boldsymbol{U} = 0$.

12.2.1.3 Force Due to a Combined Field

If both electric and magnetic fields are simultaneously present the force experienced by the moving charge is simply the sum of the forces given by Equations 12.1 and 12.2. Thus:

$$\boldsymbol{F} = Q \cdot (\boldsymbol{E} + \boldsymbol{U} \times \boldsymbol{B}) \tag{12.3}$$

This equation, known as the *Lorentz force equation,* was noted earlier in Equation 5.11. Its solution is required for determining the electron orbits in magnetron, proton paths in cyclotron, or, in general, charged-particle motion in combined electric and magnetic fields.

12.2.1.4 Force Density

The force exerted on individual charged particles ultimately gets transferred to the matter which is subjected to the influence of field(s). If a unit volume of such an affected matter is taken and the force is measured on the basis of per unit volume the resulting quantity is referred to as *force density.* It is measured in Newtons per cubic meter.

The *Lorentz force equation* in point form is given as:

$$\mathcal{P} = \rho \cdot \boldsymbol{E} + \boldsymbol{J} \times \boldsymbol{B} \;\; (\text{N/m}^3) \tag{12.3a}$$

Its unit is Newtons per cubic meter. For line and surface currents:

$$\mathcal{P}_L = \rho_L \cdot \boldsymbol{E} + (Id\ell) \times \boldsymbol{B} \;\; (\text{N/m}) \tag{12.3b}$$

and

$$\mathcal{P}_S = \rho_S \cdot \boldsymbol{E} + \boldsymbol{K} \times \boldsymbol{B} \; (\text{N/m}^2) \tag{12.3c}$$

12.2.2 Force on Differential Current Elements

Section 12.2.1 explored the impact of electric, magnetic, and combined field on an isolated but moving charged particle. The movement of charges in a conducting wire, over a surface or in a volume constitutes a current. Thus the current too is likely to be influenced by the field(s). This subsection, therefore, is devoted to studying the influence (particularly) of magnetic field on the line, surface, and volume currents.

12.2.2.1 Force on a Line Current

Figure 12.1 shows a long thin wire carrying a dc current (I). This wire is located in a steady (time-invariant) magnetic field (B).

In Equation 12.2, the force due to this magnetic field on the incremental length (dl) of the wire can be written as:

$$\boldsymbol{dF} = dq \cdot (\boldsymbol{U} \times \boldsymbol{B}) \tag{12.4a}$$

On replacement of dq by $\rho_L \cdot d\ell$ and using Equation 8.5, this equation becomes:

$$\boldsymbol{dF} = \rho_L \cdot d\ell \cdot (\boldsymbol{U} \times \boldsymbol{B}) = \tag{12.4b}$$

or,

$$\boldsymbol{dF} = (I \cdot \boldsymbol{d\ell}) \times \boldsymbol{B} \tag{12.4c}$$

On integrating Equation 12.4b, we get:

$$\boldsymbol{F} = \int (\boldsymbol{U} \times \boldsymbol{B}) \cdot \rho_L \cdot d\ell = \int ([\boldsymbol{U} \cdot \rho_L] \times \boldsymbol{B}) \cdot d\ell \tag{12.5a}$$

On substituting $\boldsymbol{U} \cdot \rho_L \cdot d\ell = \boldsymbol{Id\ell} \overset{\text{def}}{=} Id\ell \cdot \boldsymbol{a}_\ell$, this relation becomes:

$$\boldsymbol{F} = \int (\boldsymbol{a}_\ell \times \boldsymbol{B}) \cdot Id\ell \tag{12.5b}$$

where $\boldsymbol{a}_\ell$ is a unit vector in the direction of the current

In Equation 12.5b, I, $\boldsymbol{U}$ (or υ), and $d\ell$ are in the same direction.

In the case of thin wires the current is constant along the entire length; Equation 12.5b can thus be modified to:

$$\boldsymbol{F} = I \cdot \int (\boldsymbol{a}_\ell \times \boldsymbol{B}) \cdot d\ell = I \cdot \int (\boldsymbol{d\ell} \times \boldsymbol{B}) \tag{12.5c}$$

12.2.2.2 Force on a Surface Current

Figure 12.2 shows the movement of charges with surface charge density ρ_s over an incremental surface (ds). If the velocity of these charges is taken to be $\boldsymbol{U}$ the resulting

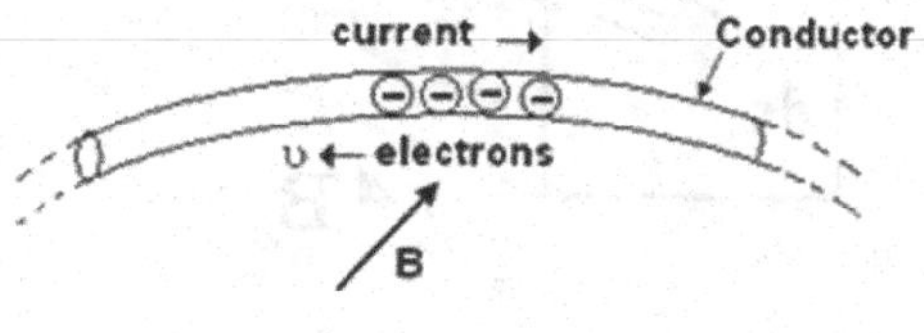

FIGURE 12.1 Current I in B-field.

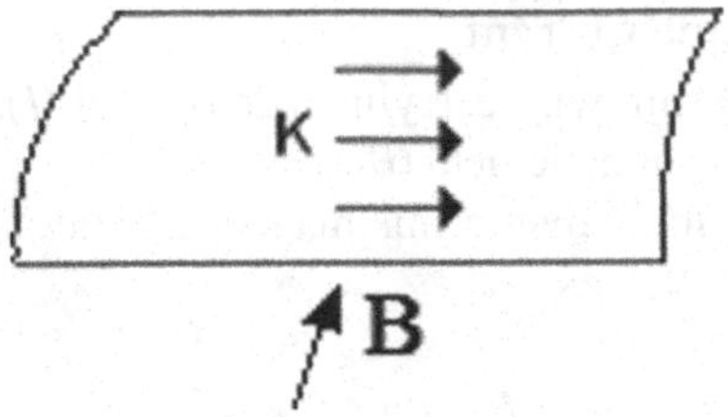

FIGURE 12.2 Surface current density (K) in B-field.

surface current will have a surface current density $\boldsymbol{K}$ which can be given as $\boldsymbol{K} = \rho_s \boldsymbol{U}$. The force exerted by the $\boldsymbol{B}$-field on these moving charges can be written as:

$$\boldsymbol{F} = \int (\boldsymbol{U} \times \boldsymbol{B}) \cdot \rho_s \cdot ds \tag{12.6a}$$

On substituting $\boldsymbol{K} = \rho_s \boldsymbol{U}$ this relation becomes:

$$\boldsymbol{F} = \int (\boldsymbol{K} \times \boldsymbol{B}) \cdot ds \tag{12.6b}$$

12.2.2.3 Force on a Volume Current

Figure 12.3 shows the movement of charges with volume charge density ρ_v (or ρ) in an incremental volume (dv). If the velocity of these charges is taken to be U the resulting volume current will have a volume current density $\boldsymbol{J}$ which can be given as $\boldsymbol{J} = \rho \cdot \boldsymbol{U}$. The force exerted by the flux density B-field on these moving charges can be written as:

$$\boldsymbol{F} = \int (\boldsymbol{U} \times \boldsymbol{B}) \cdot \rho \cdot dv \tag{12.7a}$$

On substituting $\boldsymbol{J} = \rho \cdot \boldsymbol{U}$, this relation becomes:

$$\boldsymbol{F} = \int (\boldsymbol{J} \times \boldsymbol{B}) \cdot dv \tag{12.7b}$$

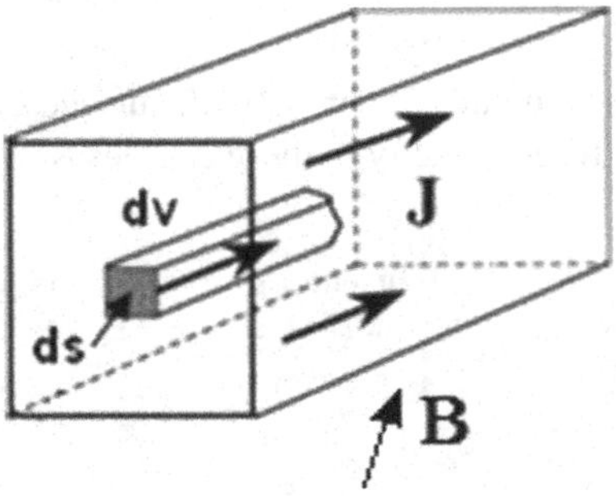

FIGURE 12.3 Volume current density in B-field.

12.2.3 Force between Differential Current Elements

This section deals with the force (i) between two differential current elements, (ii) between two straight long parallel current filaments, and (iii) the condition for equality of electric and magnetic forces

12.2.3.1 Force between Two Differential Current Elements

Figure 12.4 shows two infinitely long, parallel (filamentary) wires (or thin current-carrying conductors) spaced *d* distance apart. In Figure 12.4a the currents in the two wires are in the same direction. Such a configuration results in attractive force. In Figure 12.4b the direction of the currents are in opposition or anti parallel. This configuration results in a repulsive force between the two wires.

Like point charges, current-carrying conductors of elementary lengths exert force on each other. However, there are two differences: (i) like charges repel and unlike charges attract one another; while unidirectional current elements attract and those in opposite directions repel one another; (ii) the forces in the former case are due to the electric field, while in the latter case are due to the magnetic field. In the problem of identification of distance, the force is obtained between elements of filamentary conductors with differential lengths (say dL_1 and dL_2). This force is obtained by using Biot-Savart's law. Like charges, currents (I_1 and I_2) flowing in conductors (1 and 2) are scalar quantities. The elemental lengths dL_1 and dL_2 can, however, be assigned directions to make these vectors in accordance with the directions of flow of currents. Thus in view of the products of currents and elemental lengths, two new quantities (referred to as filamentary current elements $I_1\boldsymbol{dL}_1$ and $I_2\boldsymbol{dL}_2$) emerge in the vectorial form. The force is to be obtained between these two parallel (or anti-parallel) filamentary current elements separated by a (scalar) distance *d*. This distance can be replaced by a vector distance $\boldsymbol{R}_{12}$ or $\boldsymbol{R}_{21}$ in accordance with the direction of force to indicate the direction of the distance vectors. In view of Biot-Savart's law, this force is directly proportional to the product of these filamentary currents and inversely proportional to the square of the distance between these. In view of the constant of proportionality this force is also directly proportional to the media parameter ($\mu = \mu_o$ for free space). As this force is the cross product of two *vector*

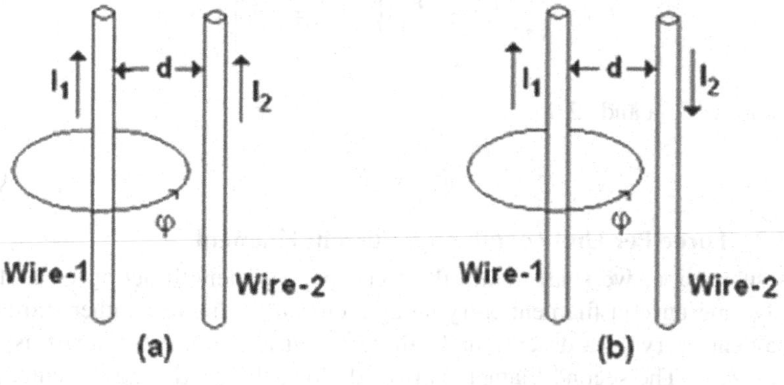

FIGURE 12.4 Two current filaments with (a) parallel; and (b) anti-parallel currents.

filamentary currents it is also a vector which is perpendicular to both the vectors involved. The final expression of vector force ($\boldsymbol{F}_{12}$ or $\boldsymbol{F}_{21}$) exerted by filament 1 on filament 2 or vice versa can be obtained by taking the following steps:

- Use Equation 11.4 to obtain $\boldsymbol{H}_2$ due to ($I_1\boldsymbol{dL}_1$) at point 2, i.e. the location of ($I_2\boldsymbol{dL}_2$) as:

$$\boldsymbol{H}_2 = \oint_{C_1} \frac{(I_1\boldsymbol{dL}_1)\times \boldsymbol{a}_{R12}}{4\pi \cdot R_{12}^2} \tag{12.8a}$$

  where C_1 is the closed path where the current $\boldsymbol{I}_1$ is flowing.
- Use Equation 11.1 to get the magnetic induction ($\boldsymbol{B}_2$):

$$\boldsymbol{B}_2 = \mu_o \cdot \oint_{C_1} \frac{(I_1\boldsymbol{dL}_1)\times \boldsymbol{a}_{R12}}{4\pi \cdot R_{12}^2} = \frac{\mu_o \cdot I_1}{4\pi} \cdot \oint_{C_1} \frac{\boldsymbol{dL}_1 \times \boldsymbol{a}_{R12}}{R_{12}^2} \tag{12.8b}$$

- Use Equation 12.4c to get the force ($\boldsymbol{F}_{12}$) exerted on $I_2\boldsymbol{dL}_2$ integrated over C_2:

$$\boldsymbol{F}_{12} = \oint_{C_2} \left[(I_2\boldsymbol{dL}_2)\times \boldsymbol{B}_2\right] = \oint_{C_2} \left[(I_2\boldsymbol{dL}_2)\times \left\{\frac{\mu_o \cdot I_1}{4\pi} \cdot \oint_{C_1} \frac{\boldsymbol{dL}_1 \times \boldsymbol{a}_{R12}}{R_{12}^2}\right\}\right] \tag{12.8c}$$

- Rearrange the terms to get:

$$\boldsymbol{F}_{12} = \frac{\mu_o}{4\pi} \cdot I_1 \cdot I_2 \cdot \oint_{C_2} \left[\oint_{C_1} \frac{\boldsymbol{a}_{R12} \times \boldsymbol{dL}_1}{R_{12}^2}\right] \times \boldsymbol{dL}_2 \tag{12.9a}$$

- Replace suffixes 1 by 2 and 2 by 1 as and where they appear to get the expression for the force exerted on element 1 by element 2. Thus we get:

$$\boldsymbol{F}_{21} = \frac{\mu_o}{4\pi} \cdot I_1 \cdot I_2 \cdot \oint_{C_1} \left[\oint_{C_2} \frac{\boldsymbol{a}_{R21} \times \boldsymbol{dL}_2}{R_{21}^2}\right] \times \boldsymbol{dL}_1 \tag{12.9b}$$

In Equations 12.9a and 12.9b:

$$\boldsymbol{F}_{21} = -\boldsymbol{F}_{12} \tag{12.9c}$$

12.2.3.2 Force Per Unit Length on a Current Filament

In this subsection, we shall obtain the force per unit length acting on a straight infinitely long current filament carrying a dc current I_1 due to another similar current filament carrying a dc current I_2; these are at a distance "*d*" apart as shown in Figure 12.4. The second filament is placed along the *z*-axis and the current I_2 is flowing in the positive *z*-direction. The two current filaments are laid parallel to each other with a separating distance R. Take a point P on the current filaments carrying the current I_1. The magnetic field intensity at the point P due to the other infinitely long current filament can be readily found using Ampère's circuital law. It is given as:

$$H_\varphi = \frac{I_2}{2\pi d} \tag{12.10}$$

Therefore, the flux density in free space at the point P is:

$$\boldsymbol{B} = \mu_o \cdot \frac{I_2}{2\pi d} \boldsymbol{a}_\varphi \tag{12.11}$$

Now, *force per unit length* $\mathcal{F}_L$ acting on the filament carrying the current I_1 due to the magnetic field is found from Equation 12.3b:

$$\mathcal{F}_L = \left(I_1 \boldsymbol{a}_1\right) \times \boldsymbol{B} \ \left(\text{N/m}\right) \tag{12.12}$$

where the unit vector $\boldsymbol{a}_1$ is in the direction of the current I_1, thus:

$$\boldsymbol{a}_1 = \begin{cases} +\boldsymbol{a}_z & \text{for } \mathrm{I}_1 \text{ parallel to } I_2 \\ -\boldsymbol{a}_z & I_1 \text{ antiparallel to } I_2 \end{cases} \tag{12.13}$$

Thus, from Equations 10.11 and 10.12, we get:

$$\mathcal{F}_L = \mu_o \cdot \frac{I_1 \cdot I_2}{2\pi d} \left(\boldsymbol{a}_1 \times \boldsymbol{a}_\varphi\right) \ (\text{N/m}) \tag{12.14}$$

or,

$$\mathcal{F}_L = \mu_o \cdot \frac{I_1 \cdot I_2}{2\pi d} \boldsymbol{a}_o \ (\text{N/m}) \tag{12.15a}$$

where

$$\boldsymbol{a}_o = \begin{cases} -\boldsymbol{a}_\rho & \text{for } I_1 \text{ parallel to } I_2 \\ +\boldsymbol{a}_\rho & I_1 \text{ antiparallel to } I_2 \end{cases} \tag{12.15b}$$

In Equations 12.15a and 12.15b, the two conductors experience attractive force with parallel currents and repulsive force of equal amount with anti-parallel currents.

12.2.3.3 Condition for Equality of Electric and Magnetic Forces

Figure 12.5 shows two long infinitesimally thin line charge distributions located "a" distance apart. Both of these carry uniform line charge density ρ_L C/m. Assuming that these line charges are placed parallel to the z-axis and both are moving with the velocity υ in the same (say positive z) direction. Thus, these filamentary charges can be regarded as current filaments with constant convection currents I_1 and I_2 flowing in the positive z-direction.

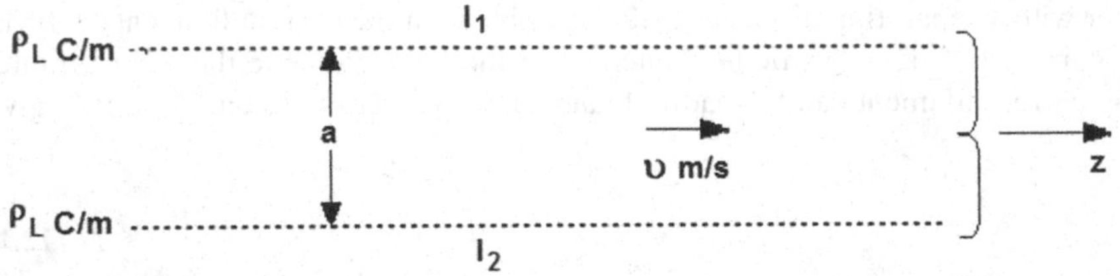

FIGURE 12.5 Two moving parallel line charges.

In Equations 11.1 and 11.60, the magnetic induction (***B***), in free space, caused at some arbitrary distance (say r) due to a long straight wire carrying a current (I), placed along the z-axis is given as:

$$\boldsymbol{B} = \frac{\mu_o \cdot I}{2\pi \cdot r} \cdot \boldsymbol{a}_\varphi \tag{12.16a}$$

Since the currents are assumed to be flowing in the (positive) z direction, ***B*** has only B_φ component. Let the wire 1 carrying current I_1 is placed along the z-axis. Thus, the field ($\boldsymbol{B}_2$) at wire 2 due to current (I_1) in wire 1 is:

$$\boldsymbol{B_2} = \frac{\mu_o \cdot I_1}{2\pi \cdot a} \cdot \boldsymbol{a}_\varphi \tag{12.16b}$$

where "a" indicates the distance between the two moving parallel line charges.

If a current filament ($I_2 \cdot \boldsymbol{d\ell_2}$) is taken on the line charge 2, the incremental force ($d\boldsymbol{F}_2$) exerted by $\boldsymbol{B_2}$ (due to the current I_1) can be given as:

$$d\boldsymbol{F}_2 = I_2 \cdot \boldsymbol{d\ell_2} \times \boldsymbol{B_2} \tag{12.17a}$$

Since $I_2 \cdot \boldsymbol{d\ell_2} = I_2 \cdot dz\boldsymbol{a}_z$ and $\boldsymbol{B_2} = B_{2\varphi} \cdot \boldsymbol{a}_\varphi$ we get:

$$I_2 \cdot dz\boldsymbol{a}_z \times B_{2\varphi} \cdot \boldsymbol{a}_\varphi = -I_2 \cdot dz \cdot B_{2\varphi} \cdot \boldsymbol{a}_\rho \tag{12.17b}$$

In Equations 12.17a and 12.16b the resulting force on the second conductor carrying current I_2 is:

$$d\boldsymbol{F}_2 = -\left(\frac{\mu_o \cdot I_1 \cdot I_2}{2\pi \cdot a}\right) \cdot dz \cdot \boldsymbol{a}_\rho \tag{12.17c}$$

In this relation the negative sign indicates that wire 1exerts an attractive force on wire 2. Thus, the magnetic force per unit length ($\mathcal{F}_{Lm} = |\boldsymbol{F}_2|$) can be obtained by integrating Equation 12.17c over a unit length.

$$\mathcal{F}_{Lm} = \frac{\mu_o \cdot I_1 \cdot I_2}{2\pi \cdot a} \tag{12.18}$$

Since both the line charges are assumed to be equal the currents are also equal. In terms of the line charge density and the velocity ($U = \upsilon$) of charges these currents can be given as:

$$I_1 = I_2 = \rho_L \cdot U = \rho_L \cdot \upsilon \tag{12.19}$$

Equation 12.18 can thus be written as:

$$\mathcal{F}_{Lm} = \frac{\mu_o \cdot (\rho_L \cdot \upsilon)^2}{2\pi \cdot a} \tag{12.20}$$

The electric field (E) of line of charge given by Equation 5.24 is reproduced from Chapter 5 here:

$$E = \frac{\rho_L}{2\pi\varepsilon_o \cdot a} \tag{5.24}$$

The electric repulsion or electrostatic force ($\mathcal{F}_{Le}$) of one wire, on each unit length of the other wire is given as:

$$\mathcal{F}_{Le} = \frac{\rho_L^2}{2\pi\varepsilon_o \cdot a} \tag{12.21}$$

On equating the magnetic and electric forces given by Equations 12.20 and 12.21, i.e. for $\mathcal{F}_{Lm} = \mathcal{F}_{Le}$, we get:

$$\frac{\mu_o \cdot (\rho_L \cdot \upsilon)^2}{2\pi \cdot a} = \frac{\rho_L^2}{2\pi\varepsilon_o \cdot a} \tag{12.22}$$

This gives:

$$\upsilon = 1/\sqrt{\mu_o \cdot \varepsilon_o} = c = 2.998 \times 10^8 \text{ m/s} \tag{12.23}$$

Therefore, the electric and magnetic forces become equal when the velocity of the two line charges equals the velocity of light (i.e. $\upsilon = c$). Here the comparison is between the Coulombian electric force and the magnetic force of *convection currents* of moving charges. The velocity of the *conduction current* in a conductor is comparable to that of light, even though the drift velocity is typically of the order of mm/s to cm/s.

12.3 A CURRENT-CARRYING LOOP IN A UNIFORM MAGNETIC FIELD

This section discusses the effect of uniform magnetic field on a closed current-carrying loop. The first likely effect, the exertion of force on this closed circuit, is described below.

12.3.1 Force on a Closed Circuit

In the case of a closed circuit, Equation 12.5c can be written in the following modified form:

$$F = -\oint\left(B\times\left[I\cdot d\ell\right]\right) \tag{12.24}$$

or,

$$F = -I\cdot\oint\left(B\times d\ell\right) \tag{12.24a}$$

If B is assumed to be uniform this equation can further be modified to:

$$F = -I\cdot B\times\oint\left(d\ell\right) \tag{12.24b}$$

In the zero value of the closed line integral the force exerted on a closed filamentary circuit by the uniform magnetic field is zero. In the case of non-uniform B, the total force can be evaluated from Equation 12.24a, which in general will not yield 0.

Equation 11.8, which relates different types of current distributions, is reproduced from Chapter 11 here:

$$I\cdot d\ell = K\cdot ds = J\cdot dv \tag{11.8}$$

In Equation 11.8, if B is taken to be uniform, Equation 12.5c can also be modified to:

$$F = -B\times\int K\cdot ds \tag{12.25a}$$

$$F = -B\times\int J\cdot dv \tag{12.25b}$$

12.3.2 Torque on a Closed Circuit

After establishing that the total vector force exerted by a uniform magnetic field on any real closed circuit carrying direct current is zero we divert our attention toward another important quantity referred to as torque.

The torque (also called moment of force) is a vector quantity and is denoted by the English capital letter T. Its magnitude $|T|$ is the product of the magnitudes of the vector force $|F|$, the vector lever arm $|R|$, and the sine of the angle between these two vectors. Its direction is normal to both the force and lever arm and is in the direction of progress of a right handed screw as the lever arm is rotated into the force vector through the smaller angle. The above description of the torque leads to the following mathematical expression:

$$T = R\times F \tag{12.26}$$

In this relation, the problem again boils down to the evaluation of force on the closed circuit. For such evaluation let us consider Figure 12.6, which shows a differential current loop located in a uniform $\boldsymbol{B}$-field. Let $\boldsymbol{B} = \boldsymbol{B}_o$ at the center of the loop ($x = y = 0$) where $\boldsymbol{B}_o$ can be written as:

$$\boldsymbol{B}_o = B_{ox}\boldsymbol{a}_x + B_{oy}\boldsymbol{a}_y + B_{oz}\boldsymbol{a}_z \tag{12.27}$$

This loop lies in the x–y plane and carries a current of I ampères. It encloses an elemental area $\boldsymbol{ds}$ ($= dxdy{\cdot}\boldsymbol{a}_z$) with its sides dx and dy lying along x and y axes respectively. This figure also shows the vector lever arm $\boldsymbol{R}$. This vector arm ($\boldsymbol{R}$) can be assigned suffixes in accordance with its orientation. Thus $\boldsymbol{R}_1$, $\boldsymbol{R}_2$, $\boldsymbol{R}_3$, and $\boldsymbol{R}_4$ will indicate $\boldsymbol{R}$ oriented toward side 1, 2, 3, and 4, respectively. Similarly, the forces exerted on sides 1, 2, 3, and 4, by $\boldsymbol{B}$-field can be indicated by $\boldsymbol{F}_1$, $\boldsymbol{F}_2$, $\boldsymbol{F}_3$, and $\boldsymbol{F}_4$, respectively. The vector forces on different differential lengths of sides can now be written in view of Equation 12.4c.

The vector force on side 1 is:

$$\boldsymbol{dF}_1 = \left(I \cdot dx \cdot \boldsymbol{a}_x\right) \times \boldsymbol{B}_o = I \cdot dx \cdot \boldsymbol{a}_x \times \left(B_{ox}\boldsymbol{a}_x + B_{oy}\boldsymbol{a}_y + B_{oz}\boldsymbol{a}_z\right)$$

or,

$$\boldsymbol{dF}_1 = I \cdot dx \cdot \left(B_{oy}\boldsymbol{a}_z - B_{oz}\boldsymbol{a}_y\right) \tag{12.28a}$$

The vector forces on side arms 2, 3, and 4 can be similarly obtained as:

$$\boldsymbol{dF}_2 = I \cdot dy \cdot \left(B_{oz}\boldsymbol{a}_x - B_{ox}\boldsymbol{a}_z\right) \tag{12.28b}$$

$$\boldsymbol{dF}_3 = -I \cdot dx \cdot \left(B_{oy}\boldsymbol{a}_z - B_{oz}\boldsymbol{a}_y\right) \tag{12.28c}$$

$$\boldsymbol{dF}_4 = -I \cdot dy \cdot \left(B_{oz}\boldsymbol{a}_x - B_{ox}\boldsymbol{a}_z\right) \tag{12.28d}$$

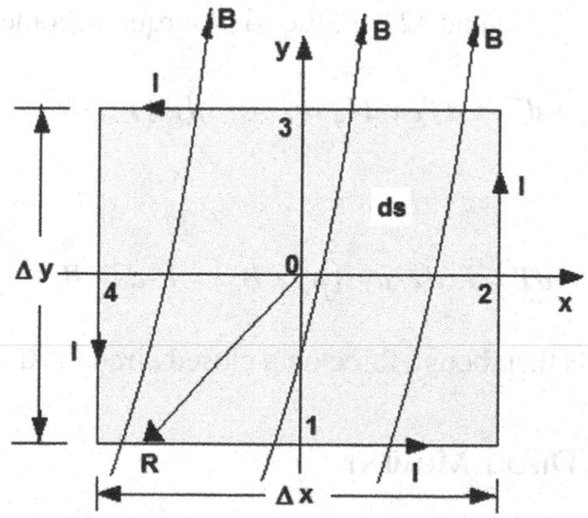

FIGURE 12.6 A closed circuit in a uniform magnetic field.

Addition of all components of Equations 12.28 a–d gives:

$$\boldsymbol{dF_1} + \boldsymbol{dF_2} + \boldsymbol{dF_3} + \boldsymbol{dF_4} = 0 \tag{12.28e}$$

Thus the total force on the closed loop is found to be zero.

For evaluating the torque, the vector arms $\boldsymbol{R_1}$, $\boldsymbol{R_2}$, $\boldsymbol{R_3}$, and $\boldsymbol{R_4}$ can be written by replacing Δx by dx, etc. shown in the Figure 12.6, as:

$$\left.\begin{aligned} \boldsymbol{R_1} &= \left(-\frac{1}{2}\right) dy \cdot \boldsymbol{a_y} \quad (\text{a}) \qquad \boldsymbol{R_2} = \left(+\frac{1}{2}\right) dx \cdot \boldsymbol{a_x} \quad (\text{b}) \\ \boldsymbol{R_3} &= \left(+\frac{1}{2}\right) dy \cdot \boldsymbol{a_y} \quad (\text{c}) \qquad \boldsymbol{R_4} = \left(-\frac{1}{2}\right) dx \cdot \boldsymbol{a_x} \quad (\text{d}) \end{aligned}\right\} \tag{12.29}$$

In view of Equations 12.26, 12.29a, and 12.28a, we get:

$$\begin{aligned} \boldsymbol{dT_1} = \boldsymbol{R_1} \times \boldsymbol{dF_1} &= (-1/2)\, dy \cdot \boldsymbol{a_y} \times I \cdot dx \cdot \left(B_{oy}\boldsymbol{a_z} - B_{oz}\boldsymbol{a_y}\right) \\ &= (-1/2)\, I \cdot B_{oy} dx \cdot dy \cdot \boldsymbol{a_x} \end{aligned} \tag{12.30a}$$

Similarly, in view of Equations 12.26, 12.29c, and 12.28c, we get:

$$\boldsymbol{dT_3} = (-1/2)\, I \cdot B_{oy} dx \cdot dy \cdot \boldsymbol{a_x} \tag{12.30b}$$

Thus,

$$\boldsymbol{dT_1} + \boldsymbol{dT_3} = -I \cdot B_{oy} dx \cdot dy \cdot \boldsymbol{a_x} \tag{12.31a}$$

Similarly, Equations 12.26, 12.29b, and 12.28b give the value of $\boldsymbol{dT_2}$ and Equations 12.26, 12.29d, and 12.28d give $\boldsymbol{dT_4}$. In view of these values:

$$\boldsymbol{dT_2} + \boldsymbol{dT_4} = +I \cdot B_{ox} dx \cdot dy \cdot \boldsymbol{a_y} \tag{12.31b}$$

In view of Equations 12.31a and 12.31b, the total torque becomes:

$$\boldsymbol{dT} = \boldsymbol{dT_1} + \boldsymbol{dT_2} + \boldsymbol{dT_3} + \boldsymbol{dT_4} = I \cdot dx \cdot dy \cdot \left(B_{ox} \cdot \boldsymbol{a_y} - B_{oy} \cdot \boldsymbol{a_x}\right)$$

or,

$$\boldsymbol{dT} = I \cdot dx \cdot dy \cdot \left(\boldsymbol{a_z} \times \boldsymbol{B_o}\right) = I \cdot \boldsymbol{ds} \times \boldsymbol{B_o} \tag{12.32}$$

Equation 12.32 reveals that though force on a closed circuit is 0, but the torque is not.

12.3.3 Magnetic Dipole Moment

In Equation 12.32 $\boldsymbol{ds}$ is the differential vector surface. On dropping the subscript from $\boldsymbol{B_o}$ the differential torque becomes:

$$dT = I \cdot ds \times B \tag{12.33a}$$

In Equation 12.33a, we can take:

$$dm = I \cdot ds \tag{12.33b}$$

Thus:

$$dT = dm \times B \tag{12.33c}$$

The quantity $\boldsymbol{dm}$ is defined as the differential dipole moment.

Earlier the electrical dipole moment $\boldsymbol{p}$ was defined in Section 7.5. Its differential version ($\boldsymbol{dp} = \boldsymbol{dq{\cdot}d}$) can be obtained in view of Equation 7.31c. In view of this differential electrical dipole moment a relation similar to that given by Equation 12.33c can readily be written as:

$$dT = dp \times E \tag{12.33d}$$

The integration of both the sides of Equation 12.33c gives:

$$T = m \times B \tag{12.34a}$$

where m represents the total dipole moment, which is related to the current and the total surface area as:

$$m = I \cdot S \tag{12.34b}$$

A simple look at Equations 12.33a and 12.33c reveals the general nature of these relations. These are true for differential surface of any shape.

12.3.4 Implications of Magnetic Dipole Moment

As noted earlier, all matter is comprised of tiny particles called molecules and atoms. The atom itself contains electrons and a nucleus containing protons and neutrons. The electrons not only revolve around this nucleus but also spin on their own axes. Such revolving and spinning electrons are shown in Figure 12.7a.

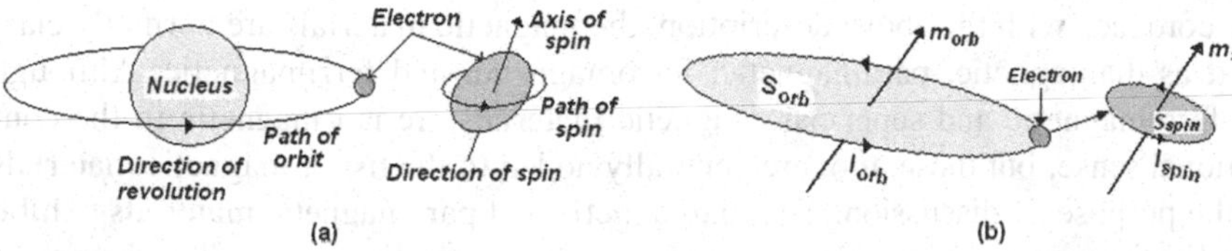

FIGURE 12.7 (a) Revolving and spinning electrons in an atom; (b) orbiting and spinning dipole moments.

A revolving or spinning electron can be considered as the flow of current along the closed path it encompasses. Let I_{orb} and I_{spin} are currents constituted by the revolving and spinning electrons and $\mathbf{S}_{\text{orb}}$ and $\mathbf{S}_{\text{spin}}$ are the differential surface areas enclosed by these two currents. The products of respective currents and enclosed areas yield the magnetic dipole moments described earlier. These can be written as:

$$\left.\boldsymbol{m}_{\text{orb}} = I_{\text{orb}} \cdot \mathbf{S}_{\text{orb}} \quad (\text{a}) \quad \text{and} \quad \boldsymbol{m}_{\text{spin}} = I_{\text{spin}} \cdot \mathbf{S}_{\text{spin}} \quad (\text{b})\right\} \tag{12.35}$$

The currents I_{orb} and I_{spin} the areas enclosed $\mathbf{S}_{\text{orb}}$ and $\mathbf{S}_{\text{spin}}$ and the dipole moments $\boldsymbol{m}_{\text{orb}}$ and $\boldsymbol{m}_{\text{spin}}$ are shown in Figure 12.7b. These dipole moments and their interaction with other electrons form the basis of magnetism. The studies have revealed that the influence of $\boldsymbol{m}_{\text{spin}}$ is more dominant on magnetic properties than that of $\boldsymbol{m}_{\text{orb}}$. As a result of the influence of dipole moments the internal magnetic field ($\boldsymbol{B}_{\text{int}}$) of some of the matters greatly change with the application of external magnetic field ($\boldsymbol{B}_{\text{app}}$).

12.4 MAGNETIC MATERIALS

In general all matters can be regarded as magnetic but some of them are more magnetic than others. In some of these there is no collective interaction of atomic magnetic moments, whereas in others it is quite strong. Thus the magnetic materials can be classified in terms of the degree of interaction.

Different materials react differently to the presence of an external magnetic field. This reaction depends on their atomic and molecular structures and the net magnetic field associated with the atoms. As described earlier, the magnetic moments associated with atoms originate from the electrons' motion and electrons' spin. The change in motion of these particles due to the presence of an external magnetic field may also result in magnetic moments.

In some of the matters the electrons occur in pairs which spin in opposite directions. The magnetic fields due to the opposite spin of these paired electrons get canceled. In such materials the net magnetic field is zero. Such materials rarely react in the presence of an external magnetic field. Some other materials which contain some unpaired electrons there is no perfect cancelation and their net magnetic field is not zero. This latter class of materials is more susceptible to an external magnetic field. In these the magnetic forces of the material's electrons get affected in accordance with Faraday's law of magnetic induction (discussed in Chapter 15).

12.4.1 Classification of Magnetic Materials

In accordance with the above description, the magnetic materials are normally classified as diamagnetic, paramagnetic, ferromagnetic, and ferrimagnetic. Although anti-ferromagnetic and super-paramagnetic materials are not magnetic in the conventional sense, but these, too, are generally added to the list of magnetic materials for the purpose of discussion. The diamagnetic and paramagnetic materials exhibit no collective magnetic interactions and are not magnetically ordered. The ferromagnetic, ferrimagnetic, and anti-ferromagnetic materials exhibit long-range magnetic order below a certain critical temperature. The ferromagnetic and ferrimagnetic are the only materials that are taken as magnetic, whereas all others are considered as weakly magnetic or non-magnetic. The ferromagnetic, ferrimagnetic, and super-paramagnetic materials are the only materials which are of practical importance.

12.4.1.1 Diamagnetic Materials

In these materials all the electron are paired so there is no net magnetic moment per atom. In other words, the orbiting and spinning electrons collectively result in zero dipole moment (i.e. $\boldsymbol{m}_{\text{orb}} + \boldsymbol{m}_{\text{spin}} = 0$). Diamagnetic properties arise from the realignment of the electron paths under the influence of an external magnetic field. These materials have a weak and negative susceptibility (see Section 12.4.3) to the magnetic fields, thus with the application of external field a negative magnetization is produced. As a result the internal magnetic field ($\boldsymbol{B}_{\text{int}}$) becomes slightly less than the applied field ($\boldsymbol{B}_{\text{app}}$). This reduction is so small that the internal and external magnetic fields can be presumed to be almost equal. In these materials the susceptibility is temperature independent. These materials do not retain the magnetic properties when the external magnetic field is removed. These materials include hydrogen, helium and other inert gases, sodium chloride, copper, gold, silicon, germanium, graphite and sulfur and metallic bismuth. The last one shows the most diamagnetic effect than others.

12.4.1.2 Paramagnetic Materials

In these materials the sum of orbiting and spinning dipole moments is almost zero (i.e. $\boldsymbol{m}_{\text{orb}} + \boldsymbol{m}_{\text{spin}} \approx 0$). These materials have a small positive susceptibility to the magnetic fields. Thus with the application of external field the internal field slightly enhances. Since this increase is almost negligible the internal and external magnetic fields can be regarded to be the same. Like diamagnetic materials, these materials also do not retain the magnetic properties when the external field is removed. The properties of these materials arise due to the presence of some unpaired electrons, and from the realignment of the electron paths caused by the external magnetic field. These materials include potassium, oxygen, tungsten, and other rare earth materials and many of their salts such as erbium chloride, neodymium, oxide, and yttrium oxide.

12.4.1.3 Ferromagnetic Materials

In paramagnetic materials atoms and hence spinning electrons are far apart not to influence each other. In ferromagnetic materials these are close enough to reinforce the effect of each other and the resulting magnetic field is stronger. In ferromagnetic materials the dipole moments due to spin electrons are much larger than those due to orbiting ones. These materials contain regions of varied shapes and sizes which are referred to as *domains* which are shown in Figure 12.8. As shown in Figure 12.8a, the dipoles in these domains are oriented in arbitrary directions. On the application of external magnetic field the dipoles in these domains try to align in the same direction. Figure 12.8b shows such an alignment for ferromagnetic case. As a result, the internal magnetic field becomes much greater than the applied field, i.e. $\boldsymbol{B}_{\text{int}} \gg \boldsymbol{B}_{\text{app}}$. Thus these materials have a large positive susceptibility to an external magnetic field. These materials retain their magnetic property even after the removal

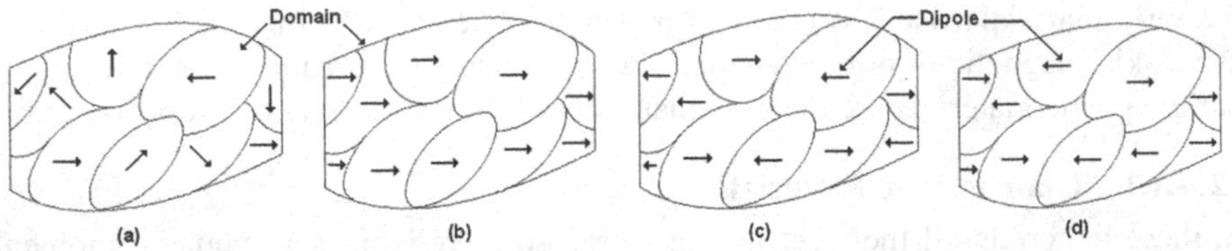

FIGURE 12.8 Domains and dipole orientation for (a) non-magnetized; (b) ferromagnetic; (c) anti-ferromagnetic; and (d) ferrimagnetic materials.

of external field. Ferromagnetic materials have fairly large μ and fairly high σ resulting in large eddy currents and high power loss at higher frequencies. Iron, nickel, and cobalt are the only ferromagnetic materials at room temperature. Some alloys, e.g. alnico (aluminum-nickel-cobalt alloy with a small amount of copper), and rare earth elements (at lower temperature), e.g. gadolinium and dysprosium, and some alloys of non-ferromagnetic metals, e.g. bismuth-manganese and copper-manganese-tin, are ferromagnetic.

12.4.1.4 Anti-Ferromagnetic Materials

In these materials the contribution of spinning electrons toward the dipole moments is much greater than that of orbiting electrons, i.e. $|\boldsymbol{m}_{\text{spin}}| \gg |\boldsymbol{mB}_{\text{orb}}|$. These materials also contain domains. The dipole moments in the adjacent domains are equal and aligned in anti-parallel directions. This results in total cancellation of dipole moments and the net internal magnetic field remains zero. With the application of external field there is not much change and the internal magnetic field remains almost equal to the applied field, i.e. $\boldsymbol{B}_{\text{int}} \approx \boldsymbol{B}_{\text{app}}$. The anti-ferromagnetic effect shown in Figure 12.8c is observed in manganese oxide.

12.4.1.5 Ferrimagnetic Materials

These materials are also referred to as ferrites. In these materials the contribution of spinning electrons toward dipole moments is more than that of orbiting electrons i.e. $|\boldsymbol{m}_{\text{spin}}| < |\boldsymbol{mB}_{\text{orb}}|$. These materials also contain domains wherein the dipole moments in adjacent domains are unequal and in opposition; thus, there is no total cancellation. With the application of an external field, the internal magnetic field increases, i.e. $\boldsymbol{B}_{\text{int}} > \boldsymbol{B}_{\text{app}}$. This situation is shown in Figure 12.8d. Ferrites have a fairly low conductivity (about 10^{-14} times of the metals) and thus at high frequencies (eddy) currents are greatly reduced resulting in less power loss. Ferrites have high μ (of the order of 5×10^3) with low $\boldsymbol{\varepsilon} \cong$ a few tens. This class of materials includes iron oxide magnetite, nickel-zink ferrite, and nickel ferrite.

12.4.1.6 Super-paramagnetic Materials

In these materials also, the contribution of spinning electrons toward the dipole moments is much larger than that of orbiting electrons. These materials are not magnetic in the conventional sense. These materials are obtained by combining magnetic and non-magnetic materials. The magnetic materials contain domains. The presence of non-magnetic materials prevents one domain to encroach into the territory of other nearby domain(s). With the application of applied field the internal magnetic field increases, i.e. $B_{int} > B_{app}$.

Some of the properties described earlier, of magnetic materials, are tabulated in Table 12.1.

12.4.2 Hysteresis Loop

Equation 11.1 gives a simple relation ($B = \mu H$) between the magnetic flux density and the magnetizing force. The media parameter called permeability (μ) is constant for all categories of magnetic materials except the ferromagnetic class wherein it exhibits non-linear behavior. As shown in Figure 12.9, this non-linear behavior results in a curve referred to as a B–H curve and a loop called the hysteresis (or B–H) loop. These can be obtained by varying H from $-H$ to $+H$ and on plotting the corresponding values of B which vary between $-B$ to $+B$. This section explains the formation of B–H curve and the B–H loop through the variation of B with H. As shown in Figure 12.9a the entire loop is divided into a number of segments, viz. 0–a, a–b, b–c, c–d, d–e, e–f, and f–a.

12.4.2.1 *B–H Curve* and Saturation

The curve shown by segment 0–a is called the *B–H curve*. It is obtained by plotting the values of B corresponding to the values of H. The origin of this curve is taken at $B = H = 0$. As H is increased from zero to $+H$, B increases from zero to $+B$. A virgin

TABLE 12.1
Some Salient Features of Magnetic Materials Shown in Figure 12.8 a, b, c, and d

	(a)	(b)	(c)	(d)
Status	**Non-magnetized**		**Magnetized**	
Material	All (b to d)	Ferromagnetic	Anti-ferromagnetic	Ferrimagnetic
Dipole Orientation	Random	Parallel	Equal and anti-parallel	Unequal and anti-parallel
Magnetic Field	Weak	Very strong	Very weak	Moderately strong

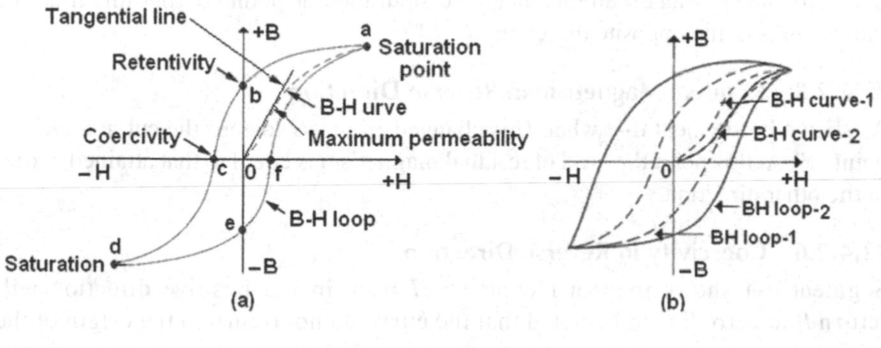

FIGURE 12.9 (a) Detail of hysteresis loop; (b) two hysteresis loops.

or perfectly demagnetized ferromagnetic material will follow a (non-linear) path shown by a dashed line between points "0 and a" in Figure 12.9a. At point "a" H attains such a value that almost all the magnetic dipoles in domains are aligned and B becomes maximum. A further increase in H results in no increase in B and the material is said to have reached the *magnetic saturation.*

12.4.2.2 Retentivity

Segment a–b shows that when the value of H is decreased from that at point "a" B does not decrease in accordance with the increasing path. At $H = 0$, B moves to point "b". At point "b" the value of B is much higher than the (original) zero value. This indicates that some magnetic flux is retained by the material. Thus the point "b" is referred to as *retentivity.* This indicates that some of the magnetic dipoles have lost their alignment. The term retentivity refers to the measure of the residual flux density corresponding to the saturation induction of a magnetic material. It spells the ability of material to retain a certain amount of residual magnetic field when the magnetizing force is removed after attaining saturation.

The *residual magnetism* or *residual flux* is another term associated with this segment. It is the magnetic flux density retained by a material at zero magnetizing force. The residual magnetism and retentivity are the same when the material is magnetized to the saturation point. In case the magnetizing force has not reach the saturation the level of residual magnetism may be lower than the retentivity value.

12.4.2.3 Coercivity

In segment b–c H is further reduced with the reversal of the polarity of H or the current. The curve moves to point "c" at which B reduces to zero but H has negative non-zero value. This point is indicated as the *coercivity* on the curve. At this point, the reversed H has flipped enough of the domains to reduce the net flux within the material to zero. Thus the *coercive force* or coercivity of the material is the force required to remove the residual magnetism from the material. Alternatively the *coercive force* is the amount of reverse magnetic field applied to a magnetic material to make the magnetic flux return to zero.

12.4.2.4 Saturation in Reverse Direction

In segment c–d the H is increased from point "c" to point 'd' in the negative direction. The material again attains magnetic saturation at point "d" but this time the saturation is in the opposite direction.

12.4.2.5 Residual Magnetism in Reverse Direction

As shown by segment d–e when H is changed from $-H$ to zero the curve travels to point "e". At this point the level of residual magnetism is equal to that attained earlier in the other direction.

12.4.2.6 Coercivity in Reverse Direction

Segment e–f shows that on increasing H back in the positive direction will return B to zero. It is to be noted that the curve do not return to the origin of the graph. This return will require some force to remove the residual magnetism. As such no name is assigned to point "f" however it indicates the coercivity in the negative direction.

12.4.2.7 Return to Saturation Point

In segment f–a, it can be seen that the curve takes a different path from point "f" back to the saturation point at "a", where it completes the loop.

12.4.2.8 Maximum Permeability

As noted in Chapter 11 and in the beginning of this section, permeability (μ) is the property of material that describes the ease with which a magnetic flux is established. It is the ratio of flux density (B) created within a material to the magnetizing field (H). The relation ($B = \mu H$) describes the slope of the curve at any point on the hysteresis loop. It is normally the maximum permeability or the maximum relative permeability which is found in the literature. The maximum permeability is the point where slope of the B–H curve for the non-magnetized material is the maximum. At this point the straight line from the origin is tangential to the B–H curve. The relative permeability (μ_r) is the ratio of the material's permeability (μ) to the permeability in free space or air ($\mu_0 = 4\pi \times 10^{-7}$) i.e. $\mu_r = \mu/\mu_0$.

12.4.2.9 Shape of Hysteresis Loop

The shape of hysteresis loop indicates the magnetization properties of the material. Figure 12.9b shows hysteresis loops for two different materials. One of these has a narrower and the other a wider configuration. A material with a wider loop has lower permeability, higher retentivity, higher coercivity, higher reluctance and higher residual magnetism. Similarly, a material with the narrower loop configuration has higher permeability, lower retentivity, lower coercivity, lower reluctance and lower residual magnetism. The permeability is related to the carbon content and alloying of the material. A material with high carbon content has low permeability and retains more magnetic flux than that with low carbon content.

12.4.3 Magnetization, Susceptibility, and Relative Permeability

This section deals with the mathematical aspects of three related quantities referred to as magnetization, susceptibility, and permeability.

12.4.3.1 Magnetization

In Equation 12.33b a new quantity named as vector differential dipole moment $\boldsymbol{dm}$ ($\boldsymbol{dm} = I\boldsymbol{ds}$) was defined, wherein I is the current and $\boldsymbol{ds}$ is the differential surface vector. All magnetic materials contain large number of such dipoles which may be identical or otherwise. For a magnetic material containing n such magnetic dipoles another quantity called magnetization and denoted by $\boldsymbol{M}$ can be defined. The magnetization for n identical magnetic dipoles can be written as:

$$\boldsymbol{M} = nI\boldsymbol{ds} \tag{12.36a}$$

For non-identical magnetic dipoles magnetization $\boldsymbol{M}$ can be obtained by averaging process over some given volume. Thus:

$$\boldsymbol{M} = \frac{1}{v}\sum_{i=1}^{m} I_i \boldsymbol{ds}_i \tag{12.36b}$$

The magnetization ($\boldsymbol{M}$) is a measure of the extent to which an object is magnetized. It is a measure of the magnetic dipole moment per unit volume of the object. It is also a vector quantity and carries the same units as a magnetic field intensity ($\boldsymbol{H}$) i.e. ampères per meter.

Figure 12.10 shows a segment dL of a closed path. It also shows a number of dipole moments along this length which are making an angle "θ_1" with this differential length. Each of these dipole moments is formed due to the flow of a current I along a closed path which encloses an area ds. In view of the configuration shown in Figure 12.10 the differential volume can be written as:

$$dv = ds\, dL \cos\theta_i = \boldsymbol{ds} \bullet \boldsymbol{dL} \tag{12.37a}$$

Since there are n dipole moments per unit volume the total dipoles in this volume are:

$$ndv = n\boldsymbol{ds} \bullet \boldsymbol{dL} \tag{12.37b}$$

If a magnetic field is applied across $\boldsymbol{dL}$ it will try to align the dipoles within the volume dv. Depending on the strength of this field these dipoles may partially (or fully) align along the differential length. This figure shows the partial alignment wherein the dipoles are twisted downward and make angle θ_2 with dL.

The concept of magnetization is similar to that of polarization introduced in Chapter 7. As polarization is related to the bound charge density the magnetization is related to the bound current density (J_b). The bound charges (Q_b) get elongated and cross the surface of the enclosed volume under the influence of electric field. Similarly, the bound current (dI_b) crosses the surface in the process of reorientation of dipole moments under the influence of magnetic field. As the magnetic

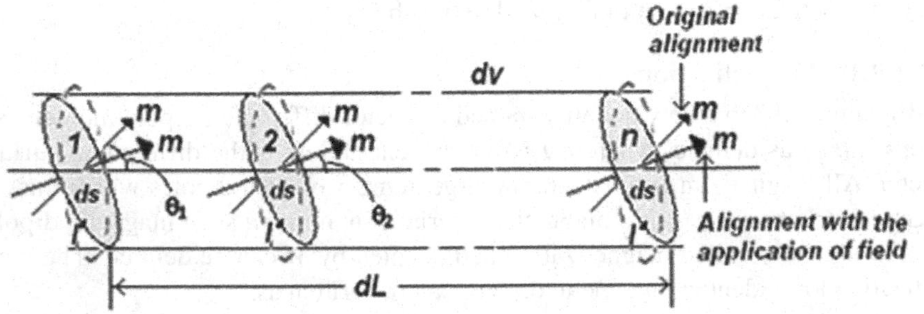

FIGURE 12.10 Dipole moments aligned along differential length before and after application of a magnetic field.

dipole moments try to fall in line with the applied magnetic field the bound current increases for each of the $\boldsymbol{nds} \bullet \boldsymbol{dL}$ dipoles and crosses the enclosed surface of the differential volume. It can be written as:

$$dI_b = nIds \bullet dL = \boldsymbol{M} \bullet \boldsymbol{dL} \tag{12.38a}$$

In view of Equation 12.36a, we get the total bound current I_b on integration:

$$I_b = \oint \boldsymbol{M} \bullet \boldsymbol{dL} \tag{12.38b}$$

If the bound current (I_b) is expressed in terms of bound current density (J_b), the RHS of Equation 12.38b can be written as:

$$\oint \boldsymbol{M} \bullet \boldsymbol{dL} = \int_s \boldsymbol{J}_b \bullet \boldsymbol{ds} \tag{12.38c}$$

The application of Stokes' theorem to the LHS of Equation 12.38c gives:

$$\int_s (\nabla \times \boldsymbol{M}) \bullet \boldsymbol{ds} = \int_s \boldsymbol{J}_b \bullet \boldsymbol{ds} \tag{12.38d}$$

In view of Equation 12.38, one gets:

$$\nabla \times \boldsymbol{M} = J_b \tag{12.39}$$

Here we reproduce a relation given by Equation 11.25:

$$\nabla \times \boldsymbol{H} = \boldsymbol{J} \ (\text{in free space}) \tag{11.25}$$

This relation is quite similar to that of Equation 12.39. Equation 11.25 was the result of the movement of free charges in vacuum (or free space) whereas Equation 12.39 arises from the bound current which produces the magnetization field. Since bound current is singled out in Equation 12.39 the current density $\boldsymbol{J}$ involved in Equation 11.25 must include the current density due to free and bound charges. Thus we can write:

$$\nabla \times \boldsymbol{H} = \boldsymbol{J}_f + \boldsymbol{J}_b \ \ (\text{in free space}) \tag{12.40a}$$

In terms of B, Equation 12.40a can be written as:

$$\nabla \times \frac{\boldsymbol{B}}{\mu_0} = \boldsymbol{J}_f + \boldsymbol{J}_b \tag{12.40b}$$

Furthermore, in Equation 12.39 we get:

$$\nabla \times \frac{\boldsymbol{B}}{\mu_0} = \boldsymbol{J}_f + \nabla \times \boldsymbol{M} \tag{12.41a}$$

or

$$\nabla \times \left(\frac{B}{\mu_0} - M \right) = J_f \tag{12.41b}$$

Equation 12.41b avoids direct consideration of bound current by merging M in the left side and defines B and H relations in more general form. Thus these can be written as:

$$\nabla \times H = J_f \tag{12.42}$$

where

$$\left. H = \frac{B}{\mu_0} - M \quad (a) \quad \text{or} \quad B = \mu_0 (H + M) \quad (b) \right\} \tag{12.43}$$

Equation 12.42 is the improved version of Maxwell's equation for steady fields when magnetic materials are present. The LHS of this equation explicitly spells about the movement of free charges while the effect of bound current is included in H on its RHS. The inclusion can be visualized in Equations 12.43a and 12.43b.

12.4.3.2 Susceptibility

Equation 12.43b can be simplified by defining a new dimensionless quantity called magnetic susceptibility denoted by a Greek symbol χ_m. It relates the magnetization (M) to the magnetic field intensity (H) as:

$$M = \chi_m H \tag{12.44a}$$

The susceptibility for diamagnetic materials is time independent whereas it is inversely proportional to the time for paramagnets. Figure 12.11 illustrates the variation of χ_m for some magnetic materials described earlier.

Thus, Equation 12.41b can be written as:

$$B = \mu_0 (H + \chi_m H) = \mu_0 (1 + \chi_m) H \tag{12.44b}$$

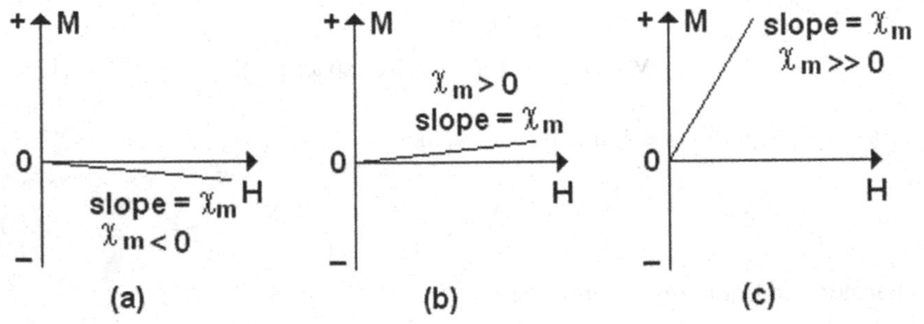

FIGURE 12.11 Magnetic susceptibility for (a) diamagnetic; (b) paramagnetic; and (c) for ferromagnetic materials.

12.4.3.3 Relative Permeability

In Equation 12.44b one can replace:

$$\mu_r = \left(1 + \chi_m\right) \tag{12.45a}$$

where μ_r represents the relative permeability of the material described earlier. It is also a dimensionless quantity. Thus Equation 12.44b reduces to:

$$\boldsymbol{B} = \mu_0 \mu_r \boldsymbol{H} \tag{12.45b}$$

In Equation 12.45b, μ_0 and μ_r are merged together to get:

$$\mu = \mu_0 \mu_r \tag{12.45c}$$

where μ is called the permeability of the material. It bears the same units as that of μ_0. Thus Equation 12.44b becomes:

$$\boldsymbol{B} = \mu \boldsymbol{H} \tag{12.46}$$

In anisotropic media, such as single ferromagnetic crystal, μ becomes a tensor. In view of Equation 12.46 the components of $\boldsymbol{B}$ can be written as:

$$B_x = \mu_{xx} H_x + \mu_{xy} H_y + \mu_{xz} H_z \tag{12.47a}$$

$$B_y = \mu_{yx} H_x + \mu_{yy} H_y + \mu_{yz} H_z \tag{12.47b}$$

$$B_z = \mu_{zx} H_x + \mu_{zy} H_y + \mu_{zz} H_z \tag{12.47c}$$

It is to mention that the given definitions of susceptibility and permeability are based on the assumption of linearity. This assumption holds good only for paramagnetic and diamagnetic materials for which the relative permeability rarely deviates from unity. The values of susceptibility for some of the diamagnetic and paramagnetic materials are given in Table 12.2.

TABLE 12.2
Values of Susceptibility for Some Diamagnetic and Paramagnetic Materials

Diamagnetic Material	Susceptibility χ_m	Paramagnetic Material	Susceptibility χ_m
Graphite	-12×10^{-5}	Oxygen	2×10^{-6}
Hydrogen	-2×10^{-5}	Yttrium oxide	0.53×10^{-6}
Copper	-0.9×10^{-5}	Tungsten	6.8×10^{-5}
Germanium	-0.8×10^{-5}	Ferric oxide	1.4×10^{-3}
Silicon	-0.3×10^{-5}		

The values of relative permeabilities for ferromagnetic materials may range from 10 to 100,000 while those for super-paramagnetic materials may vary between 1 and 10.

12.5 BOUNDARY CONDITIONS

In Section 8.5, it was mentioned that in many applications a field quantity may abruptly come across materials with different characterizing parameters. The involvement of materials with differing magnetic properties is quite common in electrical machine and other equipment and devices with static or rotating metallic parts separated by air-gaps or other non-magnetic materials. In such cases the behavior of magnetic field quantities attains significance when they cross from one medium to another. Such behavior depends on the relative permeabilities of the media. Like electrostatic field the study of magnetic boundary conditions also revolves around the behavior of Maxwell's equations in the regions in the vicinity of the interface.

In the case of an electrostatic field, it was noted that the evaluation of boundary conditions involves two common configurations and two common forms of equations. There it was also mentioned that the common configurations (Section 8.5.1) and the common form of equations (Section 8.5.2) are equally valid for electric and magnetic fields. It was further noted that the evaluation of boundary conditions for an $\boldsymbol{E}$- or $\boldsymbol{H}$-field involves a line integral whereas that for a $\boldsymbol{D}$- or $\boldsymbol{B}$-field a surface integral. The boundary conditions on dielectric-dielectric interface were described in Section 8.5.4, wherein Figure 8.10a was used to evaluate the line integral and Figure 8.10b to evaluate surface integral. Since the evaluation of boundary conditions at the interface between two magnetic materials also involve line and surface integrals, Figures 8.10a and b are equally good for magnetic field. In these figures the symbols k_1 and k_2 are to be replaced by μ_1 and μ_2 respectively; these represent the permeabilities of Region 1 and Region 2, respectively. Furthermore, in Figure 8.10a, $\boldsymbol{A}_1, \boldsymbol{A}_2$ are to be replaced by $\boldsymbol{H}_1$ and $\boldsymbol{H}_2$, and in Figure 8.10b by $\boldsymbol{B}_1$ and $\boldsymbol{B}_2$.

12.5.1 Boundary Conditions for H-Field

The boundary conditions to be satisfied by the magnetic field intensity between these two magnetic materials can be derived in view of the modified form of Figure 8.10a and Ampère's circuital law:

$$\oint \boldsymbol{H} \cdot d\boldsymbol{\ell} = I \tag{11.19}$$

In view of the description given earlier with Δh shrinking to zero Equation 11.17 gives:

$$H_{t1} \cdot \Delta w - H_{t2} \cdot \Delta w = K \cdot \Delta w \tag{12.48a}$$

This reduces to:

$$H_{t1} - H_{t2} = K \tag{12.48b}$$

In Equations 12.48a and 12.48b, K represents the component of surface current normal to the plane of the closed path. In view of Equation 12.48b the tangential components of $\boldsymbol{B}$ can simply be written as:

$$\frac{B_{t1}}{\mu_1} - \frac{B_{t2}}{\mu_2} = K \qquad (12.48c)$$

12.5.2 Boundary Conditions for B-Field

The boundary conditions to be satisfied by the magnetic flux density $\boldsymbol{B}$ at the interface between two magnetic materials can be derived in view of the modified form of Figure 8.10b and the relation:

$$\oiint \boldsymbol{B} \bullet \boldsymbol{ds} = 0 \qquad (11.29c)$$

In view of the earlier description with Δh shrinking to 0, Equation 8.29c becomes:

$$B_{n1} \cdot \Delta S - B_{n2} \cdot \Delta S = 0 \qquad (12.49a)$$

This gives:

$$B_{n1} - B_{n2} = 0 \qquad (12.49b)$$

In view of Equation 12.49b, the normal components of H are related as:

$$\mu_1 \cdot H_{n1} - \mu_2 \cdot H_{n2} = 0 \qquad (12.49c)$$

12.5.3 Boundary Conditions for M-Field

For a magnetic material, the magnetization $\boldsymbol{M}$ is defined from the following equation:

$$\boldsymbol{B} \overset{\text{def}}{=} \mu_o \cdot (\boldsymbol{H} + \boldsymbol{M}) \overset{\text{def}}{=} \mu_r \cdot \boldsymbol{H} \qquad (12.50a)$$

The magnetic susceptibility χ_m relates magnetization $\boldsymbol{M}$ to the magnetic field intensity $\boldsymbol{H}$ as:

$$\boldsymbol{M} \overset{\text{def}}{=} \chi_m \cdot \boldsymbol{H} \qquad (12.50b)$$

Therefore, the relative permeability μ_r of the material is described as:

$$\mu_r = 1 + \chi_m \qquad (12.50c)$$

The tangential components of magnetization $\boldsymbol{M}$ in Regions 1 and 2 are related as:

$$\frac{M_{t2}}{\chi_{m2}} = \frac{M_{t1}}{\chi_{m1}} - K \qquad (12.51a)$$

The normal components of magnetization $\boldsymbol{M}$ in Regions 1 and 2 are related as:

$$\frac{M_{n2}}{\chi_{m2}} \cdot \mu_2 = \frac{M_{n1}}{\chi_{m1}} \cdot \mu_1 \qquad (12.51b)$$

Equations 12.51a and 12.51b are obtained from Equations 12.48b, 12.49b, 12.50a, and 12.50b.

12.5.4 Additional Boundary Conditions

In this case, Figure 8.11 can also be used. In this configuration consider a regular point P on the boundary surface between two regions with different values of permeability, say, μ_1 and μ_2 for Region 1 and 2, respectively. Let on this boundary point there is a surface current density with its components K_x and K_y. The z-direction is normal to the boundary, directed from Region 2 to Region 1. At this point on the boundary, the following conditions are well known:

$$H_{1x}\big|_P - H_{2x}\big|_P = K_y\big|_P \qquad (12.52a)$$

$$H_{1y}\big|_P - H_{2y}\big|_P = -K_x\big|_P \qquad (12.52b)$$

We therefore have two cases, based on the tangential derivatives of Equation 10.48b on the boundary.

Case I:

$$\left.\frac{\partial H_{1x}}{\partial x}\right|_P - \left.\frac{\partial H_{2x}}{\partial x}\right|_P = \left.\frac{\partial K_y}{\partial x}\right|_P \qquad (12.53a)$$

$$\left.\frac{\partial H_{1y}}{\partial y}\right|_P - \left.\frac{\partial H_{2y}}{\partial y}\right|_P = -\left.\frac{\partial K_x}{\partial y}\right|_P \qquad (12.53b)$$

Thus, on adding we get:

$$\left.\left(\frac{\partial H_{1x}}{\partial x} + \frac{\partial H_{1y}}{\partial y}\right)\right|_P - \left.\left(\frac{\partial H_{2x}}{\partial x} + \frac{\partial H_{2y}}{\partial y}\right)\right|_P = \left.\left(\frac{\partial K_y}{\partial x} - \frac{\partial K_x}{\partial y}\right)\right|_P \qquad (12.53c)$$

These boundary conditions are valid for linear, non-linear, homogeneous, or inhomogeneous media. Let us assume that at least near the boundary surface both regions are linear and homogeneous. Thus, near the boundary surface:

$$\nabla \bullet \boldsymbol{H}_1 = \nabla \bullet \boldsymbol{H}_2 = 0 \qquad (12.54)$$

Therefore, Equation 12.53c may be written as:

$$\left.\frac{\partial H_{1z}}{\partial z}\right|_P - \left.\frac{\partial H_{2z}}{\partial z}\right|_P = -\left.\left[\nabla \times \boldsymbol{K}\right]_z\right|_P \tag{12.55}$$

Thus if the surface current density, or the normal component of its curl is absent on the boundary surface, the normal derivative of the normal component of magnetic field intensity will be continuous across the boundary between the two regions. This boundary condition has been used to solve a boundary value problem involving magnetostatic field.

Case II:

$$\left.\frac{\partial H_{1x}}{\partial y}\right|_P - \left.\frac{\partial H_{2x}}{\partial y}\right|_P = \left.\frac{\partial K_y}{\partial y}\right|_P \tag{12.56a}$$

$$\left.\frac{\partial H_{1y}}{\partial x}\right|_P - \left.\frac{\partial H_{2y}}{\partial x}\right|_P = -\left.\frac{\partial K_x}{\partial x}\right|_P \tag{12.56b}$$

For time-invariant fields, the divergence of the surface current density being zero, we have:

$$\left.\frac{\partial H_{1x}}{\partial y}\right|_P - \left.\frac{\partial H_{1y}}{\partial x}\right|_P = \left.\frac{\partial H_{2x}}{\partial y}\right|_P - \left.\frac{\partial H_{2y}}{\partial x}\right|_P$$

Or,

$$\left.\left[\nabla \times \boldsymbol{H}_1\right]_z\right|_P = \left.\left[\nabla \times \boldsymbol{H}_2\right]_z\right|_P \tag{12.57a}$$

Thus, the normal component of the curl of magnetic field intensity is continuous across the boundary between the two regions. This implies that if there is a distribution of the time-invariant source current density $\boldsymbol{J}$, its normal component must be continuous across the boundary, i.e.:

$$\left.J_{1z}\right|_P = \left.J_{2z}\right|_P \tag{12.57b}$$

The boundary conditions given by Equations 12.57a and 12.57b are valid for linear, non-linear, homogeneous, or inhomogeneous media.

In the above relations H_{1x}, H_{1y}, and H_{1z} are the x, y, and z components of H in Region 1 and H_{2x}, H_{2y}, and H_{2z} are x, y, and z components in Region 2.

Example 12.1

A point charge of 1.5 C has the velocity $5\boldsymbol{a}_x + 2\boldsymbol{a}_y - 3\boldsymbol{a}_z$ m/s. Find the magnitude of force exerted on it in the field of (a) $\boldsymbol{E} = 10\boldsymbol{a}_x + 16\boldsymbol{a}_y - 6\boldsymbol{a}_z$ V/m; (b) $\boldsymbol{B} = -2\boldsymbol{a}_x + 3\boldsymbol{a}_y + 5\boldsymbol{a}_z$ wb/m²; (c) both $\boldsymbol{E}$ and $\boldsymbol{B}$.

SOLUTION

(a) $F = QE = 1.5(10a_x + 16a_y - 6a_z) = 15a_x + 24a_y - 9a_z$

$$|F| = \sqrt{15^2 + 24^2 + 9^2} = \sqrt{225 + 576 + 81} = \sqrt{882} = 29.7 \text{ Newton}$$

(b) $F = Q(\upsilon \times B) = 1.5(5a_x + 2a_y - 3a_z) \times (-2a_x + 3a_y + 5a_z)$

$$= 1.5(19a_x - 19a_y + 19a_z) = 28.5(a_x - a_y + a_z)$$

$$|F| = 28.5 \times \sqrt{3} = 49.36 \text{ Newton}$$

(c) $F = Q(E + \upsilon \times B) = QE + Q(\upsilon \times B)$

$$= (15a_x + 24a_y - 9a_z) + 28.5(a_x - a_y + a_z) = 43.5a_x - 4.5a_y + 19.5a_z$$

$$|F| = \sqrt{43.5^2 + 4.5^2 + 19.5^2} = 47.88 \text{ Newton}$$

Example 12.2

An infinitely long filamentary conductor carrying a current of 5 A in the a_z direction lies on the z-axis. Find the magnitude of on 1 mm length of the conductor in the field of (a) $B = 0.2a_x - 0.3a_z$ wb/m²; (b) $B = 0.3a_x + 0.4a_y$ wb/m²; (c) $B = 0.5a_x$ wb/m², $E = 0.4a_y$ mV/m.

SOLUTION

$$dL = 10^{-3}a_z \qquad IdL = 5 \times 10^{-3}a_z \qquad dF = IdL \times B$$

(a) $dF = (5 \times 10^{-3}a_z) \times (0.2a_x - 0.3a_z) = 10^{-3}a_y N$

(b) $dF = (5 \times 10^{-3}a_z) \times (0.3a_x + 0.4a_y) = (1.5 \times 10^{-3}a_y - 2 \times 10^{-3}a_x) N$

(c) $dF = (5 \times 10^{-3}a_z) \times (0.5a_x) = 2.5 \times 10^{-3}a_y N$

Example 12.3

A filamentary current of 5 A is passing through point $P_1(-2,3,0)$ in a_x direction and a second filament with 6 A current passes through $P_2(3,-4,0)$ in a_y direction. Find (a) the vector force exerted on an incremental length ΔL_2 of the second conductor located at P_2 by incremental length ΔL_1 of the first conductor at P_1 (b) the vector force on ΔL_1 at P_1 by ΔL_2 at P_2.

SOLUTION

Figure 12.12 shows the parameters involved in the problem.

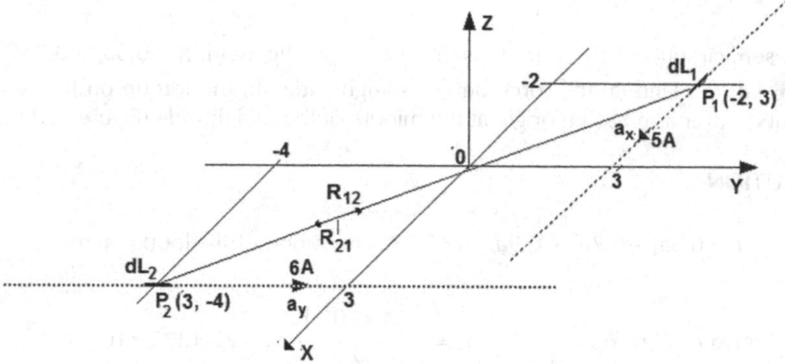

FIGURE 12.12 Two filamentary current loops lying parallel to x- and y-axes.

(a) In view of Equation 11.6: $\boldsymbol{dH_2} = \dfrac{I_1\boldsymbol{dL_1} \times \mathbf{a}_{R12}}{4\pi R_{12}^2}$

$$I_1\boldsymbol{dL_1} = 5\Delta L_1\mathbf{a}_x \qquad R_{12} = (-2-3)\,\mathbf{a}_x + (3-(-4))\,\mathbf{a}_y = -5\mathbf{a}_x + 7\mathbf{a}_y$$

$$|R_{12}| = \sqrt{25+49} = \sqrt{74} = 8.6 \qquad \mathbf{a}_{R12} = \frac{-5\mathbf{a}_x + 7\mathbf{a}_y}{8.6} = -0.581\mathbf{a}_x + 0.814\mathbf{a}_y$$

$$dH_2 = \frac{5\Delta L_1\mathbf{a}_x \times (-0.58124\mathbf{a}_x + 0.8139\mathbf{a}_y)}{4\pi \times 74} = \frac{4.07\Delta L_1\mathbf{a}_z}{4\pi \times 74}$$

$$dB_2 = \mu dH_2 = \mu_0 dH_2 = 4\pi \times 10^{-7}\left[\frac{4.07\Delta L_1\mathbf{a}_z}{4\pi \times 74}\right] = 5.5 \times 10^{-9}\,\Delta L_1\mathbf{a}_z$$

$$d(dF_2) = I_2\Delta L_2\mathbf{a}_y \times \boldsymbol{dB_2} = 6\Delta L_2\mathbf{a}_y \times (5.5 \times 10^{-9}\,\Delta L_1\mathbf{a}_z) = 33\Delta L_1\Delta L_2\mathbf{a}_x\ nN$$

(b) $\boldsymbol{dH_1} = \dfrac{I_2\boldsymbol{dL_2} \times \mathbf{a}_{R21}}{4\pi R_{21}^2} \qquad I_2\boldsymbol{dL_2} = 6\Delta L_2\mathbf{a}_y \qquad R_{21} = -R_{12} = 5\mathbf{a}_x - 7\mathbf{a}_y$

$$|R_{21}| = |R_{12}| = \sqrt{74} = 8.6 \qquad \mathbf{a}_{R21} = -\mathbf{a}_{R12} = 0.58124\mathbf{a}_x - 0.8139\mathbf{a}_y$$

$$dH_1 = \frac{6\Delta L_2\mathbf{a}_y \times (0.58124\mathbf{a}_x - 0.8139\mathbf{a}_y)}{4\pi \times 74} = \frac{-3.487\Delta L_2\mathbf{a}_z}{4\pi \times 74}$$

$$dB_1 = \mu dH_1 = \mu_0 dH_1 = 4\pi \times 10^{-7}\left[\frac{-3.487\Delta L_2\mathbf{a}_z}{4\pi \times 74}\right] = -4.712 \times 10^{-9}\,\Delta L_2\mathbf{a}_z$$

$$d(dF_1) = I_1\Delta L_1\mathbf{a}_x \times \boldsymbol{dB_1} = 5\Delta L_1\mathbf{a}_x \times (-4.712 \times 10^{-9}\,\Delta L_2\mathbf{a}_z)$$

$$= 23.56\Delta L_1\Delta L_2\mathbf{a}_y\ nN$$

Example 12.4

The semicircular current loop is in the magnetic field $\boldsymbol{B} = 0.5\mathbf{a}_x - 0.7\mathbf{a}_y + 0.9\mathbf{a}_z$ wb/m^2. Find (a) the force on the straight side; (b) the torque on the loop about a leaver arm having origin at the middle of the straight side (Figure 12.13).

SOLUTION

$$\boldsymbol{B} = 0.5\mathbf{a}_x - 0.7\mathbf{a}_y + 0.9\mathbf{a}_z \quad I = 50\text{ A} \quad r = \text{radius of the loop} = 4\text{cm}$$

$$\text{Area of the loop} = \boldsymbol{S} = \frac{\pi r^2}{2}\mathbf{a}_z = \frac{\pi\left(4\times10^{-2}\right)^2}{2}\mathbf{a}_z = 25.1327\times10^{-4}\mathbf{a}_z$$

$$I\boldsymbol{S} = 50\times25.1327\times10^{-4}\mathbf{a}_z = 12.566\times10^{-2}\mathbf{a}_z$$

$$\boldsymbol{F} = -I\int \boldsymbol{B}\times d\boldsymbol{L} = 50\int_{-0.04}^{0.04}\left(0.5\mathbf{a}_x - 0.7\mathbf{a}_y + 0.9\mathbf{a}_z\right)\times dy\mathbf{a}_y$$

(a) $$= 50\int_{-0.04}^{0.04}\left(0.5\mathbf{a}_z - 0.9\mathbf{a}_x\right)dy = \int_{-0.04}^{0.04}\left(25\mathbf{a}_z - 45\mathbf{a}_x\right)dy$$

$$= \left(25\mathbf{a}_z - 45\mathbf{a}_x\right)y\Big|_{-0.04}^{0.04} = 0.08\left(25\mathbf{a}_z - 45\mathbf{a}_x\right) = 2\mathbf{a}_z - 3.6\mathbf{a}_x\ N$$

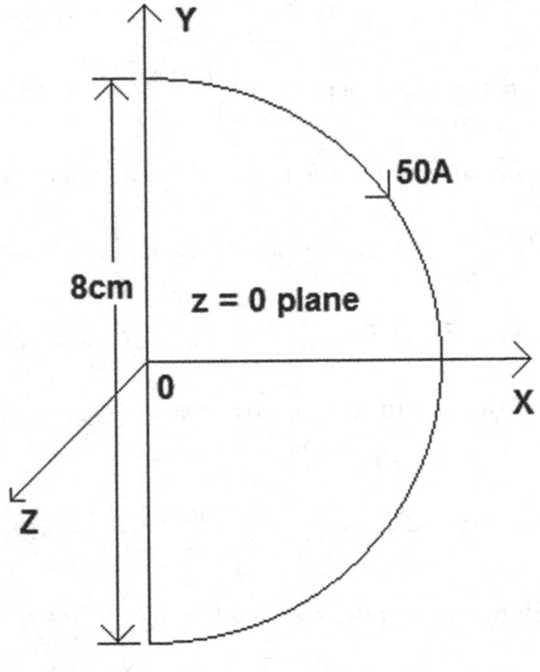

FIGURE 12.13 A semicircular current loop.

(b)
$$T = IS \times B = 12.566\times10^{-2} \boldsymbol{a}_z \times \left(0.5\boldsymbol{a}_x - 0.7\boldsymbol{a}_y + 0.9\boldsymbol{a}_z\right)$$
$$= \left(0.08796\boldsymbol{a}_x + 0.0628\pi\boldsymbol{a}_y\right)\text{Nm}$$

Example 12.5

Find the magnitude of magnetic field intensity $|H|$ in a material wherein (a) the magnetic flux density $B = 5$ mwb/m^2 and the relative permeability $\mu_r = 1.01$; (b) the magnetic susceptibility $\chi_m = -0.005$ magnetization $M = 20$ A/m; (c) there are 8.1×10^{28} atoms/m^3 each atom has a dipole moment $m = 4 \times 10^{-30}$ A m^2 and $\chi_m = 10^{-4}$.

SOLUTION

$$B = 5\ \text{mwb/m}^2 \qquad \mu_r = 1.01 \qquad B = \mu H = \mu_0\mu_r H$$

(a) $$H = \frac{B}{\mu_0\mu_r} = \frac{5\times10^{-3}}{4\pi\times10^{-7}\times1.01} = 3939\ \text{A/m}$$

(b) $$\chi_m = -0.005 \quad M = 20\ \text{A/m} \quad H = \frac{M}{\chi_m} = \frac{20}{-0.005} = -4000 \quad |H| = 4000\ \text{A/m}$$

(c) N = Number of atoms/m^3 = 8.1×10^{28} $\quad m = 4 \times 10^{-30}$ A m^2 $\quad \chi_m = 10^{-4}$

$$M = N\cdot m = 8.1\times10^{28}\times4\times10^{-30} = 32.4\times10^{-2}$$

$$H = \frac{M}{\chi_m} = \frac{32.4\times10^{-2}}{10^{-4}} = 3240\ \text{A/m}$$

Example 12.6

A magnetic material has $\boldsymbol{H} = 5\rho^4\boldsymbol{a}_\varphi\ \frac{\text{A}}{\text{m}}$ and $\mu = 5\times10^{-6}\ \frac{\text{H}}{\text{m}}$. Find at $\rho = 3$ (a) J; (b) J_T; and (c) J_b.

SOLUTION

(a) $$\boldsymbol{J} = \nabla\times\boldsymbol{H} = \frac{1}{\rho}\frac{\partial(\rho H_\varphi)}{\partial\rho}\boldsymbol{a}_z = \frac{1}{\rho}\frac{\partial(5\rho^5)}{\partial\rho}\boldsymbol{a}_z = 25\rho^3\Big|_{\rho=3}\boldsymbol{a}_z = 675\boldsymbol{a}_z\ \text{A/m}^2$$

(b) $$\mu = 5\times10^{-6} = \mu_0\mu_r \qquad \mu_r = \frac{\mu}{\mu_0} = \frac{5\times10^{-6}}{4\pi\times10^{-7}} = 3.9788 \quad \boldsymbol{B} = \mu\boldsymbol{H} = \mu_0\mu_r\boldsymbol{H}$$

$$J_T = \frac{\nabla\times\boldsymbol{B}}{\mu_0} = \nabla\times(\mu_r\boldsymbol{H}) = \mu_r\boldsymbol{J} = 3.9788\times675\boldsymbol{a}_z = 2685.74\boldsymbol{a}_z\ \text{A/m}^2$$

(c) $$\boldsymbol{B} = \mu_0(\boldsymbol{H}+\boldsymbol{M}) = \mu_0\boldsymbol{H} + \mu_0\boldsymbol{M} \qquad \boldsymbol{M} = \frac{(\boldsymbol{B}-\mu_0\boldsymbol{H})}{\mu_0} = \frac{\boldsymbol{B}}{\mu_0} - \boldsymbol{H}$$

$$J_b = \nabla \times M = \nabla \times \frac{B}{\mu_0} - \nabla \times H = J_T - J$$

$$J_b = (2685.74 - 675)\mathbf{a}_z = 2010.74\mathbf{a}_z \text{ A/m}^2$$

Example 12.7

There is an interface between the surfaces of two homogeneous, linear, isotropic materials at $x = 0$ which carries a surface current of density $K = 100\mathbf{a}_z$. For $x < 0$ (region 1) $\mu_{R1} = 3$, $\mathbf{H}_1 = 100\mathbf{a}_x - 200\mathbf{a}_y + 300\mathbf{a}_z$ and for $x > 0$ (region 2) $\mu_{R2} = 5$. Find (a) H_2; (b) $|B_1|$; and (c) $|B_2|$.

SOLUTION

$$\mathbf{H}_1 = 100\mathbf{a}_x - 200\mathbf{a}_y + 300\mathbf{a}_z \qquad \mathbf{H}_{n1} = 100\mathbf{a}_x \qquad \mathbf{H}_{t1} = -200\mathbf{a}_y + 300\mathbf{a}_z$$

$$\mathbf{H}_{t1} - \mathbf{H}_{t2} = \mathbf{K} \qquad \mathbf{K} = 100\mathbf{a}_z$$

$$\mathbf{H}_{t2} = \mathbf{H}_{t1} - \mathbf{K} = -200\mathbf{a}_y + 300\mathbf{a}_z - 100\mathbf{a}_z = -200\mathbf{a}_y + 200\mathbf{a}_z$$

$$B_{n1} = B_{n2} \qquad \mu_1 H_{n1} = \mu_2 H_{n2}$$

$$H_{n2} = \frac{\mu_1}{\mu_2} H_{n1} = \frac{\mu_{R1}\mu_0}{\mu_{R2}\mu_0} H_{n1} = \frac{3}{5} 100a_x = 60a_x$$

$$\mathbf{H}_2 = 60\mathbf{a}_x - 200\mathbf{a}_y + 200\mathbf{a}_z$$

$$\mathbf{B}_1 = \mu_{R1}\mu_0\mathbf{H}_1 = 3 \times 4\pi \times 10^{-7}(100\mathbf{a}_x - 200\mathbf{a}_y + 300\mathbf{a}_z)$$

$$= (0.377\mathbf{a}_x - 0.754\mathbf{a}_y + 1.131\mathbf{a}_z) \text{ mwb/m}^2$$

$$|\mathbf{B}_1| \approx 1.41 \text{ mwb/m}^2$$

$$\mathbf{B}_2 = \mu_{R2}\mu_0\mathbf{H}_2 = 5 \times 4\pi \times 10^{-7}(60\mathbf{a}_x - 200\mathbf{a}_y + 200\mathbf{a}_z)$$

$$= (0.377\mathbf{a}_x - 1.2566\mathbf{a}_y + 1.2566\mathbf{a}_z) \text{ mwb/m}^2$$

$$|\mathbf{B}_2| \approx 2 \text{ mwb/m}^2$$

PROBLEMS

P12.1 Find the magnitude of force exerted on a point charge of 5 C traveling with the velocity of $\boldsymbol{v} = 5\boldsymbol{a}_x - 2\boldsymbol{a}_y + 3\boldsymbol{a}_z$ m/s in the field of: (a) $\boldsymbol{E} = 20\boldsymbol{a}_x - 10\boldsymbol{a}_y - 6\boldsymbol{a}_z$ V/m; (b) $\boldsymbol{B} = 2\boldsymbol{a}_x - 5\boldsymbol{a}_y - 5\boldsymbol{a}_z$ wb/m²; (c) both $\boldsymbol{E}$ and $\boldsymbol{B}$.

P12.2 An infinitely long filamentary conductor carrying a current of 2 A in the $\boldsymbol{a}_z$ direction lies on the z-axis. Find the magnitude of on 1 mm length of the conductor in the field of (a) $=0.2\boldsymbol{a}_x + 0.5\boldsymbol{a}_z$ wb/m²; (b) $\boldsymbol{B} = 0.3\boldsymbol{a}_x - 0.2\boldsymbol{a}_y$ wb/m²; (c) $\boldsymbol{B} = -0.4\boldsymbol{a}_x$ wb/m², $\boldsymbol{E} = 0.5\boldsymbol{a}_y$ mV/m.

P12.3 Two filamentary conductors are located in $z = 0$ plane. The first conductor carrying 3 A current is passing through point $P_1(-2,1,0)$ in a_x direction whereas the second carrying 4 A current passes through $P_2(3,4,0)$ in a_y direction. Find (a) the vector force exerted on an incremental length ΔL_2 of the second conductor located at P_2 by incremental length ΔL_1 of the first conductor at P_1 (b) the vector force on ΔL_1 at P_1 by ΔL_2 at P_2.

P12.4 A semicircular loop of 10 cm diameter carrying a current of 20 A is located in $z = 0$ plane having a field $= 0.6\boldsymbol{a}_x + 0.4\boldsymbol{a}_y - 0.2\boldsymbol{a}_z$ wb/m². Find (a) the force on the straight side; (b) the torque on the loop about a leaver arm having origin at the mid of straight side.

P12.5 A square loop, of 6 cm sides, carries 10 A current. It is centered in the $z =$ 0 plane with sides parallel to x and y axes. The current at 3,0,0 cm is in the $\boldsymbol{a}_y$ direction. Find the vector torque (**T**) produced about an axis through the origin by the field: (a) $\boldsymbol{B} = 2\boldsymbol{a}_x$ wb/m² everywhere; (b) $\boldsymbol{B} = 3\boldsymbol{a}_x$ wb/m² for $x \geq 3$ cm, and zero elsewhere; (c) $\boldsymbol{B} = 5\boldsymbol{a}_x$ wb/m³ for $x \geq 0, y \geq 3$ and zero elsewhere.

P12.6 Find the magnitude of magnetic field intensity $|H|$ in a material wherein (a) the magnetic flux density $B = 3$ mwb/m² and the relative permeability $\mu_r = 2.26$; (b) the magnetic susceptibility $\chi_m = 0.004$ magnetization $M =$ 10 A/m; (c) there are 5.1×10^{26} atoms/m³ each atom has a dipole moment $m = 3 \times 10^{-30}$ A m² and $\chi_m = 2 \times 10^{-4}$.

P12.7 A magnetic material has $\boldsymbol{H} = 10\rho^3\boldsymbol{a}_\varphi \dfrac{\text{A}}{\text{m}}$ and $\mu = 3\times10^{-6}\dfrac{\text{H}}{\text{m}}$. Find at $\rho =$ 5 (a) $\boldsymbol{J}$; (b) $\boldsymbol{J}_T$; and (c) $\boldsymbol{J}_b$.

P12.8 At $x = 0$ an interface between the surfaces of two homogeneous, linear, isotropic materials carries a surface current of density $\boldsymbol{K} = 50\boldsymbol{a}_z$. For $x < 0$ (region 1) $\mu_{R1} = 2$, $\boldsymbol{H}_1 = 10\boldsymbol{a}_x + 20\boldsymbol{a}_y - 30\boldsymbol{a}_z$ and for $x > 0$ (region 2) $\mu_{R2} = 3$ Find (a) $\boldsymbol{H}_2$; (b) $|\boldsymbol{B}_1|$; and (c) $|\boldsymbol{B}_2|$.

P12.9 A B–H curve for a material is represented by two line segments. The first segment corresponds to $H = 0$, $B = 0$ to $H = 1000$ AT/m, $B = 1$ wb/m² and the second segment from $H = 1000$ AT/m, $B = 1$ wb/m² to $H = 3000$ AT/m, $B = 1.4$ wb/m² A magnetic core made-up of this material has 10 cm length of 1 cm² cross-section and 10 cm length of 0.5 cm². Find the flux produced by a 1000 turn in this magnetic circuit if is excited by a coil carrying a current of (a) 0.5 A; and (b) 2.5 A.

DESCRIPTIVE QUESTIONS

Q12.1 Derive the expressions for force due to magnetic field on a line current, surface current density and volume current density.

Q12.2 Derive the relation to represent the force between two differential current elements.

Q12.3 Obtain the condition for equality of electric and magnetic forces.

Q12.4 Derive the relations for the force and torque on a close current-carrying loop. How these relations are related to the dipole moment?

Q12.5 Discuss magnetic materials with reference to the magnetic dipole moments.

Q12.6 Describe the formation of B–H loop. Discuss its salient features and implications on the magnetic circuits.

Q12.7 Derive the relations between the magnetization, susceptibility, and permeability.

Q12.8 Discuss the boundary conditions satisfied by H and B fields at the interface between two magnetic materials.

BIBLIOGRAPHY

David J. Griffiths, *Introduction to Electrodynamics*, 3rd ed., PHI Learning Private Limited, New Delhi, 2012, pp. 477–525.

FURTHER READING

Given at the end of Chapter 15.

13 Magnetic Circuits

13.1 INTRODUCTION

Chapter 11 introduced the concepts, quantities, and laws relating to steady magnetic fields. Chapter 12 described the magnetostatic forces, torque, dipole moment, and the magnetic materials. This chapter also included the properties and boundary conditions related to magnetic materials. All of these aspects are commonly involved in the analysis, design, and fabrication of the various devices mentioned in Chapter 1. Devices composed of magnetic materials are often analyzed by modeling these in the form of magnetic circuits. This chapter therefore deals with magnetic circuits that closely resemble electrical circuits. In Chapter 8, some circuit parameters (viz. R and C) were introduced. This chapter introduces another circuit parameter, called *inductance*. This chapter also includes a discussion of the energy stored in magnetostatic fields.

13.2 MAGNETIC CIRCUITS

As such, there are no magnetic circuits in the conventional sense. The equivalent circuits depicting the magnetic phenomena involved in some devices and systems can be referred to as magnetic circuits. Such circuits can be easily drawn by noting the similarities between the relations and quantities involved in electrostatic and magnetostatic fields. These similarities are noted below.

13.2.1 Similarities between Field Quantities and Relations

The similarities between the relations of electric and magnetic fields are given in Table 13.1.

In the table, H and E, $\mathcal{V}$ and V, $\mathcal{V}_{AB}$ and V_{AB}, B and J, μ and σ, and Φ and I are analogous quantities in magnetostatic and electrostatic fields.

The electrical circuits operating with dc sources involve EMF or voltage (V), current (I), and the resistance (R). In Ohm's law, these quantities are related as:

$$V = I \cdot R \tag{8.19a}$$

where R is related to the length (L), area of cross-section (S), and the conductivity (σ) of the material of the conductor. This relation is given here:

$$R = \frac{L}{\sigma S} \tag{8.19b}$$

TABLE 13.1
Similarities between Electric and Magnetic Field Relations

Electrostatic Field		Magnetostatic Field	
$E = -\nabla V$	(7.17)	$H \overset{\text{def}}{=} -\nabla \mathcal{V}$	(11.58)
$V_{AB} = \int_A^B E \cdot dl$	(7.10b)	$\mathcal{V}_{AB} = \int_A^B H \cdot d\ell$	(11.63)
$J = \sigma E$	(8.12b)	$B = \mu H$	(11.1)
$I = \int_s J \cdot ds$	(8.1)	$\Phi = \int_s B \cdot ds$	(11.29a)
$\oint E \cdot dl = 0$	(7.13b)	$\oint H \cdot dl = I$ (for single turn coil) $= NI$ (for N turn coil)	(11.17)

In magnetic circuits, these components (V, I, and R) are replaced by the MMF ($\mathcal{V}$), magnetic flux (Φ), and the *reluctance* ($\Re$), respectively. Thus, Equation 8.19a takes the following form:

$$\mathcal{V} = \Phi\Re \tag{13.1a}$$

With the analogy of Ohm's law given by Equation 8.19a, the relation of Equation 13.1a is referred to as Hopkinson's or Rowland's Law. In Equation 13.1a, the MMF ($\mathcal{V}$) and magnetic flux (Φ) are analogous to the EMF (V) and the current (I) of electrical circuit. Similarly, with the analogy of Equation 8.19b, the expression for *reluctance* ($\Re$) can be written as:

$$\Re = \frac{L}{\mu S} \tag{13.1b}$$

The *reluctance* opposes the establishment of magnetic field in circuits comprising magnetic materials in the same way as resistance opposes the flow of current in electrical circuits.

The series reluctances in magnetic circuits are added in the same way as the series resistances in an electrical circuit. This addition is facilitated by Ampère's law, which is analogous to Kirchhoff's voltage law (KVL). Also, the sum of magnetic fluxes into any node is always zero. This follows from Gauss' law and is analogous to Kirchhoff's current law (KCL) for analyzing electrical circuits. In KVL the voltage excitation applied to a loop is equal to the sum of the voltage drops ($V = IR$) around the loop. In its magnetic analog, the MMF $\mathcal{V}$ ($= I$ or NI) is equal to the sum of MMF drops ($\mathcal{V} = \Phi\Re$) across the rest of the loop.

Like electrical circuits the magnetic circuits may also have branched and loops. In the case of a multiple loop system, the current in each branch can be solved through a matrix equation. These matrices are formed in the same way as obtained for solving the mesh circuit branch currents in loop analysis. Thereafter the individual branch currents are obtained by adding and/or subtracting the constituent loop currents. In accordance with the Ampère's law, the MMF ($\mathcal{V}$) is the excitation in magnetic circuits. It is the product of the current and the number of complete loops made (NI) and is measured in Ampère-turns.

According to Stokes' theorem, the closed line integral of H dot dl around a contour is equal to the open surface integral of curl H dot $\boldsymbol{ds}$ across the surface bounded by the closed contour. This statement given by Equation 11.56b is reproduced below:

$$\oint_l H \bullet dl = \iint_s (\nabla \times H) \cdot ds \tag{13.2}$$

This closed line integral of $H{\cdot}dl$ evaluates the total current passing through the surface and is equal to the excitation, NI.

The analogous quantities along with their units involved in electrical and magnetic circuits are summarized in Table 13.2 for ready reference.

13.2.2 Limitations of the Analogy

The similarities between electric and magnetic circuits shown in Table 13.1 can be termed as superficial in the limitations noted in Table 13.3. These limitations require due consideration in the construction of a magnetic circuit.

13.2.3 Configuration of a Magnetic Circuit

Figure 13.1a shows a core type structure made up of magnetic material of permeability μ. It contains four vertical legs each with length l_1. The cross-sectional areas

TABLE 13.2
Analogous Quantities (with Units) in Electrical and Magnetic Circuits

Electric Quantity	Unit	Magnetic Quantity	Unit
EMF ($V = \int \boldsymbol{E} \bullet \boldsymbol{dl}$)	Volts (V)	MMF ($\mathcal{V} = \oint \boldsymbol{H} \bullet \boldsymbol{dl}) = I$ or $= NI$	Ampère-turn
Electric field intensity (E)	Volt/meter	Magnetic field intensity (H)	Ampère/meter
Electric current (I)	Ampère (A)	Magnetic flux (Φ)	Weber
Resistance (R)	Ohms	Reluctance ($\mathfrak{R}$)	1/Henry
Conductance $\left(G = \frac{I}{R}\right)$	Mhos	Permanence $\left(\wp = \frac{1}{\mathfrak{R}}\right)$	Henry
Ohm's law ($V = IR$)		Hopkinson's or Rowland's law ($\mathcal{V} = \Phi\mathfrak{R}$)	
Microscopic Ohm's law ($\boldsymbol{J} = \sigma\boldsymbol{E}$)		B–H relation ($\boldsymbol{B} = \mu\boldsymbol{H}$)	
Current density ($\boldsymbol{J}$)	Amp/m^2	Magnetic flux density ($\boldsymbol{B}$)	Tesla
Conductivity (σ)	Mhos/m	Permeability (μ)	Henry/m

TABLE 13.3
Limitations of the Analogy

Electric Circuits	Magnetic Circuits
Electric currents represent flow of electrons and carry power.	Magnetic fields don't represent flow of anything.
Power is dissipated as heat in resistances.	No power is dissipated in reluctances.
Current remains confined within the circuit with little or no leakage.	All the magnetic field does not remain confined to the magnetic circuit. There is significant leakage flux in the space outside the magnetic cores.
Electrical circuits, in general, are linear as the resistances in these are constant.	The magnetic circuits are non-linear as the reluctance varies with magnetic field.
As such there is no concept of saturation in electrical circuits.	The magnetic circuits with ferromagnetic materials saturate at high magnetic fluxes. Above this level the reluctance increases rapidly.
As such there no hysteresis like phenomenon in electrical circuits.	The ferromagnetic materials (the key component of magnetic circuits) suffer from hysteresis so the flux in them depends not just on the instantaneous MMF but also on the past history of MMF. After the source of magnetic flux is removed, remnant magnetism is left creating a flux with no MMF.

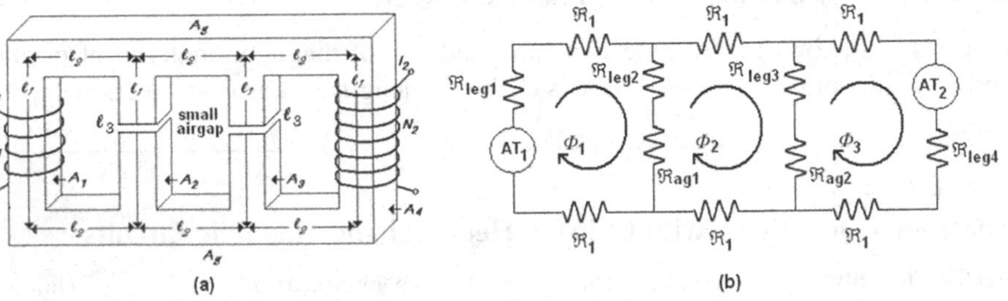

FIGURE 13.1 (a) A magnetic circuit; (b) equivalent circuit.

of legs 1, 2, 3, and 4 are given as A_1, A_2, A_3, and A_4, respectively. These legs are joined to two horizontal bars on top and bottom each with cross-sectional area A_5 and effective length $3l_2$. The two mid legs are provided with air gaps of length l_3. Leg 1 contains a coil with N_1 turns and current I_1. Similarly, leg 2 contains a coil with N_2 turns and current I_2.

13.2.4 Equivalent Circuit

The equivalent circuit for the core type structure of Figure 13.1a is shown in Figure 13.1b. This equivalent circuit can be obtained through the following steps.

13.2.4.1 Identification of Exciting Sources

As shown in Figure 13.1a this circuit contains two sources which can be written as:

$$AT_1 = V_1 = N_1 I_1 \quad (\text{in leg-1}) \tag{13.2a}$$

$$AT_2 = V_2 = N_2 I_2 \quad (\text{in leg-2}) \tag{13.2b}$$

13.2.4.2 Evaluation of Reluctances

The reluctance for different segments shown in Figure 13.1a can be calculated in the relation given by Equation 13.1b. In their equal lengths the reluctances for vertical legs can be written in a general form as:

$$\Re_{\text{legn}} = \frac{l_1}{\mu S_n} \tag{13.3a}$$

where n may take values 1, 2, 3, and 4.

It is to be noted that legs 2 and 4 contain air gaps both of lengths l_3. This length needs to be subtracted from length l_1 to get the exact value of the reluctances for legs 2 and 4. Generally the airgap lengths are quite small and subtraction of the same may be ignored.

Since each horizontal segment of top and bottom bars is of length l_2 and cross-section S_5, the reluctance for these can also be written as:

$$\Re_{\text{horseg}} = \Re_1 = \frac{l_2}{\mu S_5} \tag{13.3b}$$

The reluctances for the two air gaps both with lengths l_3 but with areas S_2 and S_3 can be written as:

$$\Re_{ag1} = \frac{l_3}{\mu_0 S_2} \tag{13.3c}$$

$$\Re_{ag2} = \frac{l_3}{\mu_0 S_4} \tag{13.3d}$$

13.2.4.3 Formation of Equivalent Circuit

The equivalent circuit of Figure 13.1b can be drawn by adopting the following sequence of steps:

- Evaluate the excitations AT_1 and AT_2 shown in legs 1 and 4 from Equations 13.2a and 13.2b.
- Evaluate the reluctances involved in different legs of core in Equations 13.3a–d.

- Connect three reluctances each for length l_1 in series in each of the top and bottom legs.
- Connect one reluctance in each of legs 1 and 4.
- Each of legs 2 and 3 contains two reluctances one of which represents the magnetic core and other the airgap.
- Obtain the resulting equivalent circuit of Figure 13.1b.
- Note that it appears like an electric circuit with two sources (AT_1 and AT_2), a number of series and parallel reluctances and the currents (or fluxes) flowing in different loops.

13.2.4.4 Evaluation of Fluxes

To evaluate fluxes in different arms, the loop equations can be written as:

$$\Phi_1 A_{11} - \Phi_2 A_{12} = AT_1 \tag{13.4a}$$

$$-\Phi_1 A_{21} + \Phi_2 A_{22} - \Phi_3 A_{23} = 0 \tag{13.4b}$$

$$-\Phi_2 A_{32} + \Phi_3 A_{33} = AT_2 \tag{13.4c}$$

where

$$A_{11} = \Re_{leg1} + 2\Re_1 + \Re_{leg2} + \Re_{ag1} \qquad A_{12} = \Re_{leg2} + \Re_{ag1} \qquad A_{13} = 0$$

$$A_{21} = \Re_{leg2} + \Re_{ag1} \quad A_{22} = 2\Re_1 + \Re_{ag1} + \Re_{leg2} + \Re_{leg3} + \Re_{ag2} \quad \boxed{A_{23} = \Re_{ag3} + \Re_{leg2}}$$

$$A_{31} = 0 \qquad A_{32} = \Re_{leg3} + \Re_{ag2} \qquad A_{33} = 2\Re_1 + \Re_{ag2} + \Re_{leg3} + \Re_{leg4}$$

These equations can be written as:

$$[A][X] = [B] \tag{13.5}$$

where

$$\left.[A] = \begin{bmatrix} A_{11} & A_{12} & 0 \\ A_{21} & A_{22} & A_{23} \\ 0 & A_{32} & A_{33} \end{bmatrix} \text{(a)} \quad [X] = \begin{bmatrix} \Phi_1 \\ \Phi_2 \\ \Phi_3 \end{bmatrix} \text{(b)} \quad [B] = \begin{bmatrix} AT_1 \\ 0 \\ AT_2 \end{bmatrix} \text{(c)}\right\} \tag{13.6}$$

The unknown values of fluxes can be obtained from the following equation:

$$[X] = [A]^{-1}[B] \tag{13.7}$$

where $[A]^{-1}$ stands for the inverse of matrix $[A]$.

13.3 INDUCTANCE

Any electrical or electronic device or component can be characterized in terms of three passive components referred to as resistance, capacitance, and inductance. The resistance (R) was defined as the ratio of potential difference to the current (i.e. V/I) in Section 8.4. Its value was given by Equation 8.19b as a function of geometrical parameters (viz. length and cross-sectional area) of the conductor and the conductivity of its material. The capacitance (C) was defined as the ratio of the total charge to the potential difference (i.e. Q/V) in Section 8.6. Its value was given by Equation 8.66 as a function of geometrical parameters of the conducting equipotential surfaces (viz. surface area and separation) and the permittivity of the material in or around them. This section deals with the inductance, the third characterizing component.

13.3.1 Defining Principle of Inductance

The inductance can be defined by using the duality between electrical and magnetic field quantities. The duality principle can be applied in the following two ways:

(a) The dual of electric charge (Q) is the magnetic flux (Φ) and that of potential difference (V) is the current (I). Also, Equation 8.66 defines the capacitance as $C = Q/V$. Thus, in the relationship of Q, V, and C, the inductance (L) can be defined as the ratio of magnetic flux to the current or as $L = \Phi/I$.

(b) The duality principle can also be applied in the energy stored in an electrostatic field or in a capacitor (W_E) and the energy stored in a magnetostatic field or in an inductor (W_H).

13.3.2 Mathematical Relations

Based on the duality principle identified in (a) and (b) the mathematical relations for inductance can be obtained as below.

13.3.2.1 Relation I

In view of the duality of ($C = Q/V$) the inductance (L) can be given as:

$$\left. L = \frac{\Phi}{I} \; (\text{for a single turn coil}) \quad \text{and} \quad L = \frac{N\Phi}{I} \; (\text{for an N turn coil}) \right\} \tag{13.8}$$

In these expressions, Φ is that flux which links (or encloses) the current I.

13.3.2.2 Relation II

Equation 8.67b gives the energy stored in an electrostatic field in terms of the capacitance as:

$$W_E = \frac{1}{2} \cdot C \cdot V_o^2 \tag{8.67b}$$

Thus, the capacitance (C) can be written as:

$$C = \frac{2W_E}{V^2} \tag{13.9a}$$

Similarly, in view of the duality principle the inductance can also be defined as:

$$L = \frac{2W_H}{I^2} \tag{13.9b}$$

where I is the total current flowing in the closed path and W_E is the energy in the magnetic field produced by this current.

Furthermore, the energy stored in the electrostatic field was given by Equation 7.60b as:

$$W_E = \iiint_{v\to\infty} \left(\frac{1}{2}\cdot \boldsymbol{D} \bullet \boldsymbol{E}\right)\cdot dv \tag{13.10a}$$

Using the duality principle, let us write:

$$W_H = \iiint_{v\to\infty} \left(\frac{1}{2}\cdot \boldsymbol{B} \bullet \boldsymbol{H}\right) dv \tag{13.10b}$$

Thus, in view of Equations 13.9b and 13.10b, inductance can be written as:

$$L = \frac{\iiint_v (\boldsymbol{B} \bullet \boldsymbol{H}) dv}{I^2} \tag{13.11a}$$

Substitution of $\boldsymbol{B} = \nabla \times \boldsymbol{A}$ (Equation 11.65a), Equation 13.11a modifies to:

$$L = \frac{1}{I^2} \iiint_v \{\boldsymbol{H} \bullet (\nabla \times \mathbf{A})\} dv \tag{13.11b}$$

Then the identity given by Equation 4.40c is:

$$\nabla \bullet (\boldsymbol{A} \times \boldsymbol{H}) \equiv \boldsymbol{H} \bullet (\nabla \times \boldsymbol{A}) - \boldsymbol{A} \bullet (\nabla \times \boldsymbol{H}) \tag{13.12a}$$

or

$$\boldsymbol{H} \bullet (\nabla \times \boldsymbol{A}) \equiv \nabla \bullet (\boldsymbol{A} \times \boldsymbol{H}) - \boldsymbol{A} \bullet (\nabla \times \boldsymbol{H}) \tag{13.12b}$$

Equation 13.11b becomes:

$$L = \frac{1}{I^2}\left[\iiint_v \{\nabla \bullet (\boldsymbol{A} \times \boldsymbol{H})\} dv + \iiint_v \{\boldsymbol{A} \bullet (\nabla \times \boldsymbol{H})\} dv\right] \tag{13.13a}$$

Applying the divergence theorem to the first integral and on substituting $\nabla \times \boldsymbol{H} = \boldsymbol{J}$ in the second integral Equation 13.13a becomes:

$$L = \frac{1}{I^2}\left[\oiint_s (\boldsymbol{A} \times \boldsymbol{H}) ds + \iiint_v (\boldsymbol{A} \cdot \boldsymbol{J}) dv\right] \tag{13.13b}$$

In the first segment of Equation 13.13b, the integral is to be carried out over the surface which encloses the entire volume containing all the magnetic energy. Since there are no energy contents on the surface, A and H must be 0 on the surface. This results in zero value of surface integral. Thus, Equation 13.13b takes the form:

$$L = \frac{1}{I^2}\iiint_v (\boldsymbol{A} \cdot \boldsymbol{J}) dv \tag{13.13c}$$

The vector magnetic potential (A) is related to the current density (J) as:

$$\boldsymbol{A} = \iiint_v \frac{\mu \boldsymbol{J}}{4\pi R} dv \tag{13.14a}$$

In view of this relation, Equation 13.13c is modified to:

$$L = \frac{1}{I^2}\iiint_v \left(\iiint_v \frac{\mu \boldsymbol{J}}{4\pi R} dv\right) \cdot \boldsymbol{J} dv \tag{13.14b}$$

In view of Equation 11.8, the differential volume with current density (*Jdv*) can be replaced by a filamentary current element (***IdL***) and the volume integral by a closed line integral. It results in the following simplified relation:

$$L = \frac{\mu}{4\pi}\frac{1}{I^2}\oint_l \left(\oint_l \frac{\boldsymbol{Idl}}{R}\right) \cdot \boldsymbol{Idl} = \frac{\mu}{4\pi}\oint_l \left(\oint_l \frac{\boldsymbol{dl}}{R}\right) \cdot \boldsymbol{dl} \tag{13.14c}$$

Equations 13.14b and 13.14c are the general forms of relation for the inductance in terms of the current density and the filamentary current. The replacement of ***Jdv*** by ***IdL*** in Equation 13.13c leads to the following relation:

$$L = \frac{1}{I}\oint_l \boldsymbol{A} \cdot \boldsymbol{dl} \tag{13.15a}$$

With the assumption of small section *dl* may be taken along the center of the filament.

Applying Stokes' theorem, Equation 13.15a becomes:

$$L = \frac{1}{I}\iint_s (\nabla \times \boldsymbol{A}) \cdot \boldsymbol{ds} \tag{13.15b}$$

Substituting $\boldsymbol{B} = \nabla \times \boldsymbol{A}$ (Equation 9.72a) we get

$$L = \frac{1}{I}\iint_s \boldsymbol{B} \cdot d\boldsymbol{s} \tag{13.15c}$$

Furthermore, in view of Equation 11.54a, the inductance can be written as:

$$L = \frac{\Phi}{I} \tag{13.16a}$$

Since Equation 13.16a is derived from Equation 13.15c, which involves surface integral, the flux Φ is that portion of the total flux that passes from any and every open surface whose perimeter is the filamentary current path. If this surface is encircled (say) N times by this filamentary current the flux will also get multiplied by N and the inductance becomes:

$$L = \frac{N\Phi}{I} \tag{13.16b}$$

As expected, the relation for inductance given by Equations 13.8 and 13.16 are the same.

13.3.3 Types of Inductance

The inductances, in general, are classified as self and mutual inductance. These can be explained using Figure 13.2.

13.3.3.1 Self-Inductance

In the above relations if the flux is produced by the same current with which it links the inductance is called the *self-inductance*. Figure 13.2a shows a straight wire carrying a current I which produces a flux Φ. This flux encloses the current which causes this flux, thus the self-inductance is given by Φ/I. Furthermore, Figure 13.2b shows two straight wires carrying currents I_1 and I_2. The flux produced by I_1 is given as Φ_1 and that by I_2 as Φ_2. If these wires are distantly located and their fields do not

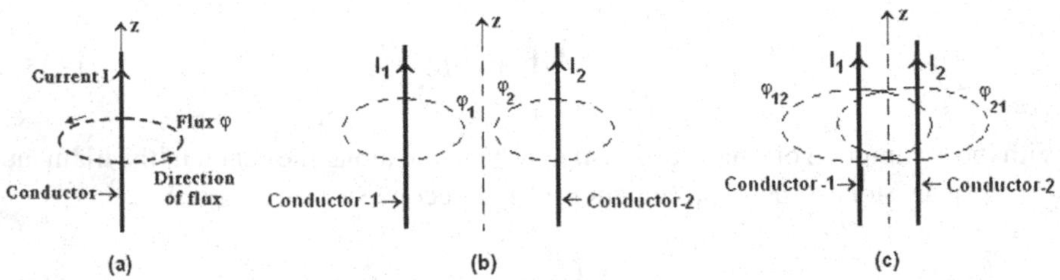

FIGURE 13.2 (a) Single conductor; (b) two distant conductors; and (c) two close conductors.

influence each other then the self-inductances of the two wires-1 and 2 can be written as:

$$\left.L_1 = \frac{\Phi_1}{I_1} \quad (a) \quad \text{and} \quad L_2 = \frac{\Phi_2}{I_2} \quad (b)\right\} \tag{13.17a}$$

Equation 13.13c gives the inductance in terms of the volume integral of the dot product of $\boldsymbol{A}$ and $\boldsymbol{J}$. The current density $\boldsymbol{J}$ can exist only within the conductor and the $\boldsymbol{A}$ caused due to this J may exist both inside and outside the volume. Since $\boldsymbol{J}$ is zero outside, the integration outside the volume is zero at every point in the volume thus the determination of $\boldsymbol{A}$ outside is of no use. The inductance evaluated from Equation 13.13c can be termed as self-inductance. Thus, the relations given by Equations 13.16a and 13.16b and those by Equations 13.17a and 13.17b respectively are the same.

13.3.3.2 Mutual Inductance

If the flux produced by the current (say I_1) flowing in one wire (coil or circuit) links with the current (say I_2) flowing in another wire (coil or circuit) the resulting inductance is referred to as *mutual inductance*. The converse is also true, that is the flux caused due to current I_2 links with current I_1. Such a situation is observed in Figure 13.2c which shows two closely located straight wires carrying currents I_1 and I_2. The flux produced by one current links with the current of the second. In case the flux produced by I_1 and linking with I_2 is to be denoted as Φ_{12}. This linking results in mutual inductance M_{12}. Similarly, the flux due to I_2 denoted as Φ_{21} links with I_1 results in mutual inductance M_{21}. These two mutual inductances are given as:

$$\left.M_{12} = \frac{\Phi_{12}}{I_1} \quad (a) \quad \text{and} \quad M_{21} = \frac{\Phi_{21}}{I_2} \quad (b)\right\} \tag{13.18}$$

If conductor 1 and 2 contain N_1 and N_2 turns, respectively, Equations 13.17a and 13.17b and 13.18a and 13.18b take the following forms:

$$\left.L_1 = \frac{N_1\Phi_1}{I_1} \quad (a) \quad \text{and} \quad L_2 = \frac{N_2\Phi_2}{I_2} \quad (b)\right\} \tag{13.19}$$

$$\left.M_{12} = \frac{N_2\Phi_{12}}{I_1} \quad (a) \quad \text{and} \quad M_{21} = \frac{N_1\Phi_{21}}{I_2} \quad (b)\right\} \tag{13.20}$$

In Equations 13.20a and 13.20b, the first suffix represents the current that produces the flux and the second the current with which it links. It is to be noted that since the direction of flux is linked to the direction of current the inductance is always a positive real quantity.

From the above description, it can be concluded that the mutual inductance is the result of the magnetic interaction of two currents. In the case of only one current, the total energy stored in magnetic field can be given in terms of single (or self) inductance, whereas in the case of two currents of non-zero value, the total energy becomes a function of two self-inductances and the mutual inductance.

In terms of energy, the values of mutual inductances given by Equations 13.20a and 13.20b can be written as:

$$M_{12} = \frac{\iiint_v (B_1 \bullet H_2)dv}{I_1 I_2} = \frac{\iiint_v (\mu H_1 \bullet H_2)dv}{I_1 I_2} \tag{13.21a}$$

Similarly,

$$M_{21} = \frac{\iiint_v (B_2 \bullet H_1)dv}{I_1 I_2} = \frac{\iiint_v (\mu H_2 \bullet H_1)dv}{I_1 I_2} \tag{13.21b}$$

In view of Equations 13.21a and 13.21b:

$$M_{12} = M_{21} \tag{13.22}$$

13.3.4 Inductance of Some Common Configurations

This section derives the mathematical relations for inductance of the coaxial cable, solenoid, and toroid, which find frequent applications in conjunction with other devices and systems.

13.3.4.1 Coaxial Cable

In view of Figure 11.9a and Equation 11.48, the field H was obtained as:

$$\boldsymbol{H} = \frac{I}{2\pi\rho}\boldsymbol{a}_\varphi \tag{13.23a}$$

From relation $\boldsymbol{B} = \mu_0 \boldsymbol{H}$ we get:

$$\boldsymbol{B} = \frac{\mu_0 I}{2\pi\rho}\boldsymbol{a}_\varphi \tag{13.23b}$$

Equation 11.29a gives the flux as:

$$\begin{aligned} \Phi &= \oint_s B \bullet ds = \oint_0^l \int_a^b \frac{\mu_0 I}{2\pi\rho} a_\varphi \bullet (d\rho \cdot dz \cdot a_\varphi) \\ &= \int_0^l \int_a^b \frac{\mu_0 I}{2\pi\rho} d\rho \cdot dz = \frac{\mu_0 I l}{2\pi} \ln\left(\frac{b}{a}\right) \end{aligned} \tag{13.24a}$$

Lastly, the inductance per unit length is given as:

$$L = \frac{\Phi}{I} = \frac{\mu_0}{2\pi} \ln\left(\frac{b}{a}\right) \text{ H/m} \tag{13.24b}$$

13.3.4.2 Solenoid

In Figure 11.2c an iron cored solenoid was shown. Assuming this solenoid to have relative permeability is μ_r, number of turns N, area of core cross-section A cm^2, and length l cm, then from Equation 11.52b we get:

$$B = \mu_0 \mu_r NI = \mu NI \tag{13.25a}$$

$$\Phi = \int_s B \bullet ds = B \cdot A = \mu NIA \tag{13.25b}$$

And the inductance (L) per unit length is given as:

$$L = \frac{N\Phi}{l} = \frac{\mu N^2 A}{l} \tag{13.26}$$

It is to be noted that small solenoid inductors may be air cored, whereas large ones are iron cored. The relative permeability of magnetic iron is around 200.

13.3.4.3 Toroid

Equation 11.53b obtained in view of Figure 11.16 gives B as:

$$B = \frac{\mu NI}{2\pi\rho_0} \tag{13.28a}$$

If A is the area of cross-section of toroid, the flux can be obtained as:

$$\Phi = B \cdot A \tag{13.28b}$$

Thus, the inductance (L) of toroid is:

$$L \cong \frac{N\Phi}{I} = \frac{N \cdot B \cdot A}{I} = \frac{\mu N^2 A}{2\pi\rho_0} \tag{13.28c}$$

In this relations N is the number of turns, A the cross-sectional area in cm^2, $\mu = \mu_{0r}$ is the permeability of the core material, μ_0 is the permeability of free space, μ_r is the relative permeability of the core. The small inductors may be air cored while large inductors are iron cored. The relative permeability of magnetic iron is around 200. Equation 13.28c involves the sign of approximation because the magnetic field changes with the radius from the centerline of toroid. The use of the centerline value for B as an average introduces an error which is small if the toroid radius (ρ_0) is much larger than the coil radius (a).

13.4 ENERGY STORED IN MAGNETOSTATIC FIELDS

In Chapter 7, a detailed derivation for the density of energy stored in electrostatic fields is presented. Similar derivation for the density of energy stored in magnetostatic fields is however, not available. Keeping in view the expression for the energy density in electrostatic fields, i.e.:

$$E_{\text{elect.}} = \frac{1}{2} \cdot \boldsymbol{E} \bullet \boldsymbol{D}. \tag{7.61a}$$

A parallel expression is usually accepted for the energy density in magnetostatic fields, i.e.:

$$\mathcal{E}_{\text{magt.}} = \frac{1}{2} \cdot \boldsymbol{H} \bullet \boldsymbol{B} \tag{13.29}$$

Equation 7.61 can be written as:

$$\mathcal{E}_{\text{elect.}} = \frac{1}{2} \cdot \boldsymbol{E} \bullet \boldsymbol{D} = \frac{\varepsilon}{2} \cdot E^2 = \frac{D^2}{2\varepsilon}$$

Similarly, Equation 7.29 can also be written as:

$$\mathcal{E}_{\text{magt.}} = \frac{1}{2} \cdot \boldsymbol{H} \bullet \boldsymbol{B} = \frac{\mu}{2} \cdot H^2 = \frac{B^2}{2\mu} \tag{13.29a}$$

One can arrive at Equation 13.29 either by using: (1) the lumped circuit approach, or (2) the magnetic charge approach. These are discussed below.

13.4.1 LUMPED CIRCUIT APPROACH

Consider a time-invariant, linear, lossless, uniformly wound long cylindrical inductor with time-varying current i at its terminals. The voltage (v), applied across its terminals opposes the induced electromotive force (EMF). The EMF (e), is caused due to the change in the magnetic flux linking the inductor winding (coil). Faraday's law of electromagnetic induction (discussed in Chapter 15), states that the EMF induced around a closed path c, is equal to the rate of decrease of the magnetic flux Φ_c linking to the closed path c. Therefore,

$$e = -\frac{d\Phi_c}{dt} \tag{13.30}$$

Thus:

$$v = -e = \frac{d\Phi_c}{dt} \tag{13.31}$$

where

$$e \overset{\text{def}}{=} \oint_c \boldsymbol{E} \bullet d\boldsymbol{\ell} \tag{13.32a}$$

and

$$\Phi_c \overset{\text{def}}{=} \iint_s \boldsymbol{B} \bullet d\boldsymbol{s} \tag{13.32b}$$

where the symbol s indicates any open surface whose boundary coincides with the closed path c and $\boldsymbol{B}$ indicates the flux density on this surface.

The instantaneous power input p, to the inductor of inductance L, is therefore given as:

$$p = i \cdot v = i \cdot \frac{d\Phi_c}{dt} = i \cdot \frac{d(L \cdot i)}{dt} = L \cdot i \cdot \frac{di}{dt} = \frac{1}{2} \cdot L \cdot \frac{di^2}{dt} = \frac{d}{dt}\left(\frac{1}{2} \cdot L \cdot i^2\right) \tag{13.33}$$

where the inductance is defined as the flux linkage per unit current, i.e.:

$$L \overset{\text{def}}{=} \Phi_c / i \tag{13.33a}$$

Since power is the rate of change of energy, the instantaneous energy stored in the inductor, found from Equation 13.33, is:

$$W_L = \frac{1}{2} \cdot L \cdot i^2 \tag{13.34}$$

The time-invariant current I is a special case of time-varying current i, so the energy stored in the inductor for dc currents can be given as:

$$W_L = \frac{1}{2} \cdot L \cdot I^2 \tag{13.35}$$

and from Equation 13.33a:

$$L \overset{\text{def}}{=} \Phi_c / I \tag{13.35a}$$

Under the steady state condition with dc excitation current the electric field will be zero. Thus, there will be no energy radiation, no terminal voltage and no power input to the ideal inductor. Therefore, Equation 13.35 gives the energy stored in the magnetostatic field in the inductor. For uniformly wound long cylindrical inductors with dc excitation, the core flux is almost axial thus the leakage flux can be neglected and the flux density B_z, is the same at every point in the inductor. Therefore, in view of Equation 13.32b:

$$\Phi_c / N = B_z \cdot a \tag{13.36}$$

where the symbols "a" and N indicate, respectively, the cross-sectional area and the number of turns of the inductor coil.

Now, Equations 13.35, 13.35a, and 13.36 give:

$$W_L = \frac{1}{2}\cdot \Phi_c \cdot I = \frac{1}{2}\cdot \left(B_z \cdot a \cdot N\right)\cdot I \tag{13.37}$$

The dc current in a uniformly wound long inductor coil can be replaced by a surface current sheet with uniform density K_φ, such that:

$$N \cdot I = K_\varphi \cdot \ell \tag{13.38}$$

where the symbol ℓ indicates the axial length of the inductor coil.

Since the magnetic field of a long, uniformly wound inductor is negligible outside the inductor, we have:

$$H_z = K_\varphi \tag{13.39}$$

Therefore, from Equations 13.37, 13.38, and 13.39:

$$W_L = \frac{1}{2}\cdot \left(B_z \cdot a\right)\cdot \left(K_\varphi \cdot \ell\right) = \frac{1}{2}\cdot \left(B_z \cdot a\right)\cdot \left(H_z \cdot \ell\right) \tag{13.40}$$

or,

$$W_L = \left(\frac{1}{2}\cdot H_z \cdot B_z\right)\cdot \left(a \cdot \ell\right) = \left(\frac{1}{2}\cdot H_z \cdot B_z\right)\cdot v \tag{13.40a}$$

where the symbol v indicates the cylindrical volume inside the inductor coil.

Therefore, the density of the energy stored in the magnetostatic field inside the inductor can be given as follows:

$$\mathcal{E}_{\text{magt.}} = \frac{1}{2}\cdot \boldsymbol{H} \bullet \boldsymbol{B} \tag{13.41}$$

The derivation presented above is under the restricted condition of a uniform one-dimensional magnetic field. An alternative treatment is found in Matthew N. O. Sadiku 2007 (reference 1 at the end of the chapter). A general treatment based on the Poynting theorem is given in a later chapter on time-varying electromagnetic fields. The starting point in all these treatments is the time-varying electromagnetic field.

13.4.2 Magnetic Charge Approach

In our derivation for the density of the energy stored in electrostatic fields we started with electrical point charges and their potential energy. Stationary point charges are the source for the electrostatic field. The energy stored in this field is the potential energy stored in these charges, expressed in terms of field vectors. The expression for the energy density thus found does not explicitly indicate its sources, i.e. point charges. Following a somewhat similar line we shall proceed to develop the expression for the energy density in magnetostatic fields. Our treatment is restricted to free space without the presence of any permanent magnet.

On comparing the magnetostatic field with electrostatic field two differences are readily found, namely; (i) the source for the magnetostatic field is a distribution of dc currents, and (ii) magnetic point charges (also known as magnetic monopoles) are apparently non-existent. In view of (i), the scalar magnetostatic potential can be defined only in the current-free region. Although magnetic charges may not exist, the concept of these hypothetical charges has been found useful[2] in certain situations. This is also demonstrated in this subsection.

Imagine a large, linear, homogeneous space. This space is divided into two regions. The volume for the region carrying dc currents is indicated as v_1. Whereas, the volume for the current-free region as v_2. Using Biot-Savart's law, the magnetic field intensity in the two regions, $\boldsymbol{H}_1$and $\boldsymbol{H}_2$, can be found for a given current distribution in Region 1. Region 2 is a current-free region, so for this region the scalar magnetic potential $\mathcal{V}_2$ can be defined by the following equation:

$$\boldsymbol{H}_2 \overset{\text{def}}{=} -\nabla \mathcal{V}_2 \tag{13.42}$$

Assuming zero potential at infinity, the magnetic potential $\mathcal{V}_P$, at any point P in Region 2 is found from the following equation:

$$\mathcal{V}_P = -\int_{\infty}^{P} \boldsymbol{H}_2 \bullet d\boldsymbol{\ell} \tag{13.43}$$

The point P could be anywhere in Region 2, including at the *common boundary* surface between the two regions.

Next, consider Region 1. Let us replace the current distribution by an *equivalent* distribution of the fictitious magnetic point charges $\mathfrak{M}_{(i)}$ and the corresponding charge density ϱ. The current distribution having been replaced by the distribution of magnetic charge density, the scalar magnetic potential $\mathcal{V}_1$ for this region can be defined by the following equation:

$$\tilde{\boldsymbol{H}}_1 \overset{\text{def}}{=} -\nabla \mathcal{V}_1 \tag{13.44}$$

Since the divergence of the magnetic flux density in this region is equal to the magnetic charge density, i.e.:

$$\nabla \bullet \tilde{\boldsymbol{B}}_1 = \varrho \tag{13.45}$$

where

$$\tilde{\boldsymbol{B}}_1 = \mu_o \cdot \tilde{\boldsymbol{H}}_1 \tag{13.45a}$$

Note that:

$$\nabla \bullet \boldsymbol{B}_1 = 0 \tag{13.45b}$$

Therefore, in general:

$$\begin{cases} \tilde{\boldsymbol{H}}_1 \neq \boldsymbol{H}_1 \\ \tilde{\boldsymbol{B}}_1 \neq \boldsymbol{B}_1 \end{cases} \tag{13.45c}$$

Thus, in Equations 13.44 and 13.45, the scalar magnetic potential $\mathcal{V}_1$ obeys Poisson's equation:

$$\nabla^2 \mathcal{V}_1 = -\frac{1}{\mu_o} \cdot \varrho \tag{13.46}$$

For a given charge density ϱ, unique solution for the potential $\mathcal{V}_1$ must satisfy the continuity of the potential at the *common boundary* surface between the two regions. For the *equivalent* distribution of the fictitious magnetic charge density ϱ, the normal derivative of the magnetic potential must be continuous across the *common boundary* surface between the two regions.

The potential energy W_{MD} of the '**D**iscreetly distributed **M**agnetic point charges' $\mathfrak{M}_{(i)}$ is given as:

$$W_{MD} = \frac{1}{2} \cdot \sum_i \mathfrak{M}_{(i)} \cdot \mathcal{V}_{1(i)} \tag{13.47a}$$

The potential energy W_{MC} of the '**C**ontinuously distributed **M**agnetic charge density' ϱ is given as:

$$W_{MC} = \frac{1}{2} \cdot \iiint_v \left(\varrho \cdot \mathcal{V}_1\right) \cdot dv \tag{13.47b}$$

where v indicates an arbitrary volume.

Therefore, in view of Equation 13.45:

$$W_{MC} = \frac{1}{2} \cdot \iiint_v \left\{ \left(\nabla \bullet \tilde{\boldsymbol{B}}_1 \right) \cdot \mathcal{V}_1 \right\} \cdot dv \tag{13.48}$$

Now, consider the following vector identity:

$$\nabla \bullet \left(\mathcal{V} \boldsymbol{B} \right) \equiv \mathcal{V} \cdot \left(\nabla \bullet \boldsymbol{B} \right) + \left(\nabla \mathcal{V} \right) \bullet \boldsymbol{B} \tag{13.49}$$

giving:

$$\left(\nabla \bullet \boldsymbol{B} \right) \cdot \mathcal{V} = \nabla \bullet \left(\mathcal{V} \boldsymbol{B} \right) - \left(\nabla \mathcal{V} \right) \bullet \boldsymbol{B} \tag{13.49a}$$

Therefore, Equation 13.48 can be rewritten as:

$$W_{MC} = \frac{1}{2} \cdot \iiint_v \left\{ \nabla \bullet \left(\nu_1 \tilde{\boldsymbol{B}}_1 \right) - \left(\nabla \nu_1 \right) \bullet \tilde{\boldsymbol{B}}_1 \right\} \cdot dv$$

or

$$W_{MC} = \frac{1}{2} \cdot \iiint_v \left\{ \nabla \bullet \left(\nu_1 \tilde{\boldsymbol{B}}_1 \right) \right\} \cdot dv - \frac{1}{2} \cdot \iiint_v \left\{ \left(\nabla \nu_1 \right) \bullet \tilde{\boldsymbol{B}}_1 \right\} \cdot dv \tag{13.50}$$

We get, using Equation 13.44:

$$W_{MC} = \frac{1}{2} \cdot \iiint_v \left\{ \nabla \bullet \left(\nu_1 \tilde{\boldsymbol{B}}_1 \right) \right\} \cdot dv + \frac{1}{2} \cdot \iiint_v \left\{ \tilde{\boldsymbol{H}}_1 \bullet \tilde{\boldsymbol{B}}_1 \right\} \cdot dv \tag{13.51}$$

Using the Divergence Theorem and Equation 13.51, we get:

$$W_{MC} = \frac{1}{2} \cdot \oiint_S \left(\nu_1 \tilde{\boldsymbol{B}}_1 \right) \bullet \boldsymbol{ds} + \frac{1}{2} \cdot \iiint_v \left\{ \tilde{\boldsymbol{H}}_1 \bullet \tilde{\boldsymbol{B}}_1 \right\} \cdot dv \tag{13.52}$$

where S is the surface of the arbitrary volume v.

Since this equation is independent of source distributions, it is applicable to both the regions. Therefore, we may rewrite Equations 13.51 and 13.52 as:

$$W_{MC} = \frac{1}{2} \cdot \iiint_v \left\{ \nabla \bullet \left(\nu \boldsymbol{B} \right) \right\} \cdot dv + \iiint_v \left(\frac{1}{2} \cdot \boldsymbol{H} \bullet \boldsymbol{B} \right) \cdot dv \tag{13.53a}$$

$$W_{MC} = \frac{1}{2} \cdot \oiint_S \left(\nu \boldsymbol{B} \right) \bullet \boldsymbol{ds} + \iiint_v \left(\frac{1}{2} \cdot \boldsymbol{H} \bullet \boldsymbol{B} \right) \cdot dv \tag{13.53b}$$

The two terms on the RHS of each equation jointly give the energy stored in the magnetostatic field spread over the entire infinite space covering both regions. As shown in Chapter 7, the first term gives the energy stored outside an arbitrary volume v, while the second term gives the energy stored in the magnetostatic field spread inside the arbitrary volume v. From this term, we can identify the expression for the energy density in the magnetic field as:

$$\mathcal{E}_{\text{magt.}} = \frac{1}{2} \cdot \boldsymbol{H} \bullet \boldsymbol{B} \tag{13.54}$$

This is true provided that the arbitrary volume is entirely either in the Region 1 or in the Region 2.

In view of Equation 13.52 for the source-free Region 2, we have:

$$W_{MC} = \frac{1}{2} \cdot \oiint_S \left(\nu_2 \boldsymbol{B}_2 \right) \bullet \boldsymbol{ds} + \frac{1}{2} \cdot \iiint_v \left\{ \boldsymbol{H}_2 \bullet \boldsymbol{B}_2 \right\} \cdot dv \tag{13.54a}$$

Now, for $v = v_2$ and $S = S_2$, we get:

$$W_{MC} = \frac{1}{2} \cdot \oiint_{S_2} \left(\mathcal{V}_2 \boldsymbol{B}_2 \right) \bullet \boldsymbol{ds} + \frac{1}{2} \cdot \iiint_{v_2} \left\{ \boldsymbol{H}_2 \bullet \boldsymbol{B}_2 \right\} \cdot dv \tag{13.54b}$$

Since the magnetostatic potential $\mathcal{V}_2$ and the magnetic field vectors $\boldsymbol{H}_2$ and $\boldsymbol{B}_2$, for Region 2 are accurately known for the given current distribution in Region 1, the second term on the RHS of this equation gives the magnetic energy stored in Region 2, and the first term on the RHS of this equation gives the magnetic energy stored in the Region 1. Likewise, Equation 13.52, for $v = v_1$ and $S = S_1$, reduces to:

$$W_{MC} = \frac{1}{2} \cdot \oiint_{S_1} \left(\mathcal{V}_1 \boldsymbol{B}_1 \right) \bullet \boldsymbol{ds} + \frac{1}{2} \cdot \iiint_{v_1} \left\{ \boldsymbol{H}_1 \bullet \boldsymbol{B}_1 \right\} \cdot dv. \tag{13.55}$$

Since the magnetostatic potential $\mathcal{V}_1$ at the *boundary* between the two regions and the magnetic field vectors $\boldsymbol{H}_1$ and $\boldsymbol{B}_1$, for Region 1 are accurately known for the given current distribution in Region 1, the second term on the RHS of this equation gives the magnetic energy stored in Region 1, and the first term on the RHS of this equation gives the magnetic energy stored in Region 2. Therefore, energy stored W_{MC}, for the two regions are:

$$W_{MC} = \begin{cases} \dfrac{1}{2} \cdot \iiint_{v_1} \left\{ \boldsymbol{H}_1 \bullet \boldsymbol{B}_1 \right\} \cdot dv = \dfrac{1}{2} \cdot \oiint_{S_2} \left(\mathcal{V}_2 \boldsymbol{B}_2 \right) \bullet \boldsymbol{ds} \text{ for Region 1;} \\ \dfrac{1}{2} \cdot \iiint_{v_2} \left\{ \boldsymbol{H}_2 \bullet \boldsymbol{B}_2 \right\} \cdot dv = \dfrac{1}{2} \cdot \oiint_{S_1} \left(\mathcal{V}_1 \boldsymbol{B}_1 \right) \bullet \boldsymbol{ds} \text{ for Region 2.} \end{cases} \tag{13.56}$$

Note that on the common boundary between S_1 and S_2: $\mathcal{V}_1 = \mathcal{V}_2$.

On the other hand, if the volume v is entirely in Region 1 ($v < v_1$), the energy stored W_{MC} outside the arbitrary volume v may satisfy the inequality:

$$W_{MC} \neq \frac{1}{2} \cdot \oiint_{S} \left(\mathcal{V}_1 \boldsymbol{B}_1 \right) \bullet \boldsymbol{ds} \tag{13.57a}$$

This is so because the magnetostatic potential in Region 1 $\mathcal{V}_1$, is a fictitious quantity.

If the volume v is entirely in Region 2 ($v < v_2$), the energy stored W_{MC} outside the arbitrary volume v may satisfy the inequality:

$$W_{MC} \neq \frac{1}{2} \cdot \oiint_{S} \left(\mathcal{V}_2 \boldsymbol{B}_2 \right) \bullet \boldsymbol{ds} \tag{13.57b}$$

This is because a part of the space outside the arbitrary volume v is in Region 1, and the potential field in Region 2 is unresponsive to the fictitious potential distribution (except at the boundary) in the Region 1.

These inequalities are justified because of the fact that no distribution of magnetic charges can produce identical field distribution caused by a given distribution of electric currents. The selection of suitable charge distribution only ensures that the field distribution in the current-free Region 2 remains unperturbed. The same reason leads to the following inequality regarding the energy stored W_{MC} *outside* the arbitrary volume v, where a part of the arbitrary volume $\overline{v}_1$ is in Region 1 and the remaining part $\overline{v}_2$ is in Region 2:

$$W_{MC} \neq \frac{1}{2} \cdot \oiint_{\overline{s}_1} (V_1 \boldsymbol{B}_1) \bullet \boldsymbol{ds} + \frac{1}{2} \cdot \oiint_{\overline{s}_2} (V_2 \boldsymbol{B}_2) \bullet \boldsymbol{ds} \tag{13.58}$$

where $\overline{s}_1$ and $\overline{s}_2$ indicate the bounding surface for the volumes $\overline{v}_1$ and $\overline{v}_2$, respectively.

The energy stored in this arbitrary volume v can however be given as:

$$\begin{aligned} W_{MC} &= \iiint_{v} \left(\frac{1}{2} \cdot \boldsymbol{H} \bullet \boldsymbol{B} \right) \cdot dv \\ &= \iiint_{\overline{v}_1} \left(\frac{1}{2} \cdot \boldsymbol{H}_1 \bullet \boldsymbol{B}_1 \right) \cdot dv + \iiint_{\overline{v}_2} \left(\frac{1}{2} \cdot \boldsymbol{H}_2 \bullet \boldsymbol{B}_2 \right) \cdot dv \end{aligned} \tag{13.59}$$

Equations 13.29, 13.41, and 13.54 gives the energy density from which the energy stored in the magnetostatic fields can be obtained on the analogy of Equations 7.61a and 7.61. Thus:

$$\begin{aligned} W_H = W_{MC} &= \int_{v \to \infty} \mathcal{E}_{mag} \cdot dv = \int_{v \to \infty} \left(\frac{\boldsymbol{B} \bullet \boldsymbol{H}}{2} \right) \cdot dv \\ &= \int_{v \to \infty} \frac{B^2}{2\mu} \cdot dv = \int_{v \to \infty} \frac{\mu H^2}{2} \cdot dv \end{aligned} \tag{13.60a}$$

As noted in Chapter 7 the work done is equal to the energy expended. In Equations 7.1a and 7.1b, the expressions for the work done for magnetostatic can similarly be written:

$$dW_H = FdL = \frac{B^2}{2\mu} SdL \quad \text{or} \quad F = \frac{B^2 S}{2\mu} \tag{13.60b}$$

Example 13.1

In a magnetic circuit shown in Figure 13.3 $B = 0.6$ wb/m² and $H = 150$ At/m at the midpoint of the left silicon steel leg, $L_1 = 10$ cm, $L_2 = 15$ cm, $S_1 = 6$ cm², $S_2 = 4$ cm². Find (a) V_{air} if airgap length $g = 0.5$ cm; (b) V_{steel}; and (c) the required current in a 2000 turn coil on the left leg.

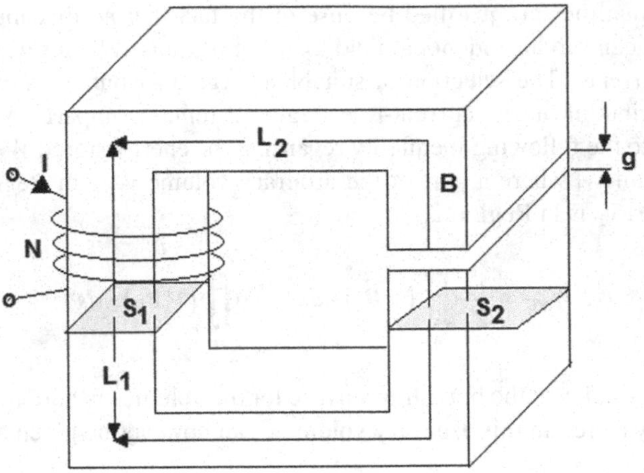

FIGURE 13.3 Magnetic circuit for Example 13.1.

SOLUTION

Since B = 0.6 wb/m² and H = 150 At/m, $\mu = \frac{B}{H} = \frac{0.6}{150} = 0.004$, thus $\mu_r = \frac{\mu}{\mu_0} = \frac{0.004}{4\pi \times 10^{-7}} = 3183$. Also given $L_1 = 0.1$ m, $L_2 = 0.15$ m, $S_1 = 6 \times 10^{-4}$ m², $S_2 = 4 \times 10^{-4}$ m².

In view of Equations 13.1b, 11.29a and 13.1a:

(a) For $g = 0.5$ cm $= 0.5 \times 10^{-2}$ m:

$$\Re_{air} = \frac{g}{\mu_0 S_2} = \frac{0.5 \times 10^{-2}}{\left(4\pi \times 10^{-7}\right) 4 \times 10^{-4}} = 9.947 \times 10^6 \text{ At/Wb}$$

$$\Phi = BS_1 = 0.6 \times 6 \times 10^{-4} = 3.6 \times 10^{-4} \text{ Wb}$$

$$\mathcal{V}_{air} = \Phi \cdot \Re_{air} = \left(3.6 \times 10^{-4}\right)\left(9.947 \times 10^6\right) = 358.1 \text{ At}$$

(b) Since Φ is the same throughout, B in the right silicon steel leg is:

$$B = \frac{\Phi}{S_2} = \frac{3.6 \times 10^{-4}}{4 \times 10^{-4}} = 0.9 \text{ Wb/m}^2$$

From the magnetization curve of silicon sheet steel B = 0.6 Wb/m² corresponds to $H \approx \frac{100 \text{ At}}{\text{m}} = H_1$ and B = 0.9 Wb/m² corresponds to $H \approx \frac{150 \text{ At}}{\text{m}} = H_2$. Thus

$$\mathcal{V}_{steel} = H_1 \times L_1 + H_2 \times L_2 = 100 \times 10 \times 10^{-2} + 150 \times 15 \times 10^{-2} = 32.5 \text{ At}$$

(c) Total At = Total mmf = $\mathcal{V}_{air} + \mathcal{V}_{steel} = 358.1 + 32.5 = 390.6$ At

$$I = \frac{\text{Total mmf}}{\text{Total turns}} = \frac{\text{Total At}}{N} = \frac{390.6}{2000} = 0.1953 \text{ A}$$

Example 13.2

Figure 13.4 shows a magnetic circuit wherein each outer leg is of length $L_1 = 10$ cm and cross-sectional area $S_1 = 1.2$ cm^{2}and the mid leg is of length $L_2 = 5$ cm and cross-sectional area $S_2 = 2.4$ cm^2. A coil with $N = 1500$ turns carrying $I = 10$ mA current is placed around the mid leg. Find B in all the three legs if the right and left outer vertical legs are provided with an airgap $g = 1$ mm. For values of B, assume the linear operation with $\mu_r = 4000$ below the knee of the magnetization curve of silicon steel.

SOLUTION

Given $S_1 = 1.2$ cm^2, $L_1 = 10$ cm, $S_2 = 2.4$ cm^2, $L_2 = 5$ cm, $g = 2$ mm, $\mu_r = 4000$, $N = 1500$, and $I = 10$ mA:

$$\mathcal{V} = NI = 1500 \times 10 \times 10^{-3} = 15 \text{ At}$$

$$\Re_{steel}(\text{left}) = \Re_{steel}(\text{right}) = \frac{L_1}{\mu_0 \mu_r S_1}$$

$$= \frac{10 \times 10^{-2}}{(4\pi \times 10^{-7}) \times 4000 \times (1.2 \times 10^{-4})} = 0.16578 \times 10^6 \text{ At/Wb}$$

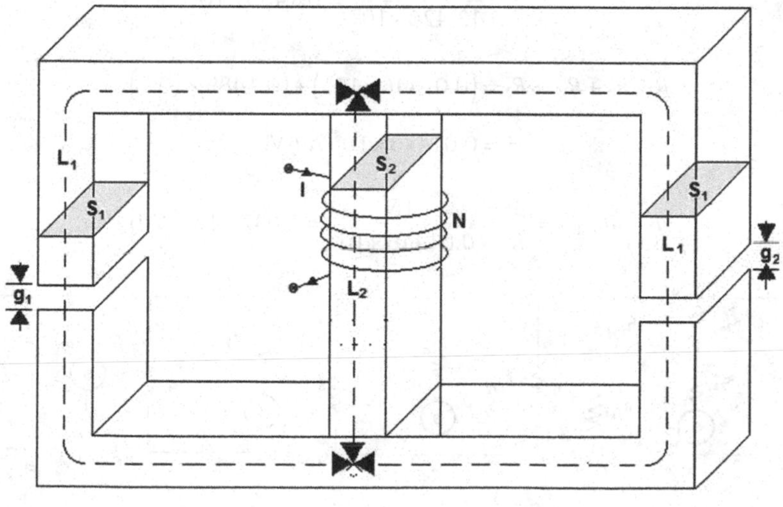

FIGURE 13.4 Magnetic circuit for Example 13.2.

$$\Re_{\text{steel}}(\text{central}) = \frac{L_2}{\mu_0 \mu_r S_2} = \frac{5\times10^{-2}}{(4\pi\times10^{-7})\times4000\times(2.4\times10^{-4})}$$

$$= 0.01446\times10^{6} \text{ At/Wb}$$

$$\Re_{\text{air}}(\text{left}) = \Re_{\text{air}}(\text{right}) = \frac{g}{\mu_0 S_1} = \frac{2\times10^{-3}}{(4\pi\times10^{-7})\times(1.2\times10^{-4})}$$

$$= 13.26\times10^{6} \text{ At/Wb}$$

In equivalent circuit shown in Figure 13.5:

$$R_{1ag} = \Re_{\text{air}}(\text{left}) = R_{3ag} = \Re_{\text{air}}(\text{right}) = 13.26\times10^{6} \text{ At/Wb}$$

$$R_{1st} = R_{3st} = \Re_{\text{steel}}(\text{left}) = \Re_{\text{steel}}(\text{right}) = 0.16578\times10^{6} \text{ At/Wb}$$

$$R_{2st} = \Re_{\text{steel}}(\text{central}) = 0.01446\times10^{6} \text{ At/Wb}$$

$$R_1 = R_{1ag} + R_{1st} = 13.26\times10^{6} + 0.16578\times10^{6} = 13.426\times10^{6} \text{ At/Wb}$$

$$R_3 = R_{3ag} + R_{3st} = 13.26\times10^{6} + 0.16578\times10^{6} = 13.426\times10^{6} \text{ At/Wb}$$

$$R_2 = R_{2st} = 0.01446\times10^{6} \text{ At/Wb}$$

$$R_{13} = \frac{R_1 + R_3}{R_1 R_3} = \frac{13.426\times10^{6} + 13.426\times10^{6}}{(13.426\times10^{6})\cdot(13.426\times10^{6})}$$

$$= \frac{2}{13.426\times10^{6}} = 0.1489\times10^{-6}$$

$$R = R_2 + R_{13} \approx R_2 = (0.01446\times10^{6}) + (0.1489\times10^{-6})$$

$$= 0.01446\times10^{6} \text{ At/Wb}$$

$$\Phi = \Phi_{\text{central}} = \frac{At}{R} = \frac{15}{0.01446\times10^{6}} = 1.037\times10^{-3} \text{ Wb}$$

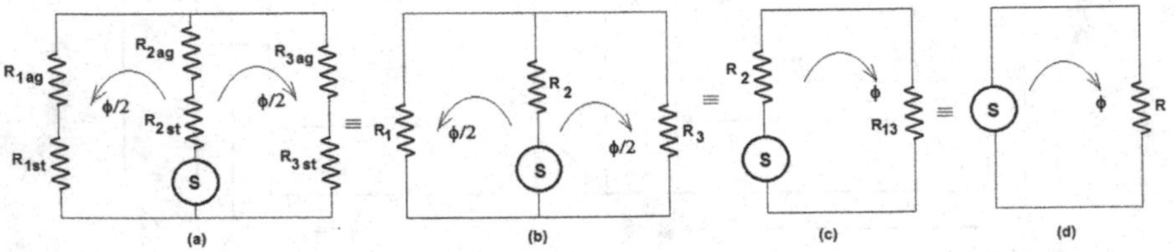

FIGURE 13.5 Equivalent circuit for Example 13.2.

Since $R_1 = R_3$, Φ will be divided into two halves, and thus we have:

$$\Phi_{\text{left}} = \Phi_{\text{right}} = \frac{\Phi}{2} = \frac{1.037 \times 10^{-3}}{2} = 0.5186 \times 10^{-3} \text{ Wb}$$

$$B_{\text{left leg}} = B_{\text{right leg}} = \frac{\Phi}{2} \times R_1 = \left(0.5186 \times 10^{-3}\right) \times \left(13.426 \times 10^{6}\right)$$

$$= 6.96 \times 10^{3} \text{ Wb/m}^2$$

$$B_{\text{cental leg}} = \Phi \times R_2 = \left(1.037 \times 10^{-3}\right) \times \left(0.01446 \times 10^{6}\right) = 14.995 \text{ Wb/m}^2$$

Example 13.3

An electromagnetic relay approximated by a magnetic circuit is shown in Figure 13.6. This circuit has an iron section of length L = 10 cm cross-sectional area S = 1 cm^2 and when the relay is open, an airgap length g of the same area. It contains N = 5000 turns carrying I = 15 mA current. For iron the relative permeability μ_r = 1500. Find the force exerted on the armature (the moving section of the magnetic circuit) when g is: (a) 1 mm (b) 0.1 mm.

SOLUTION

In the Figure L −0.1 m, $S = 10^{-4}$ m^2, N = 5000, $I = 15 \times 10^{-3}$ A, μ_r = 1500.

Thus $NI = 5000 \times 15 \times 10^{-3} = 75$ At.

In view of Equations 13.1b and 13.60b, we can proceed as below.

$$\mathfrak{R}_{\text{steel}} = \frac{L}{\mu_0 \mu_r S} = \frac{0.1}{\left(4\pi \times 10^{-7}\right) \times 1500 \times 10^{-4}} = 0.53 \times 10^{6} \text{ At/Wb}$$

$$F = \frac{B_{st}^2 S}{2\mu_0} = \frac{B_{st}^2 \times 10^{-4}}{2 \times 4\pi \times 10^{-7}} = 39.788 B_{st}^2$$

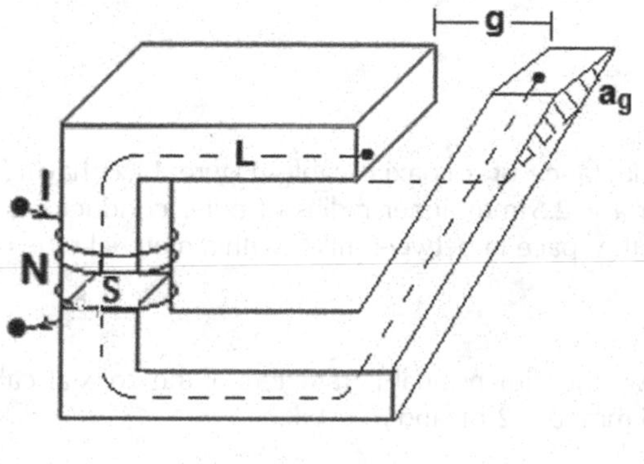

FIGURE 13.6 A magnetic circuit representing an electromagnetic relay.

(a) For g = 1 mm:

$$\Re_{gap} = \frac{g}{\mu_0 S} = \frac{10^{-3}}{\left(4\pi \times 10^{-7}\right) \times 10^{-4}} = 7.9577 \times 10^6 \text{ At/Wb}$$

$$\Re_{total} = 0.53 \times 10^6 + 7.9577 \times 10^6 = 8.488 \times 10^6 \text{ At/Wb}$$

$$\Phi_1 = \frac{NI}{\Re_{total}} = \frac{75}{8.488 \times 10^6} = 8.836 \times 10^{-6} \text{ Wb}$$

$$B_{st1} = \frac{\Phi_1}{S} = \frac{8.836 \times 10^{-6}}{10^{-4}} = 0.08836 \text{ Wb/m}^2$$

$$F_1 = 39.788 B_{st1}^2 = 39.788 (0.08836)^2 = 0.31 \text{ N}$$

(b) For g = 0.1 mm:

$$\Re_{gap} = \frac{g}{\mu_0 S} = \frac{10^{-4}}{\left(4\pi \times 10^{-7}\right) \times 10^{-4}} = 0.79577 \times 10^6 \text{ At/Wb}$$

$$\Re_{total} = 0.53 \times 10^6 + 0.79577 \times 10^6 = 1.326 \times 10^6 \text{ At/Wb}$$

$$\Phi_2 = \frac{NI}{\Re_{total}} = \frac{75}{1.326 \times 10^6} = 56.56 \times 10^{-6} \text{ Wb}$$

$$B_{st2} = \frac{\Phi_2}{S} = \frac{56.56 \times 10^{-6}}{10^{-4}} = 0.5656 \text{ Wb/m}^2$$

$$F_2 = 39.788 B_{st2}^2 = 39.788 (0.5656)^2 = 12.72 \text{ N}$$

Example 13.4

Find the self-inductance of a coaxial cable (Figure 13.7) having outer radius of inner conductor a = 2.5 mm, inner radius of outer conductor b = 7.5 mm, and length d = 2 m, the space in between filled with a material of μ_r = 150.

SOLUTION

Figure 13.7 shows the dimensional parameters of the coaxial cable wherein a = 2.5 mm, b = 7.5 mm, d = 2 m, and μ_r = 150.

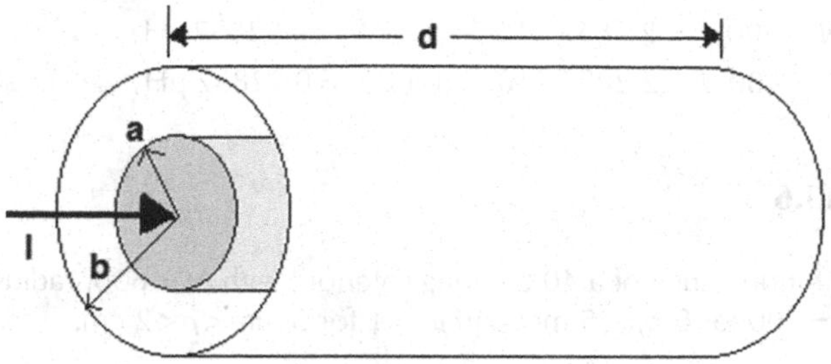

FIGURE 13.7 A coaxial cable.

In view of Equation 13.24b:

$$L = \frac{\mu d}{2\pi}\ln\left(\frac{b}{a}\right) = \frac{\mu_0 \mu_r d}{2\pi}\ln\left(\frac{b}{a}\right)$$

$$= \frac{4\pi \times 10^{-7} \times 150 \times 2}{2\pi}\ln\left(\frac{7.5}{2.5}\right) = 65.917\ \mu\text{H}$$

Example 13.5

Find the self-inductance of a square wooden cross-section toroid of 2 cm × 2 cm with inner radius $\rho = 2.5$ cm and (a) $N = 600$ turns; (b) 100 turns.

SOLUTION

In Figure 13.8 $\rho = 2.5$ cm, $2a = 2$ cm and $\mu_r = 1$ (for wood).
In $\rho_0 = \rho + a = 2.5 + 1 = 3.5$ cm.
In view of Equation 13.28:

$$L = \frac{\mu N^2 A}{2\pi\rho_0} = \frac{4\pi \times 10^{-7} \times 1 \times (N)^2 \left(4 \times 10^{-4}\right)}{2\pi \times 3.5 \times 10^{-2}} = 2.2857 \times 10^{-9} \times (N)^2$$

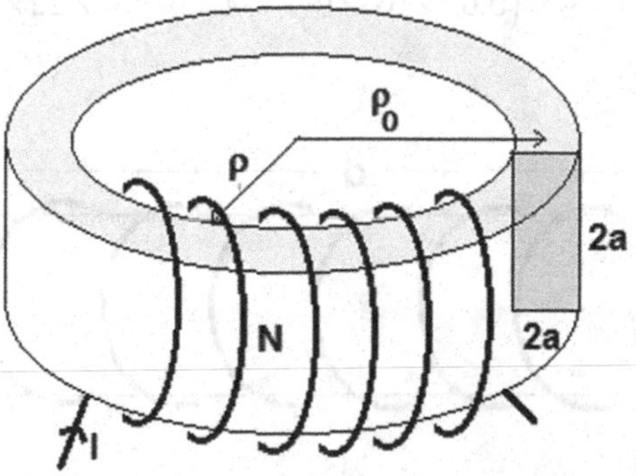

FIGURE 13.8 A square wooden toroid.

(a) For $N = 600$, $L = 2.2857 \times 10^{-9} \times (600)^2 = 8.22852\ \mu H$
(b) For $N = 100$, $L = 2.2857 \times 10^{-9} \times (100)^2 = 0.22857\ \mu H$

Example 13.6

Find the self-inductance of a 40 cm long solenoid with $N = 600$, radius $\rho = 2$ cm wherein $\mu_r = 100$ for $0<\rho<5$ mm and $\mu_r = 1$ for 5 mm$<\rho<$2 cm.

SOLUTION

In Figure 13.9, $\rho = 2$ cm, $N = 600$, $d = 40$ cm, $\mu_r = 100$ for $0<\rho<5$ mm, $\mu_r = 1$ for 5 mm$<\rho<$2 cm.

In view of Equations 13.25a–d and 13.26:

$$H = \frac{NI}{d} = \frac{600I}{40 \times 10^{-2}} = 1500I$$

$$B = \mu_0 \mu_r H = 4\pi \times 10^{-7} \times \mu_r \times 1500I = 6\pi \times 10^{-4} I \times \mu_r$$

$$B_1 = 6\pi \times 10^{-4} I \times 100 = 60\pi \times 10^{-3} I \quad \text{for } 0 < \rho < 5\text{ mm}$$

$$B_2 = 0.6\pi \times 10^{-3} I \quad \text{for } 5\text{ mm} < \rho < 2\text{ cm}$$

$$\Phi = B \cdot S \quad \Phi_1 = B_1 \cdot S_1 \quad \Phi_2 = B_2 \cdot S_2$$

For $\rho_1 = 5$ mm, $S_1 = \pi\rho_1^2 = \pi \times (5 \times 10^{-3})^2 = 25\pi \times 10^{-6}$

For $\rho_2 = 20$ mm, $S_2 = \pi\left(\rho_2^2 - \rho_1^2\right) = \pi \times (400 - 25) \times 10^{-6} = 375\pi \times 10^{-6}$

$$\Phi_1 = B_1 \cdot S_1 = \left(60\pi \times 10^{-3} I\right) \times 25\pi \times 10^{-6} = 14.8 \times 10^{-6} I$$

$$\Phi_2 = B_2 \cdot S_2 = \left(0.6\pi \times 10^{-3} I\right) \times 375\pi \times 10^{-6} = 2.22 \times 10^{-6} I$$

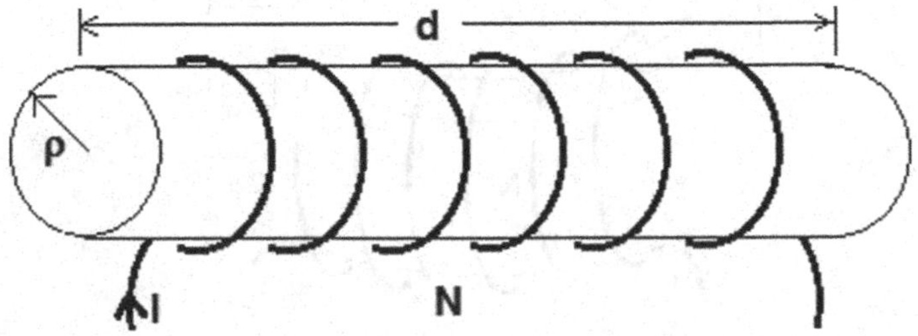

FIGURE 13.9 A long solenoid.

$$\Phi = \Phi_{\text{Total}} = \Phi_1 + \Phi_2 = 14.8I \times 10^{-6} + 2.22I \times 10^{-6} = 17.02 \times 10^{-6} I$$

$$L = \frac{N\Phi}{I} = \frac{600 \times 17.02I \times 10^{-6}}{I} = 10.213\,\text{mH}$$

Example 13.7

Figure 13.10 shows an inner and an outer concentric cylindrical solenoid, both of lengths d = 50 cm. The inner one of radius a = 0.5 cm has N_1 = 1000 turns, carrying current I_1. The outer one of radius b = 1.5 cm has N_2 = 800 turns, carrying current I_2. Find the self-inductance of each solenoid and the mutual inductance between them if the relative permeability for the material of both is μ_r = 12.

SOLUTION

In Figure 13.10, N_1 = 1000, I_1 current, d = 50 cm, a = 0.5 cm, μ_r = 12, N_2 = 800, I_2 current, d = 50 cm, b = 1.5 cm.

$$S_1 = \pi a^2 = 25\pi \times 10^{-6} \quad S_2 = (225 - 25)\pi \times 10^{-6} = 200\pi \times 10^{-6} \quad H = \frac{NI}{d}$$

$$H_1 = \frac{N_1 I_1}{d} = \frac{1000 I_1}{50 \times 10^{-2}} = 2000 I_1 \quad \text{over } 0 \le \rho \le a \text{ and } 0 \quad \text{for } \rho > a$$

$$H_2 = \frac{N_2 I_2}{d} = \frac{800 I_2}{50 \times 10^{-2}} = 1600 I_2 \quad \text{over } 0 \le \rho \le b \text{ and } 0 \quad \text{for } \rho > b$$

$$B_1 = \mu_1 H_1 = 12 \times \left(4\pi \times 10^{-7}\right) \times 2000 I_1 = 30.16 \times 10^{-3} I_1$$

$$B_2' = \mu_1 H_2 = 12 \times \left(4\pi \times 10^{-7}\right) \times 1600 I_2 = 24.13 \times 10^{-3} I_2$$

$$B_2'' = \mu_0 H_2 = \left(4\pi \times 10^{-7}\right) \times 1600 I_2 = 2.01 \times 10^{-3} I_2$$

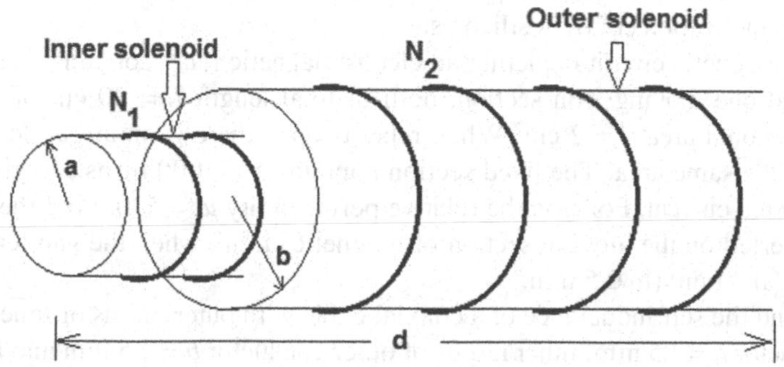

FIGURE 13.10 Two concentric cylindrical solenoids.

$$\Phi_1 = B_1 S_1 = \left(30.16\times10^{-3}\right) I_1 \left(25\pi\times10^{-6}\right) = 2.368\times10^{-6} I_1$$

$$\Phi_2' = B_2' S_1 = \left(24.13\times10^{-3}\right) I_2 \left(25\pi\times10^{-6}\right) = 1.895\times10^{-6} I_2$$

$$\Phi_2'' = B_2'' S_2 = \left(2.01\times10^{-3}\right) I_2 \left(200\pi\times10^{-6}\right) = 1.263\times10^{-6} I_2$$

$$\Phi_2 = \Phi_2' + \Phi_2'' = 3.16\times10^{-6} I_2$$

In view of Equations 13.19a, 13.19b, 13.20a, and 13.20b:

$$L_1 = \frac{N_1\varphi_1}{I_1} = \frac{1000\times2.368\times10^{-6} I_1}{I_1} = 2.368\text{ mH}$$

$$L_2 = \frac{N_2\varphi_2}{I_2} = \frac{800\times3.16\times10^{-6} I_2}{I_2} = 2.528\text{ mH}$$

$$M_{12} = \frac{N_2\varphi_{12}}{I_1} = \frac{N_2\varphi_2}{I_1} = \frac{800\times2.368\times10^{-6} I_1}{I_1} = 1.895\text{ mH}$$

$$M_{21} = \frac{N_1\varphi_{21}}{I_2} = \frac{N_1\varphi_2'}{I_2} = \frac{1000\times1.895\times10^{-6} I_2}{I_2} = 1.895\text{ mH} = M_{12}$$

PROBLEMS

P13.1 In a silicon steel core $B = 0.5$ wb/m^2 and $H = 100$ At/m at the midpoint of the left leg. In this core $L_1 = 5$ cm, $L_2 = 10$ cm, $S_1 = 4$ cm^2, and $S_2 = 2$ cm^2. Find (a) $\mathcal{V}_{air}$ if the airgap length is $g = 0.1$ cm; (b) $\mathcal{V}_{steel}$; and (c) the required current in a 1000 turn coil on the left leg.

P13.2 A silicon steel core has three legs. For each outer leg $S_1 = 1$ cm^2, $L_1 = 12$ cm and for mid leg $S_2 = 1.2$ cm^2, $L_2 = 4$ cm. The left outer leg contains a coil with N = 1200 turns carrying 10mA current. Find B in all the three legs if an airgap $g = 1.5$ mm is cut in mid and right vertical leg. For values of B, assume the linear operation with $\mu_r = 5000$ below the knee of the magnetization curve of silicon steel.

P13.3 A magnetic circuit depicting an electromagnetic relay contains one fixed and one moving iron section, both of total length $L = 20$ cm of cross-sectional area $S = 2$ cm^2 When relay is open there is an airgap length g of the same area. The fixed section contains $N = 1000$ turns carrying $I =$ 10 ma current. For iron the relative permeability $\mu_r = 500$. Find the force exerted on the moving section of magnetic circuit when the gap length g is (a) 2 mm; (b) 0.5 mm.

P13.4 Find the self-inductance of a coaxial cable with outer radius of inner conductor $a = 1.5$ mm, inner radius of outer conductor $b = 4.5$ mm and length $d = 1$ m, the space in between filled with a material of $\mu_r = 100$.

P13.5 Find the self-inductance of a square cross-section toroid of 3 cm × 3 cm with inner radius $\rho = 4.5$ cm, $N = 300$ turns. The toroid core is made of (a) soft iron, $\mu_r = 5000$; (b) cobalt, $\mu_r = 250$.

P13.6 Find the self-inductance of 60 cm long solenoid with uniformly distributed turns $N = 300$ and radius $\rho = 4$ cm wherein (a) $\mu_r = 200$ for $0 < \rho < 2$ cm and $\mu_r = 1$ for 2 cm $< \rho < d = 4$ cm; (b) $\mu_r = 1$ for $0 < \rho < 2$ cm and $\mu_r =$ 200 for 2 cm $< \rho < d = 4$ cm.

P13.7 A configuration of two concentric cylindrical solenoids is shown in Figure 13.8. The outer solenoid has $N_1 = 500$, I_1 current, $d = 25$ cm, $a = 2$ cm. The inner solenoid has $N_2 = 400$, I_2 current, $d = 30$ cm, $b = 4$ cm. For both the cores $\mu_r = 10$. Find the self-inductance of each solenoid and the mutual inductance between them.

P13.8 The two wires of a transmission line having currents $\pm I_1 \boldsymbol{a}_z$ are located at $x = 3 \mp 1$ cm. Similarly the two wires of another line with currents $\pm I_2 \boldsymbol{a}_z$ are located at $x = -3 \mp 1$ cm. Find the flux (per unit length) produced by I_1 that links with I_2 and M_{11} per unit length. Assume that both the lines are located in free space.

DESCRIPTIVE QUESTIONS

Q13.1 Describe the formation of B–H loop. Discuss its salient features and implications on the magnetic circuits.

Q13.2 Discuss the limitations of the analogy between electric and magnetic circuits.

Q13.3 In view of the duality principle, obtain the expressions for the self and mutual inductances between two circuits.

Q13.4 Derive the expressions for the self and mutual inductances between two circuits on the bases of energy stored in magnetic field.

Q13.5 Obtain the expression for inductance in a (a) coaxial cable; (b) solenoid; and (c) toroid.

BIBLIOGRAPHY

1. Matthew N. O. Sadiku, *Principles of Electromagnetics*, 4th Ed, Oxford University Press, 2007, pp. 303–306.
2. Carl T. A. Johnk, *Engineering Electromagnetic Fields and Waves*, Wiley Eastern Limited, Edition 1975, pp. 618–632.

FURTHER READING

Given at the end of Chapter 15.

14 Magnetostatic Boundary Value Problems

14.1 INTRODUCTION

In Chapters 9 and 10, a number of electrostatic boundary value problems were solved. The problems included in Chapter 9 were related to the Laplacian field, whereas those in Chapter 10 were related to Poisson's field. The present chapter is devoted to the problems of magnetostatic fields. The theory of magnetostatics and its different aspects have already been discussed in Chapters 11–13.

The electrostatic boundary value problems can easily be solved by using the boundary conditions described in Chapter 12. The method followed in the subsequent subsections is almost the same as that used to solve electrostatic field problems. These subsections include examples of magnetostatic boundary value problems. The first example, in Section 14.2, is related to the magnetostatic field in a piecewise homogenous core of given dimensions and defined magnetic properties, whereas the second example, in Section 14.3, deals with the magnetostatic field in a heterogeneous core.

14.2 MAGNETOSTATIC FIELD IN A PIECEWISE HOMOGENOUS CORE

The estimation of a magnetostatic field in a device or system or part thereof requires the understanding of its physical configuration and the governing field relation(s) and appropriate boundary conditions. If required this configuration may be divided into regions and appropriate boundary conditions for each region are to be identified. The field components in each region are to be written in view of the appropriate field relations. The arbitrary constants involved in the expressions of field components, can be evaluated in view of relevant boundary conditions. The following sections systematically describe this through some examples.

14.2.1 Physical Configuration

This problem involves a long rectangular core coaxially lying along the z-axis, as shown in Figure 14.1. It also shows the cross-section of this core extending over $(-a < x < a)$ and $(-b < y < b)$. The permeability of core varies periodically in the axial direction with the length of a period $(c + d)$. The magnetic core is piecewise homogenous such that $\mu = \mu_1$, over $c > z > 0$ and $\mu = \mu_2$, over $0 > z > -d$. The variation of the permeability in the axial direction makes it a three-dimensional problem. The process of evaluation of distribution of magnetostatic field in this core is described below.

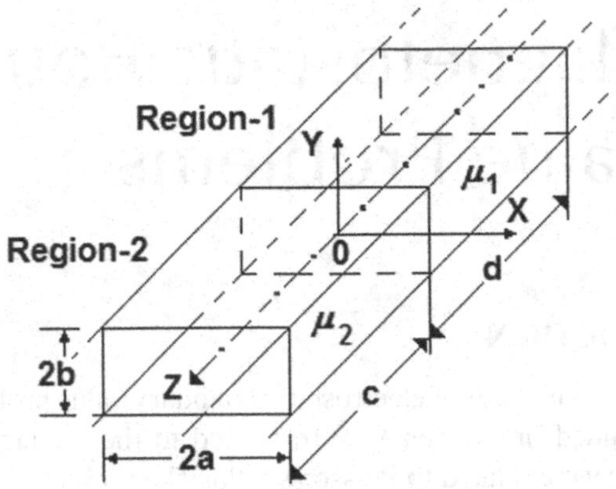

FIGURE 14.1 A piecewise homogenous long rectangular magnetic core.

14.2.2 Boundary Conditions

In view of Figure 14.1, different components of H satisfy the following boundary conditions:

$$H_z\big|_{x=\pm a} = H_o \quad \text{over } (-b < y < b) \tag{14.1a}$$

$$H_z\big|_{y=\pm b} = H_o \quad \text{over } (-a < x < a) \tag{14.1b}$$

$$H_y\big|_{x=\pm a} = 0 \quad \text{over } (-b < y < b) \tag{14.1c}$$

and

$$H_x\big|_{y=\pm b} = 0 \quad \text{over } (-a < x < a) \tag{14.1d}$$

14.2.3 Regions and Field Components

The configuration shown in Figure 14.1 is divided into two regions. Region 1 extends over ($c > z > 0$) and Region 2 over ($0 > z > -d$). The magnetic field components in Region 1 are indicated as H_{x1}, H_{y1}, and H_{z1} and those in Region 2 as H_{x2}, H_{y2}, and H_{z2}. For homogeneous current-free regions, both curl and divergence of the magnetic field intensity vector are zero, i.e.:

$$\nabla \times \boldsymbol{H} = 0 \tag{14.2a}$$

and:

$$\nabla \bullet \boldsymbol{H} = 0 \tag{14.2b}$$

In view of the identity:

$$\nabla \times (\nabla \times \boldsymbol{H}) \equiv \nabla(\nabla \bullet \boldsymbol{H}) - \nabla^2 \boldsymbol{H} \tag{14.3a}$$

$$\nabla^2 \boldsymbol{H} = 0 \tag{14.3b}$$

Thus, each component of the magnetic field expressed in a Cartesian system of space coordinates satisfies the Laplace equation. It may be noted that if $\mu_1 = \mu_2$, inside the core only the axial component of uniformly distributed magnetic field will exist. Therefore, one may tentatively take:

$$H_{z1} \stackrel{\text{def}}{=} \sum_{m\text{-odd}}^{\infty} \sum_{n\text{-odd}}^{\infty} A'_{m,n} \cdot \cos\left(\frac{m\pi}{2a} \cdot x\right) \cdot \cos\left(\frac{n\pi}{2b} \cdot y\right) \cdot \frac{\cosh \alpha_{m,n}\left(z - \frac{c}{2}\right)}{\cosh\left(\alpha_{m,n} \cdot \frac{c}{2}\right)} + H_o \tag{14.4a}$$

$$H_{z2} \stackrel{\text{def}}{=} \sum_{m\text{-odd}}^{\infty} \sum_{n\text{-odd}}^{\infty} A''_{m,n} \cdot \cos\left(\frac{m\pi}{2a} \cdot x\right) \cdot \cos\left(\frac{n\pi}{2b} \cdot y\right) \cdot \frac{\cosh \alpha_{m,n}\left(z + \frac{d}{2}\right)}{\cosh\left(\alpha_{m,n} \cdot \frac{d}{2}\right)} + H_o \tag{14.4b}$$

where

$$\alpha_{m,n} = \sqrt{\left(\frac{m\pi}{2a}\right)^2 + \left(\frac{n\pi}{2b}\right)^2} \tag{14.4c}$$

In Equations 14.4a and 14.4b, symbols $A'_{m,n}$ and $A''_{m,n}$ indicate coefficients of a double Fourier series for functions at $z = 0$.

Furthermore, in view of the relevant boundary conditions, the other field components can be written as:

$$H_{y1} \stackrel{\text{def}}{=} \sum_{m\text{-odd}}^{\infty} \sum_{n\text{-odd}}^{\infty} B'_{m,n} \cdot \cos\left(\frac{m\pi}{2a} \cdot x\right) \cdot \sin\left(\frac{n\pi}{2b} \cdot y\right) \cdot \frac{\sinh \alpha_{m,n}\left(z - \frac{c}{2}\right)}{\cosh\left(\alpha_{m,n} \cdot \frac{c}{2}\right)} \tag{14.5a}$$

$$H_{y2} \stackrel{\text{def}}{=} \sum_{m\text{-odd}}^{\infty} \sum_{n\text{-odd}}^{\infty} B''_{m,n} \cdot \cos\left(\frac{m\pi}{2a} \cdot x\right) \cdot \sin\left(\frac{n\pi}{2b} \cdot y\right) \cdot \frac{\sinh \alpha_{m,n}\left(z + \frac{d}{2}\right)}{\cosh\left(\alpha_{m,n} \cdot \frac{d}{2}\right)} \tag{14.5b}$$

$$H_{x1} \stackrel{\text{def}}{=} \sum_{m\text{-odd}}^{\infty} \sum_{n\text{-odd}}^{\infty} C'_{m,n} \cdot \sin\left(\frac{m\pi}{2a} \cdot x\right) \cdot \cos\left(\frac{n\pi}{2b} \cdot y\right) \cdot \frac{\sinh \alpha_{m,n}\left(z - \frac{c}{2}\right)}{\cosh\left(\alpha_{m,n} \cdot \frac{c}{2}\right)} \tag{14.5c}$$

and

$$H_{x2} \stackrel{\text{def}}{=} \sum_{m\text{-odd}}^{\infty} \sum_{n\text{-odd}}^{\infty} C''_{m,n} \cdot \sin\left(\frac{m\pi}{2a} \cdot x\right) \cdot \cos\left(\frac{n\pi}{2b} \cdot y\right) \cdot \frac{\sinh \alpha_{m,n}\left(z + \dfrac{d}{2}\right)}{\cosh\left(\alpha_{m,n} \cdot \dfrac{d}{2}\right)} \tag{14.5d}$$

In Equations 14.5a–d, $B'_{m,n}$, $B''_{m,n}$, $C'_{m,n}$, and $C''_{m,n}$ indicate four sets of arbitrary constants.

14.2.4 Evaluation of Arbitrary Constants

As there are no currents at the boundary surface ($z = 0$), the tangential components of magnetic field intensity are continuous at this boundary, thus:

$$H_{y1}\Big|_{z=0} = H_{y2}\Big|_{z=0} \tag{14.6a}$$

and

$$H_{x1}\Big|_{z=0} = H_{x2}\Big|_{z=0} \tag{14.6b}$$

Thus, from Equations 14.5a–d, we get:

$$-B'_{m,n} \cdot \tanh\left(\alpha_{m,n} \cdot \frac{c}{2}\right) = B''_{m,n} \cdot \tanh\left(\alpha_{m,n} \cdot \frac{d}{2}\right) \tag{14.7a}$$

and

$$-C'_{m,n} \cdot \tanh\left(\alpha_{m,n} \cdot \frac{c}{2}\right) = C''_{m,n} \cdot \tanh\left(\alpha_{m,n} \cdot \frac{d}{2}\right) \tag{14.7b}$$

Furthermore, since the normal component of magnetic flux density is continuous at $z = 0$:

$$B_{z1}\Big|_{z=0} = B_{z2}\Big|_{z=0} \tag{14.8a}$$

Therefore, using Equations 14.4a and 14.4b, we have:

$$\begin{aligned} &\mu_1 \cdot \sum_{m\text{-odd}}^{\infty} \sum_{n\text{-odd}}^{\infty} A'_{m,n} \cdot \cos\left(\frac{m\pi}{2a} \cdot x\right) \cdot \cos\left(\frac{n\pi}{2b} \cdot y\right) + \mu_1 \cdot H_o \\ &= \mu_2 \cdot \sum_{m\text{-odd}}^{\infty} \sum_{n\text{-odd}}^{\infty} A''_{m,n} . \cos\left(\frac{m\pi}{2a} \cdot x\right) \cdot \cos\left(\frac{n\pi}{2b} \cdot y\right) + \mu_2 \cdot H_o \end{aligned} \tag{14.8b}$$

On multiplying both sides by $\cos\left(\frac{p\pi}{2a}\cdot x\right)\cdot\cos\left(\frac{q\pi}{2b}\cdot y\right)$, where both p and q are odd integers, and then on integrating over $-a < x < a$ and $-b < y < b$, one gets:

$$\mu_1\cdot H_o\cdot\frac{\sin(p\pi/2)\cdot\sin(q\pi/2)}{p\cdot q}\cdot\left(\frac{4}{\pi}\right)^2+\mu_1\cdot A'_{p,q}\cdot(a\cdot b)$$

$$=\mu_2\cdot H_o\cdot\frac{\sin(p\pi/2)\cdot\sin(q\pi/2)}{p\cdot q}\cdot\left(\frac{4}{\pi}\right)^2+\mu_2\cdot A''_{p,q}\cdot(a\cdot b)$$

Furthermore, the substitution of m and n for p and q, respectively, gives:

$$\mu_1\cdot A'_{m,n}-\mu_2\cdot A''_{m,n}=(\mu_2-\mu_1)\cdot H_o\cdot\frac{\sin(m\pi/2)\cdot\sin(n\pi/2)}{(m\cdot n)\cdot(a\cdot b)}\cdot\left(\frac{4}{\pi}\right)^2 \tag{14.9}$$

for $m, n = 1, 3, 5, \ldots$

We, thus have three sets of equations between six sets of arbitrary constants. We are running short by three sets of equations. Since there is no current distribution on the surface $z = 0$, the following boundary condition is valid:

$$\left.\frac{\partial H_{z1}}{\partial z}\right|_{z=0}=\left.\frac{\partial H_{z2}}{\partial z}\right|_{z=0} \tag{14.10}$$

Therefore, use of Equations 14.4a and 14.4b gives:

$$-A'_{m,n}\cdot\tanh\left(\alpha_{m,n}\cdot\frac{c}{2}\right)=A''_{m,n}\cdot\tanh\left(\alpha_{m,n}\cdot\frac{d}{2}\right) \tag{14.11}$$

for $m, n = 1, 3, 5, \ldots$

Now, Equations 14.10 and 14.11 can be readily solved giving expressions for $A'_{m,n}$ and $A''_{m,n}$:

$$A'_{m,n}=H_o\cdot\frac{\sin(m\pi/2)\cdot\sin(n\pi/2)}{(m\cdot n)\cdot(a\cdot b)}\cdot\left(\frac{4}{\pi}\right)^2$$

$$\cdot\frac{(\mu_2-\mu_1)\cdot\tanh\left(\alpha_{m,n}\cdot\frac{d}{2}\right)}{\left[\mu_1\cdot\tanh\left(\alpha_{m,n}\cdot\frac{d}{2}\right)+\mu_2\cdot\tanh\left(\alpha_{m,n}\cdot\frac{c}{2}\right)\right]} \tag{14.12a}$$

$$A''_{m,n} = H_o \cdot \frac{\sin(m\pi/2)\cdot\sin(n\pi/2)}{(m\cdot n)\cdot(a\cdot b)}\cdot\left(\frac{4}{\pi}\right)^2 \cdot \frac{(\mu_1-\mu_2)\cdot\tanh\left(\alpha_{m,n}\cdot\frac{c}{2}\right)}{\left[\mu_1\cdot\tanh\left(\alpha_{m,n}\cdot\frac{d}{2}\right)+\mu_2\cdot\tanh\left(\alpha_{m,n}\cdot\frac{c}{2}\right)\right]} \quad (14.12b)$$

for $m, n = 1, 3, 5, \ldots$

We need two more sets of linearly independent equations between various arbitrary constants. These are obtained by considering x- and y-components of the curl of the magnetic field intensity, as shown below:

$$\nabla\times\boldsymbol{H}\big|_x = \frac{\partial H_{z1}}{\partial y} - \frac{\partial H_{y1}}{\partial z} = 0 \quad (14.13)$$

giving, in view of Equations 14.4a and 14.5a:

$$B'_{m,n} = -\frac{A'_{m,n}}{\alpha_{m,n}}\cdot\left(\frac{\pi}{2}\right)\cdot\left(\frac{n}{b}\right) \quad (14.14a)$$

Thus, in view of Equation 14.7a:

$$B''_{m,n} = \frac{A'_{m,n}}{\alpha_{m,n}}\cdot\left(\frac{\pi}{2}\right)\cdot\left(\frac{n}{b}\right)\cdot\frac{\tanh\left(\alpha_{m,n}\cdot\frac{c}{2}\right)}{\tanh\left(\alpha_{m,n}\cdot\frac{d}{2}\right)} \quad (14.14b)$$

for $m, n = 1, 3, 5, \ldots$

Also:

$$\nabla\times\boldsymbol{H}\big|_y = \frac{\partial H_{x1}}{\partial z} - \frac{\partial H_{z1}}{\partial x} = 0 \quad (14.15)$$

giving, in view of Equations 14.4a and 14.5c:

$$C'_{m,n} = -\frac{A'_{m,n}}{\alpha_{m,n}}\cdot\left(\frac{\pi}{2}\right)\cdot\left(\frac{m}{a}\right) \quad (14.16a)$$

In view of Equation 14.7b:

$$C''_{m,n} = \frac{A'_{m,n}}{\alpha_{m,n}}\cdot\left(\frac{\pi}{2}\right)\cdot\left(\frac{m}{a}\right)\cdot\frac{\tanh\left(\alpha_{m,n}\cdot\frac{c}{2}\right)}{\tanh\left(\alpha_{m,n}\cdot\frac{d}{2}\right)} \quad (14.16b)$$

for $m, n = 1, 3, 5, \ldots$

Since Equations 14.14a, 14.14b, 14.16a, and 14.16b relate $B'_{m,n}$, $B''_{m,n}$, $C'_{m,n}$, and $C''_{m,n}$ with $A'_{m,n}$, respectively, and $A'_{m,n}$ is given by Equation 14.12a, expressions for all these arbitrary constants occurring in the field equations can be found.

14.3 MAGNETOSTATIC FIELD IN A HETEROGENEOUS CORE

This problem can also be handled almost in a similar way as described in Section 14.2. Here, too, the physical configuration with all of its dimensional parameters is drawn, the boundary conditions are properly identified, the field components are written in view the appropriate field relations, and the arbitrary constants involved in the expressions of field components are evaluated with relevant boundary conditions. The procedure is stepwise as described below.

14.3.1 Physical Configuration

Figure 14.2 shows a long rectangular core, which is coaxially lying along the z-axis. This core section extends over $(-a < x < a)$ and $(-b < y < b)$. Its permeability varies exponentially in the axial direction. The magnetic core is therefore heterogeneous, such that: $\mu = \tilde{\mu} \cdot e^{-\nu \cdot |z|}$, where $\tilde{\mu}$ and ν are (known/given) positive real numbers. The variation of the permeability in the axial direction (over $-\infty < z < \infty$) makes it a three-dimensional problem. In view of the continuous variation of μ, the configuration is not to be divided into regions.

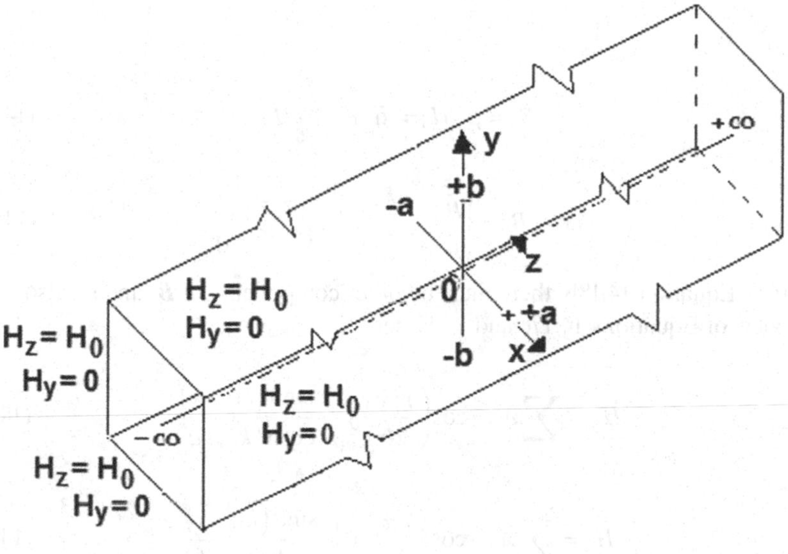

FIGURE 14.2 A heterogeneous long rectangular magnetic core.

14.3.2 Boundary Conditions

The configuration shown satisfies the following boundary conditions:

$$H_z\Big|_{x=\pm a} = H_o \quad \text{over } (-b < y < b) \tag{14.17a}$$

$$H_z\Big|_{y=\pm b} = H_o \quad \text{over } (-a < x < a) \tag{14.17b}$$

$$H_y\Big|_{x=\pm a} = 0 \quad \text{over } (-b < y < b) \tag{14.17c}$$

and

$$H_x\Big|_{y=\pm b} = 0 \quad \text{over } (-a < x < a) \tag{14.17d}$$

14.3.3 Field Components

The core section is a current-free region:

$$\nabla \times \boldsymbol{H} = 0 \tag{14.18a}$$

Also:

$$\nabla \bullet \boldsymbol{B} = 0 \tag{14.18b}$$

Let:

$$H_z = H_o \tag{14.19}$$

Thus:

$$B_z = \mu \cdot H_o = \tilde{\mu} \cdot e^{-\nu\cdot|z|} \cdot H_o \tag{14.20a}$$

$$\text{i.e.}\quad B_z = \begin{cases} \tilde{\mu} \cdot e^{-\nu\cdot z} \cdot H_o & \text{for } z \geq 0 \\ \tilde{\mu} \cdot e^{+\nu\cdot z} \cdot H_o & \text{for } z \leq 0 \end{cases} \tag{14.20b}$$

To satisfy Equation 14.18b, there must be other components of $\boldsymbol{B}$, and so also of $\boldsymbol{H}$.
In view of Equations 14.17c and 14.17, let:

$$H_x = \sum_{m\text{-odd}}^{\infty} A_{1m} \cdot \cos\left(\frac{m\pi}{2b} \cdot y\right) \cdot \frac{\sinh\left(k_{1m} \cdot x\right)}{\cosh\left(k_{1m} \cdot a\right)} \tag{14.21a}$$

$$H_y = \sum_{n\text{-odd}}^{\infty} A_{2n} \cdot \cos\left(\frac{n\pi}{2a} \cdot x\right) \cdot \frac{\sinh\left(k_{2n} \cdot y\right)}{\cosh\left(k_{2n} \cdot b\right)} \tag{14.21b}$$

where A_{1m} and A_{2n} indicate two sets of arbitrary constants.

Since:

$$\nabla\times(\nabla\times\boldsymbol{H})\equiv\nabla(\nabla\bullet\boldsymbol{H})-\nabla^2\boldsymbol{H} \tag{14.22}$$

Thus, in view of Equation 14.18a, Equation 14.22 results in:

$$\nabla^2\boldsymbol{H}=\nabla(\nabla\bullet\boldsymbol{H}) \tag{14.23a}$$

or:

$$\nabla^2\left(H_x\boldsymbol{a}_x+H_y\boldsymbol{a}_y+H_z\boldsymbol{a}_z\right)=\left(\boldsymbol{a}_x\frac{\partial}{\partial x}+\boldsymbol{a}_y\frac{\partial}{\partial y}+\boldsymbol{a}_z\frac{\partial}{\partial z}\right)\cdot(\nabla\bullet\boldsymbol{H}) \tag{14.23b}$$

where:

$$\nabla\bullet\boldsymbol{H}=\frac{\partial H_x}{\partial x}+\frac{\partial H_y}{\partial y}+\frac{\partial H_z}{\partial z} \tag{14.23c}$$

Therefore, in view of Equations 14.18b, 14.19, and 14.20b:

$$\nabla\bullet\boldsymbol{H}=\left[\frac{e^{v\cdot z}}{\tilde{\mu}}\cdot\left(\frac{\partial B_x}{\partial x}+\frac{\partial B_y}{\partial y}+\frac{\partial B_z}{\partial z}\right)+vH_o\cdot\frac{e^{v\cdot z}}{\tilde{\mu}}\cdot\left(\tilde{\mu}\cdot e^{-v\cdot z}\right)\right]+\frac{\partial H_z}{\partial z}$$

$$=vH_o,\quad(\text{a known constant})$$

Therefore, from Equation 14.23a:

$$\nabla^2\boldsymbol{H}=\nabla(\nabla\bullet\boldsymbol{H})=0 \tag{14.24}$$

Thus, from Equations 14.21a and 14.21b:

$$\left.k_{1m}=\left(\frac{m\pi}{2b}\right)\quad(\text{a})\quad\text{and}\quad k_{2n}=\left(\frac{n\pi}{2a}\right)\quad(\text{b})\right\} \tag{14.25}$$

Consider the following identities:

$$\int_{-a}^{a}\sin\left(\frac{N\pi}{2a}\cdot x\right)\cdot\sin\left(\frac{n\pi}{2a}\cdot x\right)dx\equiv\begin{cases}\dfrac{1}{a} & \text{for } N=n\\ 0 & \text{for } N\neq n\end{cases} \tag{14.26a}$$

$$\int_{-a}^{a}\sin\left(\frac{N\pi}{2a}\cdot x\right)\cdot\sinh\left(\frac{m\pi}{2b}\cdot x\right)dx\equiv\cosh\left(\frac{m\pi}{2b}\cdot a\right)\cdot\frac{2\cdot\left(\dfrac{m\pi}{2b}\right)\cdot\sin\left(\dfrac{N\pi}{2}\right)}{\left[\left(\dfrac{N\pi}{2a}\right)^2+\left(\dfrac{m\pi}{2b}\right)^2\right]} \tag{14.26b}$$

$$\int_{-b}^{b} \sin\left(\frac{M\pi}{2b}\cdot y\right)\cdot\sin\left(\frac{m\pi}{2b}\cdot y\right)dy \equiv \begin{cases} \frac{1}{b}, & \text{for } M = m \\ 0, & \text{for } M \neq m \end{cases} \tag{14.26c}$$

$$\int_{-b}^{b} \sin\left(\frac{M\pi}{2b}\cdot y\right)\cdot\sinh\left(\frac{n\pi}{2a}\cdot y\right)dy \equiv \cosh\left(\frac{n\pi}{2a}\cdot b\right)\cdot\frac{2\cdot\left(\frac{n\pi}{2a}\right)\cdot\sin\left(\frac{M\pi}{2}\right)}{\left[\left(\frac{M\pi}{2b}\right)^2+\left(\frac{n\pi}{2a}\right)^2\right]} \tag{14.26d}$$

Since:

$$\nabla\times\boldsymbol{H}\Big|_z = \frac{\partial H_y}{\partial x} - \frac{\partial H_x}{\partial y} = 0 \tag{14.27a}$$

Thus, using Equations 14.25a and 14.25b, we get from Equations 14.21a, 14.21b, and 14.27a:

$$\sum_{m\text{-odd}}^{\infty} A_{1m}\cdot\left(\frac{m\pi}{2b}\right)\cdot\sin\left(\frac{m\pi}{2b}\cdot y\right)\cdot\frac{\sinh\left(\frac{m\pi}{2b}\cdot x\right)}{\cosh\left(\frac{m\pi}{2b}\cdot a\right)}$$
$$= \sum_{n\text{-odd}}^{\infty} A_{2n}\cdot\left(\frac{n\pi}{2a}\right)\cdot\sin\left(\frac{n\pi}{2a}\cdot x\right)\cdot\frac{\sinh\left(\frac{n\pi}{2a}\cdot y\right)}{\cosh\left(\frac{n\pi}{2a}\cdot b\right)} \tag{14.27b}$$

Multiply both sides by $\left[\sin\left(\frac{N\pi}{2a}\cdot x\right)\cdot\sin\left(\frac{M\pi}{2b}\cdot y\right)\right]$ and then, on integrating over $(-a < x < a)$ and $(-b < y < b)$, we get the following equation on replacing M by m and N by n.

$$A_{1m}\cdot\frac{1}{b}\cdot\left(\frac{m\pi}{2b}\right)^2\cdot\sin\left(\frac{m\pi}{2}\right) = A_{2n}\cdot\frac{1}{a}\cdot\left(\frac{n\pi}{2a}\right)^2\cdot\sin\left(\frac{n\pi}{2}\right) \tag{14.28}$$

for $m, n = 1, 3, 5, \ldots$

Also:

$$\nabla\times\boldsymbol{H}\Big|_x = \frac{\partial H_z}{\partial y} - \frac{\partial H_y}{\partial z} = 0 \tag{14.29a}$$

and

$$\nabla\times\boldsymbol{H}\Big|_y = \frac{\partial H_x}{\partial z} - \frac{\partial H_z}{\partial x} = 0 \tag{14.29b}$$

These two equations are identically satisfied.

Consider Equations 14.18b, 14.20b, 14.21a, 14.21b, and 14.21c; for Region 1, i.e. $z \geq 0$, we have:

$$\left[\sum_{m\text{-odd}}^{\infty} k_{1m} \cdot A_{1m} \cdot \cos\left(\frac{m\pi}{2b} \cdot y\right) \cdot \frac{\cosh\left(k_{1m} \cdot x\right)}{\cosh\left(k_{1m} \cdot a\right)} + \sum_{n\text{-odd}}^{\infty} k_{2n} \cdot A_{2n} \cdot \cos\left(\frac{n\pi}{2a} \cdot x\right) \cdot \frac{\cosh\left(k_{2n} \cdot y\right)}{\cosh\left(k_{2n} \cdot b\right)}\right] - \nu \cdot H_o = 0 \tag{14.30}$$

Multiply both sides by $\left[\cos\left(\frac{N\pi}{2a} \cdot x\right) \cdot \cos\left(\frac{M\pi}{2b} \cdot y\right)\right]$ and then on integrating over $(-a < x < a)$ and $(-b < y < b)$, we get on replacing M by m and N by n:

$$2\nu \cdot H_o = A_{1m} \cdot \frac{1}{b} \cdot \left(\frac{m\pi}{2b}\right)^2 \cdot \frac{\left(\frac{n\pi}{2a}\right)^2 \cdot \sin\left(\frac{m\pi}{2}\right)}{\left[\left(\frac{n\pi}{2a}\right)^2 + \left(\frac{m\pi}{2b}\right)^2\right]} + A_{2n} \cdot \frac{1}{a} \cdot \left(\frac{n\pi}{2a}\right)^2 \cdot \frac{\left(\frac{m\pi}{2b}\right)^2 \cdot \sin\left(\frac{n\pi}{2}\right)}{\left[\left(\frac{m\pi}{2b}\right)^2 + \left(\frac{n\pi}{2a}\right)^2\right]} \tag{14.31}$$

for $m, n = 1, 3, 5, \ldots$

Solution of simultaneous Equations 14.28 and 14.31 determines the values for A_{1m} and A_{2n}.

For the Region 2, i.e. $z \leq 0$, the equation found is similar to Equation 14.31 but with a negative sign on the LHS term. On solving with Equation 14.28, this will give the same values for A_{1m} and A_{2n} but with opposite sign for the two unknowns. This indicates that the tangential components of the magnetic field intensity will be discontinuous, unless these are 0 at $z = 0$.

There being no surface current sheet at $z = 0$, we must ensure zero values for the tangential components of the magnetic field intensity at $z = 0$. To achieve this, let the modified form for Equations 14.21a and 14.21b be as follows:

$$H_x = \sum_{m\text{-odd}}^{\infty} A_{1m} \cdot \cos\left(\frac{m\pi}{2b} \cdot y\right) \cdot \frac{\sinh\left(\frac{m\pi}{2b} \cdot x\right)}{\cosh\left(\frac{m\pi}{2b} \cdot a\right)} + \sum_{m\text{-odd}}^{\infty} \sum_{n\text{-odd}}^{\infty} A'_{1mn} \cdot \sin\left(\frac{n\pi}{2a} \cdot x\right) \cdot \cos\left(\frac{m\pi}{2b} \cdot y\right) \cdot e^{-\gamma_{mn} \cdot z} \tag{14.32a}$$

and

$$H_y = \sum_{n\text{-odd}}^{\infty} A_{2n} \cdot \cos\left(\frac{n\pi}{2a} \cdot x\right) \cdot \frac{\sinh\left(\frac{n\pi}{2a} \cdot y\right)}{\cosh\left(\frac{n\pi}{2a} \cdot b\right)} + \sum_{m\text{-odd}}^{\infty} \sum_{n\text{-odd}}^{\infty} A'_{2mn} \cdot \cos\left(\frac{n\pi}{2a} \cdot x\right) \cdot \sin\left(\frac{m\pi}{2b} \cdot y\right) \cdot e^{-\gamma_{mn} \cdot z} \tag{14.32b}$$

where A'_{1mn} and A'_{2mn} are two sets of arbitrary constants and:

$$\gamma_{mn} = \sqrt{\left(\frac{m\pi}{2b}\right)^2 + \left(\frac{n\pi}{2a}\right)^2} \tag{14.32c}$$

Therefore, for the zero value of the tangential field components:

$$H_x\Big|_{z=0} = \sum_{m\text{-odd}}^{\infty} A_{1m} \cdot \cos\left(\frac{m\pi}{2b} \cdot y\right) \cdot \frac{\sinh\left(\frac{m\pi}{2b} \cdot x\right)}{\cosh\left(\frac{m\pi}{2b} \cdot a\right)} + \sum_{m\text{-odd}}^{\infty} \sum_{n\text{-odd}}^{\infty} A'_{1mn} \cdot \sin\left(\frac{n\pi}{2a} \cdot x\right) \cdot \cos\left(\frac{m\pi}{2b} \cdot y\right) = 0 \tag{14.33a}$$

and

$$H_y\Big|_{z=0} = \sum_{n\text{-odd}}^{\infty} A_{2n} \cdot \cos\left(\frac{n\pi}{2a} \cdot x\right) \cdot \frac{\sinh\left(\frac{n\pi}{2a} \cdot y\right)}{\cosh\left(\frac{n\pi}{2a} \cdot b\right)} + \sum_{m\text{-odd}}^{\infty} \sum_{n\text{-odd}}^{\infty} A'_{2mn} \cdot \cos\left(\frac{n\pi}{2a} \cdot x\right) \cdot \sin\left(\frac{m\pi}{2b} \cdot y\right) = 0 \tag{14.33b}$$

over $(-a < x < a)$ and $(-b < y < b)$.

From this we get:

$$A_{1m} \cdot \frac{\sinh\left(\frac{m\pi}{2b} \cdot x\right)}{\cosh\left(\frac{m\pi}{2b} \cdot a\right)} + \sum_{n\text{-odd}}^{\infty} A'_{1mn} \cdot \sin\left(\frac{n\pi}{2a} \cdot x\right) = 0 \tag{14.34a}$$

over $(-a < x < a)$
and

$$A_{2n} \cdot \frac{\sinh\left(\frac{n\pi}{2a} \cdot y\right)}{\cosh\left(\frac{n\pi}{2a} \cdot b\right)} + \sum_{m\text{-odd}}^{\infty} A'_{2mn} \cdot \sin\left(\frac{m\pi}{2b} \cdot y\right) = 0 \tag{14.34b}$$

over $(-b < y < b)$.

Multiply Equation 14.34a by $\sin\left(\frac{N\pi}{2a} \cdot x\right)$ and then integrate over $(-a < x < a)$; on replacing N by n, we get:

$$A'_{1mn} = -A_{1m} \cdot \frac{2a \cdot \left(\frac{m\pi}{2b}\right) \cdot \sin\left(\frac{n\pi}{2}\right)}{\left[\left(\frac{n\pi}{2a}\right)^2 + \left(\frac{m\pi}{2b}\right)^2\right]} \tag{14.35a}$$

Multiply Equation 14.34b by $\sin\left(\frac{M\pi}{2b} \cdot y\right)$ and then integrate over $(-b < y < b)$; on replacing M by m, we get:

$$A'_{2mn} = -A_{2n} \cdot \frac{2b \cdot \left(\frac{n\pi}{2a}\right) \cdot \sin\left(\frac{m\pi}{2}\right)}{\left[\left(\frac{m\pi}{2b}\right)^2 + \left(\frac{n\pi}{2a}\right)^2\right]} \tag{14.35b}$$

Since:

$$\nabla \times \boldsymbol{H}\Big|_z = \frac{\partial H_y}{\partial x} - \frac{\partial H_x}{\partial y} = 0 \tag{14.36a}$$

thus:

$$\begin{aligned}
&\sum_{m\text{-odd}}^{\infty} A_{1m} \cdot \left(\frac{m\pi}{2b}\right) \cdot \sin\left(\frac{m\pi}{2b} \cdot y\right) \cdot \frac{\sinh\left(\frac{m\pi}{2b} \cdot x\right)}{\cosh\left(\frac{m\pi}{2b} \cdot a\right)} \\
&+ \sum_{m\text{-odd}}^{\infty} \sum_{n\text{-odd}}^{\infty} A'_{1mn} \cdot \sin\left(\frac{n\pi}{2a} \cdot x\right) \cdot \left(\frac{m\pi}{2b}\right) \cdot \sin\left(\frac{m\pi}{2b} \cdot y\right) \cdot e^{-\gamma_{mn} \cdot z} \\
&= \sum_{n\text{-odd}}^{\infty} A_{2n} \cdot \left(\frac{n\pi}{2a}\right) \cdot \sin\left(\frac{n\pi}{2a} \cdot x\right) \cdot \frac{\sinh\left(\frac{n\pi}{2a} \cdot y\right)}{\cosh\left(\frac{n\pi}{2a} \cdot b\right)} \\
&+ \sum_{m\text{-odd}}^{\infty} \sum_{n\text{-odd}}^{\infty} A'_{2mn} \cdot \left(\frac{n\pi}{2a}\right) \cdot \sin\left(\frac{n\pi}{2a} \cdot x\right) \cdot \sin\left(\frac{m\pi}{2b} \cdot y\right) \cdot e^{-\gamma_{mn} \cdot z}
\end{aligned} \tag{14.36b}$$

Multiply both sides by $\left[\sin\left(\frac{N\pi}{2a}\cdot x\right)\cdot\sin\left(\frac{M\pi}{2b}\cdot y\right)\right]$ and then on integrating over ($-a < x < a$) and ($-b < y < b$), we get the following equation on replacing M by m and N by n:

$$A_{1m}\cdot\left(\frac{m\pi}{2b}\right)\cdot\frac{1}{b}\cdot\frac{\left(\frac{m\pi}{2b}\right)\cdot\sin\left(\frac{n\pi}{2}\right)}{\left[\left(\frac{n\pi}{2a}\right)^2+\left(\frac{m\pi}{2b}\right)^2\right]}+A'_{1mn}\cdot\left(\frac{m\pi}{2b}\right)\cdot\frac{1}{a}\cdot\frac{1}{b}\cdot e^{-\gamma_{mn}\cdot z}$$

$$=A_{2n}\cdot\left(\frac{n\pi}{2a}\right)\cdot\frac{1}{a}\cdot\frac{\left(\frac{n\pi}{2a}\right)\cdot\sin\left(\frac{m\pi}{2}\right)}{\left[\left(\frac{m\pi}{2b}\right)^2+\left(\frac{n\pi}{2a}\right)^2\right]}+A'_{2mn}\cdot\left(\frac{n\pi}{2a}\right)\cdot\frac{1}{a}\cdot\frac{1}{b}\cdot e^{-\gamma_{mn}\cdot z} \tag{14.37}$$

for $m, n = 1, 3, 5, \ldots$

On substituting the expressions for A'_{1mn} and A'_{2mn} from Equations 14.35a and 14.35b, we get an equation identical to Equation 14.28. On solving Equations 14.28 and 14.31:

$$2v\cdot H_o = A_{1m}\cdot\frac{1}{b}\cdot\left(\frac{m\pi}{2b}\right)^2\cdot\sin\left(\frac{m\pi}{2}\right)=A_{2n}\cdot\frac{1}{a}\cdot\left(\frac{n\pi}{2a}\right)^2\cdot\sin\left(\frac{n\pi}{2}\right) \tag{14.38}$$

Thus:

$$A_{1m}=v\cdot H_o\cdot 2b\cdot\left(\frac{2b}{m\pi}\right)^2\cdot\sin\left(\frac{m\pi}{2}\right) \tag{14.39a}$$

and

$$A_{2n}=v\cdot H_o\cdot 2a\cdot\left(\frac{2a}{n\pi}\right)^2\cdot\sin\left(\frac{n\pi}{2}\right) \tag{14.39b}$$

for $m, n = 1, 3, 5, \ldots$

From Equations 14.35a and 14.35b:

$$A'_{1mn}=-v\cdot H_o\cdot 4ab\cdot\frac{\sin\left(\frac{m\pi}{2}\right)\cdot\sin\left(\frac{n\pi}{2}\right)}{\left[\left(\frac{m\pi}{2b}\right)^2+\left(\frac{n\pi}{2a}\right)^2\right]}\cdot\left(\frac{2b}{m\pi}\right) \tag{14.40a}$$

and

$$A'_{2mn} = -\nu \cdot H_o \cdot 4ab \cdot \frac{\sin\left(\frac{m\pi}{2}\right) \cdot \sin\left(\frac{n\pi}{2}\right)}{\left[\left(\frac{m\pi}{2b}\right)^2 + \left(\frac{n\pi}{2a}\right)^2\right]} \cdot \left(\frac{2a}{n\pi}\right) \tag{14.40b}$$

for $m, n = 1, 3, 5, \ldots$

FURTHER READING

Given at the end of Chapter 15.

15 Time-Varying Fields

15.1 INTRODUCTION

In Chapters 5 to 14, the time-invariant electric and magnetic fields were described in detail. t was noted that the distribution of static electric charges, in various forms, serves as the source for electrostatic fields whereas the magnetostatic fields can be produced either by the permanent magnets or by the distribution of time-invariant currents in various forms. It also noted that the electrostatic and magnetostatic fields are independent of each other. This study resulted in the four Maxwell's equations given by Equations 6.19, 11.24, 11.25, and 11.30.

Until 1820, when Hans Christian Oersted demonstrated the effect of electric current on the compass needle, there was no concept of alternating current. On learning of this experiment Michael Faraday, a great English scientist, argued that a permanent magnet, likewise, should cause dc current to flow in a closed conducting path placed near the magnet. Although his experiments did not support this conjecture, he discovered the law of electromagnetic induction; this was the dawn for the time-varying electromagnetic fields, the subject matter of the present chapter.

15.2 FARADAY'S LAW OF ELECTROMAGNETIC INDUCTION

The theory of time-varying electromagnetic fields is based on the experimental law of Michael Faraday, formulated in 1831. This law, referred to as "Faraday's Law of Electromagnetic Induction", states that:

> Electromotive force (EMF) induced around a closed path (contour) is equal to the rate of change of magnetic flux linking to this closed path.

This law, slightly modified by Heinrich Friedrich Emil Lenz, a Russian physicist of Baltic German ethnicity in 1833 is often referred to as Faraday–Lenz's law. It states that:

> Electromotive force (EMF) induced around a closed path (contour) is equal to the rate of decrease of magnetic flux linking to this closed path.

When Oersted demonstrated that a current produces magnetic field, Faraday professed that the magnetic field should also be able to produce current. He conducted an experiment by using a simple setup. This setup, illustrated in Figure 15.1, includes two coils wound on a toroid (closed ring of magnetic material). One of these coils contains a galvanometer "G", whereas the second contains a DC source (battery "B") and a switch "S". Faraday demonstrated that by putting the switch on and off there was momentary deflection in the galvanometer in opposite directions.

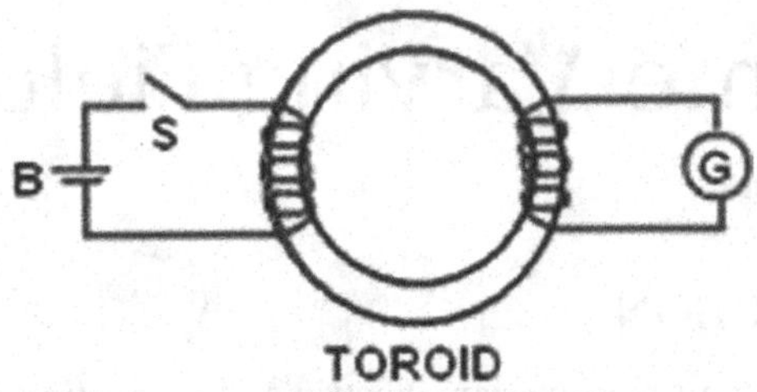

FIGURE 15.1 Faraday's experimental setup.

Figure 15.2 shows a magnet and a closed conducting loop containing a deflecting device (not shown). This magnet can be moved in this loop which is viewed from the top. When the magnet is moved up the flux linkage with the loop increases whereas when it is moved down the flux linkage with the loop decreases. Similar results are obtained when the magnet is kept stationary and the conducting loop is moved down or up. Since an electromotive force (EMF) is a voltage that arises from a conductor's motion in a magnetic field or from a changing magnetic field, the EMF or the current in the loop is reversed. The current sustains only during their relative motion. These two cases, shown in Figures 15.2a and b, demonstrate that the relative movement of a coil (or that of a magnet) will result in the alvanometer deflection.

It is worth mentioning that the (up or down) movement of the magnet or that of the coil involves time. Thus, the time-varying magnetic field produces an EMF which may establish a current in a suitable closed conducting circuit. This demonstration led to the conclusion that in time-varying fields the existence of E field is associated with H field or vice versa. Thus, for time-varying fields, the electric and magnetic fields lose their independent existence and one of these becomes dependent on the other.

Figure 15.3a shows that in an open circuited loop of conducting wire an EMF appears across the open terminals in accordance with Faraday's law. Though the EMF is distributed over the entire loop it is short circuited along the conducting part of the closed path, thus its integrated value is manifested over its open terminal. If

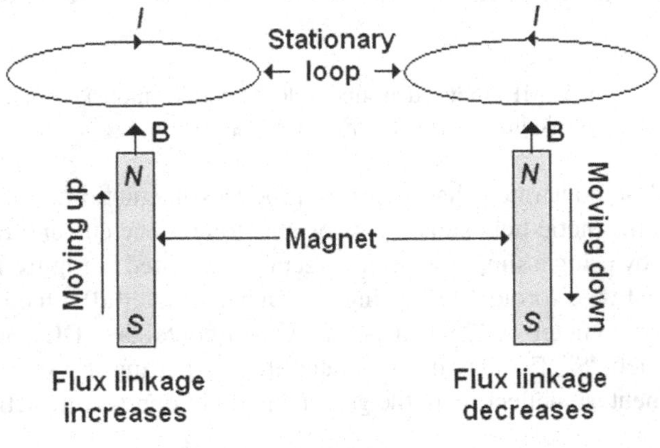

FIGURE 15.2 Movement of magnet (a) upward; (b) downward.

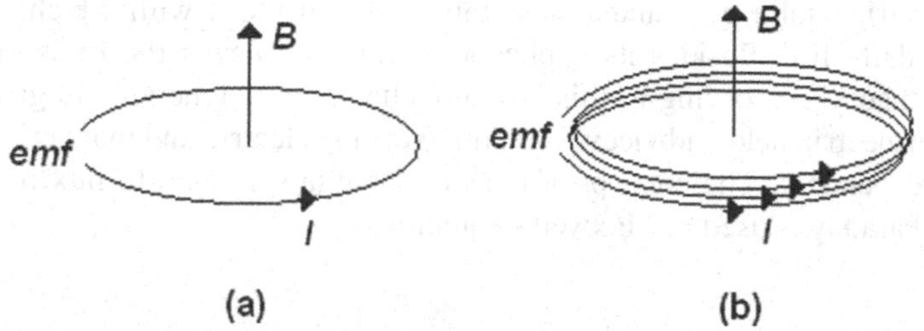

FIGURE 15.3 Current that may flow due to the induced EMF between open terminals of a conducting loop with (a) one turn; and (b) multiple turns.

the terminal is short circuited a current "i" will flow due to this EMF. If the flux linking the loop decreases with time, the induced EMF sets in a current in the closed conducting loop as shown in this figure. In mathematical terms:

$$\text{emf} = -\frac{d\phi}{dt} \tag{15.1}$$

Equation 15.1 implies induced electromotive force (EMF) around a closed path, but not necessarily a conducting closed path. It may include a series capacitor or it may purely be an imaginary closed path in space. The quantity ϕ is the magnetic flux, which passes through any and every surface whose perimeter (border) is the closed path. A negative sign indicates that EMF is in such a direction so as to produce a current in a conducting closed path such that the flux ϕ_i induced due to this current tends to oppose the change in the flux linkage. Therefore, if the original flux ϕ_o is decreasing with time the induced flux ϕ_i will be in the same direction as that of the original flux ϕ_o. If the original flux ϕ_o is increasing, the induced flux ϕ_i will be in the direction opposite to that of the original flux ϕ_o. Thus, the induced flux as if disliking any change attempts to maintain the status. The statement that induced voltage acts to produce a flux that opposes any change is known as *Lenz's law*. Equation 15.1 collectively depicts Faraday–Lenz's law.

Figure 15.3b shows a closed path containing N turns of filamentary conductor. In this case same flux is linking to each turn, thus the total flux linkage is N times the flux ϕ. The EMF induced for a fixed number of terns can therefore be written as:

$$\text{emf} = -\frac{d}{dt}(N \cdot \phi) = -N \cdot \frac{d\phi}{dt} \tag{15.2}$$

Note: The symbol ϕ (instead of Φ used in magnetostatic case) is used to indicate (time-varying magnetic flux), whereas symbol φ to indicate a coordinate in the cylindrical and spherical systems of space coordinates.

15.2.1 Applications of Faraday's Law

Faraday law is one of the most fundamental laws of electromagnetics. It forms the basis for the operation of electrical transformers, electrical motors and generators, induction cookers, electromagnetic flow meters, musical instruments (viz. electric guitar, electric violin etc.) and innumerable other devices with which one come across in daily life. Besides its applications, this law forms the basis of electromagnetic theory. According to this law, any change in magnetic field gives rise to achange in electric field and viceversa. Time-varying electric and magnetic fields are thus interdependent. The concept of lines of force (a synonym to flux lines) introduced by Faraday is used in Maxwell's equations.

15.3 CASES OF INDUCTION

It seems to be pertinent to explore the possible cases wherein EMF is induced and to assign these a suitable nomenclature so as to differentiate between the types of EMF. These cases are taken in the following subsections.

15.3.1 Stationary Loop Located in a Time-Varying Magnetic Field

When a stationary loop is located in a time-varying magnetic field an EMF is induced in the loop. This EMF can be given as:

$$\text{emf} = \oint_c \boldsymbol{E} \bullet d\boldsymbol{\ell} \tag{15.3}$$

where the symbol c indicates the closed path for the EMF.

Since the unit for the vector $\boldsymbol{E}$ is volts per meter, the unit for the EMF is volts.

In view of Stokes' Theorem, Equation 15.3 gets transformed into the following form:

$$\text{emf} = \iint_S (\nabla \times \boldsymbol{E}) \bullet d\boldsymbol{s} \tag{15.4}$$

where the symbol S is an open surface with c as its boundary.

Using Equations 15.1 and 11.29a, $\left(\phi = \iint_S \boldsymbol{B} \bullet d\boldsymbol{s}\right)$, using partial derivatives under integral sign and noting that the surface S is time-independent, i.e. not changing with time, we get:

$$\text{emf} = -\frac{d\phi}{dt} = -\frac{d}{dt}\iint_S \boldsymbol{B} \bullet d\boldsymbol{s} = -\iint_S \frac{\partial \boldsymbol{B}}{\partial t} \bullet d\boldsymbol{s} \tag{15.5}$$

On equating Equations 15.4 and 15.5, we get:

$$\iint_S (\nabla \times \boldsymbol{E}) \bullet d\boldsymbol{s} = -\iint_S \frac{\partial \boldsymbol{B}}{\partial t} \bullet d\boldsymbol{s} \tag{15.6a}$$

Equation 15.6a is true for all S, thus giving:

$$\nabla \times \boldsymbol{E} = -\frac{\partial \boldsymbol{B}}{\partial t} \quad (15.6b)$$

In view of Equations 15.3 and 15.5, we get:

$$\text{emf} = \oint_c \boldsymbol{E} \bullet d\boldsymbol{\ell} = -\iint_S \frac{\partial \boldsymbol{B}}{\partial t} \bullet \boldsymbol{ds} \quad (15.6c)$$

If $\boldsymbol{B}$ is not a function of time the, RHS of Equations 15.6b and 15.6c becomes zero and these equations reduce to:

$$\left.\nabla \times \boldsymbol{E} = 0 \quad (a) \quad \text{and} \quad \oint_c \boldsymbol{E} \bullet d\boldsymbol{\ell} = 0 \quad (b)\right\} \quad (15.7)$$

Equations 15.6b and 15.6c are Maxwell's equations in point and integral forms respectively for time-variant fields. Similarly, Equations 15.7a and 15.7b are the Maxwell's equations in point form and integral forms for time-invariant fields.

15.3.2 Varying Closed Contour in Time-Invariant Magnetic Fields

Figure15.4a illustrates a conducting bar of a fixed shape "⊏", with a voltmeter attached at the middle and a straight sliding bar conductor sliding over the two parallel sides of the "⊏" shaped conductor at a constant velocity $\boldsymbol{U}$in the Y direction, i.e. $\boldsymbol{U} = U_y \boldsymbol{a}_y$. These two jointly form a linearly time-varying closed contour, located on the x–y plane in a uniform time-invariant magnetic field oriented along the z-direction, i.e. $\boldsymbol{B} = B_z \boldsymbol{a}_z$. The conductivity for both the bars is assumed to be infinitely large. The voltmeter is assumed to be of negligible dimensions and with very high resistance drawing negligible current.

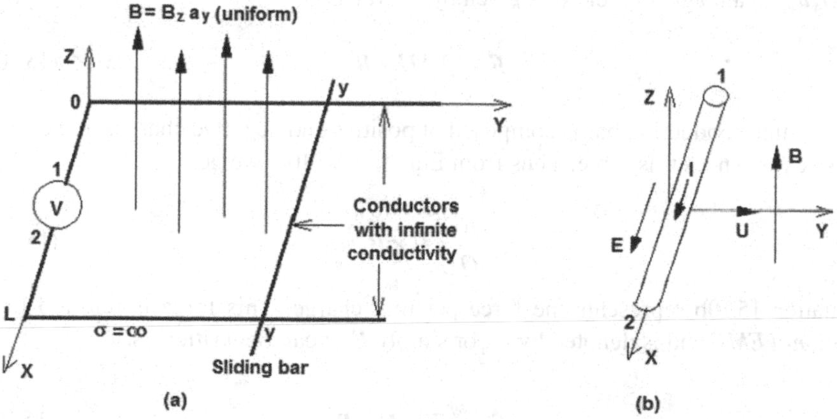

FIGURE 15.4 (a) A fixed U-shaped bar with voltmeter and a sliding bar in magnetic field B; (b) directions of I, B, and U.

The flux ϕ due to magnetic flux density $\boldsymbol{B}$ over a surface area S is given as:

$$\phi = \iint_S \boldsymbol{B} \bullet d\boldsymbol{s} = \boldsymbol{B} \bullet \boldsymbol{S} = B \cdot S \tag{15.8a}$$

In view of the parameters L and y shown in Figure 15.4a, the surface vector is:

$$\boldsymbol{S} = (L \cdot y) \cdot \boldsymbol{a}_z \overset{\text{def}}{=} S\boldsymbol{a}_z \tag{15.8b}$$

Equations 15.8a and 15.8b give:

$$\phi = B \cdot L \cdot y \tag{15.8c}$$

Since the position of sliding bar along the y-axis changes with time, its velocity can be written as:

$$\boldsymbol{U} = \frac{dy}{dt}\boldsymbol{a}_y \overset{def}{=} U_y \boldsymbol{a}_y \tag{15.8d}$$

In view of Equation 15.1, 15.3, 15.8c, and 15.8d, the EMF can be written as:

$$\text{emf} = \oint_c \boldsymbol{E} \bullet d\boldsymbol{\ell} = -\frac{d\phi}{dt} = -B \cdot L \cdot \frac{dy}{dt} = -B \cdot L \cdot U_y \tag{15.9}$$

Further, in view of the assumption $\sigma \cong \infty$, we have $E_{\tan} = 0$. It spells that the perfect conductor behaves like a short circuit and the entire closed path is short circuited except the voltmeter. Thus, the entire EMF or voltage $(= -B \cdot L \cdot U_y)$ appears across the voltmeter. As the E field is directed from terminal 2 to 1, terminal 2 is positive and 1 is negative.

The Lorentz force $\boldsymbol{F}$acting on a particle with charge Q moving with a velocity $\boldsymbol{U}$ $(= U_y \boldsymbol{a}_y)$ in a magnetic field $\boldsymbol{B}$ is given by the relation:

$$\boldsymbol{F} = Q \cdot (\boldsymbol{U} \times \boldsymbol{B}) \tag{15.10a}$$

The sliding conducting bar is composed of positive and negative charges and each of these experiences this force. Thus from Equation 15.10a, we get:

$$\frac{\boldsymbol{F}}{Q} = \boldsymbol{U} \times \boldsymbol{B} \tag{15.10b}$$

Equation 15.10b represents the force per unit charge. This force is referred to as *motional EMF* and is denoted by $\boldsymbol{E}_m$ or simply $\boldsymbol{E}$. It can be written as:

$$\boldsymbol{E}_m \overset{def}{=} \boldsymbol{E} = \boldsymbol{U} \times \boldsymbol{B} \tag{15.10c}$$

In view of Equations 15.9 and 15.10c, we get:

$$\text{emf} = \oint_c \boldsymbol{E} \bullet d\boldsymbol{\ell} = \oint_c (\boldsymbol{U} \times \boldsymbol{B}) \bullet d\boldsymbol{\ell} \tag{15.11a}$$

In Equation 15.11a, the line integral over the conducting part is negligible; therefore, almost the entire EMFis across the high-resistant voltmeter.

Thus, in view of Equation 15.9, the EMF becomes:

$$\text{emf} = -B \cdot L \cdot U_y = V \tag{15.11b}$$

where V is the voltage drop across the voltmeter, i.e. the voltmeter reading.

In view of the above, it can be concluded that if the magnetic flux density $\boldsymbol{B}$ is in Z-direction (i.e. perpendicular to the page), while the sliding conductor moves with a constant velocity U in Y direction the flux linking the loop is increasing linearly with time. The direction of the induced EMF must be such as to oppose this increase in the flux linkage. Therefore, any current that might flow, must enter the voltmeter at the terminal 2 and leave at the terminal 1. Figure 15.4b gives the direction of the current in the sliding conductor. The points marked as 1 and 2 in this figure are connected to the respective voltmeter terminals.

15.3.3 Time-Varying Closed Contour in Time-Varying Magnetic Fields

In this section, both homogeneous as well as heterogeneous space distributions of the magnetic field are considered.

15.3.3.1 Uniformly Distributed Magnetic Fields

In Section 15.3.1, a stationary loop was located in a time-varying magnetic field whereas in Section 15.3.2, the rectangular loop (or closed path) was enlarging linearly with time but the uniform magnetic field was not varying with time. In this subsection, a third possibility is to be explored. With reference to the Figure 15.4a, let the *induced EMF* is to be obtained in the closed rectangular path; one side of the rectangle is moving with a constant velocity U and the magnetic field is time-varying. This EMF will be comprised of two terms. The first of these terms will represent the EMF due to displacement of one of its sides and the second term will represent EMF due to time-varying field. The net expression of EMF can be written as:

$$\text{emf} = -\frac{d\phi}{dt} = -\frac{d}{dt}\iint_S \boldsymbol{B} \bullet d\boldsymbol{s} = -\frac{d}{dt}\left[(B_z \cdot \boldsymbol{a}_z)\cdot(L \cdot y)\boldsymbol{a}_z\right] = -\frac{d}{dt}\left[B_z \cdot L \cdot y\right] \tag{15.11c}$$

Since the flux density B_z is uniformly distributed:

$$\text{emf} = -\left[L \cdot B_z \cdot \frac{dy}{dt} + L \cdot y \cdot \frac{dB_z}{dt}\right] = -\left[L \cdot B_z \cdot U_y + S \cdot \frac{dB_z}{dt}\right] \tag{15.11d}$$

$$= -\left[\{(U_y\boldsymbol{a}_y) \times (B_z\boldsymbol{a}_z)\} \bullet (L\boldsymbol{a}_x) + (S\boldsymbol{a}_z) \bullet \frac{d(B_z\boldsymbol{a}_z)}{dt}\right]$$

$$= -\left[\{U \times B\} \bullet (La_x) + \frac{d(B_z a_z)}{dt} \bullet (Sa_z)\right]$$

or,

$$\text{emf} = -\left[\int_0^L (U \times B) \bullet d\ell + \iint_S \frac{dB}{dt} \bullet ds\right]$$

$$= \int_L^0 (U \times B) \bullet d\ell - \iint_S \frac{\partial B}{\partial t} \bullet ds$$

Since the rest of the closed path is not in motion, no motional EMF is induced in these parts.

Thus,

$$\text{emf} = \oint (U \times B) \bullet d\ell - \iint_S \frac{\partial B}{\partial t} \bullet ds \tag{15.12}$$

where

$$d\ell = d\ell a_x \quad (a) \quad \text{and} \quad ds = dSa_z \quad (b)\} \tag{15.12}$$

15.3.3.2 Arbitrarily Distributed Magnetic Fields

For an arbitrary space distribution of the flux density B, in view of Equation 15.11c we get:

$$\text{emf} = -\frac{d}{dt}\iint_S B \bullet ds = -\frac{d}{dt}\left[\int_0^y \left\{\int_0^L B_z.dx'\right\} dy'\right] \tag{15.13}$$

Let:

$$\left\{\int_0^L B_z \cdot dx'\right\} \stackrel{\text{def}}{=} \left\{\int_0^L B_z(x', y', t) \cdot dx'\right\} \stackrel{\text{def}}{=} F(y', t) \tag{15.14}$$

where x', y', and t are three independent variables; for a constant velocity U_y, the distance y is given as: $y = U_y \cdot t$. Therefore, Equation 15.13 can be rewritten as follows:

$$\text{emf} = -\frac{d}{dt}\left[\int_0^y F(y', t)\, dy'\right]$$

$$= -\frac{\partial}{\partial y}\left[\int_0^y F(y', t)\, dy'\right] \cdot \frac{dy}{dt} - \frac{\partial}{\partial t}\left[\int_0^y F(y', t)\, dy'\right] \tag{15.15}$$

Thus:

$$\text{emf} = -\left[\frac{\partial}{\partial y}\int_0^y \left\{\int_0^L B_z(x',y',t)\cdot dx'\right\}dy'\right]\cdot U_y - \left[\frac{\partial}{\partial t}\int_0^y \left\{\int_0^L B_z(x',y',t)\cdot dx'\right\}dy'\right]$$

$$= -U_y \cdot \int_0^L \left[\frac{\partial}{\partial y}\left\{\int_0^y B_z(x',y',t)dy'\right\}\right]dx' - \frac{\partial}{\partial t}\left[\int_0^y \left\{\int_0^L B_z(x',y',t).dx'\right\}dy'\right] \quad (15.16)$$

Since differentiation and integration are reverse operations, we have:

$$\frac{\partial}{\partial y}\left\{\int_0^y B_z(x',y',t)dy'\right\} \equiv B_z(x',y,t) \quad (15.17)$$

Thus:

$$\text{emf} = -U_y \cdot \int_0^L B_z(x',y,t)\Big|_{at\ y} dx' - \frac{\partial}{\partial t}\int_0^y \left\{\int_0^L B_z(x',y',t).dx'\right\}dy'$$

Since U_y is a constant, we get on taking U_y inside the integral sign:

$$\text{emf} = -\int_0^L \left\{(U_y\boldsymbol{a}_y)\times(B_z\boldsymbol{a}_z)\right\}\Big|_{at\ y} \bullet d(\ell\boldsymbol{a}_x) - \frac{\partial}{\partial t}\int_0^y\int_0^L B_z(x',y',t)\cdot dx'dy'$$

$$= -\int_0^L (\boldsymbol{U}\times\boldsymbol{B})\Big|_{at\ y} \bullet d\boldsymbol{\ell} - \iint_S \frac{d\boldsymbol{B}}{dt}\bullet d\boldsymbol{s} = \int_L^0 (\boldsymbol{U}\times\boldsymbol{B})\Big|_{at\ y} \bullet d\boldsymbol{\ell} - \iint_S \frac{\partial \boldsymbol{B}}{\partial t}\bullet d\boldsymbol{s}$$

Since the rest of the closed path is not in motion, no motional EMF is induced in these parts.

Thus:

$$\text{emf} = \oint(\boldsymbol{U}\times\boldsymbol{B})\bullet d\boldsymbol{\ell} - \iint_S \frac{\partial \boldsymbol{B}}{\partial t}\bullet d\boldsymbol{s} \quad (15.18)$$

This equation is identical to Equation 15.12.

15.3.4 Closed Contour with Two Adjacent Moving Sides in Time-Varying Magnetic Fields

Consider Figure 15.4a. Let two adjacent sides of the closed contour shown in this figure be moving in the direction normal to their length directed away from the area enclosed by the closed contour with constant velocities U_x and U_y, respectively. The size of this contour is therefore increasing with time. Equation 15.13 is accordingly revised as follows:

$$\text{emf} = -\frac{d}{dt}\iint_S \boldsymbol{B} \bullet \boldsymbol{ds} = -\frac{d}{dt}\left[\int_0^y \left\{\int_0^x B_z(x', y', t).dx'\right\} dy'\right] \tag{15.19}$$

Therefore, the induced EMF along this contour is given as:

$$\text{emf} = -\frac{\partial}{\partial x}\left[\int_0^y \left\{\int_0^x B_z(x', y', t).dx'\right\} dy'\right]\cdot\frac{dx}{dt} - \frac{\partial}{\partial y}\left[\int_0^y \left\{\int_0^x B_z(x', y', t).dx'\right\} dy'\right]\cdot\frac{dy}{dt}$$

$$-\frac{\partial}{\partial t}\left[\int_0^y \left\{\int_0^x B_z(x', y', t)\cdot dx'\right\} dy'\right] \tag{15.20}$$

or

$$\text{emf} = -\left[\int_0^y B_z(x, y', t)\Big|_{at\ x}\cdot dy'\right]\cdot U_x - \left[\int_0^x B_z(x', y, t)\Big|_{at\ y}\cdot dx'\right]\cdot U_y$$

$$-\frac{\partial}{\partial t}\left[\int_0^y \left\{\int_0^x B_z(x', y', t)\cdot dx'\right\} dy'\right] \tag{15.20a}$$

or

$$\text{emf} = \underbrace{\int_0^y \left\{(U_x\boldsymbol{a}_x)\times(B_z\boldsymbol{a}_z)\right\}\Big|_{at\ x}\bullet(d\ell\boldsymbol{a}_y) + \int_x^0 \left\{(U_y\boldsymbol{a}_y)\times(B_z\boldsymbol{a}_z)\right\}\Big|_{at\ y}\bullet(d\ell\boldsymbol{a}_x)}$$

$$-\frac{\partial}{\partial t}\iint_S (B_z\boldsymbol{a}_z)\bullet\boldsymbol{ds} \tag{15.20b}$$

where

$$\boldsymbol{ds} = dx'\cdot dy'\cdot\boldsymbol{a}_z \tag{15.20c}$$

Since there is no induced EMF over the rest of the rectangular contour, we have:

$$\text{emf} = \oint_C (\boldsymbol{U}\times\boldsymbol{B})\bullet d\boldsymbol{\ell} - \iint_S \frac{\partial\boldsymbol{B}}{\partial t}\bullet\boldsymbol{ds} \tag{15.21}$$

where the closed contour C is the boundary of the surface area S.

This equation is identical to Equations 15.12 and 15.18. Although we have derived this equation for special cases it is valid for general case as well. For instance the contour C need not be coplanar. Every elementary length of this contour may have independent constant linear velocity. The contour may vary in shape and size with time; thus it may undergo distortion and displacement. If the velocity $\boldsymbol{U}$ is same for every point on the contour; for uniform magnetic field the vector ($\boldsymbol{U} \times \boldsymbol{B}$) will be a constant. In view of Equation 15.21, we get:

$$\oint_C (\boldsymbol{U} \times \boldsymbol{B}) \bullet d\boldsymbol{\ell} = (\boldsymbol{U} \times \boldsymbol{B}) \bullet \oint_C d\boldsymbol{\ell} \tag{15.21a}$$

Since the contour integration on the RHS of this equation is zero, there will be no motional EMF. For another special case, consider an arbitrary contour with the velocity $\boldsymbol{U}$ tangential at every point on it. Let there be an arbitrary distribution of the flux density $\boldsymbol{B}$ along the contour. The vector ($\boldsymbol{U} \times \boldsymbol{B}$) will be normal at every point on the contour, thus $(\boldsymbol{U} \times \boldsymbol{B}) \bullet d\boldsymbol{\ell}$ will be zero at every point on the contour. Therefore, there will be no motional EMF in this case also (e.g. a circular path slowly rotating about its own axis). Lastly, we note that Equation 15.21 does not involve media parameters like ε, μ, and σ, therefore, this equation is valid for heterogeneous dielectric, magnetic and conducting regions as well.

15.4 DISPLACEMENT CURRENT

Maxwell's equation given by Equation 11.24 is reproduced below.

$$\nabla \times \boldsymbol{H} = \boldsymbol{J} \tag{11.24}$$

where $\boldsymbol{J} = \sigma \boldsymbol{E} + \boldsymbol{J}_o$ wherein the symbol $\boldsymbol{J}_o$ indicates source current density, if there is any.

The divergence of both sides of this equation gives:

$$\nabla \bullet (\nabla \times \boldsymbol{H}) = \nabla \bullet \boldsymbol{J} \tag{15.22}$$

As divergence of a curl is identically zero (Equation 4.40e) the LHS of Equation 15.22 is identically zero. However in view of the continuity equation, reproduced below, its RHS is not necessarily zero.

$$\nabla \bullet \boldsymbol{J} = -\frac{\partial \rho}{\partial t} \neq 0 \tag{8.23}$$

It may be noted that although electric charges are time-invariant, charge density could be time-varying.

Since LHS is not equal to RHS, Equation 11.24 does not hold good for the time-varying field and thus needs appropriate modification. In order to modify this relation let us add a new vector (say $\boldsymbol{G}$) to its RHS. Thus this equation becomes,

$$\nabla \times \boldsymbol{H} = \boldsymbol{J} + \boldsymbol{G} \tag{15.24a}$$

The divergence of both sides of Equation 15.24a gives:

$$\nabla \bullet (\nabla \times \boldsymbol{H}) = \nabla \bullet \boldsymbol{J} + \nabla \bullet \boldsymbol{G} = -\frac{\partial \rho}{\partial t} + \nabla \bullet \boldsymbol{G} = 0 \tag{15.24b}$$

Equation 15.24b for time-invariant space coordinates gives on interchanging the sequence of space and time derivatives:

$$\nabla \bullet \boldsymbol{G} = \frac{\partial \rho}{\partial t} = \frac{\partial (\nabla \bullet \boldsymbol{D})}{\partial t} = \nabla \bullet \frac{\partial \boldsymbol{D}}{\partial t} \tag{15.24c}$$

This gives:

$$\boldsymbol{G} = \frac{\partial \boldsymbol{D}}{\partial t} + \nabla \times \boldsymbol{G}'$$

where the vector $\boldsymbol{G}'$ is redundant thus arbitrarily taken as zero. Therefore:

$$\boldsymbol{G} = \frac{\partial \boldsymbol{D}}{\partial t} \stackrel{\text{def}}{=} \boldsymbol{J}_d \tag{15.24d}$$

The term $\boldsymbol{J}_d$ is called the displacement current density. Equation 11.24 now becomes:

$$\nabla \times \boldsymbol{H} = \boldsymbol{J} + \boldsymbol{J}_d = \boldsymbol{J} + \frac{\partial \boldsymbol{D}}{\partial t} \tag{15.25}$$

Equation 15.25 is the general (point) form of one of Maxwell's equations. Some special cases are as follows:

$$\nabla \times \boldsymbol{H} = \boldsymbol{J}_o \quad \text{(for time-invariant field)} \tag{15.25a}$$

where $\boldsymbol{J}_o$ indicates time-invariant source current density.

$$\nabla \times \boldsymbol{H} = \frac{\partial \boldsymbol{D}}{\partial t} \qquad (\text{if } \boldsymbol{J} = 0, \text{ e.g. in free space}) \tag{15.25b}$$

The integration of both sides of Equation 15.25 over an arbitrary stationary surface S gives:

$$\iint_S (\nabla \times \boldsymbol{H}) \bullet \boldsymbol{ds} = \iint_S \boldsymbol{J} \bullet \boldsymbol{ds} + \iint_S \frac{\partial \boldsymbol{D}}{\partial t} \bullet \boldsymbol{ds} \stackrel{\text{def}}{=} I_c + I_d \tag{15.26}$$

where

$$I_{cs} \stackrel{\text{def}}{=} \iint_S \boldsymbol{J} \bullet \boldsymbol{ds} \tag{15.26a}$$

and

$$I_d \stackrel{\text{def}}{=} \iint_S \boldsymbol{J}_d \bullet \boldsymbol{ds} = \iint_S \left(\partial \boldsymbol{D}/\partial t\right) \bullet \boldsymbol{ds} = \frac{\partial}{\partial t} \iint_S \boldsymbol{D} \bullet \boldsymbol{ds} \tag{15.26b}$$

where I_{cs} indicates the sum of the conduction current I_c and the source current I_s, if there is any; while the displacement current is indicated by the symbol I_d. These quantities are defined below:

$$I_c \stackrel{\text{def}}{=} \iint_S \sigma \boldsymbol{E} \bullet \boldsymbol{ds} \tag{15.27a}$$

and the source current I_s is defined in terms of the source current density $\boldsymbol{J_o}$ as:

$$I_s \stackrel{\text{def}}{=} \iint_S \boldsymbol{J}_o \bullet \boldsymbol{ds} \stackrel{\text{def}}{=} \iint_S \left(\boldsymbol{J} - \sigma \boldsymbol{E}\right) \bullet \boldsymbol{ds} \tag{15.27b}$$

Thus:

$$I_{cs} = I_c + I_s = \iint_S \boldsymbol{J} \bullet \boldsymbol{ds} \tag{15.27c}$$

The total current I is given as:

$$I = I_{cs} + I_d = \left(I_c + I_s\right) + I_d = \iint_S \left(\boldsymbol{J} + \boldsymbol{J}_d\right) \bullet \boldsymbol{ds} \tag{15.27d}$$

In view of Stokes' Theorem, Equation 15.25 becomes:

$$\iint_S \left(\nabla \times \boldsymbol{H}\right) \bullet \boldsymbol{ds} = \oint_C \boldsymbol{H} \bullet \boldsymbol{d\ell} = \left(I_c + I_s\right) + I_d = I$$

or,

$$\oint_C \boldsymbol{H} \bullet \boldsymbol{d\ell} = \iint_S \boldsymbol{J} \bullet \boldsymbol{ds} + \iint_S \left(\partial \boldsymbol{D}/\partial t\right) \bullet \boldsymbol{ds} \tag{15.28}$$

where the contour C is the boundary of the surface S.

Equation 15.28 is the integral form of Maxwell's equation (see Equation 15.25). It takes the following forms under specified special conditions.

$$\oint_C \boldsymbol{H} \bullet \boldsymbol{d\ell} = I_s \stackrel{\text{def}}{=} \iint_S \boldsymbol{J}_o \bullet \boldsymbol{ds} \quad \text{(For time invariant field)} \tag{15.28a}$$

Note that for (steady-state) time-invariant field: $\sigma \boldsymbol{E} \equiv \boldsymbol{0}$, for all finite values of the conductivity σ.

$$\oint_C \boldsymbol{H} \bullet \boldsymbol{d\ell} = \iint_S \left(\partial \boldsymbol{D}/\partial t\right) \bullet \boldsymbol{ds} \quad \text{(For time-varying field with } J = 0\text{)} \tag{15.28b}$$

15.4.1 Nature of Displacement Current

Figure 15.5a illustrates an electric circuit with discrete elements, which forms a closed path through a resistor R (connected between point 1 and 2) and a parallel plate capacitor C. This circuit is located in a sinusoidally time-varying magnetic field *B*. This magnetic field will result in an EMF around the closed path of the circuit. Due to the time-varying nature of *B*, the EMF and the resulting instantaneous current (*i*) flowing through the circuit will also be of sinusoidally time-varying nature and the polarities shown will keep on alternating. Let the EMF in this circuit be given as:

$$\text{emf} \overset{\text{def}}{=} v = V_o \cdot \cos(\omega t) \tag{15.29}$$

The total EMF is dropped across the resister v_R and the capacitor v_C; by the Kirchhoff's voltage law the sum of the two voltage drops must be equal to the induced EMF, thus:

$$\text{emf} \overset{\text{def}}{=} v = v_R + v_C \tag{15.30}$$

Let the induced instantaneous current *i* flowing in the circuit be defined as:

$$i = I_o \cdot \cos(\omega t + \theta) \tag{15.31}$$

where the peak current I_o and the phase angle θ are unknown quantities.

From circuit theory we have:

$$v_R = R \cdot i \quad \text{and} \quad v_C = \frac{1}{C} \cdot \int i \cdot dt$$

Therefore, from Equations 15.29, 15.30, and 15.31, we get:

$$V_o \cdot \cos(\omega t) = R \cdot I_o \cdot \cos(\omega t + \theta) + \frac{1}{C} \cdot \int I_o \cdot \cos(\omega t + \theta) \cdot dt$$

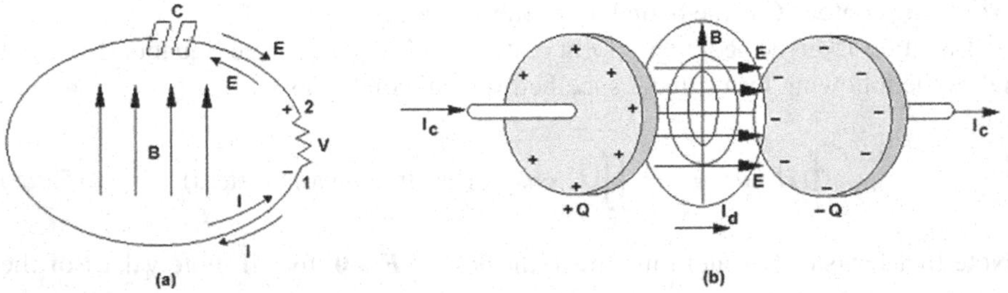

FIGURE 15.5 (a) A closed circuit in a time-varying magnetic field. (b) A capacitive current resulting in a magnetic field (note the charges on left plate appear to be outside whereas these are on the inner side of the plate).

or,

$$V_o \cdot \cos(\omega t) = R \cdot I_o \cdot \cos(\omega t + \theta) + \frac{I_o}{\omega C} \cdot \sin(\omega t + \theta) + C_o$$

Ignoring the constant of integration C_o, we get:

$$V_o \cdot \cos(\omega t) = R \cdot I_o \cdot \{\cos(\omega t) \cdot \cos(\theta) - \sin(\omega t) \cdot \sin(\theta)\}$$

$$+ \frac{I_o}{\omega C} \cdot \{\sin(\omega t) \cdot \cos(\theta) + \cos(\omega t) \cdot \sin(\theta)\}$$

or,

$$V_o \cdot \cos(\omega t) = \cos(\omega t) \cdot \left\{R \cdot I_o \cdot \cos(\theta) + \frac{I_o}{\omega C} \cdot \sin(\theta)\right\}$$

$$-\sin(\omega t)\left\{R \cdot I_o \cdot \sin(\theta) - \frac{I_o}{\omega C} \cdot \cos(\theta)\right\} \tag{15.32}$$

Since cosine and sine functions are orthogonal functions, we can equate the coefficients of these functions, thus resulting in:

$$\left\{R \cdot I_o \cdot \cos(\theta) + \frac{I_o}{\omega C} \cdot \sin(\theta)\right\} = V_o \tag{15.33a}$$

and

$$\left\{R \cdot I_o \cdot \sin(\theta) - \frac{I_o}{\omega C} \cdot \cos(\theta)\right\} = 0 \tag{15.33b}$$

On solving these equations, we get:

$$\left.\theta = \cot^{-1}(\omega CR) \quad (a) \quad \text{and} \quad I_o = \frac{V_o}{\sqrt{R^2 + \left(\frac{1}{\omega C}\right)^2}} \quad (b)\right\}$$

In the case of a perfect capacitor, the conduction current in the R–C circuit cannot cross the dielectric between the two plates of the capacitor. As a result, charges are deposited on the inner surfaces of these capacitor plates, with an average surface charge density ρ_s as given below:

$$\rho_s \cong \pm \int \frac{i}{A} \cdot dt \tag{15.35}$$

where symbol A indicates the surface area of each parallel plate of the capacitor.

For a small separation distance d between the two parallel plates, the electric flux density $\boldsymbol{D}$ between these plates will be as follows:

$$\boldsymbol{D} \cong \rho_s \boldsymbol{a}_N \tag{15.36}$$

The unit vector $\boldsymbol{a}_N$ is directed from the positive plate to the negative plate at any instant.

The displacement current density $\boldsymbol{J}_d$ in view of Equations 15.35, 15.36, and 15.24d is:

$$\boldsymbol{J}_d = \frac{d\boldsymbol{D}}{dt} = \frac{d\rho_s}{dt}\boldsymbol{a}_N = \left(\frac{d}{dt}\int \frac{i}{A}\cdot dt\right)\boldsymbol{a}_N = \frac{i}{A}\boldsymbol{a}_N \tag{15.37}$$

Therefore, the displacement current between the two capacitor plates is:

$$i_d = \iint_A \boldsymbol{J}_d \bullet \boldsymbol{ds} = i = I_o \cdot \cos(\omega t + \theta) \tag{15.38}$$

15.4.2 Conceptual Aspect of Displacement Current

In any electrical circuit the current can flow only through a completely closed path. The circuit shown in Figure 15.5a contains a capacitor with zero initial charge on its plates; it is connected in series with a resistor. As described in Section 8.6, the capacitor is composed of two parallel plate electrodes separated by a dielectric. The inclusion of dielectric midway in a circuit amounts to the discontinuity in the conducting path. Thus when a time-invariant step voltage is applied to the circuit the conduction current starts from the positive terminal of the voltage source, on reaching one of the plates of the capacitor it fails to cross the dielectric. An equal amount of return current simultaneously flows from the other capacitor plate to the negative terminal of the voltage source. These currents, however, continue to flow in the conductor joining the voltage source to the capacitor plate. As a result of the interruption to the flow of currents across the dielectric, a deposition of positive electric charges starts building up on the inner surface of the first plate and simultaneously a deposition of negative electric charges starts building up on the inner surface of the second plate. This progressively develops an electrical field between the two parallel plates. Consequently, an increasing voltage from its zero value appears across the capacitor terminals. The polarity of this voltage is in opposition to the applied constant voltage. Eventually, the two opposing voltages become equal. At this point intime, the current in the circuit stops flowing, and as a result the charges stop building up on the capacitor plates and any further growth of the plate voltage ceases. Clearly, the flow of conduction current outside the capacitor is related to the rate of increase of the charge deposited on the capacitor plate and thus the conduction current can be equated to the rate of increase in the charge deposited on a capacitor plate. This rate of increase in the charge deposited on a capacitor plate is termed as the displacement current. Therefore:

$$I_d = I_c = \frac{d}{dt}Q = \frac{d}{dt}\int_A \rho_s ds = \int_S \frac{\partial}{\partial t}\rho_s ds = \int_S \left.\frac{\partial \boldsymbol{D}}{\partial t}\right|_s \bullet \boldsymbol{ds} \tag{15.39}$$

Thus, the conduction current can be related to the rate of increase in the electric field developed between the two capacitor plates.

The simultaneous presence of these two types of currents flowing in the same direction along different parts of the closed circuit maintains continuity of the current, thus jointly fulfills the requirement that current flows in a completely closed circuit.

For small separation distance d between the two parallel plates, the displacement vector (also known as the electric flux density vector) is fairly uniform, i.e.:

$$\left.\frac{\partial \boldsymbol{D}}{\partial t}\right|_s \cong \frac{\partial \boldsymbol{D}}{\partial t} \tag{15.39a}$$

Therefore, the displacement current density, $\boldsymbol{J}_d$ is given as:

$$\boldsymbol{J}_d = \frac{\partial \boldsymbol{D}}{\partial t} \tag{15.40}$$

All this happens in view of the phenomenon of displacement explained in Section 15.2 in connection with Faraday's experiment. Faraday assumed that some sort of current is flowing between the two plates. It is surely not the conduction current but some other form which is referred to as the displacement current.

Figure 15.5b illustrates a capacitor wherein a time-varying (alternating) EMF is applied through an ac supply. When conduction current I_c arrives at the left plate it gets positively charged. Due to the phenomenon of displacement the right plate becomes negatively charged. The displacement current I_d flows between the two plates, the conduction current I_c takes over at the other end, it flows through the rest of the circuit and does the assigned job. When the polarity of EMF gets reversed the same process initiates from the right plate. The process keeps on repeating till the ac supply remains connected. During this process of alternation of EMF the capacitor keeps on charging and discharging. In view of Ampère's circuital law any flow of non-zero net current is accompanied by a magnetic field which will encircle the current. Thus there will be magnetic field encircling the conduction current. Similarly, the displacement current I_d, as the capacitor is charged, I_d can be regarded as the source of magnetic field illustrated in this figure.

In view of Ampère's law in integral form, we have:

$$\oint_c \boldsymbol{H} \bullet d\boldsymbol{\ell} = I_d \tag{15.41a}$$

where I_d is the total displacement current linking the closed contour $\boldsymbol{c}$.

The law in point form is given as:

$$\nabla \times \boldsymbol{H} = \frac{\partial \boldsymbol{D}}{\partial t} \overset{\text{def}}{=} \boldsymbol{J}_d \tag{11.41b}$$

where $\boldsymbol{D}$ and $\boldsymbol{J}_d$ indicate respectively the displacement vector (also known as the electric flux density vector) and the displacement current density, at the point where the magnetic field intensity is $\boldsymbol{H}$.

15.5 MAXWELL'S EQUATIONS

During the course of our study, we have obtained the relations for Maxwell's equations in point form and in integral form. These were obtained for time-invariant and time variant fields, and for charge-free regions and for the regions containing charges. In point form these were given by Equations 6.19a, 6.19b, 11.24, 15.6b, 15.7a, 15.25, 15.25a, and 15.25b, whereas they are given in integral form by Equations 6.21a, 6.21b, 11.19, 15.6c, 15.7b, 15.28, 15.28a, and 15.28b. The expressions of these equations are given in Table 12 of Appendix A3 in the eResources. The essence of these equations is discussed below.

Newton's (three) laws referred to as *laws of motion* describe the dynamics of physical systems. These laws spell the behavior of objects, when forces act on them but do not address the origin of forces. *Newton's law of universal gravitation* states that any objects with *rest mass* exert force on each other. This law allows us to calculate the direction and strength of the gravitational forces. These laws only hold for particles moving with speeds much less than that of light. *Maxwell's equations*, a set of four equations, hold for particles moving with any speed, are said to be relativistically more correct.

Gauss' law, which states that the electric flux through any closed surface is equal to the total charge enclosed, is the *first of Maxwell's equations*. It holds for a closed surface of any shape and size, and for any charge distribution inside that surface, moving or stationary. It is more general than *Coulomb's law*, which only holds for charges whose accelerations are small or zero. It can be derived from the first two of Maxwell's equations for static conditions, i.e. when all charges are at rest or moving with uniform velocity. The first two of Maxwell's equations, Equations 6.19 and 15.7, allow estimation of E due to any static charge distribution. Once the field is known the force that this charge distribution exerts on any charge can be estimated. If a charge distribution has a high degree of symmetry, then Gauss' law, Equation 6.21, the first Maxwell's equation, alone can be used to determine the magnitude of $\boldsymbol{E}$. The direction of $\boldsymbol{E}$ may be deduced from the symmetry of situation.

15.6 BOUNDARY CONDITIONS

As long as the fields are static or time-invariant in nature the static charge distribution remains the source of electric field intensity, electric flux density, and electric potential. Also, steady current causes the existence of magnetic field intensity, magnetic flux density, and the related quantities. The time-invariant electric and magnetic fields exist independent of each other (i.e. if the effects of the special theory of relativity are ignored). The moment the concept of time gets involved, all the field quantities referred to above become time-variant. The existence of electric and magnetic fields no more remains isolated, as the changing electric field produces magnetic field and viceversa. The pioneering work of Oersted, Faraday, Lenz, and Maxwell led to the equations that correlate the time-varying electric and magnetic field quantities. The integral form of these (Maxwell's) equations can be employed to evaluate boundary conditions obeyed by $\boldsymbol{E}$, $\boldsymbol{D}$, $\boldsymbol{H}$, and $\boldsymbol{B}$. The boundary conditions to be satisfied by the time-varying field quantities are:

$$E_{t1} = E_{t2} \tag{15.42a}$$

$$D_{n1} - D_{n2} = \rho_s \quad \text{or} \quad D_{n1} = D_{n2} \left(\text{for } \rho_s = 0\right) \tag{15.42b}$$

$$H_{t1} - H_{t2} = K \quad \text{or} \quad H_{t1} = H_{t2} \ (\text{for } K = 0) \tag{15.42c}$$

and:

$$B_{n1} = B_{n2} \tag{15.42d}$$

For the idealized physical condition ($\sigma_2 = \infty$), the time-varying current is carried on the conductor surface as a surface current density K. This is known as "*skin effect*". In this case because of zero penetration in region-2, the electromagnetic field in this region vanishes and the above conditions reduce to:

$$\left.\begin{aligned} E_{t1} &= 0 \ (a) \qquad & D_{n1} &= \rho_s \quad (b) \\ H_{t1} &= K \ (c) \ \text{and} & B_{n1} &= 0 \quad (d) \end{aligned}\right\} \tag{15.43}$$

In Equations 15.42 and 15.43, E_t, D_t, H_t, and B_t are the tangential components while E_n, D_n, H_n, and B_n are the normal components of $\boldsymbol{E}$, $\boldsymbol{D}$, $\boldsymbol{H}$, and $\boldsymbol{B}$, respectively, at the boundary surface. Suffixes 1 and 2 indicate the regions, across the boundary, to which these quantities belong. Furthermore, ρ_s represents the surface charge density and $\boldsymbol{K}$ indicates the surface current density. It needs to be mentioned that surface charge density, as well as the surface current density are considered a physical possibility for either dielectrics or conductors. These entities can either be placed or induced.

15.7 RETARDED POTENTIALS

The term *retarded potential* is frequently used in radiation problems. In order to understand its essence, consider the relations given by Equations 7.29 and 11.72a, reproduced here:

$$V = \iiint \frac{\rho \cdot dv}{4\pi \cdot \varepsilon R} \quad (7.29) \qquad \boldsymbol{A} = \iiint \frac{\mu \cdot \boldsymbol{J} \cdot dv}{4\pi \cdot R} \quad (11.72\text{a})$$

In these equations, ρ and $\boldsymbol{J}$ are the sources and V and $\boldsymbol{A}$ are the related potentials. The nature of these potentials is obviously linked to that of the sources. Thus in the case of time-variant sources, the corresponding potentials will also be time-variant. In these relations there is no involvement of space term whereas in all radiation problems the involvement of space term is imminent. A wave after leaving an antenna travels in space and if the potentials are to be estimated at some space location the above relations will yield no result. Thus, these equations are modified by using the bracketed terms $[\rho]$ and $[\boldsymbol{J}]$ in place of ρ and $\boldsymbol{J}$ in Equations 7.29 and 11.72a. The resulting equations are:

$$V = \iiint \frac{[\rho] \cdot dv}{4\pi \cdot \varepsilon R} \quad (a) \qquad A = \iiint \frac{\mu \cdot [J] \cdot dv}{4\pi \cdot R} \quad (b) \Bigg\} \tag{15.44}$$

The terms [ρ] and [J] spell much more than of their time-dependence. To elaborate it further let us take that the parameter ρ is given by the following relation:

$$\rho = e^{-t} \cdot \cos(\omega \cdot t) \tag{15.45a}$$

Its bracketed version can be written as:

$$[\rho] = e^{-t'} \cdot \cos(\omega . t') \tag{15.45b}$$

In this version, time t is replaced by t', where $t' = t - R / U$. With this substitution, Equation 15.45b becomes:

$$[\rho] = e^{-(t-R/U)} \cdot \cos\omega(t - R / U) \tag{15.45c}$$

Similarly, if t is replaced by t'', where $t'' = t + R / U$, Equation 15.45a becomes:

$$\rho = e^{-t''} \cdot \cos(\omega \cdot t'') = e^{-(t+R/U)} \cdot \cos\omega(t + R / U) \tag{15.45d}$$

The time t' is called *retarded time* and V involving $[\![\rho]\!]$ is called the *retarded potential.* Similarly the time t'' is referred to as *advanced (or accelerated) time* and V involving ρ is termed as the *advanced (or accelerated) potential.* Similarly [J] can also be written in terms of t' and $[\![J]\!]$ in terms of t''. These replacements will make "A", given by Equation 11.72a, *retarded* or *accelerated potential.*

The true meaning of retarded potential is spelled by the retarded (or delayed) time $\{t' = t - (R / U)\}$ which involves a distance term "R" and the velocity term"U". Thus retarded potential gives the instantaneous value of V (or A) at a distance R from its origination *delayed* by "R/ U", i.e. the time taken by the wave to travel distance R with velocity U. The advanced or *accelerated* potential indicates its instantaneous value at a distance R from its origination *advanced* by the time the wave takes to reach the distance R. This potential, however, is *non-causal.* It occurs in idealized situations.[1–3] (In this chapter, the superscript numbers represent the references given at the end of the chapter.)

15.8 MODIFICATION OF SOME FIELD RELATIONS

In earlier chapters, a number of relations were derived to predict the behavior of electrostatic and magnetostatic fields. Some of these relations are true only for time-invariant fields. In this section these relations are modified to make them consistent with time-varying fields.

15.8.1 Gradient Relation

The gradient relation ($\boldsymbol{E} = -\nabla V$) given by Equation 7.6 relates the electric field ($\boldsymbol{E}$) and the electric potential (V). To assess its applicability to the time-varying field let us take the curl of both its sides. This gives:

$$\nabla \times \boldsymbol{E} = \nabla \times (-\nabla V) \equiv 0 \tag{15.7a}$$

Equation 15.7a is true for a time-invariant field but for time-varying fields we have:

$$\nabla \times \boldsymbol{E} = -\frac{\partial \boldsymbol{B}}{\partial t} \tag{15.6b}$$

On comparing Equations 15.7a and 15.6b the inconsistency can easily be noticed. To remove the same let us add a new vector $\boldsymbol{N}$ to the RHS. of Equation 7.6. It becomes:

$$\boldsymbol{E} = -\nabla V + \boldsymbol{N} \tag{15.46a}$$

Taking the curl of both sides of Equation 15.46a and equating to Equation 15.6b gives:

$$\nabla \times \boldsymbol{E} = \nabla \times (-\nabla V) + \nabla \times \boldsymbol{N} \equiv \nabla \times \boldsymbol{N} = -\frac{\partial \boldsymbol{B}}{\partial t} \tag{15.46b}$$

Since $\boldsymbol{B} \overset{\text{def}}{=} \nabla \times \boldsymbol{A}$, Equation 15.46b, we get:

$$-\frac{\partial}{\partial t}\boldsymbol{B} = \underbrace{-\frac{\partial}{\partial t}(\nabla \times \boldsymbol{A}) = -\nabla \times \left(\frac{\partial}{\partial t}\boldsymbol{A}\right)} = \boxed{\nabla \times \left(-\frac{\partial \boldsymbol{A}}{\partial t}\right) = \nabla \times \boldsymbol{N}} \tag{15.46c}$$

The interchange of the sequence of the differentiation with respect to space coordinates and time is permissible in stationary systems where *space coordinates are time-invariant.*

Thus:

$$\boldsymbol{N} \triangleq -\frac{\partial \boldsymbol{A}}{\partial t} \tag{15.46d}$$

Hence, from Equation 15.46a, one gets:

$$\boldsymbol{E} = -\nabla V - \frac{\partial \boldsymbol{A}}{\partial t} \tag{15.47}$$

15.8.2 Poisson's Equation for Vector Magnetic Potential

This subsection assesses the applicability of Poisson's equation ($\nabla^2 \boldsymbol{A} = -\mu \cdot \boldsymbol{J_o}$) given by Equation 9.4b to time-varying fields. From Equations 15.25 and 15.47, we have:

$$\nabla \times \boldsymbol{H} = \boldsymbol{J} + \frac{\partial \boldsymbol{D}}{\partial t} = (\boldsymbol{J_o} + \sigma \boldsymbol{E}) + \varepsilon \cdot \frac{\partial \boldsymbol{E}}{\partial t} = \boldsymbol{J_o} + \left(\sigma + \varepsilon \cdot \frac{\partial}{\partial t}\right) \cdot \left(-\nabla V - \frac{\partial \boldsymbol{A}}{\partial t}\right) \tag{15.48a}$$

$$\nabla\times\boldsymbol{H}=\nabla\times\left(\frac{\boldsymbol{B}}{\mu}\right)=\frac{1}{\mu}\nabla\times\boldsymbol{B}=\frac{1}{\mu}\cdot\nabla\times\nabla\times\boldsymbol{A}=\frac{1}{\mu}\cdot\left[\nabla(\nabla\bullet\boldsymbol{A})-\nabla^2\boldsymbol{A}\right] \quad (15.48b)$$

Equating Equations 15.48a and 15.48b to get:

$$\frac{1}{\mu}\cdot\left[\nabla(\nabla\bullet\boldsymbol{A})-\nabla^2\boldsymbol{A}\right]=\boldsymbol{J}_o-\sigma\cdot\left(\nabla V+\frac{\partial\boldsymbol{A}}{\partial t}\right)-\varepsilon\cdot\left(\nabla\frac{\partial V}{\partial t}+\frac{\partial^2\boldsymbol{A}}{\partial t^2}\right)$$

or

$$\nabla\underbrace{(\nabla\bullet\boldsymbol{A})}-\nabla^2\boldsymbol{A}=\mu\cdot\boldsymbol{J}_o-\nabla\underbrace{\left(\mu\cdot\sigma\cdot V+\mu\cdot\varepsilon\cdot\frac{\partial V}{\partial t}\right)}-\mu\cdot\sigma\cdot\frac{\partial\boldsymbol{A}}{\partial t}-\mu\cdot\varepsilon\cdot\frac{\partial^2\boldsymbol{A}}{\partial t^2} \quad (15.48c)$$

According to the *Helmholtz Theorem*, a vector field is completely defined only when both its curl and divergence are known. The curl of **A** has already be specified by Equation 11.65a (i.e. $\boldsymbol{B}=\nabla\times\boldsymbol{A}$), but its divergence term involved in the LHS of Equation 15.48c is not yet known. This can be specified in terms of some conditions which are referred to as the *gauge condition*. The two most commonly used gauge conditions are:

$$\left.\nabla\bullet\boldsymbol{A}\overset{\text{def}}{=}-\mu\cdot\varepsilon\cdot\frac{\partial V}{\partial t}\quad(\text{a})\quad\text{and}\quad\nabla\bullet\boldsymbol{A}\overset{\text{def}}{=}0\quad(\text{b})\right\} \quad (15.49)$$

The first of these is called the "Lorentz gauge condition", whereas the second is "Coulomb's gauge condition".

For conducting regions, the "modified Lorentz gauge condition" can be defined as follows:

$$\nabla\bullet\boldsymbol{A}\overset{\text{def}}{=}-\mu\cdot\sigma\cdot V-\mu\cdot\varepsilon\cdot\frac{\partial V}{\partial t} \quad (15.49c)$$

The selection of a gauge condition depends on the application and its suitability to meet the requirement. In the present case of time-varying fields in conducting media modified Lorentz gauge condition is the most appropriate choice. For time-invariant field in non-conducting regions the *modified Lorentz gauge condition* degenerates into *Coulomb's gauge condition*. Thus in Equation 15.49c, Equation 15.48c can be rewritten as:

$$\underbrace{\nabla\left\{-\mu\cdot\sigma\cdot V-\mu\cdot\varepsilon\cdot\frac{\partial V}{\partial t}\right\}}-\nabla^2\boldsymbol{A}=\underbrace{\nabla\left\{-\mu\cdot\sigma\cdot V-\mu\cdot\varepsilon\cdot\frac{\partial V}{\partial t}\right\}} \quad (15.50a)$$

$$+\mu\cdot\boldsymbol{J}_o+-\mu\cdot\sigma\cdot\frac{\partial\boldsymbol{A}}{\partial t}-\mu\cdot\varepsilon\cdot\frac{\partial^2\boldsymbol{A}}{\partial t^2}$$

This finally gives:

$$\nabla^2 \boldsymbol{A} = -\mu \cdot \boldsymbol{J}_o + \mu \cdot \sigma \cdot \frac{\partial \boldsymbol{A}}{\partial t} + \mu \cdot \varepsilon \cdot \frac{\partial^2 \boldsymbol{A}}{\partial t^2} \tag{15.50b}$$

In regions free from source currents the vector magnetic potential satisfies the damped wave equation.

15.8.3 Poisson's Equation for Scalar Electric Potential

Poisson's equation ($\nabla^2 V = -\frac{\rho}{\varepsilon}$) satisfied by the time-invariant scalar electric potential was given by Equation 9.1a. Its modified form for the time-varying scalar electric potential can be obtained by adopting the following steps:

1. Proceed from the first Maxwell's equation given by Equation 6.19a:

$$\boxed{\nabla \bullet \boldsymbol{D} = \varepsilon\left(\nabla \bullet \boldsymbol{E}\right) = \rho} \xrightarrow{\text{yields}} \boxed{\nabla \bullet \boldsymbol{E} = \frac{\rho}{\varepsilon}} \tag{15.51a}$$

2. Substitute $\boldsymbol{E} = -\nabla V - \frac{\partial \boldsymbol{A}}{\partial t}$ from Equation 15.47 to get:

$$\nabla \bullet \left(\nabla V + \frac{\partial \boldsymbol{A}}{\partial t}\right) = \nabla \bullet \left(\nabla V\right) + \nabla \bullet \left(\frac{\partial \boldsymbol{A}}{\partial t}\right) = \nabla^2 V + \frac{\partial\left(\nabla \bullet \boldsymbol{A}\right)}{\partial t} = -\frac{\rho}{\varepsilon} \tag{15.51b}$$

3. Use the modified Lorentz gauge condition given by Equation 15.49c to get:

$$\nabla^2 V + \frac{\partial}{\partial t}\left(-\mu \cdot \sigma \cdot V - \mu \cdot \varepsilon \cdot \frac{\partial V}{\partial t}\right) = \nabla^2 V - \mu \cdot \sigma \cdot \frac{\partial V}{\partial t} - \mu \cdot \varepsilon \cdot \left(\frac{\partial^2 V}{\partial t^2}\right) = -\frac{\rho}{\varepsilon} \tag{15.52}$$

4. Find the modified Poisson's equation from Equation 15.52 as:

$$\boxed{\nabla^2 V = -\frac{\rho}{\varepsilon} + \mu \cdot \sigma \cdot \frac{\partial V}{\partial t} + \mu \cdot \varepsilon \cdot \left(\frac{\partial^2 V}{\partial t^2}\right)} \tag{15.53}$$

For charge-free regions the scalar electric potential satisfies the damped wave equation. Note that in conducting regions there can be a distribution of free electric charge ρ if the field is time-varying. For time-invariant fields ρ is zero in conducting regions.

15.9 POYNTING THEOREM

A very important contribution to the electromagnetic field theory, perhaps only second to Maxwell, is that of John Henry Poynting. Solutions of Maxwell's equations describe the nature of electromagnetic field and its distribution in space and time. Poynting theorem[1, 2] tells about the flow of electromagnetic power across closed surfaces and identifies its components. Poynting, using a vector identity brilliantly manipulated Maxwell's equations to arrive at some vital conclusions.

The Poynting vector $\mathcal{P}$ is defined as:

$$\mathcal{P} \stackrel{\text{def}}{=} \boldsymbol{E} \times \boldsymbol{H} \tag{15.54}$$

Consider the identity:

$$\nabla \bullet \mathcal{P} \stackrel{\text{def}}{=} \nabla \bullet (\boldsymbol{E} \times \boldsymbol{H}) \equiv \boldsymbol{H} \bullet (\nabla \times \boldsymbol{E}) - \boldsymbol{E} \bullet (\nabla \times \boldsymbol{H}) \tag{15.55}$$

Now, using the two curl equations of Maxwell (Equations 15.6b and 15.25), we get:

$$\nabla \bullet \mathcal{P} = -\boldsymbol{H} \bullet \left(\frac{\partial \boldsymbol{B}}{\partial t} \right) - \boldsymbol{E} \bullet \left(\boldsymbol{J} + \frac{\partial \boldsymbol{D}}{\partial t} \right) \tag{15.56}$$

Thus, in the constitutive equations, for any *linear region* with time-invariant permeability μ, permittivity ε, and conductivity σ we get:

$$\nabla \bullet \mathcal{P} = -\boldsymbol{H} \bullet \left(\mu \frac{\partial \boldsymbol{H}}{\partial t} \right) - \boldsymbol{E} \bullet \left(\overline{\sigma \boldsymbol{E} + \boldsymbol{J}_o} + \varepsilon \frac{\partial \boldsymbol{E}}{\partial t} \right) \tag{15.57}$$

where the symbol $\boldsymbol{J}_o$ indicates the density of any "source-controlled" current that might be present in the region. This current is also referred to simply as the source current; it contributes to the electromagnetic fields.

Thus:

$$\nabla \bullet \mathcal{P} = -\mu \left(\boldsymbol{H} \bullet \frac{\partial \boldsymbol{H}}{\partial t} \right) - \sigma E^2 - \boldsymbol{E} \bullet \boldsymbol{J}_o - \varepsilon \left(\boldsymbol{E} \bullet \frac{\partial \boldsymbol{E}}{\partial t} \right)$$

$$= -\mu \left(\frac{1}{2} \frac{\partial H^2}{\partial t} \right) - \varepsilon \left(\frac{1}{2} \frac{\partial E^2}{\partial t} \right) - \sigma E^2 - \boldsymbol{E} \bullet \boldsymbol{J}_o$$

$$= -\frac{\partial}{\partial t} \left(\frac{1}{2} \boldsymbol{H} \bullet \boldsymbol{B} \right) - \frac{\partial}{\partial t} \left(\frac{1}{2} \boldsymbol{E} \bullet \boldsymbol{D} \right) - \sigma E^2 - \boldsymbol{E} \bullet \boldsymbol{J}_o$$

or

$$\nabla \bullet \mathcal{P} = -\frac{\partial}{\partial t} \left(\frac{1}{2} \boldsymbol{H} \bullet \boldsymbol{B} + \frac{1}{2} \boldsymbol{E} \bullet \boldsymbol{D} \right) - \sigma E^2 - \boldsymbol{E} \bullet \boldsymbol{J}_o \tag{15.58}$$

On integrating both sides of Equation 15.58 over an arbitrary volume ν bounded by its surface s than applying the divergence theorem to the LHS of the resulting equation, one gets:

$$\oiint_s \mathcal{P} \bullet d\boldsymbol{s} = -\iiint_v \frac{\partial}{\partial t} \left(\frac{1}{2} \boldsymbol{H} \bullet \boldsymbol{B} + \frac{1}{2} \boldsymbol{E} \bullet \boldsymbol{D} \right) . dv - \iiint_v \left(\sigma E^2 + \boldsymbol{E} \bullet \boldsymbol{J}_o \right) . dv \tag{15.59a}$$

or

$$\oint\!\!\!\oint_s \mathcal{P} \bullet ds = \left[-\frac{\partial}{\partial t} \iiint_v \left(\frac{1}{2} \boldsymbol{H} \bullet \boldsymbol{B} + \frac{1}{2} \boldsymbol{E} \bullet \boldsymbol{D} \right) .dv - \iiint_v \boldsymbol{E} \bullet \boldsymbol{J}_o .dv \right] - \iiint_v \sigma E^2 .dv \quad (15.59b)$$

Consider Equation 15.59b; its RHS indicates the difference of two terms, each in a rectangular bracket. The time derivative term in the first bracket indicates the rate of decrease of a field quantity integrated over the volume. This field quantity has the dimension of energy per unit volume, i.e. energy density $\mathcal{E}$. Thus:

$$\mathcal{E} \stackrel{\text{def}}{=} \frac{1}{2} \boldsymbol{H} \bullet \boldsymbol{B} + \frac{1}{2} \boldsymbol{E} \bullet \boldsymbol{D} \quad (15.60)$$

It is believed that this expression for $\mathcal{E}$ gives the energy stored per unit volume in the electromagnetic field. We find two terms on the RHS of Equation 15.60. For time-varying electromagnetic fields, the electric and the magnetic field are dependent on each other. Therefore, these two terms jointly give energy density stored in the electromagnetic field. For time-varying fields these two terms individually do not have any identity. The volume integration gives the total energy stored in the electromagnetic field within the entire volume. So the time derivative quantity inside the first rectangular bracket gives the rate of decrease of this stored energy. The reduction of the stored energy releases power. In the absence of any conducting region in this volume; the second term on the RHS of Equation 15.59a shall be zero provided that the volume is free from any source-controlled current of density $\boldsymbol{J}_o$ (that contributes to the electromagnetic fields). Under this condition, the entire released power flows out across the closed surface of the volume. Thereby, the Poynting vector integrated over the closed surface gives the total released power from electromagnetic field in the volume flowing *out* across the closed surface of the volume. However, if inside this closed surface there are conducting regions as well as a distribution of source current of density $\boldsymbol{J}_o$; a part of the released power, by Joule's law, is dissipated in Ohmic loss in conducting regions and another part of this power is transferred to the current sources causing this distribution of the source-controlled current density $\boldsymbol{J}_o$. The remaining power released from the stored energy in the electromagnetic field distributed inside the volume, flows out across the closed surface of the volume.

Therefore, the Poynting Theorem can be stated as follows:

> The Poynting vector integrated over the closed surface containing conducting regions and current sources, gives the power flowing out across the closed surface which is the total power released from the energy stored in the electromagnetic field distributed inside this surface minus the power dissipated in the Ohmic loss and the power transferred to the current sources.

Since the vector $\boldsymbol{J}_o$ is arbitrary the term $(\boldsymbol{E} \bullet \boldsymbol{J}_o)$ could be positive, zero or a negative quantity. If it is a positive quantity, the electric field $\boldsymbol{E}$ is supporting the flow of the source current by transferring the power, i.e. volume integration of $(\boldsymbol{E} \bullet \boldsymbol{J}_o)$, into the source that drives this current. On the other hand, if $(\boldsymbol{E} \bullet \boldsymbol{J}_o)$ is a negative quantity the electric field $\boldsymbol{E}$ is opposing the flow of the source current, thus the source that drives this current must supply an extra power, equivalent to the volume integration of $(\boldsymbol{E} \cdot \boldsymbol{J}_o)$. In the event the two vectors $\boldsymbol{E}$ and $\boldsymbol{J}_o$, are mutually perpendicular, no power transfer occurs.

It may be noted that any stationary distribution of free electric charge density ρ_o that may be present inside the volume contributes to the time-invariant component of electric field; this distribution, however, does not explicitly appear in the treatment. The energy stored in the time-invariant electric (and magnetic) field in general, does not participate in the power transfer. However, there can be a transfer of power between the time-invariant electric field and the source for any time-invariant current. This is because currents constitute moving charges; these could be accelerated or retarded by the time-invariant electric field. On the other hand, the source-controlled current density $\boldsymbol{J}_o$ is not only a source for the electromagnetic fields; it appears explicitly in the derivation. The Poynting theorem in its original form may be considered as valid for source-free regions.

Corollary I

> The negative of the Poynting vector integrated over the closed surface containing conducting regions and source currents, gives the power flowing into the volume across the closed surface which increases the energy stored in the electromagnetic field distributed inside this surface, the power dissipated in the Ohmic loss therein and the power transferred to the current sources.

This corollary to the Poynting Theorem is easily realized by multiplying both sides of Equation 15.59a by minus one. The resulting equation is as follows:

$$-\oint\!\!\oint_s \boldsymbol{\mathcal{P}} \cdot \boldsymbol{ds} = \left[\frac{\partial}{\partial t} \iiint_v \left(\frac{1}{2} \boldsymbol{H} \cdot \boldsymbol{B} + \frac{1}{2} \boldsymbol{E} \cdot \boldsymbol{D} \right) \cdot dv \right]$$

$$+ \left[\iiint_v \sigma E^2 \cdot dv + \iiint_v \boldsymbol{E} \cdot \boldsymbol{J}_o \cdot dv \right] \tag{15.61}$$

Now, the LHS of Equation 15.61 gives the total power flowing *into* the volume across its closed surface. While the first term on the RHS gives the rate of increase of the energy stored in the electromagnetic field within the volume bounded by the closed surface and the second term gives the Ohmic loss therein plus the power transferred to current sources. These are the three components of the power flowing into the volume. For a source-free non-conducting volume, the entire inflow power results in an increase in the energy stored in the electromagnetic field within the volume. The question remains unanswered: Why there should be a power flowing into a source-free non-conducting volume?

Corollary II

> In a source-free region with time-invariant electromagnetic field, the integration of the Poynting vector over an arbitrary closed surface is zero.

Consider Equation 15.61. For a source-free region (i.e. $J_o = 0$) with time-invariant electromagnetic field this equation reduces to the following form:

$$-\oiint_s \mathcal{P} \bullet ds \equiv \iiint_v \sigma E^2 \cdot dv \tag{15.62a}$$

Let the volume ν inside the closed surface s be divided into two groups of volumes indicated by ν_1 and ν_2; where ν_1 indicates the combined volume for non-conducting regions in ν and ν_2 indicates the combined volume for conducting regions in ν. Therefore:

$$-\oiint_s \mathcal{P} \bullet ds \equiv \iiint_{v_1} \sigma E^2 .dv + \iiint_{v_2} \sigma E^2 \cdot dv \tag{15.62b}$$

The first term on the RHS is zero since the conductivity σ is zero in the insulating regions. The second term on the RHS is zero since the time-invariant electric field $\boldsymbol{E}$ is zero in the conducting regions. This proves the corollary.

Poynting Theorem is extensively used for the treatment of antennas, waveguides, microwave ovens, and other radiating systems. Poynting Theorem can be extended[3,4] to account for the *electromechanical power conversion* in rotating electrical machines. The mechanical power is identified as a component of the total power exchanged between the stationary and the moving part of an electrical machine.

15.9.1 Alternative Poynting Theorem

Two central features of Poynting Theorem are: (1) flow of power across a closed surface, and (2) energy stored in time-varying electromagnet field distributed within the closed surface. The expression for the density of the energy stored in electromagnetic fields, as given by Equation 15.60, precludes the possibility of negative stored energy. Poynting Theorem therefore, may not be taken as a proof for the existence of stored energy in the electromagnetic field. An alternative Poynting Theorem presented in this subsection describes the flow of power across a closed surface without invoking the existence of any stored energy in the electromagnetic field. The statement of this theorem is as follows:

> The Bedford vector $\mathfrak{B}$ integrated over the closed surface containing conducting regions and source-controlled currents, gives the power flowing out across the closed surface, which is the total power transferred from the current sources causing the source-controlled current density distributed inside this closed surface minus the power dissipated as an Ohmic loss in conducting regions therein.

The Bedford vector $\mathfrak{B}$ is uniquely specified[5] by its divergence and curl as given below:

$$\nabla \bullet \mathfrak{B} \stackrel{\text{def}}{=} \frac{\partial}{\partial t}\left(\frac{1}{2}\boldsymbol{H} \bullet \boldsymbol{B} + \frac{1}{2}\boldsymbol{E} \bullet \boldsymbol{D}\right) \tag{15.63a}$$

and

$$\nabla \times \mathfrak{B} \overset{\text{def}}{=} \frac{1}{c} \cdot \frac{\partial}{\partial t}(\boldsymbol{E} \times \boldsymbol{H}) \tag{15.63b}$$

To prove this theorem, we note in view of Equation 15.59a that:

$$\oiint_s (\mathfrak{B} + \mathcal{P}) \bullet \boldsymbol{ds} = \left[\iiint_v \{\nabla \bullet \mathfrak{B}\} - \frac{\partial}{\partial t}\left(\frac{1}{2}\boldsymbol{H} \bullet \boldsymbol{B} + \frac{1}{2}\boldsymbol{E} \bullet \boldsymbol{D}\right) \cdot dv \right]$$

$$- \left[\iiint_v \sigma E^2 \cdot dv + \iiint_v \boldsymbol{E} \bullet \boldsymbol{J}_o \cdot dv \right] \tag{15.64}$$

Now, in view of Equation 15.63b:

$$\mathcal{P} \overset{\text{def}}{=} \boldsymbol{E} \times \boldsymbol{H} = c \cdot \int (\nabla \times \mathfrak{B}) \cdot dt$$

or,

$$\mathcal{P} = \nabla \times \left(c \cdot \int \mathfrak{B} \cdot dt\right) \tag{15.65a}$$

where the velocity of propagation (c) is given as:

$$c = 1 / \sqrt{\mu \cdot \varepsilon} \tag{15.65b}$$

Therefore, Equation 15.64 can be rewritten as:

$$\oiint_s \mathfrak{B} \bullet \boldsymbol{ds} + \underbrace{\oiint_s \left\{\nabla \times \left(c \cdot \int \mathfrak{B}.dt\right)\right\} \bullet \boldsymbol{ds}}$$

$$= \left[\underbrace{\iiint_v \left\{(\nabla \bullet \mathfrak{B}) - \frac{\partial}{\partial t}\left(\frac{1}{2}\boldsymbol{H} \bullet \boldsymbol{B} + \frac{1}{2}\boldsymbol{E} \bullet \boldsymbol{D}\right)\right\} \cdot dv} \right] - \left[\iiint_v \sigma E^2 \cdot dv + \iiint_v \boldsymbol{E} \bullet \boldsymbol{J}_o \cdot dv \right] \tag{15.66}$$

Since the curl of an arbitrary vector integrated over a closed surface is identically zero, the second term on the LHS of this equation vanishes; while in view of Equation 15.63a, the first term on the RHS of this equation vanishes. Therefore, Equation 15.66 reduces to:

$$\boxed{\oiint_s \mathfrak{B} \bullet \boldsymbol{ds} = -\iiint_v \sigma E^2 \cdot dv - \iiint_v \boldsymbol{E} \bullet \boldsymbol{J}_o \cdot dv} \tag{15.67}$$

Therefore, for non-conducting source-free finite region of volume ν bounded by its surface s, we have:

$$\oiint_s \mathfrak{B} \bullet ds = 0 \tag{15.67a}$$

This theorem shows that it is possible to explain the transfer of power across a closed surface without evoking the concept of stored energy in electromagnetic fields. Evidently, no power flows across the surface of a finite non-conducting source-free region. The stored energy in electromagnetic fields is not a necessary concept in the theory of electromagnetism.

Corollary I

The negative of Bedford vector $\mathfrak{B}$ integrated over the closed surface containing conducting regions and source-controlled currents, gives the power flowing into the volume across the closed surface which is the algebraic sum of the power dissipated in the Ohmic loss and the power transferred to the current sources.

Corollary II

In an arbitrary finite region of volume ν with a time-invariant electromagnetic field, the integration of the Bedford vector over its surface bounding the region is zero.

The Bedford vector $\mathfrak{B}$ defined by Equation 15.64a and 15.64b is zero for time-invariant fields. Therefore its integration over the closed surface bounding a finite region is zero.

15.9.2 Comparison of the Two Poynting Theorems

(i) As seen from the Equation 15.65; for a finite volume, free of conducting regions and without any distribution of source-controlled currents, integration of the Bedford vector over its entire surface is zero. On the other hand, the integration of the Poynting vector over this surface, as seen from Equation 15.59a results:

$$\oiint_s \mathcal{P} \bullet ds \equiv -\left[\frac{\partial}{\partial t}\iiint_v \left(\frac{1}{2}\boldsymbol{H} \bullet \boldsymbol{B} + \frac{1}{2}\boldsymbol{E} \bullet \boldsymbol{D}\right) \cdot dv\right] \tag{15.68}$$

The LHS is in general a non-zero quantity if the stored energy in the volume is time-varying.

(ii) In view of the divergence theorem, we have:

$$-\oiint_s \mathfrak{B} \bullet ds = -\iiint_v (\nabla \bullet \mathfrak{B}) \cdot dv \tag{15.69a}$$

Thus, using Equation 15.59a, we get:

$$-\oiint_s \mathfrak{B}\bullet d\boldsymbol{s} = -\iiint_v \frac{\partial}{\partial t}\left(\frac{1}{2}\boldsymbol{H}\bullet\boldsymbol{B}+\frac{1}{2}\boldsymbol{E}\bullet\boldsymbol{D}\right)\cdot dv \qquad (15.69b)$$

On comparing Equation 15.67 and 15.69b, the volume being arbitrary, we have:

$$-\frac{\partial}{\partial t}\left(\frac{1}{2}\boldsymbol{H}\bullet\boldsymbol{B}+\frac{1}{2}\boldsymbol{E}\bullet\boldsymbol{D}\right)=\sigma E^2+\boldsymbol{E}\bullet\boldsymbol{J}_o \qquad (15.70a)$$

or

$$\sigma E^2+\left[\boldsymbol{E}\bullet\boldsymbol{J}_o+\frac{\partial}{\partial t}\left(\frac{1}{2}\boldsymbol{H}\bullet\boldsymbol{B}+\frac{1}{2}\boldsymbol{E}\bullet\boldsymbol{D}\right)\right]=0 \qquad (15.70b)$$

Since the first term on the LHS of this equation cannot be a negative quantity, the sum of the two terms within the rectangular bracket cannot be a positive quantity. For non-conducting finite regions without any source-controlled current distribution we get:

$$\frac{\partial}{\partial t}\left(\frac{1}{2}\boldsymbol{H}\bullet\boldsymbol{B}+\frac{1}{2}\boldsymbol{E}\bullet\boldsymbol{D}\right)=0 \qquad (15.70c)$$

This equation shows that for non-conducting source-free finite regions, the two terms within the parentheses often identified as the energy density in the electromagnetic field) *must* sum up to a time-invariant quantity even though each of these terms is time-varying.

This is an important conclusion. As an example; in the case of time harmonic field if $\boldsymbol{H}\bullet\boldsymbol{B}=F\cdot\cos^2(\omega\cdot t)$, then we must have: $\boldsymbol{E}\bullet\boldsymbol{D}=F\cdot\sin^2(\omega\cdot t)$, where the symbol F indicates an arbitrary scalar function of space coordinates. This implies that the magnetic field and the corresponding electric field are at a 90° time-phase apart and the following relation exists:

$$F\overset{\text{def}}{=}\mu|\boldsymbol{H}|^2=\varepsilon|\boldsymbol{E}|^2 \qquad (15.71a)$$

Thus,

$$\xi\overset{\text{def}}{=}\sqrt{\frac{\mu}{\varepsilon}}=\frac{|\boldsymbol{E}|}{|\boldsymbol{H}|}\overset{\text{def}}{=}\frac{E}{H} \qquad (15.71b)$$

where ξ indicates the characteristic impedance of the medium.

Poynting theorem, for a finite non-conducting source-free region leads to the identity:

$$\nabla\bullet\mathcal{P}\overset{\text{def}}{=}\nabla\bullet(\boldsymbol{E}\times\boldsymbol{H})\equiv-\frac{\partial}{\partial t}\left(\frac{1}{2}\boldsymbol{H}\bullet\boldsymbol{B}+\frac{1}{2}\boldsymbol{E}\bullet\boldsymbol{D}\right) \qquad (15.72)$$

This being an identity based on Maxwell's equations, the two terms within the parenthesis on the RHS of this equation (often identified as the energy density in the electromagnetic field) may not necessarily sum up to a time-invariant quantity. If this sum is time-invariant no power will flow across the surface of the region.

15.10 MODIFICATION OF FIELD RELATIONS IN VIEW OF THE RELATIVISTIC EFFECT

In Section 15.8, Equations 7.6, 9.1a, and 9.97 were modified to yield Equations 15.47, 15.53, and 15.50b, respectively, to make them consistent with time-varying fields. All of the time-invariant and time-varying relations studied to this point now are true only if the medium is assumed to be stationary with respective to an observer. If the medium becomes non-stationary, then all of these relations will be further modified in view of the relativistic effects. This section explores these modifications.

The concept of relativistic effect germane from the theory of relativity based on empirical discovery that leads to understanding the general characteristics of natural processes. The theory of relativity is generally divided into the categories of the special theory of relativity and the general theory of relativity. The special theory of relativity applies to all physical phenomena except the gravity. The general theory provides the law of gravitation, and its relation to other forces of nature. In the following subsections our study shall remain confined only to the special theory of relativity.

15.10.1 Special Theory of Relativity

The special theory of relativity is a theory of the structure of space-time.[6–8] It was introduced in Einstein's 1905 paper "On the Electrodynamics of Moving Bodies". This theory is based on two postulates:[9]

1. The laws of physics are the same for all observers in unaccelerated motion relative to one another. Therefore, when properly formulated the laws of physics are invariant to a transformation from one reference system to another moving with a linear, uniform relative velocity.
2. The speed of light in vacuum is the same for all observers, regardless of their relative motion or of the motion of the light source. Thus the velocity of propagation of an electromagnetic disturbance in free space is a universal constant c, which is independent of the reference system.

The theory based on the above postulates agrees with experiments better than classical mechanics. This theory leads to many interesting consequences. Some of these are:

- *Relativity of simultaneity*: Two events, simultaneous for one observer, may not be simultaneous for another observer if the observers are in relative motion.
- *Time dilation*: Moving clocks run slower than an observer's "stationary" clock.

- *Relativistic mass*: The mass of a moving object is larger than its value at rest.
- *Length contraction*: Objects are shortened in the direction that they are moving with respect to the observer.
- *Mass–energy equivalence*: $E = mc^2$, energy and mass are equivalent and mutually convertible.
- *Maximum speed is finite*: No physical object, message or field line can travel faster than the speed of light in a vacuum.

The defining feature of special theory of relativity is the replacement of the Galilean transformations of classical mechanics by the Lorentz transformations.

15.10.1.1 Maxwell's View

The foundation of classical electromagnetism describes light as a wave which moves with a characteristic velocity. The modern view is that light needs no medium of transmission, but Maxwell and his contemporaries were convinced that light waves were propagated in a medium,[10] analogous to sound propagating in air, and ripples propagating on the surface of a pond. This hypothetical medium was called the luminiferous ether, or simply the *ether*, through which the Earth moves.

The Michelson–Morley experiment was designed to detect second order effects of the "ether wind", i.e. the motion of the ether relative to the Earth. Michelson designed an instrument called the Michelson interferometer to accomplish this. The apparatus was more than accurate enough to detect the expected effects, but he obtained a null result when the first experiment was conducted in 1881,[11] and again in 1887.[12] Although the failure to detect an ether wind was a disappointment, the results were accepted by the scientific community.[13] In an attempt to salvage the ether paradigm, Fitzgerald and Lorentz independently proposed that the length of material bodies changes according to their motion through the ether,[13] as given below:

$$l = l_o \cdot \sqrt{1 - \beta^2} \tag{15.73}$$

where

$$\beta = u / c \tag{15.73a}$$

The symbol u, indicate the velocity of a body in the direction of its length l. The velocity of light in vacuum is given as c. While l_o is the length the body while it is at rest.

This was the origin of the Fitzgerald–Lorentz contraction, and their hypothesis had no theoretical basis. The theoretical basis for the Fitzgerald–Lorentz contraction is provided by the Einstein's special theory of relativity.

The Michelson–Morley experiment showed that the velocity of light is isotropic (same in all directions), it said nothing about how the magnitude of the velocity changed (if at all) in different inertial frames. The Kennedy–Thorndike experiment was designed to do that, and was first performed in 1932 by Roy Kennedy and

Edward Thorndike.[14] They obtained a null result, so it was concluded that the round-trip time for light is the same in all inertial (unaccelerated) reference frames.

The Ives–Stilwell experiment was carried out by Herbert Ives and G.R. Stilwell first in 1938[15] and, with better accuracy, in 1941.[16] It was designed to test the transverse "Doppler Effect", which is the redshift of light from a moving source in a direction perpendicular to its velocity predicted by Einstein in 1905. The strategy was to compare observed Doppler shifts with what was predicted by classical theory, and look for a Lorentz factor correction. Such a correction was observed, from which was concluded that the frequency of a moving atomic clock is altered according to special relativity.[15, 16] The time dilation (slowing down) of a moving clock is indicated by the following equation:

$$t = \frac{t_o}{\sqrt{1-\beta^2}} \tag{15.74}$$

where t_o is the duration for an event measured by a stationary clock, and t is the duration for the same event measured by a moving clock.

The operating principle for both, nuclear energy and nuclear bombs is based on the Einstein's equation stating that the rest mass m_o can be converted into energy E_o as per the following relation:

$$E_o = m_o \cdot c^2 \tag{15.75}$$

where c indicates the velocity of light in free space.

For the mass of a body m moving with a constant linear velocity u, the total energy E can likewise be given as follows:

$$E = m \cdot c^2 \tag{15.75a}$$

This total energy E must include its *kinetic energy* as well. Therefore,

$$m_o \cdot c^2 = m \cdot c^2 - \frac{1}{2} \cdot m \cdot u^2 \cong m \cdot c^2 \cdot \sqrt{1-(u/c)^2} \tag{15.75b}$$

Thus:

$$m = \frac{m_o}{\sqrt{1-(u/c)^2}} \tag{15.76}$$

Therefore, the mass of a moving body m, in the light of the special theory of relativity depends on its rest mass m_o as well as its constant linear velocity u. It increases with its velocity and reaches to infinity as its velocity approaches the velocity of light. It is therefore concluded that a body with non-zero rest mass cannot be moved at the velocity of light. The rest mass for a *photon* is zero, so it always moves with the velocity of light. The moving photons are identified as electromagnetic waves with non-zero electromagnetic momentum. Equation 15.76 shows that a moving particle with imaginary rest mass may exist provided that its velocity is *always* more than the velocity of light. No such particle has been discovered so far.

Classic experiments stated above have been repeated many times with increased precision. Other experiments include, for instance, relativistic energy and momentum increase at high velocities, and time dilation of moving particles.

Currently, it can be said that far from being simply of theoretical scientific interest or requiring experimental verification, the analysis of relativistic effects on time measurement is an important practical engineering concern in the operation of the global positioning systems such as GPS, GLONASS, and the forthcoming Galileo, as well as in the high precision dissemination of time.[7, 17] Instruments ranging from electron microscopes to particle accelerators simply will not work if relativistic considerations are omitted.

15.10.2 Lorentz Transformation

Imagine that a light source is placed at the origin O of a reference frame $\mathcal{R}(x, y, z, t)$ at the instant $t = 0$. An observer at the point x, y, z in the reference frame $\mathcal{R}$ will receive the light signal at the instant t, such that:

$$t = \frac{\sqrt{x^2 + y^2 + z^2}}{c} \tag{15.77}$$

Thus,

$$x^2 + y^2 + z^2 - c^2 \cdot t^2 = 0 \tag{15.77a}$$

The same observer measures position and time with respect to a second reference frame $\mathcal{R}'(x', y', z', t')$ thatis moving along the z-axis (or z'-axis) with a constant velocity u relative to O. Let us assume that the origin O′ coincides with O at the instant $t = 0$ (or $t' = 0$). Since the velocity of light is independent of the relative velocity u between the two reference frames, we have:

$$(x')^2 + (y')^2 + (z')^2 - c^2 \cdot (t')^2 = 0 \tag{15.77b}$$

In view of the Fitzgerald–Lorentz contraction, we get:[18]

$$\begin{cases} x' = x, & y' = y \\ z' = \dfrac{1}{\sqrt{1-\beta^2}} \cdot (z - u \cdot t), & t' = \dfrac{1}{\sqrt{1-\beta^2}} \cdot \left(t - u \cdot z / c^2\right) \\ \end{cases} \tag{15.78}$$

This equation is known as the *Lorentz transformation*. Maxwell's equations are invariant under this transformation. For small values of the relative velocity, i.e. $u \ll c$, (or $c \to \infty$) this equation reduces to:

$$\begin{cases} x' = x, & y' = y \\ z' = z - u \cdot t, & t' = t \end{cases} \tag{15.79}$$

The resulting equation is identified as the *Galilean transformation*. Maxwell's equations are not invariant under the Galilean transformation.

15.10.3 Maxwell's Equation for Moving Media

Maxwell's equations for a medium stationary with respective to an observer are given as:

$$\begin{cases} \nabla \times \boldsymbol{E} + \dfrac{\partial \boldsymbol{B}}{\partial t} = 0 & \nabla \bullet \boldsymbol{B} = 0 \\ \nabla \times \boldsymbol{H} - \dfrac{\partial \boldsymbol{D}}{\partial t} = \boldsymbol{J} & \nabla \bullet \boldsymbol{D} = \rho \end{cases} \tag{15.80a}$$

In view of Einstein's first postulate, let Maxwell's equations for a medium moving with a constant linear velocity u with respective to the same observer are given as follows:[13]

$$\begin{cases} \nabla' \times \boldsymbol{E}' + \dfrac{\partial \boldsymbol{B}'}{\partial t'} = 0 & \nabla' \bullet \boldsymbol{B}' = 0 \\ \nabla' \times \boldsymbol{H}' - \dfrac{\partial \boldsymbol{D}'}{\partial t'} = \boldsymbol{J}' & \nabla' \bullet \boldsymbol{D}' = \rho' \end{cases} \tag{15.80b}$$

In view of Equation 15.78, we have[13] the *Lorentz transform* for the components of the current density (including the convection currents) as shown here:

$$\begin{cases} J'_x = J_x, & J'_y = J_y \\ J'_z = \dfrac{1}{\sqrt{1-\beta^2}} \cdot (J_z - u \cdot \rho), & \rho' = \dfrac{1}{\sqrt{1-\beta^2}} \cdot \left(\rho - u \cdot J_z / c^2\right) \end{cases} \tag{15.81}$$

For small values of the relative velocity, i.e. $u \ll c$, this equation reduces to:

$$\begin{cases} J'_x = J_x, & J'_y = J_y \\ J'_z \cong J_z - u \cdot \rho, & \rho' \cong \rho \end{cases} \tag{15.81a}$$

The *Lorentz transform* for the components of the field vectors[13] $\boldsymbol{B}$ and $\boldsymbol{E}$ are as follows:

$$\begin{cases} B'_x = \dfrac{1}{\sqrt{1-\beta^2}} \cdot \left(B_x + u \cdot E_y / c^2\right), & E'_x = \dfrac{1}{\sqrt{1-\beta^2}} \cdot \left(E_x - u \cdot B_y\right) \\ B'_y = \dfrac{1}{\sqrt{1-\beta^2}} \cdot \left(B_y - u.E_x / c^2\right), & E'_y = \dfrac{1}{\sqrt{1-\beta^2}} \cdot \left(E_y + u \cdot B_x\right) \\ B'_z = B_z, & E'_z = E_z \end{cases} \tag{15.82}$$

The restriction to the motion along the z-axis can be removed by writing $\boldsymbol{u}$ as a vector:

$$\begin{cases} \boldsymbol{B}'_\perp = \dfrac{1}{\sqrt{1-\beta^2}} \cdot \left(\boldsymbol{B} - \boldsymbol{u} \times \boldsymbol{E} / c^2\right), & \boldsymbol{E}'_\perp = \dfrac{1}{\sqrt{1-\beta^2}} \cdot \left(\boldsymbol{E} + \boldsymbol{u} \times \boldsymbol{B}\right)_\perp \\ \boldsymbol{B}'_\parallel = \boldsymbol{B}_\parallel, & \boldsymbol{E}'_\parallel = \boldsymbol{E}_\parallel \end{cases} \tag{15.83}$$

For small values of the relative velocity, i.e. $u \ll c$, this equation reduces to:

$$\begin{cases} \boldsymbol{B}'_\perp \cong \boldsymbol{B}_\perp, & \boldsymbol{E}'_\perp \cong \left(\boldsymbol{E} + \boldsymbol{u} \times \boldsymbol{B}\right)_\perp \\ \boldsymbol{B}'_\parallel = \boldsymbol{B}_\parallel, & \boldsymbol{E}'_\parallel = \boldsymbol{E}_\parallel \end{cases} \tag{15.83a}$$

The *Lorentz transform* for the components of the field vectors[13] $\boldsymbol{H}$ and $\boldsymbol{D}$ are as follows:

$$\begin{cases} H'_x = \dfrac{1}{\sqrt{1-\beta^2}} \cdot \left(H_x + u \cdot D_y\right), & D'_x = \dfrac{1}{\sqrt{1-\beta^2}} \cdot \left(D_x - u \cdot H_y / c^2\right) \\ H'_y = \dfrac{1}{\sqrt{1-\beta^2}} \cdot \left(H_y - u \cdot D_x\right), & D'_y = \dfrac{1}{\sqrt{1-\beta^2}} \cdot \left(D_y + u \cdot H_x / c^2\right) \\ H'_z = H_z, & D'_z = D_z \end{cases} \tag{15.84}$$

The restriction to the motion along the z-axis can be removed by writing $\boldsymbol{u}$ as a vector:

$$\begin{cases} \boldsymbol{H}'_\perp = \dfrac{1}{\sqrt{1-\beta^2}} \cdot \left(\boldsymbol{H} - \boldsymbol{u} \times \boldsymbol{D}\right)_\perp, & \boldsymbol{D}'_\perp = \dfrac{1}{\sqrt{1-\beta^2}} \cdot \left(\boldsymbol{D} + \boldsymbol{u} \times \boldsymbol{H} / c^2\right)_\perp \\ \boldsymbol{H}'_\parallel = \boldsymbol{H}_\parallel, & \boldsymbol{D}'_\parallel = \boldsymbol{D}_\parallel \end{cases} \tag{15.85}$$

For small values of the relative velocity, i.e. $u \ll c$, this equation reduces to:

$$\begin{cases} \boldsymbol{H}'_\perp \cong \left(\boldsymbol{H} - \boldsymbol{u} \times \boldsymbol{D}\right)_\perp, & \boldsymbol{D}'_\perp \cong \boldsymbol{D}_\perp \\ \boldsymbol{H}'_\parallel = \boldsymbol{H}_\parallel, & \boldsymbol{D}'_\parallel = \boldsymbol{D}_\parallel \end{cases} \tag{15.85a}$$

The symbols $\boldsymbol{A}_\parallel$ and $\boldsymbol{A}_\perp$ indicate, respectively, the vectors in parallel and perpendicular directions to the velocity vector $\boldsymbol{u}$. The operator ∇' involves differentiation with respect to the primed space coordinates.

15.10.4 Force on Stationary Charges Due to Steady Conduction Currents

This section deals with the relativistic effects of electric field due to dc currents flowing in conductors.

An atom consists of positively charged protons in the nucleus and equal number of negatively charged electrons orbiting the nucleus. The net charge being zero, an atom is electrically neutral. Atoms in solid conducting materials are stationary and have some loosely bound electrons. These electrons often leave the atom and move at random in the inter-atomic space with zero average velocity. Thus in a solid conductor we find positively charged stationary atoms and a cloud of negatively charged free electrons in random motion. The conductor as a whole is however, electrically neutral. If a steady electric field is applied along the length of a long conducting wire the electron cloud drifts with a non-zero average velocity in the direction of the higher electric potential. This constitutes a current in the conductor, conventionally flowing in the direction opposite to the direction the electron cloud is moving. Since the free electrons are responsible for the flow of the current in a solid conductor, these free electrons are also called the conduction electrons.

Einstein's Special Theory of Relativity is applicable to unaccelerated motion. It is based on two postulates.[1] As a consequence to the second postulate the length of moving objects are shortened in the direction of its motion. This phenomenon is called the *Lorentz contraction*. This is equally applicable to an imaginary line segment of length ℓ_o moving with a constant velocity u along its length as given below:

$$\ell = \ell_o \cdot \sqrt{1-(u/c)^2} \tag{15.86}$$

where ℓ is the length of the moving line segment, and the symbol c indicates the velocity of light in free space.

Consider a long thin uncharged conducting wire. Let the average distance between two adjacent positively charged atoms in the wire be ℓ_+. For an uncharged conductor with zero drift velocity the average distance between two adjacent free electrons will be the same, so that the number of positively charged atoms per unit length n_+ is equal to the number of free electrons per unit length. Now, if a dc current I flows in the conducting wire the drift velocity u will be proportional to this current:

$$u \propto I \overset{\text{def}}{=} k \cdot I \tag{15.87}$$

where the symbol k is a constant for the given conductor material.

As the free electrons are moving with an average velocity u, the distance between adjacent electrons ℓ_- is reduced:

$$\ell_- = \ell_+ \cdot \sqrt{1-(u/c)^2} \tag{15.88}$$

Therefore, the number of free electrons per unit length n_- is increased to:

$$n_- = n_+ / \sqrt{1-(u/c)^2} \tag{15.89}$$

Thus, the line charge density ρ_L of the current carrying conductor is:

$$\rho_L = e \cdot n_+ - e \cdot n_- \tag{15.90}$$

where the symbol e indicates the charge of a proton; while the charge of an electron is negative of this value.

Using Equations 15.89 and 15.90, since $u \ll c$ we get:

$$\rho_L = e \cdot n_+ \cdot \left[1 - \frac{1}{\sqrt{1-(u/c)^2}}\right] = e \cdot n_+ \cdot \left[\frac{\sqrt{1-(u/c)^2} - 1}{\sqrt{1-(u/c)^2}}\right] \tag{15.91}$$

or

$$\rho_L \cong -e \cdot n_+ \cdot \frac{1}{2} \cdot \left(\frac{u}{c}\right)^2 \tag{15.92}$$

Thus, in view of Equation 15.87:

$$\rho_L \cong -e \cdot n_+ \cdot \frac{1}{2} \cdot \left(\frac{k}{c}\right)^2 \cdot I^2 \tag{15.92a}$$

or,

$$\rho_L \cong -e \cdot n_+ \cdot \frac{k^2}{2} \cdot \mu_o \cdot \varepsilon_o \cdot I^2 \tag{15.92b}$$

The electric field intensity at a distance ρ from a long line charge is given as:

$$\boldsymbol{E} = E_\rho \cdot \boldsymbol{a}_\rho = \frac{\rho_L}{2\pi\rho.\varepsilon_o} \cdot \boldsymbol{a}_\rho \tag{15.93}$$

where $\boldsymbol{a}_\rho$ indicates the unit vector pointing away from the line.

Using Equations 15.92b and 15.93, we get:

$$\boldsymbol{E} = -K \cdot \frac{I^2}{\rho} \cdot \boldsymbol{a}_\rho \tag{15.94}$$

where the constant K is given below:

$$K \stackrel{\text{def}}{=} e \cdot n_+ \cdot \frac{k^2}{4\pi} \cdot \mu_o \tag{15.94a}$$

Thus, the force F on a point charge Q located at a distance ρ from the conducting wire carrying a dc current I is given as:

$$\boldsymbol{F} = Q \cdot \boldsymbol{E} = -Q \cdot K \cdot \frac{I^2}{\rho} \cdot \boldsymbol{a}_\rho \tag{15.95}$$

It is interesting to note that the sources for the time-invariant electric field usually are stationary electric charges; this equation indicates that dc currents flowing in a conductor can as well be considered as the sources for the electrostatic field. The drift velocity is usually a small quantity thus the approximation leading to the Equation 15.92 is quite justified. For good conductors, the number of the positively charged atoms per unit length of the conductor n_+ being a large quantity the constant K in Equation 15.94a is not negligible.

15.10.5 The Second Observer Paradox

Imagine two identical parallel long straight line charge similar to that of Figure 12.5), each with uniform line charge density ρ_L. Let an experiment is performed in an imaginary laboratory with these line charges moving in the same axial direction with a constant velocity u and a separation distance "a" between them. Let an observer is also moving in the same direction with the same velocity. Relative to this observer both line charges are stationary. Using Coulomb's law, he finds the electrostatic force of repulsion per unit length $\mathcal{F}_{Le}$ between these line charges as given by Equation 12.21. The same is reproduced below:

$$\mathcal{F}_L \stackrel{\text{def}}{=} \mathcal{F}_{Le} = \frac{\rho_L^{\;2}}{2\pi\varepsilon_o \cdot a} \tag{12.21}$$

This observer calculates the numerical value of the force and finds it to be same as indicated on a meter that measures the force per unit length on one of the two line charges.

Now, suppose a second observer who is sitting in the laboratory notices a magnetic force of attraction between the two *moving line charges*, in addition to the repulsive electric force. The net repulsive force per unit length $\mathcal{F}_L$ in Equations 12.20 and 12.21 is:

$$\mathcal{F}_L = \mathcal{F}_{Le} + \mathcal{F}_{Lm} = \frac{\rho_L^{\;2}}{2\pi\varepsilon_o \cdot a} - \frac{\mu_o \cdot (\rho_L \cdot u)^2}{2\pi \cdot a} \tag{15.96}$$

The second observer also calculates the numerical value of the repulsive force per unit length $\mathcal{F}_L$, and finds it to be different from that indicated on the meter that measures the force on one of the two line charges. This discrepancy is because of the second term on the RHS of this equation. Then, which one is correct? The fault lies with the second observer who has used Coulomb's law for the calculation of force on moving charges. Instead of the line charge density ρ_L he should have used an *equivalent stationary* line charge density ρ_L' such that both observers find the same force:

$$\frac{\rho_L'^2}{2\pi\varepsilon_o \cdot a} - \frac{\mu_o \cdot (\rho_L \cdot u)^2}{2\pi \cdot a} = \frac{\rho_L^{\;2}}{2\pi\varepsilon_o \cdot a} \tag{15.97}$$

Solving for ρ_L' we get:

$$\rho_L' = \rho_L \cdot \sqrt{1 + \mu_o \cdot \varepsilon_o \cdot u^2} = \rho_L \cdot \sqrt{1 + (u/c)^2} \stackrel{\text{def}}{=} \rho_L \cdot \sqrt{1 + \beta^2} \tag{15.98a}$$

If $u \ll c$,

$$\rho_L' = \frac{\rho_L}{\sqrt{1-\beta^2}} \tag{15.98b}$$

This gives the *Lorentz transformation* for the moving line charge density. This transformation is based on the fact that the velocity of light in free space is independent of the velocity of the source of the light. Note that the two observers moving at a constant relative velocity is finding the same force between the two line charges. The treatment presented in this section is consistent with Einstein's special theory of relativity.

15.10.6 Magnetostatic Field as the Relativistic Effect of the Electrostatic Field

This section deals with the study of the relativistic effects of electric field due to dc currents flowing in an identical pair of thin, straight, parallel conductors of infinite length (similar to that shown in Figure 12.4). In Section 15.10.5, we have seen that an electrically neutral conductor due to the *Lorentz contraction* gets negatively charged if dc currents flow in it. Consider a straight infinitely long conductor carrying a dc current I_1 flowing in the positive of the z-direction and another identical conductor carrying a dc current I_2. The second conductor is placed along the z-axis. The two current-carrying conductors are laid parallel to each other with a separating distance R between them.

Let the average distance between two adjacent positively charged atoms in each of the two identical conductors be ℓ_+. For uncharged conductors with zero drift velocity, the average distance between two adjacent-free electrons will be the same, so that the number of free electrons per unit length is equal to the number of positively charged atoms per unit length, n_+. Let us assume that if a dc current I_2 flows in the second conductor the drift velocity u_2 will be proportional to this current:

$$u_2 \propto I_2 \overset{\text{def}}{=} k \cdot I_2 \tag{15.99}$$

where the symbol k is a constant for a given conductor.

The average distance between two adjacent free electrons in the second conductor as seen by the stationary positive atoms of the first conductor is thus given as:

$$\ell_{-2} = \ell_+ \cdot \sqrt{1-(u_2/c)^2} = \ell_+ \cdot \sqrt{1-(k \cdot I_2/c)^2} \tag{15.100}$$

Therefore, the number of negatively charged electrons per unit length n_{-2}^{+1} of the second conductor as seen by the *positively charged atom* of the *first conductor* is:

$$n_{-2}^{+1} = n_+ / \sqrt{1-(k \cdot I_2/c)^2} \tag{15.101}$$

The stationary positively charged atoms of the first conductor are subjected to a repulsive force due to the stationary positively charged atoms and an attractive force due to the moving negatively charged electrons; both of the second conductor. These charged particles of the second conductor can be represented as line charge distributions as perceived by the positively charged atoms of the first conductor:

$$\rho_{L2+}^{+1} = e \cdot n_+ \tag{15.102a}$$

and,

$$\rho_{L2-}^{+1} = -e \cdot n_{-2}^{+1} = -e \cdot n_+ / \sqrt{1-(k \cdot I_2 / c)^2} \tag{15.102b}$$

therefore, the net line charge distributions, ρ_{L+2}^{+} as perceived by each positively charged atom of the first conductor is given as:

$$\rho_{L2}^{+1} = \rho_{L2+}^{+1} + \rho_{L2-}^{+1} = e \cdot n_+ \cdot \left[1 - \frac{1}{\sqrt{1-(k \cdot I_2 / c)^2}} \right] = e \cdot n_+ \cdot \left[\frac{\sqrt{1-(k \cdot I_2 / c)^2} - 1}{\sqrt{1-(k \cdot I_2 / c)^2}} \right] \tag{15.103a}$$

or,

$$\rho_{L2}^{+1} \cong -e \cdot n_+ \cdot \frac{1}{2} \cdot \left(\frac{k \cdot I_2}{c} \right)^2 \tag{15.103b}$$

Therefore, the force on each positively charged atom of the *first conductor* will be:

$$\boldsymbol{F}_+ = e \cdot \frac{\rho_{L+2}^{+1}}{2\pi R} \boldsymbol{a}_R = -\left[e^2 \cdot n_+ \cdot \frac{1}{2} \cdot \left(\frac{k}{c} \right)^2 \right] \cdot \frac{(I_2)^2}{2\pi R} \boldsymbol{a}_\rho \tag{15.104}$$

Since the second conductor is placed along the *z*-axis, the vector $\boldsymbol{a}_\rho$ is directed from the second conductor towards the first conductor. The minus sign implies that the first conductor is experiencing a force of attraction towards the second conductor.

Thus, force per unit length perceived by the *positively charged* atoms of the *first conductor* $\mathcal{F}_{L1+}$ is:

$$\mathcal{F}_{L1+} = n_+ \cdot \boldsymbol{F}_+ = -K \cdot \frac{(I_2)^2}{4\pi R} \boldsymbol{a}_\rho \tag{15.105}$$

where

$$K \stackrel{\text{def}}{=} \left(\frac{e}{c} \cdot k \cdot n_+ \right)^2 \tag{15.105a}$$

Next, consider a negatively charged electron in the first conductor carrying the dc current I_1. The drift velocity u_1 is assumed to be proportional to this current:

$$u_1 \propto I_1 \overset{\text{def}}{=} k \cdot I_1 \tag{15.106}$$

The average distance between two adjacent *positively charged* atoms in the *second conductor* as seen by the negatively charged electron in the first conductor is thus given as:

$$\ell_{+2} = \ell_+ \cdot \sqrt{1-(u_1 / c)^2} = \ell_+ \cdot \sqrt{1-(k \cdot I_1 / c)^2} \tag{15.107}$$

Therefore, the number of positively charged atoms per unit length n_{+2}^{-1} of the second conductor as seen by the *negatively charged* free electrons in the *first conductor*is:

$$n_{+2}^{-1} = n_+ / \sqrt{1-(k \cdot I_1 / c)^2} \tag{15.108}$$

The average distance between two adjacent negatively charged electrons in the second conductor as seen by the negatively charged free electron in the first conductor is given as:

$$\ell_{-2} = \ell_+ \cdot \sqrt{1-(u_\pm / c)^2} \tag{15.109}$$

where $u_\pm$ is the drift velocity of the electrons in the second conductor with respect to the free electrons in the first conductor. Einstein's formula[1] for the relative velocity is as follows:

$$u_\pm = \mp \frac{u_1 \mp u_2}{1 \mp (u_1 \cdot u_1) / c^2} \tag{15.110}$$

where u_+ is for parallel velocities and u_- for anti-parallel velocities.

Therefore, using Equations 15.99 and 15.106:

$$u_\pm = \mp k \cdot \frac{I_1 \mp I_2}{1 \mp (I_2 \cdot I_1) \cdot (k / c)^2} \overset{\text{def}}{=} k \cdot I_{o\pm} \tag{15.111}$$

where

$$I_{o\pm} \overset{\text{def}}{=} \mp \cdot \frac{(I_1 \mp I_2)}{1 \mp \underbrace{(I_2 \cdot I_1) \cdot (k / c)^2}} \tag{15.111a}$$

Since $\underbrace{(I_2 \cdot I_1) \cdot (k / c)^2} \ll 1$, we have:

$$I_{o\pm} \cong \mp(I_1 \mp I_2) \cong I_2 \mp I_1 \tag{15.111b}$$

Thus,

$$u_{\pm} = k \cdot I_{o\pm} \cong k \cdot (I_2 \mp I_1) \tag{15.112}$$

Therefore, from Equation 15.75:

$$\ell_{-2} = \ell_{+} \cdot \sqrt{1 - \left[k \cdot (I_2 \mp I_1)/c\right]^2} \tag{15.113}$$

Therefore, the number of negatively charged electrons per unit length of the second conductor n_{-2}^{-1}, as seen by the negatively charged electrons in the first conductor is:

$$n_{-2}^{-1} = n_{+} / \sqrt{1 - \left[k \cdot (I_2 \mp I_1)/c\right]^2} \tag{15.114}$$

The negatively charged electrons of the first conductor are subjected to an attractive force due to the positively charged atoms and a repulsive force due to the negatively charged electrons; both of the *second conductor.* These charged particles of the second conductor can be represented as line charge distributions as perceived by the negatively charged electrons of the first conductor. Equivalent line charge densities ρ_{L2+}^{-1} and ρ_{L2-}^{-1} are:

$$\rho_{L2+}^{-1} = e \cdot n_{+2}^{-1} = e \cdot n_{+} / \sqrt{1 - (k \cdot I_1/c)^2} \tag{15.115a}$$

and,

$$\rho_{L2-}^{-1} = -e \cdot n_{-2}^{-1} = -e \cdot n_{+} / \sqrt{1 - \left[k \cdot (I_2 \mp I_1)/c\right]^2} \tag{15.115b}$$

Therefore, the net line charge density, ρ_{L2}^{-1} as perceived by negatively charged electrons of the first conductor is given as:

$$\rho_{L2}^{-1} = \rho_{L2+}^{-1} + \rho_{L2-}^{-1} = e \cdot n_{+} \cdot \left[\frac{1}{\sqrt{1 - (k \cdot I_1/c)^2}} - \frac{1}{\sqrt{1 - \left[k \cdot (I_2 \mp I_1)/c\right]^2}}\right] \tag{15.116a}$$

or,

$$\rho_{L+2}^{-1} \cong e \cdot n_{+} \cdot \frac{1}{2} \cdot \left(\frac{k}{c}\right)^2 \cdot \left\{(I_1)^2 - (I_2 \mp I_1)^2\right\} \tag{15.116b}$$

Therefore, the force on each negatively charged electron of the first conductor will be:

$$\boldsymbol{F}_{-} = -e \cdot \frac{\rho_{L+2}^{-1}}{2\pi R} \boldsymbol{a}_{\rho} = -\left[e^2 \cdot n_{+} \cdot \frac{1}{2} \cdot \left(\frac{k}{c}\right)^2\right] \cdot \frac{\left\{(I_1)^2 - (I_2 \mp I_1)^2\right\}}{2\pi R} \boldsymbol{a}_{\rho} \tag{15.117}$$

Thus, force per unit length perceived by the negatively charged electrons of the first conductor $\mathcal{F}_{L1-}$ is:

$$\mathcal{F}_{L1-} = n_+ \cdot \boldsymbol{F}_- = -K \cdot \frac{\left\{(I_1)^2 - (I_2 \mp I_1)^2\right\}}{4\pi R} \boldsymbol{a}_\rho \quad (15.118)$$

where the constant K is defined by Equation 15.94a.

Thus, force per unit length $\mathcal{F}_{L1}$ perceived by the first conductor:

$$\mathcal{F}_{L1} = \mathcal{F}_{L1+} + \mathcal{F}_{L1-} = -K.\frac{(I_2)^2}{2\pi R}\boldsymbol{a}_\rho - K \cdot \frac{\left\{(I_1)^2 - (I_2 \mp I_1)^2\right\}}{2\pi R}\boldsymbol{a}_\rho \quad (15.119)$$

or,

$$\mathcal{F}_{L1} = -\frac{K}{4\pi R} \cdot \left[(I_2)^2 + (I_1)^2 - (I_2 \mp I_1)^2\right]\boldsymbol{a}_\rho = \mp \frac{K}{2\pi R} \cdot (I_2 \cdot I_1)\boldsymbol{a}_\rho \quad (15.120)$$

On comparing Equation 15.120 with Equation 12.14b, we get:

$$K = \mu_o \quad (15.121a)$$

Therefore, from Equation 15.105a, we have:

Thus,

$$k \cdot n_+ = \frac{1}{e \cdot \sqrt{\varepsilon_o}} = \frac{c}{e} \cdot \sqrt{\mu_o} \quad (15.121b)$$

Therefore, Equation 15.120 can be rewritten as follows: Similarly:

$$\mathcal{F}_{L1} = \mp \frac{\mu_o}{2\pi R} \cdot (I_2 \cdot I_1)\boldsymbol{a}_\rho \quad (15.122a)$$

Also,

$$\mathcal{F}_{L2} = \pm \frac{\mu_o}{2\pi R} \cdot (I_2 \cdot I_1)\boldsymbol{a}_\rho \quad (15.122b)$$

Two conclusions can be made: (1) Equation 15.121b shows that the quantity $(k \cdot n_+)$ is a universal constant, and (2) in view of Equations 12.14b and 15.122a, a magnetostatic field may be identified as the relativistic effect of an electrostatic field.

15.10.7 Relativistic Effects on Time-Independent Fields

Maxwell's equations show that electrostatic and magnetostatic fields are not interdependent. As we saw earlier, this notion of ours has undergone modifications with the advent of the Special Theory of Relativity. Now these two fields appear to be related to each other. Even though under certain situations the link that joins the two could be weak. In the following subsections, we shall consider two cases showing the relationship between the energy densities in the electrostatic and the magnetostatic fields due to dc currents flowing in conductors.

15.10.7.1 Energy Density of Line Currents

Consider a long thin straight conducting wire placed along the z-axis. It is electrically neutral when it carries no current. Let a line charge density ρ_L^+ along the wire be defined so that:

$$\rho_L^+ \overset{\text{def}}{=} N \cdot e \tag{15.123}$$

where e is the charge of an atom deficient of an electron, and N is the number of such atoms per unit length of the wire.

With zero wire current, the cloud of free electrons has zero drift velocity. The wire being electrically neutral, same line charge density with negative sign can be taken to represent the distribution of free electrons. Now, if a dc current I flows in the conducting wire, the electron cloud drifts with a velocity u (= u_z). The modified line charge density for the free electrons can be given as follows:

$$\rho_L^- = -\frac{\rho_L^+}{\sqrt{1-\beta^2}} \tag{15.124}$$

where

$$\beta \overset{\text{def}}{=} u / c \tag{15.124a}$$

the symbol c indicates the velocity of light in free space.

Therefore the net line charge density ρ_L is given as:

$$\rho_L = \rho_L^+ + \rho_L^- = -\rho_L^+ \cdot \left(\frac{1}{\sqrt{1-\beta^2}} - 1 \right) \tag{15.125}$$

Since:

$$I = -\rho_L^- \cdot u = \frac{\rho_L^+ \cdot u}{\sqrt{1-\beta^2}} \tag{15.126a}$$

therefore, at a distance R from the line charge, the electric field intensity is given as:

$$E_R = \frac{\rho_L}{2\pi R \cdot \varepsilon_o} = -\frac{\rho_L^+}{2\pi R \cdot \varepsilon_o} \cdot \left(\frac{1}{\sqrt{1-\beta^2}} - 1 \right) = -\frac{\sqrt{1-\beta^2}}{2\pi R \cdot \varepsilon_o} \cdot \left(\frac{1}{\sqrt{1-\beta^2}} - 1 \right) \cdot \frac{I}{u} \tag{14.126b}$$

Ampère's circuital law gives the magnetic field intensity H_φ:

$$I = 2\pi R \cdot H_\varphi \tag{15.127}$$

Thus,

$$E_R = -\frac{\sqrt{1-\beta^2}}{2\pi R.\varepsilon_o}\cdot\left(\frac{1}{\sqrt{1-\beta^2}}-1\right)\cdot\frac{2\pi R\cdot H_\varphi}{u} = -\frac{\sqrt{1-\beta^2}}{\mu_o\cdot\varepsilon_o}\cdot\left(\frac{1}{\sqrt{1-\beta^2}}-1\right)\cdot\frac{B_\varphi}{u} \tag{15.128}$$

Therefore, from Equation 15.78:

$$D_R = E_R\cdot\varepsilon_o = -\sqrt{1-\beta^2}\cdot\left(\frac{1}{\sqrt{1-\beta^2}}-1\right)\cdot\frac{H_\varphi}{u} \tag{15.129}$$

Thus, from Equations 15.78 and 15.79, we get:

$$\left[\frac{1}{2}\cdot E_R\cdot D_R\right] = \left\{\frac{\left(1-\beta^2\right)}{\beta^2}\cdot\left(\frac{1}{\sqrt{1-\beta^2}}-1\right)^2\right\}\cdot\left[\frac{1}{2}\cdot H_\varphi\cdot B_\varphi\right] \tag{15.130}$$

Therefore:

$$\boxed{\frac{\mathcal{E}_{\text{elect.}}}{\mathcal{E}_{\text{magt.}}} = \left(\frac{1-\sqrt{1-\beta^2}}{\beta}\right)^2} = \begin{cases} 0 & \text{for} \quad \beta = 0 \\ 1 & \text{for} \quad \beta = 1 \end{cases} \tag{15.131}$$

This equation shows that the energy density in the electrostatic field produced due to the relativistic effects is a fraction of the energy density in the magnetostatic field produced by dc current in a conducting wire.

15.10.7.2 Energy Density of Surface Currents

Imagine a large thin conducting sheet placed on the x–y-plane. For each unit length in the x- and y-directions, let there be N number of atoms, each deficient of one electron. Thus, the surface charge density ρ_s^+ due to the ionized atoms is:

$$\rho_s^+ = N\cdot e \tag{15.132a}$$

where e indicates the charge of a proton.

The conducting sheet will be electrically neutral when no current flows in it. Thus the same number of free electrons must be present for each unit length along the x- and y-directions. Therefore, the surface charge density ρ_s^- due to the free electrons is the negative of ρ_s^+. When a dc current flows in the conducting sheet, the cloud of free electrons drifting, say in the y-direction attain an average velocity, say u (= u_y). The modified surface charge density due to the free electrons is given as:

$$\rho_s^- = -\frac{\rho_s^+}{\sqrt{1-\beta^2}} \tag{15.132b}$$

Now, let the uniform surface current of density K_y is given as:

$$K_y \overset{\text{def}}{=} -\rho_s^- \cdot u = \frac{\rho_s^+}{\sqrt{1-\beta^2}} \cdot u \tag{15.133}$$

The net surface charge density ρ_s, will be:

$$\rho_s = \rho_s^+ + \rho_s^- = \rho_s^+ - \frac{\rho_s^+}{\sqrt{1-\beta^2}} = -\rho_s^+ \cdot \left(\frac{1}{\sqrt{1-\beta^2}} - 1 \right) \tag{15.134a}$$

or,

$$\rho_s = -\frac{\sqrt{1-\beta^2}}{u} \cdot \left(\frac{1}{\sqrt{1-\beta^2}} - 1 \right) \cdot K_y \tag{15.134b}$$

Now, for $z > 0$, we have:

$$H_x = \frac{1}{2} \cdot K_y \tag{15.135}$$

Therefore:

$$\rho_s = -\frac{2}{u} \cdot \sqrt{1-\beta^2} \cdot \left(\frac{1}{\sqrt{1-\beta^2}} - 1 \right) \cdot H_x \tag{15.136}$$

Furthermore, for $z > 0$, we have:

$$D_z = \frac{1}{2} \cdot \rho_s \tag{15.137}$$

Therefore:

$$D_z = -\frac{1}{u} \cdot \sqrt{1-\beta^2} \cdot \left(\frac{1}{\sqrt{1-\beta^2}} - 1 \right) \cdot H_x = -\frac{1}{u} \cdot \left(1 - \sqrt{1-\beta^2} \right) \cdot H_x \tag{15.138}$$

and

$$E_z = -\frac{1}{u \cdot \varepsilon_o \cdot \mu_o} \left(1 - \sqrt{1-\beta^2} \right) \cdot \mu_o \cdot H_x = -\frac{1}{u \cdot \varepsilon_o \cdot \mu_o} \left(1 - \sqrt{1-\beta^2} \right) \cdot B_x \tag{15.139}$$

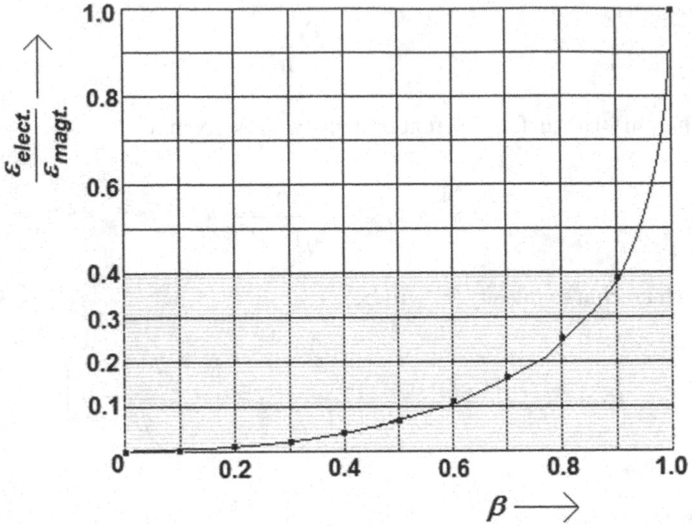

FIGURE15.6 Variation of the ratio of two energy densities with per unit velocity β.

Thus, from Equations 15.87 and 15.88, we get:

$$\left[\frac{1}{2}\cdot E_z \cdot D_z\right] = \frac{\left(1-\sqrt{1-\beta^2}\right)^2}{\beta^2}\cdot\left[\frac{1}{2}\cdot H_x \cdot B_x\right] \tag{15.140a}$$

or

$$\boxed{\frac{\mathcal{E}_{\text{elect.}}}{\mathcal{E}_{\text{magt.}}} = \left(\frac{1-\sqrt{1-\beta^2}}{\beta}\right)^2} = \begin{cases} 0 & \text{for } \beta = 0 \\ 1 & \text{for } \beta = 1 \end{cases} \tag{15.140b}$$

Therefore, the energy density in the electrostatic field produced due to the relativistic effects is a fraction of the energy density in the magnetostatic field produced by uniform dc surface current in a conducting surface.

Figure 15.6 shows the variation of the ratio of the two energy densities plotted against the per unit velocity $\beta\left(=u/c\right)$.

It is interesting to note that the relation between the energy densities in the two cases, in view of Equations 15.131 and 15.140b, is identical. This relation in its modified form is of importance in the case of convection currents caused by high speed charged particles found in particle accelerates, and those found in the magneto hydrodynamic (MHD) generators.

Example 15.1

Figure 15.7 shows a configuration wherein $d = 0.15$ m and $\boldsymbol{B} = 0.2\boldsymbol{a}_z$. If $\boldsymbol{\upsilon} = 150y\boldsymbol{a}_y$ m/s and $y = 0.1$ m at $t = 0$, find at $t = 10$ ms (a) υ_y; (b) V_{12}; (c) the current entering terminal 2 of the voltmeter if the resistance (R) of the voltmeter is 25 $K\Omega$.

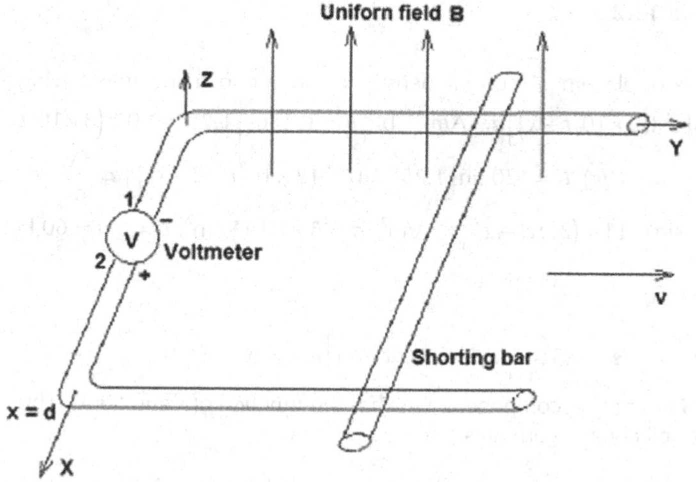

FIGURE 15.7 A U-shaped fixed bar with a sliding bar.

SOLUTION

$d = 0.15$ m, $\boldsymbol{B} = 0.2\mathbf{a}_z$, $\boldsymbol{\upsilon} = 150y\mathbf{a}_y$ m/s, and $y = 0.1$ m at $t = 0$.

(a) Given $\upsilon = 150y\mathbf{a}_y = \upsilon_y\mathbf{a}_y$, $\upsilon_y = \dfrac{dy}{dt} = 150y \quad \dfrac{dy}{y} = 150dt$

$$\int \frac{dy}{y} = \int 150dt \quad \log y + k = 150t$$

Let the constant of integration be $k = \log A$.

Thus: $\log y + \log A = \log Ay = 150t$

$$Ay = e^{150t} \quad y = \frac{e^{150t}}{A}$$

At $t = 0$ $y = 0.1m$ $\quad 0.1 = \dfrac{1}{A}$ or $A = 10$ or $y = \dfrac{e^{150t}}{10}$

At $t = 10ms$ $\quad y = \dfrac{e^{150t}}{10} = \dfrac{e^{1.50}}{10} = 0.448$

$$\upsilon = 150y\mathbf{a}_y = 150 \times 0.448\mathbf{a}_y = 67.2\mathbf{a}_y \text{ m/s}$$

(b) In view of Equation 15.9:

$$V_{12} = -B \cdot d \cdot \upsilon = -0.2 \times 0.15 \times 67.2 = -2.016 \text{ Volts}$$

(c) $I_2 = \dfrac{V_{12}}{R} = \dfrac{2.016}{25 \times 10^3} = 80.64\ \mu\Omega$

Example 15.2

Find the displacement current density $\boldsymbol{D}$ nearer to a location having (a) $\boldsymbol{H} = 0.5\cos\left[2.1\left(3\times10^8 t - x\right)\right]\boldsymbol{a}_z$ A/m; (b) $\boldsymbol{B} = 1.5\cos\left[1.257\times10^{-6}\left(3\times10^8 t - y\right)\right]\boldsymbol{a}_x$ Wb/m^2; $\mu_r = 1$ (c) $\boldsymbol{E} = 120\sin\left[1.257\times10^{-6}\left(3\times10^8 t - 2.45z\right)\right]\boldsymbol{a}_x \dfrac{\text{KV}}{\text{m}}$, $\varepsilon_r = 6$; (d) $\boldsymbol{J} = 10^6\sin\left[111.7(3.22t - z)\right]\boldsymbol{a}_x$ A/m^2, $\sigma = 3\times10^7 \mho/\text{m}^2$, $\varepsilon_r = 1$, $f = 60$ Hz.

SOLUTION

(a) Given $\boldsymbol{H} = 0.5\cos\left[2.1\left(3\times10^8 t - x\right)\right]\boldsymbol{a}_z$ A/m.

$\boldsymbol{H}$ has only $\boldsymbol{a}_z$ component, which is the function of x and t only. Thus, the displacement current is:

$$\frac{\partial \boldsymbol{D}}{\partial t} = \nabla\times\boldsymbol{H} = -\frac{\partial H_z}{\partial x}\boldsymbol{a}_y$$

$$= -\frac{\partial\left[0.5cos\left\{2.1\left(3\times10^8 t - x\right)\right\}\right]}{\partial x}\boldsymbol{a}_y = -1.05\cos\left\{2.1\left(3\times10^8 t - x\right)\right\}\boldsymbol{a}_y \text{ A/m}^2$$

(b) Given $\boldsymbol{B} = 1.5\cos\left[1.257\times10^{-6}\left(3\times10^8 t - y\right)\right]\boldsymbol{a}_x$ Wb/m^2 and $\mu_r = 1$.

$\boldsymbol{B}$ has only $\boldsymbol{a}_x$ component, which is a function of y and t only. Also $\boldsymbol{H} = \dfrac{\boldsymbol{B}}{\mu_0}$. Thus, the displacement current is:

$$\frac{\partial \boldsymbol{D}}{\partial t} = \nabla\times\boldsymbol{H} = -\frac{\partial H_x}{\partial y}\boldsymbol{a}_z = -\frac{\partial \dfrac{1.5\cos\left[1.257\times10^{-6}\left(3\times10^8 t - y\right)\right]}{\mu_0}}{\partial y}\boldsymbol{a}_z$$

$$= -\frac{1.5\times1.257\times10^{-6}\sin\left[1.257\times10^{-6}\left(3\times10^8 t - y\right)\right]}{4\pi\times10^{-7}}\boldsymbol{a}_z$$

$$= -1.5\sin\left[1.257\times10^{-6}\left(3\times10^8 t - y\right)\right]\boldsymbol{a}_z \text{ A/m}^2$$

(c) Given $\boldsymbol{E} = 120\sin\left[1.257\times10^{-6}\left(3\times10^8 t - 2.45z\right)\right]\boldsymbol{a}_x \dfrac{\text{KV}}{\text{m}}$ and $\varepsilon_r = 6$:

$$\boldsymbol{D} = \varepsilon\boldsymbol{E} = \varepsilon_r\varepsilon_0\boldsymbol{E} = 6\times\frac{10^{-9}}{36\pi}\boldsymbol{E} = \frac{10^{-9}}{6\pi}\boldsymbol{E}$$

The displacement current is:

$$\frac{\partial \boldsymbol{D}}{\partial t} = \varepsilon\frac{\partial \boldsymbol{E}}{\partial t} = \frac{10^{-9}}{6\pi}\frac{\partial\left[120\sin\left\{1.257\times10^{-6}\left(3\times10^8 t - 2.45z\right)\right\}\right]}{\partial t}\boldsymbol{a}_x$$

$$= \frac{10^{-9}}{6\pi}\left[120 \times 1.257 \times 10^{-6} \times 3 \times 10^{8} \cos\left\{1.257 \times 10^{-6}\left(3 \times 10^{8} t - 2.45z\right)\right\}\right] \boldsymbol{a_x}$$

$$= 2.4\left[\cos\left\{1.257 \times 10^{-6}\left(3 \times 10^{8} t - 2.45z\right)\right\}\right] \boldsymbol{a_x}\ \mu\text{A/m}^2$$

(d) Given $\boldsymbol{J} = 10^{6} \sin\left[111.7(3.22t - z)\right] \boldsymbol{a_x} \frac{\text{A}}{\text{m}^2}$, $\sigma = 3 \times 10^{7}$ ℧/m², $\varepsilon_r = 1$, $f =$ 60 Hz.

$$\boldsymbol{J} = \sigma \boldsymbol{E} \quad \boldsymbol{E} = \frac{\boldsymbol{J}}{\sigma} \quad \boldsymbol{D} = \varepsilon \boldsymbol{E} = \varepsilon_0 \varepsilon_r \boldsymbol{E} = \varepsilon_0 \boldsymbol{E} = \frac{\varepsilon_0}{\sigma} \boldsymbol{J}$$

$$\frac{\partial \boldsymbol{D}}{\partial t} = \frac{\varepsilon_0}{\sigma} \frac{\partial \boldsymbol{J}}{\partial t} \quad \frac{\varepsilon_0}{\sigma} = \frac{10^{-9}}{36\pi} \times \frac{1}{3 \times 10^{7}} = \frac{10^{-16}}{108\pi}$$

The displacement current is:

$$\boldsymbol{J_d} = \frac{\varepsilon_0}{\sigma} \frac{\partial \boldsymbol{J}}{\partial t} = \frac{10^{-16}}{108\pi} \frac{\partial\left[10^{6} \sin\left\{111.7(3.22t - z)\right\}\right]}{\partial t} \boldsymbol{a_x}$$

$$= \frac{10^{-16}}{108\pi} \times 10^{6} \times 111.7 \times 3.22 \times \cos\left\{111.7\left(3.22t - z\right)\right\} \boldsymbol{a_x}$$

$$= 106 \cos\left\{111.7\left(3.22t - z\right)\right\} \boldsymbol{a_x}\ \text{mmA/m}^2$$

Example 15.3

Determine whether or not the given pair of fields satisfy Maxwell's equations if $\sigma = 0$, $\mu_r = 10$, $\varepsilon_r = 1$, and (a) $\boldsymbol{E} = 2y\boldsymbol{a_y}$, $\boldsymbol{H} = 5x\boldsymbol{a_x}$; (b) $\boldsymbol{E} = 100 \sin(6 \times 10^{7} \boldsymbol{t}) \sin z\, \boldsymbol{a_y}$ $\boldsymbol{H} = -0.1328 \cos(6 \times 10^{7} t) \cos z\, \boldsymbol{a_x}$; (c) $\boldsymbol{D} = (z + 6 \times 10^{7} t)\boldsymbol{a_x}$, $\boldsymbol{B} = (-754z + 4.52 \times 10^{10} t)\boldsymbol{a_y}$.

SOLUTION

Given $\sigma = 0$, $\mu_r = 10$, $\varepsilon_r = 2.5$.

(a) Given $\boldsymbol{E} = 2y\boldsymbol{a_y}$ and $\boldsymbol{H} = 5x\boldsymbol{a_x}$
The two Maxwell's equations involving curl of ***E*** or ***H*** are:

$$\nabla \times \boldsymbol{H} = J + \frac{\partial \boldsymbol{D}}{\partial t} = \frac{\partial \boldsymbol{D}}{\partial t} = \varepsilon \frac{\partial \boldsymbol{E}}{\partial t} \quad (\text{for } \sigma = 0, J = \sigma, E = 0)$$

$$\nabla \times \boldsymbol{E} = -\frac{\partial \boldsymbol{B}}{\partial t} = -\mu \frac{\partial \boldsymbol{H}}{\partial t}$$

The given values of ***E*** and ***H*** are not functions of time; thus, both of these equations become $\nabla \times \boldsymbol{H} = 0$ and $\nabla \times \boldsymbol{E} = 0$.

Since $\boldsymbol{H} = H_x\boldsymbol{a}_x$, the components of curl involving H_x are $\left[\dfrac{\partial H_x}{\partial z}\right]\boldsymbol{a}_y$ and $-\left[\dfrac{\partial H_x}{\partial y}\right]\boldsymbol{a}_z$. Since H_x is a function of x alone, both the derivatives involved are zero. Similarly, $\boldsymbol{E} = E_y\boldsymbol{a}_y$ the components of curl involving E_y are $-\left[\dfrac{\partial E_y}{\partial z}\right]\boldsymbol{a}_x$ and $\left[\dfrac{\partial E_y}{\partial x}\right]\boldsymbol{a}_z$. Since E_y is a function of y alone, both these derivatives are zero. Thus, the curls of both $\boldsymbol{E}$ and $\boldsymbol{H}$ are identically zero. Therefore, this pair does not satisfy Maxwell's equations.

(b) Given $\boldsymbol{E} = 100\sin(6\times10^7 t)\sin z\,\boldsymbol{a}_y = E_y\boldsymbol{a}_y$

and $\boldsymbol{H} = -0.1328\cos(6\times10^7 t)\cos z\,\boldsymbol{a}_x = H_x\boldsymbol{a}_x$

both $\boldsymbol{E}$ and $\boldsymbol{H}$ are functions of z and t only. Furthermore, $\boldsymbol{E}$ only has a y component, whereas $\boldsymbol{H}$ has an x component.

Maxwell's equation: $\nabla\times\boldsymbol{E} = -\dfrac{\partial\boldsymbol{B}}{\partial t}$

$$\text{RHS} = -\mu\frac{\partial\boldsymbol{H}}{\partial t} = -\mu\frac{\partial\left\{-0.1328\cos\left(6\times10^7 t\right)\cos z\right\}}{\partial t}\boldsymbol{a}_x$$

$$= -0.1328\times6\times10^7\,\mu_0\mu_r\left\{\sin\left(6\times10^7 t\right)\cos z\right\}\boldsymbol{a}_x$$

$$= -0.1328\times6\times10^7\times4\pi\times10^{-7}\times10\times\left\{\sin\left(6\times10^7 t\right)\cos z\right\}\boldsymbol{a}_x$$

$$= -100\sin\left(6\times10^7 t\right)\cos z\boldsymbol{a}_x$$

$$\text{LHS} = \nabla\times\boldsymbol{E} = -\frac{\partial E_y}{\partial z}\boldsymbol{a}_x = -\frac{\partial\left\{100\sin\left(6\times10^7 \boldsymbol{t}\right)\sin z\right\}}{\partial z}\boldsymbol{a}_x = -100\sin\left(6\times10^7 \boldsymbol{t}\right)\cos z\boldsymbol{a}_x$$

The given pair satisfies this Maxwell's equation as RHS = LHS.

Maxwell's equation: $\nabla\times\boldsymbol{H} = \varepsilon\dfrac{\partial\boldsymbol{E}}{\partial t}$

$$\text{RHS} = \varepsilon\frac{\partial\boldsymbol{E}}{\partial t} = \varepsilon\frac{\partial\left\{100\sin\left(6\times10^7 t\right)\sin z\right\}}{\partial t}\boldsymbol{a}_y$$

$$= \varepsilon_r\varepsilon_0\times100\times6\times10^7\times\cos\left(6\times10^7 t\right)\sin z\,\boldsymbol{a}_y$$

$$= 2.5\times\frac{10^{-9}}{36\pi}\times100\times6\times10^7\times\cos\left(6\times10^7 t\right)\sin z\,\boldsymbol{a}_y$$

$$= 0.1326\cos\left(6\times10^7 t\right)\sin z\,\boldsymbol{a}_y$$

$$\text{LHS} = \frac{\partial H_x}{\partial z}\boldsymbol{a}_y = \frac{\partial\left(-0.1328\cos\left(6\times10^7 t\right)\cos z\right)}{\partial z}\boldsymbol{a}_y$$

$$= 0.1328\cos\left(6\times10^7 t\right)\sin z\,\boldsymbol{a}_y$$

The given pair satisfies this Maxwell's equation as RHS = LHS.

(c) Given $\boldsymbol{D} = (z + 6\times10^7 t)\boldsymbol{a}_x = D_x\boldsymbol{a}_x$ and $\boldsymbol{B} = (-754z - 4.52\times10^{10} t)\boldsymbol{a}_y = B_y\boldsymbol{a}_y$ Both $\boldsymbol{D}$ and $\boldsymbol{B}$ are functions of z and t only. Furthermore, $\boldsymbol{D}$ only has an x component whereas $\boldsymbol{B}$ only has a y component.

$$\boldsymbol{E} = \frac{\boldsymbol{D}}{\varepsilon} = \frac{\boldsymbol{D}}{\varepsilon_0\varepsilon_r} = \frac{\boldsymbol{D}}{2.5\varepsilon_0} = \frac{\left(z + 6\times10^7 t\right)}{2.5\varepsilon_0}\boldsymbol{a}_x$$

$$= \frac{\left(z + 6\times10^7 t\right)}{2.5\times8.854\times10^{-12}}\boldsymbol{a}_x = 45.2\times10^9\left(z + 6\times10^7 t\right)\boldsymbol{a}_x$$

Maxwell's equation: $\nabla\times\boldsymbol{E} = -\dfrac{\partial\boldsymbol{B}}{\partial t}$

$$\text{LHS} = \nabla\times\boldsymbol{E} = \frac{\partial E_x}{\partial z}\boldsymbol{a}_y = \frac{\partial\left\{45.2\times10^9\left(z + 6\times10^7 t\right)\right\}}{\partial z}\boldsymbol{a}_y = 45.2\times10^9\boldsymbol{a}_y$$

$$\text{RHS} = -\frac{\partial\boldsymbol{B}}{\partial t} = -\frac{\partial(-754z - 4.52\times10^{10} t)\boldsymbol{a}_y}{\partial t} = 45.2\times10^9\boldsymbol{a}_y$$

The given pair satisfies this Maxwell's equation as RHS = LHS.

Maxwell's equation: $\nabla\times\boldsymbol{H} = \dfrac{\partial\boldsymbol{D}}{\partial t}$

$$\boldsymbol{H} = \frac{\boldsymbol{B}}{\mu} = \frac{\boldsymbol{B}}{\mu_0\mu_r} = \frac{\boldsymbol{B}}{10\mu_0} = \frac{\left(-754z - 4.52\times10^{10}\boldsymbol{t}\right)}{10\times4.7\times10^{-7}}\boldsymbol{a}_y$$

$$= \left(-160.42\times10^6\times z - 0.9617\times10^{16}\boldsymbol{t}\right)\boldsymbol{a}_y$$

$$\text{LHS} = \nabla\times\boldsymbol{H} = -\frac{\partial H_y}{\partial z}\boldsymbol{a}_y = -\frac{\partial\left(-160.42\times10^6\times z - 0.9617\times10^{16}\boldsymbol{t}\right)}{\partial z}\boldsymbol{a}_y$$

$$= 160.42\times10^6\boldsymbol{a}_y$$

$$\text{RHS} = \frac{\partial\boldsymbol{D}}{\partial t} = \frac{\partial\boldsymbol{D}_x}{\partial t} = \frac{\partial\left(z + 6\times10^7 t\right)}{\partial t}\boldsymbol{a}_x = 6\times10^7\,\boldsymbol{a}_x$$

The given pair does not satisfy this Maxwell's equation as RHS ≠ LHS.

Example 15.4

The magnetic flux density in cylindrical coordinates for $\rho < 0.2$ m is given as $B_z = \dfrac{1}{1+100\rho^2}\sin(1000\pi t)\,\text{mWb/m}^2$. Find (a) the magnetic flux passing through the surface $\rho \le 0.1$ m, $z = 0$, in the a_z direction; (b) E at $\rho = 0.1$, $\varphi = \dfrac{\pi}{4}$, $z = 0$; (c) the current in the conducting wire of 1 Ω/cm resistance in a_φ direction, if it forms a circular path $\rho = 0.1$, $z = 0$.

SOLUTION

(a) $\boldsymbol{B} = B_z\boldsymbol{a}_z$ $B_z = \dfrac{1}{1+100\rho^2}\sin(1000\pi t)$ $\rho < 0.2$ m $\boldsymbol{ds} = \rho d\rho d\varphi \boldsymbol{a}_z$

$$\Phi = \int_s \boldsymbol{B}\bullet\boldsymbol{ds} = \int_{\varphi=0}^{2\pi}\int_{\rho=0}^{0.1} B_z\boldsymbol{a}_z \bullet \rho d\rho d\varphi \boldsymbol{a}_z = \int_{\varphi=0}^{2\pi}\int_{\rho=0}^{0.1} \frac{\rho d\rho}{1+100\rho^2}\sin(1000\pi t)d\varphi$$

$$\Phi = 2\pi\cdot\sin(1000\pi t)\int_{\rho=0}^{0.1}\frac{\rho d\rho}{1+100\rho^2}$$

Put $1 + 100\rho^2 = p$ $200\rho d^2 = dp$ $\rho d\rho = \dfrac{dp}{200}$ when $\rho = 0$, $p = 1$, and when $\rho = 0.1$, $p = 2$.

$$\Phi = 2\pi\cdot\sin(1000\pi t)\int_{p=1}^{2}\frac{dp}{200p}$$

$$= \frac{2\pi}{200}\cdot\sin(1000\pi t)\cdot\ln p\big|_1^2$$

$$= 0.01\pi\cdot\sin(1000\pi t)\cdot\ln 2$$

$$\Phi = 21.776\cdot\sin(1000\pi t)\,\mu\text{Wb}$$

(b) $$\nabla\times\boldsymbol{E} = -\frac{\partial \boldsymbol{B}}{\partial t} = -\frac{\partial\left[\dfrac{1}{1+100\rho^2}\sin(1000\pi t)\boldsymbol{a}_z\right]}{\partial t} = -\frac{1000\pi}{1+100\rho^2}\cos(1000\pi t)\boldsymbol{a}_z$$

To match both sides of the above relation, $\nabla\times\boldsymbol{E}$ will have only the $\boldsymbol{a}_z$ component, which is:

$$(\nabla\times\boldsymbol{E})\big|_{z\text{-comp}} = \left[\left(\frac{1}{\rho}\right)\cdot\left\{\frac{\partial(\rho E_\varphi)}{\partial\rho} - \frac{\partial E_\rho}{\partial\varphi}\right\}\right]$$

Furthermore, since B (or H) is a function of ρ and t, $\boldsymbol{E}$ will also be a function of ρ and t or will have a E_φ component. Thus:

$$\frac{1}{\rho}\frac{\partial(\rho E_\varphi)}{\partial\rho} = -\frac{1000\pi}{1+100\rho^2}\cos(1000\pi t) \quad \frac{\partial(\rho E_\varphi)}{\partial\rho} = -\frac{1000\pi\rho}{1+100\rho^2}\cos(1000\pi t)$$

$$\rho E_\varphi = -1000\pi\cdot\cos(1000\pi t)\int_{\rho=0}^{0.1}\frac{\rho d\rho}{1+100\rho^2}$$

In view of the earlier substitution:

$$\rho E_\varphi = -1000\pi\cdot\cos(1000\pi t)\int_{p=1}^{2}\frac{dp}{200p} = -\frac{1000\pi}{200}\cdot\cos(1000\pi t)\ln p\Big|_{p=1}^{p=2}$$

$$= -5\pi\cos(1000\pi t)\ln 2$$

$$E_\varphi = -\frac{5\pi}{\rho}\cos(1000\pi t)\ln 2 = -\frac{5\pi}{0.1}\cos(1000\pi t)\cdot\ln 2 = -108.879\cos(1000\pi t)$$

$$\boldsymbol{E} = E_\varphi \boldsymbol{a}_\varphi = -108.879\cos(1000\pi t)\boldsymbol{a}_\varphi$$

(c) From part (a): $\Phi = 21.776\cdot\sin(1000\pi t)\,\mu\text{Wb}$

$$\text{emf} = -\frac{\partial\Phi}{\partial t} = -\frac{\partial\{21.776\cdot\sin(1000\pi t)\}}{\partial t} = 21.776\cdot 1000\pi\cdot\cos(1000\pi t)\,\mu\text{V}$$

Total resistance in the path = length of the conducting wire or the circumference of the circle × 1 $\frac{\Omega}{\text{cm}}$

$$\text{Length of the conducting wire} = 2\pi\rho = 2\pi\times 0.1\,\text{m} = 2\pi\times 10\text{ cm}$$

$$\text{Total resistance} = 20\pi\times 1 = 20\pi\Omega$$

$$\frac{\text{emf}}{R} = \frac{21.776\cdot 1000\pi\cdot\cos(1000\pi t)}{20\pi} = 1088.8\cos(1000\pi t)\,\mu\text{A}$$

Example 15.5

A unit vector $\boldsymbol{a}_N = 0.48\boldsymbol{a}_x - 0.6\boldsymbol{a}_y + 0.64\boldsymbol{a}_z$ is directed from region 2 ($\sigma_2 = 0$, $\varepsilon_{r2} = 2.5$, $\mu_{r2} = 2$) toward region 1 ($\sigma_1 = 0$, $\varepsilon_{r1} = 5$, $\mu_{r1} = 10$). The boundary surface carries no charge density. If $\boldsymbol{E} = (-10\boldsymbol{a}_x - 5\boldsymbol{a}_y + 20\boldsymbol{a}_z)\sin 400t$ at point P in region 1 adjacent to the boundary, find the magnitude of (a) E_{t1}; (b) E_{n2}; (c) E_2.

SOLUTION

$$\boldsymbol{a}_N = 0.48\boldsymbol{a}_x - 0.6\boldsymbol{a}_y + 0.64\boldsymbol{a}_z$$

$$\boldsymbol{E}_1 = \left(-10\boldsymbol{a}_x - 5\boldsymbol{a}_y + 20\boldsymbol{a}_z\right)\cdot\sin 400t \quad \left|\boldsymbol{E}_1\right| = \sqrt{10^2 + 5^2 + 20^2} = 22.91$$

$$\boldsymbol{E}_1 = 22.91\left(-0.436\boldsymbol{a}_x - 0.218\boldsymbol{a}_y + 0.872\boldsymbol{a}_z\right)\cdot\sin 400t = 22.91\sin 400t\,\boldsymbol{a}_E$$

where $\boldsymbol{a}_{E_1} = -0.436\boldsymbol{a}_x - 0.218\boldsymbol{a}_y + 0.872\boldsymbol{a}_z$

$$\boldsymbol{a}_N \bullet \boldsymbol{a}_{E_1} = \left(0.48\boldsymbol{a}_x - 0.6\boldsymbol{a}_y + 0.64\boldsymbol{a}_z\right)\bullet\left(-0.436\boldsymbol{a}_x - 0.218\boldsymbol{a}_y + 0.872\boldsymbol{a}_z\right) = 0.218 = \cos\theta$$

$$\theta = \cos^{-1}(0.218) = 77.4^\circ \quad \sin\theta = 0.976$$

(a) $\left|E_{t1}\right| = \left|E\right|\sin\theta = 22.91\times 0.976 = 22.359V/m = \left|E_{t2}\right|$

(b) $\left|E_{n1}\right| = \left|E\right|\cos\theta = 22.91\times 0.218 = 4.994\text{ V/m}$ 499.92

$$D_{n1} = D_{n2} \quad \varepsilon_1 E_{n1} = \varepsilon_2 E_{n2} \quad E_{n2} = \frac{\varepsilon_1}{\varepsilon_2}E_{n1} = \frac{5}{2.5}\times 4.994 = 9.988\text{ V/m}$$

$$\left|\boldsymbol{E}_2\right| = \sqrt{\left(E_{t2}\right)^2 + \left(E_{n2}\right)^2} = \sqrt{(22.359)^2 + (9.988)^2} = 24.488\text{ V/m}$$

Example 15.6

Figure 15.8 shows two perfectly conducting planes that are located at $y = 2$ m and $y = 2.1$ m and are separated by a material having $\sigma = 0$, $\varepsilon_r = 9$, $\mu_r = 1$. If $\boldsymbol{E} = 300\cos\left(10^8 t - z\right)\boldsymbol{a}_y$V/m between the planes, find (a) $|H|$ at (5, 2.06, 1.1) at $t =$ 2ns; (b) $|K|$ at (0.7, 2, 0) at $t = 0$.

SOLUTION

$$\nabla\times\boldsymbol{E} = -\frac{\partial\boldsymbol{B}}{\partial t} = -\mu_0\frac{\partial\boldsymbol{H}}{\partial t}$$

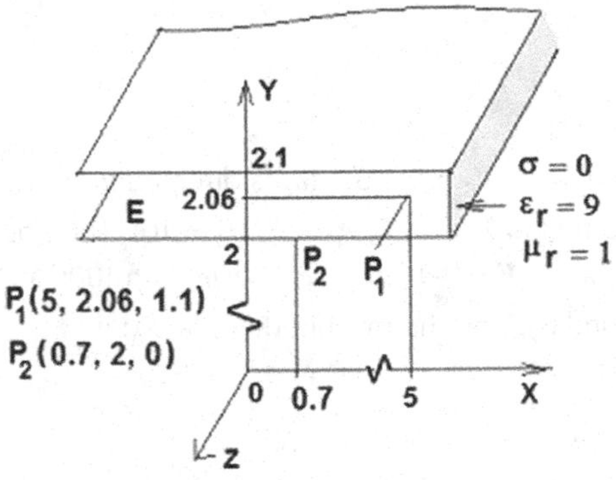

FIGURE 15.8 Location of two perfectly conducting planes.

Since $\boldsymbol{E}$ only has an $\boldsymbol{a}_y$ component, which is a function of z and t only, from Equation 4.34a we get:

$$\nabla\times\boldsymbol{E} = -\frac{\partial E_y}{\partial z}\boldsymbol{a}_x = -\frac{\partial\left(300\cos\left(10^8t-z\right)\right)}{\partial z}\boldsymbol{a}_x$$

$$= 300\sin\left(10^8t-z\right)\boldsymbol{a}_x$$

$$= -\frac{\partial\boldsymbol{B}}{\partial t}$$

$$\frac{\partial\boldsymbol{H}}{\partial t} = \frac{1}{\mu_0}\frac{\partial\boldsymbol{B}}{\partial t} \text{ or } \boldsymbol{H} = -\frac{1}{\mu_0}\int(\nabla\times\boldsymbol{E})\,dt = \frac{300}{\mu_0}\times10^{-8}\cos\left(10^8t-z\right)\boldsymbol{a}_x + C$$

(a) $$|\boldsymbol{H}|\Big|_{\substack{z=1.1\\t=2ns}} = \frac{300}{4\pi\times10^{-7}}\times10^{-8}\cos\left(10^8\times2\times10^{-9}\times-1.1\right)$$

$$= \frac{30}{4\pi}\times\cos(-0.9) = 1.484\text{ A/m}$$

(b) At (0.7, 2, 0) and $t = 0$: $|\boldsymbol{H}| = |\boldsymbol{H}_t| = |\boldsymbol{K}|$

$$|\boldsymbol{K}|\Big|_{\substack{z=0\\t=0}} = \frac{300}{4\pi\times10^{-7}}\times10^{-8}\cos(0-0) = \frac{30}{4\pi} = 2.387\text{ A/m}$$

Example 15.7

A point charge of $-4\cos10^8\pi t\ \mu C$ is located at (0, 0, –1.5), while $4\cos10^8\pi t\ \mu C$ is located at (0, 0, 1.5), both in free space. Find V at: (a) (0, 0, 2998.5) at $t = 0$; (b) (0, 0, 2999) at $t = 0$; and (c) a point on the x-axis that is 2998.5 m from each charge as a function of t.

SOLUTION

Given $Q_1 = -4\cos10^8\pi t\ \mu C$ and $Q_2 = 4\cos10^8\pi t\ \mu C$.

On replacing t by $t' = \left(t-\frac{R}{\upsilon}\right)$, where $\upsilon = 3\times10^8$ m/s for free space, we get:

$$Q_1 = -4\cos10^8\pi\left(t-\frac{R_1}{\upsilon}\right)\mu C$$

$$Q_2 = 4\cos10^8\pi\left(t-\frac{R_2}{\upsilon}\right)\mu C$$

At $t = 0$ $Q_1 = -4\cos10^8\pi\left(-\frac{R_1}{3\times10^8}\right) = -4\cos\pi\left(\frac{\pi R_1}{3}\right)$ and $Q_2 = 4\cos10^8\pi$ $\left(\frac{R_2}{3\times10^8}\right) = 4\cos\left(\frac{\pi R_2}{3}\right)$

(a) In view of Figure 15.9, the distance from Q_1 to the field point $R_1 = 2998.5 - 1.5 = 2997$ and from Q_2, $R_2 = 2998.5 + 1.5 = 3000$. Thus, at $t = 0$:

$$Q_1 = -4\cos\left(\frac{\pi R_1}{3}\right) = -4\cos\left(\frac{2997\pi}{3}\right) = -4\cos(999\pi) = 4\mu C$$

$$Q_2 = 4\cos\left(\frac{\pi R_2}{3}\right) = 4\cos\left(\frac{3000\pi}{3}\right) = 4\cos(1000\pi) = 4\mu C$$

$$\begin{aligned} V &= \frac{Q_1}{4\pi\varepsilon_0 R_1} + \frac{Q_2}{4\pi\varepsilon_0 R_2} \\ &= \frac{4\times 10^{-6}}{4\pi\varepsilon_0}\left(\frac{1}{R_1} + \frac{1}{R_2}\right) \\ &= \frac{10^{-6}}{\pi\dfrac{10^{-9}}{36\pi}}\left(\frac{1}{2997} + \frac{1}{3000}\right) \\ &= 36\times 10^3 \times 6.67\times 10^{-4} \\ &= 24 \text{ volts} \end{aligned}$$

(b) In view ofFigure 15.9, the distance from Q_1 to the field point $R_1 = 2999 - 1.5 = 2997.5$ and from Q_2, $R_2 = 2999 - 1.5 = 3000.5$. Thus, at $t = 0$:

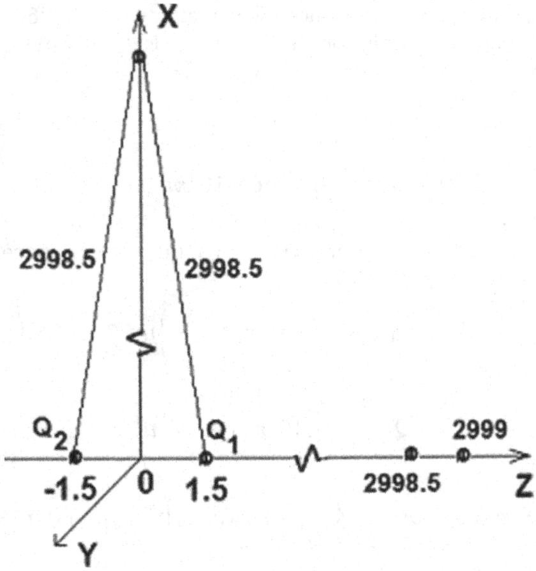

FIGURE 15.9 Location of two point charges.

$$Q_1 = -4\cos\left(\frac{\pi R_1}{3}\right) = -4\cos\left(\frac{2997.5\pi}{3}\right) = -4\cos(999.1667\pi) = 3.464\mu C$$

$$Q_2 = 4\cos\left(\frac{\pi R_2}{3}\right) = 4\cos\left(\frac{3000.5\pi}{3}\right) = 4\cos(1000.1667\pi) = 3.464\mu C$$

$$V = \frac{3.464\times 10^{-6}}{4\pi\varepsilon_0}\left(\frac{1}{R_1}+\frac{1}{R_2}\right) = \frac{3.464\times 10^{-6}}{4\pi\dfrac{10^{-9}}{36\pi}}\left(\frac{1}{2997.5}+\frac{1}{3000.5}\right) = 10.395 \text{ volts}$$

(c) $R_1 = R_2 = R = 2998.5$

$$V = \frac{Q_1}{4\pi\varepsilon_0 R}+\frac{Q_2}{4\pi\varepsilon_0 R} = \frac{1}{4\pi\varepsilon_0 R}(Q_1+Q_2) = \frac{1}{4\pi\dfrac{10^{-9}}{36\pi}2998.5}(Q_1+Q_2) = \frac{9\times 10^9}{2998.5}(Q_1+Q_2)$$

$$Q_1 = -4\cos 10^8\pi\left(t-\frac{R_1}{\upsilon}\right)\mu C$$

$$Q_2 = 4\cos 10^8\pi\left(t-\frac{R_2}{\upsilon}\right)\mu C$$

$$Q_1+Q_2 = -4\cos 10^8\pi\left(t-\frac{R_1}{\upsilon}\right)+4\cos 10^8\pi\left(t-\frac{R_2}{\upsilon}\right) = 0 \text{ for } R_1 = R_2.$$

Thus, $V = 0$ volts.

PROBLEMS

P15.1 Find the EMF developed along the path $r = 0.5$, $z = 0$ at $t = 0$, if (a) $\boldsymbol{B} = 0.1\boldsymbol{a}_z \sin 377t$; (b) $\boldsymbol{B} = 0.1\boldsymbol{a}_r \dfrac{\sin 377t}{r}$; (c) $\boldsymbol{B} = 10^{-7}\boldsymbol{a}_z \sin\left(2\pi 10^8 t - \dfrac{2\pi r}{3}\right)$.

P15.2 In Figure 15.1, let $\boldsymbol{B}$ be constant with time but not necessarily in free space. Find the voltmeter reading at $t = 0.2$ sec if $d = 0.4$ m, $U = \dfrac{dy}{dt}$ and (a) $y = 10t$ m, $B = 0.5y\boldsymbol{a}_z$ wb/m^2; (b) $y = 50t^2$ m, $B = 0.5y\boldsymbol{a}_z$ wb/m^2; (c) $y = 50t^2$ m, $B = 0.5(x+y)\boldsymbol{a}_z$ wb/m^2.

P15.3 Find the displacement current flowing at $t = 0$: (a) between the plates of $10pf$ capacitor if the voltage across this capacitor is 0.1 sin120 πt volts; (b) in a copper conductor of circular cross-section having a total resistance of 0.1Ω and a voltage of $200\sin 5\times 10^5\pi t$ volts across it; (c) in a polystyrene cylinder of 3 mm radius if $H|_{t=0} = H_\varphi a_\varphi = 5\times10^{-9} a_\varphi$ at its surface. For copper $\varepsilon = \varepsilon_0 = \dfrac{10^{-9}}{36\pi}$ and $\sigma = 5.8\times10^7$

$$\frac{\varepsilon_0}{\sigma} = \frac{\frac{10^{-9}}{36\pi}}{5.8\times10^7} = \frac{10^{-9}}{36\pi\times5.8\times10^7} = \frac{10^{-9}}{6.56\times10^9} = 0.152\times10^{-18}.$$

P15.4 Determine whether or not the following pair of fields satisfy Maxwell's equations in a linear, isotropic, homogeneous medium with $\sigma = 0, \mu_r = 2, \varepsilon_r = 8$ (a) $\boldsymbol{E} = 2\boldsymbol{a}_y, \boldsymbol{B} = 4x\boldsymbol{a}_x$; (b) $\boldsymbol{E} = 60\cos(3\times10^8 t - 4x)\boldsymbol{a}_y$, $\boldsymbol{H} = \dfrac{1}{\pi}\cos(3\times10^8 t - 4x)\boldsymbol{a}_z$; (c) $\boldsymbol{E} = 9\times10^{10}\, yt\,\boldsymbol{a}_x, \boldsymbol{B} = (80\times10^{-6} y^2 + 4\times 10^9 t^2)\boldsymbol{a}_z$.

P15.5 Two non-conducting materials with $\varepsilon_{r1} = 8$, $\varepsilon_{r2} = 3, \mu_{r1} = 2$ and $\mu_{r2} = 3$ have an interface at which the normal from Region 1 to Region 2 is given by the unit vector $\boldsymbol{a}_{n1} = \left(-2\boldsymbol{a}_x - \boldsymbol{a}_y + 2\boldsymbol{a}_z\right)/3$. At the boundary in Region 1 $\boldsymbol{B}_1 = \left(2\boldsymbol{a}_x - 3\boldsymbol{a}_y + \boldsymbol{a}_z\right)\cos 1000t$. Find (a) $|B_{n2}|$; (b) $|B_{t1}|$; and (c) $|B_{t2}|$, at $t = 0$.

P15.6 The field inside an air-filled rectangular box with perfectly conducting walls are $E_y = 1130\sin 2x \sin 3z \cos\omega t$ V/m, $H_x = 3\sin 2x\cos 3z\sin\omega t$ A/m, and $H_z = -2\cos 2x \sin 3z \sin\omega t$ A/m. Show that the possible wall locations of this cavity are at $x = 0, x = \dfrac{\pi}{2}, y = 0, y = 0.7316, z = 0,$ and $z = \dfrac{\pi}{3}$. Also find the peak amplitude of the surface current density at: (a) $x = 0, y = 0.216, z = \dfrac{\pi}{10}$; (b) $x = \dfrac{\pi}{8}, y = 0.1247, z = 0$ (a) $x = \dfrac{\pi}{8}, y = 0.7316, z = \dfrac{\pi}{12}$.

P15.7 A coaxial cable section of length L= 0.2 m is bounded by conducting surfaces at its two ends. The radii of inner and outer conductors of coaxial cable are a = 2 mm and b = 8 mm. The region $a < \rho < b$ is occupied by a material having $\varepsilon_r = 6.25, \mu_r = 1, \sigma = 0$. If $\boldsymbol{E} = \dfrac{100}{\rho}\sin 5\pi z \sin\omega t\, \boldsymbol{a}_r \dfrac{\text{V}}{\text{m}}$ find (a) ω; (b) total charge on the inner conductor at $t = \dfrac{\pi}{2\omega}$; (c) the surface current density at $\rho = 3\text{mm}, \varphi = 0, z = 0$ and at $t = 0$.

P15.8 A homogeneous medium of infinite extent is specified by $\varepsilon = 2\times10^{-10}$ F/m, $\mu = 1.25\times10^{-5}$ H/m and $\sigma = 0$. If $\boldsymbol{E} = 400\cos\left(10^9 t - kz\right)$ $\boldsymbol{a}_x$V/m find $\boldsymbol{D}$, $\boldsymbol{B}$, $\boldsymbol{H}$, and $\boldsymbol{k}$.

DESCRIPTIVE QUESTIONS

Q15.1 State and explain Faraday's law of electromagnetic induction.

Q15.2 Discuss different types of EMF.

Q15.3 Explain the meaning, nature, and the conceptual aspects of displacement current.

Q15.4 Write Maxwell's equations in point and integral forms and identify the laws from which these equations emerge.

Q15.5 In what respect the retarded potential differs from the simple time-varying one?

Q15.6 State and explain the Poynting Theorem in detail.

Q15.7 Why is the gradient relation E = −∇V not good for a time-varying field? Modify this relation to make it equally valid for time-varying and time-invariant cases.

Q15.8 Prove the relation $\nabla^2 A = -\mu J + \mu\varepsilon(\partial^2 A/\partial t^2)$ and explain the meaning of each term.

Q15.9 Modify Poisson's equation $\nabla^2 V = -\rho/\varepsilon$ to meet the requirement of time variation.

Q15.10 Prove that the average power flow ($P_{z,av}$) through any surface normal to the z-axis, in case of a lossy dielectric with attenuation constant α and θ_η as the phase angle between E_x and H_y is given by: $P_{z,av} = \dfrac{1}{2}\dfrac{E_{x0}^{\,2}}{\eta_m} e^{-\alpha z} \cos\theta_\eta$.

REFERENCES

1. William H. Jr., *Engineering Electromagnetics*, Fifth Edition, Tata McGraw-Hill Publishing Company Limited, New Delhi, pp. 353–357, (1997).
2. Sadiku, Matthew N.O., *Principles of Electromagnetics*, Fourth Edition, Oxford University Press, New Delhi and Oxford, pp. 390–394, (2009).
3. Mukerji, S.K., Goel, S.K., Bhooshan, S., and Basu, K.P. "Electromagnetic fields theory of electrical machines Part I: Poynting theorem for electromechanical energy conversion", *International Journal of Electrical Engineering Education* **41** (2): 137–145, 2004.
4. Mukerji, S.K., Khan, A.S., and Singh, Y.P., *Electromagnetics for Electrical Machines*, First Edition, CRC Press, Taylor & Francis Group, pp. 49–57, (2015).
5. ibid, pp. 4548 and pp. 298–310, (2015).
6. Einstein, Albert, "Time, space, and gravitation", *The Times*, (November 28, 1919).
7. Bergmann, Peter G., *Introduction to the Theory of Relativity*, Dover Publications, ISBN 0486632822, (1976).
8. Taylor, Edwin F., and Wheeler, John Archibald, ed., *Spacetime Physics: Introduction to Special Relativity*, Second Edition, New York: W.H. Freeman, 84–88, ISBN 0716723271, (1992).
9. Robertson, H.P., "Postulate versus observation in the special theory of relativity", *Reviews of Modern Physics* **21** (3): 378–382, (July 1949).
10. Maxwell, James Clerk, "On a possible mode of detecting a motion of the solar system through the luminiferous ether", *Nature* **21**: 314–315, (1880).
11. Michelson, Albert A., "The relative motion of the Earth and the luminiferous ether", *American Journal of Science* **22**: 120–129, (1881).
12. Michelson, Albert A., and Morley, Edward W., "On the relative motion of the Earth and the luminiferous ether", *American Journal of Science* **34**: 333–345, (1887).
13. Pais, Abraham, *Subtle is the Lord...: The Science and the Life of Albert Einstein*, First Edition, Oxford University Press, pp. 111–113, ISBN 0192806726, (1982).
14. Kennedy, R. J., and Thorndike, E. M., "Experimental establishment of the relativity of time", *Physical Review* **42** (3): 400–418, (1932).

15. Ives, H. E., and Stilwell, G. R., "An experimental study of the rate of a moving atomic clock", *Journal of the Optical Society of America* **28** (7): 215, (1938).
16. Ives, H. E., and Stilwell, G. R., "An experimental study of the rate of a moving atomic clock. II", *Journal of the Optical Society of America* **31** (5): 369, (1941).
17. Francis, S., Ramsey, B., Stein, S., Leitner, J., Moreau, J.M., Burns, R., Nelson, R.A., Bartholomew, T.R., and Gifford, A., "Timekeeping and Time Dissemination in a Distributed Space-Based Clock Ensemble", *Proceedings 34th Annual Precise Time and Time Interval (PTTI) Systems and Applications Meeting*: 201–214, (2002). Retrieved on: 14 April 2013.
18. Stratton, J. A., *Electromagnetic Theory*, McGraw-Hill Book Company, Inc., pp. 74–80, ISBN 2223242526, (1941).

FURTHER READING

Bo, Thide, *Electromagnetic Field Theory*, Second Edition, Dover Pubns, Mineola, New York, 2011, ISBN 13: 978-0-486-4773-2.

Cheng, D.K., *Field and Wave Electromagnetics*, Second Edition, Addison-Wesley, Reading, MA, (1989).

Chow, Tai L., *Introduction to Electromagnetic Theory*, Jones & Bartlett, ISBN 0-7637-3827-1, (2006).

Demarest, K.R., *Engineering Electromagnetics*, Prentice-Hall, Upper Saddle River, NJ, (1998).

Fleisch, D., *A Student's Guide to Maxwell's Equations*, Cambridge University Press, New York, ISBN13: 978-0-521-87761-9, e-ISBN: 052187761X, (2008).

Grant, I.S., and Phillips, W.R., *Electromagnetism*, Second Edition, Manchester Physics Series, ISBN 0-471-92712-0, (2008).

Griffiths, David J., *Introduction to Electrodynamics*, Third Edition, PHI Learning Private Limited, New Delhi, pp. 477525,(2012).

Harrington, R.F., *Introduction to Electromagnetic Engineering*, McGraw Hill, New York, (1968).

Ida, N., *Engineering Electromagnetics*, Second Edition, Springer, New York, (2004).

Inan, U.S., and Inan, A.S., *Engineering Electromagnetics*, Addison-Wesley-Longman, Menlo Park, CA, (1999).

Iskander, M.F., *Electromagnetic Fields and Waves*, Waveland Press, Prospect Hills, IL, (2000).

Jackson, John D., *Classical Electrodynamics*, Third Edition, Wiley, ISBN 0–471-30392-X, (1999).

Johnk, C.T.A., *Engineering Electromagnetic Fields and Waves*, Second Edition, Wiley, NY, (1988).

Kraus, J.D., and Carver, K.R., *Electromagnetics*, Third Edition, McGraw-Hill, New York, (1984).

Kraus, J.D., and Fleisch, D.A., *Electromagnetics with Applications*, Fifth Edition, McGraw-Hill, NY, (1999).

Liao, S.Y., *Engineering Applications of Electromagnetic Theory*, West Publishing Company, St. Paul, MN, (1988).

Lonngren, K. E., Savov, S. V., and Jost, R. J., *Fundamentals of Electromagnetics with MATLAB*, Second Edition, Raleigh, NC: SciTech, (2007).

Mukerji, S. K., Manoj Bhardwaj, Moley kutty George and G. K. Sharma, "Asynchronous operation of hysteresis machines", *International Journal of Applied Engineering and Computer Science* **1**(1): 1–12, (January 2014).

Mukerji, S.K., Goel, S.K., Bhooshan, Sunil, and Basu, K.P., "Electromagnetic fields theory of electrical machines Part I: Poynting theorem for electromechanical energy conversion", *International Journal of Electrical Engineering Education* **41**(2): 137–145, (2004).

Mukerji, S.K., Singh, G.K., Goel, S.K., and Manuja, S., "A theoretical study of electromagnetic transients in a large conducting plate due to current impact excitation", *Progress In Electromagnetics Research* **76**: 15–29, (2007).

Mukerji, S.K., Singh, G.K., Goel, S.K., and Manuja, S., "A theoretical study of electromagnetic transients in a large conducting plate due to voltage impact excitation", *Progress In Electromagnetics Research* **78**: 377–392, (2008).

Notaroš, B. M., *Electromagnetics*, Pearson Prentice-Hall, Upper Saddle River, NJ, (2010).

Paris, Demetrius T., and Hurd, K. Kenneth, *Basic Electromagnetic Theory*, McGraw-Hill, Physical and Quantum Electronics Series, (1969).

Paul, C. R., *Electromagnetics for Engineers with Applications*, Wiley, NY, (2004).

Plonsey, R., and Collins, R.E., *Principles and Applications of Electromagnetic Fields*, McGraw Hill, New York, (1961).

Rao, N.N., *Elements of Engineering Electromagnetics*, Sixth Edition, Prentice-Hall, Upper Saddle River, NJ, (2004).

Schelkunoff, S.A., *Electromagnetic Fields*, Blaidsell Publishing Co., Walthham, Mass., (1963).

Seely, S., and Poularikas, A.D., *Electromagnetics, Classical and Modern Theory and Applications*, Marcel Dekker, Inc., New York, (1979).

Siegel, Daniel M., *Innovation in Maxwell's Electromagnetic Theory*, Cambridge University Press, ISBN 0-521-53329-5, (2003).

Skitek, G.G., and Marshall, S.V., *Electromagnetic Concepts and Applications*, Prentice-Hall, Englewood Cliffs, NJ, (1987).

Slater, J.C., and Frank, N.H., *Electromagnetism*, (Reprint of 1947 edition, ed.). Courier Dover Publications, ISBN 0-486-62263-0, (1969).

Ulaby, F.T., *Electromagnetics for Engineers*, Prentice-Hall, Upper Saddle River, NJ, (2005).

Ulaby, F.T., *Fundamentals of Applied Electromagnetics*, Fifth Edition, Prentice-Hall, Upper Saddle River, NJ, (2006).

Wentworth, S. M., *Fundamentals of Electromagnetics with Engineering Applications*, Wiley, NY, (2005).

Zaher, F.A.A., "An analytical solution for the field of a hysteresis motor based on complex permeability", *IEEE Transactions on Energy Conversion* **5**(1); 156–163, (March 1990).

16 Electromagnetic Waves

16.1 INTRODUCTION

Electromagnetic waves can be conceived in terms of electric and magnetic field vectors, which obey some mathematical relations referred to as *wave equations*. These equations can be obtained by manipulating Maxwell's equations described earlier. These waves can be divided into two broad categories, referred to as unguided and guided waves. An unguided wave is one that travels through space after emerging from an antenna. In this mode, the energy radiated by an antenna adopts its course in accordance with the characteristics of both the media and the antenna. In a guided wave, the energy is fed to and travels through a transmitting structure. These structures may include wire pair lines, coaxial cables, conducting and dielectric waveguides, planar lines, and optical fibers. Some of these structures are discussed in Chapter 18.

The unguided waves can further be categorized as plane and non-plane waves. The plane wave in itself can be uniform or non-uniform. The uniform plane wave is the key player in almost all cases wherein communication is to be established through unguided waves. Depending on the mode of propagation, the unguided waves can also be classified as ground, space, and sky waves. The antennas are discussed in Chapter 20.

Whether guided or unguided a wave may come across different types of environments and hindrances with varied geometry and media parameters. In view of the nature, a wave may get reflected, refracted, diffracted, or absorbed. Some of the aspects of unguided waves, particularly those belonging to uniform plane waves, need careful consideration and thorough analysis. This chapter is devoted to the uniform plane waves traveling through the lossless and lossy media whereas Chapter 17 deals with the reflection and refraction phenomena.

16.2 FIELD RELATIONS AND ASSUMPTIONS

This section deals with Maxwell's equations and some assumptions made to simplify the wave analysis.

16.2.1 MAXWELL'S EQUATIONS

The electromagnetic wave phenomenon revolves around the four Maxwell's equations. The genesis of these equations was thoroughly discussed in Chapter 15. These equations in their general form are reproduced below:

$$\left.\begin{array}{llll}\nabla\times\boldsymbol{H}=J+\dfrac{\partial\boldsymbol{D}}{\partial t} & \text{(a)} & \nabla\times\boldsymbol{E}=-\dfrac{\partial\boldsymbol{B}}{\partial t} & \text{(b)}\\ \nabla\bullet\boldsymbol{D}=\rho & \text{(c)} & \nabla\bullet\boldsymbol{B}=0 & \text{(d)}\end{array}\right\}\qquad(16.1)$$

The vector field quantities involved in the above equations are electric field intensity (***E***), magnetic field intensity (***H***), electric current density (***J***), electric flux density (***D***), and magnetic flux density (***B***). The equations which relate these vector quantities through "media parameters" are referred to as constitutive relation. These relations are given below:

$$\boldsymbol{D} = \varepsilon \boldsymbol{E} \quad (a) \quad \boldsymbol{B} = \mu \boldsymbol{H} \quad (b) \quad \boldsymbol{J} = \sigma \boldsymbol{E} + J_o \quad (c)\} \tag{16.2}$$

In these relations J_o is the source controlled current density; ε, μ and σ are respectively the permittivity, permeability, and conductivity of the medium.

16.2.2 Basic Assumptions

It is to be noted that all the field quantities (viz. ***E***, ***H***, ***J***, ***D***, and ***B***) involved in Maxwell's equations are in general, time-varying. In addition, the electric and magnetic fields have to travel through the space in the form of waves. It is, therefore, necessary to ascertain the nature of these field quantities in relation to their time and space variation.

Fourier series analysis reveals that any "periodic wave" of square, triangular, trapezoidal, or of any other regular or irregular shape can be produced by summing a number of harmonics. Thus any desired waveform may be obtained by a suitable combination of the solutions to the wave equation obtained by considering sinusoidal time variation. For *non-periodic waves*, similar results can be obtained by employing *Fourier integrals*. Thus time variation can be accounted for by assuming the nature of all field quantities as sinusoidal. In view of this assumption, all the field quantities are to be multiplied by a factor $e^{j\omega t}$ where "t" indicates *time* and "ω" the *angular frequency*.

The expressions in Equation 16.1 involve either curl or divergence of a vector field. Since curl, divergence, and gradient operations involve space derivatives, it is necessary to specify the nature of variation of the field quantities with space parameters. In general, a wave may travel in any arbitrary direction (say r). The *variation* in such a direction needs to be appropriately specified. This purpose is fully served if all the field quantities are multiplied by a factor $e^{-(\gamma \cdot r)}$ where γ is the *propagation constant*.

The propagation constant (γ) is composed of two quantities α and β and is related as:

$$\gamma = \alpha + j\beta \tag{16.3}$$

The two parameters (α and β) involved in Equation 16.3 spell the following meaning.

When α is positive the wave decays in amplitude with distance and it is referred to as the *attenuation constant*. When α is negative the wave grows in amplitude with distance and α is termed as *gain coefficient*. The parameter α is measured in nepers/meter (nep./m) in both cases. In relation to the wave propagation in space, negative value of α never exists.

The second parameter β involved in Equation 16.3 is referred to as the *phase constant* and is a measure of phase-shift per unit distance. It is measured in radians/meter (rad./m).

16.3 MEDIA CLASSIFICATION

The study of wave propagation requires a complete understanding of the nature and characteristic of the media with which a wave may come across during the course of its journey. The materials involved in the media may be in solid, liquid, gaseous, or ionized state and may have any physical form e.g. flat or curved surfaces.

An electromagnetic wave is generally visualized in terms of the variations of $\boldsymbol{E}$ and $\boldsymbol{H}$ with space and time. These quantities are susceptible to the change of the media parameter (viz. σ, ε, and μ). The shape or geometry of the obstructing objects may also play a role. The paths to be followed by the electromagnetic waves to arrive at the destination may also differ from situation to situation including height and location of transmitting and receiving antennas, angle of launch of electromagnetic energy into the space, frequency of operation, polarization etc. Thus the media require an in-depth study and must be classified so as to spell the true nature of variations in space and time. The following terminology is normally used to define the media.

16.3.1 Linear/Non-Linear Media

A media is referred to as linear if $\boldsymbol{E}$ and $\boldsymbol{D}$ (or $\boldsymbol{B}$ and $\boldsymbol{H}$) bear a linear relation. If variation of $\boldsymbol{E}$ is non-linearly related with $\boldsymbol{D}$ (or that of $\boldsymbol{H}$ with $\boldsymbol{B}$) the medium is termed as non-linear. In a non-linear medium, its parameters (e.g. ε and μ) are field-dependent. As an example, $\boldsymbol{B}$ and $\boldsymbol{H}$ in ferromagnetic materials are not linearly related.

16.3.2 Homogeneous/Non-Homogeneous Media

A homogeneous medium (or region) is one for which the parameters ε, μ, and σ are constant throughout the medium (or region). If either of these parameters has a different value at different locations, the media (or region) is termed as heterogeneous or non-homogeneous.

16.3.3 Isotropic/Anisotropic Media

A medium is isotropic if ε (or μ) is a scalar constant, so that $\boldsymbol{D}$ and $\boldsymbol{E}$ (or $\boldsymbol{B}$ and $\boldsymbol{H}$) vectors have the same direction everywhere. If the value of either of the media parameter (ε or μ) is such that the vectors $\boldsymbol{D}$ and $\boldsymbol{E}$ (or $\boldsymbol{B}$ and $\boldsymbol{H}$) have different orientations, the media is termed as anisotropic or non-isotropic. The ionosphere is an example of anisotropic media.

16.3.4 Source-Free Media

A source-free medium or region is one in which no impressed voltages or current i.e. no charge and or controlled current distributions) are present.

16.3.5 Bilateral/Unilateral Media

A bilateral media is one in which the wave on its forward and backward journey behaves identically. In the case of unilateral media, such as in magnetized ferrites, the behavior of forward and backward waves will not be the same.

16.3.6 Transparent Media

A transparent media is one through which a wave passes without change in its characteristics at any space location. The free space or vacuum falls in the category.

16.3.7 Translucent Media

A translucent media is one wherein a wave loses part of its energy due to absorption during the course of its journey. In this case the phase and amplitude characteristics of the wave get partially modified. All dielectrics with a bit of conductivity fall under this category.

16.3.8 Non-Transparent or Opaque Media

An opaque media is one wherein the amplitude and phase characteristics of a wave significantly change, a traveling wave transforms into a standing wave due to reflections caused by electrical properties of the surface. All good conductors can be classified as opaque.

16.3.9 Lossless Media

A media is referred to as *lossless* if $\alpha = 0$ and $\gamma \cong j\beta$. In such a media the wave will keep on traveling without attenuation. The free space is regarded as the lossless media.

16.3.10 Lossy Media

The lossy medium is one which is characterized by the simultaneous presence of α and β. As a result, the wave will propagate as well as attenuate during the course of its journey. The rate of attenuation will depend on the value of α. There will be no propagation if β is zero.

The following two sections describe the wave propagation for both lossless and lossy media. These sections also identify the factors, which influence the wave characteristics. In both the lossless and the lossy cases, the media is assumed to be homogeneous, isotropic, source-free, bilateral, and linear.

16.4 WAVE IN LOSSLESS MEDIA

In this section, the media is assumed to be *lossless*, which is identified with $\alpha = 0$ and $\gamma \cong j\beta$. As noted earlier, *free space* is regarded as the lossless medium. It contains neither charges nor source currents. For such a medium, Equation 16.1 can be written in the following modified form.

$$\left.\begin{array}{llll}\nabla\times\boldsymbol{H}=\dfrac{\partial\boldsymbol{D}}{\partial t}=\varepsilon_o\dfrac{\partial\boldsymbol{E}}{\partial t} & \text{(a)} & \nabla\times\boldsymbol{E}=-\dfrac{\partial\boldsymbol{B}}{\partial t}=-\mu_o\dfrac{\partial\boldsymbol{H}}{\partial t} & \text{(b)}\\ \nabla\bullet\boldsymbol{D}=0 & \text{(c)} & \nabla\bullet\boldsymbol{B}=0 & \text{(d)}\end{array}\right\}\tag{16.4}$$

16.4.1 Wave Equations

The equations which govern the behavior of electromagnetic waves are referred to as wave equations. These equations can be obtained from Equation 16.4 be using the following steps.

1. Differentiate both the sides of Equation 16.4a with respect to the time and change the order of differentiation with time and space to get:

$$\frac{\partial}{\partial t}\left(\nabla\times\boldsymbol{H}\right)=\nabla\times\frac{\partial\boldsymbol{H}}{\partial t}=\varepsilon_o\frac{\partial^2\boldsymbol{E}}{\partial t^2}\tag{16.5}$$

2. Take curl of both the sides of Equation 16.4b:

$$\nabla\times\nabla\times\boldsymbol{E}=\nabla\times\left(-\mu_o\frac{\partial\boldsymbol{H}}{\partial t}\right)=-\mu_o\left(\nabla\times\frac{\partial\boldsymbol{H}}{\partial t}\right)\tag{16.6a}$$

3. Substitute Equation 16.5 into Equation 16.6a to get:

$$\nabla\times\nabla\times\boldsymbol{E}=-\mu_o\varepsilon_o\cdot\frac{\partial^2\boldsymbol{E}}{\partial t^2}\tag{16.6b}$$

4. Use the vector identity of Equation 4.41d for the LHS of Equation 16.6b:

$$\nabla\times\nabla\times\boldsymbol{E}\equiv\nabla\left(\nabla\bullet\boldsymbol{E}\right)-\nabla^2\boldsymbol{E}\tag{16.7a}$$

5. Write the first term of the RHS of Equation 16.7a for charge-free regions, in view of Equation 16.4c:

$$\nabla\left(\nabla\bullet\boldsymbol{E}\right)=\nabla\left(\nabla\bullet\left[\boldsymbol{D}/\varepsilon_o\right]\right)=\nabla\left(\frac{\nabla\bullet\boldsymbol{D}}{\varepsilon_o}\right)=0\tag{16.7b}$$

6. Rewrite Equation 16.7a in view of Equation 16.7b:

$$\nabla\times\nabla\times\boldsymbol{E}=-\nabla^2\boldsymbol{E}\tag{16.7c}$$

7. Modify Equation 16.6b in view of Equation 16.7c to get Equation 16.8a:

$$\nabla^2\boldsymbol{E}=\mu_o\varepsilon_o\cdot\frac{\partial^2\boldsymbol{E}}{\partial t^2}\tag{16.8a}$$

8. Use similar steps to get:

$$\nabla^2 \boldsymbol{H} = \mu_o \varepsilon_o \cdot \frac{\partial^2 \boldsymbol{H}}{\partial t^2} \tag{16.8b}$$

Equations 16.8a and 16.8b are referred to as a *wave equation*. These equations are obeyed by both $\boldsymbol{E}$ and $\boldsymbol{H}$ fields in free space. In view of sinusoidal time variation with the angular frequency ω, the second partial derivative with respect to time of all the field quantities is found by multiplying the field quantity with $(-\omega^2)$. Thus, Equations 16.8a and 16.8b can be rewritten as:

$$\nabla^2 \boldsymbol{E} = -\omega^2 \cdot \mu_o \varepsilon_o \cdot \boldsymbol{E} \overset{\text{def}}{=} -\beta_o^2 \cdot \boldsymbol{E} \overset{\text{yields}}{\rightarrow} \boxed{\nabla^2 \boldsymbol{E} + \beta_o^2 \cdot \boldsymbol{E} = 0} \tag{16.9a}$$

$$\nabla^2 \boldsymbol{H} = -\omega^2 \cdot \mu_o \varepsilon_o \cdot \boldsymbol{H} \overset{\text{def}}{=} -\beta_o^2 \cdot \boldsymbol{H} \overset{\text{yields}}{\rightarrow} \boxed{\nabla^2 \boldsymbol{H} + \beta_o^2 \cdot \boldsymbol{H} = 0} \tag{16.9b}$$

where the *phase constant* for free space is defined as:

$$\beta_o \overset{\text{def}}{=} \omega \cdot \sqrt{\mu_o \varepsilon_o} = \frac{\omega}{c} \tag{16.10a}$$

and the symbol c indicates the velocity of light in free space.

The quantity β_o is also referred to as *free space wave number* or the *spatial frequency*. It is in fact the phase-shift constant for uniform plane wave in free space.

For other charge-free linear media (with $\mu \neq \mu_o$ and $\varepsilon \neq \varepsilon_o$), β_o can be replaced by β as:

$$\beta \overset{\text{def}}{=} \omega \sqrt{\mu \varepsilon} \tag{16.10b}$$

The quantity β gives the *wave number* for any media. As before, it is also a measure of phase-shift per unit distance in the given media. The relations given by Equations 16.10a and 16.10b indicate that the *wave number is a function of media parameters and the frequency.*

16.4.2 Solution of Wave Equations

As stated earlier, the wave phenomenon occurs due to the variation of $\boldsymbol{E}$ and $\boldsymbol{H}$ in space and time. To account for the space variation, a coordinate system is required wherein the field variations can conveniently be represented. The three most commonly used systems include Cartesian, cylindrical, and spherical coordinate systems. The selection of a particular coordinate system depends on physical configuration wherein the field is distributed. In view of its simplicity, the present study is confined to the Cartesian coordinate system. It needs to be noted that the $\boldsymbol{E}$ and $\boldsymbol{H}$ fields may vary with one, two, or all the three coordinates and both $\boldsymbol{E}$ and $\boldsymbol{H}$ field vectors may have only one, two, or all the three components. If $\boldsymbol{E}$ (or $\boldsymbol{H}$) is taken to be the function of only one space coordinate (say x), Equations 16.8a and 16.8b can be written as:

$$\left.\frac{\partial^2 \boldsymbol{E}}{\partial x^2} = \mu_o \varepsilon_o \cdot \frac{\partial^2 \boldsymbol{E}}{\partial t^2} \quad \text{(a)} \qquad \frac{\partial^2 \boldsymbol{H}}{\partial x^2} = \mu_o \varepsilon_o \cdot \frac{\partial^2 \boldsymbol{H}}{\partial t^2} \quad \text{(b)}\right\} \tag{16.11}$$

Equation 16.11a or 16.11b in essence is equivalent to three scalar equations, one for each of the scalar components of $\boldsymbol{E}$ (or $\boldsymbol{H}$). Since both $\boldsymbol{E}$ and $\boldsymbol{H}$ follow the wave equation, a solution for only one of them needs to be obtained. From this point on, we will be focused on the solution only for $\boldsymbol{E}$.

Assume that $\boldsymbol{E} = \boldsymbol{E_y a_y}$ only. Equation 16.11a takes the form:

$$\frac{\partial^2 E_y}{\partial x^2} = \mu_o \varepsilon_o \cdot \frac{\partial^2 E_y}{\partial t^2} \tag{16.12a}$$

Solutions to Equation 16.12a can be written in a general form:

$$E_y = f_1\left(x - \upsilon_o t\right) + f_2\left(x + \upsilon_o t\right) \tag{16.12b}$$

where υ_o represents the *velocity of wave propagation*. For free space, it is given as:

$$\upsilon_o = \frac{1}{\sqrt{\mu_o \varepsilon_o}} \overset{\text{def}}{=} c \tag{16.12c}$$

where c indicates the velocity of light in vacuum or free space.

The solution given by Equation 16.12b is illustrated in Figure 16.1.

The first term f_1 involved in E_y represents a wave traveling in the positive x-direction. This wave is called the *forward wave*, which normally originates from a source and travels in the assigned forward direction. It is shown in Figure 16.1a. The second term f_2 involved in E_y represents a wave traveling in the negative x-direction. This is termed as the *backward wave* which in general is a *reflected wave*. This is shown in Figure 16.1b. The *backward wave* results when a forward wave strikes a reflecting surface. Depending on the nature and the characteristics of the reflecting surface or the surface at which the wave strikes, the reflection may be total or partial. The phenomenon of reflection along with the forward and backward waves is shown in Figure 16.1c.

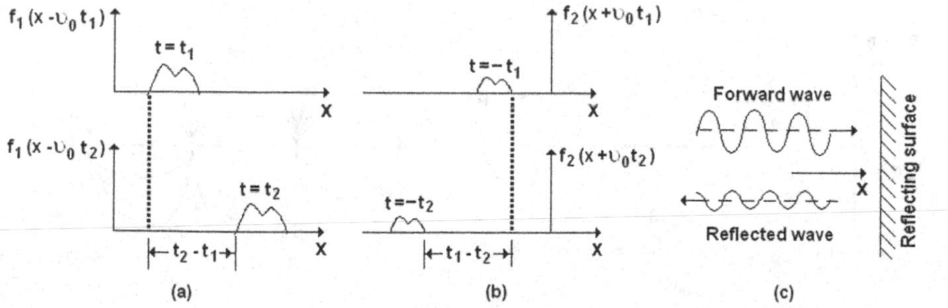

FIGURE 16.1 Components of E_y given by Equation 16.12b representing (a) forward wave; (b) backward wave; (c) direction reversal of these waves.

When a wave is progressing in a particular direction and no reflections are added to it, such a progressing wave is called a *traveling wave*. A wave represented by the equation $E_y = C_1 \cdot \cos(\omega t - \beta x)$ is a traveling wave progressing in the positive x direction. $E_y = C_2 \cdot \cos(\omega t + \beta x)$ is also a traveling wave but progressing in the negative x direction. A traveling wave is illustrated in Figure 16.2a wherein $T = 1/f = 2\pi/\omega$ (f is the frequency of the wave in Hz).

When both the forward and the backward waves are simultaneously present in equal magnitude, they combine to result in a wave called *standing wave* (or stationary wave). Such a wave does not progress and its maxima and minima occur at the same space locations. As shown in Figure 16.2b a standing wave appears to be simply expanding and contracting (i.e. increasing and decreasing in magnitude) with time. The nature of traveling and standing waves is further described in Section 16.4.5.

16.4.3 A Critical Study of the Solution

Assume that E is a function of x and t only. Equation 16.8a, which is equivalent to three scalar equations, one for each of the scalar components of E, can be written as:

$$\left.\frac{\partial^2 E_x}{\partial x^2} = \mu_o\varepsilon_o \cdot \frac{\partial^2 E_x}{\partial t^2} \quad (a) \quad \frac{\partial^2 E_y}{\partial x^2} = \mu_o\varepsilon_o \cdot \frac{\partial^2 E_y}{\partial t^2} \quad (b) \quad \frac{\partial^2 E_z}{\partial x^2} = \mu_o\varepsilon_o \cdot \frac{\partial^2 E_z}{\partial t^2} \quad (c)\right\} \tag{16.13}$$

Equation 16.1c for charge-free regions can be written as:

$$\frac{\partial E_x}{\partial x} + \frac{\partial E_y}{\partial y} + \frac{\partial E_z}{\partial z} = 0 \tag{16.14a}$$

Since E is not a function of y and z, the derivatives $\partial E_y/\partial y$ and $\partial E_z/\partial z$ will vanish and Equation 16.14a reduces merely to $\partial E_x/\partial x = 0$, representing no variation of E_x along the x-direction. Its substitution into Equation 16.13a gives:

$$\frac{\partial^2 E_x}{\partial x^2} = 0 = \mu_o\varepsilon_o \cdot \frac{\partial^2 E_x}{\partial t^2} \xrightarrow{\text{yields}} \boxed{\frac{\partial^2 E_x}{\partial t^2} = 0} \tag{16.14b}$$

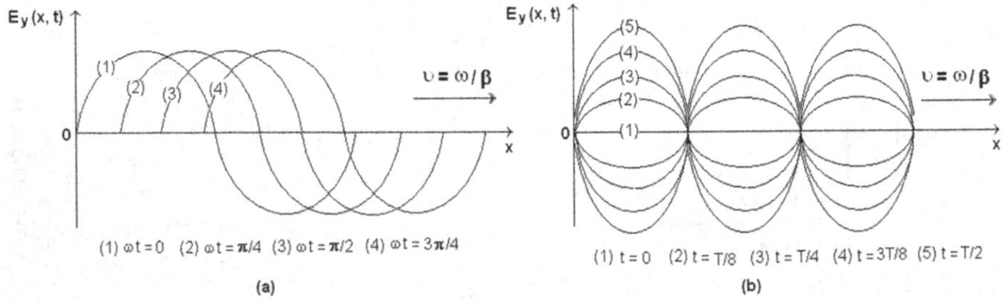

FIGURE 16.2 (a) Traveling wave; (b) standing wave.

Thus E_x is either zero or constant or uniformly (linearly) varies with time. In mathematical terms these three possibilities can be written as:

(i) $E_x = 0$; or (ii) $E_x = k$; or (iii) $E_x = k{\cdot}t$ where k is a constant.

Out of the three possible options, (ii) and (iii) do not satisfy wave equation and option (i), i.e. $E_x = 0$, is the only possibility left. Thus a wave progressing in the x-direction has no x-component of E. Similarly, in view of Equation 16.1d it can be shown that a wave progressing in the x-direction will have no x component of H. In general it can be stated that a wave traveling along a particular axis will have no components of E or H belonging to that axis.

Figure 16.2a shows a traveling wave wherein $E(y,t)$ for different time instants attains maxima at different space (x) locations. If the space locations of equiphase points are marked and joined together, and if the joining of such points results in a planar configuration such a configuration is referred to as an *equiphase plane*. Similarly, if the space locations of equiamplitude points are also marked and joined together, and if joining of such points results in a planar configuration such a configuration is referred to as an *equiamplitude plane*. The formation of equiphase and equiamplitude planes can be used for further categorization of waves as noted below.

A wave characterized by the formation of equiphase planes is termed a *plane wave*. The wave wherein equiphase and equiamplitude planes are the same is called a *uniform plane wave*. In uniform plane wave amplitude of E is uniform throughout a plane (say z = constant).

Since both electric and magnetic fields are perpendicular to the direction of propagation and both lie in a plane that is transverse to the direction of propagation a uniform plane wave is also called a *transverse electromagnetic wave*.

A uniform plane wave cannot physically exist, for it extends to infinity in (at least) two dimensions and represents an infinite amount of energy. However, the distant field of a transmitting antenna can be considered as a uniform plane wave in a limited sense; and a radar signal impinging upon a distant target may closely resemble a uniform plane wave.

16.4.4 Characteristic Impedance

Assume that a wave is traveling in x direction. It will have no E_x and H_x components. Also E and H are not functions of y and z. Equation 16.1b can be written in the expanded form as:

$$\nabla \times \boldsymbol{E} = \begin{bmatrix} \boldsymbol{a}_x & \boldsymbol{a}_y & \boldsymbol{a}_z \\ \partial/\partial x & 0 & 0 \\ 0 & E_y & E_z \end{bmatrix} = -\frac{\partial E_z}{\partial x}\boldsymbol{a}_y + \frac{\partial E_y}{\partial x}\boldsymbol{a}_z = -\frac{\partial \boldsymbol{B}}{\partial t} = -\mu_o \frac{\partial \boldsymbol{H}}{\partial t} \qquad (16.15a)$$

Also from Equation 16.1a:

$$\nabla \times \boldsymbol{H} = \begin{bmatrix} \boldsymbol{a}_x & \boldsymbol{a}_y & \boldsymbol{a}_z \\ \partial/\partial x & 0 & 0 \\ 0 & H_y & H_z \end{bmatrix} = -\frac{\partial H_z}{\partial x}\boldsymbol{a}_y + \frac{\partial H_y}{\partial x}\boldsymbol{a}_z = \frac{\partial \boldsymbol{D}}{\partial t} = \varepsilon_o \frac{\partial \boldsymbol{E}}{\partial t} \tag{16.15b}$$

In view of Equations 16.15a and 16.15b:

$$-\frac{\partial E_z}{\partial x}\boldsymbol{a}_y + \frac{\partial E_y}{\partial x}\boldsymbol{a}_z = -\mu_o \frac{\partial H_y}{\partial t}\boldsymbol{a}_y - \mu_o \frac{\partial H_z}{\partial t}\boldsymbol{a}_z \tag{16.16a}$$

$$-\frac{\partial H_z}{\partial x}\boldsymbol{a}_y + \frac{\partial H_y}{\partial x}\boldsymbol{a}_z = \varepsilon_o \frac{\partial E_y}{\partial t}\boldsymbol{a}_y + \varepsilon_o \frac{\partial E_z}{\partial t}\boldsymbol{a}_z \tag{16.16b}$$

Equating coefficients of a_y and a_z in Equations 16.16a and 16.16b to get:

$$\left.\begin{aligned} -\frac{\partial H_z}{\partial x} &= \varepsilon_o \frac{\partial E_y}{\partial t} \quad \text{(a)} & \frac{\partial E_z}{\partial x} &= \mu_o \frac{\partial H_y}{\partial t} \quad \text{(b)} \\ \frac{\partial H_y}{\partial x} &= \varepsilon_o \frac{\partial E_z}{\partial t} \quad \text{(c)} & \frac{\partial E_y}{\partial x} &= -\mu_o \frac{\partial H_z}{\partial t} \quad \text{(d)} \end{aligned}\right\}. \tag{16.17}$$

Let us consider only the forward wave component of Equation 16.12b, which is given as:

$$E_y = f_1\left(x - \upsilon_o t\right) \overset{\text{def}}{=} f_1 \tag{16.18}$$

Thus:

$$\frac{\partial E_y}{\partial t} = \frac{\partial f_1\left(x-\upsilon_o t\right)}{\partial t} = \frac{\partial f_1\left(x-\upsilon_o t\right)}{\partial\left(x-\upsilon_o t\right)} \cdot \frac{\partial\left(x-\upsilon_o t\right)}{\partial t} = f_1'\left(x-\upsilon_o t\right)\cdot\left(-\upsilon_o\right)$$

or

$$\frac{\partial E_y}{\partial t} = -\upsilon_o \cdot f_1' \tag{16.18a}$$

where

$$f_1' = f_1'\left(x-\upsilon_o t\right) = \frac{\partial f_1\left(x-\upsilon_o t\right)}{\partial\left(x-\upsilon_o t\right)} \tag{16.18b}$$

From Equation 17.16a and Equation 16.18a:

$$\frac{\partial H_z}{\partial x} = \upsilon_o \cdot \varepsilon_o \cdot f_1' = \frac{1}{\sqrt{\mu_o \cdot \varepsilon_o}} \cdot \varepsilon_o \cdot f_1' = \sqrt{\varepsilon_o / \mu_o} \cdot f_1' \tag{16.19a}$$

Thus,

$$H_z = \sqrt{\varepsilon_o / \mu_o} \cdot \int f_1' \cdot dx + k \tag{16.19b}$$

Since,

$$\frac{\partial f_1}{\partial x} = \frac{\partial f_1 (x - \upsilon_o \cdot t)}{\partial (x - \upsilon_o \cdot t)} \cdot \frac{\partial (x - \upsilon_o \cdot t)}{\partial x} = f_1' \tag{16.20}$$

Since k is constant and not a function of time (t), it does not form part of the wave motion and can be neglected. Thus, from Equation 16.19b:

$$\begin{aligned} H_z &= \sqrt{\varepsilon_o / \mu_o} \cdot \int f_1' \cdot dx = \sqrt{\varepsilon_o / \mu_o} \cdot \int \frac{\partial f_1}{\partial x} \cdot dx \\ &= \sqrt{\varepsilon_o / \mu_o} \cdot f_1 = \sqrt{\varepsilon_o / \mu_o} \cdot E_y \end{aligned} \tag{16.21}$$

Since k is constant and not a function of time (t), it does not form part of the wave motion and can be neglected. Thus, from Equation 16.21:

$$E_y / H_z = \sqrt{\mu_o / \varepsilon_o} \quad (\text{for free space or air}) \tag{16.22a}$$

$$E_y / H_z = \sqrt{\mu / \varepsilon} \quad (\text{for any linear medium}) \tag{16.22b}$$

Another relation for the ratio of E_z/H_y obtained by following the above procedure is:

$$E_z / H_y = -\sqrt{\mu / \varepsilon} \quad (\text{for any linear medium}) \tag{16.22c}$$

Now since $E_x = H_x = 0$:

$$\boldsymbol{E} = |\boldsymbol{E}| = \sqrt{E_y^2 + E_z^2} = \sqrt{\left(H_z \cdot \sqrt{\mu / \varepsilon}\right)^2 + \left(-H_y \cdot \sqrt{\mu / \varepsilon}\right)^2}$$

or

$$\boldsymbol{E} = \left[\sqrt{H_z^2 + H_y^2}\right] \cdot \sqrt{\mu / \varepsilon} \tag{16.23a}$$

and

$$\boldsymbol{H} = \left[\sqrt{H_z^2 + H_y^2}\right] \tag{16.23b}$$

Thus:

$$\boldsymbol{E} / \boldsymbol{H} = \sqrt{\mu / \varepsilon} \stackrel{\text{def}}{=} \eta \quad (\text{for any linear medium}) \tag{16.24a}$$

and

$$E/H = \sqrt{\mu_o / \varepsilon_o} \overset{\text{def}}{=} \eta_o \quad (\text{for free space}) \tag{16.24b}$$

Equation 16.24a gives the ratio of electric and magnetic field intensities. It has the unit of resistance (ohm) and is represented by a symbol η. This ratio is called the *characteristic (or intrinsic) impedance* of the medium.

The intrinsic impedanceη_o is a universal constant for free space. In Equation 16.24b, ε is replaced by $\varepsilon_o = 8.854 \times 10^{-12} = \dfrac{10^{-9}}{36\pi}$ F/m) and μ is replaced by $\mu_o = 4\pi \times 10^{-7}$ H/m the parameter η can be replaced by η_o. This gives:

$$\eta_o = 120\pi \approx 377\,\Omega \tag{16.25}$$

16.4.5 Nature of Waves

In order to further understand the nature of a wave, consider Equation 16.9a given below.

$$\nabla^2 \boldsymbol{E} + \beta^2 \boldsymbol{E} = 0 \tag{16.9a}$$

Let us assume that E is a function of x and t only and contains only $E_y$component (i.e. $E = E_y a_y$) Equation 16.10a can be written as:

$$\frac{\partial^2 E_y}{\partial x^2} + \beta^2 E_y = 0 \tag{16.26}$$

The solution of Equation 16.26 can be written as:

$$E_y(x) = C_1 e^{-j\beta x} + C_2 e^{+j\beta x} \tag{16.27a}$$

Further with the inclusion of time variation term ($e^{+j\omega t}$) Equation 16.27a becomes:

$$E_y(x,t) = C_1 e^{-j(\beta x - \omega t)} + C_2 e^{+j(\beta x + \omega t)} \tag{16.27b}$$

As observed in Equation 16.12b, Equation 16.27b contains two solutions in the complex form) representing forward and backward waves. The real (Re) and imaginary (Im) parts of $E_y(x,t)$ given by Equation 16.27b are:

$$\text{Re}\left[E_y(x,t)\right] = C_1 \cdot \cos(\omega t - \beta x) + C_2 \cdot \cos(\omega t + \beta x) \tag{16.28a}$$

$$\text{Im}\left[E_y(x,t)\right] = C_1 \cdot \sin(\omega t - \beta x) + C_2 \cdot \sin(\omega t + \beta x) \tag{16.28b}$$

The two segments in both the real and imaginary parts of $E_y(x, t)$ given by Equations 16.28a and 16.28b too represent forward and backward waves. The real part of the solution that represents only the forward wave is:

$$\mathrm{Re}\left[E_y(x,t)\right] = C_1 \cdot \cos(\omega t - \beta x) \tag{16.29}$$

Equation 16.29 reveals that when the wave completes its travel by one wavelength in the x-direction (i.e. $x = \lambda$) its phase changes by 2π radians. Thus:

$$\beta x = \beta\lambda = 2\pi \tag{16.29a}$$

Also in view of the well-known relation $\upsilon = f\lambda$ thus $\lambda = \upsilon/f$. When we multiply both sides of Equation 16.29a, (i.e. $\beta\lambda = 2\pi$) by frequency "f" we get $\beta\lambda f = \beta\upsilon = 2\pi f = \omega$. From this relation:

$$\upsilon = \frac{\omega}{\beta} \tag{16.29b}$$

This relation can also be obtained from the argument of the cosine term $(\omega t - \beta x)$by equating it to a constant (conveniently to zero). Thus:

$$\left.\begin{matrix}\omega t - \beta x = 0 \quad (\mathrm{a}) & \dfrac{x}{t} = \upsilon = \dfrac{\omega}{\beta} \quad (\mathrm{b})\end{matrix}\right\} \tag{16.30}$$

The velocity "υ" is called the *phase velocity.* Sometimes υ is used to represent the velocity of wave propagation which in vacuum or free space becomes υ_0 (= c, where c is the velocity of light). In such cases the phase velocity is denoted by υ_p.

Substitute different values of ωt in Equation 16.29 to get the following resulting expressions:

$$\mathrm{Re}\left[E_y(x,t)\right]\Big|_{\omega t=0} = C_1 \cdot \cos\beta x \tag{16.31a}$$

$$\mathrm{Re}\left[E_y(x,t)\right]\Big|_{\omega t=\pi/4} = C_1\cos(\pi/4 - \beta x) = \frac{C_1}{\sqrt{2}}(\cos\beta x - \sin\beta x) \tag{16.31b}$$

$$\mathrm{Re}\left[E_y(x,t)\right]\Big|_{\omega t=\pi/2} = C_1\cos(\pi/2 - \beta x) = C_1\sin\beta x \tag{16.31c}$$

Some more expressions for values of ωt can be obtained in a similar fashion.

Plots of E_y from the expressions given by Equations 16.31a–c are the same as were shown in Figure 16.2a, which represent the pattern of a *traveling wave.* As evident from Figure 16.2a the crests of the wave shift towards the right for subsequent higher values of ωt. Figure 16.2b illustrates the *standing wave* in order to make a comparison between the two waves wherein the locations of maximas or minimas remain unaltered.

16.5 WAVE IN LOSSY MEDIA

In lossy media, the conductivity (σ) is not zero. The presence of small conductivity in the media results in the presence of some conduction current.

16.5.1 Wave Equations

In lossy media $\sigma \neq 0$, thus the Maxwell's equations given by Equations 16.1a and 16.1b can be written as:

$$\left.\nabla \times \boldsymbol{H} = \boldsymbol{J} + \frac{\partial \boldsymbol{D}}{\partial t} = \sigma \boldsymbol{E} + \varepsilon \frac{\partial \boldsymbol{E}}{\partial t} \quad (a) \qquad \nabla \times \boldsymbol{E} = -\frac{\partial \boldsymbol{B}}{\partial t} = -\mu \frac{\partial \boldsymbol{H}}{\partial t} \quad (b)\right\} \tag{16.3}$$

The wave equations for this case can be obtained by following steps similar to those used for the lossless media. These steps are given below.

1. Differentiate Equation 16.3a with respect to time and change the order of differentiation to get:

$$\frac{\partial}{\partial t}(\nabla \times \boldsymbol{H}) = \nabla \times \frac{\partial \boldsymbol{H}}{\partial t} = \sigma \frac{\partial \boldsymbol{E}}{\partial t} + \varepsilon \frac{\partial^2 \boldsymbol{E}}{\partial t^2} \tag{16.32}$$

2. Take the curl of Equation 16.3b and use Equation 16.32 to get:

$$\nabla \times \nabla \times \boldsymbol{E} = -\nabla^2 \boldsymbol{E} = -\mu \nabla \times \frac{\partial \boldsymbol{H}}{\partial t} = -\mu\left(\sigma \frac{\partial \boldsymbol{E}}{\partial t} + \varepsilon \frac{\partial^2 \boldsymbol{E}}{\partial t^2}\right)$$

or

$$\nabla^2 \boldsymbol{E} = \left(\mu\sigma \frac{\partial \boldsymbol{E}}{\partial t} + \mu\varepsilon \frac{\partial^2 \boldsymbol{E}}{\partial t^2}\right)$$

Thus

$$\left.\nabla^2 \boldsymbol{E} - \mu\sigma \frac{\partial \boldsymbol{E}}{\partial t} - \mu\varepsilon \frac{\partial^2 \boldsymbol{E}}{\partial t^2} = 0 \quad (a) \qquad \nabla^2 \boldsymbol{H} - \mu\sigma \frac{\partial \boldsymbol{H}}{\partial t} - \mu\varepsilon \frac{\partial^2 \boldsymbol{H}}{\partial t^2} = 0 \quad (b)\right\} \tag{16.33}$$

Equations 16.33a and 16.33b are the wave equations obeyed by $\boldsymbol{E}$ and $\boldsymbol{H}$ in lossy media. In view of sinusoidal time variation the wave equation for $\boldsymbol{E}$ can be rewritten as:

$$\begin{aligned}\nabla^2 \boldsymbol{E} - j\omega\mu\sigma \boldsymbol{E} + \omega^2 \mu\varepsilon \boldsymbol{E} &= \nabla^2 \boldsymbol{E} + \left(\omega^2 \mu\varepsilon - j\omega\mu\sigma\right) \boldsymbol{E} \\ &= \nabla^2 \boldsymbol{E} - j\omega\mu\left(\sigma + j\omega\varepsilon\right) \boldsymbol{E} = 0\end{aligned}$$

From the above and similar steps, we get:

$$\nabla^2 \boldsymbol{E} - \gamma^2 \boldsymbol{E} = 0 \tag{16.34a}$$

$$\nabla^2 \boldsymbol{H} - \gamma^2 \boldsymbol{H} = 0 \tag{16.34b}$$

where

$$\gamma = \sqrt{j\omega\mu\left(\sigma + j\omega\varepsilon\right)} = \alpha + j\beta \tag{16.34c}$$

The parameters α and β have already been explained earlier. As before β, the phase shift constant, is a measure of alternation of phase of the wave. The values of the β_0 and β were given by Equations 16.10a and 16.10b, which were referred to as wave numbers for free space and general media, respectively.

16.5.2 Solution to Wave Equation

The solution to Equation 16.34a can be written as:

$$E_y\left(x\right) = C_1 e^{-\gamma x} + C_2 e^{+\gamma x} \tag{16.35a}$$

With the inclusion of time variation Equation 16.35a becomes:

$$E_y\left(x,t\right) = \left[C_1 e^{-(\alpha+j\beta)x} + C_2 e^{(\alpha+j\beta)x}\right] e^{j\omega t} \tag{16.35b}$$

As before, Equation 16.35balso includes forward and backward waves, and both of these contain real and imaginary parts. Consider only the real part of the forward wave, which is:

$$\text{Re}\left[E_y\left(x,t\right)\right] = C_1 e^{-\alpha x}\cos\left(\omega t - \beta x\right) \tag{16.36}$$

Substitution of different values of ωt in Equation 16.36 results in the following expressions:

$$\text{Re}\left[E_y\left(x,t\right)\right]_{\omega t=0} = C_1 e^{-\alpha x}\cos\beta x \tag{16.36a}$$

$$\text{Re}\left[E_y\left(x,t\right)\right]_{\omega t=\pi/4} = C_1 e^{-\alpha x}\cos\left(\frac{\pi}{4} - \beta x\right) = \frac{C_1}{\sqrt{2}} e^{-\alpha x}\left[\cos\beta x + \sin\beta x\right] \tag{16.36b}$$

$$\text{Re}\left[E_y\left(x,t\right)\right]_{\omega t=\pi/2} = C_1 e^{-\alpha x}\cos\left(\frac{\pi}{2} - \beta x\right) = C_1 e^{-\alpha x}\sin\beta x \tag{16.36c}$$

Some more expressions for the values of ωt can be obtained in similar fashion. The nature of waves obtained from these is discussed in the following subsection.

The characteristic impedance for a lossy media obtained by using the procedure of Section 16.4.4 can be written as:

$$\frac{\mathbf{E}}{\mathbf{H}} = \eta = \sqrt{\frac{j\omega\mu}{\sigma + j\omega\varepsilon}} \tag{16.37}$$

The plot obtained for $\boldsymbol{E}_y$ from the expression given by Equation 16.36 is shown in Figure 16.3. This represents a traveling wave pattern but with a difference from that

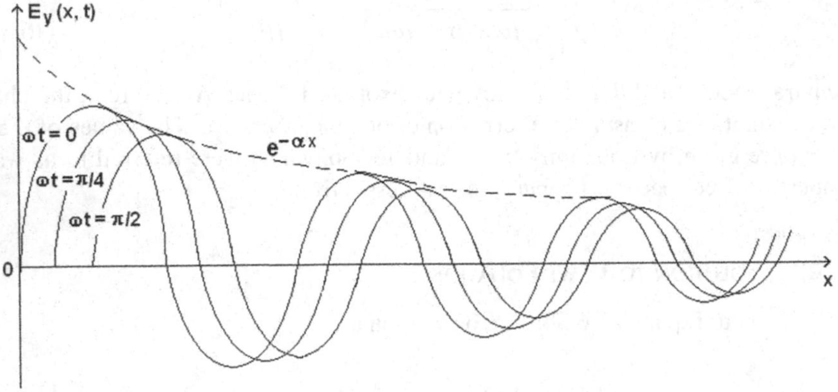

FIGURE16.3 Traveling waves in lossy media.

of Figure 16.2a. In this case the presence of α results in gradual reduction in magnitude of the wave as it progresses along the x-axis.

The time average power density for a wave traveling in the z-direction can be obtained by using the Poynting Theorem. If $\boldsymbol{E} = E_x\boldsymbol{a_x}$ then $\boldsymbol{H} = H_y\boldsymbol{a_y}$. In view of the Poynting Theorem:

$$\boldsymbol{P} = \boldsymbol{E} \times \boldsymbol{H} = E_x\boldsymbol{a_x} \times H_y\boldsymbol{a_y} = P_z\boldsymbol{a_z}$$

Let $E_x = E_{x0}\cos(\omega t - \beta z)$, thus: $H_y = \dfrac{E_{x0}}{\eta}\cos(\omega t - \beta z)$
and

$$P_z = \frac{E_{x0}^2}{\eta}\cos^2(\omega t - \beta z) \tag{16.38a}$$

$$P_{z,av} = f\int_0^{1/f}\frac{E_{x0}^2}{\eta}\cos^2(\omega t - \beta z)\,dt = \frac{f}{2}\frac{E_{x0}^2}{\eta}\int_0^{1/f}\left[1 + \cos(2\omega t - 2\beta z)\right]dt$$

$$= \frac{f}{2}\frac{E_{x0}^2}{\eta}\left[t + \frac{1}{2\omega}\sin(2\omega t - 2\beta z)\right]_0^{1/f} = \frac{f}{2}\frac{E_{x0}^2}{\eta}\frac{1}{f} = \frac{1}{2}\frac{E_{x0}^2}{\eta} \tag{16.38b}$$

$$P_{z,av}(\text{rms}) = \frac{E_{x0}^2}{\eta} \quad \text{for lossless media} \tag{16.38c}$$

$$P_{z,av} = \frac{E_{x0}^2}{2\eta}e^{-2ax}\cos\theta_\eta \quad \text{where } \eta = \eta_m\angle\theta_\eta \tag{16.38d}$$

16.6 CONDUCTORS AND DIELECTRICS

The conducting and dielectric materials were discussed in Chapter 8 in detail. Since at that stage the picture of all of Maxwell's equations did not emerge, the criterion to distinguish between these was not explained. This criterion is based on the Maxwell's equation given by Equation 16.1a. This equation is reproduced below in the modified form:

$$\nabla \times \boldsymbol{H} = \boldsymbol{J} + \frac{\partial \boldsymbol{D}}{\partial t} = \boldsymbol{J} + \boldsymbol{J}_d \tag{16.1a}$$

Its RHS comprises two terms. The first term represents the conduction current density (J) and the second the displacement or capacitive current density (J_d). The magnitude of the ratio of these two current densities (for sinusoidal time variation) can be written as:

$$\left|\frac{\boldsymbol{J}}{\boldsymbol{J}_d}\right| = \frac{|\sigma \boldsymbol{E}|}{|\omega\varepsilon \boldsymbol{E}|} = \frac{\sigma}{\omega\varepsilon} \tag{16.39}$$

A critical look at the RHS of Equation 16.39 reveals that (a) the conductivity σ may assume values between "0" and "∞"; (b) the angular frequency ω may also assume values between "0" and "∞"; (c) in the case of permittivity ($\varepsilon = \varepsilon_0\varepsilon_r$), ε_r is never less than unity (for a vacuum) and not more than 80 or so (for distilled water); and (d) none of the above parameters has a negative value.

Depending on the above noted limits the RHS of Equation 16.39 can have any value between "0" and "∞" including unity. The unity value of this ratio is normally taken as the dividing line for the identification of conducting and dielectric materials. Thus, for $\frac{\sigma}{\omega\varepsilon} > 1$ the material can be taken as a conductor and for $\frac{\sigma}{\omega\varepsilon} < 1$ it is considered as a dielectrics. Furthermore, the farther the value of this ratio from this dividing value, the more perfect will be the dielectric, i.e. $\frac{\sigma}{\omega\varepsilon} \gg 1$, or the conductor, i.e. $\frac{\sigma}{\omega\varepsilon} \ll 1$, as the case may be. This classification of materials is depicted through an illustration given by Figure 16.4.

16.6.1 Parameters Related to the Dielectrics

There are some special parameters, which further spell the properties of a dielectric. These include the dissipation factor, power factor, and the loss tangent. These are spelled below in terms of ε and σ.

The *dissipation factor* of the dielectric is denoted by $Đ$ and is given as:

$$Đ = \frac{\sigma}{\omega\varepsilon} \tag{16.39a}$$

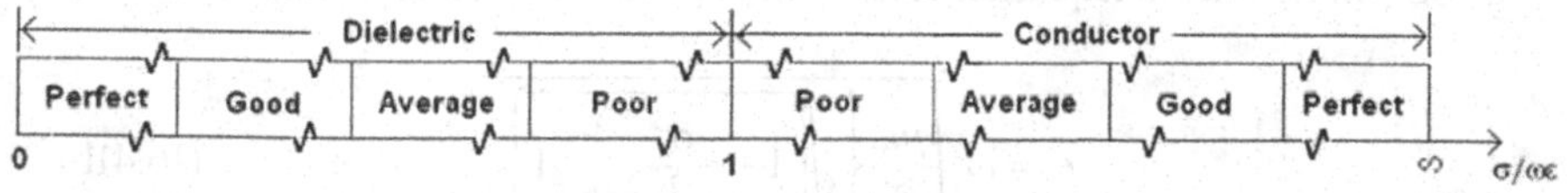

FIGURE 16.4 Relative values of $\sigma/\omega\varepsilon$ for conductors and dielectrics.

The *power factor* is given as sin φ where the angle φ is related with the $Đ$ as:

$$\varphi = \tan^{-1} Đ \tag{16.39b}$$

The values of the dissipation factor ($Đ$) and the power factor ($\sin\varphi$) differ by about 1% when $Đ$ is less than 0.15.

The term *loss tangent* is defined as:

$$\tan\varphi = Đ \tag{16.39c}$$

For small values: $\varphi \approx Đ$

It may be noted that for mica the value of $\sigma/\omega\varepsilon$ even at audio frequencies is approximately 0.002 whereas for copper it is about 3.5×10^8 at 30 GHz. Furthermore, for most of the dielectrics, σ and ε are functions of frequency, whereas for good conductors, σ and ε are frequency-independent.

16.6.2 Values of α and β in Terms of Media Parameters

The values of α and β can be obtained by taking the following steps:

1. Write γ from Equation 16.34b:

$$\gamma = \sqrt{j\omega\mu(\sigma + j\omega\varepsilon)} = \alpha + j\beta \tag{16.34b}$$

2. Square both sides of Equation 16.34b:

$$(\alpha + j\beta)^2 = \alpha^2 - \beta^2 + 2j\alpha\beta = -\omega^2\mu\varepsilon + j\omega\mu\sigma \tag{16.40a}$$

3. Separate the real and imaginary parts of Equation 16.40a:

$$\left.\alpha^2 - \beta^2 = -\omega^2\mu\varepsilon \quad (b) \qquad 2\alpha\beta = \omega\mu\sigma \quad (c)\right\} \tag{16.40}$$

4. Obtain α and β from Equation 16.40c as:

$$\left.\alpha = \frac{\omega\mu\sigma}{2\beta} \quad (d) \qquad \beta = \frac{\omega\mu\sigma}{2\alpha} \quad (e)\right\} \tag{16.40}$$

5. Substitute α from Equation 16.40d in Equation 16.40b and simplify to get:

$$\beta = \omega\sqrt{\frac{\mu\varepsilon}{2}\left\{\sqrt{\left(1 + \frac{\sigma^2}{\omega^2\varepsilon^2}\right)} + 1\right\}} \tag{16.41a}$$

6. Substitute β from Equation 16.40e, in Equation 16.40b and simplify to get:

$$\alpha = \omega\sqrt{\frac{\mu\varepsilon}{2}\left\{\sqrt{\left(1 + \frac{\sigma^2}{\omega^2\varepsilon^2}\right)} - 1\right\}} \tag{16.41b}$$

16.6.3 Impact of Conductivity on α, β, υ, and η

For a *good dielectric,* $\frac{\sigma}{\omega\varepsilon} \ll 1$ the inner square root terms in Equations 16.41a and 6.41b get simplified as:

$$\sqrt{\left(1+\frac{\sigma^2}{\omega^2\varepsilon^2}\right)} \approx 1+\frac{\sigma^2}{2\omega^2\varepsilon^2} \quad (16.42a)$$

In view of Equation 16.42a, the expressions of α and β given by Equation 16.41b and Equation 16.41a, respectively, reduce to:

$$\left.\alpha = \frac{\sigma}{2}\sqrt{\frac{\mu}{\varepsilon}} \quad (b) \qquad \beta = \omega\sqrt{\mu\varepsilon}\left(1+\frac{\sigma^2}{8\omega^2\varepsilon^2}\right) \quad (c)\right\} \quad (16.42)$$

In view of Equation 16.42c, the velocity (υ) given by Equation 16.29b becomes:

$$\upsilon = \frac{\omega}{\beta} \approx \upsilon_0\left(1-\frac{\sigma^2}{8\omega^2\varepsilon^2}\right) \quad (16.43)$$

Also, the expression of characteristic impedance (η) given by Equation 16.25 reduces to:

$$\eta = \sqrt{\frac{j\omega\mu}{\sigma + j\omega\varepsilon}} \approx \sqrt{\frac{\mu}{\varepsilon}}\left(1+\frac{j\sigma}{2\omega\varepsilon}\right) \quad (16.44a)$$

$$\left.\eta \approx \sqrt{\frac{\mu}{\varepsilon}} = \eta_0 \quad (\text{for } \sigma = 0) \quad (b) \qquad \eta \approx \eta_0\left(1+\frac{j\sigma}{2\omega\varepsilon}\right) \quad (\text{for } \sigma \neq 0) \quad (c)\right\} \quad (16.44)$$

In view of Equation 16.43 it can be observed that in good dielectrics the presence of small σ results in slight reduction in the velocity of propagation of the wave. Similarly, Equation 16.44c explains that the presence of small σ results in the addition of a small reactive component to the characteristic impedance.

For *good conductors,* σ is large, the ratio $\omega\varepsilon/\sigma$ approaches 0, and Equation 16.34b gets modified as:

$$\gamma = \alpha + j\beta = \sqrt{j\omega\mu(\sigma + j\omega\varepsilon)} = \sqrt{j\omega\mu\sigma\left(1+\frac{j\omega\varepsilon}{\sigma}\right)} \approx \sqrt{j\omega\mu\sigma} \quad (16.45a)$$

Thus:

$$\alpha = \beta = \sqrt{\omega\mu\sigma}\angle 45° = \sqrt{\frac{\omega\mu\sigma}{2}} \quad (16.45b)$$

$$\left.\upsilon = \frac{\omega}{\beta} = \sqrt{\frac{2\omega}{\mu\sigma}} \quad (a) \quad \eta = \sqrt{\frac{j\omega\mu}{\sigma + j\omega\varepsilon}} = \sqrt{\frac{j\omega\mu}{\sigma\left(1 + j\omega\varepsilon/\sigma\right)}} \approx \sqrt{\frac{\omega\mu}{\sigma}}\angle 45^\circ \quad (b)\right\} \tag{16.46}$$

From Equations 16.45 and 16.46, it can be noted that both α and β are large. Thus due to large $\alpha$the wave greatly attenuates as it progresses through the conductor. As β is also large it results in the large reduction of the wave velocity almost to the level of sound waves in air. The characteristic impedance is also very small and contains a reactive component. As can be seen from Equation 16.46b, the angle of this impedance is always 45° for good conductors.

16.7 POLARIZATION

Polarization may be defined in terms of the orientation of the ***E*** vector with reference to the ground (or the surface of the Earth, assumed to be perfectly smooth and flat). To elaborate its meaning consider the flat surface of the Earth over which an antenna is installed. If at the instant of the emergence of the wave its ***E*** field is perpendicular to the Earth's surface the wave is termed as *vertically polarized* and if it is parallel it is said to be the *horizontally polarized.* A wave emerging from a vertical antenna is expected to be *vertically polarized* whereas that from horizontal antenna *horizontally polarized.* However, the above is true in the case of a horizontal antenna whereas in the case of a vertical antenna it is not perfectly vertically polarized. The above description relates to only one time instant. During the course of the journey, the orientation of ***E*** changes, mainly due to the variations of media parameters. Thus the polarization of a wave is normally defined in terms of the direction of alignment of ***E*** vector during the passage of at least one complete cycle. Depending on the variation of magnitude and/or the orientation of the ***E*** vector during each cycle such a vector may mapout an ellipse, a circle, or a line in a plane normal to the direction of propagation. Due to such mappings, a wave may be termed as *elliptic, circular,* or *linearly* polarized. In fact the elliptic polarization is the general form and circular and linear polarizations are its special cases.

To further elaborate let us consider a wave traveling in an arbitrary (say, for simplicity, positive *z*) direction. Thus, the perpendicular component(s) of its ***E*** vector may lie along the *x*- or *y*-axis or in between *x* and *y*. This ***E*** vector can be written as $\boldsymbol{E} = E_x\boldsymbol{a_x} + E_y\boldsymbol{a_y}$. In view of this expression of ***E*** vector one can conclude the following:

1. If $E_y = 0$ and E_x is present, the wave is said to be *polarized in the x-direction.*
2. If $E_x = 0$ and E_y is present, the wave is said to be *polarized in the y-direction.*
3. If both E_x and E_y are present and are in phase, the resultant ***E*** vector will change in magnitude but not in direction. Such a wave is called a *linearly polarized wave.*
4. If both E_x and E_y are present and have the phase difference of 180° the wave is again linearly polarized but the vector direction will be opposite to that of an in-phase case.

5. If both E_x and E_y are present and have a phase difference of 90°, the resultant **E** vector at different time instants will have the same magnitude but will assume different directions. An envelope joining tips of this vector at different time instants will appear to be a circle. Such a wave is called *circularly polarized.*
6. If the phase difference between E_x and E_y is 270° the wave is again circularly polarized but the sense of polarization will get reversed, i.e. the directions of **E** vectors will be opposite to that in the case of the 90° phase difference.
7. If E_x and E_y components have a phase difference of other than 0°, 90°, 180°, or 270°, the resultant **E** vector will change in magnitude as well in direction. If an envelope joins tips of this vector at different time instants, it will appear to be an ellipse. Such a wave is termed as an *elliptically polarized wave.* The sense of polarization will reverse for angles between 180° and 360° from that obtained for angles between 0° and 180°.

In an ellipse the ratio of its minor to major axis (called axial ratio) is referred to as *ellipticity*. It is expressed in dBs and is always less than unity for an ellipse. The direction of the major axis of an ellipse is referred to as the *polarization orientation*. The polarization of a wave is completely specified by the *axial ratio, orientation*, and the *sense* of rotation and is defined at a point in a homogeneous isotropic medium. The linear, circular, and elliptical polarizations are further elaborated below.

16.7.1 Linear Polarization

A linearly polarized wave is defined as a transverse electromagnetic (TEM) wave for which the electric field vector at all times lies along a fixed line. It can be considered as a special case of elliptical polarization when its ellipticity becomes infinity. Figure 16.5a shows the orientation of the **E** vector at different time instants (T_1, T_2, etc.). In this case the magnitude of the **E** vector keeps on changing whereas its orientation remains unaltered.

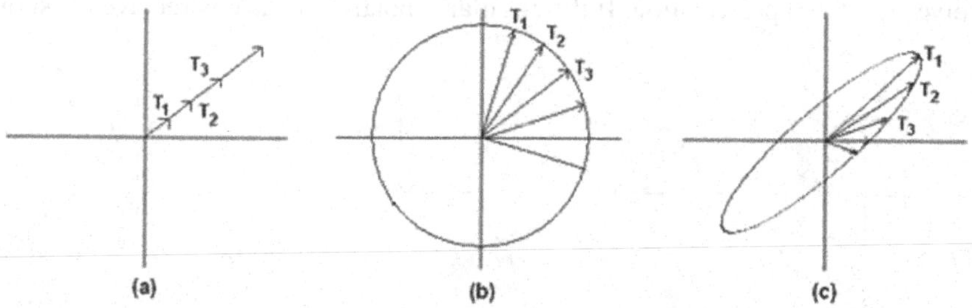

FIGURE 16.5 Orientation of an E vector for (a) linear; (b) circular; and (c) elliptic polarization.

16.7.2 Circular Polarization

A *circularly polarized wave* is defined as a TEM wave for which the electric field vector at a point describes a circle. It can be considered as a special case of elliptical polarization when its ellipticity becomes zero. For circular polarization, the orientation of the **E** vector at T_1, T_2, etc. is shown in Figure 16.5b. In this case, the magnitude of the **E** vector remains the same whereas its orientation keeps on changing.

16.7.3 Elliptic Polarization

As stated earlier, this is the most basic form of polarization. In this case, the vector **E** maps out an ellipse in the plane normal to the direction of propagation. Figure 16.5c shows the orientation of the **E** vector at different time instants in the case of elliptical polarization. Here the magnitude as well as direction of **E** keeps on changing at T_1, T_2, etc.

In most of the practical applications, the linearly polarized wave plays the major role. The circularly polarized wave, however, provides a response to a linearly polarized wave of arbitrary orientation. Such a situation may become necessary where the media causes the rotation of the plane of polarization. In addition, it helps in the suppression of precipitation clutter in radar services.

16.7.4 Sense of Polarization

The sense of polarization may be identified in reference to an observer. For an observer looking in the direction of propagation, the rotation of the electric field vector in a transverse plane *is clockwise for right-hand polarization* and *counterclockwise for left-hand polarization.* The elliptically and circularly polarized waves may have a *left-hand or right-hand sense.* Figures 16.6a and b illustrate the rotation of the **E** vector for right-hand and left-hand senses in circular polarization case. Similarly, Figures 16.6c and d illustrate the rotation of the **E** vector for right- and left-hand senses in the case of elliptic polarization.

The sense of rotation of a circularly polarized wave can be determined by using two helical antennas of opposite senses. A right-hand helical antenna transmits or receives right-hand polarization while a left-hand helical antenna transmits or receives left-hand polarization. If the circularly polarized wave is received first on a

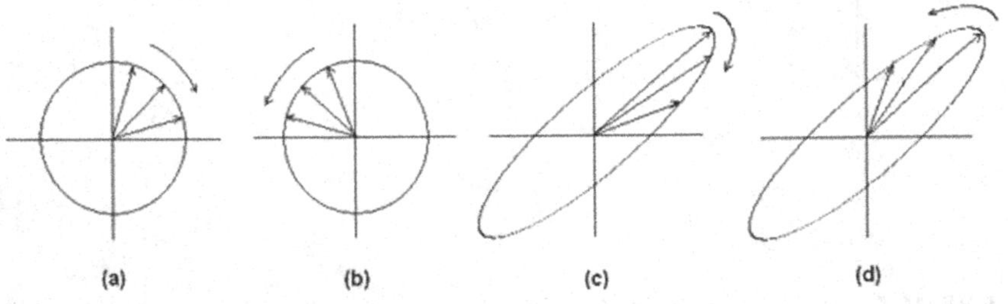

FIGURE 16.6 Rotation of an E vector for circular polarization (a) clockwise and (b) counterclockwise, and for elliptic polarization (c) clockwise and (d) counterclockwise.

right-hand helical antenna and then on the left-hand helical antenna, and the signal received by the right-handed helix is more than that of the left-hand helix, the wave will have right-hand sense. Thus the antenna receiving the greater amount of signal will have sense corresponding to the sense of the received wave. In the case of an elliptically polarized wave, the sense will be taken to be the same as that of the predominant circular component.

The polarizations of the two antennas are said to be *orthogonal* if one of these antennas does not respond to the wave emanating from the other antenna. A circularly polarized antenna is entirely unable to see its own image in any symmetrical reflecting surface. It is due to the fact that the reflected wave has its sense reversed and is thus orthogonal to the polarization of the antenna from which it has originated.

16.7.5 Mathematical Representation of Polarization

The mathematical expressions to represent polarization can be obtained by considering an un-attenuated traveling wave. Let such a wave traveling (say) in positive z-direction has E_x and E_y components with θ degree phase shift between them. In view of Equation 16.29, the real parts for the E_x and E_y components of the solution can be rewritten as:

$$E_x(z,t) = E_{0x}\cos(\omega t - \beta z) \quad (a) \quad E_y(z,t) = E_{0y}\cos(\omega t - \beta z + \theta) \quad (b)\} \qquad (16.47)$$

where E_{0x} and E_{0y} are the magnitudes of $E_x(z,t)$ and $E_y(z,t)$, respectively.

At some fixed space location (say $z = 0$) these equations reduce to:

$$E_x(t) = E_{0x}\cos(\omega t) \quad (a) \quad E_y(t) = E_{0y}\cos(\omega t + \theta) \quad (b)\} \qquad (16.48)$$

Since $E_x(t)$ and $E_y(t)$ are orthogonal to each other, vector $\boldsymbol{E}$ can now be written as:

$$\boldsymbol{E} = E_x(t) + jE_y(t) = E_{0x}\cos(\omega t) + jE_{0y}\cos(\omega t + \theta) \qquad (16.49)$$

From Equation 16.49 the magnitude $|\boldsymbol{E}|$ and phase (φ) of $\boldsymbol{E}$ can be obtained as:

$$\left.\begin{aligned} |\boldsymbol{E}| &= \sqrt{\{E_{0x}\cos(\omega t)\}^2 + \{E_{0y}\cos(\omega t + \theta)\}^2} \quad (a) \\ \varphi &= \tan^{-1}\left\{\frac{E_{0y}\cos(\omega t + \theta)}{E_{0x}\cos(\omega t)}\right\} \quad (b) \end{aligned}\right\} \qquad (16.50a)$$

The polarization plots can be obtained either from Equation 16.49 or Equation 16.50 by assigning suitable values to the parameters involved in the relations.

16.8 DEPTH OF PENETRATION AND SURFACE IMPEDANCE

This section deals with important aspects of plane wave viz. depth of penetration and the surface impedance. These are described below.

16.8.1 Depth of Penetration

When a wave travels it may come across different types of hindrances on its way. The moment it impinges on a surface it may get reflected back or penetrate the striking surface. This reflection or penetration will obviously depend on the material properties and the geometry of the surface. In addition, the amount of penetration will depend on the conductivity of the surface and the frequency of the wave. The amplitude of the penetrating wave will decrease in accordance with the factor $e^{-\alpha x}$ where x is the amount of penetration. If we assume a unit amplitude of a penetrating wave at $x = 0$, it will reduce by $e^{-\alpha\delta}$ of its original value at $x = \delta$. In reference to the above, the term "depth of penetration (δ)" can be defined as the measure of penetration at which the amplitude of a wave reduces by a factor e^{-1} of its original value. This gives $\alpha\delta = 1$ or $\delta = 1/\alpha$.

Thus, Equation 16.45b is reproduced here:

$$\alpha = \beta \approx \frac{\sqrt{\omega\mu\sigma}}{2} \tag{16.45b}$$

$$\delta = \frac{1}{\alpha} \approx \sqrt{\frac{2}{\omega\mu\sigma}} \tag{16.51}$$

In view of Equation 16.51, depth of penetration has an inverse relation with the frequency. At low and very low frequencies the wave can penetrate much deeper into land or water and thus these frequencies can be used for mine and marine communication.

16.8.2 Surface Impedance

It is a well-known fact that a conductor offers more resistance to the flow of ac current than to dc. This difference in resistance is attributed to the skin effect. The cause of this effect is the presence of inductance, which comes into existence the moment a current flows in a conductor irrespective of its ac or dc nature. As inductance is defined in terms of flux linkage per unit ampère the current well inside the conductor has more flux linkage or more inductance than at its boundary. As long as the frequency remains zero the inductance has no role to play. The moment that frequency starts increasing the reactance or impedance starts increasing and disturbs the uniformity of current flow. As per the inherent nature of current, it chooses the path of least resistance. At higher frequencies, more current starts flowing in the outer periphery than in the interior of the conductor. As illustrated in Figure 16.7 at very high frequencies the current is almost entirely confined to a very thin sheet at the surface of the conductor. The impedance offered by the surface to the flow of current is referred to as surface impedance (Z_s). It is given as:

$$Z_s = \frac{E_{\tan}}{J_s} \tag{16.52}$$

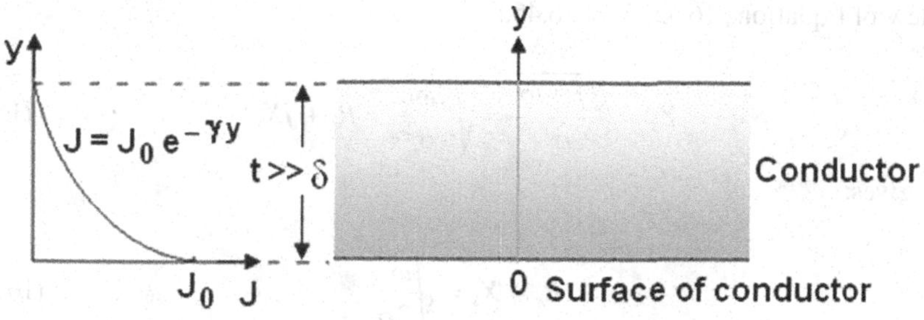

FIGURE 16.7 Current distribution in thick (conducting) flat plate.

where $E_{\tan}$ is the electric field parallel to and at the surface and J_s is the linear current density that flows as a result of $E_{\tan}$. It represents the total conduction current per meter width flowing in a thin sheet.

Assuming the conductor to be absolutely flat with one of its surfaces lying along the $y = 0$ plane, the current in the conductor in the y-direction can be given by:

$$J = J_0 e^{-\gamma y} \tag{16.53}$$

where J_0 is the current density at the surface.

With the assumption that the thickness of conductor (t) is much greater than the depth of penetration (δ) and thus there is no reflection from the back surface of the conductor, the surface current (J_s) can be written as:

$$J_s = \int_0^\infty J dy = J_0 \int_0^\infty e^{-\gamma y}\, dy = \left(-\frac{J_0}{\gamma} e^{-\gamma y} \right)_0^\infty = \frac{J_0}{\gamma} \tag{16.54a}$$

But the current density at the surface is:

$$J_0 = \sigma E_{\tan} \tag{16.54b}$$

Thus:

$$J_s = \sigma \frac{E_{\tan}}{\gamma} \tag{16.54c}$$

or

$$\frac{\gamma}{\sigma} = \frac{E_{\tan}}{J_s} = Z_s \tag{16.54d}$$

The relation for γ in good conductors is given as:

$$\gamma \approx \sqrt{j\omega\mu\sigma} \tag{16.45a}$$

In view of Equations 16.45a and 16.45d, we get:

$$Z_s = \frac{\sqrt{j\omega\mu\sigma}}{\sigma} = \sqrt{\frac{j\omega\mu}{\sigma}} = R_s + jX_s \tag{16.55}$$

This gives:

$$R_s = X_s = \sqrt{\frac{\omega\mu}{2\sigma}} \tag{16.56a}$$

Since

$$\delta \approx \sqrt{\frac{2}{\omega\mu\sigma}} \tag{16.51}$$

$$R_s = X_s = \frac{1}{\sigma\delta} \tag{16.56b}$$

The concept of surface impedance becomes relevant in many applications such as propagation of ground waves. As the wave glides over the Earth's surface it finds different conductivities and permittivities during the course of its journey. Accordingly, it may travel only along the surface or may penetrate. The charges appear and disappear and the current flows along or beneath the Earth's surface. The moment it penetrates, impedance is offered by the Earth. This impedance is referred to as surface impedance and is a function of the depth of penetration (δ) and the Earth's parameters σ and ε.

Example 16.1

Find the wave number for a plane wave with a frequency of 5 MHz traveling through (a) free space; (b) a medium having $\mu_r = 2$ and $\varepsilon_r = 3$.

SOLUTION

For $f = 5$ MHz, $\omega = 2\pi f = \pi \times 10^7$.

(a) For free space, $\mu = \mu_0$ and $\varepsilon = \varepsilon_0$.
In view of Equation 16.10a:

$$\beta_o \overset{\text{def}}{=} \omega \cdot \sqrt{\mu_o\varepsilon_o} = \frac{\omega}{c} = \frac{\pi \times 10^7}{3 \times 10^8} = 0.1045 \text{ rad/m}$$

(b) For medium having $\mu_r = 2$ and $\varepsilon_r = 3$, $\mu = \mu_0\mu_r = 2\mu_0$ and $\varepsilon = \varepsilon_0\varepsilon_r = 3\varepsilon_0$.
In view of Equation 16.10b:

$$\beta \overset{\text{def}}{=} \omega\sqrt{\mu\varepsilon} = \pi \times 10^7 \sqrt{2\mu_0 3\varepsilon_0} = \pi\sqrt{6} \times 10^7 \sqrt{\mu_0\varepsilon_0} = \frac{\pi\sqrt{6} \times 10^7}{3 \times 10^8} = 0.26 \text{ rad/m}$$

Example 16.2

A 50 MHz plane wave with an average Poynting vector of 10 W/m^2 travels through a lossless medium with relative permeability $\mu_r = 1$ and relative permittivity $\varepsilon_r = 4$. Find (a) the velocity of propagation υ; (b) wavelength λ; (c) impedance of medium η; and (d) the RMS value of electric field E_{rms}.

SOLUTION

Given $f = 5$ MHz, $P = 10$ W/m^2, $\mu_r = 1$, $\varepsilon_r = 4$:

$$\text{(a)}\ \upsilon = \frac{1}{\sqrt{\mu\varepsilon}} = \frac{1}{\sqrt{\mu_0\mu_r\varepsilon_0\varepsilon_r}} = \frac{1}{\sqrt{\mu_r\varepsilon_r}}\frac{1}{\sqrt{\mu_0\varepsilon_0}} = \frac{c}{\sqrt{\mu_r\varepsilon_r}} = \frac{3\cdot 10^8}{\sqrt{1\times 4}} = 1.5\times 10^8 \text{ m/s}$$

$$\text{(b)}\ \lambda = \frac{\upsilon}{f} = \frac{1.5\times 10^8}{50\times 10^6} = 3 \text{ m}$$

$$\text{(c)}\ \eta = \sqrt{\frac{\mu}{\varepsilon}} = \sqrt{\frac{\mu_0\mu_r}{\varepsilon_0\varepsilon_r}} = \sqrt{\frac{\mu_0}{\varepsilon_0}}\sqrt{\frac{\mu_r}{\varepsilon_r}} = 120\pi\sqrt{\frac{1}{4}} = 60\pi = 188.5\ \Omega$$

$$P = E\cdot H = \frac{(E_{rms})^2}{\eta} \quad \text{thus } (E_{rms})^2 = \eta\cdot P = 10\times 188.5 = 1885$$

$$E_{rms} = 43.416 \text{ V/m}$$

Example 16.3

The real part of E vector for a plane wave traveling in free space at a velocity of 1.5×10^8 m/s is given as $E_y(x,t) = 10\cos(10^8 t - \beta x)$. Find β, λ, and $\mathbf{H}$.

SOLUTION

In view of Equation 16.29 $\text{Re}\left[E_y(x,t)\right] = C_1\cdot\cos(\omega t - \beta x)$.

On comparing the given expression $E_y(x,t) = 10\cos(10^8 t - \beta x)$ with that of Equation 16.29, we get:

$\omega = 10^8 = 2\pi f$, thus: $f = \dfrac{\omega}{2\pi} = \dfrac{10^8}{2\pi} = 15.915$ MHz

Given $\upsilon = 1.5 \times 10^8 = f\lambda$, thus: $\lambda = \dfrac{\upsilon}{f} = \dfrac{1.5\times 10^8}{15.915\times 10^6} = 0.9425$ m,

$$\beta = \frac{\omega}{\upsilon} = \frac{10^8}{1.5\times 10^8} = 0.6666 \text{ rad/m}$$

$$H = \frac{E}{\eta_0} = \frac{10\cos(10^8 t - \beta x)}{377} = 0.0265\cos(10^8 t - \beta x) \text{ A/m}$$

Example 16.4

Find the time-average power density (PW/m^2) at $x = 1$ associated with a uniform plane wave with $\mathbf{E} = 30e^{-\alpha x}\cos\left(10^8 t - \beta x\right)\mathbf{a_z}$ V/m propagating through the medium characterized by: (a) μ_0, ε_0; (b) $\varepsilon_r = 2.26$, $\sigma = 0$.

SOLUTION

Since power is flowing in the positive x-direction and E has E_z component (or $E = E_z a_z$), H will have a $-H_y$ component (or $H = -H_y a_y$):

$$\mathbf{P} = \mathbf{E} \times \mathbf{H} = E_z \mathbf{a_z} \times \left(-H_y \mathbf{a_y}\right) = E_z H_y \mathbf{a_x} = P_x \mathbf{a_x}$$

$$E_z = 30e^{-\alpha x}\cos\left(10^8 t - \beta x\right) \qquad E_{z0} = 30$$

(a) For free space $\alpha = 0$, $\mu_r = \varepsilon_r = 1$, $\eta = \eta_0 = 377\ \Omega$.
In view of Equation 16.38b:

$$P_{x,av} = \frac{1}{2}\frac{E_{z0}^2}{\eta} = \frac{(30)^2}{2\eta_0} = \frac{450}{377} = 1.19 \qquad \mathbf{P} = P_x \mathbf{a_x} = P_{x,av}\mathbf{a_x} = 1.19\mathbf{a_x}$$

(b) For $\sigma = 0$, $\varepsilon_r = 2.26$:

$$\eta = \sqrt{\frac{\mu}{\varepsilon}} = \sqrt{\frac{\mu_0}{\varepsilon_0 \varepsilon_r}} = \sqrt{\frac{\mu_0}{\varepsilon_0}}\sqrt{\frac{1}{\varepsilon_r}} = \eta_0\sqrt{\frac{1}{\varepsilon_r}} = 377\sqrt{\frac{1}{2.26}} = 250.777$$

$$P_{x,av} = \frac{1}{2}\frac{E_{z0}^2}{\eta} = \frac{(30)^2}{2 \times 250.7772} = \frac{900}{501.55} = 1.7944 \qquad \mathbf{P} = 1.7977\mathbf{a_x}$$

Example 16.5

Using $\frac{\sigma}{\omega\varepsilon} = 1$ as the demarcation, identify (a) porcelain; and (b) copper as a good conductor or a good dielectric at different frequencies.

SOLUTION

In view of Equation 16.38, if $\frac{\sigma}{\omega\varepsilon} \ll 1$ the material is a good dielectric and if $\frac{\sigma}{\omega\varepsilon} \gg 1$ the material is a good conductor.

(a) From Appendix A2 in the eResources, for porcelain $\sigma = 2 \times 10^{-13}$ mhos/m and $\varepsilon_r = 6$.

For $f = 10$ KHz, $\omega = 2\pi \times 10^4$, $\frac{\sigma}{\omega\varepsilon} = \frac{2 \times 10^{-13}}{2\pi \times 10^4 \times 6} = 0.05305 \times 10^{-17} \ll 1$

For $f = 10$ MHz, $\omega = 2\pi \times 10^7$, $\frac{\sigma}{\omega\varepsilon} = \frac{2 \times 10^{-13}}{2\pi \times 10^7 \times 6} = 0.05305 \times 10^{-20} \ll 1$

For $f = 10$ GHz, $\omega = 2\pi \times 10^{10}$, $\dfrac{\sigma}{\omega\varepsilon} = \dfrac{2\times10^{-13}}{2\pi\times10^{10}\times6} = 0.05305\times10^{-23} \ll 1$

Thus, the porcelain is a good dielectric over $\dfrac{\sigma}{\omega\varepsilon} = \dfrac{5.8\times10^{7}}{2\pi\times10^{4}\times8.854\times10^{-12}}$

$= 0.104258\times10^{15} \gg 1$ a wide range of frequency as $\dfrac{\sigma}{\omega\varepsilon} \gg 1$.

(b) From Appendix A2, for copper $\sigma = 5.8\times 10^7$ mhos/m and $\varepsilon_r = 1$.

For $f = 10$ KHz

For $f = 10$ MHz $\dfrac{\sigma}{\omega\varepsilon} = \dfrac{5.8\times10^{7}}{2\pi\times10^{7}\times8.854\times10^{-12}} = 0.104258\times10^{12} \gg 1$

For $f = 10$ GHz $\dfrac{\sigma}{\omega\varepsilon} = \dfrac{5.8\times10^{7}}{2\pi\times10^{10}\times8.854\times10^{-12}} = 0.104258\times10^{9} \gg 1$

Thus, copper is a good conductor over a wide range of frequency as $\dfrac{\sigma}{\omega\varepsilon} \gg 1$.

Example 16.6

Find the type of polarization for a wave traveling in positive z-direction with field components $E_x = E_{0x}\angle\theta_1 a$ and $E_y = E_{0y}\angle\theta_2$ if (a) $E_{0x} = E_{0y} = 5$, $\theta_1 = \theta_2 = 45°$; (b) $E_{0x} = 5$, $E_{0y} = 10$, $\theta_1 = \theta_2 = 45°$; (c) $E_{0x} = E_{0y} = 5$, $\theta_1 = 0°$, $\theta_2 = 90°$; (d) $E_{0x} = 5$, $E_{0y} = 10$, $\theta_1 - 15°$, $\theta_2 = 75°$.

SOLUTION

At some fixed point in space location (say $z = 0$), the field components for different cases can be written as:

(a) $E_x(t) = 5\cos(\omega t + 45°)$ and $E_y(t) = 5\cos(\omega t + 45°) = 1\times E_x(t)$

Since $E_y(t)$ is linearly related to $E_x(t)$, the wave is linearly polarized.

(b) $E_x(t) = 5\cos(\omega t+45°)$ and $E_y(t) = 10\cos(\omega t+45°) = 2 \times E_x(t)$
Since $E_y(t)$ is linearly related to $E_x(t)$, the wave is linearly polarized.

(c) $E_x(t) = 5\cos(\omega t + 0°) = 5\cos\omega t$ and $E_y(t) = 5\cos(\omega t + 90°) = -5\sin\omega t$

$$E_x^2(t) + E_y^2(t) = 50,$$

This is the equation of a circle; thus, the wave is circularly polarized.

(d) $E_x(t) = 5\cos(\omega t + 15°) = 5\cos\omega t\cos 15° - 5\sin\omega t \sin 15°$

$$= 4.83\cos\omega t - 1.294\sin\omega t$$

$$E_y(t) = 10\cos(\omega t + 75°) = (10\cos\omega t\cos 75° - 10\sin\omega t\sin 75°)$$
$$= 2.588\cos\omega t - 9.66\sin\omega t$$

The polarization plots can be obtained by evaluating $|E|$ and φ in view of Equations 16.50a and 16.50b, respectively, by assigning suitable values to ωt in these relations. This will result in an ellipse, thus the wave is elliptically polarized.

Example 16.7

Find the depth of penetration for a wave in copper at (a) 20 MHz; (b) 5 GHz; and (c) 50 GHz.

SOLUTION

For copper $\sigma = 5.8 \times 10^7$ mhos/m, $\mu = \mu_0 = 4\pi \times 10^{-7}$.
In view of Equation 16.51:

$$\delta = \sqrt{\frac{2}{\omega\mu\sigma}} = \sqrt{\frac{1}{\omega}}\sqrt{\frac{2}{\mu\sigma}} = \sqrt{\frac{1}{\omega}}\sqrt{\frac{2}{(4\pi\times 10^{-7})(5.8\times 10^{7})}} = \frac{0.16565}{\sqrt{\omega}}$$

(a) At $f = 20\,\text{MHz}, \omega = 40\pi\cdot 10^6, \sqrt{\omega} = 11.21\times 10^3, \delta = \dfrac{0.16565}{11.21\times 10^3} = 14.777\ \mu\text{m}$

(b) At $f = 5$ GHz, $\omega = \pi\cdot 10^{10}$, $\sqrt{\omega} = 1.772\times 10^5, \delta = \dfrac{0.16565}{1.772\times 10^5} = 0.9346\ \mu\text{m}$

(c) At $f = 50$ GHz, $\omega = 10\pi\cdot 10^{10}$, $\sqrt{\omega} = 5.6\times 10^5$, $\delta = \dfrac{0.16565}{5.6\times 10^5} = 0.2955\ \mu\text{m}$

Example 16.8

Find the surface resistance and surface reactance for the sandy soil having $\sigma = 2 \times 10^{-5}$ mhos/m, $\varepsilon_r = 3$ and $\mu_r = 1$ at: (a) 5 kHz; (b) 50 kHz; and (c) 500 kHz.

SOLUTION

Given $\sigma = 10^{-5}$ mhos/m, $\varepsilon_r = 3$, thus $\varepsilon = 3\varepsilon_0$ and $\mu_r = 1$, $\mu = \mu_0$.

$$\sqrt{\frac{\mu}{2\sigma}} = \sqrt{\frac{\mu_0}{2\sigma}} = \sqrt{\frac{4\pi\times 10^{-7}}{4\times 10^{-5}}} = \sqrt{\frac{\pi}{100}} = 0.1772$$

In view of Equation 16.56a:

$$R_s = X_s = \sqrt{\frac{\omega\mu}{2\sigma}} = \sqrt{\omega}\sqrt{\frac{\mu}{2\sigma}} = 0.1772\times\sqrt{\omega}$$

(a) $f = 5$ KHz, $\omega = 2\pi f = \pi \times 10^4$, $\sqrt{\omega} = 100\sqrt{\pi} = 177.2$

$$R_s = X_s = 0.1772 \times \sqrt{\omega} = 0.1772 \times 177.2 = 31.4\ \Omega$$

(b) $f = 50$ KHz, $\omega = 2\pi f = \pi \times 10^5$, $\sqrt{\omega} = 100\sqrt{10\pi} = 560.5$

$$R_s = X_s = 0.1772 \times \sqrt{\omega} = 0.1772 \times 560.5 = 99.32\ \Omega$$

(c) $f = 500$ KHz, $\omega = 2\pi f = \pi \times 10^6$, $\sqrt{\omega} = 1000\sqrt{\pi} = 1772$

$$R_s = X_s = 0.1772 \times \sqrt{\omega} = 0.1772 \times 1772 = 314\ \Omega$$

PROBLEMS

P16.1 Find the wave number for a plane wave of frequency of 20 MHz traveling through a medium having (a) $\mu_r = 1$ and $\varepsilon_r = 3$; (b) $\mu_r = 2$ and $\varepsilon_r = 4$.

P16.2 Find (a) the velocity of propagation v; (b) wavelength λ; (c) impedance of medium η; and (d) the RMS value of electric field E_{rms} for a 150 MHz plane wave having an average Poynting vector of 1 W/m^2 traveling through a lossless medium with relative permeability $\mu_r = 2$ and relative permittivity $\varepsilon_r = 3$.

P16.3 Find the relative dielectric constant (ε_r) for a non-magnetic media, which carries a wave characterized by (a) characteristic impedance $\eta = 200\ \Omega$; (b) phase shift constant $\beta = 0.01$, frequency $f = 200$ kHz; and (c) wavelength $\lambda = 2$ cm at $f = 10$ GHz.

P16.4 Find time-average power density (P W/m^2) at $x = 1$ associated with a uniform plane wave with $\boldsymbol{E} = 20e^{-\alpha x}\cos\left(10^6 t - \beta x\right)\boldsymbol{a}_x$ V/m propagating in a medium having (a) $\varepsilon_r = 4.9$, $\sigma = 0$; (b) $\varepsilon_r = 3.4$, $\dfrac{\sigma}{\omega\varepsilon} = 0.2$.

P16.5 Find the values of α, β, v, and η for copper at 5 MHz.

P16.6 Find the values of α, β, v, and η for porcelain at 5 MHz.

P16.7 Find the type of polarization for a wave traveling in the positive z-direction with the field components $E_x = E_{0x}\angle\theta_1$ and $E_y = E_{0y}\angle\theta_2$ if: (a) $E_{0x} = 5$, $E_{0y} = 10$, $\theta_1 = \theta_2 = 30°$; (b) $E_{0x} = 5$, $E_{0y} = 25$, $\theta_1 = \theta_2 = 60°$.

P16.8 Find the type of polarization for a wave traveling in positive z direction with field components $E_x = E_{0x}\angle\theta_1$ and $E_y = E_{0y}\angle\theta_2$ if: (a) $E_{0x} = E_{0y} = 10$, $\theta_1 = 15°$, $\theta_2 = 105°$; (b) $E_{0x} = 5$, $E_{0y} = 10$, $\theta_1 = 40°$, $\theta_2 = 80°$.

P16.9 How deep a 100 Hz wave will penetrate before it reduces to 1/e times of its original value at the surface in (a) silver ($\sigma = 6.17 \times 10^7$ ℧/m); (b) aluminum ($\sigma = 3.72 \times 10^7$ ℧/m); and (c) brass ($\sigma = 1.5 \times 10^7$ ℧/m).

P16.10 Find the surface resistance for the slabs for which the conductivity (σ) and depth of penetration (δ) at 1000 Hz are: (a) silver $\sigma = 6.17 \times 10^7$ ℧/m, $\delta = 0.002$ m; (b) copper $\sigma = 5.8 \times 10^7$ ℧/m, $\delta = 0.0021$ m; (c) aluminum $\sigma = 3.72 \times 10^7$ ℧/m, $\delta = 0.0026$ m; and (d) brass $\sigma = 1.5 \times 10^7$ ℧/m, $\delta = 0.0041$ m.

DESCRIPTIVE QUESTIONS

Q16.1 What is the logic of assuming a sinusoidal time variation, while the actual signals may not be sinusoidal?

Q16.2 Discuss types of media that may be encountered by an electromagnetic wave.

Q16.3 Obtain the solution of wave equation obeyed by an $\boldsymbol{E}$ field in a lossless media. How does it lead to a composite solution involving forward and backward waves?

Q16.4 Explain the terms plane wave, uniform plane wave, traveling wave, and standing wave.

Q16.5 Obtain the relation that describes the characteristic impedance of the medium.

Q16.6 Obtain the solution of the wave equation in a lossy media. How does it differ from that of lossless media?

Q16.7 Discuss the criteria through which the conductors and dielectrics can be distinguished.

Q16.8 Discuss the impact on α, β, v, and η of small conductivity in dielectrics.

Q16.9 Explain the meaning and type of polarization of a wave.

Q16.10 Explain the meaning of depth of penetration and surface impedance.

FURTHER READING

Given at the end of Chapter 17.

17 Reflection and Refraction of Electromagnetic Waves

17.1 INTRODUCTION

Reflection and refraction of electromagnetic waves are a key concern in any communication system and must be properly addressed. Whether guided or unguided, a wave may come across different type of environments and hindrances with varied geometries and media parameters. As a consequence, it may get reflected, refracted, diffracted, or absorbed. The wave behavior at the interface of different types of media requires the understanding of certain preliminaries, viz. the direction cosines, wave and phase velocities, nature of medium, boundary conditions to be satisfied, and the nature of striking waves, etc. In this chapter, reflection and refraction cases are classified on the basis of type of material, type of surfaces, mode of incidence, and the type of polarization of the wave. In the first case, a material may be a perfect conductor, a perfect dielectric, or a mixture of both. In the second case, a surface may be perfectly flat (without any aberrations and irregularities) or of an irregular shape. In the third case, a wave may strike a surface perpendicularly or obliquely. Finally, in the fourth case, a wave may be vertically or horizontally polarized.

17.2 DESCRIPTION OF SOME BASIC TERMS

In electromagnetic terminology a wave is conceived in terms of the periodic variation of E and H fields with time and space. Thus, the term *striking wave* refers to E and H impinging upon the boundary or on an interface between two media. This striking wave may cause induction of charges on the surface or flow of current in the material to which this surface belongs. The charges so induced remain on the surface, however, the current may either remain confined to the skin or may penetrate a little deeper. As observed in Chapter 16 the depth of penetration depends on the characteristics of the material and the wave frequency.

The surfaces which are perpendicular to each other are referred to as *orthogonal surfaces*. These surfaces, referred to as *boundary surface* and *plane of incidence*, are shown in Figure 17.1.

A noted in Chapter 16, a *ray* is the perpendicular drawn to an equiphase plane. A ray striking on a surface is called the *incident ray*, a ray rebounding from a surface is the *reflected ray*, and a ray passing into the second media with the changed angle is called the *refracted ray*. An angle between incident ray and the normal is called the *angle of incidence*, angle between reflected ray and the normal *angle of reflection* and the angle between refracted ray and the normal *angle of refraction*. *Normal* is the perpendicular drawn to the boundary surface at the point of incidence.

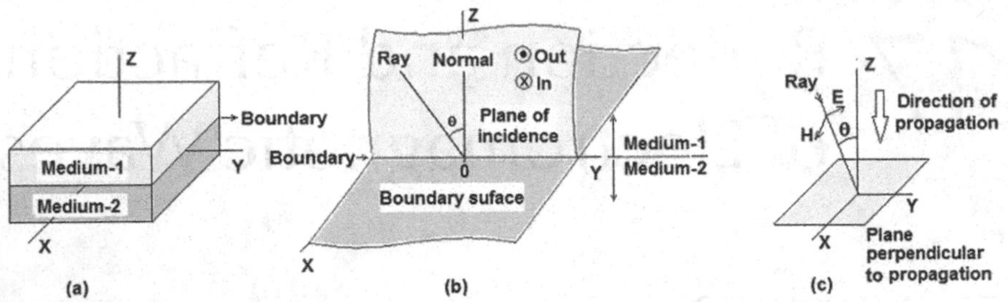

FIGURE 17.1 (a) Boundary between media 1 and 2; (b) boundary surface, plane of incidence, ray, and normal; (c) direction of wave propagation and a perpendicular plane.

Figure 17.1a shows two slabs located in an X–Y–Z space coordinate system. These slabs have different characteristic parameters and are called Medium 1 and Medium 2. Figure 17.1b shows two perpendicular surfaces referred to as the boundary surface and the plane of incidence. This also shows the incidence angle (θ), which the incident ray makes with the normal. When a wave strikes the boundary surface such that the ray lies along the normal the wave it is referred to as *perpendicular incidence.* When a wave strikes the boundary at an angle between 0° and 90° it is said to be *oblique incidence.* Figure 17.1c the direction of propagation of a wave, a ray striking a plane perpendicular to the direction of propagation. A ray represented by a straight line does not indicate the direction of the *E* or *H* fields. Figure 17.1c also shows E and H components. In Figure 17.1b, the *E* field perpendicular to the plane of incidence, is represented by a cross in a circle while entering the plane and by a dot in a circle while emerging from the plane. Some of these terms are further illustrated in Figure 17.2.

The term *reflection* indicates rebounding of electric field (or light, heat or sound) from a surface through the same medium in which the incident field exists. The

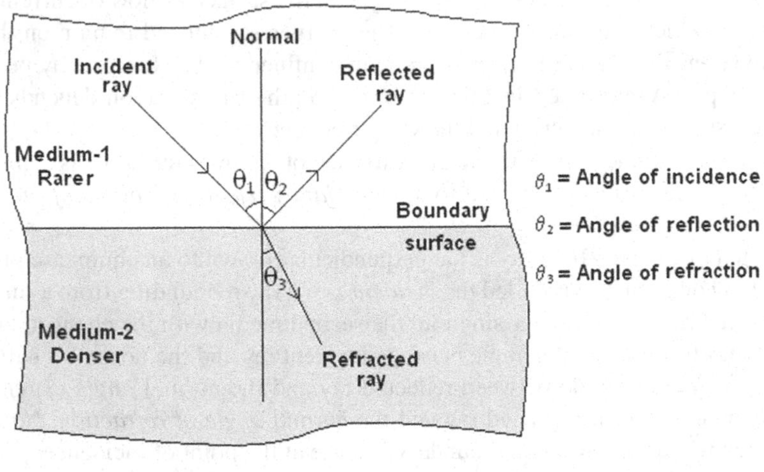

FIGURE 17.2 Parameters related with reflection and refraction.

term *refraction* refers to the change of direction of a ray which is obliquely incident and passes through a surface bounding the two media in which the ray has different velocity. In connection with the reflection and refraction one can observe that: (a) in isotropic media when a ray passes from rarer to denser media it refracts toward the normal and when it passes from denser to rarer media away from the normal; and (b) the sines of incident and refracted angles bear a constant ratio to each other for any given media. If the first medium is air (or a vacuum), this ratio is called the *refractive index* or *index of refraction*; (c) the incident, reflected, and refracted rays lie in the same plane; and (d) the incident and reflected angles are always equal.

There are some more terms commonly referred to in connection with the reflection, refraction, and the transmission of waves. These include: (a) the ratio of the field contents of E (or H) in reflected and incident waves, referred to as *reflection coefficient*; (b) the ratio of the field contents of E (or H) in refracted and incident waves, referred to as *refraction coefficient*; and (c) the ratio of field contents of E (or H) in transmitted and incident waves, as the *transmission coefficient*, all for E (or H).

As the striking ray may also be referred to as striking wave thus this must represent both the E and H fields. This representation leads to the identification of a wave vis-à-vis its polarization. Thus, when E is perpendicular to the plane of incidence and H perpendicular to boundary surface the wave is referred to as *perpendicularly* or *horizontally polarized*; when E is parallel to the plane of incidence and H parallel to the boundary surface the wave is called *parallelly* or *vertically polarized*. These two cases are illustrated in Figures 17.3a and b. In the following sections these two cases are separately discussed for conductors and dielectrics.

17.3 BOUNDARY CONDITIONS AND THEIR IMPLICATIONS

The idea of reflection and refraction emanates from the conditions at the boundary, the wave is impinging upon, being satisfied by the electric and magnetic fields. For time-varying E, D, H, and B fields, the boundary conditions have already been discussed in Chapter 15.

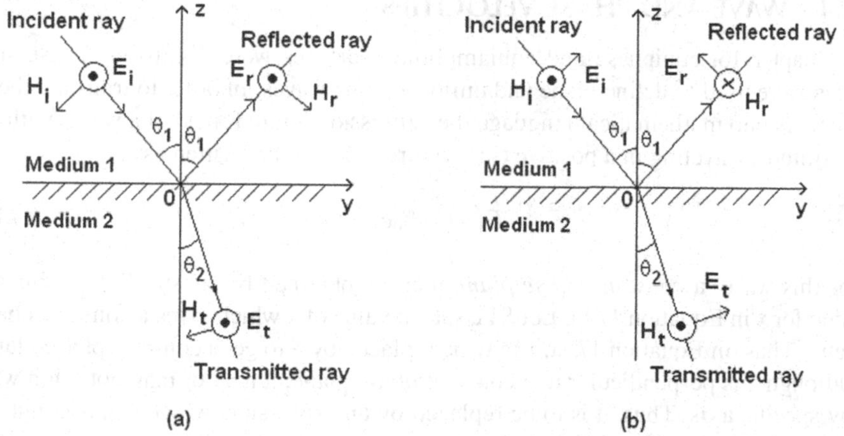

FIGURE 17.3 (a) Perpendicular (horizontal); and (b) parallel (vertical) polarizations.

In Chapter 8, it was noted that no time-invariant E field can exist inside a good conductor. There it was further noted that if a charge is somehow placed inside the conductor it will appear on the surface within a very short time. This time, referred to as the *relaxation time*, for copper was found to be of the order of 10^{-19} seconds. Thus, the moment a ray strikes on the surface of a good conductor it is bound to reflect as the E field cannot penetrate inside. If this incidence is perpendicular the rebound will be perpendicular and if oblique the rebound will also be oblique. Further the reflected ray will carry the energy contents in the direction of its travel. Thus in perpendicular incidence the direction of reflected ray will be just opposite to that of the incident ray whereas in oblique case the reflected ray will make an angle which will be equal to that of incident angle but on the other side of normal to the surface. Similarly, the current flow shall remain confined to the skin of the surface of perfect conductor due to the non-penetration of the field. The reflection will be total in the case of infinite conductivity. When conductivity reduces some field may penetrate inside and the reflection will be partial. In the case of perfect dielectrics, the wave penetrates and crosses the medium through refraction. This refraction toward or away from normal will depend on the relative permittivities of the two media.

If the reflection is assumed to be from a smooth surface acting as an interface, one can observe that (a) when a *vertically polarized wave* is reflected there is no change in its characteristics; (b) when a *horizontally polarized wave* is reflected there is 180° phase change, because of the coordinate system reversal in space when one looks in the reversed direction of propagation; (c) when a *circularly polarized wave* is reflected the phase of its horizontal component gets altered by 180°. Since the magnitudes of vertical and horizontal components are the same the sense of polarization gets reversed; (d) in the case of the reflection of an *elliptically polarized* wave, the phase of its horizontal component changes by 180° and the sense of polarization again gets reversed. But since the magnitudes of vertical and horizontal components are not the same the new polarization ellipse may be obtained by vectorially combining the vertical and phase reversed horizontal components.

17.4 WAVE AND PHASE VELOCITIES

In Chapter 16, equiphase and equiamplitude surfaces were discussed. These surfaces were used to define plane and uniform plane waves. In order to translate these concepts into mathematical language the expression for an E field of a wave, with E_0 magnitude, traveling in a positive (say, x) direction can be written as:

$$E(x) = E_0 e^{-j\beta x} \tag{17.1a}$$

For this wave, a *constant phase plane* may be obtained by substituting a constant value for x in Equation 17.1a. Let S be such a value of x which gives a constant phase plane. Thus, in Equation 17.1a, x may be replaced by S to get a constant phase plane. Although S is perpendicular to a constant phase plane, it may or may not align with any specific axis. Thus, it is to be replaced by an expression, which can account for any or all the three axes of the selected coordinate system. If such an expression is equated to a constant it will give an equiphase surface.

Let the equation of a plane is given by $\tilde{n} \cdot \check{r} = a$ where a is a constant. In this relation ř represents a vector with components x, y, z, and $\tilde{n}$ is a unit vector (lying along S) making angles A, B, and C with these axes, respectively. Thus, cos A, cos B, and cos C are the components of the unit vector ($\tilde{n}$) along the x-, y-, and z-axes. These vectors and angles are shown in Figure 17.4. In view of this description we can write:

$$\tilde{n} \cdot \check{r} = x\cos A + y\cos B + z\cos C \tag{17.1b}$$

In Equation 17.1b, cos A, cos B, and cos C are referred to as direction cosines. In terms of these a plane wave traveling in the direction $\check{r}$, with ñ as the normal to the plane of constant phase, can be written as:

$$E(r) = E_0 e^{-j\beta\tilde{n}\cdot\check{r}} = E_0 e^{-j\beta(x\cos A + y\cos B + z\cos C)} \tag{17.2a}$$

In time-varying form with the assumption of E_0 (= E_r + jE_i) being complex and incorporation of the factor $e^{j\omega t}$ for time variation of Equation 17.1a, Equation 17.2a can be written as:

$$E(r,t) = \text{Re}\left[E_0 e^{-j(\beta n\cdot r \ \ \omega t)}\right] = E_r \cos(\beta \boldsymbol{n}\cdot\boldsymbol{r} - \omega t) + E_i \sin(\beta \boldsymbol{n}\cdot\boldsymbol{r} - \omega t) \tag{17.2b}$$

In Equation 16.12c, the term wave velocity was defined, indicated by v_0. In lossless media it was taken equal to the *velocity of light.* Equation 16.30 defined another velocity, i.e. phase velocity, which is the velocity with which the phase of progressing wave changes (with space). In free space the wave and phase velocities have the same value. If there is any slight imperfection of the dielectric, the phase velocity gets slightly reduced. Equation 16.43 gives the difference between the two velocities. For conductors with high conductivities, the value of *phase velocity* was given by Equation 16.46a.

In general, wave propagation in space is a three-dimensional phenomenon. Assuming that a wave of a specified frequency is traveling in an arbitrary

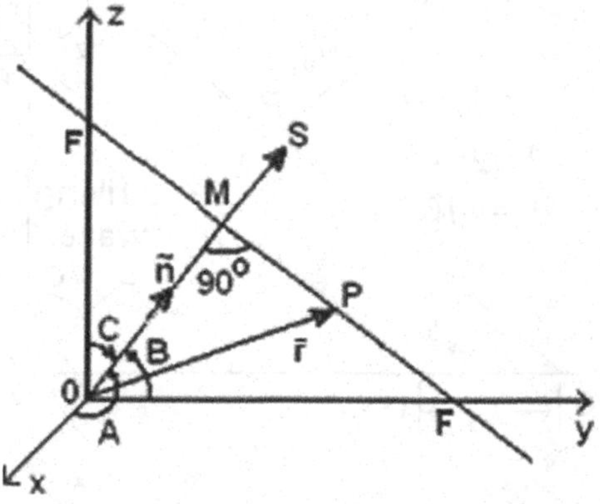

FIGURE 17.4 Illustration of different vectors and angles involved in Equations 17.1 and 17.2.

direction with a velocity (say υ). In accordance with the frequency, it will have some wavelength (say λ). If one wishes to find velocity and wavelength of this wave at any location, it will obviously be different. In the case these are to be obtained along different axes, one has to draw projections on the different coordinates.

Figure 17.5 illustrates such a situation where a wave strikes at the y-axis with a velocity υ_0 (or c). As before the wave is making angles A, B, and C with x-, y-, and z-axes. In this figure the components of wavelength λ and velocity υ along different axes are:

$$\left.\lambda_x = \frac{2\pi}{\beta\cos A} = \frac{\lambda}{\cos A} \quad \text{(a)} \quad \lambda_y = \frac{2\pi}{\beta\cos B} = \frac{\lambda}{\cos B} \quad \text{(b)} \quad \lambda_z = \frac{2\pi}{\beta\cos C} = \frac{\lambda}{\cos C} \quad \text{(c)}\right\} \tag{17.3}$$

$$\left.\upsilon_x = \frac{\omega}{\beta\cos A} = \frac{\upsilon}{\cos A} \quad \text{(a)} \quad \upsilon_y = \frac{\omega}{\beta\cos B} = \frac{\upsilon}{\cos B} \quad \text{(b)} \quad \upsilon_z = \frac{\omega}{\beta\cos C} = \frac{\upsilon}{\cos C} \quad \text{(c)}\right\} \tag{17.4}$$

As long as angles A, B, and C are not 0, the wavelengths and phase velocities along the respective axes are greater than when measured along the wave normal (i.e. λ_y, $\lambda_z > \lambda$). Figure 17.5 shows the wavelengths (λ_y, λ_z) and phase velocities (υ_y, υ_z) along the y- and z-axes. These values depend on the striking angle. It is to be noted that wavelengths λ_x, λ_y, and λ_z are simply the projections of λ on x-, y-, and z-axes and not the vector components like υ_x, υ_y, and υ_z.

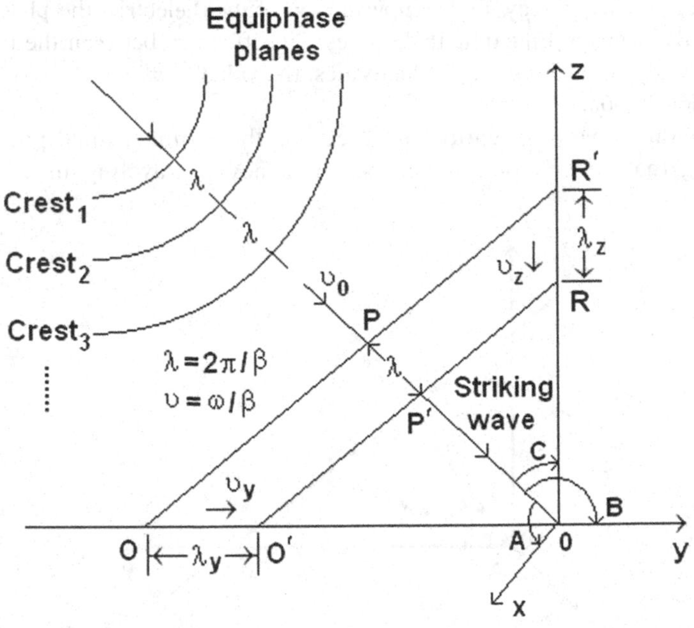

FIGURE 17.5 Wavelength and velocity components.

17.5 REFLECTION FROM PERFECT CONDUCTOR

In this case the surface is assumed to be perfectly conducting and flat without any aberrations. Since a ray may strike a surface perpendicularly or obliquely, the two cases are separately considered.

17.5.1 Normal Incidence

Let us consider that a wave strikes at the surface of a perfect conductor. In the boundary condition a perfect conductor does not allow electric field to penetrate or exist inside and the striking energy is bound to reflect. This situation is depicted in Figure 17.6 wherein a forward (or incident) and a backward (or reflected) waves are shown.

Let us assume that E_i and E_r be the electric field intensities of incident and reflected waves at surface ($x = 0$). Since there is total reflection and the two (incident and reflected) waves travel in opposite directions the continuity of tangential components of E demands that the electric field intensities of incident and reflected waves must bear the relation $E_i = -E_r$. In the Poynting theorem ($\boldsymbol{P} = \boldsymbol{E} \times \boldsymbol{H}$), the reversal of the direction of power flow requires change of phase either of $\boldsymbol{E}$ or $\boldsymbol{H}$ vector but not of both. Thus, in the case of magnetic field intensity ($\boldsymbol{H}$) there will be no phase reversal at $x = 0$ and $H_i = H_r$. With the assumption of lossless medium the expressions of E for incident and reflected waves can be given by $E_i e^{-j\beta x}$ and $E_r e^{j\beta x}$, respectively. Thus, the total electric field $\{E_t(x)\}$ at any point along the x-axis can be written as:

$$E_t\left(x\right) = E_i e^{-j\beta x} + E_r e^{j\beta x} = E_i\left(e^{-j\beta x} - e^{j\beta x}\right) = -2jE_i \sin\left(\beta x\right) \tag{17.5a}$$

$$E_t\left(x,t\right) = \text{Re}\left\{-2jE_i \sin\left(\beta x\right) e^{j\omega t}\right\} = 2E_i \sin\left(\beta x\right)\sin\left(t\right) \tag{17.5b}$$

The total magnetic field $H_t(x)$ obtained in a similar manner can be written as:

$$H_t\left(x\right) = H_i e^{-j\beta x} + H_r e^{j\beta x} = H_i\left(e^{-j\beta x} + e^{j\beta x}\right) = 2H_i \cos\left(\beta x\right) \tag{17.6a}$$

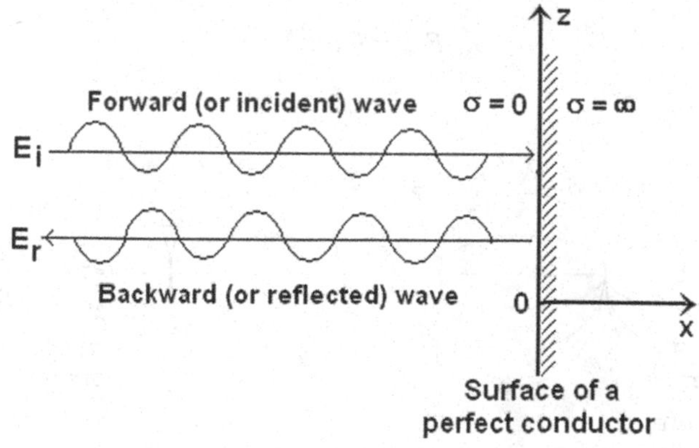

FIGURE 17.6 Normal incidence – perfect conductor.

$$H_t(x,t) = \text{Re}\left\{2H_i \cos(\beta x) e^{j\omega t}\right\} = 2H_i \cos(\beta x)\cos(\omega t) \tag{17.6b}$$

Study of Equations 17.5b and 17.6b shows that both E and H exhibit standing wave patterns. As per the illustration of Figure 17.7 neither E nor H wave is progressing and their minima and maxima are stagnated at the same physical location of x. The magnitudes of maxima of both E_t and H_t are, however, changing at different instants of time. In Equations 17.5b and 17.6b and Figure 17.7 the orthogonal nature of E and H can clearly be observed.

17.5.2 Oblique Incidence

The oblique incidence cases of perpendicular and parallel polarization are considered separately.

17.5.2.1 Perpendicular Polarization

This case is shown in Figure 17.8 wherein E is perpendicular to the plane of incidence. In this figure E_{ref} and E_{inc} can be written as below.

$$E_{\text{ref}} = E_r e^{-j\beta \boldsymbol{n}\cdot\boldsymbol{r}} = E_r e^{-j\beta(x\cos A + y\cos B + z\cos C)} \tag{17.7a}$$

where

$$\boldsymbol{n} \bullet \boldsymbol{r} = \left[x\cos\frac{\pi}{2} + y\cos\left(\frac{\pi}{2} - \theta\right) + z\cos\theta \right] = y\sin\theta + z\cos\theta \tag{17.7b}$$

Thus:

$$E_{\text{ref}} = E_r e^{-j\beta(y\sin\theta + z\cos\theta)} \tag{17.7c}$$

Similarly:

$$E_{\text{inc}} = E_i e^{-j\beta \boldsymbol{n}\cdot\boldsymbol{r}} \tag{17.8a}$$

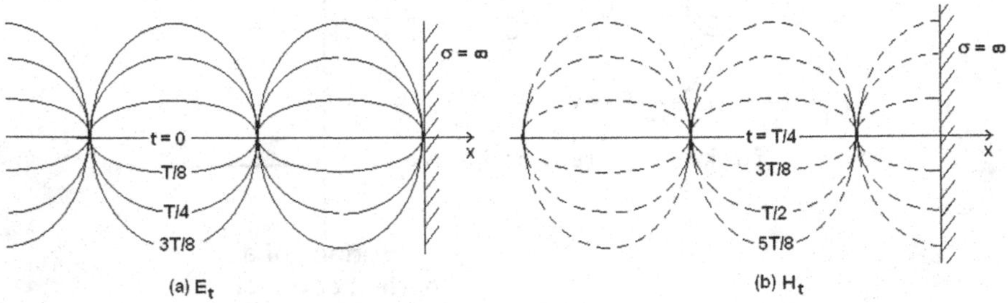

FIGURE 17.7 Standing waves for E_t and H_t fields.

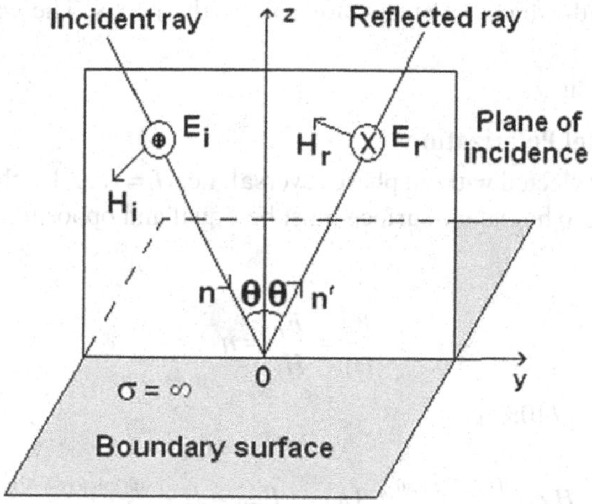

FIGURE 17.8 Perpendicular polarization case.

where

$$\boldsymbol{n} \bullet \boldsymbol{r} = \left[x\cos\frac{\pi}{2} + y\cos\left(\frac{\pi}{2} - \theta\right) + z\cos(\pi - \theta) \right] = y\sin\theta - z\cos\theta \qquad (17.8b)$$

Thus:

$$E_{\text{inc}} = E_i e^{-j\beta(y\sin\theta - z\cos\theta)} \qquad (17.8c)$$

At any point along the z-axis, the total field E_t will be the sum of incident and reflected field contents, which can be obtained by summing E_{ref} and E_{inc}. Since at $z = 0$, $E_t = 0$, it leads to the relation $E_i = -E_r$. Thus:

$$E_t = E_i\left[e^{-j\beta(y\sin\theta - z\cos\theta)} - e^{-j\beta(y\sin\theta + z\cos\theta)} \right] = 2jE_i\left[\sin\beta_z z\, e^{-j\beta_y y} \right] \qquad (17.9a)$$

where

$$\left.\beta_z = \beta\cos\theta \ \ (\text{b}) \quad \text{and} \quad \beta_y = \beta\sin\theta \ \ (\text{c})\right\} \qquad (17.9)$$

β_z and β_y, given by Equations 17.9b and 17.9c indicate the phase shift constants along the z and y directions, respectively. Equation 17.9a shows a standing wave distribution of E along the z-axis. The wavelength (twice the distance between nodal points) measured along the z-axis (λ_z) is greater than λ of the incident wave.

As illustrated by Figure 17.9 planes of zero E occur at even multiples of $\lambda_z/4$ from the reflecting surface, whereas planes of maximum E occur at odd multiples of $\lambda_z/4$ from the reflecting surface. The whole standing wave distribution of E is seen to be traveling in υ direction with a velocity $\upsilon_y = \dfrac{\omega}{\beta_y} = \dfrac{\omega}{\beta\sin\theta} = \dfrac{\upsilon}{\sin\theta}$. This is the velocity with which the crests move along the y-axis. The wavelength in this direction is $\lambda_y = \dfrac{\lambda}{\sin\theta}$.

17.5.2.2 Parallel Polarization

In this case H is reflected without phase reversal, i.e. $H_i = H_r$. Also the components of E_i and E_r parallel to boundary surface must be equal and opposite.

Also:

$$\frac{E_i}{H_i} = \frac{E_r}{H_r} = \eta \tag{17.10}$$

In view of Figure 17.10:

$$\left.H_{\text{inc}} = H_i e^{-j\beta(y\sin\theta - z\cos\theta)} \quad (\text{a}) \qquad E_{\text{ref}} = H_r e^{-j\beta(y\sin\theta + z\cos\theta)} \quad (\text{b})\right\} \tag{17.11}$$

Thus:

$$H_t = H_i + H_r = 2H_i \cos\beta_z z e^{-j\beta_y y} \tag{17.12}$$

where again:

$$\left.\beta_z = \beta\cos\theta \quad (\text{a}) \qquad \beta_y = \beta\sin\theta \quad (\text{b})\right\} \tag{17.13}$$

In the above it can be observed that H has a standing wave distribution in the z direction with plane of maximum H located at $z = 0$ and even multiple of $\lambda_z = 4$ from the

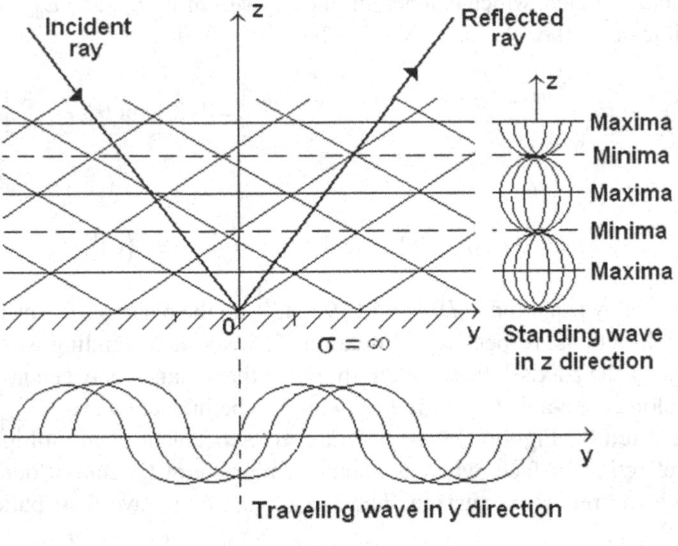

FIGURE 17.9 Standing wave pattern along the z-axis and traveling wave along the y-axis.

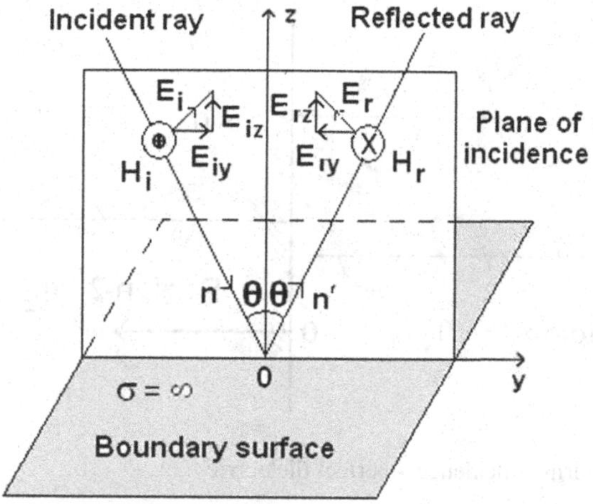

FIGURE 17.10 E in parallel polarization.

surface. The planes of zero H occur at odd multiples of $\lambda_z = 4$ from the surface. Also the total E_z and E_y can be written as:

$$E_z = 2\eta \sin\theta H_i \left[\cos\beta_z z\, e^{-j\beta_y y}\right] \tag{17.14a}$$

$$E_y = 2j\eta \cos\theta H_i \left[\sin\beta_z z\, e^{-j\beta_y y}\right] \tag{17.14b}$$

Thus both E_z and E_y have standing wave distribution above the reflecting surface. Furthermore, the maxima for E_z occur at the plane and odd multiples of $\lambda_z = 4$ from the plane whereas for E_y minima occur at the plane and even multiples $\lambda_z = 4$ of from the plane.

17.6 REFLECTION FROM PERFECT DIELECTRIC

In this case the dielectric surface is assumed to be flat without any aberrations. Since a ray may strike a surface perpendicularly or obliquely, the two cases are separately considered.

17.6.1 Normal Incidence

In this case there will be only partial reflection of incident energy. A part of this energy will penetrate the dielectric it will pass (or transmitted) through Region 2. The amount of reflection and transmission will depend on the relative dielectric constants of the two media. This situation is shown in Figure 17.11.

In this case, the relation $\eta = E/H$ is taken as the starting point. If the characteristic impedance of Medium 1 is taken as η_1 and that of Medium 2 as η_2 then E and H can be related as:

$$E_i = \eta_1 H_i \quad (\text{a}) \qquad E_r = -\eta_1 H_r \quad (\text{b}) \qquad E_t = \eta_2 H_t \quad (\text{c})\} \tag{17.15}$$

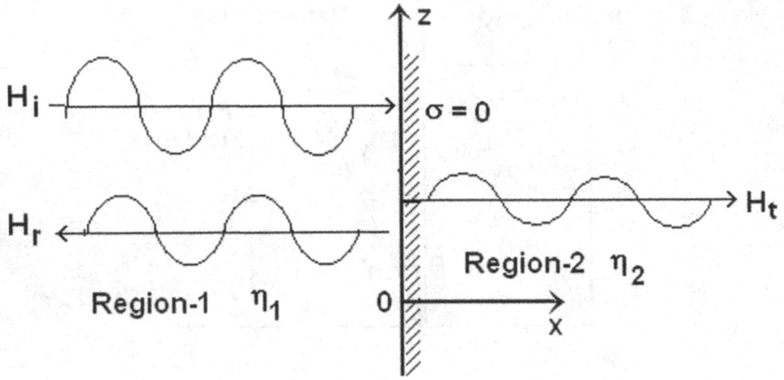

FIGURE 17.11 Normal incidence – perfect dielectric.

Since both E_{tan} and H_{norm} at $x = 0$ are continuous, we get:

$$H_i + H_r = H_t \quad (a) \qquad E_i + E_r = E_t \quad (b)\} \tag{17.16}$$

$$H_i + H_r = \frac{E_i - E_r}{\eta_1} = \frac{E_i + E_r}{\eta_2} \tag{17.17a}$$

$$\eta_2 (E_i - E_r) = \eta_1 (E_i + E_r) \tag{17.17b}$$

$$(\eta_2 - \eta_1) = E_r (\eta_2 + \eta_1) \tag{17.17c}$$

Finally, we get:

$$\frac{E_r}{E_i} = \frac{\eta_2 - \eta_1}{\eta_2 + \eta_1} \tag{17.18}$$

Equation 17.18 represents the reflection coefficient for E in terms of characteristic impedances η_1 and η_2. It can also be expressed in terms of media parameters. Furthermore, the reflection coefficient for H and transmission coefficient for E and H can also be obtained from the reflection coefficient for E. These are given below.

Reflection coefficient (Γ_E) for E:

$$\Gamma_E = \frac{E_r}{E_i} = \frac{\eta_2 - \eta_1}{\eta_2 + \eta_1} = \frac{\sqrt{\mu_0/\varepsilon_2} - \sqrt{\mu_0/\varepsilon_1}}{\sqrt{\mu_0/\varepsilon_2} + \sqrt{\mu_0/\varepsilon_1}} = \frac{\sqrt{\varepsilon_1} - \sqrt{\varepsilon_2}}{\sqrt{\varepsilon_1} + \sqrt{\varepsilon_2}} \tag{17.19a}$$

Transmission coefficient (T_E) for E:

$$T_E = \frac{E_t}{E_i} = \frac{E_r + E_i}{E_i} = 1 + \frac{E_r}{E_i} = \frac{2\eta_2}{\eta_2 + \eta_1} = \frac{2\sqrt{\varepsilon_1}}{\sqrt{\varepsilon_1} + \sqrt{\varepsilon_2}} \tag{17.19b}$$

Reflection coefficient (Γ_H) for H:

$$\Gamma_H = \frac{H_r}{H_i} = -\frac{E_r}{E_i} = \frac{\eta_1 - \eta_2}{\eta_1 + \eta_2} = \frac{\sqrt{\varepsilon_2} - \sqrt{\varepsilon_1}}{\sqrt{\varepsilon_1} + \sqrt{\varepsilon_2}} \tag{17.20a}$$

Transmission coefficient (T_H) for H:

$$T_H = \frac{H_t}{H_i} = \frac{\eta_1 E_t}{\eta_2 E_i} = \frac{2\eta_1}{\eta_1 + \eta_2} = \frac{2\sqrt{\varepsilon_2}}{\sqrt{\varepsilon_1} + \sqrt{\varepsilon_2}} \tag{17.20b}$$

17.6.2 Oblique Incidence

Like perfect conductors, perfect dielectrics also require separate consideration for perpendicular and parallel polarizations. These two cases are discussed after developing the following required relations.

Figure 17.12 illustrates a boundary surface between two media, which lies along the y-axis. Medium 1 is characterized by μ_0, ε_1 whereas Medium 2 by μ_0, ε_2. Two incident rays are striking this surface at points A and B. The energy contents of these rays are partly reflected and remain in Medium 1. The remaining part passes through Medium 2 through refraction. Since the velocity of a wave depends on media parameters it will obviously be different in two media. Let these velocities be υ_1 and υ_2 for Media 1 and 2, respectively. With these differing velocities, at the same time the incident ray travels from C to B and the transmitted ray travels from A to D.

From this figure and the above description, one can write:

$$CB = AB \sin\theta_1 \quad (a) \qquad \text{and} \qquad AD = AB \sin\theta_2 \quad (b)\} \tag{17.21}$$

Thus:

$$\frac{CB}{AD} = \frac{\sin\theta_1}{\sin\theta_2} = \frac{\upsilon_1}{\upsilon_2} \tag{17.22}$$

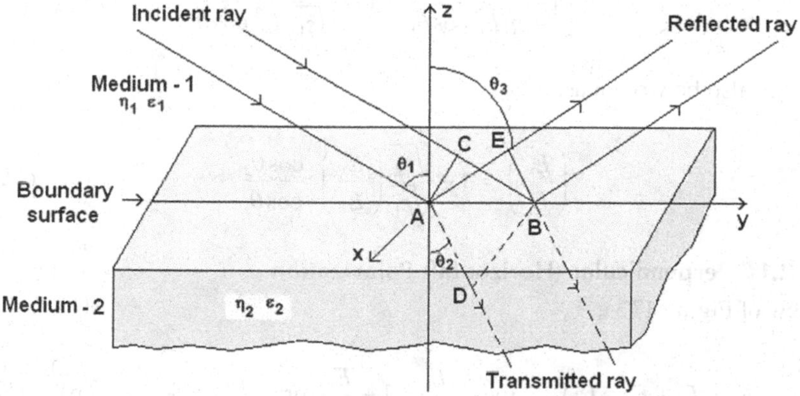

FIGURE 17.12 Incident, reflected, and transmitted rays on perfect dielectric.

where:

$$\left.\begin{aligned}
&\upsilon_1 = \frac{1}{\sqrt{\mu_1\varepsilon_1}} = \frac{1}{\sqrt{\mu_0\varepsilon_1}} \quad (a) \quad \upsilon_2 = \frac{1}{\sqrt{\mu_1\varepsilon_2}} = \frac{1}{\sqrt{\mu_0\varepsilon_2}} \quad (b)\\
&\text{Thus,} \quad \frac{\sin\theta_1}{\sin\theta_2} = \sqrt{\frac{\varepsilon_2}{\varepsilon_1}} \quad (c) \quad \text{or} \quad \sin\theta_2 = \sqrt{\frac{\varepsilon_1}{\varepsilon_2}}\sin\theta_1 \quad (d)
\end{aligned}\right\} \tag{17.23}$$

Also, since $AE = CB$, it leads to the relation:

$$\sin\theta_1 = \sin\theta_3 \tag{17.24}$$

The relation is called the *law of sines* or *Snell's law.*

In view of the Poynting theorem ($P = E \times H = E^2/\eta$), the three components of power contents can be identified as:

1. Incident wave striking AB is proportional to $\dfrac{E_i^2\cos\theta_1}{\eta_1}$
2. Reflected wave across AB is proportional to $\dfrac{E_r^2\cos\theta_1}{\eta_1}$
3. Transmitted wave striking AB is proportional to $\dfrac{E_t^2\cos\theta_2}{\eta_2}$

These power contents can be equated to get the following relation:

$$\frac{E_i^2\cos\theta_1}{\eta_1} = \frac{E_r^2\cos\theta_1}{\eta_1} + \frac{E_t^2\cos\theta_2}{\eta_2} \tag{17.25a}$$

Equation 17.25a finally leads to:

$$\frac{E_r^2}{E_i^2} = 1 - \frac{\eta_1 E_t^2\cos\theta_2}{\eta_2 E_i^2\cos\theta_1} = 1 - \frac{\sqrt{\varepsilon_2}}{\sqrt{\varepsilon_1}}\frac{E_t^2\cos\theta_2}{E_i^2\cos\theta_1} \tag{17.25b}$$

This can also be written as:

$$\left(\frac{E_r}{E_i}\right)^2 = 1 - \frac{\sqrt{\varepsilon_2}}{\sqrt{\varepsilon_1}}\left(\frac{E_t}{E_i}\right)^2\frac{\cos\theta_2}{\cos\theta_1} \tag{17.25c}$$

17.6.2.1 Perpendicular (Horizontal) Polarization

In view of Figure 17.3a:

$$\left. E_t = E_i + E_r \quad (a), \quad \text{thus} \quad \frac{E_t}{E_i} = 1 + \frac{E_r}{E_i} \quad \text{or} \quad -\frac{E_r}{E_i} = 1 - \frac{E_t}{E_i} \quad (b)\right\} \tag{17.26}$$

In view of Equation 17.25b:

$$\left(\frac{E_r}{E_i}\right)^2 = \left(1-\frac{E_t}{E_i}\right)^2 = 1-\sqrt{\frac{\varepsilon_2}{\varepsilon_1}}\left(1+\frac{E_r}{E_i}\right)^2\left(\frac{\cos\theta_2}{\cos\theta_1}\right) \tag{17.27a}$$

Equation 17.27a can be manipulated as:

$$1-\left(\frac{E_r}{E_i}\right)^2 = \left(1-\frac{E_r}{E_i}\right)\left(1+\frac{E_r}{E_i}\right) = \sqrt{\frac{\varepsilon_2}{\varepsilon_1}}\left(1+\frac{E_r}{E_i}\right)\left(1+\frac{E_r}{E_i}\right)\left(\frac{\cos\theta_2}{\cos\theta_1}\right)$$

$$\left(1-\frac{E_r}{E_i}\right) = \sqrt{\frac{\varepsilon_2}{\varepsilon_1}}\left(1+\frac{E_r}{E_i}\right)\left(\frac{\cos\theta_2}{\cos\theta_1}\right)$$

$$\frac{E_r}{E_i} = \frac{\sqrt{\varepsilon_1}\cos\theta_1 - \sqrt{\varepsilon_2}\cos\theta_2}{\sqrt{\varepsilon_1}\cos\theta_1 + \sqrt{\varepsilon_2}\cos\theta_2} \tag{17.27b}$$

Using Equation 17.23d, $\sin\theta_2 = \sqrt{\frac{\varepsilon_1}{\varepsilon_2}}\sin\theta_1$, we get:

$$\frac{E_r}{E_i} = \frac{\sqrt{\varepsilon_1}\cos\theta_1 - \sqrt{\varepsilon_2 - \varepsilon_1\sin^2\theta_1}}{\sqrt{\varepsilon_1}\cos\theta_1 + \sqrt{\varepsilon_2 - \varepsilon_1\sin^2\theta_1}} \tag{17.28}$$

Finally, we get:

$$\frac{E_r}{E_i} = \frac{\cos\theta_1 - \sqrt{\left(\frac{\varepsilon_2}{\varepsilon_1}\right) - \sin^2\theta_1}}{\cos\theta_1 + \sqrt{\left(\frac{\varepsilon_2}{\varepsilon_1}\right) - \sin^2\theta_1}} \tag{17.29}$$

17.6.2.2 Parallel (Vertical) Polarization

In view of Figures 17.3b and 17.12:

$$\left(E_i - E_r\right)\cos\theta_1 = E_t\cos\theta_2 \tag{17.30}$$

$$\frac{E_t}{E_i} = \left\{1-\frac{E_r}{E_i}\right\}\frac{\cos\theta_1}{\cos\theta_2} \quad \text{or} \quad \left(\frac{E_t}{E_i}\right)^2 = \left\{1-\frac{E_r}{E_i}\right\}^2\frac{\cos^2\theta_1}{\cos^2\theta_2}$$

Substitution of equivalent expression of $\left(\frac{E_t}{E_i}\right)^2$ in Equation 17.25c gives:

$$\left(\frac{E_r}{E_i}\right)^2 = 1 - \frac{\sqrt{\varepsilon_2}}{\sqrt{\varepsilon_1}}\left(\frac{E_t}{E_i}\right)^2 \frac{\cos\theta_2}{\cos\theta_1} = 1 - \frac{\sqrt{\varepsilon_2}}{\sqrt{\varepsilon_1}}\left\{1 - \frac{E_r}{E_i}\right\}^2 \frac{\cos^2\theta_1}{\cos^2\theta_2}\frac{\cos\theta_2}{\cos\theta_1}$$

$$= 1 - \frac{\sqrt{\varepsilon_2}}{\sqrt{\varepsilon_1}}\left\{1 - \frac{E_r}{E_i}\right\}^2 \frac{\cos\theta_1}{\cos\theta_2}$$

$$1 - \left(\frac{E_r}{E_i}\right)^2 = \left\{1 - \left(\frac{E_r}{E_i}\right)\right\}\left\{1 + \left(\frac{E_r}{E_i}\right)\right\} = \frac{\sqrt{\varepsilon_2}}{\sqrt{\varepsilon_1}}\left\{1 - \frac{E_r}{E_i}\right\}^2 \frac{\cos\theta_1}{\cos\theta_2}$$

$$1 + \left(\frac{E_r}{E_i}\right) = \frac{\sqrt{\varepsilon_2}}{\sqrt{\varepsilon_1}}\left\{1 - \left(\frac{E_r}{E_i}\right)\right\}\frac{\cos\theta_1}{\cos\theta_2} = \frac{\sqrt{\varepsilon_2}}{\sqrt{\varepsilon_1}}\frac{\cos\theta_1}{\cos\theta_2} - \frac{\sqrt{\varepsilon_2}}{\sqrt{\varepsilon_1}}\frac{\cos\theta_1}{\cos\theta_2}\left(\frac{E_r}{E_i}\right)$$

$$\frac{\sqrt{\varepsilon_2}}{\sqrt{\varepsilon_1}}\frac{\cos\theta_1}{\cos\theta_2} - 1 = \left(\frac{E_r}{E_i}\right)\left\{\frac{\sqrt{\varepsilon_2}}{\sqrt{\varepsilon_1}}\frac{\cos\theta_1}{\cos\theta_2} + 1\right\}$$

$$\frac{\sqrt{\varepsilon_2}\cos\theta_1 - \sqrt{\varepsilon_1}\cos\theta_2}{\sqrt{\varepsilon_1}\cos\theta_2} = \left(\frac{E_r}{E_i}\right)\frac{\sqrt{\varepsilon_2}\cos\theta_1 + \sqrt{\varepsilon_1}\cos\theta_2}{\sqrt{\varepsilon_1}\cos\theta_2}$$

or

$$\left(\frac{E_r}{E_i}\right) = \frac{\sqrt{\varepsilon_2}\cos\theta_1 - \sqrt{\varepsilon_1}\cos\theta_2}{\sqrt{\varepsilon_2}\cos\theta_1 + \sqrt{\varepsilon_1}\cos\theta_2}$$

The replacement of θ_2 by θ_1 using Equation 17.23d finally gives:

$$\frac{E_r}{E_i} = \frac{\left(\frac{\varepsilon_2}{\varepsilon_1}\right)\cos\theta_1 - \sqrt{\left(\frac{\varepsilon_2}{\varepsilon_1}\right) - \sin^2\theta_1}}{\left(\frac{\varepsilon_2}{\varepsilon_1}\right)\cos\theta_1 + \sqrt{\left(\frac{\varepsilon_2}{\varepsilon_1}\right) - \sin^2\theta_1}} \tag{17.31}$$

17.6.2.3 Brewster's Angle

From Equation 17.31 it can be noted that if its numerator becomes zero there will be no reflection. This condition is met when:

$$\sqrt{\frac{\varepsilon_2}{\varepsilon_1} - \sin^2\theta_1} = \frac{\varepsilon_2}{\varepsilon_1}\cos\theta_1 \tag{17.32}$$

Square both the sides and replace $\cos\theta_1$ by an equivalent term of $\sin\theta_1$:

$$\frac{\varepsilon_2}{\varepsilon_1} - \sin^2\theta_1 = \left(\frac{\varepsilon_2}{\varepsilon_1}\right)^2 - \left(\frac{\varepsilon_2}{\varepsilon_1}\right)^2 \sin^2\theta_1$$

Or

$$\left(\varepsilon_1^2 - \varepsilon_2^2\right)\sin^2\theta_1 = \varepsilon_2\left(\varepsilon_1 - \varepsilon_2\right) \tag{17.33}$$

From Equation 17.33:

$$\left.\sin^2\theta_1 = \frac{\varepsilon_2}{\varepsilon_1 + \varepsilon_2} \quad (a) \qquad \cos^2\theta_1 = \frac{\varepsilon_1}{\varepsilon_1 + \varepsilon_2} \quad (b)\right\} \tag{17.34}$$

Thus, we get:

$$\left.\tan\theta_1 = \sqrt{\varepsilon_2/\varepsilon_1} \quad (a) \quad \text{and} \quad \theta_1 = \tan^{-1}\sqrt{\varepsilon_2/\varepsilon_1} \quad (b)\right\} \tag{17.35}$$

Angle θ_1 is called *Brewster's angle*, at which there will be no reflection at all. As can be seen from Equation 17.29, no such condition exists for perpendicular polarization.

17.6.3 Total Internal Reflection

The concept of total internal reflection is the basis of propagation of light in the optical fibers.

When $\varepsilon_1 > \varepsilon_2$, $\sin^2\theta_1 > \sqrt{\varepsilon_2/\varepsilon_1}$, Equations 17.29 and 17.31 can be written as:

$$\frac{E_r}{E_i} = \frac{\cos\theta_1 - j\sqrt{\sin^2\theta_1 - \left(\dfrac{\varepsilon_2}{\varepsilon_1}\right)}}{\cos\theta_1 + j\sqrt{\sin^2\theta_1 - \left(\dfrac{\varepsilon_2}{\varepsilon_1}\right)}} \tag{17.36}$$

$$\frac{E_r}{E_i} = \frac{\left(\dfrac{\varepsilon_2}{\varepsilon_1}\right)\cos\theta_1 - j\sqrt{\sin^2\theta_1 - \left(\dfrac{\varepsilon_2}{\varepsilon_1}\right)}}{\left(\dfrac{\varepsilon_2}{\varepsilon_1}\right)\cos\theta_1 + j\sqrt{\sin^2\theta_1 - \left(\dfrac{\varepsilon_2}{\varepsilon_1}\right) - \sin^2\theta_1}} \tag{17.37}$$

Equations 17.36 and 17.37 are of the form:

$$\left.\frac{E_r}{E_i} = \frac{a - jb}{a + jb} \quad (a) \quad \text{or} \quad \left|\frac{E_r}{E_i}\right| = 1 \quad (b)\right\} \tag{17.38}$$

The unity magnitude in Equation 17.38b signifies that the reflection is total. Thus, if θ_1 is large enough and Medium 1 is denser than Medium 2, there will be total internal reflection of wave when it strikes a dielectric obliquely. The mode of internal reflection is shown in Figure 17.13.

The wave propagation in rod waveguides and optical fibers follows the principle of total internal reflection. In these the energy travels in a zig-zag fashion and

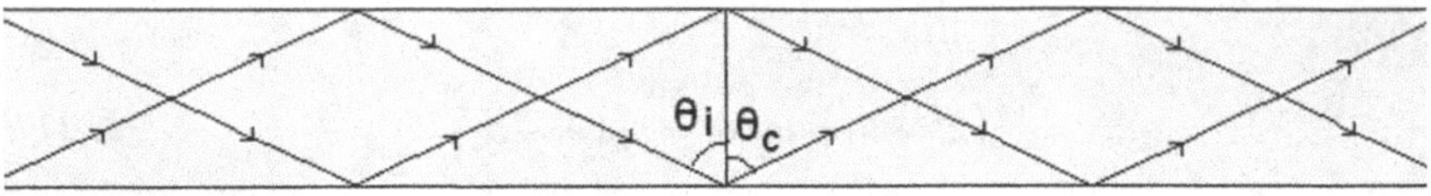

FIGURE 17.13 Illustration of total internal reflection.

remains confined within the rod or the fiber. To abide by this principle, the angle of incidence θ_i (also called angle of launch) has to be greater than the critical angle (θ_c) which is given as:

$$\theta_c = \tan^{-1}\sqrt{1/(\mu_r\varepsilon_e)} \tag{17.39}$$

17.7 REFLECTION FROM IMPERFECT MATERIALS

Sections 17.5 and 17.6 were devoted to the reflection cases wherein the surfaces were assumed to be either perfect conductors or perfect dielectrics. In practice such situations are rarely found. A propagating wave may strike a wall, a tree, a tower, a hill, or simply the Earth. The material of none of these may be a perfect conductor or a perfect dielectric. This situation calls for reconsideration of the reflection cases for imperfect materials. This section will assess the impact of such imperfections on the reflection coefficients, particularly for oblique incidence. Here too the surface is assumed to be a perfect flat plane.

The relations for reflection coefficients for oblique incidence on perfect dielectric were given by Equations 17.29 and 17.31 for horizontal and vertical polarizations, respectively. For imperfect materials, these equations require a relook at one of the basic Maxwell's equations, which is reproduced here:

$$\nabla \times \boldsymbol{H} = \boldsymbol{J} + \frac{\partial \boldsymbol{D}}{\partial t} = \sigma \boldsymbol{E} + \varepsilon\frac{\partial \boldsymbol{E}}{\partial t} \tag{17.40a}$$

Since

$$\boldsymbol{E} = E_0 e^{j\omega t} \qquad \frac{\partial \boldsymbol{E}}{\partial t} = j\omega E_0 e^{j\omega t} = j\omega \boldsymbol{E} \qquad \boldsymbol{E} = \frac{1}{j\omega}\frac{\partial \boldsymbol{E}}{\partial t} \tag{17.40b}$$

In view of Equation 17.40b, Equation 17.40a can be written as:

$$\nabla \times \boldsymbol{H} = \sigma \boldsymbol{E} + \varepsilon\frac{\partial \boldsymbol{E}}{\partial t} = (\sigma + j\omega\varepsilon)\boldsymbol{E} \tag{17.40c}$$

Substitution of E from Equation 17.40b in Equation 17.40c gives:

$$\nabla \times \boldsymbol{H} = (\sigma + j\omega\varepsilon)\boldsymbol{E} = \frac{\sigma + j\omega\varepsilon}{j\omega}\frac{\partial \boldsymbol{E}}{\partial t} = \left(\varepsilon + \frac{\sigma}{j\omega}\right)\frac{\partial \boldsymbol{E}}{\partial t} = \varepsilon'\frac{\partial \boldsymbol{E}}{\partial t} \tag{17.40d}$$

where

$$\varepsilon' = \varepsilon + \frac{\sigma}{j\omega} = \varepsilon\left(1 + \frac{\sigma}{j\omega\varepsilon}\right) = \varepsilon\left(1 - j\frac{\sigma}{\omega\varepsilon}\right) \tag{17.40e}$$

In Equation 17.40e, ε' represents the complex permittivity.

Figure 17.14 illustrates an incident ray, which makes an angle θ_1 with the normal drawn on the surface of a flat surface (say of Earth). This is the same angle as shown in Figure 17.12 and used in Equations 17.29 and 17.31. This ray also makes an angle ψ_2 with this surface. In this figure $\psi_2 = 90 - \theta_1$ or $\theta_1 = 90 - \psi_2$. In view of the relation of θ_1 and ψ_2, $\cos\theta_1 = \sin\psi_2$ and $\sin\theta_1 = \cos\psi_2$. Thus, if ε_1 is replaced by ε_0 and ε_2 by ε', $\cos\theta_1$ by $\sin\psi_2$ and $\sin\theta_1$ by $\cos\psi_2$. Equation 17.29 takes a new form:

$$\frac{E_r}{E_i} = \frac{\cos\theta_1 - \sqrt{(\varepsilon_2/\varepsilon_1)\sin^2\theta_1}}{\cos\theta_1 + \sqrt{(\varepsilon_2/\varepsilon_1)\sin^2\theta_1}} = \frac{\sin\psi_2 - \sqrt{(\varepsilon'/\varepsilon_0) - \cos^2\psi_2}}{\sin\psi_2 + \sqrt{(\varepsilon'/\varepsilon_0) - \cos^2\psi_2}} \tag{17.41a}$$

In view of Equation 17.40e:

$$\frac{\varepsilon'}{\varepsilon_0} = \frac{\varepsilon}{\varepsilon_0}\left(1 + \frac{\sigma}{j\omega\varepsilon}\right) = \varepsilon_r\left(1 + \frac{\sigma}{j\omega\varepsilon}\right) = \left(\varepsilon_r + \frac{\sigma\varepsilon_r}{j\omega\varepsilon}\right) = \left(\varepsilon_r + \frac{\sigma(\varepsilon/\varepsilon_0)}{j\omega\varepsilon}\right)$$
$$= \left(\varepsilon_r + \frac{\sigma}{j\omega\varepsilon_0}\right) = \left(\varepsilon_r - j\frac{\sigma}{\omega\varepsilon_0}\right) = (\varepsilon_r - jx) \tag{17.41b}$$

where

$$x = \frac{\sigma}{\omega\varepsilon_0} = \frac{\sigma}{2\pi f\varepsilon_0} = \frac{\sigma}{2\pi f\left(10^{-9}/36\pi\right)} = \frac{(18\times10^9)\sigma}{f} = \frac{(18\times10^3)\sigma}{f_{MHZ}} \tag{17.41c}$$

In view of Equation 17.41b, Equation 17.41a can be rewritten as:

$$R_h = \frac{E_r}{E_i} = \frac{\sin\psi_2 - \sqrt{(\varepsilon_r - jx) - \cos^2\psi_2}}{\sin\psi_2 + \sqrt{(\varepsilon_r - jx) - \cos^2\psi_2}} \tag{17.42}$$

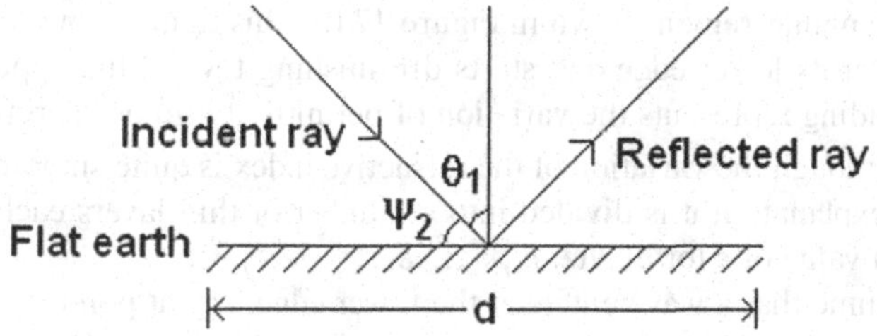

FIGURE 17.14 Incident and reflected waves.

Following the above procedure, Equation 17.31 can also be modified to the following form:

$$R_v = \frac{E_r}{E_i} = \frac{(\varepsilon_r - jx)\cdot \sin\psi_2 - \sqrt{(\varepsilon_r - jx) - \cos^2\psi_2}}{(\varepsilon_r - jx)\cdot \sin\psi_2 + \sqrt{(\varepsilon_r - jx) - \cos^2\psi_2}} \tag{17.43}$$

Equations 17.42 and 17.43 give the reflection coefficients for horizontally and vertically polarized waves, respectively. From these equations, it can be noted that both R_h and R_v have become complex quantities due to imperfection of the material. Thus, R_h and R_v can be written as:

$$R_h = |R_h|\angle R_h \quad (a) \quad \text{and} \quad R_v = |R_v|\angle R_v \quad (b)\} \tag{17.44}$$

where $|R_h|$ and $|R_v|$ indicate the magnitudes and $\angle R_h$ and $\angle R_v$ indicate the phases of R_h and R_v, respectively.

These parameters (R_h and R_v) are the key to understand the ground wave propagation. These are also required in the case of space waves wherein part of energy from an elevated transmitting antenna strikes the ground and then arrives at the receiving antenna after reflection.

17.8 REFRACTION IN THE IONIZED MATERIAL

So far, our study has revolved mainly around the reflection phenomenon. In a few places, the refraction of electromagnetic waves was only tangentially touched. Since this phenomenon is deeply involved in wave propagations, this section is devoted to understanding this.

In Section 17.2, the meanings of some basic terms including that of the refraction were included. Thereat it was noted that a wave gets refracted when it passes from rarer to denser or from denser to rarer media. In the first case it refracts toward the normal and in the second away from normal. With this statement one should not conclude that the rarer and denser media are only in the form of two slabs with two distinct dielectric constants or refractive indices. There may be a gradual change of the refractive index through which the wave will gradually refract. If it meets suitable conditions it may even change its direction of travel. Such situations are frequently met when waves travel through the ionospheric region.

To understand the process of gradual bending through refraction, let us consider a shaded rectangular region shown in Figure 17.15. This figure shows that the shade is quite dark at its lower edge and starts diminishing toward the upper edge. This change in shading represents the variation of permittivity (ε) or the refractive index $\{n = \sqrt{\varepsilon}\}$. Although the variation of the refractive index is quite smooth but for convenience of explanation it is divided into a number of thin layers each presumably with constant value of ε (or n), viz. $\varepsilon_1, \varepsilon_2, \ldots \varepsilon_n$.

Let us assume that a wave enters at the lower edge, say at point A. When it progresses upward, it turns away from the normal (shown in the figure at the lower edge of each layer) since the second layer is rarer than the first (i.e. $\varepsilon_1 > \varepsilon_2$). As it

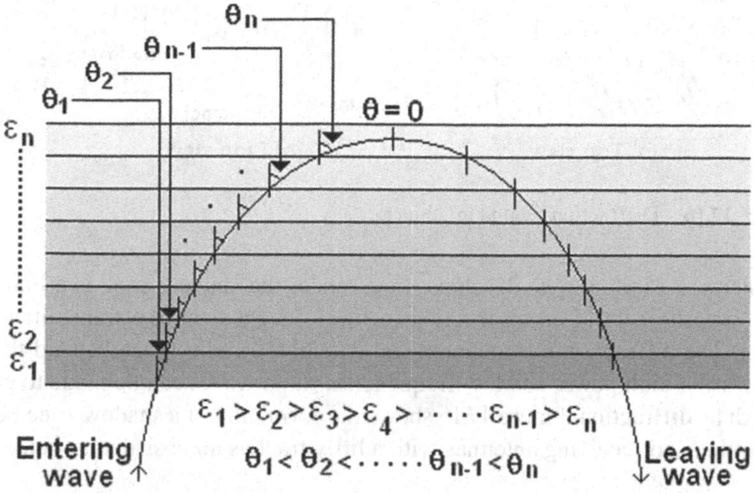

FIGURE 17.15 Process of gradual bending of a wave.

progresses further to the third layer it further turns away from the normal as $\varepsilon_2 > \varepsilon_3$. The process continues until it arrives at point *B*. As a result of this drifting away from the normal, the angle between the normal and the direction of wave travel gradually keeps on increasing until it becomes 90° at point *B*. It is worthwhile mentioning that point *B* does not necessarily lie on the topmost layer.

If this wave travels further and finds the same refractive index there will be no further bending in its path and the wave will keep on traveling, unhindered, along a straight line in the forward direction. If it somehow turns downward after attaining the maximum height in some layer, it will find conditions that are just opposite to those met in the upward travel. This time it will find the refractive indices, of successive layers, in the increasing order. As a result, at each of its crossing, from one layer to another, its direction of travel will bend toward the normal. Ultimately, the wave will appear at point *C*.

17.9 DIFFRACTION OF ELECTROMAGNETIC WAVES

Diffraction is another phenomenon with which an electromagnetic wave comes across during the course of its journey. Particularly a ground wave may have to travel through trees, buildings, hills, and other natural or manmade structures. It may strike sharp corners, sharp edges, flat surfaces, and objects of various shapes and sizes. In the case of flat surfaces, a wave may get reflected or refracted in accordance with the material properties of the obstructing object whereas, when a wave negotiates with sharp corners or bends, its energy contents may get scattered or the direction of part of energy around these may change. A wave sees the obstruction in accordance with its wavelength. Thus, a wave with larger wavelength than the size of the obstructing object can easily propagate around it. As the wavelength decreases more and more energy is obstructed and there is more and more attenuation. At high

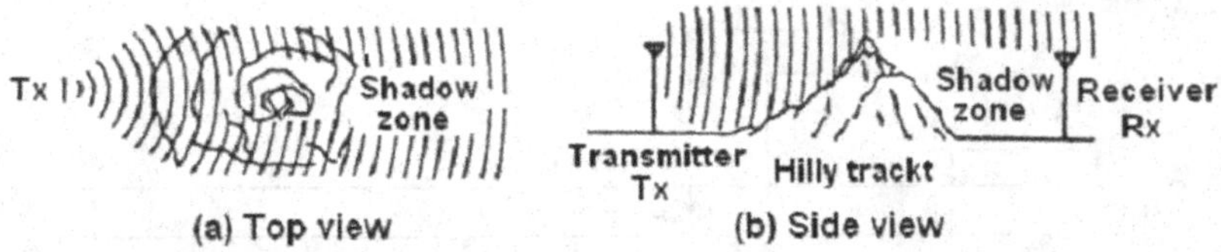

FIGURE 17.16 Diffraction around an object.

frequencies, a *shadow zone* develops. In essence, the shadow zone is a blank area on the opposite side of an obstruction in line-of-sight from the transmitter to the receiver. The diffraction phenomenon can extend the radio range beyond the horizon. By using high power and low-frequencies, radio waves can be made to encircle the earth by diffraction. Figure 17.16 shows the formation of a shadow zone between transmitting and receiving antennas with a hilly tract as an obstructing object.

Example 17.1

Find the direction cosines for a position vector $\check{\boldsymbol{r}}$ (1, 2, 3).

SOLUTION

Vector $\check{\boldsymbol{r}}$ can be written as $\check{\boldsymbol{r}} = \mathbf{a}_x + 2\mathbf{a}_y + 3\mathbf{a}_z$ $\quad \left|\check{r}\right| = \sqrt{1+4+9} = \sqrt{14}$

$$\mathbf{a}_r = \frac{\check{\boldsymbol{r}}}{\left|\check{\boldsymbol{r}}\right|} = \frac{\mathbf{a}_x + 2\mathbf{a}_y + 3\mathbf{a}_z}{\sqrt{14}} = 0.267\mathbf{a}_x + 0.534\mathbf{a}_y + 0.802\mathbf{a}_z$$

The direction cosines are the cosines of angles which the unit vectors make with the three axes. In Equation 17.1b:

$$\check{\boldsymbol{n}} \cdot \check{\boldsymbol{r}} = x\cos A + y\cos B + z\cos C$$

Thus, the direction of cosines are $\cos A = 0.267$, $\cos B = 0.534$, and $\cos C = 0.802$.

The angles that a position vector makes are: $A = \cos^{-1}0.267 = 74.51°$, $B = \cos^{-1}0.534 = 57.72°$, and $C = \cos^{-1}0.802 = 36.7°$.

Example 17.2

Find the wavelength and the velocity components for a wave with $\lambda = 3$ m traveling in free space if the direction cosines are given as $\cos A = 0.3$, $\cos B = 0.6$, and $\cos C = 0.8$.

SOLUTION

In free space, $\upsilon = c = 3 \times 10^8$. From Equations 17.3 and 17.4, we get:

$$\lambda_x = \frac{\lambda}{\cos A} = \frac{3}{0.3} = 10 \text{ m} \qquad \upsilon_x = \frac{\upsilon}{\cos A} = \frac{3\times10^8}{0.3} = 10\times10^8 \text{ m/s}$$

$$\lambda_y = \frac{\lambda}{\cos B} = \frac{3}{0.6} = 5\,\text{m} \qquad \upsilon_y = \frac{\upsilon}{\cos B} = \frac{3\times 10^8}{0.6} = 5\times 10^8\ \text{m/s}$$

$$\lambda_z = \frac{\lambda}{\cos C} = \frac{3}{0.8} = 3.75\,\text{m} \qquad \upsilon_z = \frac{\upsilon}{\cos C} = \frac{3\times 10^8}{0.8} = 3.75\times 10^8\ \text{m/s}$$

Example 17.3

A uniform plane wave is traveling in a region with dielectric constant $\boldsymbol{\varepsilon_1}$. It perpendicularly strikes at the surface of another region with dielectric constant $\boldsymbol{\varepsilon_2}$. Find (a) the reflection and transmission coefficients for ***E*** and ***H*** if $\varepsilon_1 = 2.53$ and $\varepsilon_2 = 1$; (b) the percentage of reflected and transmitted power.

SOLUTION

In view of Equations 17.19a and 17.19b and 17.20a and 17.20b, we get:

Reflection coefficient for *E*:

$$\Gamma_E = \frac{\sqrt{\varepsilon_1} - \sqrt{\varepsilon_2}}{\sqrt{\varepsilon_1} + \sqrt{\varepsilon_2}} = \frac{\sqrt{2.53} - \sqrt{1}}{\sqrt{2.53} + \sqrt{1}} = \frac{0.59}{2.59} = 0.228$$

Transmission coefficient for *E*:

$$T_E = \frac{2\sqrt{\varepsilon_1}}{\sqrt{\varepsilon_1} + \sqrt{\varepsilon_2}} = \frac{2\sqrt{2.53}}{\sqrt{2.53} + \sqrt{1}} = \frac{3.18}{2.59} = 1.228$$

Reflection coefficient for *H*:

$$\Gamma_H = \frac{\sqrt{\varepsilon_2} - \sqrt{\varepsilon_1}}{\sqrt{\varepsilon_1} + \sqrt{\varepsilon_2}} = -0.228$$

Transmission coefficient for *H*:

$$T_E = \frac{2\sqrt{\varepsilon_2}}{\sqrt{\varepsilon_1} + \sqrt{\varepsilon_2}} = \frac{2}{2.59} = 0.772$$

In view of the Poynting theorem, power is the product of *E* and *H*. Thus, the reflected and transmitted powers will be the product of reflection and transmission coefficients of *E* and *H*, respectively.

Thus, the reflected power $P_r = \Gamma_E \cdot \Gamma_H = 0.228 \times -0.228 = -0.052$ *or* 5.2%.

Also, the transmitted power $P_t = T_E \cdot T_H = 1.228 \times 0.772 = 0.948$ *or* 94.8%.

Example 17.4

A horizontally polarized wave obliquely incidents on the interface between two regions. It makes 30° angle with the perpendicular drawn at the boundary surface. Find the reflection coefficient (Γ) if the relative dielectric constants of region 1 and 2 are $\varepsilon_1 = 2.53$ and $\varepsilon_2 = 2.26$.

SOLUTION

Since $\theta = \theta_1 = 30°$, $\cos 30° = 0.866$, $\sin 30° = 0.5$, and $\frac{\varepsilon_2}{\varepsilon_1} = \frac{2.26}{2.53} = 0.893$

In view of Equation 17.29: $\Gamma_E = \frac{E_r}{E_i} = \frac{\cos\theta_1 - \sqrt{\left(\frac{\varepsilon_2}{\varepsilon_1}\right) - \sin^2\theta_1}}{\cos\theta_1 + \sqrt{\left(\frac{\varepsilon_2}{\varepsilon_1}\right) - \sin^2\theta_1}}$

$$\Gamma_E = \frac{0.866 - \sqrt{0.893 - 0.25}}{0.866 + \sqrt{0.893 - 0.25}} = \frac{0.866 - \sqrt{0.616}}{0.866 + \sqrt{0.616}}$$

$$= \frac{0.866 - 0.7848}{0.866 + 0.7848} = \frac{0.0811}{1.65} = 0.04913$$

Example 17.5

Regions 1 and 2 are characterized the relative dielectric constants ε_1 and ε_2, respectively. Find Brewster's angle (θ_1) if (a) $\varepsilon_1 = 2.53$, $\varepsilon_2 = 1$; (b) $\varepsilon_1 = 1$, $\varepsilon_2 = 3.4$.

SOLUTION

In view of Equation 17.35: $\theta_1 = \tan^{-1}\sqrt{\varepsilon_2/\varepsilon_1}$

(a) $\theta_1 = \tan^{-1}\sqrt{1/2.53} = \tan^{-1} 0.628 = 32.16°$

(b) $\theta_1 = \tan^{-1}\sqrt{3.4/1} = \tan^{-1} 1.84 = 61.52°$

Example 17.6

Estimate parameter x for flat earth with $\sigma = 4 \times 10^{-5}$ at (a) f = 300 KHz; (b) 3 MHz.

SOLUTION

In view of Equation 17.41c: $x = \frac{(18 \times 10^3)\sigma}{f_{MHZ}}$

(a) $x = \frac{(18 \times 10^3) \times 4 \times 10^{-5}}{0.3} = 2.4$

(b) $x = \frac{(18 \times 10^3) \times 4 \times 10^{-5}}{3} = 0.24$

Example 17.7

Find R_h, R_v, $\angle R_h$ and $\angle R_v$ for an average earth at 3 MHz with the following data: $\varepsilon_r = 15$ and $x = 0.24$ and $\psi_2 = 45°$.

SOLUTION

In view of Equations 17.42 and 17.43, $\psi_2 = 45°$, $\varepsilon_r = 15$, and $x = 0.24$. Thus, $\sin\psi_2 = \sin 45° = 0.707$, $\cos\psi_2 = \cos 45° = 0.707$, $\cos^2\psi_2 = 0.5$, $\varepsilon_r - jx = 15 - j0.24$.

$$R_h = \frac{\sin\psi_2 - \sqrt{(\varepsilon_r - jx) - \cos^2\psi_2}}{\sin\psi_2 + \sqrt{(\varepsilon_r - jx) - \cos^2\psi_2}} = \frac{0.707 - \sqrt{(15 - j0.24) - 0.5}}{0.707 + \sqrt{(15 - j0.24) - 0.5}}$$

$$= \frac{0.707 - \sqrt{(14.5 - j0.24)}}{0.707 + \sqrt{(14.5 - j0.24)}} = \frac{0.707 - 3.808\sqrt{(1 - j0.01655)}}{0.707 + 3.808\sqrt{(1 - j0.01655)}}$$

$$= \frac{0.707 - 3.808 \times (1 - j0.008277)}{0.707 + 3.808 \times (1 - j0.008277)} = -\frac{3.101 + j0.0315}{4.515 - j0.0374}$$

$$= -\frac{3.101 + j0.0315}{4.515 - j0.0374} \cdot \frac{4.515 + j0.0374}{4.515 + j0.0374}$$

$$\approx -\frac{14 + j0.258}{20.3814} \approx -0.687 + j0.01266$$

$$|R_h| = \sqrt{0.472 - 0.01266} = 0.6777 \qquad \theta = \tan^{-1}\left(-\frac{0.01266}{0.687}\right) = -1.055°$$

$$\angle R_h = \theta = -1.055°$$

$$R_v = \frac{(\varepsilon_r - jx)\cdot\sin\psi_2 - \sqrt{(\varepsilon_r - jx) - \cos^2\psi_2}}{(\varepsilon_r - jx)\cdot\sin\psi_2 + \sqrt{(\varepsilon_r - jx) - \cos^2\psi_2}}$$

$$= \frac{(15 - j0.24).0.707 - \sqrt{(15 - j0.24) - 0.5}}{(15 - j0.24).0.707 + \sqrt{(15 - j0.24) - 0.5}}$$

$$= \frac{(10.605 - j0.167) - \sqrt{(14.5 - j0.24)}}{(10.605 - j0.167) + \sqrt{(14.5 - j0.24)}}$$

$$= \frac{(10.605 - j0.167) - 3.808\sqrt{(1 - j0.01655)}}{(10.605 - j0.167) + 3.808\sqrt{(1 - j0.01655)}}$$

$$= \frac{(10.605 - j0.167) - (3.808 - j0.0315)}{(10.605 - j0.167) + (3.808 - j0.0315)} \approx \frac{(6.797 - j0.1355)}{(14.413 - j0.1985)}$$

$$= \frac{(6.797 - j0.1355)}{(14.413 - j0.1985)} \cdot \frac{(14.413 + j0.1985)}{(14.413 + j0.1985)}$$

$$= \frac{98 - j0.604}{207.74} = 0.4717 - j0.0029$$

$$|R_v| \approx 0.4717$$

$$\angle R_v = \theta = \tan^{-1} 0.0061 = 0.35°$$

PROBLEMS

P17.1 Find the position vector which makes 70°, 50°, and 30° angles with the *x*-, *y*-, and *z*-axes, respectively. Its magnitude is estimated to be 5.

P17.2 A wave of wavelength $\lambda = 30$ m traveling in free space makes 30°, 40°, and 50° angles with the *x*-, *y*-, and *z*-axes, respectively. Find its wavelength components along these axes.

P17.3 A wave traveling with velocity of light in free space makes 30°, 45°, and 60° angles with the *x*-, *y*-, and *z*-axes, respectively. Find its velocity components along these axes.

P17.4 Find the reflection and transmission coefficients for ***E*** and ***H*** of a uniform plane wave traveling in a region and perpendicularly striking at the surface of another region. The relative dielectric constants of these regions are $\varepsilon_1 = 3.4$ and $\varepsilon_2 = 2.53$.

P17.5 Calculate the percentage of reflected and transmitted power if the reflection coefficients for E (Γ_E), reflection coefficients for H (Γ_H), transmission coefficients for E (T_E) and transmission coefficient for H (T_H) are 0.06, −0.06, 0.9, and 1.1, respectively.

P17.6 A vertically polarized wave obliquely incidents on an interface between two regions. It makes 60° angle with the perpendicular drawn at the boundary surface. Find the reflection coefficient (Γ) if the relative dielectric constant for region 1 is $\varepsilon_1 = 2.26$, and for region 2 it is $\varepsilon_2 = 3.4$.

P17.7 Two regions referred to as 1 and 2 are characterized by the relative dielectric constants ε_1 and ε_2, respectively. Find Brewster's angle (θ_1) if (a) $\varepsilon_1 = 2.53$, $\varepsilon_2 = 2.26$; (b) $\varepsilon_1 = 2.26$, $\varepsilon_2 = 3.4$.

P17.8 Estimate parameter x for flat earth with $\sigma = 2 \times 10^{-4}$ at (a) $f = 250$ KHz; (b) 6 MHz.

P17.9 Find R_h and $\angle R_h$ at 5 MHz for an average earth with $\varepsilon_r = 10$, $x = 3$, and $\psi_2 = 30°$.

P17.10 Find R_h and R_v at 5 MHz for an average earth with $\varepsilon_r = 10$, $x = 3$, and $\psi_2 = 30°$.

DESCRIPTIVE QUESTIONS

Q17.1 Explain the terms wave velocity, phase velocity, and direction cosines.

Q17.2 Explain the meaning of perpendicular and parallel polarizations.

Q17.3 Explain the concept of complex dielectric constant vis-à-vis the wave phenomenon.

Q17.4 Describe the phenomenon of refraction from ionosphere.

Q17.5 Discuss the significance of diffraction phenomenon vis-à-vis the wave propagation.

FURTHER READING

A. S. Khan, *Microwave Engineering: Concepts and Fundamentals*, CRC Press, New York, Mar. 2014.

M. F. Iskander, *Electromagnetic Fields and Waves*, Waveland Press, Prospect Hills, IL, 2000.

Y. Ito, *Radio Wave Propagation Handbook*, ed. Y. Hosoya, Realize Inc., Japan, 1999, pp. 171–199.

E. C. Jordon, K. G. Balmain, *Electromagnetic Waves and Radiating Systems*, Prentice Hall Ltd., New Delhi, 1987.

J. R. Wait, *Electromagnetic Wave Theory*, Harper and Row, New York, 1985.

Richard Arthur Waldron, *Theory of Guided Electromagnetic Waves*, Van Nostrand Reinhold, London, New York, 1970.

S. Ramo, J. R. Whinnery, T. Van Duzer, *Fields and Waves in Communication Electronics*, John Wiley & Sons, Inc., New York, 1965.

Raghunath Shevgoankar, *Electromagnetic Waves*, McGraw Hill Education India, 2005. ISBN13 978-0-07-059116-5.

18 Transmission Lines

18.1 INTRODUCTION

The guidance and confinement of energy within some specified boundaries require some form of structure. Such structures may have different shapes and sizes. The waves carried by these are referred to as guided waves and the structure in itself is termed as a transmission line. In these, a wave may travel within, along or over the surface of the structure. In some of these (e.g. in wire pairs) the field does not fully confine whereas in others (e.g. in coaxial cables and waveguides) it remains well within. The understanding of the guiding principles of the waves in any of these structures solely rely on the basic Maxwell's equations.

As it is a material medium, the transmission line provides a path for energy to travel from one place to another. These lines may be used over a very wide range of frequencies. At lower frequencies, these are used for power transmission and distribution. At higher frequencies, these are employed for connecting radio transmitters and receivers with antennas, distributing television signals, and for computer networking. At low frequencies, these serve the purpose in efficient and effective manner but at higher frequencies the energy tends to radiate and RF currents to reflect from the discontinuities. These factors result in loss and prevent power from reaching its destination. In order to minimize losses and reflections, high-frequency transmission lines employ structures with precise conductor dimensions and spacing in conjunction with impedance-matching techniques. Such transmission lines may include ladder lines, coaxial cables, dielectric slabs, striplines, waveguides, and optical fibers.

In general, the wavelength at transmission frequency must be of the order of line length. At higher frequencies such as for millimetric and optical waves, wavelengths become much smaller than the dimensions of the guiding structures. At these frequencies, the method of optics takes over as conventional techniques become inadequate and inefficient.

In low-frequency electric circuits, the length of the wires connecting the components can be ignored provided the voltage on the wire at a given time is assumed to be the same at all points. The length assumes significance if this assumption of constancy of voltage does not hold. In other words, the length of the line becomes important when the signal includes frequency components with corresponding wavelengths comparable to or less than the length of the line. In general, a cable or wire is treated as a transmission line if its length is greater than $\lambda/10$. At this length, the phase delay and interference due to reflections assume significance and can lead to unpredictable behavior of the systems that involve transmission lines. Depending on the applications, the lengths and types of transmission line may greatly vary. The lengths of a transmission path may vary from millimeters or even less in processors to hundreds and thousands of kilometers in the interconnection of long-haul networks. With regard to its type, the transmission lines may take many forms, some of which are mentioned in this chapter.

Transmission lines usually consist of at least two conductors between which a voltage exists. Waveguides, on the other hand, contain single hollow conductors of different shapes and sizes that are in accordance with frequency and applications. In these transmission structures, a part of the power is lost due to the presence of resistance. It is referred to as *Ohmic* or *resistive* loss. At higher frequencies, *dielectric loss* due to absorption of energy by the insulating material within the structure also becomes significant. A transmission line is usually modeled in terms of its resistance (R) and inductance (L) in series and capacitance (C) and conductance (G) in parallel. Resistance and conductance contribute toward the losses usually depend on the signal frequency and are specified in dB/m. Through the ordinary transmission line theory, a distributed parameter configuration can be represented by a lumped parameter configuration. Such configurations not only facilitate the application of circuit theory rules and theorems, but also bridge the gap between Maxwell's equations and circuit theory.

In transmission structures, E and H fields of electromagnetic waves may distribute in a variety of ways. These distributions lead to another classification, referred to as Transverse Electric (TE), Transverse Magnetic (TM), and Transverse Electro Magnetic (TEM) waves. For TEM waves the field is entirely transverse between the parallel planes. This mode has the lowest cutoff frequency and thus can carry all of the frequencies down to zero. Its velocity of propagation remains of the order of free space velocity as long as the space between the planes is a perfect dielectric. In the case of imperfection, the wave velocity slightly reduces. All transmission lines used to carry power or signals at lower frequencies use TEM waves. It needs to mention that TEM wave does not exist in hollow pipe-like structures commonly known as waveguides which are meant to carry TE and TM waves.

18.2 MODELING OF TRANSMISSION LINE USING CIRCUIT APPROACH

The transmission line model, developed by Oliver Heaviside, is based on Maxwell's equations. This model contains a set of equations referred to as the telegraph or *telegrapher's equations*. To arrive at these equations the transmission line is represented by an infinite series of infinitesimally short segments, as shown in Figure 18.1a. Each of these segments is represented by a two-port network comprising a combination of elementary components shown in Figure 18.1b.

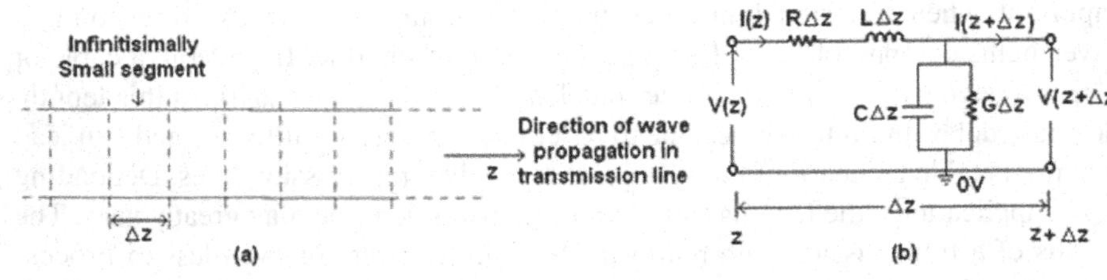

FIGURE 18.1 (a) Small segments of transmission line; (b) equivalent circuit.

The networks involved in this model contain the following distributed parameters:

- The resistance (R in Ohms) of conductors as a series resistor.
- The self-inductance (L in Henries) due to the magnetic field around the wires as a series inductor.
- The capacitance (C in Farads) between the two conductors as a shunt capacitor.
- The conductance (G in Siemens) of the dielectric material separating the two conductors as a shunt resistor between the signal wire and the return wire.

The values of distributed parameters R, L, C, and G are specified per unit length.

The *telegrapher's equations* contain a pair of linear differential equations. For a line running along the z-axis, these equations describe the voltage and the current on an electrical transmission line with distance (z) and time (t). The line voltage $V(z)$ and the current $I(z)$can be expressed in the frequency domain as:

$$\left.\frac{\partial V(z)}{\partial z} = -(R + j\omega L)I(z) \quad (\mathrm{a}) \qquad \frac{\partial I(z)}{\partial z} = -(G + j\omega C)V(z) \quad (\mathrm{b})\right\} \tag{18.1}$$

Equation 18.1 with the involvement R and G represents a lossy line. When R and G are negligibly small the transmission line is considered as a lossless structure. In the lossless case, the model depends only on L and C elements. As a consequence, the line analysis gets greatly simplified. For lossless line the second-order steady-state telegraph equations are:

$$\left.\frac{\partial^2 V(z)}{\partial z^2} + \omega^2 LC\,V(z) = 0 \quad (\mathrm{a}) \qquad \frac{\partial^2 I(z)}{\partial z^2} + \omega^2 LC\,I(z) = 0 \quad (\mathrm{b})\right\} \tag{18.2}$$

Equations 18.2a and 18.2b are two wave equations. The solutions of these equations represent two plane waves, which have equal propagation speeds in forward and reverse directions. The forward wave propagates down the transmission line and the reflected wave travels backward interferes with the original signal. These are the fundamental equations of the transmission line theory. If R and G are not neglected, the telegrapher's equations become:

$$\left.\frac{\partial^2 V(z)}{\partial z^2} = \gamma^2 V(z) \quad (\mathrm{a}) \qquad \frac{\partial^2 I(z)}{\partial z^2} = \gamma^2 I(z) \quad (\mathrm{b})\right\} \tag{18.3}$$

Equation 18.3 leads to two analogous relations to that of Equations 16.34c and 16.37, representing the propagation constant (γ) and the characteristic impedance (Z_0), respectively. These relations are:

$$\left.\gamma = \sqrt{(R + j\omega L)(G + j\omega C)} \quad (\mathrm{a}) \qquad Z_0 = \sqrt{\frac{R + j\omega L}{G + j\omega C}} \quad (\mathrm{b})\right\} \tag{18.4}$$

18.3 MODELING OF TRANSMISSION LINE USING FIELD APPROACH

Equations 18.1 to 18.4 can also be obtained by adopting the field approach as any system of conductors, carrying a low-frequency TEM wave, may be referred to as a transmission line. Such a line is illustrated in Figure 18.2a, which contains two narrow strips of width *b*, thickness *t*, and separation *a*. In view of the lowest cutoff frequency TEM mode is suitable for all practical transmission lines, for all powers and for all frequencies below 200 or 300 MHz. Since the propagation of this mode requires much smaller separation between parallel planes in comparison to wavelength, the basic transmission line equations can be derived from such a configuration.

As shown in Figure 18.2a, the parallel strips run along the *z*-axis, which is also the assumed direction of wave propagation. Figure 18.2b illustrates the field distribution between the two parallel strips and Figure 18.2c shows the field viewed from the top of the strip. In view of these figures, the transmission line equations and parameters involved therein can be obtained by applying electromotive force (EMF) and magnetomotive force (MMF) concepts. These concepts emerge from Maxwell's equations, which govern the behavior of the transmission line. These equations in point and integral form are:

$$\left.\nabla\times\boldsymbol{E} = -\frac{\partial \boldsymbol{B}}{\partial t} = -j\omega\boldsymbol{B} \quad (\text{a}) \qquad \oint \boldsymbol{E}\bullet dl = -j\omega\int \boldsymbol{B}\bullet ds \quad (\text{b})\right\} \tag{18.5}$$

$$\left.\nabla\times\boldsymbol{H} = \frac{\partial \boldsymbol{D}}{\partial t} = j\omega\boldsymbol{D} \quad (\text{a}) \qquad \oint \boldsymbol{H}\bullet dl = j\omega\int \boldsymbol{D}\bullet ds \quad (\text{b})\right\} \tag{18.6}$$

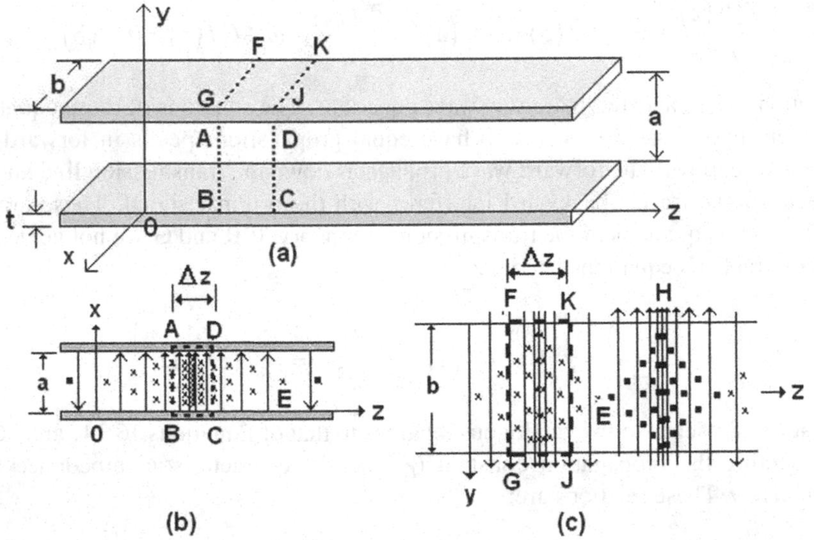

FIGURE 18.2 (a) Two parallel strips alongwith some geometrical parameters; (b) loop ABCDA in E field; (c) loop FGJK in H field.

Equations 18.5a and 18.5b are referred to as EMF equations and Equations 18.6a and 18.6b as MMF equations. The procedure to arrive at transmission line equations through these equations revolve around two loops, ABCDA and FGJKF, shown in Figure 18.2a, and again in Figure 18.2b and 18.2c. In the next two subsections, both the EMF and MMF equations are solved for lossless and lossy lines.

18.3.1 EMF Equations

Consider the loop ABCDA shown in Figure 18.2b. Equation 18.5b around this loop gives:

$$V_{AB} + V_{BC} - V_{CD} - V_{DA} = -j\omega B_y a\Delta z \tag{18.7a}$$

For *lossless line*, the conductivity (σ) of the strips is assumed to be infinite, $V_{BC} = V_{DA} = 0$ and Equation 18.7a reduces to:

$$V_{AB} - V_{CD} = \Delta V = -j\omega B_y a\Delta z \tag{18.7b}$$

Substitution of $B_y = \mu H_y = \mu J_{sz} = \dfrac{\mu I}{b}$ in Equation 18.7b leads to:

$$\left.\frac{dV}{dz} = -j\omega\left(\frac{\mu a}{b}\right)I = -j\omega LI \quad (\text{c}) \qquad \text{where } L = \frac{\mu a}{b} \quad (\text{d})\right\} \tag{18.7}$$

In Equation 18.7d, L is the inductance per unit length, which is simply the permeability μ of the transmission medium multiplied by a factor a/b. If $a = b = 1$, $L = \mu$.

For a *lossy line* having imperfect conductor, $\sigma \neq \infty$, $V_{BC} = V_{DA} = J_{sz}Z_s\Delta z$, and Equation 18.7a becomes:

$$V_{AB} - V_{CD} = \Delta V = -j\omega B_y a\Delta z - 2J_{sz}Z_s\Delta z \tag{18.8a}$$

On substitution of $B_y = \dfrac{\mu I}{b}$ Equation 18.8a gives:

$$\left.\begin{aligned}&\frac{\Delta V}{\Delta z} = -j\omega LI - \left(\frac{2Z_s}{b}\right)I = -j\omega LI - Z_s^* I \quad (\text{b}),\\ &Z_s^* = \left(\frac{2Z_s}{b}\right) = R^* + j\omega L^* \qquad (\text{c})\end{aligned}\right\} \tag{18.8}$$

Thus, $\dfrac{\Delta V}{\Delta z} = -j\omega LI - \left(R^* + j\omega L^*\right)I$. If R^* is replaced by R and $L+L^*$ by L, then:

$$\left.\frac{\Delta V}{\Delta z} = -(R + j\omega L)I = ZI \quad (\text{d}), \quad Z = R + j\omega L \quad (\text{e})\right\} \tag{18.8}$$

In Equation 18.8e, Z (the impedance) and R (the resistance) are both in per unit length.

18.3.2 MMF Equation

Similarly, from loop FGHKF in Figure 18.2c and the Equation 18.6b:

$$bH_{FG} + \Delta z H_{GH} - bH_{HK} - \Delta z H_{KF} = -j\omega\varepsilon E_x b\Delta z \quad (18.9a)$$

For a *lossless line* with perfect dielectric $\sigma = 0$, $H_{GH} = H_{KF} = 0$, Equation 18.9a becomes:

$$-b\left(H_{HK} - H_{FG}\right) = j\omega\varepsilon E_x b\Delta z \quad (18.9b)$$

Since $E_x = \frac{V}{a}$, $-b(H_{HK} - H_{FG}) = \Delta\left(bH_y\right) = \Delta I = j\omega\varepsilon Vb\Delta z/a$, Equation 18.5b gets modified to:

$$\left.\frac{dI}{dz} = j\omega\left(\frac{\varepsilon b}{a}\right)V = -j\omega CV \quad (\mathrm{c}), \quad C = \frac{\varepsilon b}{a} \quad (\mathrm{d})\right\} \quad (18.9)$$

In Equation 18.9d, C is the capacitance per unit length, which is simply the permittivity ε of the transmission medium multiplied by a factor b/a. If $a = b = 1$, $C = \varepsilon$.

For a *lossy line* with imperfect dielectric $\sigma \neq 0$, $H_{GH} = H_{KF} = \sigma E_x b$, thus:

$$-b(H_{HK} - H_{FG}) = \left(\sigma E_x + j\omega\varepsilon E_x\right)b\Delta z \quad (18.10a)$$

or

$$\frac{d\left(bH_y\right)}{dz} = -b(\sigma + j\omega\varepsilon)E_x \quad (18.10b)$$

Replacement of bH_y by $bH_y = I$ and E_x by V/a gives:

$$\frac{dI}{dz} = -\left(\frac{b\sigma}{a} + \frac{j\omega\varepsilon b}{a}\right)V = -\left(G + j\omega C\right)V = -YV \quad (18.10c)$$

where

$$\left.Y = G + j\omega C \quad (\mathrm{d}) \quad G = \frac{b\sigma}{a} \quad (\mathrm{e})\right\} \quad (18.10)$$

In Equation 18.10d, Y is the admittance per unit length. Also in Equation 18.10e, G is the conductance per unit length, which is simply the conductivity σ of the transmission medium multiplied by a factor b/a. If $a = b = 1$, $G = \sigma$.

Equations 18.7d and 18.9d lead to the following useful relations:

$$\left. \upsilon = \frac{1}{\sqrt{LC}} = \frac{1}{\sqrt{\mu\varepsilon}} \quad (a) \qquad Z_0 = \sqrt{\frac{L}{C}} = \sqrt{\frac{\mu}{\varepsilon}} \quad (b) \right\} \tag{18.11}$$

18.4 VOLTAGE AND CURRENT RELATIONS

Equations 18.8d and 18.10d obtained above are reproduced here:

$$\left. \frac{dV}{dz} = -(R + j\omega L)I \quad (18.8d), \qquad \frac{dI}{dz} = -(G + j\omega C)V \quad (18.10c) \right\}$$

These are the same telegrapher's equations as given earlier by Equations 18.1a and 18.1b:

The differentiation of Equations 18.8d and 18.10c with respect to z and some manipulation of the resulting equations leads to:

$$\left. \begin{aligned} \frac{d^2V}{dz^2} &= -(R + j\omega L)\frac{dI}{dz} = (R + j\omega L)(G + j\omega C)V = \gamma^2 V \quad (a) \\ \frac{d^2I}{dz^2} &= -(G + j\omega C)\frac{dV}{dz} = (R + j\omega L)(G + j\omega C)I = \gamma^2 I \quad (b) \end{aligned} \right\} \tag{18.12}$$

Equations 18.12a and 18.12b and Equations 18.2a and 18.2b are the same. In both of these equations, γ (the propagation constant) is a complex quantity, and we have a similar relation to that given by Equation 18.4c:

$$\gamma = \sqrt{(R + j\omega L)(G + j\omega C)} = \alpha + j\beta \tag{18.12c}$$

In Equation 18.12c, γ is comprised of attenuation constant (α) and phase shift constant (β).

As described in the following two subsections, Equations 18.12a and 18.12b can have two forms of solution.

18.4.1 Exponential Form

$$\left. V = V'e^{\gamma z} + V''e^{-\gamma z} \quad (a) \qquad I = I'e^{\gamma z} + I''e^{-\gamma z} \quad (b) \right\} \tag{18.13}$$

Both Equations 18.13a and 18.13b contain two segments wherein $V'e^{\gamma z}$ and $I'e^{\gamma z}$ represent forward waves and $V''e^{-\gamma z}$ and $I''e^{-\gamma z}$ represent backward waves. In these, V' and V'' are the magnitudes of voltages of forward waves and backward (or reflected) waves, I' and I'' are the magnitudes of currents of forward and backward (or reflected) waves respectively. The ratio of different quantities involved in Equations 18.13a and 18.13b lead to three important quantities viz. characteristic impedance, terminating impedance, and reflection coefficient.

Characteristic impedance is denoted by Z_0. It is the ratio of magnitudes of voltage and current either of a forward wave or that of a backward wave. This is given as:

$$\left.\frac{V'}{I'} = Z_0 \quad (a), \quad \frac{V''}{I''} = -Z_0 \quad (b)\right\} \tag{18.14}$$

Where Z_0 was given earlier by Equation 18.4b. This single parameter describes the behaviorof a line. The termination of a line in its characteristic impedance ensures the transfer of maximum power to the load and minimum reflections toward the source.

Terminating impedance is denoted by Z_R or Z_L. It is the ratio of voltage and current at the terminating end. This is given as:

$$Z_R = Z_L = \left.\frac{V}{I}\right]_{z=0} = \frac{V'+V''}{I'+I''} = Z_0\frac{I'-I''}{I'+I''} = Z_0\frac{V'+V''}{V'+V''} \tag{18.15}$$

A transmission line meant to carry power or signal is ultimately connected to some load or terminated in some impedance wherein this power or signal is utilized or absorbed. It is this load which is referred to as terminating impedance.

The *reflection coefficient* is defined as the ratio of magnitudes of voltages or currents of forward and reflected waves. It is denoted by a Greek letter Γ. This is given as:

$$\left.\frac{V''}{V'} = \Gamma \quad (\text{for voltage}) \quad (a) \quad \frac{I''}{I'} = -\Gamma \quad (\text{for current}) \quad (b)\right\} \tag{18.16}$$

Manipulation of Equation 18.16a and 181.6b gives:

$$\left.\Gamma = \frac{Z_R - Z_0}{Z_R + Z_0} \quad (\text{for voltage}) \quad (c) \quad \Gamma = \frac{Z_0 - Z_R}{Z_R + Z_0} \quad (\text{for current}) \quad (d)\right\} \tag{18.16}$$

As noted earlier, if a line is terminated in impedance other than its characteristic impedance the wave will reflect back. Reflections may also occur due to the presence of discontinuities on the line. It is the reflection coefficient which accounts for the amount of reflection.

18.4.2 Hyperbolic Form

The solutions of Equations 18.13a and 18.13b in hyperbolic form can be written as:

$$\left.V = A_1\cosh(\gamma z) + B_1\sinh(\gamma z) \quad (a) \quad I = A_2\cosh(\gamma z) + B_2\sinh(\gamma z) \quad (b)\right\} \tag{18.17}$$

In a transmission line, let V_S and I_S be the voltage and current at the sending end (at $z = -l$) and V_R and I_R be the voltage and current at the receiving end (at $z = 0$).

Substitution of $z = 0$ into Equations 18.17a and 18.17b gives $A_1 = V_R$ and $A_2 = I_R$. Thus, these equations become:

$$\left.\begin{aligned} V &= V_R \cosh(\gamma z) + B_1 \sinh(\gamma z) \quad (\text{a}) \\ I &= I_R \cosh(\gamma z) + B_2 \sinh(\gamma z) \quad (\text{b}) \end{aligned}\right\} \tag{18.18}$$

Differentiation of Equations 18.18a and 18.18b with respect to z and substitution of Equations 18.8d and 18.10c gives:

$$\left.\begin{aligned} \frac{dV}{dz} &= \gamma V_R \sinh(\gamma z) + \gamma B_1 \cosh(\gamma z) = -(R + j\omega L)I \quad (\text{a}) \\ \frac{dI}{dz} &= \gamma I_R \sinh(\gamma z) + \gamma B_2 \cosh(\gamma z) = -(G + j\omega C)V \quad (\text{b}) \end{aligned}\right\} \tag{18.19}$$

Use of the conditions at $z = 0$ again gives $V = V_R$ and $I = I_R$. Substitution of these values into Equations 18.19a and 18.19b gives $B_1 = -Z_0 I_R$ and $B_2 = -V_R/Z_0$. We finally get:

$$\left.\begin{aligned} V &= V_S = V_R \cosh(\gamma z) - Z_0 I_R \sinh(\gamma z) \quad (\text{a}) \\ I &= I_S = I_R \cosh(\gamma z) - \frac{V_R}{Z_0} \sinh(\gamma z) \quad (\text{b}) \end{aligned}\right\} \tag{18.20}$$

In view of Figure 18.3, $z = -l$ represents the sending end. Substitution of $z = -l$ into Equations 18.20a and 18.20b gives:

$$\left.\begin{aligned} V_S &= V_R \cosh(\gamma l) + Z_0 I_R \sinh(\gamma l) \quad (\text{a}) \\ I_S &= I_R \cosh(\gamma l) + \frac{V_R}{Z_0} \sinh(\gamma l) \quad (\text{b}) \end{aligned}\right\} \tag{18.21}$$

Equations 18.21a and 18.21b are the general transmission line equations that relate the voltage and current at the transmitting and receiving ends.

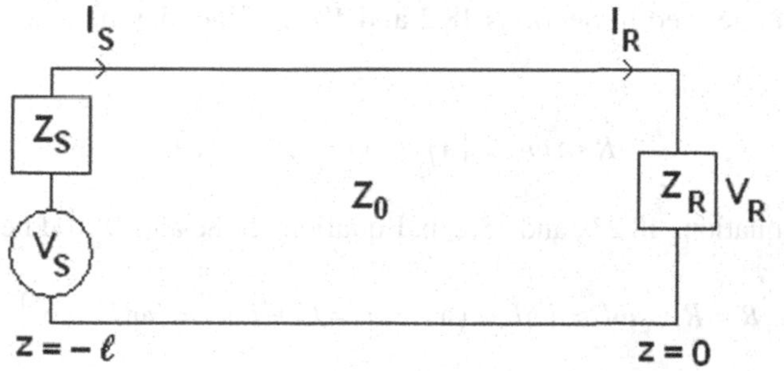

FIGURE 18.3 Circuit representation of transmission line.

18.4.2.1 Input Impedance

Equations 18.21a and 18.21b lead to another important quantity referred to as the *input impedance*. It is the ratio of voltage and current at sending end of the transmission line. Denoted by Z_{in} or Z_S, it is given by:

$$Z_{in} = Z_S = \frac{V_S}{I_S} = \frac{V_R \cosh(\gamma l) + Z_0 I_R \sinh(\gamma l)}{I_R \cosh(\gamma l) + \frac{V_R}{Z_0} \sinh(\gamma l)} \tag{18.22a}$$

The value of input impedance will obviously depend on the termination of the transmission line. The input impedance obtained from Equation 18.22a when the receiving end is short-circuited and open-circuited are:

In a *short-circuit* case, $Z_R = 0$, $V_R = 0$, the input impedance is:

$$Z_{in} = Z_{sc} = Z_0 \tanh(\gamma l) \tag{18.22b}$$

In an *open-circuit* case, $Z_R = \infty$, $I_R = 0$, the input impedance is:

$$Z_{in} = Z_{oc} = Z_0 \coth(\gamma l) \tag{18.22c}$$

The product of Equations 18.22b and 18.22c gives:

$$Z_{sc} Z_{oc} = Z_0^2 \tag{18.22d}$$

Equation 18.22d is one of the most important relations of transmission line and is used for impedance and voltage transformations in line sections.

18.5 LOSSLESS LINES AT HIGHER FREQUENCIES

In Sections 18.2 and 18.3, two types of transmission lines, namely lossless and lossy, were referred. A line that does not fall perfectly under any of the above categories may be referred to as a low loss line. A line is called low loss if its attenuation constant becomes much smaller than its phase shift constant (i.e. $\alpha \ll \beta$). This situation arises in most of the practical cases at higher frequencies. In terms of parameters derived in Sections 18.2 and 18.3, a line may be considered as low loss if:

$$\left. R \ll \omega L \quad (a) \qquad G \ll \omega C \quad (b) \right\} \tag{18.23}$$

In view of Equations 18.23a and 18.23b, Equations 18.8e and 18.10d become:

$$\left. R = R + j\omega L \approx j\omega L \quad (a) \qquad Y = G + j\omega C \approx j\omega C \quad (b) \right\} \tag{18.24}$$

In view of Equations 18.24a and 18.24b, Equations 18.12c and 18.14c get modified to:

$$\left.\begin{aligned} \gamma &= \sqrt{(R+j\omega L)(G+j\omega C)} \approx j\omega\sqrt{LC} \quad &\text{(a)} \\ Z_0 &= \sqrt{\frac{R+j\omega L}{G+j\omega C}} = \sqrt{\frac{L}{C}} \quad &\text{(b)} \end{aligned}\right\} \tag{18.25}$$

Since $\gamma = \alpha + j\beta$, Equation 18.25a gives:

$$\left.\alpha = 0 \quad \text{(a)} \qquad \beta \approx \omega\sqrt{LC} \quad \text{(b)}\right\} \tag{18.26}$$

18.5.1 Approximation for α

The above approximation for α holds well, in most of the cases, but sometimes $\alpha = 0$ may not be valid. A closer approximation for α is obtained as given here:

$$\gamma = \sqrt{(R+j\omega L)(G+j\omega C)} = \sqrt{(j\omega L)(j\omega C)\left(1+\frac{R}{j\omega L}\right)\left(1+\frac{G}{j\omega C}\right)}$$

$$\approx j\omega\sqrt{LC}\left\{\left(1+\frac{R}{j2\omega L}\right)\left(1+\frac{G}{j2\omega C}\right)\right\} \approx j\omega\sqrt{LC}\left\{1+\left(\frac{R}{j2\omega L}\right)+\left(\frac{G}{j2\omega C}\right)\right\}$$

$$\approx\left\{j\omega\sqrt{LC} + j\omega\sqrt{LC}\left(\frac{R}{j2\omega L}\right) + j\omega\sqrt{LC}\left(\frac{G}{j2\omega C}\right)\right\}$$

$$\approx \frac{R}{2}\sqrt{\frac{C}{L}} + \frac{G}{2}\sqrt{\frac{L}{C}} + j\omega\sqrt{LC} = \alpha + j\beta$$

Thus:

$$\left.\alpha \approx \frac{R}{2}\sqrt{\frac{C}{L}} + \frac{G}{2}\sqrt{\frac{L}{C}} = \frac{1}{2}\left(\frac{R}{Z_0} + GZ_0\right) \quad \text{(c)} \qquad \alpha \approx \frac{R}{2Z_0}\,(\text{for small G}) \quad \text{(d)}\right\} \tag{18.26}$$

18.5.2 Modified Voltage and Current Relations

In view of the above, Equations 18.21a and 18.21b reduce to:

$$\left.\begin{aligned} V_S &= V_R\cos(\beta l) + jZ_0 I_R\sin(\beta l) \quad &\text{(a)} \\ I_S &= I_R\cos(\beta l) + j\frac{V_R}{Z_0}\sin(\beta l) \quad &\text{(b)} \end{aligned}\right\} \tag{18.27}$$

Equation 18.22a also gets modified to:

$$Z_S = \frac{V_R\cos(\beta l) + jZ_0 I_R \sin(\beta l)}{I_R\cos(\beta l) + j\dfrac{V_R}{Z_0}\sin(\beta l)} = Z_R \frac{\cos(\beta l) + j\left(\dfrac{Z_0}{Z_R}\right)\sin(\beta l)}{\cos(\beta l) + j\left(\dfrac{Z_R}{Z_0}\right)\sin(\beta l)} \quad (18.27c)$$

$$= Z_0 \frac{Z_R\cos(\beta l) + jZ_0\sin(\beta l)}{Z_0\cos(\beta l) + jZ_R\sin(\beta l)}$$

18.5.3 Voltage and Current at an Arbitrary Location

If the voltage and current are to be obtained at an arbitrary location (x) on the line, this can be done by simply replacing l by x. Equations 18.27a and 18.27b then become:

$$\left.\begin{aligned} V_x &= V_R\cos(\beta x) + jZ_0 I_R\sin(\beta x) \quad (a) \\ I_x &= I_R\cos(\beta x) + j\frac{V_R}{Z_0}\sin(\beta x) \quad (b) \end{aligned}\right\} \quad (18.28)$$

Furthermore, if this line is terminated in pure resistance (i.e. $Z_R = R$) and its characteristic impedance is purely resistive (i.e. $Z_0 = R_0$), Equations 18.28a and 18.28b can be written as:

$$|V_x| = V_R\sqrt{\cos^2(\beta x) + \left(\frac{R_0}{R}\right)^2 \sin^2(\beta x)} \quad (18.29a)$$

$$|I_x| = I_R\sqrt{\cos^2(\beta x) + \left(\frac{R}{R_0}\right)^2 \sin^2(\beta x)} \quad (18.29b)$$

Equations 18.28 and 18.29 get further modified in case of different types of terminations. These cases are given in the following subsections.

(a) In a short-circuited line, $V_R = 0$, $I_R R_0 = V_m$, and $I_R = I_m$, thus:

$$\left.|V_x| = |V_m\sin(\beta x)| \quad (a) \qquad |I_x| = |I_m\cos(\beta x)| \quad (b)\right\} \quad (18.30)$$

In this case, voltage is minimum and the current is maximum at the point of termination. Both V and I follow the standing wave pattern as shown in Figure 18.4.

(b) In an open-circuited line, $I_R = 0$, $V_R/R_0 = I_m$, and $V_R = V_m$, thus:

$$\left.|V_x| = |V_m\cos(\beta x)| \quad (a) \qquad |I_x| = |I_m\sin(\beta x)| \quad (b)\right\} \quad (18.31)$$

In this case, voltage is maximum and the current is minimum at the point of termination. Both V and I follow the standing wave pattern as shown in Figure 18.5.

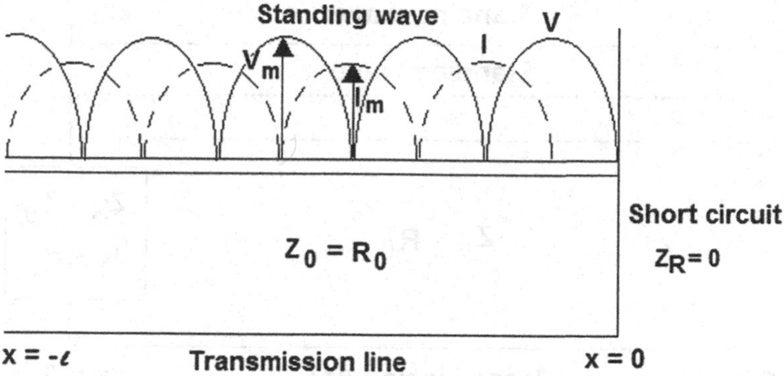

FIGURE 18.4 A short-circuited line with current and voltage standing waves.

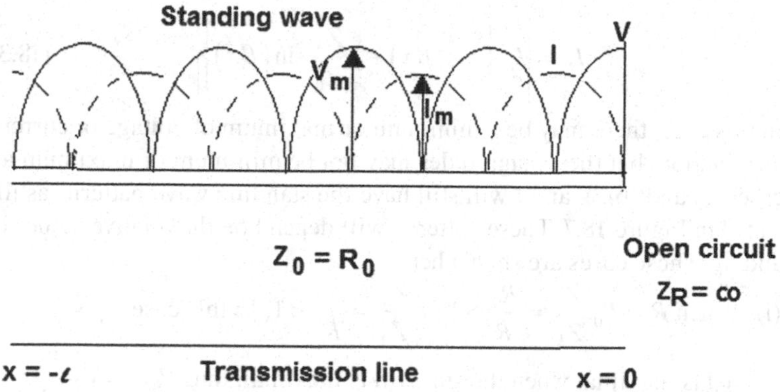

FIGURE 18.5 An open-circuited line with current and voltage standing waves.

(c) If the line is terminated in characteristic impedance, which is purely resistive, $Z_R = Z_0 = R_0$, we get:

$$|V_x| = V_R\sqrt{\cos^2(\beta x) + \sin^2(\beta x)} = V_R \tag{18.32a}$$

$$|I_x| = I_R\sqrt{\cos^2(\beta x) + \sin^2(\beta x)} = I_R \tag{18.32b}$$

In this case, voltage and current are shown by straight lines in Figure 18.6. As there is no reflection, there are no standing waves on the line.

(d) If the line is not terminated in characteristic impedance, which is purely resistive, $Z_R \neq Z_0(=R_0)$, and we get:

$$|V_x| = \left|V_R\left\{\cos(\beta x) + j\frac{Z_0}{Z_R}\sin(\beta x)\right\}\right| \tag{18.33a}$$

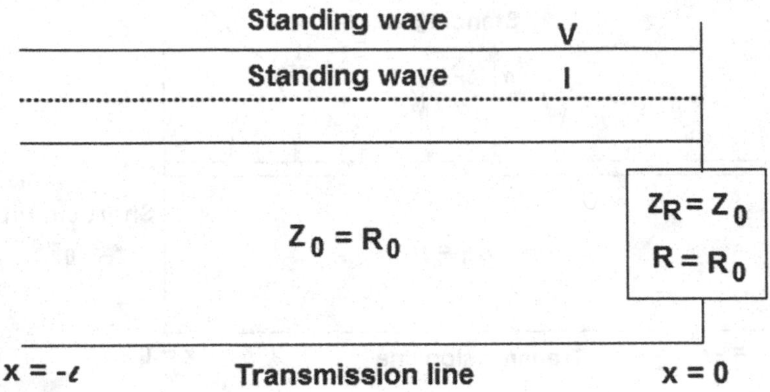

FIGURE 18.6 Line terminated in its characteristic impedance.

$$|I_x| = \left| I_R \left\{ \cos(\beta x) + j\frac{Z_R}{Z_0}\sin(\beta x) \right\} \right| \tag{18.33b}$$

In this case, there may be a minimum or maximum of voltage or current at termination, but their magnitudes may not be minimum or maximum as in cases (a) and (b). V and I will still have the standing wave patterns as illustrated in Figure 18.7. These patterns will depend on the relative values of R and R_0. These cases are shown here.

(i) When $R < R_0, \frac{Z_0}{Z_R} = \frac{R_0}{R} > 1$ or $\frac{Z_R}{Z_0} = \frac{R}{R_0} < 1$. In this case:

$|V_x|$ is maximal when the sin term is maximal, thus $V_{max} = V_R \frac{R_0}{R}$.

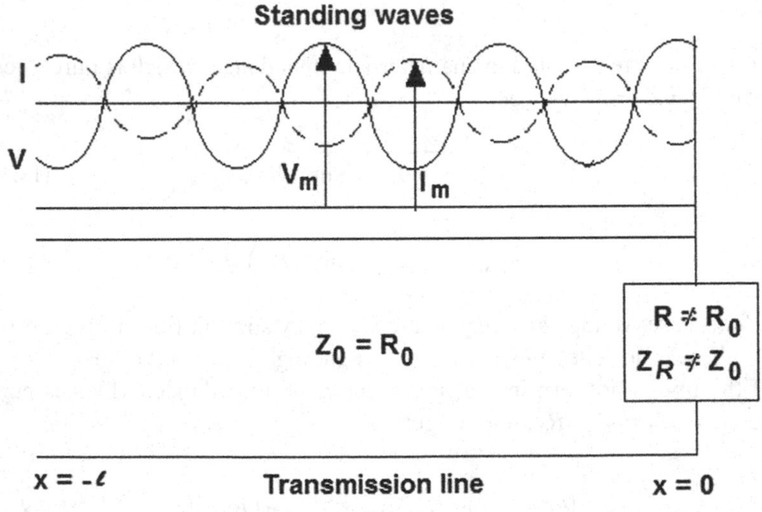

FIGURE 18.7 Terminating impedance is not equal to the characteristic impedance of the line.

$|I_x|$ is maximal when the cos term is maximal, thus $I_{max} = I_R$.
$|V_x|$ is minimal when the cos term is minimal, thus $V_{min} = V_R$.

$|I_x|$ is minimal when the sin term is minimal, thus $I_{min} = I_R \dfrac{R}{R_0}$.
These conditions lead to:

$$\frac{V_{max}}{V_{min}} = \frac{R_0}{R} \quad \text{and} \quad \frac{I_{max}}{I_{min}} = \frac{R_0}{R} \tag{18.34}$$

(ii) When $R > R_0, \dfrac{Z_0}{Z_R} = \dfrac{R_0}{R} < 1$ or $\dfrac{Z_R}{Z_0} = \dfrac{R}{R_0} > 1$. In this case:

$|V_x|$ is maximal when the cos term is maximal, thus $V_{max} = V_R \dfrac{R}{R_0}$.

$|I_x|$ is maximal when the sin term is maximal, thus $I_{max} = I_R$.
$|V_x|$ is minimal when the sin term is minimal, thus $V_{min} = V_R$.

$|I_x|$ is minimal when the cos term is minimual, thus $I_{min} = I_R \dfrac{R}{R_0}$.
These conditions give:

$$\frac{V_{max}}{V_{min}} = \frac{R}{R_0} \quad \text{and} \quad \frac{I_{max}}{I_{min}} = \frac{R}{R_0} \tag{18.35}$$

In Equations 18.34 and 18.35, the ratios V_{max}/V_{min} and I_{max}/I_{min} are very important quantities called *voltage* and *current standing wave ratios*, respectively. The voltage standing wave ratio (VSWR) and current standing wave ratio (CSWR) are denoted by ρ or *s*. The value of VSWR is always greater than 1. These are given as:

$$\left.\rho = \frac{R}{R_0} \;\; (\text{for } R > R_0) \;\; (a) \qquad \rho = \frac{R_0}{R} \;\; (\text{for } R < R_0) \;\; (b)\right\} \tag{18.36}$$

A relation between the standing wave ratio and the reflection coefficient can be obtained in view of Equation 18.16 and the conditions described by Equation 18.35.

$$\rho = \frac{1+\Gamma}{1-\Gamma} \tag{18.36c}$$

(e) If the line is terminated in an impedance, which is a complex quantity and is not equal to the characteristic impedance, i.e. $Z_R = R \pm jX \neq R_0$ there will be neither minima nor maxima of voltage or current at the location of termination. The minima or maxima will always be displaced to either side from the terminating end. The direction and location of these displacements along with the standing wave measurements leads to the estimation of the value of reactance. The sign of reactance will spell its type.

Figure 18.8a illustrates a transmission line ($Z_0 = R_0$) terminated in an unknown impedance ($R \pm X$) at location *c*. To know the values of *R* and *X*let the line be

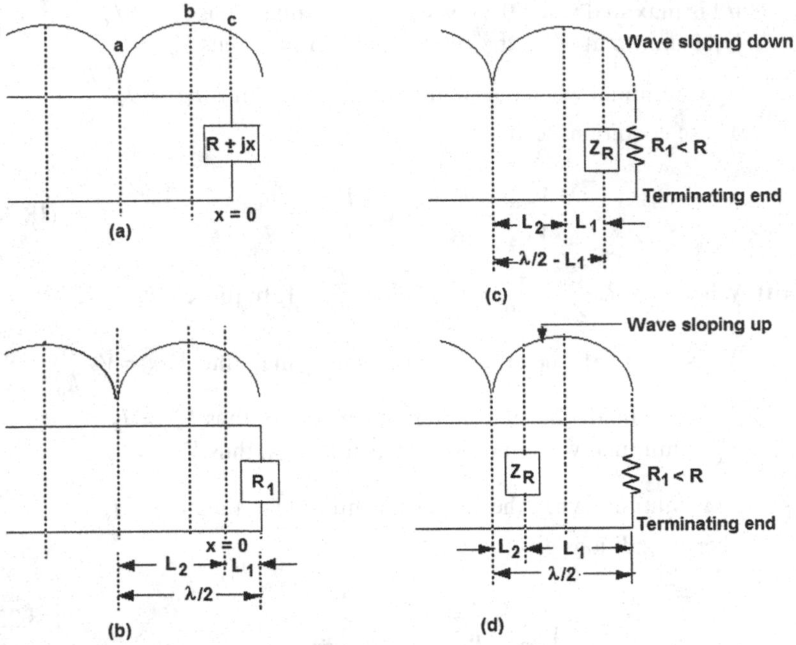

FIGURE 18.8 Voltage and current distributions with (a) complex termination Z = R ± jX; (b) line terminated in pure resistance (R_1) at hypothetically extended distance (L_1); (c) wave sloping up; (d) wave sloping down.

terminated in a pure resistance (R_1) at a hypothetically extended distance (L_1) shown in Figure 18.8b. At this location voltage (or current) becomes minimum (or maximum) or vice-versa. The value of R_1 is such that it appears at location *c* as $R \pm JX$. It is to be noted it is referred to as Z_R as it the terminating impedance but since it is the input impedance at location c for the hypothetically extended length L_1which terminated in R_1it can be regarded as input impedance Z_s at c. Thus in view of Equation 18.27c, Z_S (now Z_R) can be written in the modified form:

$$Z_R = R \pm JX = R_0 \frac{R_1 \cos(\beta L_1) + jR_0 \sin(\beta L_1)}{R_0 \cos(\beta L_1) + jR_R \sin(\beta L_1)} \tag{18.37a}$$

Simplification of Equation 18.33a gives:

$$R \pm JX = R_0 \frac{\{R_1 \cos(\beta L_1) + jR_0 \sin(\beta L_1)\}\{R_0 \cos(\beta L_1) - jR_1 \sin(\beta L_1)\}}{\{R_0 \cos(\beta L_1) + jR_1 \sin(\beta L_1)\}\{R_0 \cos(\beta L_1) - jR_1 \sin(\beta L_1)\}}$$

$$R \pm JR_0 \frac{\begin{array}{l}\{R_0 R_1 \cos^2(\beta L_1) + R_0 R_1 \sin^2(\beta L_1)\} \\ \quad + j\{R_0^2 \sin(\beta L_1)\cos(\beta L_1) - R_1^2 \sin(\beta L_1)\cos(\beta L_1)\}\end{array}}{R_0^2 \cos^2(\beta L_1) + R_1^2 \sin^2(\beta L_1)}$$

Thus

$$\left.\begin{aligned} R &= \frac{R_0^2 R_1}{R_0^2\cos^2(\beta L_1) + R_1^2\sin^2(\beta L_1)} \quad &\text{(b)} \\ X &= \frac{R_0\left(R_0^2 - R_1^2\right)\sin(\beta L_1)\cos(\beta L_1)}{R_0^2\cos^2(\beta L_1) + R_1^2\sin^2(\beta L_1)} \quad &\text{(c)} \end{aligned}\right\} \quad (18.37)$$

In Equations 18.37b and 18.37c, the length L_1 cannot be measured and is to be replaced by a measurable length. In view of Figure 18.8b, $L_1 = \frac{\lambda}{2} - L_2$. Thus:

$$\sin(\beta L_1) = \sin\left\{\beta\left(\frac{\lambda}{2} - L_2\right)\right\} = \sin(\beta L_2)$$

and

$$\cos(\beta L_1) = \cos\left\{\beta\left(\frac{\lambda}{2} - L_2\right)\right\} = -\cos(\beta L_2)$$

In view of Equations 18.36b, 18.37b, 18.37c, and the above relations, we get:

$$\left.\begin{aligned} R &= \frac{\rho R_0}{\rho^2\cos^2(\beta L_2) + \sin^2(\beta L_2)} \quad &\text{(d)} \\ X &= \frac{-R_0\left(\rho^2 - 1\right)\sin(\beta L_2)\cos(\beta L_2)}{\rho^2\cos^2(\beta L_2) + \sin^2(\beta L_2)} \quad &\text{(e)} \end{aligned}\right\} \quad (18.37)$$

In view of Equations 18.37d and 18.37e, it can be seen for $R_1 < R_0$ that (a) when $L_2 < \frac{\lambda}{4}$, the wave is sloping up, X is positive or inductive; and (b) when $\frac{\lambda}{4} < L_2 < \frac{\lambda}{2}$, the wave is sloping down, X is negative or capacitive. Cases a and b are shown in Figures 18.8d and 18.8c, respectively.

18.6 LINE SECTIONS AS CIRCUIT ELEMENTS

As noted earlier, at low frequencies transmission lines are used to transfer power between distant points whereas at higher frequencies these are employed to transfer signals. Similarly at lower frequencies circuits can be formed by lumped elements whereas at higher frequencies limped elements lose their significance. After certain frequency construction of lumped elements becomes difficult and the required physical size of transmission line sections becomes sufficiently small. Above 150 MHz ($\lambda < 2$ m), these line sections become good contenders to replace lumped circuit elements. Such line sections are conveniently employed up to about 3000 MHz, beyond which the size becomes too small to be handled and waveguide technology begins to take over. So, what do these line sections represent? Obviously, the elements represented by a lossless line will be different from those represented by a lossy line. These cases are therefore considered separately in the following subsections.

18.6.1 Lossless Line

For a lossless line, the input impedance given by Equation 18.27c is reproduced here:

$$Z_S = Z_0 \frac{Z_R \cos(\beta l) + jZ_0 \sin(\beta l)}{Z_0 \cos(\beta l) + jZ_R \sin(\beta l)} \tag{18.27c}$$

When this line is *short-circuited*, $Z_R = 0$ and its input impedance becomes:

$$Z_S = jZ_0 \tan(\beta l) \tag{18.38a}$$

From Equation 18.38a, the input impedance for $0 < l < \frac{\lambda}{4}$ or for $0 < \beta l < \frac{\pi}{2}$ is positive (or inductive), and for $\frac{\lambda}{4} < l < \frac{\lambda}{2}, \frac{\pi}{2} < \beta l < \pi$, negative (or capacitive).

When this line is *open-circuited*, $Z_R = \infty$ and its input impedance becomes:

$$Z_S = jZ_0 \cot(\beta l) \tag{18.38b}$$

In Equation 18.38b the input impedance for $0 < l < \frac{\lambda}{4}$ or for $0 < \beta l < \frac{\pi}{2}$ is negative (or capacitive) and for $\frac{\lambda}{4} < l < \frac{\lambda}{2}$ or for $\frac{\pi}{2} < \beta l < \pi$, it is positive (or inductive).

18.6.2 Lossy Lines

For lossy lines, the input impedance given by Equation 18.22a can be written in the following modified form:

$$Z_S = Z_0 \frac{Z_R \cosh(\gamma l) + Z_0 \sinh(\gamma l)}{Z_0 \cosh(\gamma l) + Z_R \sinh(\gamma l)} \tag{18.22a}$$

In this equation, γ is to be replaced by $\alpha + j\beta$. It will result in terms like sin(βl), cos(βl), cosh(αl), and sinh(αl). Since α is assumed to be small its approximate value $\left(\alpha \approx \frac{R}{2Z_0}\right)$ given by Equation 18.26d is to be used. For small α, cosh(αl) $\approx$ 1 and sinh(αl) $\approx$ αl. Since we are considering the equivalence of line sections with $l = \frac{\lambda}{4}$ and $l = \frac{\lambda}{2}$ thus for $l = \frac{\lambda}{4}$, $\beta l = \frac{\pi}{2}$, sin(βl) = 1, and cos(βl) = 0, and for $= \frac{\lambda}{2}$, $\beta l = \pi$, sin(βl) = 0, and cos(βl) = –1. Equation 18.22a may take the following forms under different terminating conditions.

When the line is *short-circuited*, $Z_R = 0$ and Equation 18.22a gives:

$$Z_S = Z_0 \frac{\sinh(\gamma l)}{\cosh(\gamma l)} = Z_0 \frac{\sinh\{(\alpha + j\beta)l\}}{\cosh\{(\alpha + j\beta)l\}}$$
$$= Z_0 \frac{\sinh \alpha l \cosh \beta l + j \cosh \alpha l \sinh \beta l}{\cosh \alpha l \cosh \beta l + j \sinh \alpha l \sinh \beta l} \quad (18.39a)$$

The input impedance from Equation 18.39a for *quarter wave* and *half wave sections* are given by Equations 18.39b and 18.39c, respectively.

$$\left. Z_S = \frac{2Z_0^2}{Rl} \ (\text{b}) \ \text{for } l = \frac{\lambda}{4} \quad Z_S = \frac{Rl}{2} \ (\text{c}) \ \text{for } l = \lambda/2 \right\} \quad (18.39)$$

This expression of Equation 18.39b is also valid for a short-circuited line whose length is an odd multiple of $\lambda/4$.

When this line is *open-circuited*, $Z_R = \infty$ and Equation 18.22a gives:

$$Z_S = Z_0 \frac{\cosh\{(\alpha + j\beta)l\}}{\sinh\{(\alpha + j\beta)l\}} = Z_0 \frac{\cosh \alpha l \cosh \beta l + j \sinh \alpha l \sinh \beta l}{\sinh \alpha l \cosh \beta l + j \cosh \alpha l \sinh \beta l} \quad (18.40a)$$

The input impedance from Equation 18.40a for *quarter wave* and *half wave sections* are given by Equations 18.40b and 18.40c, respectively:

$$\left. Z_S = \frac{Rl}{2} \ (\text{b}) \ \text{for } l = \frac{\lambda}{4}, \quad Z_S = \frac{2Z_0^2}{Rl} \ (\text{c}) \ \text{for } l = \lambda/2 \right\} \quad (18.40)$$

The equivalent elements given by Equations 18.39 and 18.40 are illustrated in Figure 18.9.

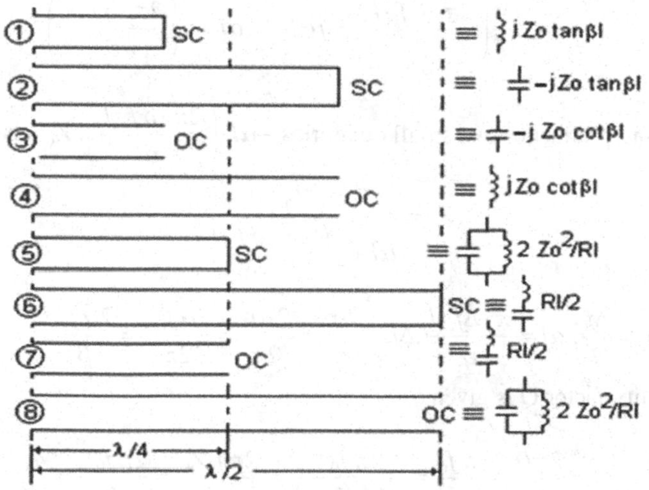

FIGURE 18.9 Lines sections and their equivalent elements for different lengths and terminations.

18.6.3 Quality Factor of the Resonant Section

The input impedance of a shorted line is given by Equation 18.39a. At resonant frequency $f = f_0$, $\beta l = \frac{n\pi}{2}$, where n is an odd integer.

This gives $\cos \beta l = 0$, $\sin \beta l = \pm 1$, thus Equation 18.39a reduces to:

$$Z_s = Z_0 \frac{\cosh \alpha l}{\sinh \alpha l} = \frac{Z_0}{\alpha l} \tag{18.41a}$$

Let there be an incremental change in the frequency, i.e. $f = f_0 + \delta f$.

Thus $\beta l = \frac{2\pi fl}{\upsilon} = \frac{2\pi(f_0 + \delta f)l}{\upsilon} = \frac{n\pi}{2} + \frac{2\pi\delta fl}{\upsilon}$ $\quad \cos \beta l = -\sin\left(\frac{2\pi \cdot \delta f \cdot l}{\upsilon}\right)$

$\sin \beta l = \cos\left(\frac{2\pi \cdot \delta f \cdot l}{\upsilon}\right)$

and

$$Z_s = Z_0 \frac{-\sinh \alpha l \cdot \sin\left(\frac{2\pi \cdot \delta f \cdot l}{\upsilon}\right) + j\cosh \alpha l \cdot \cos\left(\frac{2\pi \cdot \delta f \cdot l}{\upsilon}\right)}{\cosh \alpha l \cdot \sin\left(\frac{2\pi \cdot \delta f \cdot l}{\upsilon}\right) + j\sinh \alpha l \cdot \cos\left(\frac{2\pi \cdot \delta f \cdot l}{\upsilon}\right)} \tag{18.41b}$$

Also since, $\sinh \alpha l \approx \alpha l$, $\cosh \alpha l \approx 1$, $\cos\left(\frac{2\pi \cdot \delta f \cdot l}{\upsilon}\right) \approx 1$ $\sin\left(\frac{2\pi \cdot \delta f \cdot l}{\upsilon}\right) \approx \frac{2\pi \cdot \delta f \cdot l}{\upsilon}$

$$Z_s = Z_0 \frac{-\alpha l \cdot \left(\frac{2\pi \cdot \delta f \cdot l}{\upsilon}\right) + j1}{-\left(\frac{2\pi \cdot \delta f \cdot l}{\upsilon}\right) + j\alpha l} = \frac{Z_0}{\alpha l + j\left(\frac{2\pi \cdot \delta f \cdot l}{\upsilon}\right)} \tag{18.41c}$$

On neglecting product of two small quantities $-\alpha l \cdot \left(\frac{2\pi \cdot \delta f \cdot l}{\upsilon}\right)$, Z_s becomes:

$$Z_s = \frac{Z_0}{\alpha l + j\left(\frac{2\pi \cdot \delta f \cdot l}{\upsilon}\right)} \tag{18.41d}$$

If we let $\delta f = \frac{\Delta f}{2}$, $\alpha l = \frac{\pi \cdot \Delta f \cdot l}{\upsilon}$, $\Delta f = \frac{\alpha \upsilon}{\pi} = \frac{2\alpha \upsilon}{2\pi} = \frac{2\alpha f_{0\lambda}}{2\pi} = \frac{2\alpha f_0}{\beta}$,

The quality factor Q is given as:

$$Q = \frac{f_0}{\Delta f} = \frac{\beta}{2\alpha} = \frac{2\pi/\lambda}{2\alpha} = \frac{\pi f_0}{\upsilon \alpha} = \frac{\pi f_0}{\upsilon(R/2Z_0)} = \frac{2\pi f_0 Z_0}{\upsilon R} = \frac{2\pi f_0 L}{R} = \frac{\omega_0 L}{R} \tag{18.42a}$$

or simply:

$$Q = \frac{\omega L}{R} \tag{18.42b}$$

In above equations, $\upsilon = f\lambda \quad \lambda = \frac{\upsilon}{f}$ and $\beta = \frac{2\pi}{\lambda} \quad Z_0 = \sqrt{\frac{L}{C}} \quad \upsilon = \frac{1}{\sqrt{LC}} \quad \frac{Z_0}{\upsilon} = L$

18.7 STUB MATCHING

In order to prevent reflections from reaching the generator (or sending end) transmission lines are often matched by using some auxiliary or stub lines. The reflections can be prevented by using one or two auxiliary lines called single or double stub matching. The working of stub lines is briefly described below.

18.7.1 Single Stub Matching

The single stub matching uses only one auxiliary line of (variable) length (says). This line is made to slide over the main line. Its location (say d) from the load end determines the admittance it introduces. Figure 18.10 shows the single stub matching and the related phasor diagram. According to this diagram, the load admittance Y_R is transformed by length d into admittance $Y_A = G_A \pm JB_A$. Also Y_{SC} is transformed by length s to Y_S. If $Y_S = JB_A$ and $G_A = Y_C$, the sum of Y_A and Y_S is simply the characteristic admittance Y_C and the line will be flat towards the generator or there will be no reflection from the location of stub. The transmission line is said to be matched (or flat) if there are no reflections.

18.7.2 Double Stub Matching

The double stub matching uses two auxiliary lines both of variable lengths (say s_1 and s_2). The lengths of these stubs determine the admittances they introduce. Both of these stubs have fixed locations (l and $l + d$) from the terminating end, over the main line. The double stub matching and the related phasor diagram are shown in Figure 18.11. According to this diagram, the load admittance Y_R is transformed by

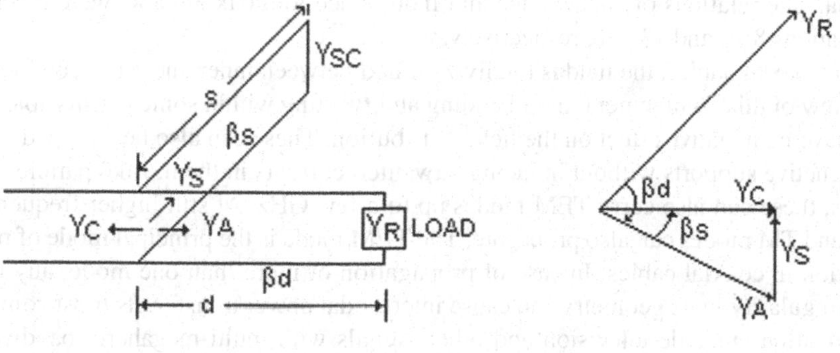

FIGURE 18.10 Single stub matching.

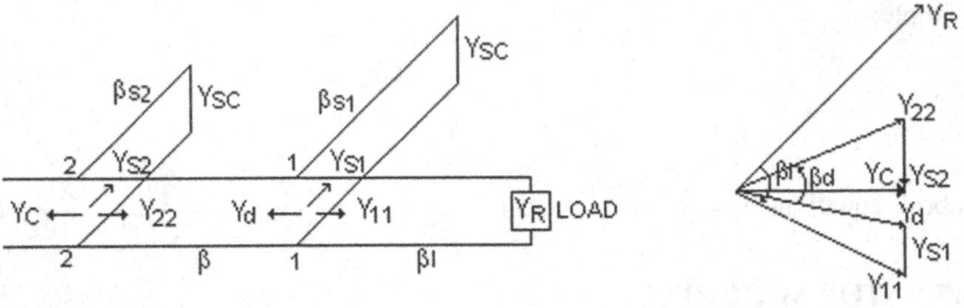

FIGURE 18.11 Double stub matching.

length l into an admittance $Y_{11} = G_{11} \pm JB_{11}$. In addition, the Y_{SC} of stub 1 is translated by length s_1 to a pure susceptance $Y_{S1} = jb_1$ and Y_{SC} of stub 2 to $Y_{S2} = jb_2$. At location 1, Y_{S1} gets added to Y_{11} and partially neutralizes the susceptance of Y_{11}, hence the net admittance becomes Y_d. This Y_d again gets transformed by length d to a new value $Y_{22} = G_{22} \pm JB_{22}$. If the addition of Y_{S2} to Y_{22} neutralizes its susceptance part $\pm JB_{22}$, the matching is accomplished provided $G_{22} = Y_C$. Thus, $\pm JB_{22}$ and jb_2 must add to result in a zero susceptance.

In the single stub case, matching is accomplished through two variables s and d. The process of matching by double stub system also requires two variables. The stub lengths s_1 and s_2 serve this purpose, which can be conveniently varied by shorting plungers. In single stub matching system variation of d is not as convenient as variation of stub length. Thus double stub matching is normally preferred over single stub matching.

18.8 SOME TRANSMISSION LINE STRUCTURES

This section briefly deals with two types of transmission line structures. These include the coaxial cables and the planar lines.

18.8.1 Coaxial Cable

Coaxial cables were discussed earlier in Sections 11.4.2.1 and 13.3.4.1. Its configurations were shown earlier in Figures 8.15a, 9.5, 11.9a, 11.14, 11.19b, 11.20, and 13.8a. The relations of capacitance and inductance for coaxial cable were given by Equations 8.70 and 13.24b, respectively.

In coaxial cables, the field is totally confined between inner and outer conductors. In view of this confinement, their bending and twisting within some permissible limits have no negative effect on the field distribution. These can also be strapped to the conductive supports without inducing unwanted currents in them. Like parallel wire lines, these can also carry TEM modes up to a few GHz. At still higher frequencies TE and TM modes can also propagate. The TEM mode is the principal mode of propagation in coaxial cables. In case of propagation of more than one mode, any bend or irregularity in its geometry can cause intermodal power transfer. Its most common applications include television and other signals with multi-megahertz bandwidth. The characteristic impedance of a coaxial cable may be of the order of 50 to 75 ohms.

18.8.2 Planar Lines

The planar lines are also forms of transmission lines that are used to guide electromagnetic waves in centimetric to the submillimetric range. These lines have the same characterizing parameters (viz. Z_0, α, β, v_p, etc.) as applicable to other transmission structures. In planar lines, these parameters are controlled by their geometry wherein the dimensions are defined in a single plane. In view of this unique feature, a complete transmission line circuit can be fabricated in one step by using thin film and photolithography techniques. These lines have a number of forms such as striplines; microstrip, slotline, coplanar lines, and fin lines, and each of these has a number of versions. Each of these is also characterized by their nature of field distributions, achievable levels of characteristic impedances, phase velocities, and the modes supported by these structures. These lines can be combined to produce delay lines, crossovers, resonators, transitions, power splitters/combiners, and filters. Microstrips and other planar lines can be used to realize lumped parameters, viz. resistors, inductors, capacitors, and even transformers. The design, analysis, and circuit implementation of these lines are well established. With the addition of normal metals; terminations and broadband absorbers can also be realized. The circuit responses of these lines can easily be tailored as per the requirements.

18.8.2.1 Stripline

Stripline, one of the forms of planar lines, is illustrated in Figure 18.13. In this, a flat conducting strip is symmetrically placed between two large ground planes (GPs). The space between these planes is filled with a homogeneous dielectric material having no air pockets. A conventional stripline is a balanced line. In this structure, almost all the field remains confined nearer to the central conductor. This field rapidly decays away from this strip. This property of stripline allows its termination in the transverse direction without affecting its transmission characteristics.

A stripline is well suited for microwave-integrated circuits and photolithographic fabrication. It usually carries a pure TEM mode. This line can also support higher-order TE and TM modes, but in practice, all such modes are avoided. The striplines can have many versions including (a) centered stripline, (b) off-centered stripline, and (c) dual-orthogonal stripline. Each of these has its own merits. Figure 18.12 shows the most commonly used centered stripline. It illustrates the separation (b) between outer conductors, thickness (t), and the width (w) of the central conductor. The width of the strip is generally much larger than its thickness. The thickness is of the order of 1.4 to 2.8 mils.

18.8.2.2 Microstrip

When the upper ground plane of a stripline is removed the resulting configuration is referred to as microstrip line or simply microstrip. It is one of the most popular lines which can be fabricated by photolithographic process. Its versions include (i) a common microstrip; (ii) an embedded microstrip; and (iii) a covered microstrip.

Figure 18.13 illustrates a microstrip (also called open-strip line) which contains a ground plane at the bottom and a conducting strip. Due to the absence of a top ground plane the strip can easily be accessed. This makes it convenient to connect

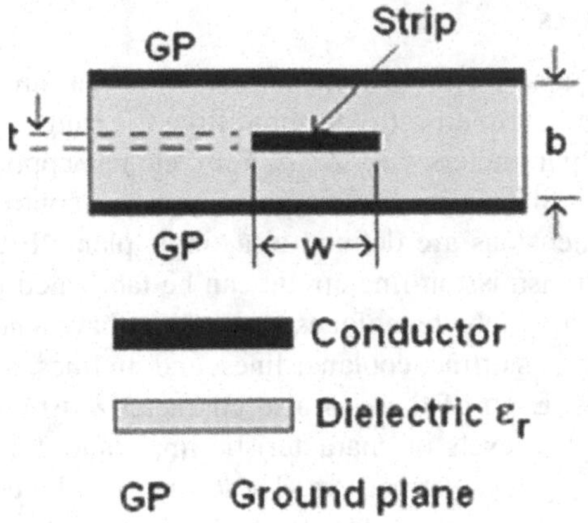

FIGURE 18.12 Geometrical configuration of stripline.

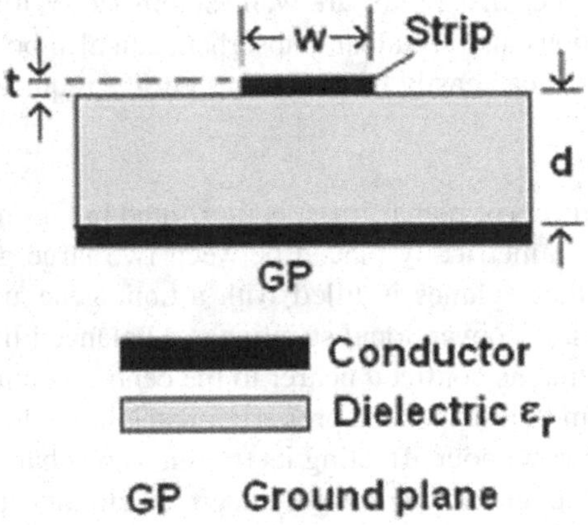

FIGURE 18.13 Geometrical configuration of microstrip.

the strip with the discrete passive components and active microwave devices. Also, the signals are readily available for probing. Besides, minor adjustments can also be made after the fabrication of the circuit.

The openness of this strip structure also has some negative aspects. These include (i) coupling of some of the transmitted energy to space or the adjacent traces; (ii) likelihood of radiation loss or interference from nearby conductors. These losses are likely to become more significant at discontinuities viz. short-circuit posts, corners etc. and calls for some remedial measures for their reduction. The use of high permittivity substrate is one such remedial measure which results in the confinement of the field nearer to the strip. As a consequence, there is reduction of phase velocity, guide wavelength, and the circuit dimensions. The penetration of field into the space makes the microstrip configuration a mixed dielectric transmission structure due to which the analysis becomes complicated.

Due to the advances in IC technology microwave semiconductor devices are usually fabricated as semiconductor chips. In view of their small volumes (on the order of 0.008 to 0.08 mm³) microstrips are commonly used for feeding into or extracting energy from, these chips. In most practical applications, the thickness (d) of dielectric substrate is kept much smaller than the electrical wavelength ($d << \lambda$). Due to this thinness microstrip does not support a pure TEM wave and the modes which propagate in these are often referred to as quasi-TEM. Thus the theory of TEM-coupled lines is only approximately valid in the case of microstrip lines.

18.8.2.3 Slotline

Figure 18.14 shows a slotline containing a pair of ground planes with a narrow slot between them. In slotline, the signal propagates in TE mode making it a non-ideal general purpose transmission medium. The TE mode, however, makes it useful for balanced mixer and amplifier circuits requiring push-pull operation. In the cross-section of a slot line, there exists a location where the magnetic field is circularly polarized. This feature makes it useful for the design of several components such as ferrite isolators. A slotline with two conductors in one plane makes shunt mounting of active or passive components quite easy.

The comparison of a slot line with other planar lines reveals the following aspects.

1. It yields high characteristic impedance with large slot width, whereas a microstrip requires a very narrow strip. The fabrication of narrow strip imposes a technological limitation.
2. A coplanar line has quasi-TEM mode and a slot line has TE mode. The combination of these two leads to the formation of useful hybrid junctions and transition circuits.
3. The coplanar line to slotline transition is one of the most popular formations which has demonstrated balun operation over more than two octaves of bandwidth and has been used in miniaturized uniplanar mixers and amplifiers.
4. The coaxial to microstrip transition is relatively easier to fabricate than that from coaxial to slot line.
5. A slot line is more dispersive than microstrips i.e. the variations of Z_0 and l_0/l_g with frequency are larger.

18.8.2.4 Coplanar Line

Figure 18.15 shows a configuration of a coplanar line (or coplanar waveguide). It contains a conductor located between two ground planes. In these lines, quasi-TEM

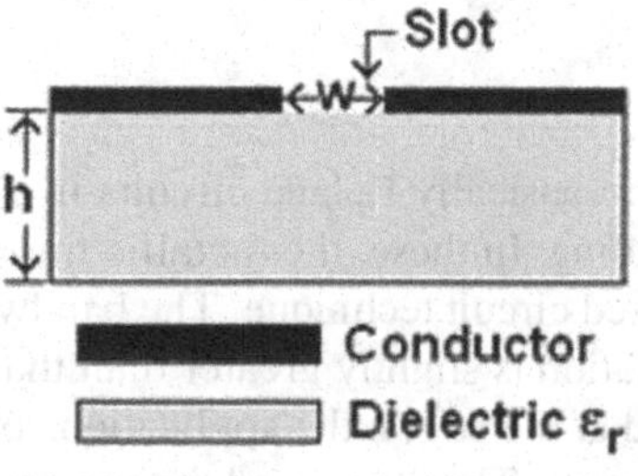

FIGURE 18.14 Geometrical configuration of slotline.

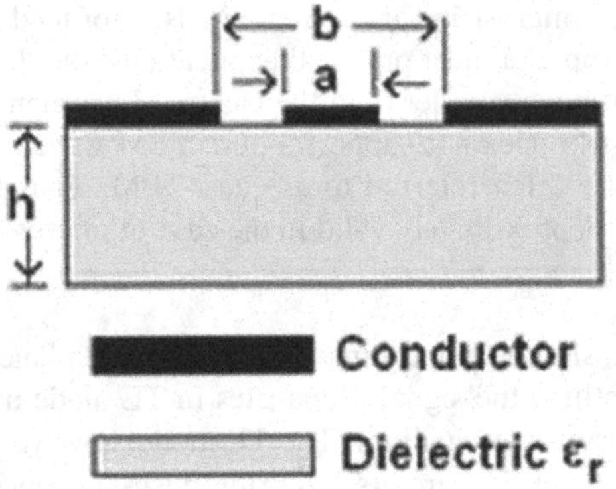

FIGURE 18.15 Geometrical configuration of coplanarline.

is the dominant mode and there is no low frequency cutoff. Coplanar line exhibits some of the advantageous features of microstrips and slot lines. Microstrip geometry is convenient for series mounting of components across the gap in the strip conductor whereas slot lines are suitable for shunt mounting of components across the slots. In coplanar lines, both series and shunt mounted components can be easily incorporated.

The transition from coplanar line to coaxialline and microstrip line can be conveniently achieved. In coplanar line the dimensions can be tapered, for uniform impedance, to connect to components of large to a minute dimension. The coplanar line finds extensive applications in balanced mixers, balanced modulators etc. where both balanced and unbalanced signals are to be carried. The advantage of circularly polarized magnetic field region of slot line is also available in coplanar lines. The versions of coplanar line configurations include coplanar waveguides with or without the ground.

The major attraction of coplanar lines includes that no via-holes through substrate for grounding are needed. A coplanar line has low dispersion and its chip components exhibit less parasitic capacitance. Tapers with constant Z_0 can be realized by varying the track width and gap combination. Its ground planes may provide some shielding between lines. One of its major drawbacks is that its mode of propagation gets easily degenerated from quasi-TEM into a balanced coupled-slotline mode, particularly at discontinuities. The degeneration can be prevented by incorporating grounding straps between the ground planes either through bond-wires or by using a second metallic layer.

18.8.2.5 Fin-Lines

The integrated fin lines are basically Eplane circuits in a rectangular waveguide in the form of capacitive loading. In these, the metallic fins are incorporated on to the dielectric slabs using printed circuit technique. The bandwidth of a fin-line is greater than an octave and attenuation is slightly greater than that of microstrip. Due to the concentration of high field at the edges, the applications of fin-lines are restricted to low and medium power ranges. These are used for constructing low cost millimetric

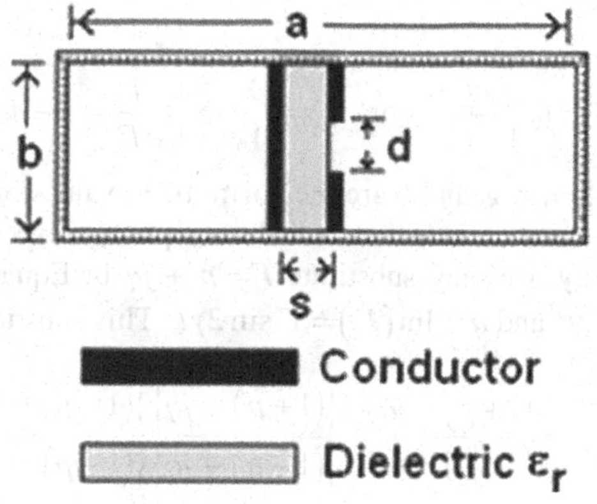

FIGURE 18.16 Integrated unilateral fin-line.

wave circuits. Like other planar lines, fin-lines too have many versions referred to as unilateral, bilateral, and insulated. The unilateral version of a fin-line is shown in Figure 18.16.

18.9 SMITH CHART

A Smith chart is a tool used to solve transmission line problems. It is a circular plot containing a large number of interlaced circles. Although it is obtained from the relations of two-wire transmission lines, it is equally applicable to waveguides, coaxiallines, planar lines, and other structures used to transmit or guide electromagnetic energy. It contains a pair of coordinates for impedance and admittance and represents all possible (real, imaginary, or complex) values of these. It provides an easy method for the conversion of impedance to admittance and vice versa. It is a graphical method for determining the impedance and admittance transformation due to the length of the transmission line.

18.9.1 Formation of a Smith Chart

A Smith chart is obtained from the exponential form of solution given by Equations 18.12 to 18.16. On replacing Z_R by Z in Equation 18.16, the expressions for the reflection coefficient (Γ) for voltage and current can be written as:

$$\left.\Gamma = \frac{Z - Z_0}{Z + Z_0} \;(\text{for voltage})\; (\text{c}) \quad \Gamma = \frac{Z_0 - Z}{Z + Z_0} \;(\text{for current})\; (\text{d})\right\} \tag{18.43}$$

Equation 18.43 can be manipulated to yield:

$$Z = Z_0 \frac{(1+\Gamma)}{(1-\Gamma)} \quad \text{or} \quad z = \frac{Z}{Z_0} = \frac{1+\Gamma}{1-\Gamma} = r + jx \tag{18.44a}$$

Similarly we can get:

$$Y = Y_0 \frac{1-\Gamma}{1+\Gamma} \quad \text{or} \quad y = \frac{Y}{Y_0} = \frac{1-\Gamma}{1+\Gamma} = g + jb \tag{18.44b}$$

In Equation 18.44 *z*, *y*, *r*, *x*, *g*, and *b* are the normalized values of impedance, admittance, resistance, reactance, conductance, and susceptance respectively. Since Γ may be a complex quantity, we may substitute $\Gamma = p + jq$ in Equation 18.44a. In this $p = \mathrm{Re}(\Gamma) = \Gamma\cos 2\gamma l$ and $q = \mathrm{Im}(\Gamma) = \Gamma\sin 2\gamma l$. This substitution gives:

$$r + jx = \frac{1+\Gamma}{1-\Gamma} = \frac{(1+p)+jq}{(1-p)-jq} = \frac{\{(1+p)+jq\}\{(1-p)+jq\}}{\{(1-p)-jq\}\{(1-p)+jq\}}$$

$$= \frac{1-p^2-q^2}{1+p^2+q^2-2p} + j\frac{-2q}{1+p^2+q^2-2p}$$

Thus, we get:

$$\left. r = \frac{1-p^2-q^2}{1+p^2+q^2-2p} \quad (\mathrm{a}) \qquad x = \frac{-2q}{1+p^2+q^2-2p} \quad (\mathrm{b}) \right\} \tag{18.45}$$

Equation 18.45 can be manipulated to yield:

$$\left. \left[p - \left(\frac{r}{1+r}\right)\right]^2 + q^2 = \left(\frac{1}{1+r}\right)^2 \quad (\mathrm{a}) \quad (p-1)^2 + \left(q + \frac{1}{x}\right)^2 = \left(\frac{1}{x}\right)^2 \quad (\mathrm{b}) \right\} \tag{18.46}$$

Equation 18.46a is an equation of a circle with $p = \mathrm{Re}(\Gamma)$ as the horizontal axis and $q = \mathrm{Im}(\Gamma)$ as the vertical axis, radius $R = \frac{1}{1+r}$ and center at $p = \frac{r}{1+r}$, $q = 0$. The locus of centers of constant *r* circles lies on the $q = 0$ plane or *p*-axis. These centers move from $p = 0$ to $p = 1$ as *r* changes from 0 to ∞. The center of the largest constant *r* circle lies at $p = 0$, $q = 0$ with unit radius, and corresponds to $r = 0$. The center of the smallest constant *r* circle lies at $p = 1$, $q = 0$ with zero radius, and corresponds to $r = \infty$. All constant *r* circles pass through an infinite impedance point located at $p = 1$, $q = 0$. The horizontal line is the $x = 0$ line. It is the resistance axis of impedance coordinates with values from $r = 0$ to ∞.

Similarly, Equation 18.46b is also an equation of a circle with *p* as the horizontal axis and *q* vertical axis, radius $R = \frac{1}{|x|}$ and center at $p = 1$, $q = -\frac{1}{x}$. Since *x* is reactance, when *x* is inductive (or positive) $q = -\frac{1}{x}$, when *x* is capacitive (or negative) $q = \frac{1}{x}$. The locus of centers of constant *x* circles lies on the $p = 1$ axis. As normalized *x* may assume positive or negative values the centers of constant *x* circles will gradually move away from $p = 1$, $q = 0$. Movement of centers along *q* axis will be in the direction opposite to the sign of *x*. The center of the largest constant *x* circle will lie at $p = 1$, $q = \pm\infty$ with infinite radius, and will correspond to $x = 0$. Centers of smallest constant x circles will lie at $p = 1$, $q = 0$ with zero radius and will correspond to $x = \infty$. All impedance values on the outermost ($r = 0$) circle are purely reactive ranging from 0 to ∞. This agrees that for pure reactance (i.e. at the outermost circle) $|\Gamma| = 1$.

This is to be noted that ρ is related to r (for $R > R_0$), to $1/r$ (for $R < R_0$) and to Γ by Equations 18.36a, 18.36b, and 18.36c, respectively. Table 18.1, drawn from Equations 18.45a and 18.45b, shows the same.

In view of the above description and Table 18.1, a number of figures are drawn. Figure 18.17a shows constant r circles, whereas Figure 18.17b shows constant x circles.

If the constant r and constant x circles of Figure 18.17a and b are superimposed, the resulting configuration is referred to as a Smith chart. This Smith chart shown in Figure 18.18a is also called the impedance chart as it based on normalized impedance given by Equation 18.44a. Another Smith chart of Figure 18.18b, called an admittance chart, can also be obtained in a similar manner, by using the expression of normalized admittance given by Equation 18.44b.

TABLE 18.1
Resistance Circles and Reactance Circles

Resistance Circles				Reactance Circles			
Normalized Resistance	**Coordinates of Center**		**Radius**	**Normalized Reactance**	**Coordinates of the Center**		**Radius**
r	q	$p=\frac{r}{1+r}$	$R=\frac{1}{1+r}$	x	p	$q=-\frac{1}{x}$	$R=\frac{1}{\lvert x\rvert}$
0.0	0	0	1	±0	1	∓∞	∞
0.25	0	0.2	0.8	±0.1	1	∓10	10
0.5	0	0.333	0.667	±0.5	1	∓2	2
1.0	0	0.5	0.5	±1	1	∓1	1
4.0	0	0.8	0.2	±5	1	∓0.2	0.2
9.0	0	0.9	0.1	±10	1	∓0.1	0.1
∞	0	1.0	0.0	±∞	1	0.0	0

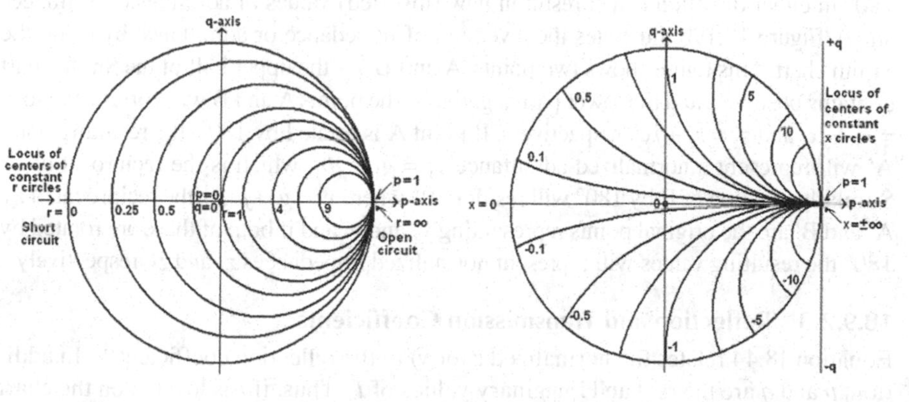

FIGURE 18.17 (a) Constant r circles; (b) constant x circles.

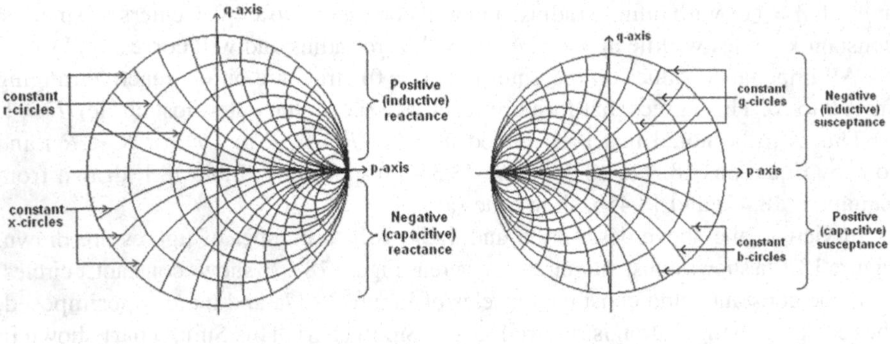

FIGURE 18.18 (a) Impedance chart; (b) admittance chart.

As can be seen from Figure 18.18a and b, it is the same Smith chart which is labeled as an impedance or admittance chart. It contains the same axes, labeled "inductive reactance" and "capacitive". This is due to the fact that the short circuit location in a Smith chart will be replaced by the open circuit in admittance cart and vice versa.

18.9.2 Operations Facilitated by the Smith Chart

As noted earlier the solutions to transmission line problems can be obtained through Smith charts with much ease. Some of the operations which can be performed through Smith charts, with special reference to the impedance chart, are noted below.

18.9.2.1 Impedance Mapping

All (real, imaginary, and complex) values of normalized impedances can be mapped on the smith chart. Figure 18.19a shows mapping of some normalized impedances on the Smith chart. It is to be noted that the impedances on the real axis are always real, on the outermost circle always imaginary, and on intersections of constant r and constant x circles are always complex.

18.9.2.2 Impedance Inversion

If a point representing impedance (or admittance) plotted on Smith chart is rotated by 180° in either direction it will result in new (inverted) values of admittance (or impedance). Figure 18.19b illustrates the inversion of impedance or admittance by using the Smith chart. This figure shows two points, A and B. As the upper half of the Smith chart contains positive x and the lower part negative x, the points A and B will correspond to $z_1 = r_1 + jx_1$ and $z_2 = r_2 - jx_2$, respectively. If point A is rotated by 180°, the resulting point A′ will represent a normalized admittance $y_1 = g_1 - jb_1$, which is the reciprocal of z_1. Similarly, rotation of B by 180° will result in B′ where $y_2 = g_2 + jb_2$ is the reciprocal of z_2. A′ and B′ are the original points representing y_1 and y_2 and if both of these are rotated by 180° the resulting values will represent normalized impedances z_1 and z_2, respectively.

18.9.2.3 Reflection and Transmission Coefficients

Equation 18.44 relates the normalized z (or y) to the reflection coefficient Γ. In addition, p and q are the real and imaginary values of Γ. Thus, if z is located on the chart the length oz will represent the magnitude of the reflection coefficient $|\Gamma|$ and length

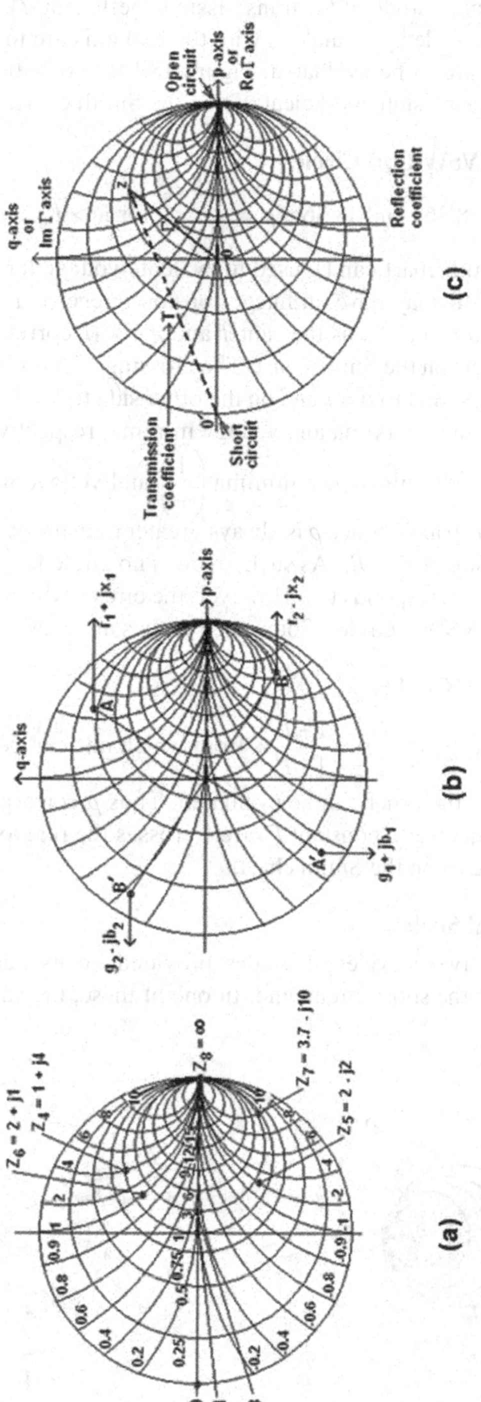

FIGURE 18.19 (a) Normalized impedances; (b) inversion of impedance or admittance; and (c) reflection and transmission.

$o'z$ will represent the magnitude of the transmission coefficient $|T|$. For complex values of Γ and T, angles made by oz and $o'z$ with the real axis are to be measured and the respective values are to be evaluated. Figure 18.19c shows the reflection coefficient (Γ) and the transmission coefficient (T) on the Smith chart.

18.9.2.4 Constant VSWR (ρ) Circles

In view of Equations 18.36a and 18.36b, $\rho = \dfrac{R}{R_0} = r$ for $R > R_0$ and $\rho = \dfrac{R_0}{R} = \dfrac{1}{r}$ (for $R < R_0$). Thus, the Smith chart can be used to evaluate voltage (or current) standing wave ratios. The VSWR may have an integer or non-integer real value. To find its value take the $p = 0$, $q = 0$ point as the center and $p = k$ (k corresponds to r) as the radius, and draw a circle on the Smith chart. The crossing of this circle on one side of the real axis will correspond to $p = r$ and on the other side to $p = 1/r$. These locations will correspond to voltage maxima and voltage minima, respectively. Thus voltage minima correspond to the impedance minima $\left(\dfrac{1}{r}\right)$ and voltage maxima to impedance maxima (r) on the p axis. Since ρ is always greater than unity, it will correspond to r with the assumption of $R > R_0$. As such, there is no circle for ρ since it has only one value and that too corresponds to r. However, the circles which correspond to the values of r are called VSWR circles. Some such circles are shown in Figure 18.20a.

18.9.2.5 Constant Γ Circles

Equation 18.36c relates ρ to $\Gamma \left\{ \rho = \dfrac{1+\Gamma}{1-\Gamma} \right\}$ from which one can get $\left\{ \Gamma = \dfrac{\rho - 1}{\rho + 1} \right\}$. If one of these is known the other can be evaluated. Thus ρ corresponds to the value of r at the location wherever a constant Γ circle crosses the real axis. Figure 18.20b illustrates some Γ circles on the Smith chart.

18.9.2.6 Peripheral Scales

Figure 18.21a shows two wavelength scales provided on its periphery. Both the scales start at zero on the short-circuit end. In one of these, the value of wavelength

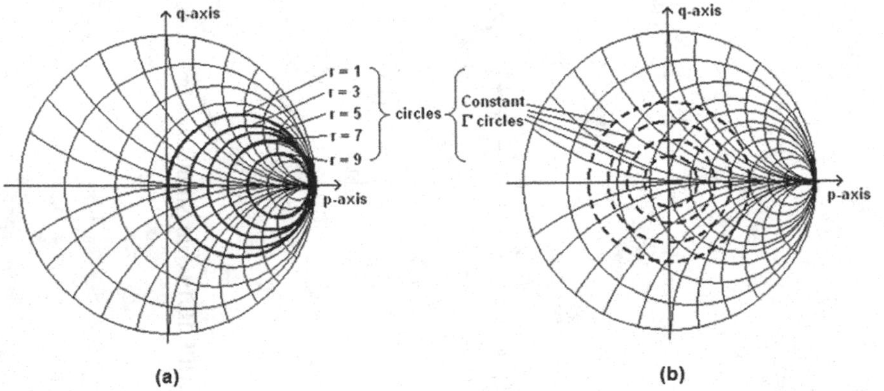

FIGURE 18.20 (a) VSWR circles; and (b) Γ circles on the Smith chart.

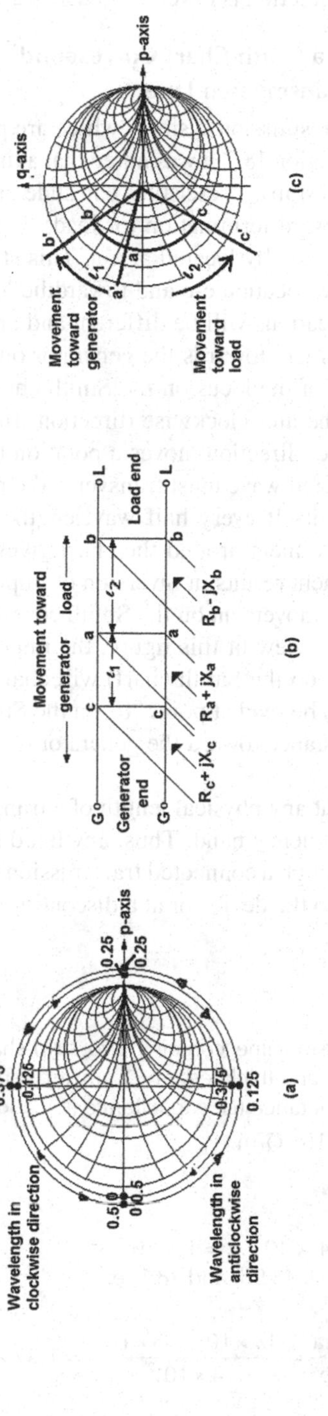

FIGURE 18.21 (a) Wavelength scales; (b) transmission line with stubs; (c) equivalent of movement on transmission line.

increases in clockwise direction whereas in the second in an anticlockwise direction. Both of these scales coincide at the real axis with a value of quarter wavelength. The entire periphery in either direction is covered by half wavelength.

18.9.2.7 Movement on a Smith Chart Corresponding to that on a Transmission Line

As discussed earlier, all transmission systems which are prone to reflections require impedance matching. In Section 18.7, it was noted that a line can be matched by using some auxiliary lines, called stubs. These stubs can slide on the main line towards the source (or generator) or toward termination (or load). Figure 18.21b shows a transmission line with a stub originally located at a-a. This stub is moved by distance l_1 toward the generator to new location c-c and towardthe load by l_2 to b-b. Obviously the impedances at these locations will be different and are to be found out by using the Smith chart. The movement towards the generator on the transmission line corresponds to the movement of the locus on the Smith chart in a clockwise direction and that towards load to the anti-clockwise direction. In addition, θ degree movement along the line in either direction moves a point on the locus by 2θ degrees on the chart because the reflected wave must transverse the round-trip distance moved. As the impedance repeats itself every half wavelength along a uniform transmission line, the one-time movement around the chart gives the same impedance and quarter-wavelength movement results in inversion of impedance.

Figure 18.21c shows the movement on the Smith chart equivalent to that of a stub on the transmission line. In view of this figure, the impedance $r_a + jx_a$ seen across a-a (to be read at point "a" on the Smith chart) will change to $r_b - jx_b$ at b-b, i.e. at l_2 distance towards load (to be read at point "b" on the Smith chart) and will change to $r_c + jx_c$ at c-c, i.e. at l_1 distance toward the generator (to be read at point "c" on the Smith chart).

It is to be mentioned that any physical length of a transmission becomes variable electrical length over a frequency band. Thus, any fixed impedance will spread out to a locus when viewed through a connected transmission line. This aspect facilitates wideband matching close to the device or at a discontinuity.

Example 18.1

A parallel wire transmission line contains two strip-shaped conductors. These strips of width b (= 4 cm) and thickness t (= 1 cm) are separated by a distance a (= 5 cm). Calculate the inductance (L), capacitance (C), and conductance (G), if $\mu = \mu_0$, $\varepsilon = \varepsilon_0$ and $\sigma = 3.72\times10^7$ ℧/m.

SOLUTION

Given $a = 5 \times 10^{-2}$, $b = 4 \times 10^{-2}$, $t = 1 \times 10^{-2}$.
In view of Equations 18.7d, 18.9d, and 18.10e:

$$L = \frac{\mu a}{b} = \frac{4\pi\times10^{-7}\times5\times10^{-2}}{4\times10^{-2}} = 1.57\times10^{-6}$$

$$C = \frac{\varepsilon b}{a} = \frac{8.854\times10^{-12}\times4\times10^{-2}}{5\times10^{-2}} = 7.0832\times10^{-12}\text{ F}$$

$$G = \frac{\sigma b}{a} = \frac{3.72\times10^{7}\times4\times10^{-2}}{5\times10^{-2}} = 2.976\times10^{7}\ \text{℧/m}$$

Example 18.2

Find the velocity of propagation (υ) in a transmission line which carries a sinusoidal signal of 100 MHz. This line is characterized by an inductance L (= 0.5 μH/m) a capacitance C (= 50 pF/m) and (a) $R = 0, G = 0$; (b) $R = 0.3\ \Omega/\text{m}, G = 10^{-5}$ ℧/m; and (c) $R = 200\ \Omega/\text{m}, G = 0$.

SOLUTION

Given $f = 100 \times 10^6 = 10^8$, $\omega = 2\pi f = 6.283 \times 10^8$

$L = 0.5\times10^{-6} \qquad C = 50\times10^{-12}$

$\omega L = 6.283\times10^{8}\times0.5\times10^{-6} = 3.1415\times10^{2}$

$\omega C = 6.283\times10^{8}\times50\times10^{-12} = 3.1415\times10^{-2}$

(a) $R = 0,\ G = 0,\quad \gamma = \alpha + j\beta$

$$(R + j\omega L) = j\omega L = j3.1415\times10^{2}$$

$$(G + j\omega C) = j\omega C = j3.1415\times10^{-2}$$

From Equation 18.12c $\gamma = \sqrt{(R + j\omega L)(G + j\omega C)} = \alpha + j\beta$

Thus $\gamma = \sqrt{(j3.1415\times10^{2})(j3.1415\times10^{-2})} = j3.1415$

$\alpha = 0,\quad \beta = 3.1416$

Also $\upsilon = \frac{\omega}{\beta} = \frac{6.283\times10^{8}}{3.1415} = 2\times10^{8}$ m/s

(b) $R = 0.1\ \Omega/\text{m},\ G = 10^{-5}$ ℧/m

$$R + j\omega L = 0.3 + j3.1415\times10^{2} \qquad G + j\omega C = 10^{-5} + j3.1415\times10^{-2}$$

$$\gamma = \sqrt{(0.3 + j3.1415\times10^{2})(10^{-5} + j3.1415\times10^{-2})}$$

$$= \sqrt{3\times10^{-6} - 9.87 + j3.1415(3\times10^{-3} + 1\times10^{-3})}$$

$$\approx \sqrt{-9.87 + j12.566\times10^{-3}} = j3.1416(1 - j1.273\times10^{-3})^{1/2}$$

$$= j3.1416(1 - j0.636\times10^{-3}) = 2\times10^{-3} + j3.1416$$

$$\alpha = 2\times10^{-3} \quad \beta = 3.1416 \quad \upsilon = \frac{\omega}{\beta} = \frac{6.283\times10^{8}}{3.1416} = 2\times10^{8}\text{ m/s}$$

(c) $R = 200$, $G = 0$

$$R + j\omega L = 200 + j3.1415\times10^{2} \qquad G + j\omega C = j3.1415\times10^{-2}$$

$$\gamma = \alpha + j\beta = \sqrt{\left(200 + j3.1415\times10^{2}\right)\left(j3.1415\times10^{-2}\right)} = \sqrt{-9.87 + j6.283}$$

$$= j3.1416\left(1 + j0.6365\right)^{1/2} = j3.1416\left(1 + j0.3183\right) = -1 + j3.1416$$

$$\alpha = -1 \quad \beta = 3.1416 \quad \upsilon = \frac{\omega}{\beta} = \frac{6.283\times10^{8}}{3.1416} = 2\times10^{8}\text{ m/s}$$

Example 18.3

A 10 cm long lossless transmission line has a characteristic impedance of 50 Ω. Find its input impedance at 50 MHz if it is terminated in (a) open circuit; (b) short circuit; and (c) 10pF capacitor.

SOLUTION

Given = 50×10^6 Hz, $Z_0 = 50\ \Omega$, $l = 0.1$ m, $\omega = 100\pi \times 10^6$.

$$\upsilon = 3\times10^{8}\ \frac{\text{m}}{\text{s}} \qquad \beta = \frac{\omega}{\upsilon} = \frac{100\pi\times10^{6}}{3\times10^{8}} = \frac{\pi}{3} \qquad \beta L = \frac{\pi}{3}\times0.1 = 0.1047$$

$$\sin\beta l = \sin 0.1047 = 0.1045 \qquad \cos\beta l = \cos 0.1047 = 0.9945$$

$$\tan\beta l = \tan 0.1047 = 0.1051 \qquad \cot\beta l = \cot 0.1047 = 9.5161$$

(a) Line is open-circuited or $Z_R = \infty$.

$$Z_{in} = -jZ_0 \cot\beta L = -j50\times9.5161 = -j475.81\,\Omega$$

(b) Line is short-circuited or $Z_R = 0$.

$$Z_{in} = jZ_0 \tan\beta L = j50\times0.1051 = j5.254\,\Omega$$

(c) Line is terminated in 10 pF capacitor.

$$X_C = \omega C = 100\pi\times10^{6}\times10\times10^{-12} = \pi\times10^{-3}$$

$$Z_R = \frac{-j}{\omega C} = \frac{-j}{\pi\times10^{-3}} = -j318.31\,\Omega$$

$$Z_{in} = Z_0 \frac{Z_R \cos\beta l + jZ_0 \sin\beta l}{Z_0 \cos\beta l + jZ_R \sin\beta l} = 50\frac{-j318.31\times 0.9945 + j50\times 0.1045}{50\times 0.9945 + 318.31\times 0.1045}$$

$$= -j187.561\,\Omega$$

Example 18.4

Find the propagation velocity (υ), characteristic impedance (Z_0), propagation constant (γ), attenuation constant (α), and phase shift constant (β) for a line having inductance ($L = 0.5$ μH/m), capacitance ($C = 30$ pF/m), resistance ($R = 0.25$ Ω/m), and conductance ($G = 210^{-5}$ ℧/m) at (a) 20 MHz; (b) 500 MHz.

SOLUTION

Given $R = 0.25$ Ω/m, $G = 2\times 10^{-5}$ ℧/m, $L = 0.5$ μH/m, $C = 30$ pF/m.

(a) $f = 20$ MHz, $\omega = 2\pi \times 20\times 10^6 = 4\pi \times 10^7$

$$Z = R + j\omega L = 0.25 + j\left(4\pi\times 10^7\right)\times\left(0.5\times 10^{-6}\right) = 0.25 + j20\pi$$

$$Y = G + j\omega C = 2\times 10^{-5} + j\left(4\pi\times 10^7\right)\times\left(30\times 10^{-12}\right) = 2\times 10^{-5} + j12\pi\times 10^{-4}$$

$$\gamma = \alpha + j\beta = \sqrt{(R + j\omega L)(G + j\omega C)} = \sqrt{(0.25 + j20\pi)\left(2\times 10^{-5} + j12\pi\times 10^{-4}\right)}$$

$$\approx \sqrt{-0.23687 + j0.0022} = j0.4867(1 - j0.0039)^{1/2}$$

$$= j0.4867(1 - 0.002) = 0.00097 + j0.4867$$

$$\alpha = 0.00097 \text{ nepers/m} \quad \beta = 0.4867 \text{ rad/m} \quad \upsilon = \frac{\omega}{\beta} = \frac{4\pi\times 10^7}{0.4867} = 2.58125\times 10^8 \text{ m/s}$$

$$Z_0 = \sqrt{\frac{(R + j\omega L)}{(G + j\omega C)}} = \sqrt{\frac{0.25 + j20\pi}{2\times 10^{-5} + j12\pi\times 10^{-4}}}$$

$$= \sqrt{\frac{(0.25 + j20\pi)\left(2\times 10^{-5} - j12\pi\times 10^{-4}\right)}{\left(2\times 10^{-5} + j12\pi\times 10^{-4}\right)\left(2\times 10^{-5} - j12\pi\times 10^{-4}\right)}}$$

$$\approx \sqrt{\frac{0.23687 + j\pi\times 10^{-4}}{1421.227\times 10^{-8}}} = \sqrt{\left(1.6667\times 10^4 + j22.1\right)}$$

$$= 1.291\times 10^2\left(1 + j13.25\times 10^{-4}\right)^{1/2}$$

$$= 1.291\times 10^2\left(1 + j6.625\times 10^{-4}\right) = 129.1 + j0.085523\,\Omega$$

(b) $f = 500$ MHz $\omega = 2\pi \times 500 \times 10^6 = \pi \times 10^9$

$$Z = R + j\omega L = 0.25 + j\left(\pi \times 10^9\right) \times \left(0.5 \times 10^{-6}\right) = 0.25 + j1570.8$$

$$Y = G + j\omega C = 2\times 10^{-5} + j\left(\pi \times 10^9\right) \times \left(30 \times 10^{-12}\right) = 210^{-5} + j9.425 \times 10^{-2}$$

$$\gamma = \alpha + j\beta = \sqrt{(R + j\omega L)(G + j\omega C)} = \sqrt{(0.25 + j500\pi)\left(2\times 10^{-5} + j3\pi \times 10^{-2}\right)}$$

$$\approx \sqrt{-148.044 + j0.055} = j12.167\sqrt{1 - j3.7136 \times 10^{-4}}$$
$$= j12.167\left(1 - j1.8568 \times 10^{-4}\right)$$

$$\gamma = j12.167\left(1 - j1.8568 \times 10^{-4}\right) = 22.5916 \times 10^{-4} + j12.167$$

$$\alpha = 22.5916 \times 10^{-4} \quad \beta = j12.167 \quad \upsilon = \frac{\omega}{\beta} = \frac{\pi \times 10^9}{12.167} = 2.582 \times 10^8 \text{ m/s}$$

$$Z_0 = \sqrt{\frac{(R + j\omega L)}{(G + j\omega C)}} = \sqrt{\frac{0.25 + j1570.8}{2\times 10^{-5} + j9.425 \times 10^{-2}}}$$

$$= \sqrt{\frac{(0.25 + j1570.8)\left(2\times 10^{-5} - j9.425 \times 10^{-2}\right)}{\left(2\times 10^{-5} + j9.425 \times 10^{-2}\right)\left(2\times 10^{-5} - j9.425 \times 10^{-2}\right)}}$$

$$= 129.1 + j0.0342\ \Omega$$

Example 18.5

A 100 MHz transmission line has inductance L = 0.4 μH/m, capacitance C = 40 pF/m, resistance R = 0.1 Ω/m, and conductance $G = 10^{-5}$ ℧/m. Find propagation velocity (υ), phase shift constant (β), characteristic impedance (Z_0), and attenuation constant (α).

SOLUTION

Given $f = 10^8$ Hz, $L = 0.4 \times 10^{-6}$ H/m, $C = 40 \times 10^{-12}$ F/m, $R = 0.1$ Ω/m, $G = 10^{-5}$ ℧/m

$$\omega = 2\pi f = 2\pi \times 10^8 \quad \upsilon = \frac{1}{\sqrt{LC}} = \frac{1}{\sqrt{0.4 \times 10^{-6} \times 40 \times 10^{-12}}} = 2.5 \times 10^8 \text{ m/s}$$

$$\beta = \frac{\omega}{\upsilon} = \frac{2\pi \times 10^8}{2.5 \times 10^8} = 2.51327 \text{ rad/m}$$

$$Z = R + j\omega L = 0.1 + j2\pi \times 10^8 \times 0.4 \times 10^{-6} = 0.1 + j0.8\pi \times 10^2\ \Omega$$

$$Y = G + j\omega C = 10^{-5} + j2\pi \times 10^8 \times 40 \times 10^{-12} = 10^{-5} + j0.8\pi \times 10^{-2}\ \Omega$$

$$Z_0 = \sqrt{\frac{Z}{Y}} = \sqrt{\frac{0.1 + j0.8\pi \times 10^2}{10^{-5} + j0.8\pi \times 10^{-2}}}$$

$$= \sqrt{\frac{\left(0.1 + j0.8\pi \times 10^2\right)\left(10^{-5} - j0.8\pi \times 10^{-2}\right)}{\left(10^{-5} + j0.8\pi \times 10^{-2}\right)\left(10^{-5} - j0.8\pi \times 10^{-2}\right)}} = 100\ \Omega$$

$$\alpha = \frac{R}{2}\sqrt{\frac{C}{L}} + \frac{G}{2}\sqrt{\frac{L}{C}} = \frac{0.1}{2}\sqrt{\frac{40 \times 10^{-12}}{0.4 \times 10^{-6}}} + \frac{10^{-5}}{2}\sqrt{\frac{0.4 \times 10^{-6}}{40 \times 10^{-12}}}$$

$$= 0.5 \times 10^{-3} + 0.5 \times 10^{-3} = 10^{-3}\ \text{nep/m}$$

Example 18.6

A 50 MHz transmission line with characteristic impedance of 50 Ω is terminated in 200 Ω. Find its reflection coefficient (Γ) and VSWR (ρ). Also find its quality factor (Q) if this line is a resonant section and its inductance is $L = 0.4$ μH/m, capacitance $C = 40$ pF/m, resistance $R = 0.1$ Ω, and conductance $G = 10^{-5}$ ℧/m.

SOLUTION

Given $Z_R = 200$ Ω, $Z_0 = 50$ Ω, $L = 0.4 \times 10^{-6}$ H/m, $C = 40 \times 10^{-12}$ F/m, $R = 0.1$ Ω, $G = 10^{-5}$ ℧/m, $f = 50 \times 10^6$ Hz, $\omega = 2\pi f = \pi \times 10^8$.

$$\text{Reflection coefficient } \Gamma = \frac{Z_R - Z_0}{Z_R + Z_0} = \frac{200 - 50}{200 + 50} = 0.6$$

$$\text{VSWR } \rho = \frac{1 + \Gamma}{1 - \Gamma} = \frac{1 + 0.6}{1 - 0.6} = 4$$

$$\text{Quality factor } Q = \frac{\omega L}{R} = \frac{\pi \times 10^8 \times 0.4 \times 10^{-6}}{0.1} = 1256.6$$

Example 18.7

A 300 MHz transmission line of characteristic impedance 50 Ω is terminated in an unknown impedance $R + jX$. The voltage standing wave ratio on the line is 3. Find R and X if the distance (L_2) between the first minima and the terminating end is 0.7 m.

SOLUTION

Given $f = 3 \times 10^8$ Hz, $\omega = 2\pi f = 6\pi \times 10^8$, $\upsilon = 3 \times 10^8$, $\beta = \dfrac{\omega}{\upsilon} = \dfrac{6\pi \times 10^8}{3 \times 10^8} = 2\pi$

$Z_0 = R_0 = 50$ Ω, $\rho = 3$, $\rho^2 = 9$, $\rho^2 - 1 = 8$, $L_2 = 0.7$ m, $\beta L_2 = 1.4\pi$.

$$\cos(\beta L_2) = \cos(1.4\pi) = -0.309 \qquad \cos^2(\beta L_2) = 0.0955$$

$$\sin(\beta L_2) = \sin(1.4\pi) = -0.951 \qquad \sin^2(\beta L_2) = 0.9044$$

In view of Equations 18.37d and 18.37e:

$$R = \frac{\rho R_0}{\rho^2\cos^2(\beta L_2) + \sin^2(\beta L_2)} = \frac{3\times 50}{9\times 0.0955 + 0.9044} = \frac{150}{0.95} = 157.9\ \Omega$$

$$X = \frac{-R_0(\rho^2-1)\sin(\beta L_2)\cos(\beta L_2)}{\rho^2\cos^2(\beta L_2) + \sin^2(\beta L_2)} = \frac{-50\times(8)\times 0.951\times 0.309}{0.95} = -123.72\ \Omega$$

Example 18.8

Calculate the inductance (L), capacitance (C), velocity of propagation (υ), wavelength (λ), and the characteristic impedance (Z_0) for a lossless coaxialline operating at 3 GHz and having: (a) $a = 1$ mm, $b = 3$ mm, $\mu_r = \varepsilon_r = 1$; (b) $a = 1$ mm, $b = 3$ mm, $\mu_r = 1$, $\varepsilon_r = 3$; (c) $a = 1$ mm, $b = 5$ mm, $\mu_r = 1$, $\varepsilon_r = 3$.

SOLUTION

Given $f = 3 \times 10^9$, $\mu_0 = 4\pi \times 10^{-7}$, $\varepsilon_0 = 8.854 \times 10^{-12}$

$\upsilon_0 = 3 \times 10^8$, $\lambda_0 = \frac{\upsilon_0}{f} = \frac{3\times 10^8}{3\times 10^9} = 0.1\,\text{m}$

(a) $a = 10^{-3}$, $b = 3 \times 10^{-3}$, $d = 1$ m, $\mu_r = 1$, $\varepsilon_r = 1$, $\mu = \mu_0 = 4\pi \times 10^{-7}$, $\varepsilon = \varepsilon_0 = 8.854 \times 10^{-12}$.

From Equation 13.24b:

$$L = \frac{\mu d}{2\pi}\ln\left(\frac{b}{a}\right) = \frac{\mu_0}{2\pi}\ln\left(\frac{b}{a}\right) = \frac{4\pi\times 10^{-7}}{2\pi}\ln\left(\frac{3}{1}\right) = 0.2197\ \mu\text{H/m}$$

From Equation 8.70:

$$C = \frac{2\pi\cdot\varepsilon_0\cdot\varepsilon_r}{\log\left(\frac{b}{a}\right)} = \frac{2\pi\times 8.854\times 10^{-12}}{\log\left(\frac{3}{1}\right)} = 50.666\ \text{pF/m}$$

In Equation 18.11a:

$$\upsilon = \frac{1}{\sqrt{LC}} = \frac{1}{\sqrt{0.2197\times 10^{-6}\times 50.666\times 10^{-12}}} = 2.997\times 10^8\ \text{m/s}$$

$$\lambda = \frac{\upsilon}{f} = \frac{2.997\times 10^8}{3\times 10^9} = 10\ \text{cm}$$

In Equation 18.11b:

$$Z_0 = \sqrt{\frac{L}{C}} = \sqrt{\frac{0.2197\times 10^{-6}}{50.666\times 10^{-12}}} = 65.85\ \Omega$$

(b) $a = 10^{-3}$, $b = 3 \times 10^{-3}$, $\mu_r = 1$, $\varepsilon_r = 3$, $\mu = \mu_0 = 4\pi \times 10^{-7}$, $\varepsilon = 3\varepsilon_0 = 3 \times 8.854 \times 10^{-12}$

$$L = \frac{\mu d}{2\pi} \ln\left(\frac{b}{a}\right) = \frac{4\pi \times 10^{-7} \times 1}{2\pi} \ln\left(\frac{3}{1}\right) = 0.2197\ \mu\text{H/m}$$

$$C = \frac{2\pi \cdot \varepsilon_0 \cdot \varepsilon_r}{\log\left(\frac{b}{a}\right)} = \frac{2\pi \times 3 \times 8.854 \times 10^{-12}}{\log\left(\frac{3}{1}\right)} = 151.998\ \text{pF/m}$$

$$\upsilon = \frac{1}{\sqrt{LC}} = \frac{1}{\sqrt{0.2197 \times 10^{-6} \times 151.998 \times 10^{-12}}} = 1.73 \times 10^8\ \text{m/s}$$

$$\lambda = \frac{\upsilon}{f} = \frac{1.73 \times 10^8}{3 \times 10^9} = 5.77\ \text{cm}$$

$$Z_0 = \sqrt{\frac{L}{C}} = \sqrt{\frac{0.2197 \times 10^{-6}}{151.998 \times 10^{-12}}} = 38.02\ \Omega$$

(c) $a = 10^{-3}$, $b = 5 \times 10^{-3}$, $\mu_r = 1$, $\varepsilon_r = 3$,

$$\mu = \mu_0 = 4\pi \times 10^{-7}, \quad \varepsilon = 3\varepsilon_0 = 3 \times 8.854 \times 10^{-12}$$

$$L = \frac{\mu d}{2\pi} \ln\left(\frac{b}{a}\right) = \frac{4\pi \times 10^{-7} \times 1}{2\pi} \ln\left(\frac{5}{1}\right) = 0.3218\ \mu\text{H/m}$$

$$C = \frac{2\pi \cdot \varepsilon_0 \cdot \varepsilon_r}{\log\left(\frac{b}{a}\right)} = \frac{2\pi \times 3 \times 8.854 \times 10^{-12}}{\log\left(\frac{5}{1}\right)} = 103.7\ \text{pF/m}$$

$$\upsilon = \frac{1}{\sqrt{LC}} = \frac{1}{\sqrt{0.3218 \times 10^{-6} \times 103.7 \times 10^{-12}}} = 1.731 \times 10^8\ \text{m/s}$$

$$\lambda = \frac{\upsilon}{f} = \frac{1.731 \times 10^8}{3 \times 10^9} = 5.77\ \text{cm}$$

$$Z_0 = \sqrt{\frac{L}{C}} = \sqrt{\frac{0.3218 \times 10^{-6}}{103.7 \times 10^{-12}}} = 55.7\ \Omega$$

PROBLEMS

P18.1 A parallel wire transmission line has two strip-shaped conductors of width b = 3 cm, thickness t = 1 cm, and separation a = 2 cm. Find the inductance (L), capacitance (C), conductance (G), velocity of propagation (υ) and the characteristic impedance (Z_0) if $\mu = \mu_0$, $\varepsilon = \varepsilon_0$ and $\sigma = 5.8 \times 10^7$ ℧/m.

P18.2 Find the velocity of propagation (v) in a transmission line carrying a sinusoidal signal of 100 MHz and characterized by an inductance L = 0.3 μH/m, a capacitance C = 20 pF/m, R = 10 Ω, and G = 0.

P18.3 Find the velocity of propagation (v) in a transmission line carrying a sinusoidal signal of 100 MHz and characterized by an inductance L = 0.3 μH/m, a capacitance C = 20 pF/m, a resistance R= 100 Ω/m, and $G = 10^{-3}$ ℧/m.

P18.4 A 30 cm long lossless transmission line has a characteristic impedance of 50 Ω. Find its input impedance at 15 MHz if it is terminated in (a) an open circuit; and (b) a short circuit.

P18.5 Find the input impedance at 15 MHz of a 30 cm long lossless transmission line with 50 Ω characteristic impedance terminated in a 5 μH inductor.

P18.6 Find the propagation velocity (v), characteristic impedance (Z_0), propagation constant (γ), attenuation constant (α) and phase shift constant (β) for a line having inductance (L = 0.4 μH/m), capacitance (C = 20 pF/m), resistance (R = 0.2 Ω/m), and conductance ($G = 10^{-5}$ ℧/m) at 50 MHz.

P18.7 A 50 MHz transmission line has inductance L = 0.2 μH/m, capacitance C = 20 pF/m, resistance R = 0.25 Ω/m, and conductance $G = 10^{-4}$ ℧/m. Find propagation velocity (v), phase shift constant (β), characteristic impedance (Z_0), and attenuation constant (α).

P18.8 A transmission line with characteristic impedance of 100 Ω is operating at 50 MHz. Find its voltage and current reflection coefficients if it is terminated in 10 μH inductor.

P18.9 A 100 MHz transmission line with characteristic impedance of 30 Ω is terminated in 100 Ω. Find its reflection coefficient (Γ) and VSWR (ρ). Also find its quality factor (Q) if this line is a resonant section and its inductance is L = 0.2 μH/m, capacitance C = 50 pF/m), resistance is R = 0.2 Ω and conductance G = 0 ℧/m.

P18.10 A 100 MHz transmission line of characteristic impedance 30 Ω is terminated in an unknown impedance $R + jX$. The voltage standing wave ratio on the line is 5. Find R and X if the distance (L_2) between the first minima and the terminating end is 0.5 m.

P18.11 Calculate inductance (L), capacitance (C), velocity of propagation (v), wavelength (λ) and the characteristic impedance (Z_0) of a lossless coaxial line having a = 1 mm, b = 5 mm and operating at 1 GHz The space between the inner and outer conductors is filled with a material of $\mu_r = 1$, $\varepsilon_r = 3$.

P18.12 Find the inductance (L), capacitance (C), velocity of propagation (v), wavelength (λ), and the characteristic impedance (Z_0) of a coaxial line having a = 3 mm, b = 5 mm and operating at 3 GHz. The space between the inner and outer conductors is filled with a material of μ_r = 2, $\varepsilon_r = 4$.

DESCRIPTIVE QUESTIONS

Q18.1 Describe the telegrapher's equations and the information spelled by these.

Q18.2 Write EMF and MMF equations for lossy and lossless transmission lines

Q18.3 Derive the expressions that govern the behaviour of lossless and lossy lines.

Q18.4 Find the solutions to the transmission line equations in exponential form. Obtain characteristic impedance, terminating impedance, and reflection coefficient from these.

Q18.5 Obtain the solutions of transmission line equations in hyperbolic form. Write the general expression for the input impedance.

Q18.6 What is the propagation constant? How it is related to the line parameters?

Q18.7 Obtain the expressions for voltage and current distributions in transmission lines terminated with different impedances. Use these expressions to illustrate the voltage and current variations along the line.

Q18.8 Write input impedance expressions for short-circuited and open-circuited lines.

Q18.9 Discuss the applications of transmission line sections as circuit elements.

Q18.10 Illustrate the equivalent circuit elements of different transmission line sections.

Q18.11 Illustrate and explain the working of single and double stub matching.

Q18.12 What are planar lines? Discuss their relative merits and demerits.

FURTHER READING

A. S. Khan, *Microwave Engineering: Concepts and Fundamentals*, CRC Press, New York, Mar. 2014.

David M. Pozar, *Microwave Engineering*, 3rd Ed, John Wiley & Sons, Inc., New Jersey, 2005.

T. H. Lee, *Planar Microwave Engineering*, Cambridge University Press, UK, 2004.

K. Chang, I. Bahl, V. Nair, *RF and Microwave Circuits and Component Design for Wireless Systems*, John Wiley & Sons, New York, 2002.

W. Hayt, Jr., J. A. Buck, *Engineering Electromagnetics*, 4th Ed, McGraw Hill Publishing Co. Ltd., New Delhi, 2001.

I. Bahl, P. Bhartia, K. C. Gupta, *Microstrip Lines and Slotlines*, Artech House, Norwood, MA, 1996.

T. C. Edward, *Foundation for Micro strip Circuit Design*, Wiley, Chichester, UK, 1981 (2nd Ed, 1992).

Brian C. Wadell, *Transmission Line Design Handbook*, Artech House, Norwood, MA, 1991.

F. Gardiol, *Micro Strip Circuits*, Wiley, New York, 1990.

S. Y. Liao, *Engineering Applications of Electromagnetic Theory*, West Publishing Company, St. Paul, MN, 1988.

R. K. Hoffmann, *Handbook of Microwave Integrated Circuits*, Artech House, Norwood, MA, 1987.

E. C. Jordon, K. G. Balmain, *Electromagnetic Waves and Radiating Systems*, Prentice Hall Ltd., New Delhi, 1987.

L. N. Dworsky, *Modern Transmission Line Theory and Applications*, John Wiley & Sons, New York, 1979.

W. Sinnema, *Electronic Transmission Technology: Lines and Applications*, Prentice-Hall, Inc., Englewood Cliffs, NJ, 1979.

K. C. Gupta, R. Garg, I. J. Bahl, *Micro Strip Lines and Slot Lines*, Artech House, Dedham, MA, 1979.

H. Howe, Jr., *Strip Line Circuit Design*, Artech House, Dedham, MA, 1974.

R. A. Waldron, *Theory of Guided Electromagnetic Waves*, London, New York, Van Nostrand Reinhold, 1970.

Richard B. Alder, Lan Jen Chu, Robert M. Fano, *Electromagnetic Energy Transmission and Radiation*, MIT Press, Cambridge, MA, 1969.

R. K. Moore, *Travelling-Wave Engineering*, McGraw-Hill Book Company, New York, 1960.

W. C. Johnson, *Transmission Lines and Networks*, McGraw-Hill Book Company, New York, 1950.

W. L. Everitt, *Communication Engineering*, McGraw-Hill, New York, 1937.

19 Waveguides and Cavity Resonators

19.1 INTRODUCTION

In Chapter 18, it was noted that the parallel wire/plane transmission structures can only be used to propagate low frequency signals with *TEM* mode having zero cut-off frequency. At higher frequencies, the suitability of *TEM* mode diminishes due to increased attenuation. Thus, at higher frequencies the use of *TE* and *TM* modes becomes a necessity. These modes can be supported by another form of transmission structures called waveguides. These structures have a variety of forms in terms of their shapes and materials. These can be in the form of hollow rectangular or circular metallic pipes or solid dielectric rods or slabs. Except the differing modes, many of the features of these structures resemble those of wire/plane transmission lines. These include the concept of reflection, current flow in the conductor skin, properties of quarter and half wave sections, impact of discontinuities on propagation etc.

Since waveguides are the key players for guiding electromagnetic energy at centimetric and millimetric ranges, this chapter is fully devoted to the study of the field behavior in these. This chapter encompasses the mathematical theory of rectangular and circular waveguides, describes different modes and their excitation, and includes the physical interpretation of various terms involved.

19.2 RECTANGULAR WAVEGUIDE

This is the simplest and most commonly used waveguide. It is a hollow rectangular pipe containing four metallic walls. Its physical structure along with its dimensional parameters and coordinate system is shown in Figure 19.1. In view of the presence of four metallic walls $\boldsymbol{E}$ and $\boldsymbol{H}$ fields have to satisfy appropriate boundary conditions. These include the continuity of the tangential component of $\boldsymbol{E}$ and the normal component of $\boldsymbol{H}$ for appropriate values of x and y.

19.2.1 PRELIMINARY MATHEMATICS

In order to understand the wave phenomenon in a rectangular waveguide it is necessary to describe certain basic mathematical aspects. These are given in the following subsections.

The wave is assumed to be propagating along its z-axis. As in the case of transmission lines the time variation of all the field parameters is accounted by $e^{j\omega t}$ and the space variation by $e^{-\gamma z}$. The parameter γ is again a complex quantity having a real part α and an imaginary part $j\beta$. These parameters (γ, α, and β) represent propagation, attenuation, and phase constants, respectively. Furthermore, the behavior of

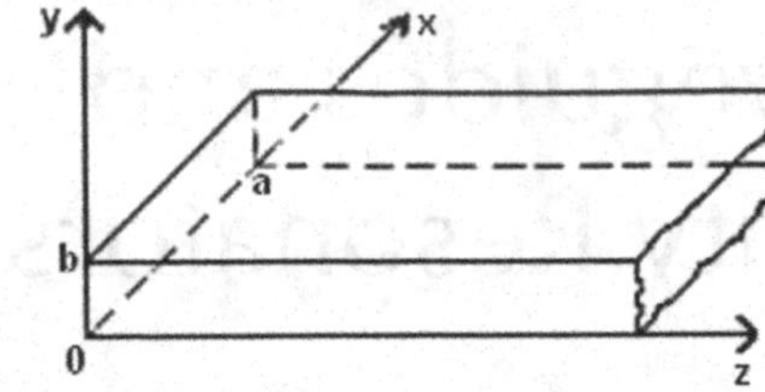

FIGURE 19.1 Rectangular waveguide.

wave propagation in this structure is solely governed by Maxwell's equations. These equations, for current-free region, given by Equations 15.25b and 15.6b, are reproduced here:

$$\nabla \times \boldsymbol{H} = \frac{\partial \boldsymbol{D}}{\partial t} = j\omega \boldsymbol{D} = j\omega\varepsilon \boldsymbol{E} \quad (a) \quad \nabla \times \boldsymbol{E} = -\frac{\partial \boldsymbol{B}}{\partial t} = -j\omega \boldsymbol{B} = -j\omega\mu \boldsymbol{H} \quad (b) \Big\} \qquad (19.1)$$

Also, the wave equations (Equations 16.34a and 16.34b) followed by $\boldsymbol{E}$ and $\boldsymbol{H}$ fields, are:

$$\nabla^2 \boldsymbol{E} = \gamma^2 \boldsymbol{E} \quad (a) \qquad \nabla^2 \boldsymbol{H} = \gamma^2 \boldsymbol{H} \quad (b) \} \qquad (19.2)$$

where γ and β are given as:

$$\left.\begin{aligned} &\gamma = \sqrt{\left[(\sigma + j\omega\varepsilon) j\omega\mu\right]} \quad (\text{general}) \quad (a) \\ &\gamma = j\beta = j\omega\sqrt{\mu\varepsilon} \quad (\text{if } \sigma = 0) \quad (b) \\ &\beta = \omega\sqrt{\mu\varepsilon} \quad (\text{for any medium}) \quad (c) \\ &\beta = \beta_0 = \omega\sqrt{\mu_0\varepsilon_0} \quad (\text{free space}) \quad (d) \end{aligned}\right\} \qquad (19.3)$$

In view of the of time and space variations the governing Maxwell's equations given by Equations 19.1a and 19.1b can be written in the following expanded form:

$$\frac{\partial H_x}{\partial z} - \frac{\partial H_z}{\partial x} = \frac{\partial H_z}{\partial x} + \gamma H_x = -j\omega\varepsilon E_y \quad (a) \quad \frac{\partial H_y}{\partial x} - \frac{\partial H_x}{\partial y} = j\omega\varepsilon E_z \quad (b) \Big\} \qquad (19.4)$$

$$\left.\begin{aligned} &\frac{\partial E_z}{\partial y} - \frac{\partial E_y}{\partial z} = \frac{\partial E_z}{\partial y} + \gamma E_y = -j\omega\mu H_x \quad (a) \\ &\frac{\partial E_x}{\partial z} - \frac{\partial E_z}{\partial x} = \frac{\partial E_z}{\partial x} + \gamma E_x = j\omega\mu H_y \quad (b) \quad \frac{\partial E_y}{\partial x} - \frac{\partial E_x}{\partial y} = -j\omega\mu H_z \quad (c) \end{aligned}\right\} \qquad (19.5)$$

Similarly, the differential forms of Equations 19.2a and 19.2b for E_z and H_z (for $\sigma = 0$) are:

$$\left.\begin{aligned}\frac{\partial^2 E_Z}{\partial x^2}+\frac{\partial^2 E_Z}{\partial y^2}+\frac{\partial^2 E_Z}{\partial z^2}=\frac{\partial^2 E_Z}{\partial x^2}+\frac{\partial^2 E_Z}{\partial y^2}+\gamma^2 E_z=-\omega^2\mu\varepsilon E_Z \quad &\text{(a)}\\ \frac{\partial^2 H_Z}{\partial x^2}+\frac{\partial^2 H_Z}{\partial y^2}+\frac{\partial^2 H_Z}{\partial z^2}=\frac{\partial^2 H_Z}{\partial x^2}+\frac{\partial^2 H_Z}{\partial y^2}+\gamma^2 H_Z=-\omega^2\mu\varepsilon E_Z \quad &\text{(b)}\end{aligned}\right\} \tag{19.6}$$

Since the wave is assumed to be progressing in the z-direction, Equations 19.4 and 19.5 are to be manipulated in such a way so as to obtain all the field components in terms of E_z and H_z. Such manipulations yield the following relations:

$$\left.\begin{aligned}E_x=-\frac{\gamma}{h^2}\frac{\partial E_z}{\partial x}-\frac{j\omega\mu}{h^2}\frac{\partial H_z}{\partial y} \quad \text{(a)} \quad & E_y=-\frac{\gamma}{h^2}\frac{\partial E_z}{\partial y}+\frac{j\omega\mu}{h^2}\frac{\partial H_z}{\partial x} \quad \text{(b)}\\ H_x=-\frac{\gamma}{h^2}\frac{\partial H_z}{\partial x}+\frac{j\omega\varepsilon}{h^2}\frac{\partial E_z}{\partial y} \quad \text{(c)} \quad & H_y=-\frac{\gamma}{h^2}\frac{\partial H_z}{\partial y}-\frac{j\omega\varepsilon}{h^2}\frac{\partial E_z}{\partial x} \quad \text{(d)}\end{aligned}\right\} \tag{19.7}$$

A cursory look at Equations 19.7a to 19.7d reveals that when both E_z and H_z are zero all the field components vanish. If one of these components is zero and the other survives, the resulting waves are assigned two different nomenclatures. Thus when $H_z \equiv 0$ the wave is called *Transverse Magnetic* (*TM*) *wave* and when $E_z \equiv 0$ it is referred to as *Transverse Electric* (*TE*) *wave*. The genesis of this terminology lies in the condition that either H or E field is entirely transverse or perpendicular to the direction of propagation. Since the mode of field distributions for *TE* and *TM* waves are likely to differ, these cases are to be separately treated.

Equation 19.7 involves a new parameter h^2, related to the angular frequency, media parameters, and the propagation constant, is given as:

$$h^2=\omega^2\mu\varepsilon+\gamma^2 \tag{19.8}$$

Further in view of the continuity of tangential components of E at the conducting surfaces the boundary conditions for different components of E can be written as:

$$E_x=E_z=0 \text{ at } y=0 \text{ \& } y=b \quad (\text{a}) \quad E_y=E_z=0 \text{ at } x=0 \text{ \& } x=a \quad (\text{b})\} \tag{19.9}$$

In view of the above description, the electromagnetic waves classified as *TM* and *TE* waves are discussed in the following two subsections.

19.2.2 Transverse Magnetic (TM) Waves

In *TM* waves, the magnetic field is entirely transverse (i.e. $H_z \equiv 0$) and the mean energy transmission takes place through E_z (*as* $E_z \neq 0$). Thus, for TM waves, the solution of Equation 19.2a is to be obtained by using the method of separation of variables discussed in Chapter 9. In general E_z, is a function of all the three coordinates (x, y, and z), and can be written as:

$$\left.\begin{aligned} &E_z(x,y,z) = E_z(x,y)e^{-\gamma z} && \text{(a)} \\ \text{where}\quad &E_z(x,y) = X(x)Y(y) \text{ or } XY && \text{(b)} \end{aligned}\right\} \tag{19.10}$$

In Equation 19.10b, *X*(*x*) indicates that *X* is a function of *x* alone and *Y*(*y*) indicates that *Y* is a function of *y* alone.

The substitution of Equation 19.10b in Equation 19.2a, for $\sigma = 0$, gives:

$$Y\left(\frac{d^2X}{dx^2}\right) + X\left(\frac{d^2Y}{dy^2}\right) + \gamma^2 XY = -\omega^2 \mu\varepsilon XY \tag{19.11}$$

Division of Equation 19.11 by *XY* and separation of *X* and *Y* terms lead to the relations:

$$\left.\frac{1}{X}\left(\frac{d^2X}{dx^2}\right) + h^2 = A^2 \quad \text{(a)} \qquad \frac{1}{Y}\left(\frac{d^2Y}{dy^2}\right) = -A^2 \quad \text{(b)}\right\} \tag{19.12}$$

Equation 19.12 can be manipulated into the following forms:

$$\frac{d^2X}{dx^2} + \left(h^2 - A^2\right)X = \frac{d^2X}{dx^2} + B^2X = 0 \tag{19.13a}$$

$$\frac{d^2Y}{dy^2} + A^2Y = 0 \tag{19.13b}$$

where

$$B^2 = h^2 - A^2 \text{ or } h^2 = A^2 + B^2 \tag{19.13c}$$

Solutions of Equations 19.13a and 19.13a can be written as:

$$X = C_1 \cos(Bx) + C_2 \sin(Bx) \tag{19.14a}$$

$$Y = C_3 \cos(Ay) + C_4 \sin(Ay) \tag{19.14b}$$

Thus:

$$E_Z = XY = \{C_1 \cos(Bx) + C_2 \sin(Bx)\}\{C_3 \cos(Ay) + C_4 \sin(Ay)\} \tag{19.14c}$$

Equation 19.14c is the solution to Equation 19.2a for *E* replaced by E_Z. This solution involves arbitrary constants C_1, C_2, C_3, and C_4. To evaluate these arbitrary constants the boundary conditions, given by Equations 19.9a and 19.9b, are to be separately applied on *X*(*x*) and *Y*(*y*) given by Equations 19.14a and 19.14b. This gives:

$$\left.E_Z = 0 \text{ at } x = 0, \text{ thus } C_1 = 0 \quad \text{(a)} \qquad E_Z = 0 \text{ at } y = 0, \text{ thus } C_3 = 0 \quad \text{(b)}\right\} \tag{19.15}$$

Substitution of $C_1 = 0$ and $C_3 = 0$ in Equation 19.14c gives:

$$\left.\begin{aligned} E_Z = C_2 \sin(Bx) C_4 \sin(Ay) = C \sin(Bx)\sin(Ay) \quad &(\mathrm{a}) \\ \text{where, } C = C_2 C_4 \quad &(\mathrm{b}) \end{aligned}\right\} \quad (19.16)$$

The remaining two boundary conditions can now be applied to Equation 19.16a. A cursory look at this equation reveals that the boundary condition $E_Z = 0$ at $x = a$, is met only when $Ba = m\pi$ and $E_Z = 0$ at $y = 0$ is met when $Ab = n\pi$. Thus, parameters B and A and field intensity E_Z can be written as:

$$B = \frac{m\pi}{a} \quad (\mathrm{a}) \quad A = \frac{n\pi}{b} \quad (\mathrm{b}) \quad E_Z = C \sin\left(\frac{m\pi}{a}x\right)\sin\left(\frac{n\pi}{b}y\right) \quad (\mathrm{c})\Bigg\} \quad (19.17)$$

In view of Equations 19.13c, 19.17a, and 19.17b:

$$h^2 = A^2 + B^2 = \left(\frac{n\pi}{b}\right)^2 + \left(\frac{m\pi}{a}\right)^2 \quad (\mathrm{a}) \quad \text{or} \quad h = \sqrt{\left(\frac{n\pi}{b}\right)^2 + \left(\frac{m\pi}{a}\right)^2} \quad (\mathrm{b})\Bigg\} \quad (19.18)$$

The parameter h (sometimes also represented as h_{nm}) is referred to as the *cutoff wave number.*

The value of γ obtained in view of Equations 19.8a and 19.8b can be written as:

$$\gamma = \sqrt{\left\{\left(\frac{m\pi}{a}\right)^2 + \left(\frac{n\pi}{b}\right)^2\right\} - \omega^2\mu\varepsilon} \quad (19.19)$$

In the above equations both m and n are positive integers and may take values 0, 1, 2, … etc.

If C is replaced by E_{0z} to represent the magnitude of E_z, Equation 19.17c can be written as:

$$E_z = E_{0z} \sin\left(\frac{m\pi}{a}x\right)\sin\left(\frac{n\pi}{b}y\right)e^{-j\beta z} = E_{0z}\sin(Bx)\sin(Ay)e^{-j\beta z} \quad (19.20\mathrm{a})$$

In view of Equation 19.20a the expression for other field components involved in Equations 19.7a to 19.7d, for free space, can be written as:

$$\left.\begin{aligned} E_x = E_{0x}\cos(Bx)\sin(Ay)e^{-j\beta z} \quad (\mathrm{b}) \quad E_y = E_{0y}\sin(Bx)\cos(Ay)e^{-j\beta z} \quad (\mathrm{c}) \\ H_x = H_{0x}\sin(Bx)\cos(Ay)e^{-j\beta z} \quad (\mathrm{d}) \quad H_y = H_{0y}\cos(Bx)\sin(Ay)e^{-j\beta z} \quad (\mathrm{e}) \end{aligned}\right\}$$

(19.20)

In Equations 19.20b to 19.20e:

$$\left.\begin{array}{llll} E_{0x} = CB\left(\dfrac{-j\beta}{h^2}\right) & \text{(f)} & E_{0y} = CA\left(\dfrac{-j\beta}{h^2}\right) & \text{(g)} \\ H_{0x} = CA\left(\dfrac{j\omega\varepsilon}{h^2}\right) & \text{(h)} & H_{0y} = CB\left(-\dfrac{j\omega\varepsilon}{h^2}\right) & \text{(i)} \end{array}\right\} \quad (19.20)$$

In view of Equation 19.17 A and B are functions of n and m, respectively. Equations 19.20a–e reveal that for m or $n = 0$ all the field components vanish. Thus the minimum value, which can be assigned to m and n is 1.

19.2.3 Transverse Electric (TE) Waves

In a TE wave, the electric field is entirely transverse (i.e. $E_z \equiv 0$) and the mean energy transmission takes place through H_z (as $H_z \neq 0$). Thus, for this case only Equation 19.2b needs to be solved. This solution can be obtained by using the steps which are used to arrive at Equation 19.20 from Equation 19.10. However, in this case the boundary conditions cannot be directly applied to H_z as it is a tangential component and in the case of H only the continuity of normal component is ensured. For the application of boundary conditions Equation 19.7 is to be used to get E_x and E_y. The boundary conditions given by Equations 19.9a and 19.9b can now be applied to these tangential components of E. After evaluation of arbitrary constants we can revert back to H_z and get its expression. The final expression for H_z obtained after incorporation of time variation is:

$$H_z = H_{0z}\cos(Bx)\cos(Ay)e^{-j\beta z} \quad (19.21a)$$

In this relation H_{0z} represents the magnitude of H_z.

In view of Equations 19.7 and 19.21a the remaining field components can be written as:

$$\left.\begin{array}{llll} E_x = E_{0x}\cos(Bx)\sin(Ay)e^{-j\beta z} & \text{(b)} & E_y = E_{0y}\sin(Bx)\cos(Ay)e^{-j\beta z} & \text{(c)} \\ H_x = H_{0x}\sin(Bx)\cos(Ay)e^{-j\beta z} & \text{(d)} & H_y = H_{0y}\cos(Bx)\sin(Ay)e^{-j\beta z} & \text{(e)} \end{array}\right\} \quad (19.21)$$

In Equations 19.21b to 19.21e:

$$\left.\begin{array}{llll} E_{0x} = CA\left(\dfrac{j\omega\mu}{h^2}\right) & \text{(f)} & E_{0y} = CB\left(\dfrac{-j\omega\mu}{h^2}\right) & \text{(g)} \\ H_{0x} = CB\left(\dfrac{j\beta}{h^2}\right) & \text{(h)} & H_{0y} = CA\left(\dfrac{j\beta}{h^2}\right) & \text{(i)} \end{array}\right\} \quad (19.21)$$

In these relations, parameters A and B have the same meaning as given by Equations 19.16c and 19.16d. In addition, Equations 19.18 and 19.19 are equally valid for both *TM* and *TE* waves.

Equations 19.21a–e reveal that for $m = 0$ and $n \neq 0$, $B = 0$, $H_x = E_y = 0$, H_z, H_y, and E_x survive, whereas for $n = 0$ and $m \neq 0$, $A = 0$, $H_y = E_x = 0$, H_z, H_x, and E_y survive. Also, when $m = n = 0$, H_z survives, but there is no surviving component of E. Thus, in the case of a TE wave, the lowest possible values that can be assigned to m and n are 0 and 1 or 1 and 0.

19.2.4 Wave Behavior with Frequency Variation

The study of Equation 19.19 leads to the following terms that indicate the wave behavior.

19.2.4.1 Attenuation Constant

When $\left\{\left(\frac{m\pi}{a}\right)^2 + \left(\frac{n\pi}{b}\right)^2\right\} > \omega^2\mu\epsilon$, γ is purely real (i.e. $\gamma = \alpha$), the wave does not progress but simply attenuates. The attenuation constant for such a wave is given as:

$$\alpha = \sqrt{\left[\left\{\left(\frac{m\pi}{a}\right)^2 + \left(\frac{n\pi}{b}\right)^2\right\} - \omega^2\mu\epsilon\right]} \quad \text{or} \quad \alpha_{nm}^2 = h_{nm}^2 - \omega^2\mu\epsilon \tag{19.22a}$$

19.2.4.2 Phase Constant

When $\left\{\left(\frac{m\pi}{a}\right)^2 + \left(\frac{n\pi}{b}\right)^2\right\} < \omega^2\mu\epsilon$, γ is purely imaginary (i.e. $\gamma = j\beta$), the wave (theoretically) progresses without attenuation. The phase shift constant for such a wave is given as:

$$\beta = \sqrt{\left[\omega^2\mu\epsilon - \left\{\left(\frac{m\pi}{a}\right)^2 + \left(\frac{n\pi}{b}\right)^2\right\}\right]} \quad \text{or} \quad \beta_{nm}^2 = \omega^2\mu\epsilon - h_{nm}^2 \tag{19.22b}$$

19.2.4.3 Cutoff Frequency

When, $\left\{\left(\frac{m\pi}{a}\right)^2 + \left(\frac{n\pi}{b}\right)^2\right\} = \omega^2\mu\epsilon\omega$ can be replaced ω_c or $2\pi f_c$ where f_c is:

$$f_c = \frac{1}{2\sqrt{\mu\varepsilon}}\sqrt{\left\{\left(\frac{m}{a}\right)^2 + \left(\frac{n}{b}\right)^2\right\}} \tag{19.22c}$$

Equation 19.22c indicates a frequency (f_c referred to as cutoff, or critical) frequency. This frequency is the demarcation line below which the wave propagation ceases to exist.

19.2.4.4 Cutoff Wavelength

In view of the value of f_c given by Equation 19.22c and the relations $\upsilon = f\lambda$ and $\lambda_c = \upsilon/f_c$:

$$\lambda_c = \frac{2}{\sqrt{(m/a)^2 + (n/b)^2}} \tag{19.22d}$$

19.2.4.5 Dispersion Phenomenon

When wave progresses along its path it suffers from a phenomenon called dispersion. This phenomenon makes the energy of a signal to spread over a larger time period. This spreading reduces the effectiveness of the signal and thus is not a desirable feature.

19.2.4.6 Usable Frequency Range

The dispersion mainly occurs between f_c and $2f_c$. Thus, the use of a waveguide should be avoided at the extremes of this range. This range is referred to as the *usable frequency range.*

19.2.4.7 Phase Velocity

The phase velocity of a wave along the walls of a waveguide is not a real velocity. This in fact gives the rate of change of phase and is always greater than *c*. For both *TE* and *TM* wave in rectangular waveguide it is given by:

$$\upsilon_p = \frac{c}{\sqrt{1-(f_c/f)^2}} \tag{19.22e}$$

19.2.4.8 Group Velocity

In a hollow waveguide having air or vacuum inside the electromagnetic wave propagates at the speed of light (*c*) but follows a zigzag path. Thus the overall speed at which the signal travels the guide is always less than c. This velocity of the signal is called group velocity and is denoted by "υ_g". In view of the relation $\upsilon_p\upsilon_g = c^2$ and the expression of υ_p the expression of υ_g can be written as:

$$\upsilon_g = c\sqrt{1-(f_c/f)^2} \tag{19.22f}$$

19.2.4.9 Guide Wavelength

Wavelength in rectangular (or circular) waveguide (λ_g) is larger than that in free space at the same frequency. For both *TE* and *TM* wave it is given by:

$$\lambda_g = \frac{\lambda}{\sqrt{1-(f_c/f)^2}} \tag{19.22g}$$

19.2.4.10 Guide Phase Constant

In terms of the cutoff frequency, the phase shift constant in the waveguide is given as:

$$\beta_g = \beta\sqrt{1-(f_c/f)^2} \quad (19.22h)$$

19.2.4.11 Wavelength Parallel to Waveguide Walls

Since $\left(\frac{\omega_c}{\omega}\right)^2 = \left(\frac{\lambda}{\lambda_c}\right)^2$, the value of λ parallel to waveguide walls can be written as:

$$\lambda = \frac{\lambda_0\lambda_c}{\sqrt{(\lambda_c^2-\lambda_0^2)}} \quad \text{or} \quad \lambda_0 = \frac{\lambda\lambda_c}{\sqrt{(\lambda_c^2+\lambda^2)}} \quad (19.22j)$$

19.2.5 Modes

The term "*mode*" represents the way of the distribution of *E* and *H* fields in a waveguide. Each mode has a cutoff frequency below which there will be no propagation. In addition, the time taken by the wave to move down the waveguide varies with the mode.

19.2.5.1 TE Mode

The transverse electric (or *TE*) mode is one wherein the components of an *E* field are at a right angle to the direction of travel. In Equations 19.20a–19.20e, parameters *A* and *B* are functions of *n* and *m*, respectively. Thus while describing a mode suffixes are assigned to *TE*. In view of such assignment, *TE* is replaced by TE_{mn}. When *m* and *n* are given numerical values the resulting abbreviations completely specify the actual propagating mode. As noted earlier the minimum values that can be assigned to *m* and *n* in *TE* case are 1,0 or 0,1. Thus $TE_{10}$and TE_{01} are the lowest *TE* modes.

19.2.5.2 TM Mode

The transverse magnetic (or *TM*) mode is one wherein the components of an *H* field are at right angles to the direction of travel. In Equations 19.21a to 19.21e, parameters *A* and *B* are functions of integer *n* and *m*, respectively. Thus, while describing a mode, suffixes are assigned to *TM*. In view of the assigned suffixes, these can be written as TM_{mn}. To specify the actual propagating mode, *m* and *n* are assigned numerical values. As noted earlier, the minimum values that can be assigned to *m* and *n* in the case of a *TM* wave are 1 and 1. Thus, in a rectangular waveguide, TM_{11} is the lowest possible mode for a *TM* wave.

19.2.5.3 Dominant Mode

The mode that has the lowest cutoff frequency is referred to as the *dominant mode*. For rectangular waveguide the TE_{10} mode has the lowest cutoff frequency and thus it is the dominant *TE* mode. It has one half-cycle along the broader dimension of the waveguide. The wavelength (λ) is related to "*a*" as "$a = \lambda_c/2$", where *c* is the velocity of light. Modes with next higher cutoff frequency are TE_{01} and TE_{20}. In the case of a *TM* wave, TM_{11} is the mode with the lowest cutoff frequency and hence is referred to as the dominant *TM* mode. In view of Equation 19.22g, the cutoff wavelengths for *TM* and *TE* modes are $\frac{2ab}{\sqrt{(a^2+b^2)}}$ and 2a, respectively.

19.2.5.4 TEM Mode

The *TEM* mode is one in which both *E* and *H* fields are perpendicular to the direction of propagation. The cut of frequency for *TEM* mode is 0. Waves in power lines travel in *TEM* mode whereas in some planar lines it may have *TEM* or quasi *TEM* modes. In the case of waveguides irrespective of their shapes of cross-section there is *no possibility of existence of TEM wave.* It is due to the fact that for the existence of TEM waves in a hollow waveguide H lines must lie in a transverse plane. Also in a non-magnetic material divergence of *B* (or *H*) must be 0. Thus *H*-lines have to be in the form of closed loops in the plane perpendicular to the direction of travel of the wave. As spelled by the first Maxwell's equation ($\nabla \times \boldsymbol{H} = \boldsymbol{J} + \frac{\partial \boldsymbol{D}}{\partial t}$) the MMF around each of the close loops must be equal to the axial current passing through the loop. This requirement is met only when either conduction current density (*J*) or the displacement current density ($\partial \boldsymbol{D}/\partial t$) is present, which in turn requires the presence of E_z. But neither conduction current nor displacement current is present thus *TEM* wave cannot exist inside a hollow waveguide of any shape.

19.2.5.5 Single Mode Propagation

If the wave in a waveguide is carried through one mode it is called single mode propagation. Single mode propagation is always desirable as it has the *lesser dispersion.*

19.2.5.6 Multimode Propagation

When more than one mode propagates through the waveguide it is referred to as *multimode propagation.* According to the velocity relation, there will be faster propagation of lower order mode and slower for higher order. Correspondingly, the wave will strike the two waveguide walls more frequently in the case of higher order modes than in the case of lower order. The multiplicity of modes results in increased dispersion.

19.2.5.7 Mode Disturbance

The term *mode disturbance* refers to the simultaneous propagation of more than one mode. It occurs when the length of broader or narrower wall of a waveguide does not correspond to the excited mode. In the case of the TE_{10} mode, the broader wall has to be of $\lambda/2$ length. If it is slightly less or more than $\lambda/2$ the propagating mode will differ from TE_{10} and will be a combination of more than one mode. This mixing results in mode disturbance.

19.2.5.8 Impact of Mode Order

As noted earlier, *m* and *n* are the order of the modes. Thus, if the frequency is kept constant and *m* and *n* are changed, all of the parameters involved in Equations 19.22a–h are affected in one way or the other.

19.2.5.9 Wave Front Movement

The *wave front* is nothing but an equiphase plane. The polarity of a wave front keeps on changing as the wave progresses in the waveguide. This polarity change will correspond to the velocity of the wave and hence to the mode order.

19.2.6 Field Patterns

In view of Equations 19.20a–e and 19.21a–e, the patterns of field distributions for the *TM* and *TE* modes can be obtained by substituting values of *m* and *n* in the following manner.

19.2.6.1 TE Wave

Substitution of $m = 1$ and $n = 0$ in Equations 19.21a–e gives $E_x = H_y = 0$. The expressions for the real parts of the surviving components E_y, H_x, and H_z with the presumption of multiplication of all these components by $e^{-j\beta z}$ are:

$$\left.\begin{aligned} H_z = H_{0z}\cos\left(\frac{\pi x}{a}\right)\cos(\beta z) \quad (a) \quad E_y = E'_{0y}\sin\left(\frac{\pi x}{a}\right)\sin(\beta z) \quad (b) \\ H_x = H'_{0x}\sin\left(\frac{\pi x}{a}\right)\sin(\beta z) \quad (c) \end{aligned}\right\} \tag{19.23}$$

where $H_{0z} = C$, $E'_{0y} = CB\left(-\frac{\omega\mu}{h^2}\right)$ and $H'_{0x} = CB\left(\frac{\beta}{h^2}\right)$ are the magnitudes of H_z, E_y, and H_x respectively.

In view of Equations 19.23a–c, *E* has only E_y component whereas *H* has H_x and H_z components. Thus, *E* and *H* can be written as:

$$\left. E = E'_{0y}\sin\left(\frac{\pi x}{a}\right)\sin(\beta z) \quad (a) \quad H = \sqrt{H_x^2 + H_z^2} \quad (b) \right\} \tag{19.24}$$

In view of Equation 19.24a, *E* field is sinusoidal along x axis (or the broader wall) and becomes zero at $x = 0$ and "*a*" and maximum at $x = a/2$. E is also sinusoidal along *z* axis. If $m = 1$ and $n = 0$, *E* becomes zero at $x = 0$, $a/2$ and "*a*" and maximum at $x = a/4$ and $3a/4$. Similarly when $m = 0$ and $n = 1$, *E* becomes zero at $y = 0$ and *b* and maximum at $y = b/2$. Thus locations of minimas and maximas depend on the values of *m* and *n*. Figure 19.2 illustrates the TE_{10} modes in x–y, z–y, and z–x planes and Figure 19.3 shows different TE modes in the x–y plane.

19.2.6.2 TM Wave

Substitution of values of *m* and *n* in Equations 19.20a–e and appropriate manipulations may lead to different *TM* modes. Since in this case both *m* and *n* take one as the minimum value all the components given by Equations 19.20a–e will survive. Thus resultant *E* will depend on E_x, E_y, and E_z and resultant *H* on H_x and H_y. From these resultant *E* and *H*, field patterns for different *m* and *n* can be obtained. Some such field patterns are given in Figure 19.4.

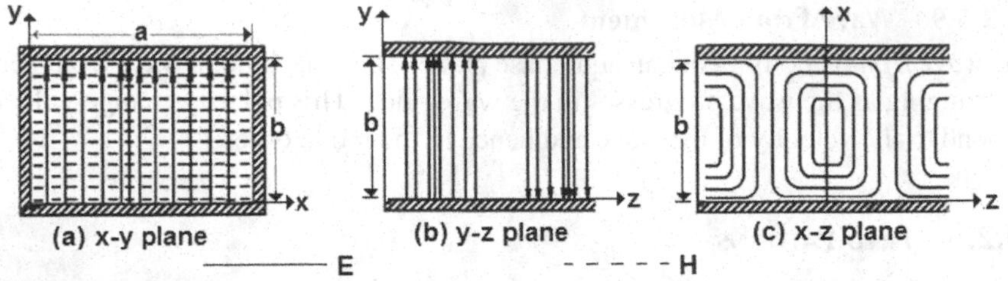

FIGURE 19.2 TE_{10} modes in x–y, z–y, and z–x planes.

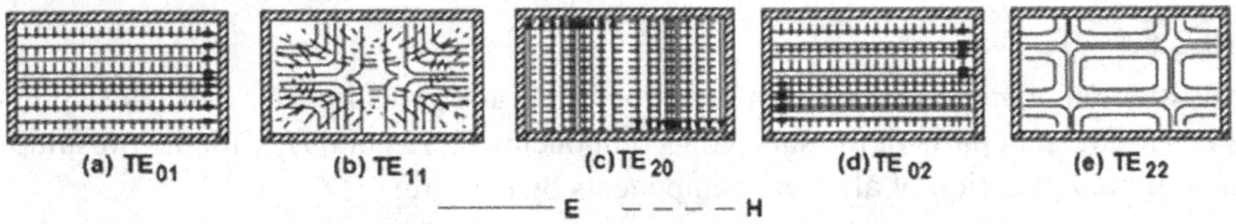

FIGURE 19.3 TE modes in x–y plane.

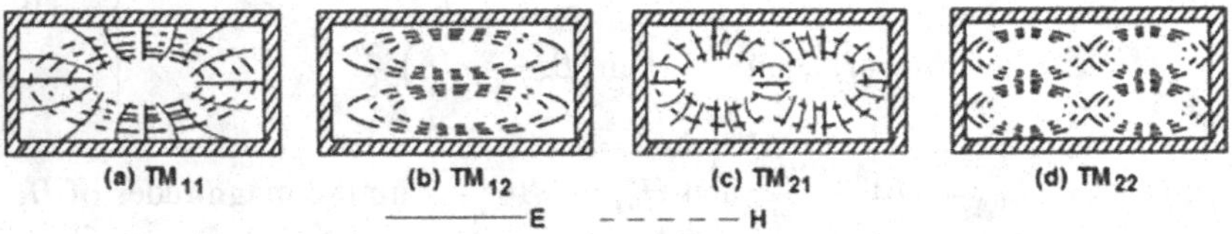

FIGURE 19.4 TM modes in x–y plane.

19.2.7 Excitation of Modes

The electromagnetic signal generated by a device is to be fed into waveguide. This feeding is referred to as excitation. Once it is fed this signal is carried by the waveguide to its destination in the form of electromagnetic wave. In general the output of a signal generator is taken out through a wire(s) or probe(s). The probe(s) is/are inserted in a waveguide are appropriate location(s) to excite *TM* and *TE* modes. This excitation is described below.

19.2.7.1 TE Modes

Figure 19.5 illustrates the excitation of different *TE* modes. The energy from a signal generator is fed into waveguide. The location(s) of probe(s) in the broader or narrower wall will results in the excitation of a particular mode. Once a mode is excited its field pattern will progress in the direction of propagation.

19.2.7.2 TM Modes

Figure 19.6 illustrates the excitation of different *TM* modes. Accordingly, different numbers of probes are inserted at different locations into the cross-section of the waveguide for exciting different modes.

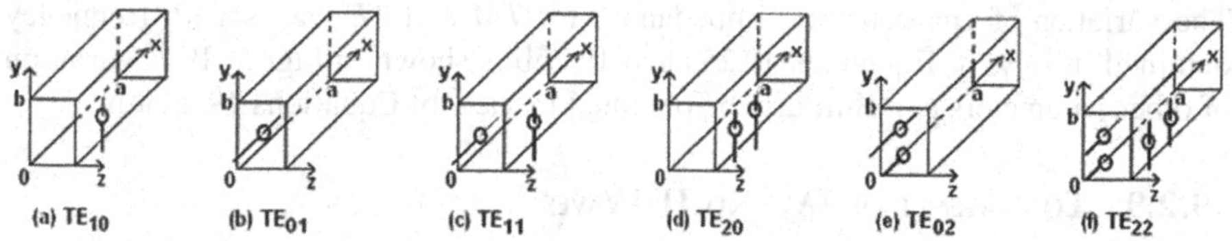

FIGURE 19.5 Excitation of different TE modes.

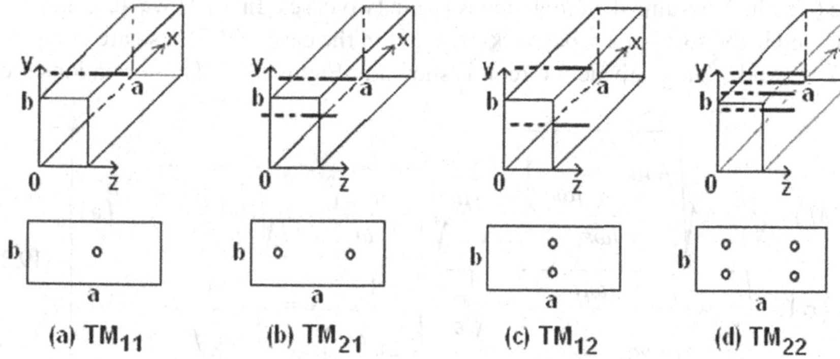

FIGURE 19.6 Excitation of different TM modes.

19.2.8 Wave Impedances

A wave traveling in different directions may come across different types of environments or obstacles. For example in the eastward direction it may come across some buildings, westward it may see a dense forest, northward the atmosphere may be wet and southward it may be quite dry. Obviously in each direction the wave will see different impedances. Thus a new concept referred to as "wave impedance" emerges. The wave impedance can thus be spelled as the impedance which is seen by the wave when it travels in a particular direction. In addition, the same environment may be differently seen at different times by different modes of *TE* and *TM* modes.

In the case of waveguide the situation is quite simplified as a wave is assumed to be traveling only along the *z*-direction. The impedance offered to this wave will be simply the ratio of orthogonal *E* and *H* components across the *x*–*y* plane. The mathematical analysis yields the following impedance relations for *TM* and *TE* waves.

$$\left.\begin{aligned} Z_z(TM) &= \frac{\beta}{\omega\varepsilon} = \eta\sqrt{\left\{1-\left(\frac{\omega_c}{\omega}\right)^2\right\}} = \eta\sqrt{\left\{1-\left(\frac{f_c}{f}\right)^2\right\}} \quad &\text{(a)} \\ Z_z(TE) &= \frac{\omega\mu}{\beta} = \frac{\eta}{\sqrt{\left\{1-\left(\frac{\omega_c}{\omega}\right)^2\right\}}} = \frac{\eta}{\sqrt{\left\{1-\left(\frac{f_c}{f}\right)^2\right\}}} \quad &\text{(b)} \end{aligned}\right\} \tag{19.25}$$

The variation of characteristic impedances for *TM* and *TE* waves with frequency obtained in view of Equations 19.25a and 19.25b is shown in Figure 19.7. Variation of other parameters can similarly be obtained in view of Equations 19.22a–h.

19.2.9 Equivalence of TM and TE Waves

In view of the mathematical analysis, a wave can be represented by an equivalent circuit. This circuit is obviously different for *TM* and *TE* waves. The final relations for both the waves are given in terms of the series impedance (*Z*) and shunt admittance (*Y*) which assume different values in the two cases. In a *TM* wave, *Z* and *Y* are connected in the form of a π-network, whereas in the case of *TE* these are connected in a T-network. The equivalent circuit is shown in Figure 19.8. These relations are:

$$\left.\begin{aligned} Z_z(TM) &= \sqrt{\frac{Z}{Y}} = \sqrt{\frac{j\omega\mu + \left(\dfrac{h^2}{j\omega\varepsilon}\right)}{j\omega\varepsilon}} = \sqrt{\frac{\mu}{\varepsilon}}\sqrt{1-\left(\frac{\omega_c}{\omega}\right)^2} = \eta\sqrt{1-\left(\frac{f_c}{f}\right)^2} \quad &(a)\\ Z_z(TE) &= \sqrt{\frac{Z}{Y}} = \frac{j\omega\mu}{\sqrt{j\omega\varepsilon + \left(\dfrac{h^2}{j\omega\mu}\right)}} = \sqrt{\frac{\mu}{\varepsilon}}\frac{1}{\sqrt{1-\left(\dfrac{\omega_c}{\omega}\right)^2}} = \frac{\eta}{\sqrt{1-\left(\dfrac{f_c}{f}\right)^2}} \quad &(b)\end{aligned}\right\} \quad (19.26)$$

In these relations, suffix *z* indicates the direction of wave travel and bracketed terms (*TM*) and (*TE*) means the type of wave. $Z_Z(TM)$ and $Z_Z(TE)$ in fact represent the

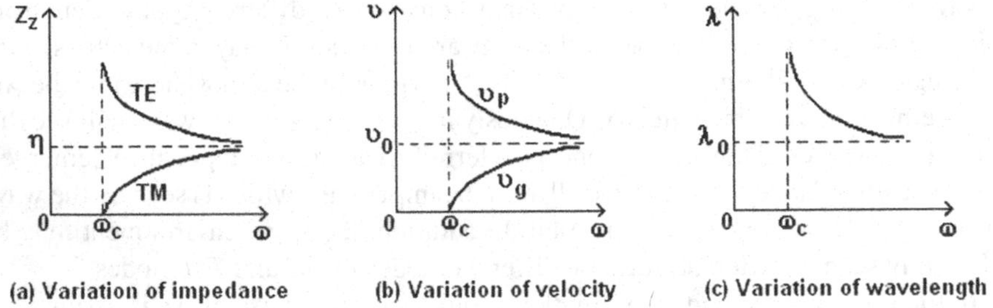

FIGURE 19.7 Variation of $Z_z(TE)$ and $Z_z(TM)$ with frequency.

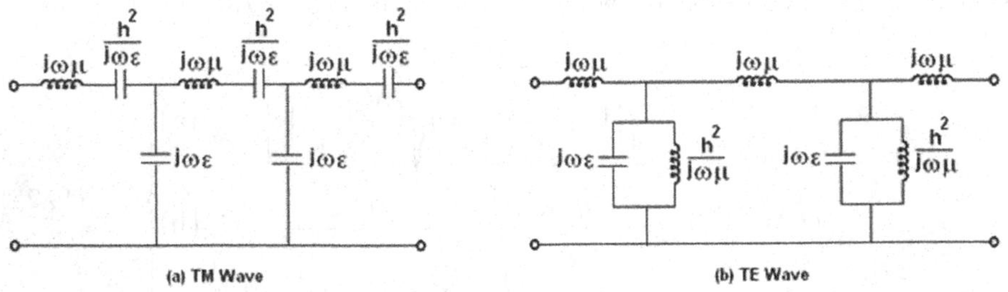

FIGURE 19.8 Equivalent circuits for (a) TM and (b) TE waves.

impedances of these two equivalent circuits. It can be noted that Equations 19.26a and 19.26b are the same as Equations 19.25a and 19.25b because it is the same environment is faced by the wave in the waveguide.

This is to be noted that if the waveguide is filled with a dielectric, other than air or vacuum c is to be replaced by v, η_0 by η etc. in equations wherever they involve.

19.3 CIRCULAR WAVEGUIDE

A circular waveguide is a hollow metallic pipe with circular cross-section for propagating the electromagnetic waves by continuous reflections from the surfaces or walls of the guide. The circular waveguides are generally avoided because (a) the frequency difference between the lowest frequency on the dominant mode and the next mode is smaller than in a rectangular waveguide, with $b/a = 0.5$; (b) the circular symmetry of the waveguide may reflect on the possibility of the wave not maintaining its polarization throughout the length of the guide and; (c) for the same operating frequency, circular waveguide is bigger in size than a rectangular waveguide. Circular waveguides are mainly used for making some microwave components viz. circulators, isolators etc. The physical structure along with its radius and the coordinate system is shown in Figure 19.9.

19.3.1 Preliminary Mathematics

As in a rectangular waveguide, in a circular waveguide the wave is assumed to be propagating in a positive z direction, the time variation is accounted by $e^{j\omega t}$, and the space variation along the z-axis by $e^{-\gamma z}$. These parameters have the same meaning as described in Section 19.2.1 It is further assumed that the waveguide is made of conducting material with $\sigma = \infty$. Thus, the continuity conditions of E_{tan} and H_{norm} hold at the surface of conductor that is for $0 \le \varphi \le 2\pi$ and $0 \le z \le \infty$. The waveguide

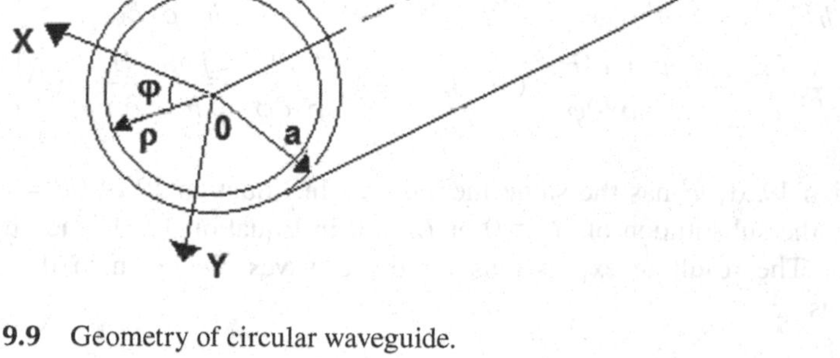

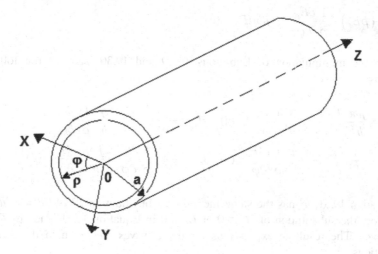

FIGURE 19.9 Geometry of circular waveguide.

is assumed to be filled with air or a dielectric for which $\sigma = 0$ and thus $\gamma = j\beta$. In view of the shape of the waveguide cylindrical coordinate system (ρ,φ,z) is employed.

The basic governing Maxwell's equations and wave equations for perfect dielectric media are the same as were given by Equations 19.1 and 19.2. In addition, the parameter γ has the same meaning as spelled by Equation 19.3. For this case the governing Equations 19.1a and 19.1b in cylindrical coordinates can be expanded into the following form:

$$\left.\begin{aligned} &\frac{1}{\rho}\frac{\partial H_z}{\partial \varphi} - \frac{\partial H_\varphi}{\partial z} = j\omega\varepsilon E_\rho \quad \text{(a)} \qquad \frac{\partial H_\rho}{\partial z} - \frac{\partial H_z}{\partial \rho} = j\omega\varepsilon E_\varphi \quad \text{(b)} \\ &-\frac{1}{\rho}\frac{\partial}{\partial \rho}(\rho H_\varphi) - \frac{1}{\rho}\frac{\partial H_\rho}{\partial \varphi} = j\omega\varepsilon E_z \qquad \text{(c)} \end{aligned}\right\} \tag{19.27}$$

$$\left.\begin{aligned} &\frac{1}{\rho}\frac{\partial E_z}{\partial \varphi} - \frac{\partial E_\varphi}{\partial z} = -j\omega\mu H_\rho \quad \text{(a)} \qquad \frac{\partial E_\rho}{\partial z} - \frac{\partial E_z}{\partial \rho} = j\omega\mu H_\varphi \quad \text{(b)} \\ &\frac{1}{\rho}\frac{\partial}{\partial \rho}(\rho E_\varphi) - \frac{1}{\rho}\frac{\partial E_\rho}{\partial \varphi} = j\omega\mu E_z \qquad \text{(c)} \end{aligned}\right\} \tag{19.28}$$

After incorporating z derivatives, Equations 19.27 and 19.28 become:

$$\left.\begin{aligned} &\frac{1}{\rho}\frac{\partial H_z}{\partial \varphi} + \gamma H_\varphi = -j\omega\varepsilon E_\rho \quad \text{(a)} \qquad -\gamma H_\rho + \frac{\partial H_z}{\partial \rho} = j\omega\varepsilon E_\varphi \quad \text{(b)} \\ &\frac{1}{\rho}\frac{\partial}{\partial \rho}(\rho H_\varphi) - \frac{1}{\rho}\frac{\partial H_\rho}{\partial \varphi} = j\omega\varepsilon E_z \quad \text{(c)} \end{aligned}\right\} \tag{19.29}$$

$$\left.\begin{aligned} &\frac{1}{\rho}\frac{\partial E_z}{\partial \varphi} + \gamma E_\varphi = -j\omega\mu H_\rho \quad \text{(a)} \qquad -\gamma E_\rho + \frac{\partial E_z}{\partial r} = -j\omega\mu H_\varphi \quad \text{(b)} \\ &\frac{1}{\rho}\frac{\partial}{\partial \rho}(\rho E_\varphi) - \frac{1}{\rho}\frac{\partial E_\rho}{\partial \varphi} = -j\omega\mu E_z \quad \text{(c)} \end{aligned}\right\} \tag{19.30}$$

Further the manipulation of Equations 19.29 and 19.30 leads to the following relations:

$$\left.\begin{aligned} &H_\rho = \frac{j\omega\epsilon}{h^2}\frac{1}{\rho}\frac{\partial E_z}{\partial \varphi} - \frac{\gamma}{h^2}\frac{\partial H_z}{\partial \rho} \quad \text{(a)} \qquad H_\varphi = -\frac{j\omega\epsilon}{h^2}\frac{\partial E_z}{\partial \rho} - \frac{\gamma}{h^2}\frac{1}{\rho}\frac{\partial H_z}{\partial \varphi} \quad \text{(b)} \\ &E_\rho = -\frac{\gamma}{h^2}\frac{\partial E_z}{\partial \rho} - \frac{j\omega\mu}{h^2}\frac{1}{\rho}\frac{\partial H_z}{\partial \varphi} \quad \text{(c)} \qquad E_\phi = -\frac{\gamma}{h^2}\frac{1}{\rho}\frac{\partial E_z}{\partial \varphi} + \frac{j\omega\mu}{h^2}\frac{\partial H_z}{\partial \rho} \quad \text{(d)} \end{aligned}\right\} \tag{19.31}$$

In Equation 19.31, h^2 has the same meaning as in Equation 19.19 ($h^2 = \omega^2\mu\varepsilon + \gamma^2$). As before the substitution of $H_z \equiv 0$ or $E_z \equiv 0$ in Equation 19.31 leads to *TM* and *TE* waves. The resulting expressions for these waves are given in the following subsections.

19.3.2 Transverse Magnetic Wave

In Equations 19.31a–d if $H_z \equiv 0$, the field components for *TM* wave become:

$$\left.\begin{aligned} E_\rho &= -\frac{\gamma}{h^2}\frac{\partial E_z}{\partial \rho} \quad (a) \qquad E_\varphi = -\frac{\gamma}{h^2}\frac{1}{\rho}\frac{\partial E_z}{\partial \varphi} \quad (b) \\ H_\rho &= \frac{j\omega\epsilon}{h^2}\frac{1}{\rho}\frac{\partial E_z}{\partial \varphi} \quad (c) \qquad H_\varphi = -\frac{j\omega\epsilon}{h^2}\frac{\partial E_z}{\partial \rho} \quad (d) \end{aligned}\right\} \tag{19.32}$$

19.3.3 Transverse Electric Wave

For *TE* wave $E_Z \equiv 0$, the field components given by Equations 19.31a–d become:

$$\left.\begin{aligned} E_\rho &= -\frac{j\omega\mu}{h^2}\frac{1}{\rho}\frac{\partial H_z}{\partial \varphi} \quad (a) \qquad E_\varphi = \frac{j\omega\mu}{h^2}\frac{\partial H_z}{\partial \rho} \quad (b) \\ H_\rho &= -\frac{\gamma}{h^2}\frac{\partial H_z}{\partial \rho} \quad (c) \qquad H_\varphi = -\frac{\gamma}{h^2}\frac{1}{\rho}\frac{\partial H_z}{\partial \varphi} \quad (d) \end{aligned}\right\} \tag{19.33}$$

19.3.4 Solution of Wave Equation

From Equations 19.31 and 19.32, it is evident that all the field components are related either to E_z or to H_z. The E_z and H_z can be obtained by solving Equations 19.2a or 19.2b for E or H. These equations can be written in differential forms in cylindrical coordinates as:

$$\left.\begin{aligned} \frac{1}{\rho}\frac{\partial}{\partial \rho}\left(\rho\frac{\partial E_z}{\partial \rho}\right) + \frac{1}{\rho^2}\frac{\partial^2 E_z}{\partial \varphi^2} + h^2 E_z = 0 \quad (a) \\ \frac{1}{\rho}\frac{\partial}{\partial \rho}\left(\rho\frac{\partial H_z}{\partial \rho}\right) + \frac{1}{\rho^2}\frac{\partial^2 H_z}{\partial \varphi^2} + h^2 H_z = 0 \quad (b) \end{aligned}\right\} \tag{19.34}$$

To obtain the solution of Equations 19.34a and 19.34b through the method of separation of variables, in cylindrical coordinates, let:

$$E_z \text{ or } H_z = R(\rho)S(\varphi) \tag{19.35}$$

In Equation 19.35 R is a function of ρ alone and S is a function of φ alone. Substitution of E_z in Equation 19.34a and of H_z in Equation 19.34b leads to the following relation:

$$S\frac{d^2 R}{d\rho^2} + \frac{1}{\rho}S\frac{dR}{d\rho} + \frac{R}{\rho^2}\frac{d^2 S}{d\varphi^2} + R\cdot S\cdot h^2 = 0 \tag{19.36}$$

Dividing Equation 19.36 by $R{\cdot}S$, adding and subtracting n^2/ρ^2 in the resulting expression, and rearranging the terms therein, results in:

$$\left[\frac{1}{R}\frac{d^2 R}{d\rho^2} + \frac{1}{R\rho}\frac{dR}{d\rho} + h^2 - \frac{n^2}{\rho^2}\right] + \left[\frac{1}{S\rho^2}\frac{d^2 S}{d\varphi^2} + \frac{n^2}{\rho^2}\right] = 0 \tag{19.37}$$

Separation of R and S terms gives:

$$\left.\begin{aligned} &\frac{1}{R}\frac{d^2R}{d\rho^2}+\frac{1}{R\rho}\frac{dR}{d\rho}+h^2-\frac{n^2}{\rho^2}=\frac{d^2R}{d\rho^2}+\frac{1}{\rho}\frac{dR}{d\rho}+Rh^2-R\frac{n^2}{\rho^2}=0 \quad \text{(a)}\\ &\frac{1}{S\rho^2}\frac{d^2S}{d\varphi^2}+\frac{n^2}{\rho^2}=0 \quad \text{(b)}\end{aligned}\right\} \quad (19.38)$$

If Equation 19.38a is divided by h^2 and h is merged with ρ, it will amount to replace ρ by ρh. Thus, this equation will take the following form:

$$\frac{d^2R}{d(\rho h)^2}+\frac{1}{(\rho h)}\frac{dR}{d(\rho h)}+\left(1-\frac{n^2}{(\rho h)^2}\right)R=0 \quad (19.39)$$

Equation 19.39 has the same form as that of Equation 9.203, which was identified as a Bessel's equation. Its solution can be written as:

$$R(\rho h)=A_{n1}J_n(\rho h)+B_{n1}Y_n(\rho h) \quad (19.40)$$

In Equation 19.40, J and Y represent Bessel's functions of the first and second kind, suffix n indicates their order, and the term (ρh) indicates their argument.

Equation 19.38b can be rewritten as:

$$\frac{d^2S}{d\varphi^2}+n^2S=0 \quad (19.41)$$

The solution of Equation 19.41 can be written as:

$$S=C_{n1}\cos(n\varphi)+D_{n1}\sin(n\varphi) \quad (19.42)$$

Substitution of Equations 19.40 and 19.42 in Equation 19.35 gives:

$$E_z \text{ or } H_z=\left[\{A_{n1}J_n(\rho h)+B_{n1}Y_n(\rho h)\}\cdot\{C_{n1}\cos(n\varphi)+D_{n1}\sin(n\varphi)\}\right] \quad (19.43)$$

Equation 19.43 represents the complete solution to both the E_z and H_z fields. This equation involves four arbitrary constants (viz. A_{n1}, B_{n1}, C_{n1}, and D_{n1}). The procedure for evaluation of these constants is described in the following subsection.

In Equation 19.43, if $\rho = 0$, $\rho h = 0$, $Y_n(\rho h) = \infty$, then Y_n does not represent a physical field. This situation is met by taking B_{n1} as zero. The constant A_{n1} can be merged with the C_{n1} and D_{n1}. This leads to the following new sets of arbitrary constants.

$$\left.A_n\,(\text{or } C_n)(=A_{n1}C_{n1}) \quad \text{(a)} \quad B_n\,(\text{or } D_n)(=A_{n1}D_{n1}) \quad \text{(b)}\right\} \quad (19.44)$$

In view of Equation 19.43, it was noted that E_z and H_z have the same form of solutions. In order to differentiate the solutions for E_z and H_z different symbols are used to represent their arbitrary constants. Thus, for the TM wave:

$$\left.\begin{aligned} E_z &= J_n(\rho h)\{A_n\cos(n\varphi)+B_n\sin(n\varphi)\} \quad (\text{for TM wave}) \quad (\text{a}) \\ H_z &= J_n(\rho h)\{C_n\cos(n\varphi)+D_n\sin(n\varphi)\} \quad (\text{for TE wave}) \quad (\text{b}) \end{aligned}\right\} \quad (19.45)$$

In expressions of Equation 19.45a, the relative values of A_n and B_n determine the orientation of field in a waveguide. In a circular waveguide φ-axis (for any value of n) can be oriented so as to make either A_n or B_n equal to zero. For $\varphi = 0, 2\pi$, $A_n = 0$ and for $\varphi = \frac{\pi}{2}, \frac{3\pi}{2}$, $B_n = 0$.

Selection of $\varphi = \pi/2$ gives $B_n = 0$. Thus, the expression for E_z becomes:

$$E_z = A_n J_n(\rho h)\cos(n\varphi) \quad (19.46\text{a})$$

Other field components for *TM* wave obtained in view of Equations 19.46a and 19.32 are:

$$\left.\begin{aligned} E_\rho &= -\frac{\gamma}{h^2}A_n J_n(\rho h)\cos(n\varphi) \quad (\text{b}) \quad E_\varphi = -\frac{\gamma}{h^2}\frac{n}{\rho}A_n J_n'(\rho h)\sin(n\varphi) \quad (\text{c}) \\ H_\rho &= \frac{-j\omega\epsilon}{h^2}\frac{n}{\rho}A_n J_n'(\rho h)\sin(n\varphi) \quad (\text{d}) \quad H_\varphi = -\frac{j\omega\epsilon}{h^2}A_n J_n(\rho h)\cos(n\varphi) \quad (\text{e}) \end{aligned}\right\}$$

(19.46)

Using a similar procedure, the expression for H_z becomes:

$$H_z = C_n J_n'(\rho h)\cos(n\varphi) \quad (19.47\text{a})$$

In view Equations 19.47a and 19.32, the other field components for *TE* wave are:

$$\left.\begin{aligned} E_\rho &= -\frac{j\omega\mu}{h^2}\frac{n}{\rho}C_n J_n(\rho h)\sin(n\varphi) \quad (\text{b}) \quad E_\varnothing = \frac{j\omega\mu}{h^2}C_n J_n'(\rho h)\cos(n\varphi) \quad (\text{c}) \\ H_\rho &= -\frac{\gamma}{h^2}C_n J_n'(\rho h)\cos(n\varphi) \quad (\text{d}) \quad H_\varnothing = -\frac{\gamma}{h^2}\frac{n}{\rho}C_n J_n(\rho h)\sin(n\varphi) \quad (\text{e}) \end{aligned}\right\} \quad (19.47)$$

In Equations 19.46c and 19.47, J_n' represents the derivative of $J_n(\rho h)$ with respect to ρ (or ρh). Also, all of the expressions in these equations are assumed to be multiplied by $e^{-j\beta z}$. The variation of $J_n(x)$ with the argument (x) is illustrated in Figure 19.10.

In the *TM* wave $E_z = 0$ at the conducting wall of waveguide (i.e. at $\rho = a$). In view of Equation 13.46a $E_z = 0$ when $J_n(ha) = 0$. Figure 19.10 shows that $J_n(ha)$ is zero infinite number of times for any integer n. The values of argument (ha) at which $J_n(ha) = 0$ are referred to as zeros of Bessel's function.

In the cylindrical waveguide case, boundary conditions cannot be directly applied to H_z in the case of *TE* wave. First, the tangential component E_φ is obtained, and then the boundary conditions are applied. This operation leads to $J_n'(ha) = 0$ where dash (′) over J indicates its derivative. Like $J_n(ha)$, $J_n'(ha)$ is also zero infinite number of

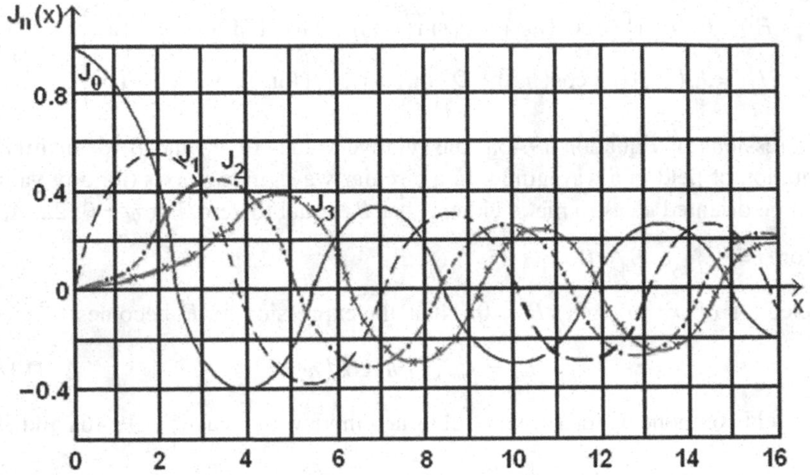

FIGURE 19.10 variation of $J_n(x)$ with the argument (x).

times for any integer n. All the values of argument (ha) at which $J'_n(ha)$ becomes zero are referred to as zeroes of $J'_n(ha)$. If ha is replaced by x, $J_n(ha)$ and $J'_n(ha)$ can be replaced by $J_n(x)$ and $J'_n(x)$ respectively. First few zeroes of $J_n(x)$ and $J'_n(x)$ for different values of order (n) are listed in Table 19.1. In this table, p is the number of zeros.

19.3.5 Mode Designation

In circular waveguides, the mode designation is not as simple as in the case of rectangular guides. The procedure of mode designation in circular waveguides is explained below.

19.3.5.1 Mode Designation in TM Case

For TM mode the wave equation is solved for E_z. At the conducting surface of the waveguide $E_z = 0$ and thus $J_n(ha) = 0$. Equation 19.22b means that in waveguide

TABLE 19.1

Zeroes of $J_n(x)$ and $J'_n(x)$ (Abramowitz and Stegun, pp. 409, 411)

pth zero of $J_n(x)$					pth zero of $J'_n(x)$				
$p \rightarrow$	1	2	3	4	$p \rightarrow$	1	2	3	4
$J_0(x)$	2.405	5.520	8.645	11.792	$J'_0(x)$	0.000	3.832	7.016	10.173
$J_1(x)$	3.832	7.106	10.173	13.324	$J'_1(x)$	1.841	5.331	8.536	11.706
$J_2(x)$	5.136	8.417	11.620	14.796	$J'_2(x)$	3.054	6.706	9.969	13.170
$J_3(x)$	6.380	9.761	13.015	16.223	$J'_3(x)$	4.201	8.015	11.346	14.586
$J_4(x)$	7.588	11.065	14.372	16.616	$J'_4(x)$	5.317	9.282	12.682	15.964

a wave will propagate only when h^2 is less than $\omega^2\mu\varepsilon$. This condition is met only when h is small or ω is large. Earlier it was noted that $J_n(ha)$ becomes zero infinite number of times and the values of "ha" at which it becomes zero are referred to as zeros (or roots) of Bessel's function. These roots are designated as $(ha)_{nm}$ wherein the first suffix (n) indicates the order of Bessel's function and the second suffix (m) represents the number of roots. Out of the zeros of the Bessel's function listed in Table 19.1, only a few of these are of practical importance. These include, $(ha)_{01}$, $(ha)_{11}$, $(ha)_{02}$, and $(ha)_{12}$. Similarly, the first few roots of practical importance of $J'_n(ha) = 0$ include $(ha)'_{01}$, $(ha)'_{11}$, $(ha)'_{02}$, and $(ha)'_{12}$. These roots are designated as $(ha)'_{nm}$. The TM modes can be designated as TM_{01}, TM_{02}, TM_{11}, and TM_{12}. The possible TM modes in a circular waveguide along with their root values are: $(ha)_{01} = 2.405$ for TM_{01}, $(ha)_{02} = 5.53$ for TM_{02},

$(ha)_{11} = 3.85$ for TM_{11}, $(ha)_{12} = 7.02$ for TM_{12}.

19.3.5.2 Mode Designation in TE Case

The description given for TM mode designation is equally valid for TE waves. In the case of a TE wave $E_z \equiv 0$ and the field components are given by Equation 20.47. In view of the zeros (or roots) of $J_n(ha) = 0$ and $J'_n(ha) = 0$ the possible TE modes can be designated as TE_{01}, TE_{02}, TE_{11}, and TE_{12}. The root values for these modes are: $(ha)_{01} = 3.85$ for TE_{01}, $(ha)_{02} = 7.02$ for TE_{02}, $(ha)_{11} = 1.841$ for TE_{11}, $(ha)_{12} = 5.53$ for TE_{12}.

19.3.5.3 TEM Mode

As mention earlier, no TEM mode can propagate in a circular waveguide.

19.3.5.4 Dominant Modes

The *dominant mode* for a circular waveguide is defined as the lowest order mode having the lowest root value. The dominant mode for TE waves in a circular waveguide is the TE_{11} because it has the lowest root value of 1.841. In TM waves, it is TM_{01} with a lowest root value of 2.405. Since the root value of TE_{11} is lower than that for TM_{01}, TE_{11} is the dominant or the lowest order mode for a circular waveguide.

19.3.5.5 Cutoff Wave Number

The designated roots $(ha)_{nm}$ and $(ha)'_{nm}$ can be related to h or h_{nm} given in Equation 19.18b by the following relations.

$$\left.\begin{aligned} h_{nm} &= \frac{(ha)_{nm}}{a} = \omega_c\sqrt{\mu\varepsilon} \quad \text{(for TM wave)} \quad \text{(a)} \\ h'_{nm} &= \frac{(ha)'_{nm}}{a} = \omega_c\sqrt{\mu\varepsilon} \quad \text{(for TE wave)} \quad \text{(b)} \end{aligned}\right\} \tag{20.48}$$

These parameters are referred to as cut off wave numbers for *TM* and *TE* waves, respectively. In view of Equation 19.48, the expressions given by Equations 19.46 and 19.47 can be written in the modified form by replacing h by h_{nm} or h'_{nm}.

19.3.5.6 Phase Shift Constant

Expressions for phase shift constant for *TM* and *TE* waves take the following form:

$$\left.\begin{aligned} \beta_{nm} &= \sqrt{\omega^2\mu\varepsilon - h_{nm}^{'2}} \quad (\text{for TE wave}) \quad (\text{a}) \\ \beta_{nm} &= \sqrt{\omega^2\mu\varepsilon - h_{nm}^{2}} \quad (\text{for TM wave}) \quad (\text{b}) \end{aligned}\right\} \tag{19.49}$$

19.3.5.7 Cutoff Frequency

The cutoff or critical frequencies can be written as:

$$\left. f_c = \frac{h'_{nm}}{2\pi\sqrt{\mu\varepsilon}} \quad (TE \text{ wave}) \quad (\text{a}) \qquad f_c = \frac{h_{nm}}{2\pi\sqrt{\mu\varepsilon}} \quad (\text{TM wave}) \quad (\text{b}) \right\} \tag{19.50}$$

19.3.5.8 Other Parameters

For circular waveguide the relations for υ_p, υ_g, λ_g, λ, β_g, $Z_z(TM)$ and $Z_z(TE)$, etc., for both *TM* and *TE* waves are the same as given by Equations 19.22 and 19.26 for a rectangular waveguide.

19.3.6 Field Distributions

In view of the analysis presented above field distributions for different *TM* and *TE* modes can be obtained. Some such distributions across the cross-section of a circular waveguide are shown in Figure 19.11.

19.3.7 Mode Excitation

Figure 19.12 illustrates the excitation of TE_{11} and TM_{01} modes in a circular waveguide.

19.3.8 Transformation of Modes

In the introduction to this chapter, it was mentioned that rectangular waveguide is more commonly used but in some applications the involvement of circular waveguide

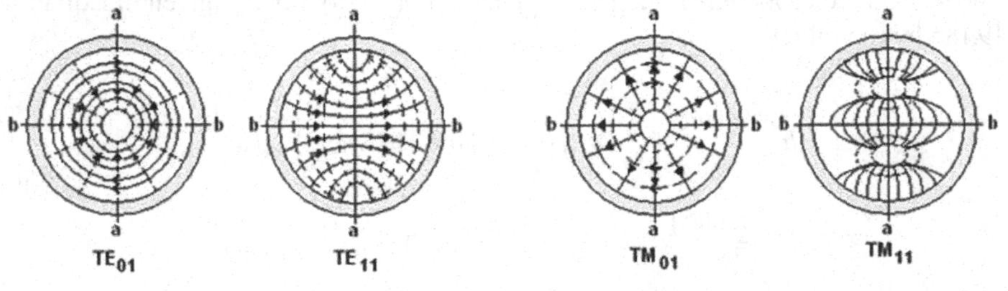

FIGURE 19.11 TE and TM modes in circular waveguide.

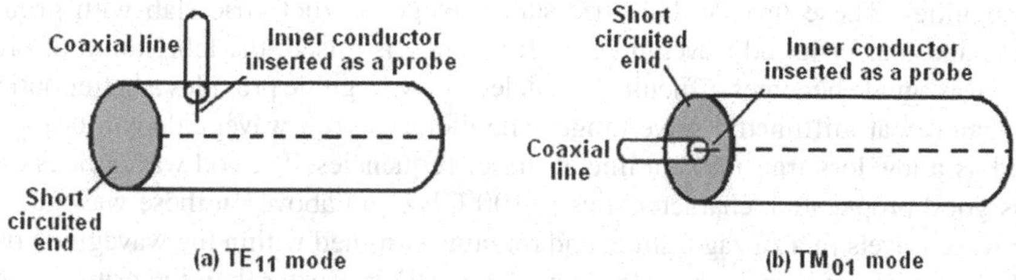

FIGURE 19.12 Excitation of TE_{11} and TM_{01} modes.

becomes unavoidable. In some cases, the simultaneous use of these two waveguides becomes a necessity and one of these may be followed by the other. The transition between two different shapes is accomplished by a component called rectangular to circular transition. During this transition TE_{10} mode of rectangular waveguide transforms into TE_{11} mode in circular waveguide or vice versa.

This transformation can be justified if a rectangular waveguide is hammered until it takes the shape of circular waveguide. Its TE_{10} mode will get transformed into the TE_{11} mode of the circular waveguide. Similarly, if a circular waveguide is hammered to take the shape of the rectangular waveguide TE_{11} mode of the circular waveguide will transform to the TE_{10} mode in the rectangular waveguide. Figure 19.13 illustrates such a transition.

19.4 DIELECTRIC WAVEGUIDES

The waveguides can also be made of dielectric materials. These waveguides may have rectangular or circular shapes. Figure 19.14 illustrates some of the dielectric

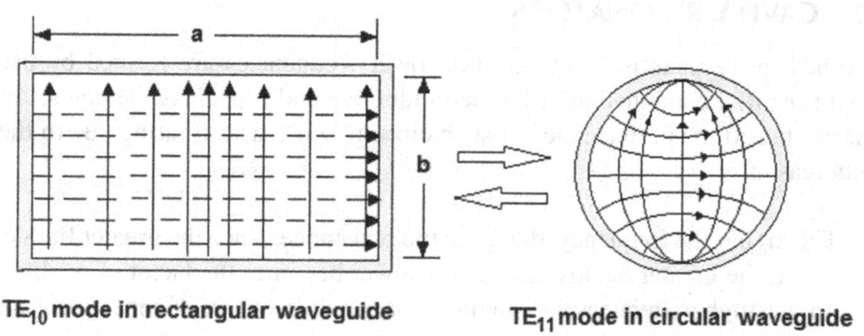

FIGURE 19.13 Transformation of modes.

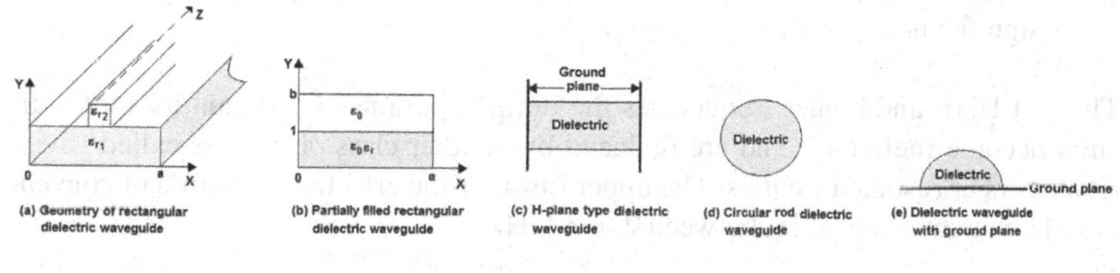

FIGURE 19.14 Dielectric waveguides of different shapes.

waveguides. These include dielectric slab waveguide, dielectric slab with ground plane and dielectric rod waveguide. As frequency is raised, the fabrication of ordinary waveguide becomes difficult. The dielectric waveguide provides a better option, particularly at millimetric wave range. The dielectric rod waveguide can be safely used as a low loss transmission line at these frequencies. The rod waveguides possess good propagation characteristics at 100 GHz and above. In these waveguides, the wave travels in a zigzag fashion and remains confined within the waveguide provided its angle of launch or angle of incidence (θ_i) is greater than the critical angle (θ_c) where the $\theta_c = \tan^{-1}\sqrt{\dfrac{1}{\mu_r\varepsilon_r}}$.

The wave in dielectric waveguide propagates in a similar way as it propagates in metallic waveguides. These waves propagate unattenuated by bouncing back and forth between two surfaces of a dielectric slab in accordance with the principle of total internal reflection. As before the wave phenomenon is governed by Maxwell's and wave equations. In this case, the wave is assumed to be progressing in the z direction and the time and space variations are represented by $e^{j\omega t}$ and $e^{-\gamma z}$. Since this waveguide is normally surrounded by air the wave equations are to be solved in two different regions referred to as outer (O or air) and inner (I or dielectric) regions. These two solutions are matched at the boundary between these two regions to yield values of arbitrary constants and other parameters involved. In these regions the propagation constant γ takes values γ_0 and γ_1 respectively. It is further assumed that there is no variation of field in the x direction; thus, all x derivatives vanish (i.e. $\dfrac{\partial}{\partial x} \equiv 0$). The variation of field along the remaining (y) axis is then evaluated.

19.5 CAVITY RESONATORS

The tuned (or resonant) circuits used at lower frequencies are formed by using a series or parallel combination of lumped inductive and capacitive elements. As frequency is raised the performance of such circuits starts deteriorating due to the following reasons:

- The higher the frequency, the more the resistance, since the greater the skin effect, the greater the losses; thereby lower becomes the factor of quality.
- To get higher frequencies, smaller and smaller values of inductance and capacitance are required. Beyond certain minimum limits, a further reduction in the values of these parameters become impractical.
- At higher frequencies, stray inductance and stray capacitance attain greater significance.

Thus, at UHF and higher frequencies the lumped parameter resonant (or tank) circuits become ineffective and are replaced by another class of devices called cavity resonators or resonant cavities. The upper limit for the effective operation of conventional resonant circuits lies between 2 to 3 GHz.

Cavity resonators are spaces that are completely enclosed by conducting walls. These are capable of containing oscillating electromagnetic fields. Such enclosures possess resonant properties and can resonate at more than one frequency. This is in contrast to the low frequency circuits which resonate at only one frequency.

In Section 18.7, it was noted that quarter-wave line-section can act as a resonant circuit. Since the basic principles of transmission lines and waveguides are the same, a quarter-wave hollow waveguide section can be considered as a resonant cavity with similar conditions, as applicable in the case of a quarter-wave line section.

The physical shape is one of the main factors which determine the resonance frequency of a cavity. A cavity may be carved-out of a waveguide, coaxial cable, and spherical structure. In spite of the variations in their physical structures all these cavities follow the same basic principle of operation. Figure 19.15 illustrates some of the commonly used cavities. These cavities may also be branded as absorption and transmission cavities. This branding is based on the capability of a cavity to transmit or absorb electromagnetic energy. In the absorption type there is maximum absorption, whereas in the transmission type there is maximum transmission of energy at resonant frequencies.

The size of the cavity is another key factor for governing the resonant frequency. As a thumb rule smaller is the cavity higher is its resonant frequency. A half-wave cavity resonates at the principal frequency (f) and its even multiples whereas a quarter-wave cavity exhibits resonance at f and its odd multiples. In the case of a spherical cavity, the first resonance frequency corresponds to the resonant wavelength (λ_r) of 2.28a, and the second to 1.4a, where a is the radius of the sphere.

19.5.1 Coaxial Cavity

This cavity is shown in Figure 19.15a. It is formed by shorting a coaxial line at its two ends and joining these at the center by a capacitor. This cavity supports an infinite number of resonant frequencies or modes of oscillation. Its resonance frequency is given by:

$$f_r = \frac{c}{2\pi\sqrt{\varepsilon_r}}\left[al\left\{\frac{a}{2d} - \frac{2}{l}\ln\frac{0.765}{\sqrt{l^2+(b-a)^2}}\right\}\ln\frac{b}{a}\right]^{-1/2} \tag{19.51}$$

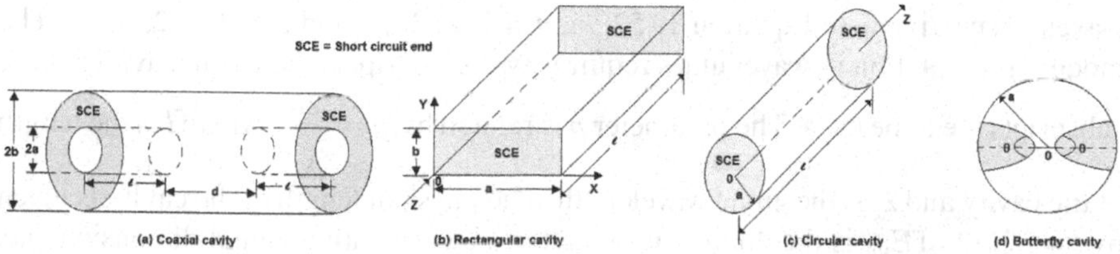

FIGURE 19.15 Some commonly used cavities.

where "a" and "b" are the radii of inner and outer conductors, "d" is the separation of two ends of the inner conductors forming the capacitance, "c" is the velocity of light in free space, "ε_r" is the relative permittivity, and l is the length of the coaxial cable. The maximum quality factor for a coaxial cavity is obtained when $b/a = 3.6$.

19.5.2 Rectangular Cavity

This cavity is shown in Figure 19.15b. It is formed by shorting both the ends of a half-wave section of a rectangular waveguide by conducting plates. The field components in this cavity can be obtained by adopting the same method as used in the case of a rectangular waveguide. These components are:

$$E_z = E_{0z}\sin\left(\frac{m\pi x}{a}\right)\sin\left(\frac{n\pi y}{b}\right)\cos\left(\frac{p\mu z}{l}\right) \quad \text{(for } TM_{mnp} \text{ mode)} \tag{19.52a}$$

$$H_z = H_{0z}\cos\left(\frac{m\pi x}{a}\right)\cos\left(\frac{n\pi y}{b}\right)\sin\left(\frac{p\mu z}{l}\right) \quad \text{(for } TE_{mnp} \text{ mode)} \tag{19.52b}$$

The phase constant for this case can be given as:

$$\beta^2 = \omega^2\mu\varepsilon - h_{nmp}^2 \quad \text{(for both } TM_{mnp} \text{ and } TE_{mnp} \text{ modes)} \tag{19.52c}$$

where

$$h_{mnp} = \sqrt{\left(\frac{m\pi}{a}\right)^2 + \left(\frac{n\pi}{b}\right)^2 + \left(\frac{p\pi}{l}\right)^2} \tag{19.52d}$$

The resonant frequency of this cavity (for TM_{mnp} and TE_{mnp} modes) is given as:

$$f_r = \frac{1}{2\sqrt{\mu\varepsilon}}\sqrt{\left(\frac{m}{a}\right)^2 + \left(\frac{n}{b}\right)^2 + \left(\frac{p}{l}\right)^2} \tag{19.52e}$$

In these relations, "a", "b", and "l" are the lengths of broader wall, narrower wall, and the length of waveguide, respectively. In a cavity, the modes TM_{mnp} and TE_{mnp} are described in terms of the fields along the x-, y-, and z-axes. Thus, the subscripts "m", "n", and "p" indicate the number of half wavelengths along the x-, y-, and z-axes, respectively. In Equation 19.52, $m, n = 1, 2, 3, \ldots$ and $p = 0, 1, 2, 3, \ldots$. The mode representation in waveguides requires two subscripts whereas in cavities three subscripts are to be used. The parameter p is related by $p = \dfrac{2l}{\lambda_g\, l}$, where l is the length of the cavity and λ_g is the guide wavelength. The physical length of the cavity is given by $l = p\lambda_g/2$. TE_{101} is the dominant mode for a cavity with unequal dimensions i.e. for $a > b > 1$.

19.5.2.1 Mode Degeneracy

If there are many modes that have the same resonant frequency, the situation is called mode degeneracy.

If $a \neq b \neq 1$, there will be twofold degeneracy.
If a = b, $\neq 1$ or $a = 1, \neq b$ or $b = 1, \neq a$, there will be fourfold degeneracy.
If $a = b = 1$, there will be twelvefold degeneracy (six each for the *TE* and *TM* mode).

The slightest manufacturing defect can cause the separation and degeneration of modes.

19.5.3 Cubic Cavity

When the lengths a, b, and l of a rectangular cavity becomes equal the cavity takes the form of a cubic cavity. The resonant frequency for this cavity obtained by substituting $a = b = 1$ in Equation 19.52e becomes:

$$f_r = \frac{1}{2a\sqrt{\mu\varepsilon}}\sqrt{(m)^2 + (n)^2 + (p)^2} \tag{19.53a}$$

From Equation 19.53a, the expression of resonant length for TE_{101} mode (i.e. $m = p = 1$, $n = 0$) can be obtained as:

$$\lambda_r = \sqrt{2}a \tag{19.53b}$$

The approximate number of finite resonances (N) from a practicable limit can be given in terms of volume of cavity (V) and the minimum wavelength (λ_m) by the following relation:

$$N = \frac{8\pi V}{3\lambda_m^3} \tag{19.54}$$

The maximum quality factor for a rectangular cavity is obtained when $a = b = 1$.

19.5.4 Circular Cavity

This cavity is shown in Figure 19.15c. It is formed by shorting the two ends of a circular waveguide by metal plates. In this cavity, the field components for the *TE* and *TM* waves are: TM_{mnp} and TE_{mnp}

$$H_z = H_{0z} J_n'\left(h_{mn}' r\right)\cos(n\phi)\sin(p\pi z/l) \quad \text{(for } TE_{mnp} \text{ mode)} \tag{19.55a}$$

$$E_z = E_{0z} J_n\left(h_{mn} r\right)\cos(n\phi)\sin(p\pi z/l) \quad \text{(for } TM_{mnp} \text{ mode)} \tag{19.55b}$$

The separation equations for the TE and TM modes are:

$$h^2 = \left(h'_{nm}\right)^2 + \left(q\pi / l\right)^2 \quad \text{(for } TE \text{ mode)} \tag{19.55c}$$

$$h^2 = \left(h_{nm}\right)^2 + \left(p\pi / l\right)^2 \quad \text{(for TM mode)} \tag{19.55d}$$

The resonant frequency for such a cavity is given by:

$$f_r = \frac{1}{2\sqrt{\mu\varepsilon}}\sqrt{\left(h'_{nm}\right)^2 + \left(q\pi / l\right)^2} \quad \text{(for } TE_{mnq} \text{ mode)} \tag{19.55e}$$

$$f_r = \frac{1}{2\sqrt{\mu\varepsilon}}\sqrt{\left(h_{nm}\right)^2 + \left(p\pi / l\right)^2} \quad \text{(for } TM_{mnp} \text{ mode)} \tag{19.55f}$$

where

$$h'_{nm} = \frac{(ha)'_{nm}}{a} \tag{19.55g}$$

and

$$h_{nm} = \frac{(ha)_{nm}}{a} \tag{19.55h}$$

In these relations, "a" is the radius and "l" is the length of a circular cavity, and n = 0,1,2, ... m = 1,2, ..., q = 1,2, ... and p = 0,1,2, ... In these equations, it can be noted that TM_{110} is the dominant mode for $2a < l$ and TE_{111} mode is dominant when $l \geq 2a$. The maximum quality factor for a circular cavity is obtained when $2a = l$.

19.5.5 Semicircular Cavity

It is formed by shorting the two ends of a semicircular waveguide of radius "a" and length "l".

The expressions for resonant frequencies for the *TE* and *TM* modes for circular and semicircular cavities are the same. The values of n, m, and p, however, differ from those for a circular cavity. Also, TE_{111} mode is dominant if $l > a$, and TM_{110} mode is dominant for $l < a$.

19.5.6 Reentrant Cavity

It is one of the most useful forms of cavities. It is used in klystrons and other microwave tubes. As shown in Figure 19.15d in this cavity the metallic boundaries are extended into its interior. It helps in the reduction of capacitance. It yields a maximum quality factor when $\theta = 33.5°$, where θ is the angle of a reentrant cone.

19.5.7 Dielectric Resonators

Like dielectric waveguides resonators are also made of dielectrics. These are called dielectric resonators or dielectric resonator oscillators (DROs). A DRO is an electronic component which exhibits resonant property. It traps energy in an extremely narrow frequency band. The resonance of this resonator closely resembles the resonance of a circular hollow metallic waveguide. These have negligible radiation losses.

These differ from the metallic resonators in terms of their boundary conditions. In a dielectric resonator the boundary is defined by a large change in permittivity, whereas in a circular waveguide it is defined by a conductor. These generally consist of a "puck" of ceramic that has a large permittivity and a low dissipation factor. Its resonance frequency is mainly governed by the physical dimensions of the puck and the dielectric constant of the material. It can be loaded by using probe (for an E field) or loop (for an H field) coupling methods. These also differ from metallic cavities in terms of field distributions. A metallic cavity has zero E and H fields outside the walls, whereas in dielectric resonators these are not zero. However, in dielectric resonators, E and H fields quickly decay away from the resonator walls. At a given resonant frequency these resonators with sufficiently high permittivity store most of the energy. These exhibit extremely high Q, comparable to a metal cavity.

Dielectric resonators are used in filters, combiners, frequency-selective limiters, etc. These are commonly employed in satellite communications, GPS antennas, and wireless internet.

Example 19.1

A TE_{10} wave propagates in a rectangular waveguide with cross-section $a = 4$ cm and $b = 2$ cm. Find the signal frequency if the distance d between the adjacent minima and maxima of the propagating wave is 5 cm.

SOLUTION

Given mode TE_{10}, $m = 1$; $n = 0$; $a = 0.04$ m, $b = 0.02$ m, $\upsilon_0 = 3\times10^8$ m/s and $d =$ 0.05 m, thus $\lambda = 4d = 0.2$ m.

From Equation 19.22f:

$$\lambda_c = \frac{2}{\sqrt{(m/a)^2 + (n/b)^2}}$$

$$\lambda_c = \frac{2}{\sqrt{(1/0.04)^2 + (0/0.02)^2}} = 0.08 \text{ m}$$

From Equation 19.22h:

$$\lambda_0 = \frac{\lambda\lambda_c}{\sqrt{(\lambda_c^2 + \lambda^2)}}$$

$$\lambda_0 = \frac{0.2 \times 0.08}{\sqrt{(0.08^2 + 0.2^2)}} = \frac{0.016}{0.2154} = 0.07428\ \text{m}$$

$$\text{The signal frequency } f_0 = \frac{\upsilon_0}{\lambda_0} = \frac{3 \times 10^8}{0.07428} = 4.03887\ \text{GHz}$$

Example 19.2

A rectangular waveguide with air dielectric and dimensions $a = 4$ cm and $b = 2$ cm carrying a TE_{10} mode signal of 5 GHz. Compute (a) cutoff frequency f_c; (b) cutoff wavelength λ_c; (c) phase velocity υ_p; (d) group velocity υ_g; (e) guide wavelength λ_g; (f) phase shift constant β_g; and (f) waveguide impedance Z_g.

SOLUTION

Given TE_{10} mode, $m = 1$, $n = 0$, $a = 0.04$ m, $b = 0.02$ m, and $f = 5 \times 10^9$ Hz, $\omega = 2\pi f = \pi \times 10^{10}$, $\mu_0 = 4\pi \times 10^{-7}$, $\varepsilon_0 = \frac{10^{-9}}{36\pi}$, $\lambda = \frac{c}{f} = \frac{3 \times 10^8}{5 \times 10^9} = 0.06$ m,

$$\upsilon_0 = \frac{1}{\sqrt{\mu_0 \varepsilon_0}} = 3 \times 10^8, \quad \frac{1}{2\sqrt{\mu\varepsilon}} = \frac{3 \times 10^8}{2}, \quad \eta = \sqrt{\frac{\mu_0}{\varepsilon_0}} = 120\pi,$$

$$\beta = \frac{\omega}{\upsilon_0} = \frac{\pi \times 10^{10}}{3 \times 10^8} = 104.72, \text{ and } \sqrt{\left\{\left(\frac{m}{a}\right)^2 + \left(\frac{n}{b}\right)^2\right\}} = \sqrt{\left\{\left(\frac{1}{0.04}\right)^2 + \left(\frac{0}{0.02}\right)^2\right\}} = 25$$

(a) From Equation 19.22c:

$$f_c = \frac{1}{2\sqrt{\mu\varepsilon}} \sqrt{\left\{\left(\frac{m}{a}\right)^2 + \left(\frac{n}{b}\right)^2\right\}} = \frac{3 \times 10^8}{2} 25 = 3.75 \times 10^9$$

(b) From Equation 19.22d:

$$\lambda_c = \frac{2}{\sqrt{(m/a)^2 + (n/b)^2}} = \frac{2}{25} = 0.08$$

$$\sqrt{1 - (f_c/f)^2} = \sqrt{1 - (3.75 \times 10^9 / 5 \times 10^9)^2} = 0.66144$$

$$\frac{1}{\sqrt{1 - (f_c/f)^2}} = 1.51186$$

(c) From Equation 19.22e:

$$\upsilon_p = \frac{c}{\sqrt{1 - (f_c/f)^2}} = \frac{3 \times 10^8}{0.66144} = 4.53557$$

(d) From Equation 19.22f:

$$\upsilon_g = c\sqrt{1-(f_c/f)^2} = 3\times10^8\times0.66144 = 1.58342\times10^8$$

(e) From Equation 19.22g:

$$\lambda_g = \frac{\lambda}{\sqrt{1-(f_c/f)^2}} = 0.06\times1.51186 = 0.0907\text{ m}$$

(f) From Equation 19.22g:

$$\beta_g = \beta\sqrt{1-(f_c/f)^2} = 104.72\times0.66144 = 69.265\text{ rad/s}$$

(g) From Equation 19.26b:

$$Z_g = Z_z(TE) = \frac{\eta_0}{\sqrt{1-(f_c/f)^2}} = \frac{120\pi}{0.66143} = 569.96$$

Example 19.3

A 6 cm × 4 cm rectangular waveguide operates at 5 MHz in TM_{11} mode. Compute (a) cutoff frequency f_c; (b) guide wavelength λ_g; (c) phase constant β_g; (d) phase velocity υ_p; (e) group velocity υ_g; and (f) impedance Z_g if the waveguide is filled with a dielectric having $\mu_r = 1$ and $\varepsilon_r = 2.1$.

SOLUTION

Given TM_{11} mode, $m = 1$, $n = 1$, $a = 0.06$ m, $b = 0.04$ m, $f = 5\times10^9$ Hz, $\omega = \pi\times10^{10}$, $\mu = \mu_0 = 4\pi\times10^{-7}$, $\varepsilon_0 = \dfrac{10^{-9}}{36\pi}$, $\varepsilon = \varepsilon_0\varepsilon_r = \dfrac{10^{-9}}{36\pi}\times2.1$

$\upsilon_0 = \dfrac{1}{\sqrt{\mu_0\varepsilon_0}} = 3\times10^8\ \dfrac{\text{m}}{\text{s}}$, $\eta_0 = 120\pi$, $\upsilon = \dfrac{1}{\sqrt{\mu\varepsilon}} = \dfrac{\upsilon_0}{\sqrt{2.1}} = 2.07\times10^8$ and

$\dfrac{\upsilon}{2} = 1.035\times10^8$,

$$\sqrt{\left(\frac{m}{a}\right)^2+\left(\frac{n}{b}\right)^2} = \sqrt{\left(\frac{1}{0.06}\right)^2+\left(\frac{1}{0.04}\right)^2} = 30.046$$

From Equation 19.22c:

(a) $$f_c = \frac{\upsilon}{2}\sqrt{\left(\frac{m}{a}\right)^2+\left(\frac{n}{b}\right)^2} = 1.035\times10^8\times30.046 = 31.1\times10^8\text{ m/s}$$

$$\sqrt{1-(f_c/f)^2} = \sqrt{1-\left(31.1\times10^8/5\times10^9\right)^2} = 0.783$$

(b) $\lambda_g = \dfrac{\upsilon / f}{\sqrt{1-(f_c/f)^2}} = \dfrac{2.07\times10^8 / 5\times10^9}{0.783} = 0.05287 \text{ m}$

(c) $\beta_g = \dfrac{\omega}{\upsilon}\sqrt{1-(f_c/f)^2} = \dfrac{\pi\times10^{10}}{2.07\times10^8}0.783 = 118.837 \text{ rad/m}$

(d) $\upsilon_p = \dfrac{\upsilon}{\sqrt{1-(f_c/f)^2}} = \dfrac{2.07\times10^8}{0.783} = \dfrac{2.07\times10^8}{0.783} = 2.644\times10^8 \text{ m/s}$

(e) $\upsilon_g = \upsilon\sqrt{1-(f_c/f)^2} = 2.07\times10^8\times0.783 = 1.621\times10^8 \text{ m/s}$

(f) $Z_g = \eta\sqrt{1-(f_c/f)^2} = \dfrac{120\pi}{\sqrt{2.1}}\times0.783 = 203.7\ \Omega$

Example 19.4

An air-filled hollow rectangular waveguide carrying a 20 GHz signal has $a = 5$ cm and $b = 2.5$ cm. Calculate the wave impedance for (a) TE_{11}; (b) TM_{11} mode.

SOLUTION

Given TE_{11} and TM_{11} modes, $m = 1$; $n = 1$; $a = 0.05$ m and $b = 0.025$ m; $f = 20 \times 10^9$ Hz for air dielectric, $\upsilon_0 = 3 \times 10^8$ m/s; $\eta_0 = 377\ \Omega$:

$$f_c = \frac{\upsilon}{2}\sqrt{\left(\frac{m}{a}\right)^2+\left(\frac{n}{b}\right)^2} = 1.5\times10^8\sqrt{\left(\frac{1}{0.05}\right)^2+\left(\frac{1}{0.025}\right)^2} = 67.08\times10^8 \text{ m/s}$$

$$\frac{f_c}{f} = \frac{67.08\times10^8}{20\times10^9} = 0.3354 \qquad \left(\frac{f_c}{f}\right)^2 = (0.3354)^2 = 0.1125$$

$$\sqrt{1-(f_c/f)^2} = \sqrt{1-0.1125} = 0.942$$

In view of Equations 19.26a and 19.26b:

(a) $Z(TE) = \dfrac{\eta_0}{\sqrt{1-\left(\dfrac{f_c}{f}\right)^2}} = \dfrac{120\pi}{0.942} = 400.2\ \Omega$

(b) $Z(TM) = \eta_0\sqrt{1-\left(\dfrac{f_c}{f}\right)^2} = 120\pi\times0.942 = 355.125\ \Omega$

Example 19.5

Calculate the values of inductive and capacitive reactance of equivalent circuits for (a) TE_{11}; and (b) TM_{11} of 10 GHz wave propagating in an air-filled hollow rectangular waveguide with $a = 5$ cm and $b = 2.5$ cm. Also calculate corresponding $Z_0(TM)$ and $Z_0(TE)$.

SOLUTION

Given, $m = 1$; $n = 1$; $a = 0.05$ m and $b = 0.025$ m; $f = 10 \times 10^9$ Hz
For air dielectric $\upsilon_0 = 3 \times 10^8$ m/s; $\eta_0 = 377\ \Omega$, $\omega = 2\pi \times 10^{10}$. From Equation 19.18a:

$$h = \sqrt{\left(\frac{m\pi}{a}\right)^2 + \left(\frac{n\pi}{b}\right)^2} = \sqrt{\left(\frac{\pi}{0.05}\right)^2 + \left(\frac{\pi}{0.025}\right)^2} = \sqrt{2000\pi^2} = 140.5$$

From Figure 19.8:

$$\text{Series inductive reactance} = \omega\mu_0 = 2\pi \times 10^{10} \times 4\pi \times 10^{-7} = 7.895 \times 10^4\ \Omega$$

$$\text{Shunt capacitive reactance} = \omega\varepsilon_0 = 2\pi \times 10^{10} \times \frac{10^{-9}}{36\pi} = \frac{5}{9} = 0.5555\ \Omega$$

$$\text{Series capacitive reactance} = \frac{h^2}{\omega\varepsilon_0} = \frac{(140.5)^2}{0.5555} = \frac{19739}{0.5555} = 3.5532 \times 10^4\ \Omega$$

$$\text{Shunt inductive reactance} = \frac{h^2}{\omega\mu_0} = \frac{(140.5)^2}{78.95 \times 10^3} = \frac{19739}{78.95 \times 10^3} = 0.25\ \Omega$$

From Equation 19.26, $Z_z(TM)$ and $Z_z(TE)$ for air-filled waveguide are:

$$Z_z(TM) = \sqrt{\frac{j\omega\mu_0 + \left(\frac{h^2}{j\omega\varepsilon_0}\right)}{j\omega\varepsilon_0}} = \sqrt{\frac{j7.895 \times 10^4 - j\left(3.5532 \times 10^4\right)}{j0.5555}} = 279.57\ \Omega$$

$$Z_z(TE) = \frac{j\omega\mu_0}{\sqrt{j\omega\varepsilon_0 + \left(\frac{h^2}{j\omega\mu_0}\right)}} = \sqrt{\frac{j7.895 \times 10^4}{j0.5555 - j0.25}} = 508\ \Omega$$

Example 19.6

An air-filled circular waveguide carries a TE_{11} mode at 5 GHz. Its cutoff frequency is 80% of the operating frequency. Calculate its (a) diameter; (b) phase velocity; (c) guide wavelength; (d) guide phase shift; and (e) wave impedance.

SOLUTION

Given $m = 1$; $n = 1$; $\upsilon_0 = \frac{1}{\sqrt{\mu_0\varepsilon_0}} = 3 \times 10^8$, $f = 5 \times 10^9$ Hz, $f_c = 0.8 f_0 = 4 \times 10^9$ Hz

$$\lambda = \frac{\upsilon_0}{f} = \frac{3 \times 10^8}{5 \times 10^9} = 0.06 \text{ m},\ \beta = \frac{\omega}{\upsilon_0} = \frac{\pi \times 10^{10}}{3 \times 10^8} = 104.72,\ \eta = \eta_0 = 377$$

and from Table 19.1 $(ha)'_{11} = 1.841$

(a) In view of Equation 20.48b radius $a = \frac{(ha)'_{11}}{2\pi f_c} \upsilon_0 = \frac{1.841}{2\pi \times 4 \times 10^9} \times 3 \times 10^8 =$

0.02197, diameter = 2a = 0.044 m

$$\sqrt{1-(f_c/f)^2} = \sqrt{1-(0.8)^2} = 0.6$$

(b) In view of Equation 20.22d $\upsilon_p = \frac{c}{\sqrt{1-(f_c/f)^2}} = \frac{3 \times 10^8}{0.6} = 5 \times 10^8$ m/s

(c) In view of Equation 20.22g $\lambda_g = \frac{\lambda}{\sqrt{1-(f_c/f)^2}} = \frac{0.06}{0.6} = 0.1$

(d) In view of Equation 20.22h $\beta_g = \beta\sqrt{1-(f_c/f)^2} = 104.72 \times 0.6 = 62.832$

(e) In view of Equation 20.26b $Z_z(TE) = \frac{\eta}{\sqrt{1-\left(\frac{f_c}{f}\right)^2}} = \frac{377}{0.6} = 628.333$

Example 19.7

A circular waveguide is operating in the dominant mode. Its cutoff frequency is 10 GHz Calculate the inner diameter of the guide if it is filled with a dielectric having $\mu_r = 1$ and $\varepsilon_r = 2.5$.

SOLUTION

Given dominant mode TE_{11}, thus $m = 1$; $n = 1$; $f_c = 10 \times 10^9$ Hz, and from Table 19.1: $(ha)'_{11} = 1.841$

Also given $\mu_r = 1$, $\varepsilon_r = 2.5$, thus $\mu = \mu_0 = 4\pi \times 10^{-7}$ and $\varepsilon = 2.5\varepsilon_0 = 2.5 \times \frac{10^{-9}}{36\pi}$,

$\upsilon = \frac{\upsilon_0}{\sqrt{2.5}} = 1.897 \times 10^8$

$$d = \frac{2\upsilon \times (ha)'_{11}}{2\pi f_c} = \frac{2 \times 1.897 \times 10^8 \times 1.841}{2\pi \times 10^{10}} = 0.01111\text{m}$$

Example 19.8

The cutoff frequency for an air-filled circular waveguide of 5 cm diameter is 6 GHz. Identify the modes that it can carry.

SOLUTION

Given a = 0.025 m, $f_c = 6 \times 10^9$ Hz, thus $\omega_c = 12\pi \times 10^9$. For air $\mu = \mu_0$, $\varepsilon = \varepsilon_0$, thus $\upsilon_0 = 3 \times 10^8$.

$$\omega_c\sqrt{\mu\varepsilon} = \frac{\omega_c}{\upsilon_0} = \frac{12\pi\times10^9}{3\times10^8} = 125.664 \quad \omega_c\sqrt{\mu\varepsilon}\,a = 125.664\times0.025 = 3.1416$$

In Equation 19.48, for a mode to propagate through waveguide, h_{nm} (for TM wave) and h'_{nm} (for TE wave) or the corresponding zero of $J_n(x)$ or $J'_n(x)$ has to be less than $\omega_c\sqrt{\mu\varepsilon}\,a(=3.1416)$

Zeroes of $J_n(x)$ and $J'_n(x)$ have to be less than $\omega_c\sqrt{\mu\varepsilon}\,a$ or 3.1416.

From Table 19.1, the values of such zeros are $(ha)_{01} = 2.405$ and $(ha)'_{11} = 1.841$, $(ha)'_{21} = 3.054$. Thus only TM_{01} and TE_{11}, TE_{21} modes can propagate in the given waveguide.

Example 19.9

Calculate (a) cutoff wave number; and (b) resonant frequency for an air-filled rectangular cavity operating in dominant TE_{101} mode. Its dimensions are given as: width a = 6 cm, height b = 4 cm, and length l = 8 cm.

SOLUTION

Given TE_{101} mode, $m = 1$, $n = 0$, $p = 1$, $a = 0.06$ m, $b = 0.04$ m, $l = 0.08$ m, $\mu = \mu_0 = 4\pi\times10^{-7}$, $\varepsilon = \varepsilon_0 = \frac{10^{-9}}{36\pi}$, $\upsilon_0 = \frac{1}{\sqrt{\mu_0\varepsilon_0}} = 3\times10^8\ \frac{\text{m}}{\text{s}}$

(a) In view of Equation 19.52d:

$$h_{mnp} = \sqrt{\left(\frac{m\pi}{a}\right)^2 + \left(\frac{n\pi}{b}\right)^2 + \left(\frac{p\pi}{l}\right)^2} = \pi\sqrt{\left(\frac{1}{0.06}\right)^2 + \left(\frac{1}{0.08}\right)^2} = 65.45$$

(b) In view of Equation 19.52e:

$$f_r = \frac{1}{2\sqrt{\mu\varepsilon}}\sqrt{\left(\frac{m}{a}\right)^2 + \left(\frac{n}{b}\right)^2 + \left(\frac{p}{l}\right)^2} = \frac{\upsilon_0}{2}\frac{h}{\pi} = \frac{3\times10^8}{2}\frac{65.45}{\pi}$$

$$= 1.5\times10^8\times20.8333 = 3.125\times10^9\ \text{Hz}$$

Example 19.10

An air-filled rectangular cavity with width a = 6 cm and height b = 4 cm operates in dominant TE_{101} mode at 12 GHz resonant frequency. Calculate its length.

SOLUTION

Given TE_{101} mode, $m = 1$, $n = 0$, $p = 1$, $a = 0.06$ m, $b = 0.04$ m, $\mu = \mu_0 = 4\pi\times10^{-7}$, $\varepsilon = \varepsilon_0 = \frac{10^{-9}}{36\pi}$, $\upsilon_0 = \frac{1}{\sqrt{\mu_0\varepsilon_0}} = 3\times10^8\ \frac{\text{m}}{\text{s}}$, $f_r = 12\times10^9$

Since $f_r = \frac{\upsilon_0}{2}\frac{h}{\pi}$, $h = \frac{2\pi f_r}{\upsilon_0} = \frac{2\pi \times 12 \times 10^9}{3 \times 10^8} = 80\pi = 251.327$, $h^2 = 63165$.

$$h = \sqrt{\left(\frac{m\pi}{a}\right)^2 + \left(\frac{n\pi}{b}\right)^2 + \left(\frac{p\pi}{l}\right)^2} = \sqrt{\left(\frac{\pi}{0.06}\right)^2 + \left(\frac{\pi}{l}\right)^2}$$

$$h^2 = 63165 = \left(\frac{\pi}{0.06}\right)^2 + \left(\frac{\pi}{l}\right)^2 = 2741.57 + \left(\frac{\pi}{l}\right)^2$$

$$\frac{\pi}{l} = \sqrt{63165 - 2741.57} = 245.811$$

$$l = \frac{\pi}{245.811} = 0.01284\ \text{m}$$

Example 19.11

Find the resonant frequency of an air-filled circular cavity operating in TE_{111} mode. It has 8 cm length and 5 cm radius.

SOLUTION

Given $a = 0.05$ m, $l = 0.08$ m, $m = 1$, $n = 1$, $q = 1$, $\upsilon_0 = 3 \times 10^8$ m/s.

From Table 19.1 $h'_{11} = \frac{1.841}{0.05} = 36.82$.

In view of Equation 20.55e:

$$f_r = \frac{\upsilon_0}{2}\sqrt{(h'_{nm})^2 + (q\pi/l)^2} = 1.5 \times 10^8 \sqrt{(36.82)^2 + (\pi/0.08)^2} = 8.07\ \text{GHz}$$

Example 19.12

Find the length of an air-filled circular cavity of 3 cm radius operating at $f = 10$ GHz in TE_{111} mode.

SOLUTION

Given $a = 0.03$ m and $f = 10$ GHz, $m = 1$, $n = 1$, $p = 1$, $\upsilon_0 = 3 \times 10^8$ m/s. In Table 19.1 and Equation 20.48b $h'_{11} = \frac{1.841}{0.03} = 61.366$.

In view of Equation 20.55e:

$$f_r = \frac{1}{2\sqrt{\mu\varepsilon}}\sqrt{(h'_{11})^2 + (p\pi/l)^2} = 1.5 \times 10^8 \sqrt{(61.366)^2 + (\pi/l)^2} = 10\ \text{GHz}$$

$$\sqrt{(61.366)^2 + (\pi/l)^2} = \frac{10^{10}}{1.5 \times 10^8} = 66.667 \quad (\pi/l)^2 = 4444.44 - 3765.78$$

$$l = \frac{\pi}{\sqrt{4444.44 - 3765.78}} = 0.121\ \text{m}$$

PROBLEMS

P19.1 A rectangular waveguide with dimensions $a = 2$ cm and $b = 1$ cm carries a TE_{10} wave. Find the signal frequency if the adjacent minima and maxima of the propagating wave are separated by a distance of 10 cm.

P19.2 A rectangular waveguide with dimensions $a = 6$ cm and $b = 3$ cm carries a TE_{01} mode signal of 10 GHz. Find (a) cutoff frequency f_c; (b) cutoff wavelength λ_g; (c) phase velocity v_p; (d) group velocity v_g; (e) guide wavelength λ_g; (f) phase shift constant β_g; and (f) waveguide impedance Z_g if $\mu_r = 1$ and $\varepsilon_r = 2.6$.

P19.3 The magnetic field intensity for a TE_{11} wave of 6 GHz traveling in z-direction in an air-filled rectangular waveguide having $\beta_g = 1.1$ rad/m is given as $H_z = H_{z0}\cos\left(\frac{\pi x}{0.06}\right)\cos\left(\frac{\pi y}{0.04}\right)$ A/m. Find its (a) cutoff frequency f_c; (b) guide wavelength λ_g; (c) phase velocity v_p; and (d) group velocity v_g. Also write the surviving E field components.

P19.4 Find $Z_0(TM)$ and $Z_0(TE)$ for (a) TE_{11}; (b) TM_{11} wave propagating in hollow rectangular waveguide with dimensions $a = 8$ cm and $b = 4$ cm at 10 GHz. The waveguide is filled with a dielectric having $\mu = \mu_0$ and $\varepsilon_r = 3.6$.

P19.5 Find the inductive and capacitive reactance of equivalent circuits for (a) TE_{11}; and (b) TM_{11} of 5 GHz wave propagating in a hollow rectangular waveguide with $a = 6$ cm and $b = 3$ cm. Also calculate corresponding $Z_0(TM)$ and $Z_0(TE)$. The waveguide is filled with a material having $\mu = \mu_0$ and $\varepsilon_r = 2.6$.

P19.6 An air-filled circular waveguide carries a TM_{11} mode at 10 GHz. Its cutoff frequency is 90% of the operating frequency. Calculate its (a) diameter; (b) phase velocity; (c) group velocity; (d) wave impedance.

P19.7 A circular waveguide carries TE_{01} mode. Its cutoff frequency is 5 GHz. Calculate its inner diameter if it is filled with air.

P19.8 Identify the modes carried by a circular waveguide of 6 cm diameter at 5 GHz cutoff frequency if it is filled by a material with $\mu = \mu_0$ and $\varepsilon_r = 2.1$.

P19.9 Calculate (a) cutoff wave number; and (b) resonant frequency for a rectangular cavity operating in TM_{111} mode. The cavity is filled with a material of $\mu_r = 1$ and $\varepsilon_r = 2.6$ and having width $a = 8$ cm, height $b = 4$ cm, and length $l = 10$ cm.

P19.10 Calculate the length of an air-filled rectangular cavity with width $a = 4$ cm and height $b = 2$ cm and operating in dominant TE_{111} mode at 10 GHz resonant frequency.

P19.11 Calculate the resonant frequency of an air-filled circular cavity operating in TM_{111} mode, if its length is 10 cm and radius 6 cm.

P19.12 Find the length of an air-filled circular cavity of 8 cm radius is operating at $f = 9$ GHz in TM_{111} mode.

DESCRIPTIVE QUESTIONS

Q19.1 Explain the meaning of TE, TM, and TEM waves. Write the wave equation and the boundary conditions to be satisfied by a field E in a hollow rectangular waveguide.

Q19.2 Discuss the behavior of an electromagnetic wave with the frequency variation in terms of its attenuation constant, phase constant, and cutoff frequency.

Q19.3 Explain the meaning of dispersion, phase velocity, group velocity, cutoff wavelength, guide wavelength, and mode.

Q19.4 Write the lowest possible TE and TM modes. Which of these is called the dominant mode and why?

Q19.5 Illustrate TE_{11}, TE_{21}, TM_{11}, and TM_{12} modes across the waveguide cross-section.

Q19.6 Illustrate the excitation of TE_{10}, TE_{02}, TM_{21}, and TM_{22} in rectangular waveguide.

Q19.7 Explain the designation of modes in a circular waveguide.

Q19.8 Discuss the mode transformation between rectangular and circular waveguides.

Q19.9 Illustrate the dielectric waveguides of different shapes.

Q19.10 Discuss cavity resonators in terms of their requirements and shapes.

Q19.11 How do the modes in cavity resonators differ from those of the waveguides?

FURTHER READING

A. S. Khan, *Microwave Engineering: Concepts and Fundamentals*, CRC Press, New York, Mar. 2014.

A. Das, S. K. Das, *Microwave Engineering*, Tata McGraw Hill, New Delhi, 2002.

Peter A. Rizzi, *Microwave Engineering – Passive Circuits*, Prentice Hall, New Delhi, 2001.

David M. Pozar, *Microwave Engineering*, 2nd Ed, John Wiley & Sons, New Delhi, 1999.

S.Y. Liao, *Microwave Devices & Circuits*, 3rd Ed, Prentice Hall of India, New Delhi, 1995.

R. Chatterjee, *Microwave Engineering*, East West Press, New Delhi, 1990.

E. C. Jordon, K. G. Balmain, *Electromagnetic Waves and Radiating Systems*, Prentice Hall Ltd., New Delhi, 1987.

O. P. Gandhi, *Microwave Engineering and Applications*, Pergamon Press, New York, 1981.

K. C. Gupta, *Microwaves*, Wiley Eastern, New Delhi, 1979.

H. J. Reich, et al. *Microwaves*, Affiliated East West Press, 1976.

T. Saad, and R. C. Hansen, *Microwave Engineer's Handbook* Vol. 1, Artech House, Dedham, MA, 1971.

S. F. Adam, *Microwave Theory and Applications*, Prentice Hall, Inc., 1969.

A. L. Lance, *Introduction to Microwave Theory and Measurements*, McGraw Hill, 1969.

R. E. Collins, *Foundation for Microwave Engineering*, McGraw-Hill, New York, 1966.

G. J. Wheeler, *Introduction to Microwaves*, Prentice Hall, Upper Saddle River, NJ, 1963.

20 Radiation Mechanism

20.1 INTRODUCTION

The radiation of electromagnetic energy is one of the most important aspects of field phenomena. This phenomenon is the soul of radar, satellite, mobile, line of sight, and troposcatter communication, and numerous other wireless systems. No wireless service can be conceived in the absence of radiation. It is, therefore, necessary to fully understand both, its qualitative and quantitative aspects. The radiation process too revolves around the basic Maxwell's equations, which were thoroughly discussed in Chapter 15. This chapter is devoted to understanding the mechanism involved in the radiation of electromagnetic energy.

20.2 RADIATION PHENOMENON

In Chapter 18 it was stated that a wave traveling along a transmission line in the forward direction gets reflected back from its receiving end if it is not terminated in its characteristic impedance. Thereat it was further observed that more is the difference between the characteristic and terminating impedances more is the reflection. In the two extreme cases wherein the terminating end of the line is either short-circuited or open-circuited there is total reflection of electromagnetic energy.

20.2.1 Process of Escape of Energy

An open-circuited line with total reflection and the formation of standing waves was shown in Figure 18.5. There the assumption of total reflection was taken to be true. But since the reflected energy has to travel in the backward direction it has to change its direction. During the course of this direction reversal, the energy takes some time as it possesses some moment of inertia like property. Though the velocity of the wave, in free space, is of the order of velocity of light, and the required time of reversal is quite small, some of the energy may still escape into the space. This escape is referred to as radiation.

Figure 20.1 shows a two parallel wire transmission line with alternation of traveling E field. At OC end theoretically there must be perfect reflection and energy must travel in the backward direction. But since energy possesses some moment of inertia-like property, it will take some non-zero time to flow back. Meanwhile a fraction of energy may escape into spacc. Besides, at the open end a sort of fringing phenomenon begins and the E-lines start bending (or bulging) outward. Due to this bending, the reversal of direction may require a bit more time than if these lines, at end, were straight. This bending results in more energy escaping into the space.

Figure 20.1b shows that if the end of the line is flared outward, the E-lines start bending outward and assume a curved shape. Thus the direction reversal of energy flow will take a longer time, which will result in more energy escaping. Figure 20.1c

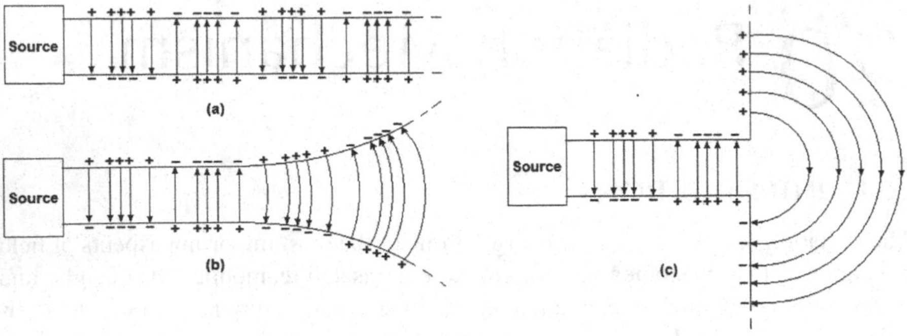

FIGURE 20.1 E field between two (a) parallel; (b) partially flared; and (c) fully flared wires.

shows that if the upper and lower λ/4 lengths at the end of this line are completely bent outward, the resulting configuration, called a half wave dipole, will result in maximum bending of E-lines and maximum escape of energy.

20.2.2 Process of Loop Formation

Figure 20.2 illustrates the process of formation of loops on a transmission line. Assume that a line is divided into a number of small segments. Figure 20.2a shows a simplified equivalent circuit for one such segment. Is contains a series resistance *R*, a series inductance *L*, and a shunt capacitance *C*. In this parameter, *G* (which is usually small) is neglected. In this circuit, the conduction current (*I*) is shown in the upper conductor from left to right and in the lower conductor from right to left. Another current referred to as the capacitive (or displacement) current (I_c) is shown to be between the conductors from upper to lower and lower to upper at the two ends of this segment. Thus, the current can be considered to be forming a closed loop within each of such segments.

Figure 20.2b illustrates the formation of E-loop. Since the conduction and displacement current densities are related to **E** as $\boldsymbol{J} = \sigma \boldsymbol{E}$ and $\boldsymbol{J}_d = \varepsilon_0 \dot{\boldsymbol{E}}$, respectively. Thus, like current, an ***E*** field also forms a loop in each segment. These loops travel along the line towards the open-circuited end one after the other. After arriving nearer to the end of the line these loops start bulging in the forward direction. The moment a loop reaches the end it gets detached. As long as the input end of the line is fed with an alternating source and there is movement of E field along the line in the forward direction, the process of formation and detachment of the loops continues.

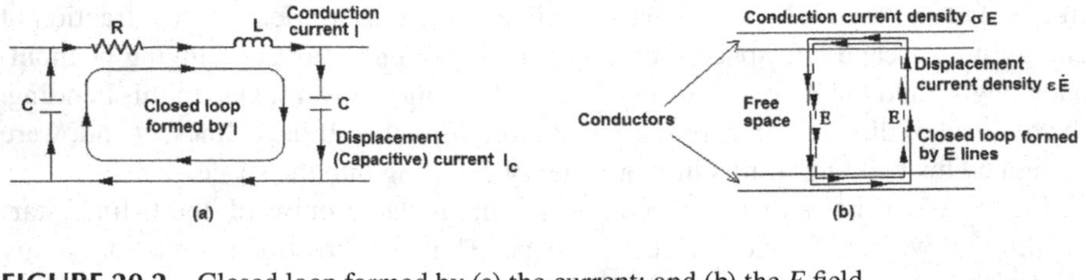

FIGURE 20.2 Closed loop formed by (a) the current; and (b) the *E* field.

20.2.3 Hertzian (Oscillating) Dipole

Figure 20.3a shows a configuration of very short and very thin conducting (alternating) current-carrying element. Figure 20.3b shows a chain of small spheres which alternately carry charges of opposite polarity. If the charge content of each sphere is the same there will be perfect cancellation of all the adjacent charges, barring the ones residing over the two extreme ends of the wire. The condition of the same content of the charges will be met only if the current in the current-carrying element is perfectly uniform. In fact, it is this condition, which is illustrated by Figure 20.3a wherein a thin wire shown to be terminated in two small spheres at both ends. This figure also shows the radius of wire (r') connecting the two spheres, the radius of each sphere (r) on which charges reside, and the length of the wire (dl). The uniformity of the current or perfect cancellation of charges will be possible only if the relative values of these parameters are such that $r' \ll r$, $r \ll dl$, and $dl \ll \lambda$. In the case of non-uniformity of the current the condition of equality of charge contents on each sphere and thus the perfect cancellation of charges will not be met. The thinness of joining wire in Figure 20.3a is also required to make the distributed capacitance between these two charged spheres negligible. Figure 20.3c shows both of the cases of uniform and non-uniform current distributions.

In the alternating nature of current in the element the polarities of charges on the two terminating spheres will keep on changing (or oscillating). Such a configuration is referred to as an *oscillating dipole* or a *Hertzian dipole.*

20.2.4 Detachment and Expansion of Loop

Figure 20.4 illustrates an oscillating dipole consisting of two charges of equal magnitude and opposite polarity. Figure 20.4a shows an E-line emerging from the positive charge and entering into the negative charge. This line can be visualized as a wavefront. Since the dipole is oscillating the two charges will keep on gradually exchanging their positions. Figure 20.4b shows that during the process of change of

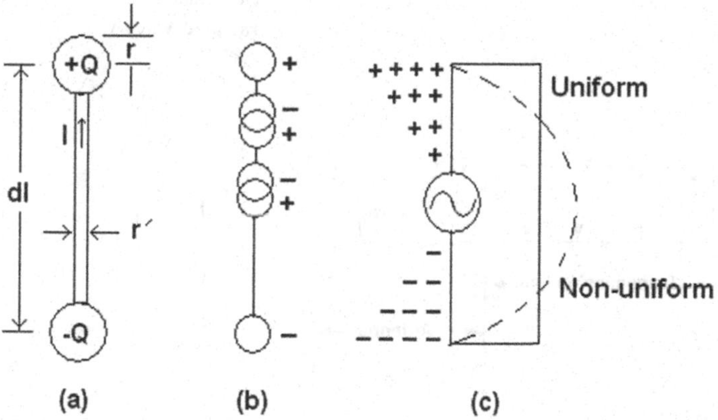

FIGURE 20.3 Hertzian dipole (a) current element terminated in two small spheres; (b) charges on adjacent spheres; and (c) the resulting current distributions.

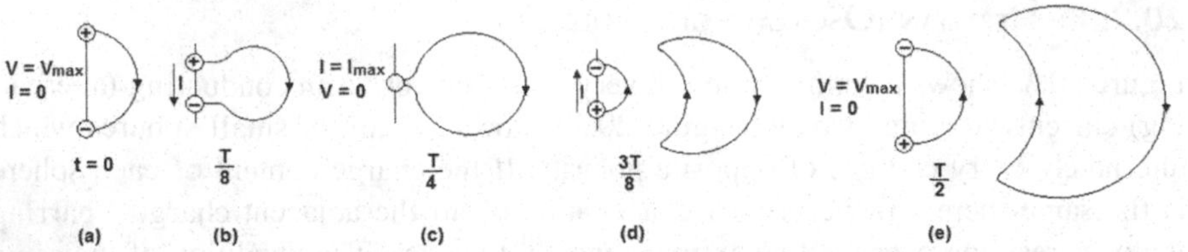

FIGURE 20.4 (a) At t = 0, E-line or wavefront with charges at the ends of dipole; (b) at t = T/8, wavefront moves out as the charges go in; (c) at t = T/4, as charges pass mid-point, the wavefront cuts loose; (d) at t = 3T/8, the detached wavefront moves out, the second wavefront forms; (e) at t = T/2, the first moving wavefront expands, the second wavefront moves out.

position, the wavefront moves out as the charges go in. When the charges pass mid-point the wavefront cut loose. This situation is shown in Figure 20.4c. Figure 20.4d illustrates that this wavefront gets detached and moves away. Meanwhile, a second wavefront forms. Figure 20.4e shows that the first wavefront moves away and expands and a second wavefront moves out. In each segment of Figure 20.4, the time of occurrence of a particular event is also given. As long as this process of formation, expansion, and detachment of loops continues, the dipole keeps on radiating electromagnetic energy into space. This detached or radiated energy keeps on traveling in the space in the forward direction until it is received somewhere or comes across some obstacle (s).

Figure 20.4 also reveals information of utmost importance which indicates the difference between a (simple) dipole and an oscillating dipole. As described in Chapter 7 the distance between the charges and the polarities in a dipole remain the same for all times whereas in oscillating dipole both of these parameters keep on changing with the time.

Having explained the detachment and the expansion of loops we can revisit the configuration of the transmission line shown in Figure 20.1. Figure 20.5 can be

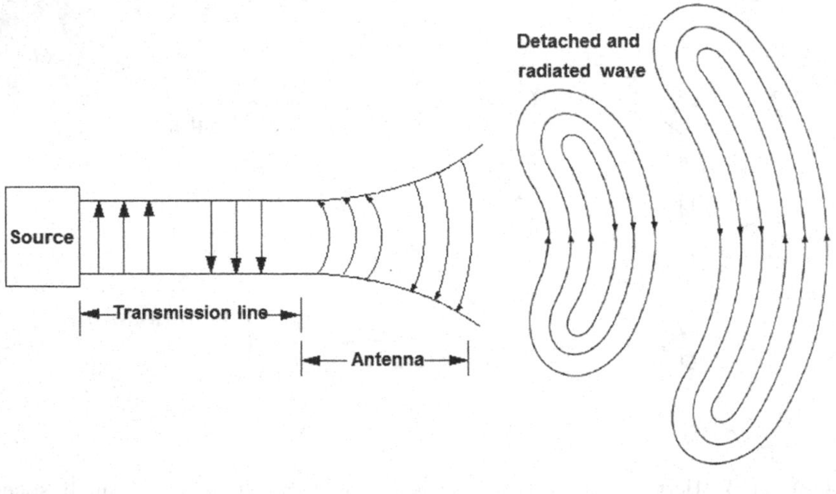

FIGURE 20.5 Movement and expansion of detached loops formed by the E field.

considered as an extension of this figure. It illustrates that while moving away in the forward direction the detached closed loops start expanding. These detached loops expand more and more as they travel farther and farther from the output end of the line. It is to be noted that the direction of E field in these closed loops not only alternate but maintain their relative distances in accordance with the pattern exhibited by the field within the transmission structure. This expansion and travel of loops can be visualized as expansion and travel of rings of smokes released by a smoker, but without any polarity therein.

20.3 RETARDED (OR TIME-VARYING) POTENTIALS

This section revisits Maxwell's equations, which form the basis for the study of the radiation phenomenon. It also briefly touches the retarded potentials, which were discussed earlier in Chapter 15.

20.3.1 Maxwell's Equations

The differential and integral forms of Maxwell's equations were given in Chapter 15. In view of the suitability of differential form for the study of radiation problems, these are reproduced below:

$$\left.\begin{aligned} \nabla \times \boldsymbol{H} &= \frac{\partial \boldsymbol{D}}{\partial t} \quad \text{(a)} \qquad \nabla \times \boldsymbol{E} = -\frac{\partial \boldsymbol{B}}{\partial t} \quad \text{(b)} \\ \nabla \cdot \boldsymbol{D} &= 0 \quad \text{(c)} \qquad \nabla \cdot \boldsymbol{B} = 0 \quad \text{(d)} \end{aligned}\right\} \tag{20.1}$$

The connecting equations for the field quantities involved in Equation 20.1 are:

$$\boldsymbol{D} = \varepsilon \boldsymbol{E} \;\; (\text{a}) \qquad \boldsymbol{B} = \mu \boldsymbol{H} \;\; (\text{b}) \qquad \boldsymbol{J} = \sigma \boldsymbol{E} \;\; (\text{c})\} \tag{20.2}$$

As before, ε is the permittivity, μ is the permeability, and σ is the conductivity of the media.

20.3.2 Assumptions

In Chapter 16 it was noted that for wave propagation the time variation of all the field quantities can be accounted by the term $e^{j\omega t}$ and the space variation by $e^{-\gamma r} = e^{-(\alpha + j\beta)r}$. There the meaning and the nature of γ, α, β, and r were thoroughly explained. These assumptions are also valid for the radiation mechanism. Since for free space $\alpha = 0$ the study of radiation revolves only around those fields that are characterized by the term $e^{-j\beta r}$.

20.3.3 Scalar and Vector Potentials

The relations for scalar electric potential (V) were given by Equations 7.25, 7.27, and 7.29 for different forms of charge distributions. Similarly, the relations for vector magnetic potential (**A**) were given by Equations 11.72a–c for different types of current distributions.

20.3.3.1 Time-Invariant Sources

In the sets of equations noted above the charge and current distributions were assumes to be time-invariant. These relations are reproduced below for ready reference:

$$\left.\begin{aligned} V &= \int_L \frac{\rho_L d\ell}{4\pi\varepsilon_o R} = \iint_s \frac{\rho_s ds}{4\pi\varepsilon_o R} = \iiint_v \frac{\rho_v}{4\pi\varepsilon_o r} dv && \text{(a)} \\ \mathbf{A} &= \int_L \frac{\mu_o \cdot \mathbf{I}}{4\pi \cdot R} \cdot d\mathbf{L} = \iint_s \frac{\mu_o \cdot \mathbf{K}}{4\pi \cdot R} \cdot ds = \iiint_v \frac{\mu_o \cdot \mathbf{J}}{4\pi \cdot R} \cdot dv && \text{(b)} \end{aligned}\right\} \quad (20.3)$$

In these relations the line charge density (ρ_L), the surface charge density (ρ_s), and the volume charge density (ρ_v) represent the sources for the scalar electric potential (V) and the current (**I**) the linear (or surface) current density (**K**) and the (volumetric) current density (**J**) indicate the sources for the vector magnetic potential (**A**). In all these relations the parameter R indicates the distance between the source and the point at which V or **A** is to be evaluated. These charge or current densities acting as sources for V or **A**, reside on or flow in, the elements which are referred to as radiating elements or antennas.

20.3.3.2 Time-Variant Sources

As long as these sources are time-independent the fields are time-invariant. But since wave propagation and the radiation mechanism are the time-variant phenomena all these sources have to be time dependant. Thus, in Equations 20.3a and 20.3b all the time-independent quantities may be replaced by time-dependent quantities. These equations can thus be rewritten as:

$$\left.\begin{aligned} V &= \int_L \frac{\rho_{L(t)}}{4\pi\varepsilon R} dl = \iint_s \frac{\rho_{s(t)}}{4\pi\varepsilon R} ds = \iiint_v \frac{\rho_v(t)}{4\pi\varepsilon r} dv && \text{(c)} \\ A &= \int_L \frac{\mu \cdot \mathbf{I}(t)}{4\pi \cdot R} \cdot d\mathbf{L} = \iint_s \frac{\mu \cdot \mathbf{K}(t)}{4\pi \cdot R} \cdot ds = \iiint_v \frac{\mu \cdot \mathbf{J}(t)}{4\pi \cdot R} \cdot dv && \text{(d)} \end{aligned}\right\} \quad (20.3)$$

Due to the involvement of time the two potentials given by the above equations become inter-dependent as is evident from the Lorentz gauge condition given by Equation 15.49a.

20.3.3.3 Modified Time

The energy after leaving the radiating element travels in the form of an electromagnetic wave. After traveling a distance (R) with a velocity (v) the values of V and **A** will assume different values. To account for the impact of this traveled distance (R) on the values of V and **A**, the time (t) in their expressions is to be replaced by another time (say t') where $t' = t - R/v$. In view of this newly defined time the quantities $\rho(t)$

and *J*(*t*) can be replaced by [ρ] and [*J*], respectively. Thus, on replacing ρ_v by ρ in Equation 20.3c the last segments of Equations 21.3c and 21.3d can be written as:

$$\left. V = \int_v \frac{[\rho]\,dv}{4\pi\varepsilon R} \quad (a) \qquad A = \int_v \frac{\mu[J]}{4\pi R} \quad (b) \right\} \tag{20.4}$$

As before if $\rho = \rho_0 \cos \omega t$, and *t* is replaced by *t*′ one gets $[\rho] = \rho_0\cos[\omega(t - R/\upsilon)]$. In this expression ρ_0 is the magnitude of ρ (at $t = 0$), *R* is the distance between the differential volume *dv* located in current-carrying conductor and the arbitrary point *P* shown in Figure 20.6 and υ is the velocity with which the field progresses or the wave travels in the forward direction. As described in Chapter 15, *V* and *A* given by Equations 20.4a and 20.4b are called the retarded potentials.

20.3.3.4 Location of Origin

The parameter *R*, as seen in Equations 20.4a and 20.4b, is the distance between *dv* and point *P*. In the case the origin lies within the elemental volume, *R* can be directly substituted. If origin lies outside the volume, distance *R* is to be evaluated afresh. In Figure 20.6 the origin is shown to be outside the current-carrying conductor. From this origin the differential volume (*dv*) is shown to be at distance *r*′ and point *P* at distance *r*. Thus, distance *R* is to be evaluated in terms of *r*′ and *r*. For such a situation Equations 20.4a and 20.4 b will get modified accordingly. If *V* is replaced by *V*(*r*,*t*) and *A* by *A*(*r*,*t*) in these equations, it will indicate that these quantities are functions of (some arbitrary distance) *r* and time *t* with reference to the given origin and initial time frame. Equations 20.4a and 20.4b can thus be written as:

$$\left. V(r,t) = \frac{1}{4\pi\varepsilon}\int_v \frac{\rho(r',t)}{R}\,dv' \quad (a) \qquad A(r,t) = \frac{\mu}{4\pi}\int_v \frac{J(r',t)}{R}\,dv' \quad (b) \right\} \tag{20.5}$$

The distance *r*′ involved in Equations 20.5a and 20.5b is shown in Figure 20.6. Since with the shift of the origin, the limits of integral will also change, *dv* is replaced

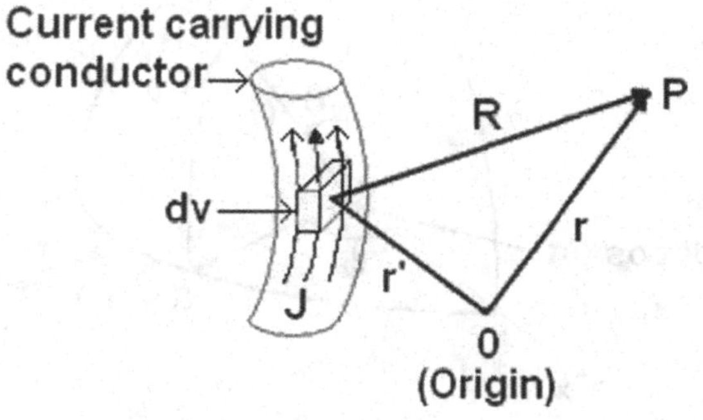

FIGURE 20.6 Configuration showing an elemental volume dv and an arbitrary point P.

by dv' to account for the same. Equations 20.5a and 20.5b can also be modified by replacing t by t'. The resulting expressions of retarded potentials become:

$$\left. V(r,t) = \frac{1}{4\pi\varepsilon}\int_v \frac{\rho(r',t-R/\upsilon)}{R}dv' \quad (a) \quad A(r,t) = \frac{\mu}{4\pi}\int_v \frac{J(r',t-R/\upsilon)}{R}dv' \quad (b) \right\} \quad (20.6)$$

Equation 20.6b can be taken as the starting point for the study of the radiation process.

20.4 FIELD DUE TO AN OSCILLATING DIPOLE

As noted earlier an oscillating dipole may be regarded as a Hertzian dipole provided it conforms to the conditions discussed in Section 20.2.3. The radiation from this infinitesimal element follows the same principle of formation of loops, their detachments, and their expansion as were described in connection with radiation phenomenon. In this section an effort has been made to assess the quantum of radiation of electromagnetic energy.

20.4.1 Components of Vector Magnetic Potential

Figure 20.7 shows a filamentary current-carrying-conducting element that lies along the z-axis. This element in fact is a very short and very thin wire of diminishing length $\boldsymbol{dL}$ ($\boldsymbol{dL}$ tends to zero). It carries a time-varying current flowing in z-direction which is given by $\boldsymbol{IdL}\cos\omega t$. Thus, the current density $\boldsymbol{J}$ has only z-component (or $\boldsymbol{J} = J_z\boldsymbol{a}_z$). Since the components of $\boldsymbol{A}$ corresponds to the components of $\boldsymbol{J}$, $\boldsymbol{A} = A_z\boldsymbol{a}_z$, and $A_\rho = A_\varphi = 0$. As vector magnetic potential $\boldsymbol{A}$ follows Poisson's equation in the current-carrying region we have:

$$\nabla^2\boldsymbol{A} = \nabla^2 A_z = -\mu\cdot\boldsymbol{J} \quad (20.7)$$

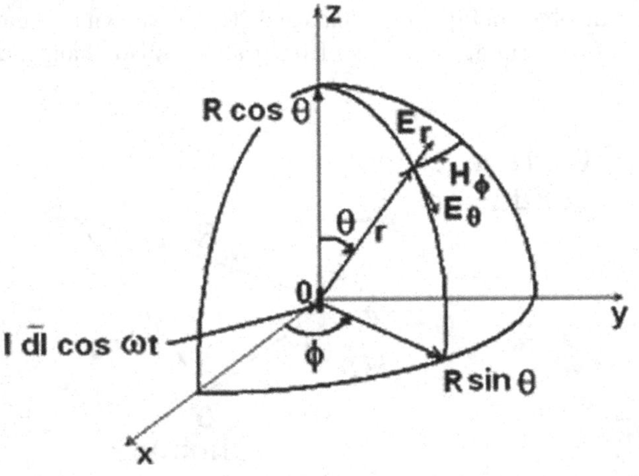

FIGURE 20.7 Configuration of filamentary current-carrying conductor.

This description shows that this configuration can easily be accommodated in a cylindrical coordinate system. But in view of the description of Hertzian dipole, this filamentary element is likely to radiate in all the possible directions. Thus, the problem needs to be tackled in a spherical coordinate system and A_z is to be transformed from a cylindrical to a spherical coordinate system. As $\boldsymbol{a}_z$ in a cylindrical coordinate system has the same meaning as in Cartesian coordinates the transformation in fact is required between rectangular and spherical systems. The relevant components of this transformation in view of Table 3.4 are:

$$\left.\begin{array}{c} \boldsymbol{a}_z \bullet \boldsymbol{a}_r = \boldsymbol{a}_r \bullet \boldsymbol{a}_z = \cos\theta \quad (\text{a}), \quad \boldsymbol{a}_z \bullet \boldsymbol{a}_\theta = \boldsymbol{a}_\theta \bullet \boldsymbol{a}_z = -\sin\theta \quad (\text{b}), \\ \boldsymbol{a}_z \bullet \boldsymbol{a}_\varphi = \boldsymbol{a}_\varphi \bullet \boldsymbol{a}_z = 0 \quad (\text{c}) \end{array}\right\} \tag{20.8}$$

In view of Equation 20.8, the two surviving components of A are obtained as:

$$\left. A_r = A_z \cos\theta \quad (\text{d}), \qquad A_\theta = -A_z \sin\theta \quad (\text{e}) \right\} \tag{20.8}$$

In view of the relation $\boldsymbol{Idl} = \boldsymbol{Kds} = \boldsymbol{J}dv$, A_z for filamentary current can be written as:

$$A_z = \frac{\mu}{4\pi} \frac{Idl \cos\omega\left(t - \dfrac{r}{\upsilon}\right)}{r} \tag{20.9}$$

In view of Equations 20.8 and 20.9:

$$\left. A_r = \frac{\mu}{4\pi} \frac{Idl\cos\omega\left(t - \dfrac{r}{\upsilon}\right)}{r} \cos\theta \quad (\text{a}), \quad A_\theta = -\frac{\mu}{4\pi} \frac{Idl\cos\omega\left(t - \dfrac{r}{\upsilon}\right)}{r} \sin\theta \quad (\text{b}) \right\} \tag{20.10}$$

20.4.2 Components of Electric and Magnetic Fields

In view of Equation 20.8c, $A_\varphi = 0$. It can also be noted that $\dfrac{\partial}{\partial\varphi} \equiv 0$ for all field components due to symmetry along φ. Thus, from the relation $\boldsymbol{B} = \nabla \times \boldsymbol{A}$ the components of $\nabla \times \boldsymbol{A}$ can be obtained as:

$$\left.\begin{array}{ll} (\nabla \times A)_r = \dfrac{1}{\sin\theta}\left[\dfrac{\partial}{\partial\theta}\left(\sin\theta A_\varphi - \dfrac{\partial A_\theta}{\partial\varphi}\right)\right] = B_r = 0 & (\text{a}), \\ (\nabla \times A)_\theta = \left[\dfrac{1}{r\sin\theta}\dfrac{\partial A_r}{\partial\varphi} - \dfrac{1}{r}\dfrac{\partial(rA_\varphi)}{\partial r}\right] = B_\theta = 0 & (\text{b}), \\ (\nabla \times A)_\varphi = \dfrac{1}{r}\left[\dfrac{\partial}{\partial r}(rA_\theta) - \dfrac{\partial A_r}{\partial\theta}\right] = B_\varphi = \mu H_\varphi & (\text{c}) \end{array}\right\} \tag{20.11}$$

In the above relations only the H_φ component survives. Thus, from Equations 20.10 and 20.11c:

$$H_\varphi = \frac{Idl\sin\theta}{4\pi}\left[-\frac{\omega}{r\upsilon}\sin\omega\left(t-\frac{r}{\upsilon}\right)+\frac{\cos\omega\left(t-r/\upsilon\right)}{r^2}\right] \tag{20.12}$$

Equation 20.1a can also be written in the following modified form:

$$E = \frac{1}{\varepsilon}\int\left(\nabla\times H\right)dt \tag{20.13}$$

From Equation 20.13, we can get the following two components of E:

$$\left.E_r = \frac{1}{\varepsilon}\int\left(\nabla\times H\right)_r dt \quad (\text{a}),\quad E_\theta = \frac{1}{\varepsilon}\int\left(\nabla\times H\right)_\theta dt \quad (\text{b})\right\} \tag{2.14}$$

From Equation 4.31c:

$$\nabla\times H = \frac{1}{r\sin\theta}\frac{\partial\left(H_\varphi\sin\theta\right)}{\partial\theta}a_r - \frac{1}{r}\frac{\partial\left(rH_\varphi\right)}{\partial r}a_\theta \tag{20.15}$$

Substitution of H_φ from Equation 20.12 into Equation 20.15 gives:

$$\left(\nabla\times H\right)_r = \frac{1}{r\sin\theta}\frac{\partial}{\partial\theta}\left[\frac{Idl}{4\pi}\sin^2\theta\left\{-\frac{\omega}{r\upsilon}\sin\omega\left(t-r/\upsilon\right)+\frac{\cos\omega\left(t-r/\upsilon\right)}{r^2}\right\}\right] \tag{20.16a}$$

$$\left(\nabla\times H\right)_\theta = \frac{1}{r}\frac{\partial}{\partial r}\left[\frac{Idl}{4\pi}\sin\theta\left\{-\frac{\omega}{\upsilon}\sin\omega\left(t-r/\upsilon\right)+\frac{\cos\omega\left(t-r/\upsilon\right)}{r}\right\}\right] \tag{20.16b}$$

In view of Equations 20.14 and 20.16 we substitute $t' = t - \frac{r}{\upsilon}$:

$$\left.\begin{aligned} E_r &= \frac{2Idl\cos\theta}{4\pi\varepsilon}\left[\frac{\cos\omega t'}{r^2\upsilon}+\frac{\sin\omega t'}{\omega r^3}\right] && (\text{a}) \\ E_\theta &= \frac{Idl\sin\theta}{4\pi\varepsilon}\left[\frac{-\omega\sin\omega t'}{r\upsilon^2}+\frac{\cos\omega t'}{r^2\upsilon}+\frac{\sin\omega t'}{\omega r^3}\right] && (\text{b}) \end{aligned}\right\} \tag{20.17}$$

In view of the substitution $t' = t - \frac{r}{\upsilon}$ expression of H_φ can also be modified to:

$$H_\varphi = \frac{Idl\sin\theta}{4\pi}\left[\frac{\cos\omega t'}{r^2}+\frac{\omega\sin\omega t'}{r\upsilon}\right] \tag{20.17c}$$

20.4.3 Classification of Fields on the Basis of Terms Involved

Inspection of Equations 20.17a, 20.17b, and 20.17c reveals that these equations involve expressions, some of which are inversely proportional to r, r^2, and r^3. This typical variation of components leads to the classification of fields into the following three categories:

1. The terms that are inversely proportional to r^3 represent the electrostatic field. Such terms are involved in expressions of E_θ and E_r given by Equations 20.17a and 20.17b.
2. The terms that are inversely proportional to r^2 give induction or near field. Such terms are involved in the expressions of E_θ, E_r, and H_φ given by Equations 20.17a, 20.17b, and 20.17c.
3. Finally, the terms that are inversely proportional to r, represent radiation, distance, or far field. These terms are involved in the expressions of E_θ and H_φ given by Equations 20.17b and 20.17c.

Electrostatic, induction, and radiation fields are explained below.

20.4.3.1 Electrostatic Field

The electric field due to dipole was discussed in Section 7.5. The configuration of a dipole comprising two equal and opposite charges was shown in Figure 7.7. In view of the geometry of this configuration and the assumption that point P is far removed, R_1 and R_2 were taken as parallel to r. The potential and the electric field intensity at point P were given by Equations 7.32 and 7.33, respectively. These are reproduced below:

$$V \cong \frac{Q}{4\pi\varepsilon_o} \cdot \frac{d \cdot \cos\theta}{r^2} \quad (a) \qquad \mathbf{E} = \frac{Qd}{4\pi\varepsilon_o \cdot r^3} \cdot \left(2\cos\theta \cdot \mathbf{a}_r + \sin\theta \cdot \mathbf{a}_\theta\right) \quad (b) \Bigg\} \tag{20.18}$$

Equation 20.18b comprises the following two components of E:

$$E_r = \frac{2Qd\cos\theta}{4\pi\varepsilon_o \cdot r^3} \quad (a) \quad \text{and} \quad E_\theta = \frac{Qd\sin\theta}{4\pi\varepsilon_o \cdot r^3} \quad (b) \Bigg\} \tag{20.19}$$

In view of the relation $\dfrac{dQ}{dt} = I\cos\omega t$:

$$Q = \frac{I\sin\omega t}{\omega} \tag{20.20}$$

On replacing Q by its equivalent term from Equation 20.20, d by the elemental length dl and t by t', the expressions of Equation 20.19 take the following form:

$$E_\theta = \frac{Idl\sin\theta\sin\omega t'}{4\pi\omega\varepsilon_o \cdot r^3} \quad (a) \quad \text{and} \quad E_r = \frac{2Idl\cos\theta\sin\omega t'}{4\pi\omega\varepsilon_o \cdot r^3} \quad (b) \Bigg\} \tag{20.21}$$

Equations 20.21a and 20.21b represent the electrostatic fields due to a dipole and are the same as involved in the expressions of E_θ and E_r given by Equations 20.17a and 20.17b.

20.4.3.2 Induction Field

The incremental magnetic field intensity given by Equation 11.4 is reproduced here:

$$dH = \frac{I \cdot \boldsymbol{dl} \times \boldsymbol{a_R}}{4\pi \cdot R^2} \tag{20.22}$$

Equation 20.22 can be modified by replacing $I{\cdot}\boldsymbol{dl}$ by $I{\cdot}\boldsymbol{dl}\cos\omega t$, R by r and $\boldsymbol{dl} \times \boldsymbol{a}_R$ by $\sin\theta$. Since filamentary current is the only current, $\boldsymbol{dH}$ can be replaced by $\boldsymbol{H}$. Also the current is flowing in the z-direction $\boldsymbol{H}$ will have only H_φ component. Lastly, t' is substituted for t. In view of these replacements and substitutions, Equation 20.22 becomes:

$$H_\varphi = \frac{Idl\sin\theta\cos\omega t'}{4\pi r^2} \tag{20.23}$$

Equation 20.23 represents the expression of induction field involved in Equation 20.17.

20.4.3.3 Dominant Field

It can be noted that the magnitudes of two bracketed terms in Equation 20.17c will become equal if the following relation is satisfied:

$$\frac{1}{r^2} = \frac{\omega}{r\upsilon} \quad \text{or} \quad r = \frac{\upsilon}{\omega} = \frac{f\lambda}{2\pi f} = \frac{\lambda}{2\pi} \quad \text{or} \quad r \approx \frac{\lambda}{6} \tag{20.24}$$

Equation 20.24 spells that when $r < \frac{\lambda}{6}$ induction field dominates whereas for $r > \frac{\lambda}{6}$ the radiation field assumes significance. In practice for radiation (or far) fields the distance r is taken to be much larger than $\lambda/6$.

20.4.4 Radiated Power

In view of Equations 20.17a, 20.17b, and 20.17c, it is noted that E has no φ component and H has only φ component. Thus, E and H can be written as:

$$\boldsymbol{E} = E_r\boldsymbol{a}_r + E_\theta\boldsymbol{a}_\theta \quad (\text{a}) \quad \text{and} \quad \boldsymbol{H} = H_\varphi\boldsymbol{a}_\varphi \quad (\text{b})\} \tag{20.25}$$

The radiated power can be expressed as view of Poynting vector $\boldsymbol{P}$:

$$\begin{aligned} \boldsymbol{P} = \boldsymbol{E} \times \boldsymbol{H} &= \left(E_r\boldsymbol{a}_r + E_\theta\boldsymbol{a}_\theta\right) \times H_\varphi\boldsymbol{a}_\varphi = -E_rH_\varphi\boldsymbol{a}_\theta + E_\theta H_\varphi\boldsymbol{a}_r \\ &= P_\theta\boldsymbol{a}_\theta + P_r\boldsymbol{a}_r \end{aligned} \tag{20.26}$$

where

$$P_\theta = -E_r H_\varphi$$

$$= -\frac{2I^2dl^2 \sin\theta\cos\theta}{16\pi^2\varepsilon}\left[-\frac{\omega\sin\omega t'\cos\omega t'}{r^3\upsilon^2}+\frac{\cos^2\omega t'}{r^4\upsilon}-\frac{\omega\sin^2\omega t'}{\omega r^4\upsilon}+\frac{\sin\omega t'\cos\omega t'}{\omega r^5}\right]$$

$$= \frac{I^2dl^2\sin2\theta}{16\pi^2\varepsilon}\left[-\frac{\cos2\omega t'}{r^4\upsilon}-\frac{\sin2\omega t'}{2\omega r^5}+\frac{\omega\sin2\omega t'}{2r^3\upsilon^2}\right]$$

(20.27a)

Similarly,

$$P_r = E_\theta H_\varphi = \frac{I^2dl^2\sin^2\theta}{16\pi^2\varepsilon}\left[\frac{\cos2\omega t'}{r^4\upsilon}-\frac{\omega\sin2\omega t'}{r^3\upsilon^2}+\frac{\sin2\omega t'}{2\omega r^5}+\frac{\omega^2(1-\cos2\omega t')}{2r^2\upsilon^3}\right]$$

(20.27b)

20.4.4.1 Average Power

To calculate the average of the radiated power, the average of the terms involved in Equations 20.27a and 20.27b is to be taken. Since the average of sin2$\omega t'$ and cos2$\omega t'$ terms over a complete cycle is zero all such terms from the expressions of P_θ and P_r may be eliminated.

In the case of P_θ (Equation 20.27a) all the three involved terms are of this form and thus the resulting average power becomes zero. This power in fact is surging back and forth during each half cycle. In other words, power emanating from oscillating dipole during the positive half cycle returns to the dipole during the negative half cycle. Thus, this component does not contribute any power toward radiation.

In the case of P_r(Equation 20.27b), the elimination of sin $2\omega t'$ and cos $2\omega t'$ gives:

$$P_r = \frac{\omega^2I^2dl^2\sin^2\theta}{32\pi^2r^2\upsilon^3\varepsilon} = \frac{1}{2\varepsilon\upsilon}\left(\frac{\omega Idl\sin\theta}{4\pi r\upsilon}\right)^2 = \frac{\eta}{2}\left(\frac{\omega Idl\sin\theta}{4\pi r\upsilon}\right)^2 \text{ watts/m}^2 \quad (20.28)$$

In view of Equation 20.28:

$$v = \frac{1}{\sqrt{\mu\varepsilon}} \qquad \eta = \sqrt{\frac{\mu}{\varepsilon}} \qquad \text{thus} \varepsilon v = \sqrt{\frac{\varepsilon}{\mu}} = \frac{1}{\eta}$$

Since P_θ does not contribute toward the radiation, $P_{av} = P_r$ is the only radiated power.

In view of Equation 20.28, it can be noted that the amplitudes of components that contribute toward the net power flow are:

$$\left.\begin{aligned} |E_\theta| &= \frac{\omega Idl\sin\theta}{4\pi r\varepsilon\upsilon^2} = \frac{\eta Idl\sin\theta}{2\pi r} = \frac{60\pi Idl\sin\theta}{\lambda r} \quad (\text{a}),\\ |H_\varphi| &= \frac{\omega Idl\sin\theta}{4\pi r\upsilon} = \frac{Idl\sin\theta}{2\lambda r} \quad (\text{b}) \end{aligned}\right\} \quad (20.29)$$

20.4.4.2 Characteristic Impedance

The ratio of components of E given by Equations 20.29a and 20.29b leads to:

$$\frac{E_\theta}{H_\varphi} = 120\pi = \eta_0 = 377\ \Omega \qquad (20.30)$$

Parameter η_0 is called the *characteristic impedance* for free space.

20.4.4.3 Total Radiated Power

Since oscillating dipole is an infinitesimally short element it can be considered a point source. Thus, this dipole acts as an omnidirectional antenna radiating in all of the possible directions. The total radiated power (P_T) can be obtained by integrating the average power (P_{av}) over the entire surface of an imaginary sphere of radius r:

$$P_T = \int_s P_{av}\, ds \qquad (20.31)$$

Where the elemental surface area (ds) of an imaginary sphere of radius r is:

$$ds = rd\varphi\, rd\theta \sin\theta = 2\pi r^2 \sin\theta d\theta \qquad (20.32)$$

In view of Equations 20.31 and 20.32 we get:

$$P_T = \int_0^\pi \frac{\eta}{2}\left(\frac{\omega Idl \sin\theta}{4\pi r\upsilon}\right)^2 2\pi r^2 \sin\theta d\theta = \frac{\eta\omega^2 I^2 dl^2}{12\pi\upsilon^2}\ \text{Watts} \qquad (20.33)$$

In Equation 20.33, I represents the peak current. If P_T is to be obtained in terms of effective current I_{eff} it can be written as:

$$P_T = \frac{20\omega^2 I_{\text{eff}}^2 dl^2}{\upsilon^2}\ \text{Watts} \qquad (20.34)$$

On replacing ω by $2\pi f$ and υ by the velocity of light c Equation 20.34 yields:

$$P_T = 20\frac{(2\pi f)^2}{c^2} I_{\text{eff}}^2 dl^2 = 80\pi^2\left(\frac{dl}{\lambda}\right)^2 I_{\text{eff}}^2 = R_{\text{rad}} I_{\text{eff}}^2\ \text{Watts} \qquad (20.35)$$

To arrive at Equation 20.33 from Equation 20.31, the following relation is employed:

$$\int_0^\pi \sin^3\theta\, d\theta = \left[\frac{1}{3}\cos 3\theta - \cos\theta\right]_0^\pi = \frac{4}{3} \qquad (20.36)$$

In Equation 20.35, the term R_{rad} is the radiation resistance, which can be written as:

$$R_{\text{rad}} = 80\pi^2\left(\frac{dl}{\lambda}\right)^2\ \Omega \qquad (21.37)$$

Radiation resistance, seen in Equations 20.35 and 20.37, is further elaborated on in the next subsection.

20.4.5 Radiation Resistance

"Radiation resistance" is the fictitious resistance that, when substituted in a series with an antenna, will consume the same power as is actually radiated by the antenna.

Figure 20.8 illustrates three different radiating elements. The radiation resistance for these elements with given current distributions can be obtained in view of Equation 20.37.

20.4.5.1 Elemental Length

Figure 20.8a shows an element of length dl with uniform current distribution. The radiation resistance expression given by Equation 20.37 can be modified to the following form:

$$R_{\text{rad}} = 80\pi^2\left(\frac{dl}{\lambda}\right)^2 \approx 800\left(\frac{dl}{\lambda}\right)^2 \ \Omega \qquad (20.38a)$$

20.4.5.2 Short Dipole

This case is shown in Figure 20.8b. As noted earlier the uniformity of time-varying current can be presumed only over a very short (or elemental) length *dl*. For a larger length (say) *L*, the current can no longer remain uniform. If we presume that this current has non-uniform (tapered) distribution the length *dl* is to be replaced by (say) *L*. In view of this tapering, the radiation resistance reduces to one fourth that of elemental length. Thus the radiation resistance for a short dipole with *dl* replaced by *L* in Equation 20.38a becomes:

$$R_{\text{rad}} = 20\pi^2\left(\frac{dl}{\lambda}\right)^2 \approx 200\left(\frac{L}{\lambda}\right)^2 \ \Omega \qquad (20.38b)$$

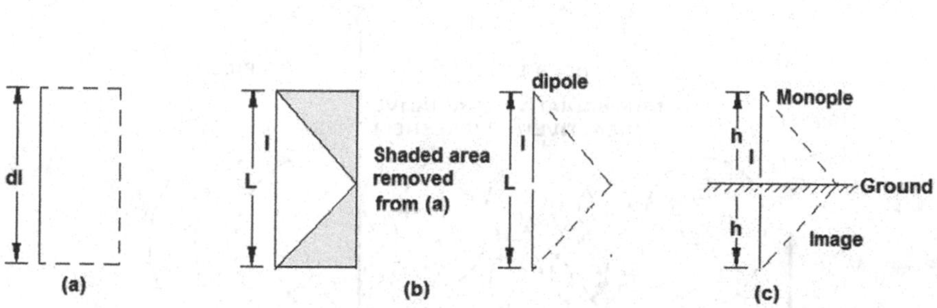

FIGURE 20.8 Current distribution on radiating element (a) uniform along elemental length *dl*; (b) tapered (non-uniform) along short dipole of length *L* (*dl* is replaced by *L*); and (c) tapered (non-uniform) along a short monopole of length *h* (here *L* is replaced 2*h*).

20.4.5.3 Short Monopole

This case is shown in Figure 20.8c. In this case again the element of length has non-uniform (tapered) current distribution. Since the length of the radiating element is only half of the dipole the radiation resistance is also half. Thus, the radiation resistance for a short monopole with length $L = 2h$ becomes:

$$R_{\text{rad}} = 10\pi^2\left(\frac{L}{\lambda}\right)^2 \approx 100\left(\frac{L}{\lambda}\right)^2 = 400\left(\frac{h}{\lambda}\right)^2 \ \Omega \tag{20.38c}$$

The relations given by Equations 20.38a, 20.38b, and 20.38c hold good for very short antennas. In general, these lengths should not exceed $\lambda/10$ (and in some cases $\lambda/8$) but not beyond for treating an antenna as a short antenna in electrical terms.

20.5 FIELDS DUE TO AN ANTENNA WITH SINUSOIDAL CURRENT DISTRIBUTION

As shown in Figure 20.9, the field around an antenna is divided into two principal regions. The region near to the antenna is called the *near field* or *Fresnel zone*, whereas the one away from the antenna is called the *far field* or *Fraunhofer zone*. These cases are discussed in the following two subsections. These fields are obtained due to half (or quarter) wave elements with sinusoidal current distribution.

20.5.1 Far Field

We have so far described the field caused due to an element of very short length. Therein the current distribution was assumed to be uniform. If the length of this element is increased or the current distribution becomes non-uniform the analysis presented earlier cannot yield the desired results and we have to look at the analysis afresh. In this section the length of the element is taken to be half of the wavelength and current distribution is assumed to be sinusoidal. Such a configuration may be referred to as a half-wave dipole. If only half of this length is considered the configuration is referred to as quarter-wave monopole. Both of the configurations are shown in Figure 20.10.

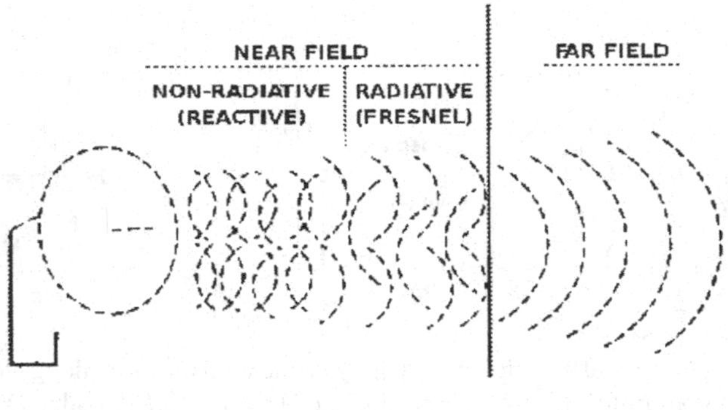

FIGURE 20.9 Near and far field of an antenna.

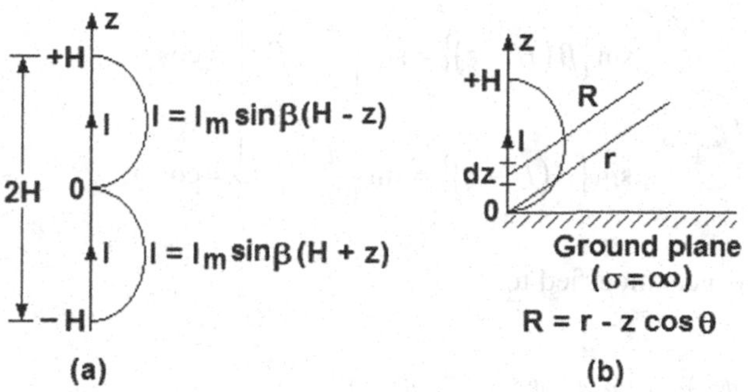

FIGURE 20.10 (a) Half-wave dipole; and (b) quarter-wave monopole.

20.5.1.1 Geometry of the Configuration

Figure 20.10a shows a half-wave dipole wherein the length of the conducting element runs from $-H$ to $+H$ along the z-axis. It contains a sinusoidally distributed current which for the segments of the element can be given as:

$$\left.\begin{aligned} I &= I_m \sin\beta(H-z) \quad \text{for } z>0 \quad (\text{a}),\\ I &= I_m \sin\beta(H+z) \quad \text{for } z<0 \quad (\text{b}) \end{aligned}\right\} \tag{20.39}$$

Figure 20.10b shows a quarter-wave monopole with the same current distribution as that in the dipole. This monopole is assumed to be located on a perfectly conducting ground.

As shown in Figure 20.10b, r and R are the distances from the origin and length dz to a point at which the field is to be evaluated. For large distances, r and R are generally taken to be parallel.

As the current is taken to be in the z-direction, $J = J_z a_z$ and $A = A_z a_z$. The differential vector magnetic potential can be written as:

$$dA_z = \frac{\mu I dz}{4\pi R} e^{-\beta R} \tag{20.40}$$

In view of Equations 20.39 and 20.40:

$$A_z = \frac{\mu}{4\pi}\int_{-H}^{0} \frac{I_m \sin\beta(H+z)}{R} e^{-j\beta R}\, dz + \frac{\mu}{4\pi}\int_{0}^{H} \frac{I_m \sin\beta(H-z)}{R} e^{-j\beta R}\, dz$$

If R is replaced by r for the estimation of magnitude and by $r - z\cos\theta$ for phase β Equation 20.40 becomes:

$$A_z = \frac{\mu I_m}{4\pi r} e^{-j\beta r}\left[\int_{-H}^{0} \sin\{\beta(H+z)\} e^{j\beta z\cos\theta}\, dz + \int_{0}^{H} \sin\{\beta(H-z)\} e^{-j\beta z\cos\theta}\, dz\right] \tag{20.41}$$

For a half-wave dipole $H = \frac{\lambda}{4}$, $\beta H = \frac{2\pi}{\lambda}\frac{\lambda}{4} = \frac{\pi}{2}$, we get

$$\sin\{\beta(H+z)\} = \sin\left\{\frac{\pi}{2}+\beta z\right\} = \cos\beta z \tag{20.42a}$$

$$\sin\{\beta(H-z)\} = \sin\left\{\frac{\pi}{2}-\beta z\right\} = \cos\beta z \tag{20.42b}$$

Equation 20.41 gets modified to:

$$\begin{aligned}
A_z &= \frac{\mu I_m}{4\pi r}e^{-j\beta r}\int_0^{\lambda/4}\cos\beta z\left(e^{j\beta z\cos\theta}+e^{-j\beta z\cos\theta}\right)dz \\
&= \frac{\mu I_m}{4\pi r}e^{-j\beta r}\,2\int_0^{\lambda/4}\cos\beta z\,\cos(\beta z\cos\theta)\,dz \\
&= \frac{\mu I_m}{2\pi r}e^{-j\beta r}\int_0^{\lambda/4}\left[\cos\beta z(1+\cos\theta)+\cos\beta z(1-\cos\theta)\right]dz \\
&= \frac{\mu I_m}{2\pi r}e^{-j\beta r}\left[\frac{\sin\beta z(1+\cos\theta)}{\beta(1+\cos\theta)}+\frac{\sin\beta z(1-\cos\theta)}{\beta(1-\cos\theta)}\right]_0^{\lambda/4} \\
&= \frac{\mu I_m}{2\pi\beta r}e^{-j\beta r}\left[\left\{(1-\cos\theta)\left(\frac{\pi}{2}\cos\theta\right)+(1+\cos\theta)\left(\frac{\pi}{2}\cos\theta\right)\right\}\frac{1}{\sin^2\theta}\right] \\
&= \frac{\mu I_m}{2\pi\beta r}e^{-j\beta r}\left[\frac{\cos\left(\frac{\pi}{2}\cos\theta\right)}{\sin^2\theta}\right]
\end{aligned} \tag{20.43}$$

20.5.1.2 Field Components

The required field components can be obtained as below:

Since $J = J_z a_z$, $B = \nabla\times A = B_\varphi a_\varphi$ only, and $E_\theta = \eta H_\phi = 120\pi H_\varphi$, thus:

$$H_\phi = -\frac{1}{\mu}\frac{\partial A_z}{\partial r}\sin\theta = \frac{jI_m e^{-j\beta r}}{2\pi r}\left[\frac{\cos\left(\frac{\pi}{2}\cos\theta\right)}{\sin\theta}\right] \tag{20.44}$$

$$E_\theta = -\frac{j60 I_m e^{-j\beta r}}{r}\left[\frac{\cos\left(\frac{\pi}{2}\cos\theta\right)}{\sin\theta}\right] \tag{20.45}$$

The magnitudes of E_θ and H_φ are:

$$\left.|E_\theta| = \frac{60I_m}{r}\left[\frac{\cos\left(\frac{\pi}{2}\cos\theta\right)}{\sin\theta}\right] \quad (a) \qquad |H_\varphi| = \frac{I_m}{2\pi r}\left[\frac{\cos\left(\frac{\pi}{2}\cos\theta\right)}{\sin\theta}\right] \quad (b)\right\} \tag{20.46}$$

20.5.1.3 Radiated Power

In view of Equation 20.46 the expressions for average (P_{av}) and total (P_T) radiated power are:

$$\left.\begin{aligned} P_{av} &= |E_\theta||H_\varphi| = \frac{\eta I_m^2}{8\pi^2 r^2}\left[\frac{\cos^2\left(\frac{\pi}{2}\cos\theta\right)}{\sin^2\theta}\right] && (a) \\ P_T &= \int P_{av}\,ds = \frac{\eta I_m^2}{4\pi}\int_0^{\frac{\pi}{2}}\left[\frac{\cos^2\left(\frac{\pi}{2}\cos\theta\right)}{\sin^2\theta}\right]d\theta && (b) \end{aligned}\right\} \tag{20.47}$$

The evaluation of Equation 20.47 gives:

$$P_T = \frac{0.609\eta I_m^2}{4\pi} = \frac{0.609\eta I_{\text{eff}}^2}{2\pi} = 36.5 I_{\text{eff}}^2 = R_{\text{rad}} I_{\text{eff}}^2 \tag{20.48}$$

20.5.1.4 Radiation Resistance

In deriving the Equation 20.48, the configuration shown in Figure 14.10b was taken as the basis. Since this configuration represents a quarter-wave monopole, the radiation resistance can be written as:

$$R_{\text{rad}} = 36.5\ \Omega \quad \text{(for quarter-wave monopole)}$$

$$R_{\text{rad}} = 73\ \Omega \quad \text{(for half-wave dipole)}$$

20.5.2 Near Field

The computation of near field in general is not of much significance except when antennas are located nearby such as at airports, seaports, etc. Near field also assumes significance in antenna arrays wherein the elements are closely located. In both of these, the field due to one antenna (or element) may influence other adjacent antenna(s) or element(s). This situation demands a critical study of the near field.

20.5.2.1 Geometry of the Configuration

Figure 20.11 shows a (sinusoidal) current-carrying dipole lying along the z-axis. It also shows the four distances to the point (P) at which the field is to be evaluated. These include the distance (r) to point P from the origin, distance (R) from elemental length dh located at distance h from the origin along the z-axis, distance (R_1) from the upper tip (at $z = +H$) and distance (R_2) from the lower tip (at $z = -H$) of the dipole. From the geometry of this configuration these four distances can be written as:

$$\left.\begin{aligned} R_1 &= \sqrt{(z-H)^2 + y^2} \quad (\text{a}), \quad R_2 = \sqrt{(z+H)^2 + y^2} \quad (\text{b}), \\ R &= \sqrt{(z-h)^2 + y^2} \quad (\text{c}), \quad r = \sqrt{z^2 + y^2} \quad (\text{d}) \end{aligned}\right\} \tag{20.49}$$

The expressions for the sinusoidal current distribution can be written as:

$$\left.\begin{aligned} I &= I_m \sin\beta(H-h) \quad \text{for } h > 0 \quad (\text{a}), \\ I &= I_m \sin\beta(H+h) \quad \text{for } h < 0 \quad (\text{b}) \end{aligned}\right\} \tag{20.50}$$

As before, the coordinates of point P can be represented in a spherical system but for the near field case the problem can be better tackled with both the rectangular and cylindrical coordinate systems. Thus, in view of Figure 20.11 and Equations 20.49 and 20.50, the z-component of A can be written as:

$$A_z = \frac{\mu I_m}{j8\pi} \begin{bmatrix} e^{j\beta H} \displaystyle\int_0^H \frac{e^{-j\beta(R+h)}}{R} dh - e^{-j\beta H} \int_0^H \frac{e^{-j\beta(R-h)}}{R} dh \\ + e^{j\beta H} \displaystyle\int_0^H \frac{e^{-j\beta(R-h)}}{R} dh - e^{-j\beta H} \int_0^H \frac{e^{-j\beta(R+h)}}{R} dh \end{bmatrix} \tag{20.51}$$

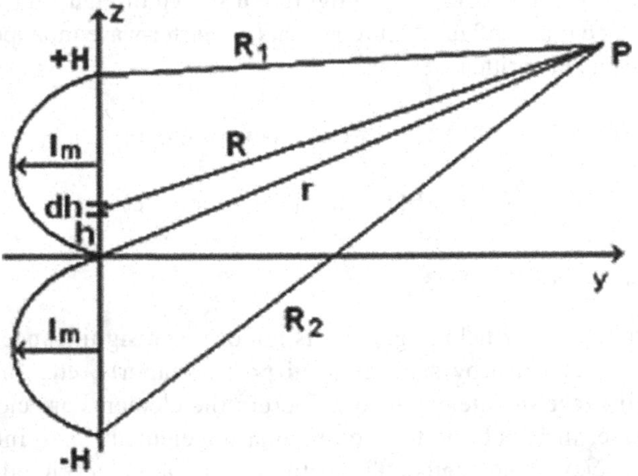

FIGURE 20.11 Dipole with sinusoidal current distribution.

20.5.2.2 Field Components

From Equation 20.51, the surviving component of magnetic field intensity can be written as:

$$H_\varphi = \frac{1}{\mu}(\nabla \times A)_\varphi = -\frac{1}{\mu}\frac{\partial A_z}{\partial \rho} = -\frac{1}{\mu}\frac{\partial A_z}{\partial y} = -H_x \tag{20.52}$$

In Equation 20.51, ρ is replaced by y. Also, from Equations 20.51 and 20.52, we get:

$$\begin{aligned} H_\varphi = &-\frac{I_m}{j8\pi}\left[e^{j\beta H}\int_0^H \frac{\partial}{\partial y}\left\{\frac{e^{-j\beta(R+h)}}{R}dh\right\} - e^{-j\beta H}\int_0^H \frac{\partial}{\partial y}\left\{\frac{e^{-j\beta(R-h)}}{R}dh\right\}\right] \\ &-\frac{I_m}{j8\pi}\left[+e^{j\beta H}\int_{-H}^0 \frac{\partial}{\partial y}\left\{\frac{e^{-j\beta(R-h)}}{R}dh\right\} - e^{-j\beta H}\int_{-H}^0 \frac{\partial}{\partial y}\left\{\frac{e^{-j\beta(R+h)}}{R}dh\right\}\right] \end{aligned} \tag{20.53}$$

Equation 20.53 involves four integral expressions. All of these integrals involve partial derivatives with y. The first of these terms is rewritten as:

$$\int_0^H \frac{\partial}{\partial y}\left\{\frac{e^{-j\beta(R+h)}}{R}dh\right\} \tag{20.54}$$

The parameter R involved in this expression is given by Equation 20.49c. This term can be differentiated by using the following standard relation involving u and v which are functions of variable x:

$$\frac{d}{dx}\left(\frac{u}{v}\right) = \frac{v\left(\frac{du}{dx}\right) - u\left(\frac{dv}{dx}\right)}{v^2} \tag{20.55}$$

Thus, Equation 20.54 becomes:

$$\int_0^H \left[\frac{e^{-j\beta y}e^{-j\beta(R+h)}}{R^2} - \frac{ye^{-j\beta(R+h)}}{R^3}\right]dh \tag{20.56}$$

It is a coincidence that the integrand of Equation 20.56 is a perfect differential. Also, the integral involved in Equation 20.56 can be evaluated by using the relation for the product terms:

$$\int uvdx = u\int vdx - \int\left\{\left(\frac{du}{dx}\right)\int vdx\right\} \tag{20.57}$$

In view of Equation 20.57, Equation 20.56 yields:

$$e^{j\beta H}\left[\frac{ye^{-j\beta(R+h)}}{R(R+h-z)}\right]_{h=0}^{h=H} = ye^{j\beta H}\left[\frac{e^{-j\beta(R+h)}}{R_1(R_1+H-z)} - \frac{(r+z)e^{-j\beta r}}{r(r-z)}\right] \tag{20.58}$$

But from Equation 20.49a:

$$R_1^2 - (H-z)^2 = r^2 - z^2 = y^2 \quad (20.59)$$

Thus, the terms of Equation 20.56 become:

$$\frac{e^{j\beta H}}{y}\left[\left(1-\frac{H-z}{R_1}\right)e^{-j\beta(R_1+h)} - \left(1-\frac{z}{r}\right)e^{-j\beta r}\right] \quad (20.60a)$$

$$\frac{e^{j\beta H}}{y}\left[\left(1+\frac{H-z}{R_1}\right)e^{-j\beta(R_1-h)} - \left(1-\frac{z}{r}\right)e^{-j\beta r}\right] \quad (20.60b)$$

$$\frac{e^{-j\beta H}}{y}\left[\left(1-\frac{H+z}{R_2}\right)e^{-j\beta(R_2+h)} - \left(1-\frac{z}{r}\right)e^{-j\beta r}\right] \quad (20.60c)$$

$$\frac{e^{-j\beta H}}{y}\left[\left(1+\frac{H+z}{R_2}\right)e^{-j\beta(R_2-h)} - \left(1-\frac{z}{r}\right)e^{-j\beta r}\right] \quad (20.60d)$$

The total magnetic field obtained by summing all four terms of Equation 20.60 is:

$$H_\varphi = \frac{-I_m}{4j\pi}\left(\frac{e^{-j\beta R_1}}{y} + \frac{e^{-j\beta R_2}}{y} - \frac{2\cos\beta H e^{-j\beta r}}{y}\right) \quad (20.61)$$

Using the first Maxwell's equation, we can write $E = \frac{1}{j\omega\varepsilon}\nabla\times H$ in free space. Now, if the $x = 0$ plane is replaced by y, we get:

$$\left.E_y = -\frac{1}{j\omega\varepsilon}\frac{\partial H_\varphi}{\partial z}\ (\text{a}), \qquad E_z = -\frac{1}{j\omega\varepsilon y}\frac{\partial(yH_\varphi)}{\partial y}\ (\text{b})\right\} \quad (20.62)$$

Since H_ϕ is the only surviving component of H the curl of H becomes:

$$\nabla\times H = \frac{\partial H_\varphi}{\partial z}a_y + \frac{1}{y}\frac{\partial(yH_\varphi)}{\partial y}a_z \quad (20.63)$$

Using Equations 20.62 and 20.63, to get:

$$\begin{aligned} E_z &= \frac{-j\beta I_m}{4\pi\varepsilon\omega y}\left(\frac{ye^{-j\beta R_1}}{R_1} + \frac{ye^{-j\beta R_2}}{R_2} - 2\cos\beta H\frac{ye^{-j\beta r}}{r}\right) \\ &= -j30I_m\left(\frac{e^{-j\beta R_1}}{R_1} + \frac{e^{-j\beta R_2}}{R_2} - 2\cos\beta H\frac{e^{-j\beta r}}{r}\right) \end{aligned} \quad (20.64a)$$

$$E_y = j30I_m\left(\frac{z-H}{y}\frac{e^{-j\beta R_1}}{R_1}+\frac{z+H}{y}\frac{e^{-j\beta R_2}}{R_2}-2\frac{z\cos\beta H}{y}\frac{e^{-j\beta r}}{r}\right) \quad (20.64b)$$

$$H_\varphi = \frac{j30I_m}{\eta y}\left(e^{-j\beta R_1}+e^{-j\beta R_2}-2\cos\beta He^{-j\beta r}\right) \quad (20.64c)$$

20.5.2.3 Interpretation of Different Terms

The terms involved in the expressions of E_z, E_y, and H_φ given by Equations 20.64a–c impart the following information:

1. Expressions involving $e^{-j\beta R_1}$ represent spherical waves, which originate at the top of the antenna.
2. Expressions involving $e^{-j\beta R_2}$ represent spherical waves, which originate from the bottom of the antenna in the case of a dipole and from the lower tip of the image in the case of a monopole.
3. Expressions involving $e^{-j\beta r}$ represent spherical waves, which originate from the center of the antenna in the case of a dipole and from the base in the case of a monopole.
4. Amplitudes of these waves will depend on length H. Since $\beta H = \frac{\pi}{2}$, cos $\beta H = 0$. Thus, for quarter-wave monopole or half-wave a dipole the amplitude will be 0.
5. All of the spherical waves are omnidirectional in nature.
6. In all expressions, numerators represent the phase factor and denominators represent the inverse distance factor.
7. Equations 20.64a and 20.64b, respectively, represent parallel (E_z) and perpendicular (E_y) components. By assuming suitable values of the involved parameters E_z and E_y can be computed from these relations in the immediate neighborhood of the antenna.

20.6 RADIATION PATTERNS

The "radiation pattern" also called directional pattern is the graphical representation of field distribution in space due to radiation from an antenna. It is a function of direction or space coordinates. The form of this pattern is often referred to as a beam. This beam may be very narrow (pencil beam), very wide (fan beam), omni, or an arbitrary shape. A radiation pattern may comprise a single beam or multiple beams. In radar, these beams may be mechanically or electronically steered.

The radiation patterns may be defined in terms of electric field (called E pattern), magnetic field (called H pattern), or power. When defined in terms of E or H fields these are termed as radiation density patterns or spatial variation of E or H. When a pattern is defined in terms of power it is called the *power pattern*. These are the trace of received power ($P = E \times H$) per unit solid angle (in watts per steradian). Since the E pattern is more commonly used, the next section deals only with the E patterns.

20.6.1 Required Relations

The expression of A_z given by Equation 20.41 is reproduced here:

$$A_z = \frac{\mu I_m}{4\pi r} e^{-j\beta r} \left[\begin{array}{l} \int_{-H}^{0} \sin\{\beta(H+z)\} e^{j\beta z\cos\theta} dz \\ + \int_{0}^{H} \sin\{\beta(H-z)\} e^{j\beta z\cos\theta} dz \end{array} \right] \qquad (20.41)$$

Since $\beta = 2\pi/\lambda$, $\beta H = \left(\frac{2\pi}{\lambda}\right) H$.

Table 20.1 includes the values of βH, $\sin \beta(H + z)$ and $\sin \beta(\beta - z)$ for some selected values of H.

In Section 20.4, the case of $H = \lambda/4$ was considered, and the limits of z were taken accordingly. In view of Equation 20.27b, E_θ is the only component that contributes towards the radiated power, thus only the E_θ component needs to be considered. The magnitude of E_θ given by Equation 20.46a is reproduced here:

$$|E_\theta| = \frac{60 I_m}{r} \left[\frac{\cos\left(\frac{\pi}{2}\cos\theta\right)}{\sin\theta} \right] \qquad (20.46a)$$

If we proceed without substitution of value of βH and corresponding values of $\sin \beta(H + z)$ and $\sin \beta(H - z)$ are substituted in Equation 20.41, we get the following relation:

$$|E_\theta| = \frac{60 I_m}{r} \left[\frac{\cos \beta H - \cos(\beta H \cos\theta)}{\sin\theta} \right] \qquad (20.65)$$

If H is assigned different values (say $\frac{\lambda}{4}, \frac{\lambda}{2}, \frac{3\lambda}{4}, \lambda$, etc.) the corresponding values of βH ($\frac{\pi}{2}, \pi, \frac{3\pi}{2}, 2\pi$ etc.) can be obtained. The substitution of these values in

TABLE 20.1
Values of βH, $\sin \beta(H + z)$, and $\sin \beta(H - z)$ for Different Values of H

$H =$	$\lambda = 4$	$\lambda = 2$	$3\lambda/4$	λ	$3\lambda/2$	2λ
$\beta H = \left(\frac{2\pi}{\lambda}\right) H$	$\pi/2$	π	$3\pi/2$	2π	3π	4π
$\sin \beta(H + z)$	$\cos(\beta z)$	$-\sin(\beta z)$	$-\cos(\beta z)$	$\sin(\beta z)$	$-\sin(\beta z)$	$\sin(\beta z)$
$\sin \beta(H - z)$	$\cos(\beta z)$	$\sin(\beta z)$	$-\cos(\beta z)$	$-\sin(\beta z)$	$\sin(\beta z)$	$-\sin(\beta z)$

Equation 20.65 will give new corresponding expressions for $|E_\theta|$. Values of $|E_\theta|$ can then be calculated by varying r and θ over a selected range. Let us assume that $I_m = I$, r varies between 0 and 1 and θ varies between 0 and π. The values so obtained can be plotted on a polar graph to give the radiation patterns. These patterns, classified as horizontal and vertical radiation patterns are discussed in the next subsections.

20.6.2 Horizontal Radiation Pattern

The horizontal radiation pattern (*HRP*) is one that is plotted in the horizontal (or $\theta = 0$) plane wherein φ varies between 0 and 2π. Since there is no involvement of φ in the expressions of $|E_\varphi|$ the pattern in the $\theta = 0$ plane is a circle for all lengths. This is shown in Figure 20.12a.

20.6.3 Vertical Radiation Pattern

The vertical radiation pattern (*VRP*) is one that is plotted in the vertical plane at some constant value of φ and θ is varied between 0 and π.

20.6.3.1 Vertical Radiation Pattern for $l < \lambda/10$

If the length of the radiating element is less than $\lambda/10$ its *VRP* can be obtained from Equation 20.29a, which is reproduced here:

$$E_\theta = \frac{60\pi I dl}{r\lambda}\sin\theta \tag{20.29a}$$

The *VRP* obtained from this relation for length dl is shown in Figure 20.12b.

20.6.3.2 Vertical Radiation Patterns for $l > \lambda/10$

For all other lengths ($l > \lambda/10$) *VRPs* are to be obtained from the $|E_\theta|$ relations as discussed above. Since θ is involved in all the expressions of $|E_\theta|$ *VRPs* obtained are of different shapes for different lengths (*H*) shown in Figure 20.9. Figure 20.13 shows *VRPs* for some selected lengths.

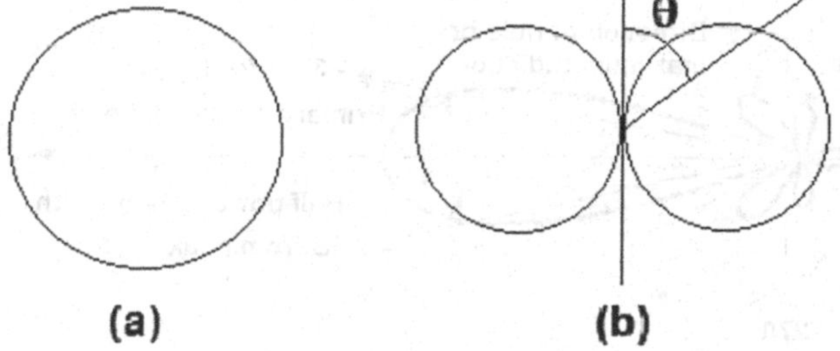

FIGURE 20.12 (a) HRP for all lengths; and (b) VRP for elemental length *dl*.

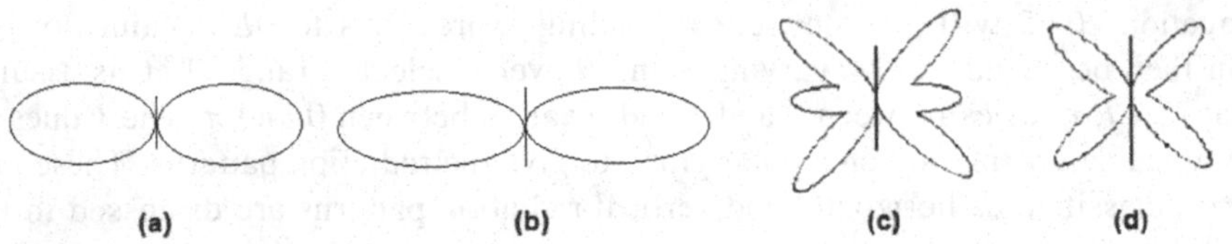

FIGURE 20.13 Vertical radiation patterns for H=: (a) $\lambda/2$; (b) λ; (c) $3\lambda/2$; (d) 2λ.

20.6.3.3 Components of Radiation Pattern

Earlier it was mentioned that radiation pattern is the graphical representation of E, H, or P distribution in the space and thus a function of direction or space coordinates. As illustrated in Figures 20.13 beam in fact is a three-dimensional configuration.

A radiation pattern may be comprised of one beam or more beams. If there is only one beam it will have an angular location in which there will be maximum radiation. If we move away from this location, in either direction, the radiation will start reducing until the zero-radiation point is reached. In the case of more than one beam, the radiation will again start increasing from this zero-radiation point until a second maximum is attained and thereafter start decreasing until it again reaches the zero-radiation point. This process goes on repeating unless all the beams are exhausted.

Figure 20.14 shows a number of constituents of a beam. It may have a main lobe, one or more side lobes, and a back lobe. The main beam will have a direction of maximum radiation, the directions in which the radiation reduces to half of the maximum and the directions of zero radiations. The angular distance between two half-power points is called the half-power beamwidth, and the angular distance between two zero-radiation points is called the beamwidth between the first nulls.

Some other parameters defined in connection with the beam are beam efficiency, gain, power gain, directive gain, and the directivity. The *beam efficiency* is the ratio of the main beam area to the total beam area. The *gain* is the measure of squeezed or concentrated radiations in a particular direction with respect to an omni-directional antenna which is normally taken to be a point source. The *power gain* is the ratio of radiation intensity in a given direction to the average total power. The *directive gain* is the ratio of power density in a particular direction at a given point to the power that

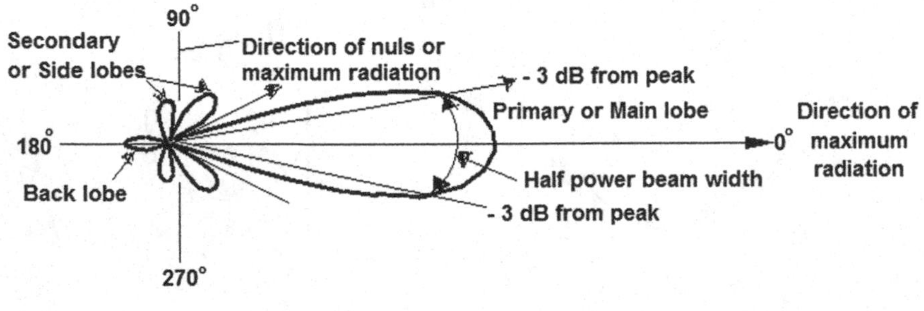

FIGURE 20.14 Components of a lobe.

would be radiated at the same distance by an omni-antenna. And, lastly, the *directivity* is the maximum directive gain.

20.6.4 Orientation of Vertical Radiation Patterns

The orientation of radiation pattern depends on the location of feed. For all the cases shown in Figure 20.13 the feed is assumed to be located in the center of the radiating element. In the case that it is shifted toward an end of the element, the *VRP* will also twist toward that end. Figures 20.15a and 20.15b illustrate this basic norm for $l = \lambda/2$ and λ, respectively. Figures 20.15a(ii) and 20.15b(ii) show the feeds to be located at the left ends of the elements and the *VRPs* also twisted towards the same ends. Similarly, in Figure 20.15a(iii) and 20.15b(iii), the feeds are located at the right ends and *VRPs* also twists towards the right ends. The same is true for other lengths whether it is small or large.

20.6.5 Effect of Termination on RPs

Besides the electrical length and location of feed, the radiation pattern of an antenna is also affected by termination. This leads to the following classification of antennas.

20.6.5.1 Unterminated Antenna

An antenna that is not terminated at its end is called an *unterminated* antenna. Such antennas have standing wave current distributions and are also called a *standing wave, periodic wave,* or *resonant antenna.* In Chapter 16, it was noted that the standing waves are formed due to the combination of forward and reflected waves. Thus, due to the presence of two waves traveling in opposite directions the radiation will also be in both the forward and backward directions. The same is evident from Figure 20.16a.

20.6.5.2 Terminated Antenna

An antenna that is terminated at its end is called a *terminated* antenna. Such antennas have traveling wave current distributions and are also called *traveling wave, aperiodic,* or *non-resonant antennas.* In this case as there are no reflections and no backward waves the energy will be radiated only in the forward direction. The resulting radiation pattern is shown in Figure 20.16b.

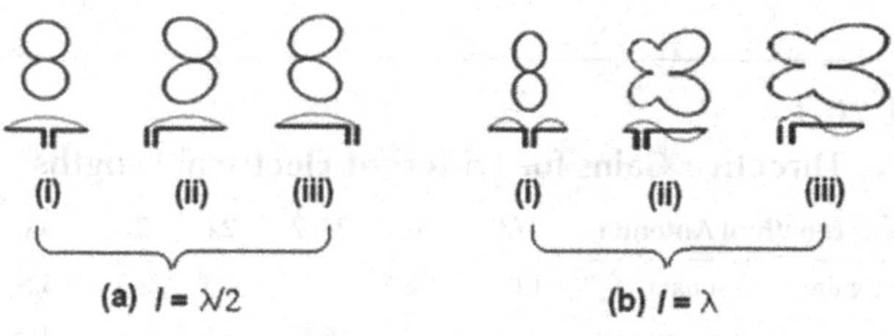

FIGURE 20.15 Vertical radiation patterns for feeds at different locations.

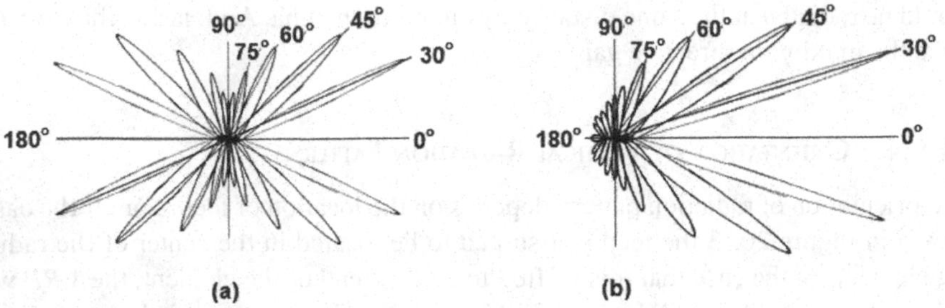

FIGURE 20.16 (a) Vertical radiation patterns of (a) resonant antenna; and (b) non-resonant antenna.

There are a large number of lobes as shown in Figures 20.16a and b. This number will depend on the electrical length of an antenna. As a rule, for every $\lambda/2$ length there will be one lobe in the radiation pattern. Thus, for an element with $n\lambda/2$ length there will be n lobes. Here, n is an integer that may take an even or odd value. If n is even, there will be two main lobes that will be symmetrically located along the axis perpendicular to the wire or element. For an odd n, there will be only one main lobe that will lie along the axis perpendicular to the wire.

20.6.5.3 Relative Directive Gains

As shown in Figure 20.16a the radiation is in both forward and backward directions whereas in Figure 20.16b it is only in the forward direction. Thus, there is more concentration of energy in a forward direction. This results in more directive gain for non-resonant antennas. Table 20.2 shows the relative directive gain for the two antennas for different electrical lengths.

20.6.6 Effect of Current Distribution on RPs

In Section 20.4, the current distribution was assumed to be sinusoidal and the field relations were derived accordingly. When the current distribution deviates from sinusoidal these relations get modified. Correspondingly the radiation patterns and related beam parameters get modified. The *Antenna Engineering Handbook* by Jesic (see references) includes the relative gain and half-power beamwidth (HPBW) for different current distributions along a wire. Antennas are not always in the form of straight wires and may take the form of folded wires, loops, or surfaces of flat or

TABLE 20.2
Relative Directive Gains for Different Electrical Lengths

Electrical Length of Antenna		$\lambda/2$	λ	$3\lambda/2$	2λ	3λ	4λ	8λ
Directive gain	Resonant	1.64	1.8	2	2.3	2.8	3.5	7.1
	Non-resonant	3.2	4.3	5.5	6.5	8.6	10.5	17.4

curved shapes. Jesic also includes current distributions for rectangular and circular flat surfaces referred to as rectangular and circular aperture distributions.

20.6.7 Effect of Ground on Radiation Patterns

The radiation pattern of an antenna with different lengths and/or current distributions may get modified due to the presence of a reflecting surface in its vicinity. The ground or earth too acts as a reflecting surface. Since an antenna may be installed on or nearer to the ground (or a flat surface such as in aircrafts or other vehicles) its radiation pattern is bound to be affected. The quantum of effect will be determined by the conductivity, relative permittivity, and the distance of the antenna from the ground or the surface. The effect will also depend on the orientation of the antenna with respect to the ground or the surface. The logic of this effect is described below.

Perfect earth or ground plane (GP) acts as a mirror to the radiating system. The antenna located above the earth sees its own image on the other side of the boundary surface. Figure 20.17 shows the vertical and horizontal radiators located above GP and the formation of their images. This figure also shows the direction of current flow and polarity of charges induced in antennas and their images. It results in the formation of images at different distances is in accordance with the locations of antennas.

Figure 20.18 shows the transmitting and receiving antennas of heights h_1 and h_2 located d distance apart. Energy radiated from antenna A will arrive at antenna B through a direct ray (DR) and a reflected ray (RR). The RR may be presumed to be coming from the lower tip of the image and will satisfy the relation AC + CB = A*C + CB. In the case of perfect GP there will be perfect reflection whereas in the case of imperfect GP part of the energy of the striking ray will get absorbed. This case can be visualized as that of a blurred mirror. Thus, the RP of an antenna gets modified. The degree of modification will depend on the degree of imperfection.

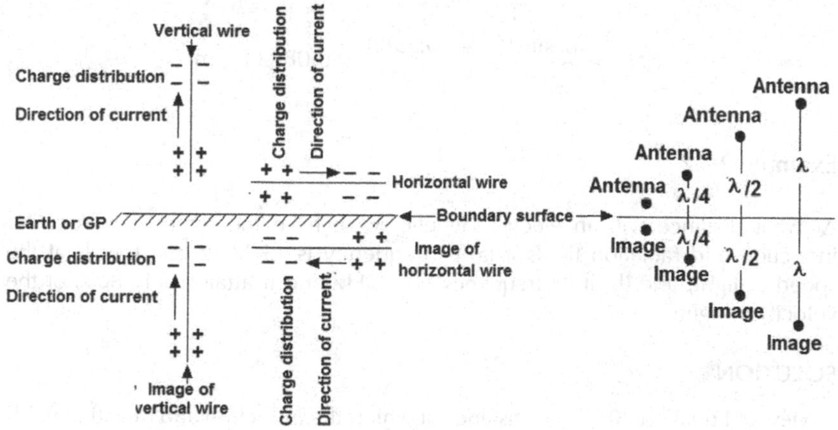

FIGURE 20.17 Antennas at different heights above the GP and formation of images.

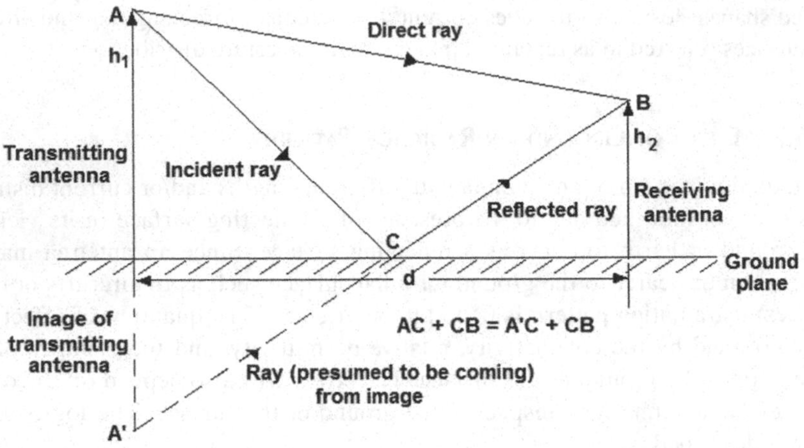

FIGURE 20.18 Direct and reflected rays between transmitting and receiving antennas.

Example 20.1

A 4 m long wire carrying 1 amp current radiates in $\theta = 15°$ direction in free space at $f = 5$ MHz. Find $|E|$ and $|H|$ at a distance of 100 km from the location of the radiating element.

SOLUTION

Given $f = 5$ MHz, thus $\lambda = 60$ m, length of the element $l = 4$ m, thus $l = \frac{\lambda}{15}$, $Idl =$ 4 A-m, $\theta = 15°$, $\sin\theta = 0.2588$, $r = 10^5$ m.

In view of Equation 20.29, the components contributing toward the radiation field are:

$$|E_\theta| = \frac{60\pi Idl \sin\theta}{\lambda r} = \frac{60\pi \times 4 \times 0.2588}{60 \times 10^5} = 19.513\ \mu\text{V/m}$$

$$|H_\varphi| = \frac{Idl \sin\theta}{2\lambda r} = \frac{4 \times 0.2588}{2 \times 60 \times 10^5} = 0.0863\ \mu\text{A/m}$$

Example 20.2

At what distance will an electromagnetic wave have the same magnitude for induction and radiation fields if (a) its frequency is 15 MHz and travels at the speed of light; and (b) if its frequency is 5 GHz and it attains only 80% of the velocity of light?

SOLUTION

In view of Equation 20.24, the distance at which the induction and radiation fields have the same magnitude is $r = \frac{\upsilon}{\omega} = \frac{\upsilon}{2\pi f}$. Thus:

(a) $r = \frac{\upsilon}{2\pi f} = \frac{3\times 10^8}{2\pi \times 15\times 10^6} = 3.1831\,\text{m}$

(b) $r = \frac{\upsilon}{2\pi f} = \frac{0.8\times 3\times 10^8}{2\pi \times 5\times 10^9} = 0.00764\,\text{m}$

Example 20.3

A radiating element of 1 cm carries an effective current of 1 Ampere. Calculate the radiated power at f = (a) 5 GHz; (b) 10 GHz.

SOLUTION

Given $I_{eff} = 1$ A, $dl = 10^{-2}$ m, $\upsilon = c \approx 3\times 10^8\ \frac{\text{m}}{\text{s}}$. Thus, in view of Equation 20.34:

(a) At 5 GHz, $\omega = 2\pi f = 2\pi \times 5 \times 10^9 = \pi \times 10^{10}$

$$P = \frac{20\omega^2 I_{eff}^2 dl^2}{\upsilon^2} = \frac{20\times\left(\pi\times 10^{10}\right)^2\times(1)^2\times\left(10^{-2}\right)^2}{\left(3\times 10^8\right)^2} = 21.93\ \text{W}$$

(b) At 10 GHz, $\omega = 2\pi f = 2\pi \times 10\times 10^9 = 2\pi\times 10^{10}$

$$P = \frac{20\omega^2 I_{eff}^2 dl^2}{\upsilon^2} = \frac{20\times\left(2\pi\times 10^{10}\right)^2\times(1)^2\times\left(10^{-2}\right)^2}{\left(3\times 10^8\right)^2} = 87.72\ \text{W}$$

Example 20.4

Evaluate the radiation resistance of a radiating element having 3 m length and operating at f =: (a) 200 KHz; (b) 20 MHz.

SOLUTION

(a) At f = 200 KHz, $\lambda = 1.5 \times 10^3$ m, $\frac{\lambda}{10} = 150$ m. Since the given length of element $L\ (= 3\ \text{m}) << \lambda/10$ the radiation resistance is to be evaluated by Equation 20.38a.

$$R_{rad} = 800\left(dl/\lambda\right)^2 = 800\left(\frac{3}{1.5\times 10^3}\right)^2 = 3.2\ \text{m}\Omega$$

(b) At f = 20 MHz, $\lambda = 15$ m, $\frac{\lambda}{10} = 1.5$ m. Since L (=3 m) > $\lambda/10$ the radiation resistance is to be obtained from Equations 20.38b and 20.38c for short dipole and short monopole, respectively.
Thus, for short dipole $R_{rad} = 200(L/\lambda)^2 = 200(3/15)^2 = 8\ \Omega$.

For a short monopole, $L = 2h$ or $h = L/2 = 1.5$ m.

$$R_{rad} \approx 400\left(h/\lambda\right)^2 = 400\left(1.5/15\right)^2 = 4\ \Omega$$

Example 20.5

Calculate the average power available in free space at 10 km distance if an element carrying a maximum 3 amp current radiates in $\theta =$: (a) 45° direction; (b) 30° direction.

SOLUTION

Given $r = 10^4$ m and $\eta = 120\pi \cdot I_m = 3$ A.

In view of Equation 20.47a:

$$P_{av} = \frac{\eta I_m^2}{8\pi^2 r^2}\left[\frac{\cos^2\left\{\frac{\pi}{2}\cos\theta\right\}}{\sin^2\theta}\right]$$

$$\frac{\eta I_m^2}{8\pi^2 r^2} = \frac{120\pi \times (3)^2}{8\pi^2 \times \left(10^4\right)^2} = \frac{3392.9}{78.95 \times 10^8} = 42.97 \times 10^{-8}$$

(a) For $\theta = 45°$, sin 45° = 0.707, $\sin^2 45° = 0.5$.

$$\cos 45° = 0.707, \quad \frac{\pi}{2}\cos\theta = 1.11 \quad \cos^2\left(\frac{\pi}{2}\cos 45\right) = 1.233,$$

$$\frac{\cos^2\left(\frac{\pi}{2}\cos\theta\right)}{\sin^2 45°} = \frac{1.233}{0.5} = 2.466$$

$$P_{av} = 42.97 \times 10^{-8} \times 2.466 = 1.0596\ \mu\text{W}$$

(b) For $\theta = 30°$, sin 30° = 0.5, $\sin^2 30° = 0.25$.

$$\cos 30° = 0.866, \quad \frac{\pi}{2}\cos 30 = 1.36, \quad \cos^2\left(\frac{\pi}{2}\cos\theta\right) = 1.8496$$

$$\frac{\cos^2\left(\frac{\pi}{2}\cos\theta\right)}{\sin^2 45°} = \frac{1.8496}{0.5} = 3.6992$$

$$P_{av} = 42.97 \times 10^{-8} \times 3.6992 = 1.5895\ \mu\text{W}$$

PROBLEMS

P20.1 A 3 m long radiating element carries a current of 3 amp. It radiates in $\theta = 25°$ direction in free space at $f = 10$ MHz. Estimate the ratio of magnitudes of E and H at a point located 50 km from the point of origination.

P20.2 Calculate the distance at which ratio of induction and radiation fields will be: (a) ¼; (b) 2; (c) 5 if the wave frequency is 5 MHz.

P20.3 If the medium of propagation allows the wave to attain only 70% of the velocity of light at what distance the induction and radiation fields will become equal in magnitude at $f = 3$ GHz.

P20.4 If an element of 1 cm length radiates 1 W at 3 GHz estimate the effective current carried by the element.

P20.5 Find the radiation resistance of an element of length $L = 2$ m at: (a) $f =$ 500 kHz; (b) $f = 50$ MHz.

P20.6 Calculate the average power available at 100 km distance if a current element with 2 amp radiates in the direction of $\theta =$: (a) 15°; (b) 25°.

DESCRIPTIVE QUESTIONS

Q20.1 What is a Hertzian dipole? How it is related to the filamentary current?

Q20.2 Explain the process of detachment of the wavefront from an oscillating dipole.

Q20.3 What is retarded potential? How is its concept relevant to the radiation mechanism?

Q20.4 Classify the fields on the basis of the terms involved in E and H relations.

Q20.5 What is radiation resistance? Give its value for a short dipole and a short monopole.

Q20.6 What are horizontal and vertical radiation patterns? Draw a vertical radiation pattern for the following the lengths: (a) $< \lambda/10$; (b) $\lambda/4$; (c) $\lambda/2$; (d) 2λ; and (e) 3λ.

Q20.7 Illustrate a lobe and show its different components.

Q20.8 Illustrate the vertical radiation patterns for feeds at different locations if the length of radiator is (a) $3\frac{\lambda}{2}$; and (b) 2λ.

Q20.9 What are terminated and unterminated antennas? Show the approximate radiation patterns for both if the wire length is (a) 5λ; and (b) 6λ.

Q20.10 What is the meaning of aperture distributions? What is their impact on radiation?

Q20.11 Discuss the effect of ground on (a) vertical radiator; and (b) horizontal radiator.

Q20.12 Derive the relations for the electric and magnetic far field components due to an oscillating dipole aligned along the z-axis.

Q20.13 Obtain the radiated power if the electric and magnetic field components are:

$$E_r = \frac{2Idl\cos\theta}{4\pi\varepsilon}\left[\frac{\cos\omega t'}{r^2\upsilon} + \frac{\sin\omega t'}{\omega r^3}\right]$$

$$E_\theta = \frac{Idl\sin\theta}{4\pi\varepsilon}\left[\frac{-\omega\sin\omega t'}{r\upsilon^2} + \frac{\cos\omega t'}{r^2\upsilon} + \frac{\sin\omega t'}{\omega r^3}\right]$$

and

$$H_\varphi = \frac{Idl \sin\theta}{4\pi}\left[\frac{\cos\omega t'}{r^2} + \frac{\omega \sin\omega t'}{r\upsilon}\right]$$

where $t' = t - \dfrac{r}{\upsilon}$ and the other parameters have their usual meaning.

Q20.14 Obtain the radiated power due to sinusoidal current distribution if the electric and magnetic far field components are:

$$|E_\theta| = \frac{60 I_m}{r}\left[\frac{\cos\left\{\cos\left(\frac{\pi}{2}\right)\cos\theta\right\}}{\sin\theta}\right] \text{ and } |H_\varphi| = \frac{I_m}{2\pi r}\left[\frac{\cos\left\{\cos\left(\frac{\pi}{2}\right)\cos\theta\right\}}{\sin\theta}\right]$$

FURTHER READING

J. D. Kraus, R. J. Marhefka, A. S. Khan, *Antennas and Wave Propagation*, 5th Ed, Tata McGraw, New Delhi, 2018.

M. L. Sisodia, V. L. Gupta, *Microwaves – Introduction to Circuits Devices and Antennas*, New Age International (Pvt) Ltd., New Delhi, 2006.

Y. Ito, *Radio Wave Propagation Handbook*, ed. Y. Hosoya, Realize Inc., Japan, 1999, pp. 171–199.

K. Chang, *Microwave Ring Circuits and Antennas*, Wiley, New York, 1996.

E. C. Jordon, K. G. Balmain, *Electromagnetic Waves and Radiating Systems*, Prentice Hall Ltd., New Delhi, 1987.

R. S. Elliot, *Antenna Theory and Design*, Prentice Hall, New Delhi, 1985.

W. L. Weeks, *Antenna Engineering*, Tata McGraw Hill, New Delhi, 1974.

R. G. Brown, et al., *Lines, Waves and Antennas*, 2nd Ed, John Wiley & Sons, New York, 1970.

R. B. Alder, et al., *Electromagnetic Energy Transmission and Radiation*, MIT Press, Cambridge, MA, 1969.

H. Jesic, *Antenna Engineering Handbook*, McGraw Hill Book Co., USA, 1961.

21 Eddy Currents

21.1 INTRODUCTION

The terms current, current density, and their types were introduced in Chapter 8. The displacement current and displacement current density were described at length in Chapter 15. Yet another type of current, referred to as an eddy current, is the subject of the present chapter. This chapter describes the genesis, meaning, importance, and the applications of eddy currents. The eddy current equation and its solution in one, two, and three dimensions also is discussed. Finally, eddy currents in anisotropic medium are described.

21.2 BASIC ASPECTS

Eddy currents are circular electric currents induced within conductors by a changing magnetic field in the conductor, in accordance with *Faraday's law of induction*. These are also referred to as *Foucault currents*. These flow in closed loops within conductors, in planes perpendicular to the magnetic field. They can be induced within nearby stationary conductors by a time-varying magnetic field created by an electromagnet or transformer, for example, by relative motion between a magnet and a nearby conductor. The magnitude of the current in a given loop is proportional to the strength of the magnetic field, the area of the loop, and the rate of change of flux, and inversely proportional to the resistivity of the material.

By *Lenz's law*, an eddy current creates a magnetic field that opposes the magnetic field that created it, and thus eddy currents react back on the source of the magnetic field. For example, a nearby conductive surface will exert a drag force on a moving magnet that opposes its motion, due to eddy currents induced in the surface by the moving magnetic field. This effect is employed in *eddy current brakes*, which are used to stop rotating power tools quickly when they are turned off. The current flowing through the resistance of the conductor also dissipates energy as heat in the material. Thus, eddy currents are a source of energy loss in alternating current (ac) *inductors*, *transformers*, *electric motors*, and *generators*, and other ac machinery, requiring special construction such as *laminated magnetic cores* to minimize them. Eddy currents are also used to heat objects in *induction heating furnaces and equipment*, and to detect cracks and flaws in metal parts using *eddy current testing instruments*.

Before dealing with the mathematical aspects of the eddy currents it appears to be necessary to understand some of its basic aspects. These are given in the following subsections.

21.2.1 Genesis and Meaning

The 25th prime minister of France, the mathematician, physicist, and astronomer Francis Arago, was the first person to observe current eddies in 1824, which he called *rotatory magnetism*. This discovery was completed and explained by Michael Faraday. As per the Lenz's law, developed by Heinrich Lenz in 1834, the direction of induced current flow in an object is such that its magnetic field opposes the change of magnetic field that caused the current flow. Eddy currents produce a secondary field that cancels a part of the external field and causes some of the external flux to avoid the conductor. In 1855, Leon Foucault, a French physicist, discovered that the force required for the rotation of a copper disc becomes greater when it is made to rotate with its rim between the poles of a magnet, the disc at the same time becoming heated by the eddy current induced in the metal. The first use of eddy current for non-destructive testing was reportedly made in 1879 when David E. Hughes used this principle to conduct metallurgical sorting tests.

The presence of a time-varying magnetic flux in a conducting medium results in an electromotive force. This force lies in the plane perpendicular to the direction of flux change and causes flow of current in the material. This current is referred to as an *eddy current*. The magnitude of this current depends on the *geometry of the medium, rate of alternation of flux*, and the *electric and magnetic properties of the materials* involved. The direction of this current is always such so as to oppose the change in flux that produces it. Such a flow of current prevents immediate penetration of flux into the interior of the matter. Thus, for continuously varying applied field the magnetizing force in the interior of the material never exceeds a small fraction of the magnetizing force at the surface.

To understand the meaning of an eddy current, let us consider Figure 21.1a. It shows a water channel with a side trench. The direction of water flow is shown to be from left to right. The water will enter in trench at A in the downward direction. After coming to a dead end of the trench at B it will turn in the right direction and then upward at C. Depending on the relative speeds of water in the main channel and the trench a part of it may join the water flow in the forward direction at D and part of it may turn toward A. In the latter case it will again turn toward B, etc. The process will continue and the part of water in the trench will keep on circulating forming what is referred to as eddies. This analogy of eddies can be extended to understand the flow of eddy current in a metallic core as described below.

Figure 21.1b shows a flat metallic sheet with a hole (such as the top of a transformer tank). A current carrying conductor with an alternating current is passing

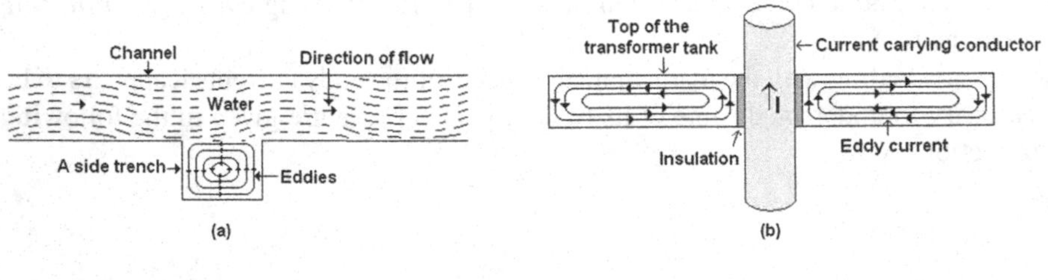

FIGURE 21.1 Eddies due to (a) water flow; (b) current flow.

through this hole. The direction and magnitude of the current and the insulation between the sheet and conductor are also shown. This current will result in an EMF which will link with the metallic sheet. This linking EMF will result in flow of current in the metallic sheet. In the finite dimensions of the sheet this current will first flow upward, turn rightward, turn downward, and then toward left. Since the flow of this current is in the form eddies produced in the water it can be referred to as eddy current. This current will also induce an EMF which will be in opposition to the EMF which produces this eddy.

Figures 21.2a and 21.2b illustrate the flow of currents in solid and laminated cores respectively.

21.2.2 Importance and Applications

The interaction of eddy currents with inducing field results in mechanical torque. Such a useful phenomenon is observed in polyphase induction machines with solid rotors. The flow of these currents also results in power loss referred to as eddy current loss. This power loss produces heat in the material and becomes the base for induction heating. There are many other situations wherein the study of eddy current phenomenon becomes essential particularly when the field and current in conducting media are time-varying. The thorough understanding of eddy current phenomenon and the proper estimation of penetration of alternating flux into the solid iron is needed in the design of *electromechanical clutches, eddy current brakes, electromagnetic coupling, solid iron rotor induction machines*, and the *self-starting synchronous motors.*

21.2.3 Eddy Current Equation

The *eddy current equation* (known in physics as the *diffusion equation*), is a useful tool for modelling the effect of eddy currents in a material. It can be derived from the differential, magnetostatic form of Ampère's law providing an expression for the magnetizing field *H* surrounding a current density *J*, reproduced here:

$$\nabla \times \boldsymbol{H} = \boldsymbol{J} \tag{11.24}$$

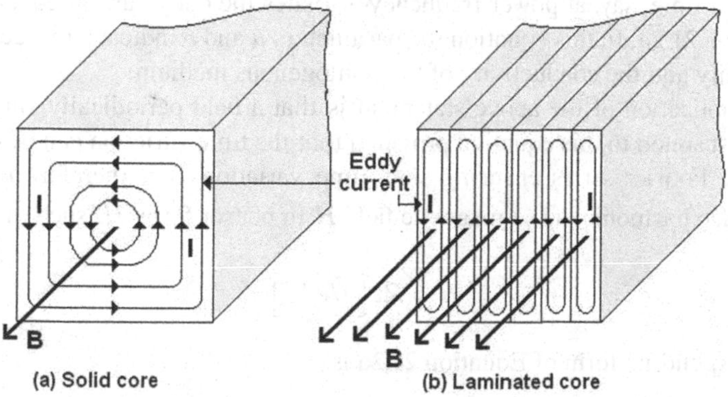

FIGURE 21.2 Flow of current in (a) solid core; (b) laminated core.

In view of the vector identity given by Equation 4.41d and Equation 11.24:

$$\nabla \times \boldsymbol{H}(\nabla \bullet \boldsymbol{H}) - \nabla^2 \boldsymbol{H} = \nabla \times \boldsymbol{J}$$

Since $\nabla \bullet \boldsymbol{H} = \boldsymbol{0}$ (Equation 11.55), we get:

$$\nabla^2 \boldsymbol{H} = -\nabla \times \boldsymbol{J} \tag{21.1a}$$

Using the relation $J = \sigma E$, for an isotropic homogeneous conductivity, the above equation becomes:

$$\nabla^2 \boldsymbol{H} = -\sigma(\nabla \times \boldsymbol{E}) \tag{21.1b}$$

Using $\nabla \times \boldsymbol{E} = -\frac{\partial \boldsymbol{B}}{\partial t}$ (from Equation 15.6b), we get:

$$\nabla^2 \boldsymbol{H} = -\sigma\left(-\frac{\partial \boldsymbol{B}}{\partial t}\right) \tag{21.1c}$$

Also substituting $\boldsymbol{B} = \mu \boldsymbol{H}$ (from Equation 11.1), we finally get:

$$\nabla^2 \boldsymbol{H} = \sigma\mu\left(\frac{\partial \boldsymbol{H}}{\partial t}\right) \tag{21.2a}$$

Equation 21.2a is referred to as the *eddy current (or diffusion) equation.* If the magnetization ($\boldsymbol{M}$) of the material involved in the relation $\mathbf{B} = \mu_0 (\mathbf{H} + \mathbf{M})$ (Equation 12.43b) is also to be considered, this equation becomes:

$$\nabla^2 \boldsymbol{H} = \sigma\mu_0\left(\frac{\partial \boldsymbol{H}}{\partial t} + \frac{\partial \boldsymbol{M}}{\partial t}\right) \tag{21.2b}$$

The slowly time-varying electromagnetic fields are often referred to as quasi-stationary fields. The magnetic field intensity $\boldsymbol{H}$, in a conducting medium, varying slowly with time, say, at power frequency, satisfies the eddy current equation given by Equation 21.2a. In this equation the parameters μ and σ indicate, respectively, the permeability and the conductivity of the homogenous medium.

The implication of the above statement is that a field periodically varying with time is best suited to this equation provided that the time variation can be expressed by a finite Fourier series ensuring slow time variation. It is therefore possible to express each harmonic of the magnetic field $\boldsymbol{H}$ in phasor form $\tilde{\boldsymbol{H}}$, such that:

$$\boldsymbol{H} = \mathcal{R}e\left[\tilde{\boldsymbol{H}} e^{j\omega t}\right] \tag{21.3}$$

The corresponding form of Equation 21.2a is:

$$\nabla^2 \tilde{\boldsymbol{H}} = (j\omega\mu\sigma)\tilde{\boldsymbol{H}} \tag{21.4}$$

21.2.4 Boundary Conditions

For the determination of the unique solution of this equation giving the field distribution in a specified region, it is necessary that any one of the following sets of boundary conditions must be prescribed and the solution must satisfy the same. These conditions are:

1. The tangential component of the magnetic field intensity may be given over the entire boundary surface bounding the region.
2. The tangential component of the electric field intensity may be given over the entire boundary surface bounding the region.
3. The components of both, the magnetic as well as electric field intensities along arbitrarily selected tangential directions may be specified over the entire boundary surface bounding the region.
4. A combination of any two or all the three conditions stated above; each valid for different parts of closed bounding surface.

21.2.5 Type of Eddy Current Problems

The electromagnetic field problems in general are classified as one-, two-, or three-dimensional boundary value problems. This feature is reflected by the expression for the Laplacian operator, ∇^2, discussed in Chapter 4. As with problems involving Laplace's and Poisson's equations discussed in Chapters 9 and 10, and those involving the wave equation discussed in Chapter 16, the problems related to the eddy current equation also belong to the family of *boundary value problems.* A unique solution for the eddy current equation given by Equation 21.4 has to satisfy one or the other boundary condition given above. The art of solving eddy current equation as one-, two-, or three-dimensional boundary value problems is described in the subsequent sections in detail.

21.3 ONE-DIMENSIONAL PROBLEM

The solution to the one-dimensional problem is illustrated by the following example.

21.3.1 Infinite Space with Two Regions

The one-dimensional boundary value problems involving eddy currents in an infinite space can be solved by using the following steps.

21.3.1.1 Physical Configuration

The configuration used for solving one-dimensional boundary value problems, involving eddy currents, is shown in Figure 21.3. In this figure the entire space is divided into two infinite-half regions. Region 1 encompasses the entire space for $z > 0$ and Region 2 covers the entire space for $z < 0$. Region 1 is taken to be the free space, whereas Region 2 is considered to be a homogeneous conducting region. The figure also shows a current sheet located at $z = 0$, which acts as the

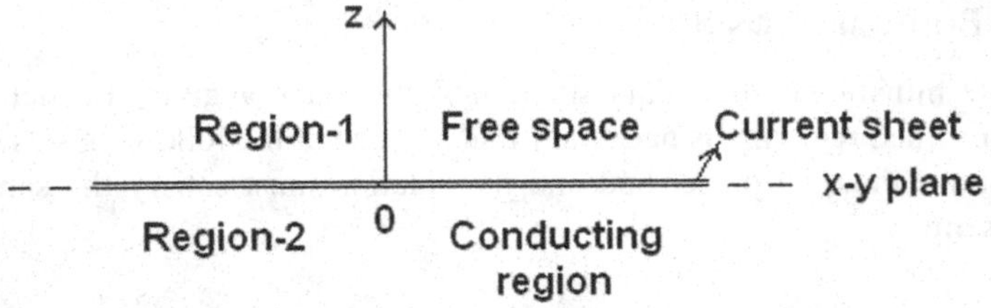

FIGURE 21.3 Physical configuration.

source for the eddy currents in the conducting sheet. It is basically the nature of variation of current along the coordinates, in the sheet which makes the problem one-, two-, or three-dimensional. In this case the variation of field is taken only along the z-axis.

21.3.1.2 Source

The source for magnetic field is a uniformly distributed, sinusoidally time-varying current sheet located at $z = 0$. The density of this surface current is given as:

$$K_x = \mathcal{R}e\left[\tilde{K}_o e^{j\omega t}\right] \tag{21.5}$$

or simply as:

$$K_x \overset{\text{def}}{=} \tilde{K}_o \tag{21.6}$$

where the constant $\left|\tilde{K}_o\right|$ indicates the RMS value for the surface current density.

21.3.1.3 Boundary Conditions

The boundary conditions to be satisfied by the field are:

$$H_{2y}\Big|_{z=0} = H_{1y}\Big|_{z=0} + K_x \tag{21.7}$$

$$H_{2y}\Big|_{z=-\infty} = \text{finite} \tag{21.8}$$

$$H_{1y}\Big|_{z=+\infty} = \text{finite} \tag{21.9}$$

Since each infinite-half region is homogenous and the source is independent of x- and y-coordinates, the magnetic field will also be independent of x- and y-coordinates. Region 1 being free space, the field component H_{1y} will satisfy the Laplace equation, i.e.:

$$\left.\nabla^2 H_{1y} = 0 \quad (\text{a}) \quad \text{or} \quad \frac{d^2}{dz^2} H_{1y} = 0 \quad (\text{b})\right\} \tag{21.10}$$

21.3.1.4 Solution

The complete solution of Equation 21.10b can be written as:

$$H_{1y} \stackrel{\text{def}}{=} \tilde{H}_{1y} = \tilde{c}_1 + \tilde{c}_1' \cdot z \tag{21.11}$$

where $\tilde{c}_1$ and $\tilde{c}_1'$ indicate two arbitrary constants. In view of Equation 21.9, the field is finite at $z = +\infty$, thus we choose:

$$\tilde{c}_1' = 0 \tag{21.12}$$

This gives:

$$\tilde{H}_{1y} = \tilde{c}_1 \tag{21.13}$$

This shows that the field in free space remains undamped, i.e. it is the same at all distances from its source, the current sheet.

The magnetic field in Region 2, which is a homogenous conducting region, satisfies the following equation:

$$\nabla^2 \tilde{H}_{2y} = \left(j\omega\mu\sigma\right)\tilde{H}_{2y} \tag{21.14}$$

Since the phasor $\tilde{H}_{2y}$ is a function of the z-coordinate only, Equation 21.14 reduces to:

$$\frac{d^2}{dz^2}\tilde{H}_{2y} = \left(j\omega\mu\sigma\right)\tilde{H}_{2y} \tag{21.15}$$

This is an ordinary second-order homogenous differential equation. Its solution is:

$$\tilde{H}_{2y} = \tilde{c}_2\, e^{\frac{1+j}{d}\cdot z} + \tilde{c}_2'\, e^{-\frac{1+j}{d}\cdot z} \tag{21.16}$$

where $\tilde{c}_2$ and $\tilde{c}_2'$ indicate two arbitrary constants.

The classical depth of penetration for a *one-dimensional field* is defined as:

$$d \stackrel{\text{def}}{=} \sqrt{\frac{2}{\omega\mu\sigma}} \tag{21.17a}$$

Using the boundary condition given by Equation 21.8, we have:

$$\tilde{c}_2' = 0 \tag{21.17b}$$

Therefore, Equation 21.16 reduces to:

$$\tilde{H}_{2y} = \tilde{c}_2\, e^{\frac{1+j}{d}\cdot z} \tag{21.18}$$

From Equation 21.7, we have the boundary condition for the magnetic fields in phasor form:

$$\tilde{H}_{2y}\Big|_{z=0} = \tilde{H}_{1y}\Big|_{z=0} + \tilde{K}_o \tag{21.19}$$

Therefore, from Equations 21.13 and 21.18:

$$\tilde{c}_2 = \tilde{c}_1 + \tilde{K}_o \tag{21.20}$$

This gives one equation between two unknowns, and apparently no more boundary conditions are available! Let us rewrite Equation 21.18 as:

$$\tilde{H}_{2y} = \tilde{c}_2 - \tilde{c}_2\left[1 - e^{\frac{1+j}{d}.z}\right] \tag{21.21}$$

The second term on the RHS vanishes as d becomes infinity when the conductivity of the region is set to zero. This term, therefore, can be identified as the damping field caused by eddy currents induced in this region. Note that the damping field is zero at the boundary (i.e. at $z = 0$), it increases with the distance from the current sheet, and at infinite distance (i.e. at $z = -\infty$) this field becomes equal to the undamped field. It can thus be concluded that the damping field has no effect on the boundary condition described by Equation 21.19. Furthermore, in the absence of eddy currents, the magnetic field intensity is asymmetrically distributed about the $z = 0$ plane. It can be given as:

$$\tilde{H}_{2y}\left(\sigma = 0\right)\Big|_{z\le 0} = -\tilde{H}_{1y}\Big|_{z\ge 0} \tag{21.22}$$

This is the boundary condition we were looking for. From this equation, we get:

$$\tilde{c}_2 = -\tilde{c}_1 \tag{21.23}$$

Solving Equation 21.20 and 21.23, we obtain:

$$\tilde{c}_2 = -\tilde{c}_1 = \frac{1}{2}\tilde{K}_o \tag{21.24}$$

The solution for the eddy current equation thus found is:

$$\tilde{H}_{2y} = \frac{1}{2}\tilde{K}_o\, e^{\frac{1+j}{d}.z} \tag{21.25}$$

In the absence of any other source, the x- and z-components of the magnetic field will be 0.

21.3.1.5 Eddy Current Density

The eddy current density $\tilde{J}_{2x}$ can be obtained from Maxwell's equation for a quasi-stationary field as:

$$\boldsymbol{J} \cong \nabla \times \boldsymbol{H} \tag{21.26}$$

Thus:

$$\tilde{J}_{2x} = -\frac{d}{dz}\tilde{H}_{2y} = -\frac{1}{2}\tilde{K}_o \cdot \frac{1+j}{d} \cdot e^{\frac{1+j}{d} \cdot z} \tag{21.27}$$

21.4 TWO-DIMENSIONAL BOUNDARY VALUE PROBLEMS

This section deals with two cases of two-dimensional eddy current problems. In both of these cases, the field is assumed to be varying along the y and z directions. The current sources for these cases are taken to be of a different nature.

21.4.1 Infinite Space with Two Regions

The eddy current problem in an infinite space with two regions is solved as below.

21.4.1.1 Physical Configuration

This case uses the configuration shown by Figure 21.3 wherein the entire space is divided into two infinite-half regions as described in Section 21.3.1.

21.4.1.2 Source

For this case, the distribution of the surface current density at $z = 0$ is given as:

$$K_x = \mathcal{R}e\left[\tilde{K}_o \cos\left(\frac{\pi}{\tau} y\right) e^{j\omega t}\right] \tag{21.28}$$

or simply as:

$$\tilde{K}_x \overset{\text{def}}{=} \tilde{K}_o \cos\left(\frac{\pi}{\tau} y\right) \tag{21.29}$$

This current sheet simulates an ideal single-phase armature winding of a rotating electrical machine, with all winding harmonics neglected.

21.4.1.3 Boundary Conditions

For this case, the boundary conditions are:

$$\tilde{H}_{2y}\Big|_{z=0} = \tilde{H}_{1y}\Big|_{z=0} + \tilde{K}_o \cos\left(\frac{\pi}{\tau} y\right) \tag{21.30}$$

$$\tilde{H}_{2y}\Big|_{z=-\infty} = \text{finite} \tag{21.31}$$

$$\tilde{H}_{1y}\Big|_{z=+\infty} = \text{finite} \tag{21.32}$$

21.4.1.4 Solution

The field in the Region 1 obeys Laplace's equation. Thus, in view of Equations 21.30 and 21.32, we may assume:

$$\tilde{H}_{1y} = \tilde{c}_1 \cos\left(\frac{\pi}{\tau} y\right) e^{-\frac{\pi}{\tau} z} \tag{21.33}$$

where $\tilde{c}_1$ indicates an arbitrary constant.

Since the field is independent of the x coordinate, and its divergence is 0, the x-component of this field is absent as there is no source to produce it, while the z-component of this field can be given in view of Equation 21.33 as:

$$\tilde{H}_{1z} = -\tilde{c}_1 \sin\left(\frac{\pi}{\tau} y\right) e^{-\frac{\pi}{\tau} z} \tag{21.34}$$

To the list of boundary conditions, we should add another boundary condition:

$$\mu \tilde{H}_{2z}\Big|_{z=0} = \mu_o \tilde{H}_{1z}\Big|_{z=0} \tag{21.35}$$

Now, consider the Region 2. In view of what we have learned so far, we choose a tentative solution for Equation 21.14 as:

$$\tilde{H}_{2y} = \tilde{c}_2 \cos\left(\frac{\pi}{\tau} y\right) e^{\tilde{\alpha} z} \tag{21.36}$$

where $\tilde{c}_2$ indicates an arbitrary constant. To determine the expression for $\tilde{\alpha}$, we substitute this expression for $\tilde{H}_{2y}$ in the Equation 21.14. Thus:

$$\left(\partial^2/\partial x^2 + \partial^2/\partial y^2 + \partial^2/\partial z^2\right)\tilde{c}_2 \cos\left(\frac{\pi}{\tau} y\right) e^{\tilde{\alpha} z} = \left(j\omega\mu\sigma\right)\tilde{c}_2 \cos\left(\frac{\pi}{\tau} y\right) e^{\tilde{\alpha} z}$$

Therefore:

$$\left[-\left(\frac{\pi}{\tau}\right)^2 + \tilde{\alpha}^2\right]\cdot \tilde{c}_2 \cos\left(\frac{\pi}{\tau} y\right) e^{\tilde{\alpha} z} = \left(j\omega\mu\sigma\right)\cdot \tilde{c}_2 \cos\left(\frac{\pi}{\tau} y\right) e^{\tilde{\alpha} z}$$

or

$$\left[-\left(\frac{\pi}{\tau}\right)^2 + \tilde{\alpha}^2\right] = \left(j\omega\mu\sigma\right) \tag{21.37}$$

From Equation 21.37:

$$\tilde{\alpha} = \sqrt{\left(\frac{\pi}{\tau}\right)^2 + j\omega\mu\sigma} \tag{21.38}$$

Noting that the field is independent of the x coordinate, and its divergence is 0, the x-component of this field is absent as there is no source to produce it, while the z-component of this field can be given, in view of Equation 21.36, as:

$$\tilde{H}_{2z} = \tilde{c}_2 \frac{\pi}{\tilde{\alpha}\tau} \sin\left(\frac{\pi}{\tau} y\right) e^{\tilde{\alpha} z} \tag{21.39}$$

In developing field expressions, boundary conditions given by Equations 21.31 and 21.32 have been taken into account. Now, we are left with two boundary conditions, as given by Equations 21.30 and 21.35. These boundary conditions must be used to determine the two arbitrary constants, $\tilde{c}_1$ and $\tilde{c}_2$. In view of Equations 21.30, 21.33, and 21.36, we get:

$$\tilde{c}_2 = \tilde{c}_1 + \tilde{K}_o \tag{21.40}$$

while from Equations 21.35, 21.34, and 21.39:

$$\mu \cdot \tilde{c}_2 \frac{\pi}{\tilde{\alpha}\tau} = -\mu_o \cdot \tilde{c}_1$$

or,

$$\tilde{c}_1 = -\mu_r \frac{\pi}{\tilde{\alpha}\tau} \cdot \tilde{c}_2 \tag{21.41}$$

where μ_r indicates the relative permeability of Region 2.

Solving Equations 21.40 and 21.41, one gets:

$$\left.\begin{aligned} \tilde{c}_2 = \frac{\tilde{K}_o}{\left[1 + \mu_r \dfrac{\pi}{\tilde{\alpha}\tau}\right]} \quad (\text{a}) \qquad \tilde{c}_1 = -\frac{\mu_r \cdot \tilde{K}_o}{\left[\mu_r + \dfrac{\tilde{\alpha}\tau}{\pi}\right]} \quad (\text{b}) \end{aligned}\right\} \tag{21.42a}$$

21.4.2 Rectangular Conducting Core

This case of a two-dimensional problem involving eddy current is solved as follows.

21.4.2.1 Physical Configuration

Figure 21.4 shows the physical configuration to be used to deal with this two-dimensional case. It contains a long, solid conducting core with a rectangular cross-section ($a \times b$).

21.4.2.2 Source

In this case, excitation is due to the alternating current in a uniformly distributed winding wound around the core. This current carrying winding is simulated by a current sheet on the core surface. Let the current density in this current sheet be defined as:

$$\tilde{\boldsymbol{K}}\Big|_{x=\pm a/2} \overset{\text{def}}{=} \pm \tilde{K}_o \boldsymbol{a}_y, \quad \text{over} \quad b/2 < y < b/2 \tag{21.43a}$$

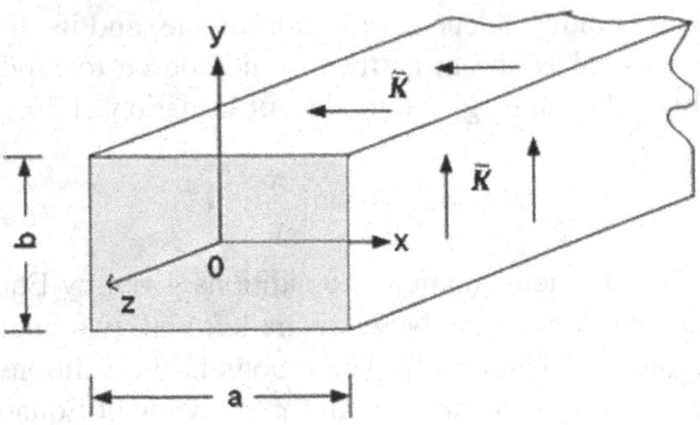

FIGURE 21.4 Rectangular solid conducting core.

and

$$\tilde{K}\Big|_{y=\pm b/2} \overset{\text{def}}{=} \mp \tilde{K}_o \boldsymbol{a}_x, \quad \text{over} \quad a/2 < x < a/2 \tag{21.43b}$$

where $\left|\tilde{K}_o\right|$ indicates the RMS value of the surface current density.

At power frequencies, the displacement currents are neglected resulting in zero fields outside the core, while inside the core only axial component of the magnetic field, $\tilde{H}_z$ exists.

21.4.2.3 Boundary Conditions

The boundary conditions are as follows:

$$\tilde{H}_z\Big|_{x=\pm a/2} = \tilde{K}_o \quad \text{over} \quad b/2 < y < b/2 \tag{21.44a}$$

$$\tilde{H}_z\Big|_{y=\pm b/2} = \tilde{K}_o \quad \text{over} \quad a/2 < x < a/2 \tag{21.44b}$$

Since the field is independent of the z-coordinate, Equation 21.14 can be rewritten as:

$$\left(\partial^2/\partial x^2 + \partial^2/\partial y^2\right)\tilde{H}_z = \left(j\omega\mu\sigma\right)\tilde{H}_z \tag{21.45}$$

Boundary conditions given by Equations 21.44a and 21.44b suggest that the field expression to satisfy the first boundary condition may be chosen as a periodic function of y, while to satisfy the second boundary condition it should be a periodic function of x.

21.4.2.4 Solution

For Equation 21.45, no simple solution emerges that is periodic in both the x- as well as y-direction and also independent of z at the same time. One of several possible solutions could be to split up the solution for Equation 21.45 into two parts, each satisfying this equation. One solution may satisfy boundary conditions given by Equation 21.44 and give zero values on the remaining boundaries, while the other solution should satisfy the boundary conditions given by Equation 21.44b and give zero values on the remaining boundaries. This way, the sum of the two solutions shall jointly satisfy all boundary conditions. Therefore, let:

$$\tilde{H}_z \overset{\text{def}}{=} \tilde{H}_z' + \tilde{H}_z'' \tag{21.46}$$

Since the field $\tilde{H}_z$ is an even function of both x and y, the tentative partial solutions are:

$$\tilde{H}_z' = \sum_{m\text{-odd}}^{\infty} F_m' \cdot \cos\left(\frac{m\pi}{b} y\right) \cdot \frac{\cosh\left(\gamma_m' \cdot x\right)}{\cosh\left(\gamma_m' \cdot a/2\right)} \tag{21.47a}$$

and

$$\tilde{H}_z'' = \sum_{n\text{-odd}}^{\infty} F_n'' \cdot \cos\left(\frac{n\pi}{a} x\right) \cdot \frac{\cosh\left(\gamma_n'' \cdot y\right)}{\cosh\left(\gamma_n' \cdot b/2\right)} \tag{21.47b}$$

where the two Fourier coefficients, viz. F_m' and F_m'', are chosen to satisfy the boundary conditions specified by Equations 21.44a and 21.44b, respectively, thus:

$$\tilde{H}_z\Big|_{x=\pm a/2} = \tilde{H}_z'\Big|_{x=\pm a/2} = \sum_{m\text{-odd}}^{\infty} F_m' \cdot \cos\left(\frac{m\pi}{b} y\right)$$

Therefore:

$$\sum_{m\text{-odd}}^{\infty} F_m' \cdot \cos\left(\frac{m\pi}{b} y\right) = \tilde{K}_o \tag{21.48}$$

To determine the Fourier coefficient F_m', multiply both sides of Equation 21.48 by $\cos\left(\frac{n\pi}{b} y\right)$ and then integrate over $-b/2 < y < b/2$. If we chose an odd integer value for n, all terms on the LHS will vanish except the one that corresponds to $m = n$. Therefore, the LHS:

$$\int_{-b/2}^{b/2} F_n' \cdot \cos^2\left(\frac{n\pi}{b} y\right) dy = \int_{-b/2}^{b/2} F_n' \cdot \frac{1}{2}\left[1 + \cos\left(\frac{2n\pi}{b} y\right)\right] dy = \frac{b}{2} \cdot F_n'$$

and for the RHS:

$$\int_{-b/2}^{b/2} \tilde{K}_o \cos\left(\frac{n\pi}{b} y\right) dy = \frac{b}{n\pi} \cdot 2\sin\left(\frac{n\pi}{2}\right) \cdot \tilde{K}_o$$

thus giving:

$$F_n' = \tilde{K}_o \cdot \frac{4}{n\pi} \cdot \sin\left(\frac{n\pi}{2}\right)$$

Finally, on replacing n by m, we get:

$$\left. F_m' = \tilde{K}_o \cdot \frac{4}{m\pi} \cdot \sin\left(\frac{m\pi}{2}\right) \quad (a) \quad \text{and} \quad F_n'' = \tilde{K}_o \cdot \frac{4}{n\pi} \cdot \sin\left(\frac{n\pi}{2}\right) \quad (b) \right\} \tag{21.49}$$

Therefore, the complete solution is given as follows:

$$\begin{aligned} \tilde{H}_z = {} & \sum_{m\text{-odd}}^{\infty} \tilde{K}_o \cdot \frac{4}{m\pi} \cdot \sin\left(\frac{m\pi}{2}\right) \cdot \cos\left(\frac{m\pi}{b} y\right) \cdot \frac{\cosh(\gamma_m' \cdot x)}{\cosh(\gamma_m' \cdot a/2)} \\ & + \sum_{n\text{-odd}}^{\infty} \tilde{K}_o \cdot \frac{4}{n\pi} \cdot \sin\left(\frac{n\pi}{2}\right) \cdot \cos\left(\frac{n\pi}{a} x\right) \cdot \frac{\cosh(\gamma_n'' \cdot y)}{\cosh(\gamma_n' \cdot b/2)} \end{aligned} \tag{21.50}$$

On substituting the two parts of this expression separately in Equation 21.45, we find:

$$\left. \gamma_m' = \sqrt{\left(\frac{m\pi}{b}\right)^2 + j\omega\mu\sigma} \quad (a) \quad \gamma_n'' = \sqrt{\left(\frac{n\pi}{a}\right)^2 + j\omega\mu\sigma} \quad (b) \right\} \tag{21.51}$$

21.5 THREE-DIMENSIONAL BOUNDARY VALUE PROBLEMS

This section deals with a three-dimensional eddy current boundary value problem.

21.5.1 RECTANGULAR CONDUCTING CORE

The steps followed to solve this problem are as below.

21.5.1.1 Physical Configuration

This case also uses the configuration of rectangular core shown in Figure 21.4.

21.5.1.2 Source

The field is produced by a poly phase winding carrying balanced poly phase currents, wrapped around the core. Here the symbol $\mathcal{R}e$ and the exponential factor (showing sinusoidal variation with time t and the axial direction z) has been suppressed.

21.5.1.3 Boundary Conditions

The configuration satisfies the following boundary conditions:

$$H_z\big|_{x=\pm a/2} = \mathcal{R}e\left[\tilde{H}_o \cdot e^{j\left(\omega t - \frac{\pi}{\tau} z\right)}\right] \quad \text{over:} \quad b/2 < y < b/2 \tag{21.52a}$$

or simply:

$$\tilde{H}_z\Big|_{x=\pm a/2} \overset{\text{def}}{=} \tilde{H}_o \qquad \text{over:} \quad b/2 < y < b/2 \tag{21.52b}$$

and

$$H_z\Big|_{y=\pm b/2} = \mathcal{R}e\left[\tilde{H}_o \cdot e^{j\left(\omega t - \frac{\pi}{\tau}z\right)}\right] \qquad \text{over:} \quad a/2 < x < a/2 \tag{21.53a}$$

or simply:

$$\tilde{H}_z\Big|_{y=\pm b/2} \overset{\text{def}}{=} \tilde{H}_o \qquad \text{over:} \quad a/2 < x < a/2 \tag{21.53b}$$

21.5.1.4 Solution

The solution to this problem is obtained as follows.

As the field is dependent on the z-coordinate as well, Equation 21.45 can be rewritten as:

$$\left(\partial^2/\partial x^2 + \partial^2/\partial y^2 + \partial^2/\partial z^2\right) = \left(j\omega\mu\sigma\right)\tilde{H}_z \tag{21.54}$$

Noting that $\tilde{H}_z$ is an even function of both x- and y-coordinates, the complete solution is given as follows:

$$\begin{aligned}\tilde{H}_z = &\sum_{m\text{-odd}}^{\infty} \tilde{H}_o \cdot \frac{4}{m\pi} \cdot \sin\left(\frac{m\pi}{2}\right) \cdot \cos\left(\frac{m\pi}{b}y\right) \cdot \frac{\cosh\left(\delta'_m . x\right)}{\cosh\left(\delta'_m \cdot a/2\right)} \\ &+ \sum_{n\text{-odd}}^{\infty} \tilde{H}_o \cdot \frac{4}{n\pi} \cdot \sin\left(\frac{n\pi}{2}\right) \cdot \cos\left(\frac{n\pi}{a}x\right) \cdot \frac{\cosh\left(\delta''_n \cdot y\right)}{\cosh\left(\delta'_n \cdot b/2\right)}\end{aligned} \tag{21.55}$$

On substituting the two parts of this expression separately in Equation 21.54, we find:

$$\left.\delta'_m = \sqrt{\left(\frac{m\pi}{b}\right)^2 + \left(\frac{\pi}{\tau}\right)^2 + j\omega\mu\sigma} \quad (\text{a}) \quad \delta''_n = \sqrt{\left(\frac{n\pi}{a}\right)^2 + \left(\frac{\pi}{\tau}\right)^2 + j\omega\mu\sigma} \quad (\text{b})\right\} \tag{21.56}$$

For the homogeneous region inside the rectangular core, the divergence of the magnetic field intensity $\tilde{\boldsymbol{H}}$ is zero. The expression for $\tilde{H}_z$ as given by Equation 21.55, shows that $\tilde{H}_z$ is a sinusoidal function of z. Therefore, inside the core, x and y components of the field should also exist such that:

$$\begin{aligned}\frac{\partial}{\partial x}\tilde{H}_x + \frac{\partial}{\partial y}\tilde{H}_y = -\frac{\partial}{\partial z}\tilde{H}_z = &\sum_{m\text{-odd}}^{\infty} \tilde{H}_o \cdot j\frac{4}{m\tau} \cdot \sin\left(\frac{m\pi}{2}\right) \cdot \cos\left(\frac{m\pi}{b}y\right) \cdot \frac{\cosh\left(\delta'_m \cdot x\right)}{\cosh\left(\delta'_m \cdot a/2\right)} \\ &+ \sum_{n\text{-odd}}^{\infty} \tilde{H}_o \cdot j\frac{4}{n\tau} \cdot \sin\left(\frac{n\pi}{2}\right) \cdot \cos\left(\frac{n\pi}{a}x\right) \cdot \frac{\cosh\left(\delta''_n \cdot y\right)}{\cosh\left(\delta'_n \cdot b/2\right)}\end{aligned} \tag{21.57}$$

For further boundary conditions, let us assume that:

$$\tilde{H}_x\Big|_{y=\pm b/2} \stackrel{\text{def}}{=} 0 \quad \text{over:} \quad a/2 < x < a/2 \tag{21.58a}$$

and

$$\tilde{H}_y\Big|_{x=\pm a/2} \stackrel{\text{def}}{=} 0 \quad \text{over:} \quad b/2 < y < b/2 \tag{21.58b}$$

From Equations 21.57 and 21.58a, we get:

$$\tilde{H}_x = \sum_{m\text{-odd}}^{\infty} \tilde{H}_o \cdot \frac{j4}{m\tau\delta'_m} \cdot \sin\left(\frac{m\pi}{2}\right) \cdot \cos\left(\frac{m\pi}{b}y\right) \cdot \frac{\sinh(\delta'_m \cdot x)}{\cosh(\delta'_m \cdot a/2)} \tag{21.59}$$

while from Equations 21.57 and 21.58b, we get:

$$\tilde{H}_y = \sum_{n\text{-odd}}^{\infty} \tilde{H}_o \cdot \frac{j4}{n\tau\delta''_n} \cdot \sin\left(\frac{n\pi}{2}\right) \cdot \cos\left(\frac{n\pi}{a}x\right) \cdot \frac{\sinh(\delta''_n \cdot y)}{\cosh(\delta'_n \cdot b/2)} \tag{21.60}$$

As the tangential components of magnetic field intensity is defined on all the four boundary surfaces (see Equations 21.47, 21.47b, 21.57a, and 21.57b), the field expressions as given by Equations 21.55, 21.59, and 21.60 are unique.

21.5.1.5 Eddy Current Density

The components of eddy current density are obtained using the relation:

$$\boldsymbol{J} = \nabla \times \boldsymbol{H}$$

Thus, we have:

$$\begin{aligned}
\tilde{J}_x &= \frac{\partial}{\partial y}\tilde{H}_z - \frac{\partial}{\partial z}\tilde{H}_y \\
&= -\sum_{m\text{-odd}}^{\infty} \tilde{H}_o \cdot \frac{4}{b} \cdot \sin\left(\frac{m\pi}{2}\right) \cdot \sin\left(\frac{m\pi}{b}y\right) \cdot \frac{\cosh(\delta'_m \cdot x)}{\cosh(\delta'_m \cdot a/2)} \\
&\quad + \sum_{n\text{-odd}}^{\infty} \tilde{H}_o \cdot \frac{4\delta''_n}{n\pi} \cdot \sin\left(\frac{n\pi}{2}\right) \cdot \cos\left(\frac{n\pi}{a}x\right) \cdot \frac{\sinh(\delta''_n . y)}{\cosh(\delta'_n \cdot b/2)} \\
&\quad - \sum_{n\text{-odd}}^{\infty} \tilde{H}_o \cdot \frac{4}{m\tau\delta''_n} \cdot \frac{\pi}{\tau} \cdot \sin\left(\frac{n\pi}{2}\right) \cdot \cos\left(\frac{n\pi}{a}x\right) \cdot \frac{\sinh(\delta''_n . y)}{\cosh(\delta'_n \cdot b/2)}
\end{aligned} \tag{21.61a}$$

$$\tilde{J}_y = \frac{\partial}{\partial z}\tilde{H}_x - \frac{\partial}{\partial x}\tilde{H}_z$$

$$= \sum_{m\text{-odd}}^{\infty} \tilde{H}_o \cdot \frac{4}{m\tau\delta'_m} \cdot \frac{\pi}{\tau} \cdot \sin\left(\frac{m\pi}{2}\right) \cdot \cos\left(\frac{m\pi}{b}y\right) \cdot \frac{\sinh\left(\delta'_m \cdot x\right)}{\cosh\left(\delta'_m \cdot a/2\right)}$$

$$- \sum_{m\text{-odd}}^{\infty} \tilde{H}_o \cdot \frac{4\delta'_m}{m\pi} \cdot \sin\left(\frac{m\pi}{2}\right) \cdot \cos\left(\frac{m\pi}{b}y\right) \cdot \frac{\sinh\left(\delta'_m \cdot x\right)}{\cosh\left(\delta'_m \cdot a/2\right)} \quad (21.61b)$$

$$+ \sum_{n\text{-odd}}^{\infty} \tilde{H}_o \cdot \frac{4}{a} \cdot \sin\left(\frac{n\pi}{2}\right) \cdot \sin\left(\frac{n\pi}{a}x\right) \cdot \frac{\cosh\left(\delta''_n \cdot y\right)}{\cosh\left(\delta'_n \cdot b/2\right)}$$

$$\tilde{J}_z = \frac{\partial}{\partial x}\tilde{H}_y - \frac{\partial}{\partial y}\tilde{H}_x$$

$$= -\sum_{n\text{-odd}}^{\infty} \tilde{H}_o \cdot \frac{j4}{\tau\delta''_n} \cdot \frac{\pi}{a} \cdot \sin\left(\frac{n\pi}{2}\right) \cdot \sin\left(\frac{n\pi}{a}x\right) \cdot \frac{\sinh\left(\delta''_n \cdot y\right)}{\cosh\left(\delta'_n \cdot b/2\right)} \quad (21.61c)$$

$$+ \sum_{m\text{-odd}}^{\infty} \tilde{H}_o \cdot \frac{j4}{\tau\delta'_m} \cdot \frac{\pi}{b} \cdot \sin\left(\frac{m\pi}{2}\right) \cdot \sin\left(\frac{m\pi}{b}y\right) \cdot \frac{\sinh\left(\delta'_m \cdot x\right)}{\cosh\left(\delta'_m \cdot a/2\right)}$$

In view of Equations 21.59 and 21.60, we note that:

$$\left.\tilde{H}_x\right|_{x=\pm a/2} \neq 0 \quad (a) \qquad \left.\tilde{H}_y\right|_{y=\pm b/2} \neq 0 \quad (b)\Big\} \quad (21.62)$$

It may therefore be concluded that in such problems magnetic field could be present outside the long rectangular core, even at low frequencies.

21.6 EDDY CURRENTS IN ANISOTROPIC MEDIA

It has been noticed that cold rolled steels exhibit anisotropic properties such that the media parameters in the direction of rolling are different from those in the direction normal to the rolling. Also, for the purpose of field analysis in the heterogeneous (or rather, piecewise homogeneous) media representing laminated cores are often treated as homogeneous but anisotropic media. This section considers the distribution of harmonic fields in homogeneous anisotropic media.

Let us assume the following constitutive relations between various field vectors:

$$\tilde{\boldsymbol{B}} \overset{\text{def}}{=} \mu_x \tilde{H}_x \boldsymbol{a_x} + \mu_y \tilde{H}_y \boldsymbol{a_y} + \mu_z \tilde{H}_z \boldsymbol{a_z} \quad (21.63a)$$

$$\tilde{\boldsymbol{D}} \overset{\text{def}}{=} \varepsilon_x \tilde{E}_x \boldsymbol{a_x} + \varepsilon_y \tilde{E}_y \boldsymbol{a_y} + \varepsilon_z \tilde{E}_z \boldsymbol{a_z} \quad (21.63b)$$

$$\tilde{J} \stackrel{\text{def}}{=} \sigma_x \tilde{E}_x \boldsymbol{a}_x + \sigma_y \tilde{E}_y \boldsymbol{a}_y + \sigma_z \tilde{E}_z \boldsymbol{a}_z \tag{21.63c}$$

where the parameters (μ_x, μ_y, μ_z), (ε_x, ε_y, ε_z), and (σ_x, σ_y, σ_z) are constants for a given region.

Consider Maxwell's equation for harmonic fields:

$$\nabla \times \tilde{\boldsymbol{E}} = -j\omega \tilde{\boldsymbol{B}} \tag{21.64a}$$

On equating the coefficients of the unit vector $\boldsymbol{a}_x$ on both sides of this equation, one gets:

$$\frac{\partial}{\partial y}\tilde{E}_z - \frac{\partial}{\partial z}\tilde{E}_y = -j\omega\mu_x \tilde{H}_x \tag{21.64b}$$

Thus:

$$\tilde{H}_x = \frac{j}{\omega\mu_x}\left[\frac{\partial}{\partial y}\tilde{E}_z - \frac{\partial}{\partial z}\tilde{E}_y\right] = \frac{j}{\omega\mu_x}\left[\frac{1}{\sigma_z}\cdot\frac{\partial}{\partial y}\tilde{J}_z - \frac{1}{\sigma_y}\cdot\frac{\partial}{\partial z}\tilde{J}_y\right] \tag{21.65}$$

Similarly:

$$\tilde{H}_y = \frac{j}{\omega\mu_y}\left[\frac{\partial}{\partial z}\tilde{E}_x - \frac{\partial}{\partial x}\tilde{E}_z\right] = \frac{j}{\omega\mu_y}\left[\frac{1}{\sigma_x}\cdot\frac{\partial}{\partial z}\tilde{J}_x - \frac{1}{\sigma_z}\cdot\frac{\partial}{\partial x}\tilde{J}_z\right] \tag{21.66}$$

and

$$\tilde{H}_z = \frac{j}{\omega\mu_z}\left[\frac{\partial}{\partial x}\tilde{E}_y - \frac{\partial}{\partial y}\tilde{E}_x\right] = \frac{j}{\omega\mu_z}\left[\frac{1}{\sigma_y}\cdot\frac{\partial}{\partial x}\tilde{J}_y - \frac{1}{\sigma_x}\cdot\frac{\partial}{\partial y}\tilde{J}_x\right] \tag{21.67}$$

Next, consider Maxwell's equation for the quasi-stationary field in a source-free region:

$$\nabla \times \tilde{\boldsymbol{H}} = \tilde{\boldsymbol{J}} \tag{21.68}$$

We get:

$$\left.\begin{aligned} &\tilde{J}_x = \frac{\partial}{\partial y}\tilde{H}_z - \frac{\partial}{\partial z}\tilde{H}_y \quad (\text{a}) \qquad \tilde{J}_y = \frac{\partial}{\partial z}\tilde{H}_x - \frac{\partial}{\partial x}\tilde{H}_z \quad (\text{b}) \\ &\tilde{J}_z = \frac{\partial}{\partial x}\tilde{H}_y - \frac{\partial}{\partial y}\tilde{H}_x \quad (\text{c}) \end{aligned}\right\} \tag{21.69}$$

On substituting the components of current density thus obtained into Equations 21.65, 21.66, and 21.67, one obtains equations exclusively between components of the magnetic field intensity:

$$\tilde{H}_x = \frac{j}{\omega\mu_x}\left[\frac{1}{\sigma_z}\cdot\frac{\partial}{\partial y}\left\{\frac{\partial}{\partial x}\tilde{H}_y - \frac{\partial}{\partial y}\tilde{H}_x\right\} - \frac{1}{\sigma_y}\cdot\frac{\partial}{\partial z}\left\{\frac{\partial}{\partial z}\tilde{H}_x - \frac{\partial}{\partial x}\tilde{H}_z\right\}\right] \quad (21.70)$$

$$\tilde{H}_y = \frac{j}{\omega\mu_y}\left[\frac{1}{\sigma_x}\cdot\frac{\partial}{\partial z}\left\{\frac{\partial}{\partial y}\tilde{H}_z - \frac{\partial}{\partial z}\tilde{H}_y\right\} - \frac{1}{\sigma_z}\cdot\frac{\partial}{\partial x}\left\{\frac{\partial}{\partial x}\tilde{H}_y - \frac{\partial}{\partial y}\tilde{H}_x\right\}\right] \quad (21.71)$$

and

$$\tilde{H}_z = \frac{j}{\omega\mu_z}\left[\frac{1}{\sigma_y}\cdot\frac{\partial}{\partial x}\left\{\frac{\partial}{\partial z}\tilde{H}_x - \frac{\partial}{\partial x}\tilde{H}_z\right\} - \frac{1}{\sigma_x}\cdot\frac{\partial}{\partial y}\left\{\frac{\partial}{\partial y}\tilde{H}_z - \frac{\partial}{\partial z}\tilde{H}_y\right\}\right] \quad (21.72)$$

Likewise, on substituting the components of the magnetic field intensity from Equations 21.65, 21.66, and 21.67, into Equations 21.69a, 21.69b, and 21.69c, one obtains equations exclusively between components of the current density:

$$\tilde{J}_x = \frac{j}{\omega\mu_z}\frac{\partial}{\partial y}\left[\frac{1}{\sigma_y}\cdot\frac{\partial}{\partial x}\tilde{J}_y - \frac{1}{\sigma_x}\cdot\frac{\partial}{\partial y}\tilde{J}_x\right] - \frac{j}{\omega\mu_y}\frac{\partial}{\partial z}\left[\frac{1}{\sigma_x}\cdot\frac{\partial}{\partial z}\tilde{J}_x - \frac{1}{\sigma_z}\cdot\frac{\partial}{\partial x}\tilde{J}_z\right] \quad (21.73)$$

$$\tilde{J}_y = \frac{j}{\omega\mu_x}\frac{\partial}{\partial z}\left[\frac{1}{\sigma_z}\cdot\frac{\partial}{\partial y}\tilde{J}_z - \frac{1}{\sigma_y}\cdot\frac{\partial}{\partial z}\tilde{J}_y\right] - \frac{j}{\omega\mu_z}\frac{\partial}{\partial x}\left[\frac{1}{\sigma_y}\cdot\frac{\partial}{\partial x}\tilde{J}_y - \frac{1}{\sigma_x}\cdot\frac{\partial}{\partial y}\tilde{J}_x\right] \quad (21.74)$$

and

$$\tilde{J}_z = \frac{j}{\omega\mu_y}\frac{\partial}{\partial x}\left[\frac{1}{\sigma_x}\cdot\frac{\partial}{\partial z}\tilde{J}_x - \frac{1}{\sigma_z}\cdot\frac{\partial}{\partial x}\tilde{J}_z\right] - \frac{j}{\omega\mu_x}\frac{\partial}{\partial y}\left[\frac{1}{\sigma_z}\cdot\frac{\partial}{\partial y}\tilde{J}_z - \frac{1}{\sigma_y}\cdot\frac{\partial}{\partial z}\tilde{J}_y\right] \quad (21.75)$$

The applications of these field equations are demonstrated through two different cases. The first of these cases relates to a simplified model for an induction machine whereas the second deals with a long rectangular laminated core. These cases are discussed here.

21.6.1 Induction Machine with Solid Rotor

Figure 21.5 shows a highly simplified model for an induction machine with solid rotor made of cold rolled steel. In this figure the infinite-half space for $z > 0$ represents the solid rotor and the infinite-half space for $z < 0$ represents the highly permeable stator core. There is negligible airgap length between the two.

The stator winding carrying balanced poly phase currents is simulated by the current sheet, K_x, located at $z = 0$, where:

$$K_x = \mathcal{R}e\left[\tilde{K}_o \cdot e^{j\left(\omega t - \frac{\pi}{\tau}y\right)}\right] \quad (21.76a)$$

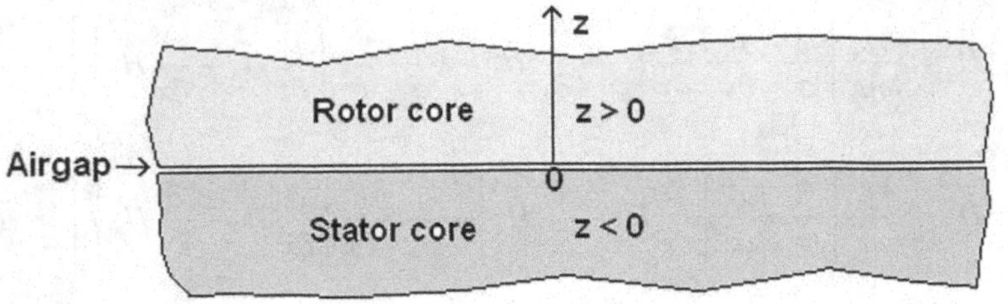

FIGURE 21.5 Simplified model for an induction machine.

or simply:

$$\tilde{K}_x \overset{\text{def}}{=} \tilde{K}_o \tag{21.76b}$$

The symbol $|\tilde{K}_o|$ indicates the RMS value of the surface current density.

It is further assumed that the field variation in the axial direction (i.e. the x-direction) is negligible. The field is varying sinusoidally in the peripheral direction (i.e. the y-direction) and also in time t.

Neglecting the field variation in the x-direction, the field equations developed above simplifies to:

$$\tilde{H}_x = \frac{j}{\omega\mu_x}\left[\frac{1}{\sigma_z}\cdot\left(\frac{\pi}{\tau}\right)^2 \tilde{H}_x - \frac{1}{\sigma_y}\cdot\frac{d^2}{dz^2}\tilde{H}_x\right] \tag{21.77}$$

$$\tilde{H}_y = \frac{j}{\omega\mu_y}\cdot\frac{1}{\sigma_x}\left[-j\left(\frac{\pi}{\tau}\right)\frac{d}{dz}\tilde{H}_z - \frac{d^2}{dz^2}\tilde{H}_y\right] \tag{21.78}$$

and

$$\tilde{H}_z = \frac{j}{\omega\mu_z}\cdot\frac{1}{\sigma_x}\left[-j\left(\frac{\pi}{\tau}\right)\frac{d}{dz}\tilde{H}_y + \left(\frac{\pi}{\tau}\right)^2 \tilde{H}_z\right] \tag{21.79}$$

Similarly:

$$\tilde{J}_x = \frac{j}{\omega\sigma_x}\left[\frac{1}{\mu_z}\cdot\left(\frac{\pi}{\tau}\right)^2 \tilde{J}_x - \frac{1}{\mu_y}\cdot\frac{d^2}{dz^2}\tilde{J}_x\right] \tag{21.80}$$

$$\tilde{J}_y = \frac{j}{\omega\mu_x}\left[-\frac{1}{\sigma_y}\cdot\frac{d^2}{dz^2}\tilde{J}_y - \frac{1}{\sigma_x}\cdot j\left(\frac{\pi}{\tau}\right)\frac{d}{dz}\tilde{J}_z\right] \tag{21.81}$$

$$\tilde{J}_z = \frac{j}{\omega\mu_x}\left[\frac{1}{\sigma_x}\cdot\left(\frac{\pi}{\tau}\right)^2 \tilde{J}_z - \frac{1}{\sigma_y}\cdot j\left(\frac{\pi}{\tau}\right)\frac{d}{dz}\tilde{J}_y\right] \tag{21.82}$$

Note that all the field components in phasor form are functions of the z-coordinate only.

The source of the field is the surface current sheet located at $z = 0$; the electromagnetic field must vanish at infinite distance from the source, i.e. at $z = \infty$. The second boundary condition is as follows:

$$\left.\tilde{H}_y\right|_{z=0} = -\tilde{K}_o \tag{21.83}$$

Now, since there is no source for the field component $\tilde{H}_x$, it must be 0 throughout. Therefore, from Equations 21.69b and 21.69c, we have:

$$\tilde{J}_y = \tilde{J}_z = 0 \tag{21.84}$$

Hence, from Equations 21.66 and 21.67:

$$\tilde{H}_y = \frac{j}{\omega\mu_y} \cdot \frac{1}{\sigma_x} \cdot \frac{d}{dz}\tilde{J}_x \tag{21.85}$$

and

$$\tilde{H}_z = -\frac{1}{\omega\mu_z} \cdot \frac{1}{\sigma_x} \cdot \frac{\pi}{\tau} \cdot \tilde{J}_x \tag{21.86}$$

Now, from Equation 21.80:

$$\frac{d^2}{dz^2}\tilde{J}_x = \left[\frac{\mu_y}{\mu_z} \cdot \left(\frac{\pi}{\tau}\right)^2 + j\omega\sigma_x\mu_y\right] \cdot \tilde{J}_x \tag{21.87}$$

The tentative solution for this equation may be taken as:

$$\tilde{J}_x = \tilde{J}_o \cdot e^{-\tilde{k} \cdot z} \tag{21.88}$$

where $\tilde{J}_o$ indicates an arbitrary constant, and $\tilde{k}$ is the following complex quantity with positive real part (since the field must vanish at $z = \infty$):

$$\tilde{k} = \sqrt{\frac{\mu_y}{\mu_z} \cdot \left(\frac{\pi}{\tau}\right)^2 + j\omega\sigma_x\mu_y}, \tag{21.89}$$

To find the arbitrary constant $\tilde{J}_o$ we substitute the expression for $\tilde{J}_x$ from Equation 21.88 into Equation 21.86, giving:

$$\tilde{H}_y = -\frac{j}{\omega\mu_y} \cdot \frac{1}{\sigma_x} \cdot \tilde{k} \cdot \tilde{J}_o \cdot e^{-\tilde{k} \cdot z} \tag{21.90}$$

Now, using the boundary condition given by Equation 21.83, we get:

$$\tilde{J}_o = -j\omega\mu_y\sigma_x \cdot \frac{1}{\tilde{k}} \cdot \tilde{K}_o \tag{21.91}$$

Thus, from Equations 21.85, 21.86, and 21.88, the distributions of electromagnetic fields found are:

$$\tilde{H}_y = -\tilde{K}_o \cdot e^{-\tilde{k}\cdot z} \tag{21.92}$$

$$\tilde{H}_z = j\frac{\mu_y}{\mu_z} \cdot \frac{\pi}{\tau} \cdot \frac{1}{\tilde{k}} \cdot \tilde{K}_o \cdot e^{-\tilde{k}\cdot z} \tag{21.93}$$

and

$$\tilde{J}_x = -j\omega\mu_y\sigma_x \cdot \frac{1}{\tilde{k}} \cdot \tilde{K}_o \cdot e^{-\tilde{k}\cdot z} \tag{21.94}$$

21.6.2 Long Rectangular Laminated Core

Figure 21.6 shows a long rectangular laminated core with its surfaces located at $x = \pm a/2$ (over $-b/2 < y < b/2$), and at $y = \pm b/2$ (over $-a/2 < x < a/2$). These surfaces carry uniformly distributed current sheets simulating the excitation coil with alternating current.

The surface current density is given as:

$$\boldsymbol{K}\Big|_{x=\pm a/2} = \pm \mathcal{R}\boldsymbol{e}\left[\tilde{K}_o \boldsymbol{e}^{j\omega t}\right]\boldsymbol{a}_y \quad \text{over} \quad b/2 < y < b/2 \tag{21.95a}$$

or simply

$$\tilde{K}_y\Big|_{x=\pm a/2} = \pm\tilde{K}_o \quad \text{over} \quad b/2 < y < b/2 \tag{21.95b}$$

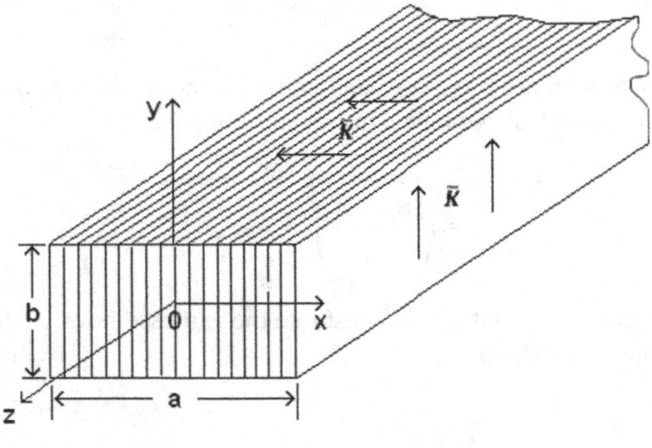

FIGURE 21.6 A long rectangular laminated core.

and

$$\boldsymbol{K}\Big|_{y=\pm b/2} = \mp \mathcal{Re}\left[\tilde{K}_o\, e^{j\omega t}\right]\boldsymbol{a}_x \quad \text{over} \quad a/2 < x < a/2 \tag{21.96a}$$

or simply

$$\tilde{K}_x\Big|_{y=\pm b/2} = \mp \tilde{K}_o \quad \text{over} \quad a/2 < x < a/2 \tag{21.96b}$$

Let the heterogeneous laminated core be simulated by solid homogeneous anisotropic medium. For a long core with uniformly distributed surface currents of low frequencies, the field outside the core will be negligible. Furthermore, there will be no variation of fields inside the core in its axial direction, i.e. z-direction. Therefore, from Equation 21.72, we have:

$$\tilde{H}_z = -\frac{j}{\omega\mu_z}\left[\frac{1}{\sigma_y}\cdot\frac{\partial^2}{\partial x^2}\tilde{H}_z + \frac{1}{\sigma_x}\cdot\frac{\partial^2}{\partial y^2}\tilde{H}_z\right]$$

or

$$\frac{1}{\sigma_y}\cdot\frac{\partial^2}{\partial x^2}\tilde{H}_z + \frac{1}{\sigma_x}\cdot\frac{\partial^2}{\partial y^2}\tilde{H}_z = j\omega\mu_z\cdot\tilde{H}_z \tag{21.97}$$

This equation to be solved is subjected to the following boundary conditions:

$$\tilde{H}_z\Big|_{x=\pm a/2} = \tilde{K}_o \quad \text{over} \quad b/2 < y < b/2 \tag{21.98a}$$

$$\tilde{H}_z\Big|_{y=\pm b/2} = \tilde{K}_o \quad \text{over} \quad a/2 < x < a/2 \tag{21.98b}$$

Let us assume that $\tilde{H}_z$ can be expressed as the product of $\tilde{\mathbb{X}}(x)$ and $\tilde{\mathbb{Y}}(y)$; the former is a function x and the latter is a function of y. Therefore:

$$\tilde{H}_z \overset{\text{def}}{=} \tilde{\mathbb{X}}(x)\cdot\tilde{\mathbb{Y}}(y) \tag{21.99}$$

Therefore, from Equations 21.97 and 21.99, we get:

$$\frac{1}{\sigma_y\cdot\tilde{\mathbb{X}}}\cdot\frac{d^2}{dx^2}\tilde{\mathbb{X}} + \frac{1}{\sigma_x\cdot\tilde{\mathbb{Y}}}\cdot\frac{d^2}{dy^2}\tilde{\mathbb{Y}} = j\omega\mu_z \tag{21.100}$$

From Equation 21.100, we conclude that each term on its LHS must be a constant quantity, say α_m^2 and β_n^2, such that:

$$\left(\alpha^2/\sigma_y\right) + \left(\beta^2/\sigma_x\right) = j\omega\mu_z \tag{21.101}$$

Therefore:

$$\left.\frac{d^2}{dx^2}\tilde{\mathbb{X}} = \alpha^2 \cdot \tilde{\mathbb{X}} \quad (a) \qquad \frac{d^2}{dy^2}\tilde{\mathbb{Y}} = \beta^2 \cdot \tilde{\mathbb{Y}} \quad (b)\right\} \tag{21.102}$$

This way, we have converted a partial differential equation into two separate ordinary differential equations. This method of solution is called the method of separation of variables. The coefficients α and β are called the constants of separation.

Therefore, on solving these equations, we get:

$$\left.\tilde{\mathbb{X}} = c_1\, e^{\alpha x} + c_2\, e^{-\alpha x} \quad (a) \qquad \tilde{\mathbb{Y}} = c_3\, e^{\beta x} + c_4\, e^{-\beta x} \quad (b)\right\} \tag{21.103}$$

Our next step involves the determination of the constants of separation α and β. Consider the boundary conditions given by Equations 21.98a and 21.98b. Let the solution for Equation 21.97 be given as the sum of two partial solutions;

$$\tilde{H}_z = \tilde{H}_{1z} + \tilde{H}_{2z} \tag{21.104}$$

such that one partial solution, $\tilde{H}_{1z}$ satisfies Equation 21.98a and gives zero value on the remaining boundaries, whereas the other partial solution, $\tilde{H}_{2z}$ satisfies Equation 21.98b and sets zero values on the remaining boundaries. Since the field component $\tilde{H}_z$ is an even function of x and y, let the first partial solution provisionally be given as follows:

$$\tilde{H}_{1z} = \sum_{m\text{-odd}}^{\infty} \tilde{d}_{1m} \cdot \cos\left(\frac{m\pi}{b} \cdot y\right) \cdot \frac{\cosh(\gamma_m \cdot x)}{\cosh(\gamma_m \cdot a/2)} \tag{21.105}$$

In this expression, $\tilde{d}_{1m}$ indicates a set of arbitrary constants. The separation constant β is taken as $\left(j\dfrac{m\pi}{b}\right)$, while the separation constant α found from Equation 21.101 is as follows:

$$\alpha \stackrel{\text{def}}{=} \gamma_m = \sqrt{\frac{\sigma_y}{\sigma_x} \cdot \left(\frac{m\pi}{b}\right)^2 + j\omega\mu_z\sigma_y} \tag{21.106}$$

Likewise, let the second partial solution provisionally be given as follows:

$$\tilde{H}_{2z} = \sum_{n\text{-odd}}^{\infty} \tilde{d}_{2n} \cdot \cos\left(\frac{n\pi}{a} \cdot x\right) \cdot \frac{\cosh(\delta_n \cdot y)}{\cosh(\delta_n \cdot b/2)} \tag{21.107}$$

In this expression, $\tilde{d}_{2n}$ indicates a set of arbitrary constants. The separation constant α is taken as $\left(j\dfrac{n\pi}{a}\right)$, while the separation constant β found from Equation 21.101 is as follows:

$$\beta \stackrel{\text{def}}{=} \delta_n = \sqrt{\frac{\sigma_x}{\sigma_y} \cdot \left(\frac{n\pi}{a}\right)^2 + j\omega\mu_z\sigma_x} \tag{21.108}$$

Therefore, the complete solution for Equation 21.97, found in view of Equation 21.104 is as follows:

$$\begin{aligned} \tilde{H}_z = &\sum_{m\text{-odd}}^{\infty} \tilde{d}_{1m} \cdot \cos\left(\frac{m\pi}{b} \cdot y\right) \cdot \frac{\cosh(\gamma_m \cdot x)}{\cosh(\gamma_m \cdot a/2)} \\ &+ \sum_{n\text{-odd}}^{\infty} \tilde{d}_{2n} \cdot \cos\left(\frac{n\pi}{a} \cdot x\right) \cdot \frac{\cosh(\delta_n \cdot y)}{\cosh(\delta_n \cdot b/2)} \end{aligned} \tag{21.109}$$

Now, to determine the arbitrary constants involved in this equation, we identify $\tilde{d}_{1m}$ and $\tilde{d}_{2n}$ as Fourier coefficients in the Fourier expansion of the RHS of Equations 21.98a and 21.98b, respectively. Thus:

$$\left.\tilde{d}_{1m} = \tilde{K}_o \cdot \frac{4}{m\pi} \cdot \sin\left(\frac{m\pi}{2}\right) \quad (\text{a}) \qquad \tilde{d}_{2n} = \tilde{K}_o \cdot \frac{4}{n\pi} \cdot \sin\left(\frac{n\pi}{2}\right) \quad (\text{b})\right\} \tag{21.110a}$$

DESCRIPTIVE QUESTIONS

Q21.1 What do you understand by eddy current and how it is different from other currents?

Q21.2 Describe the conditions that give rise to eddy currents.

Q21.3 Discuss the importance and applications of eddy currents.

Q21.4 Write the eddy current equation and describe the boundary conditions for obtaining its unique solution.

Q21.5 Why it is necessary to classify the boundary value problems in accordance with dimensions, particularly those related with the eddy currents?

FURTHER READING

S. K. Mukerji, A. S. Khan, Y. P. Singh, *Electromagnetics for Electrical Machines*, CRC Press, New York, Mar. 2015.

S. K. Mukerji, Y. P. Singh, M. George, P. Gautam, "Eddy currents in cores with triangular cross-sections", *International Journal of Applied Engineering and Computer Science* Vol. 1, No. 1, pp. 29–32, Jan. 2014.

S. K. Mukerji, Y. P. Singh, "Eddy currents in long solid conducting cores with triangular cross-sections", National Conference on Recent Advances in Technology and Engineering (RATE 2013), Mangalayatan University, Aligarh, India, 2013, pp. 1–3.

S. K. Mukerji, D. S. Srivastava, Y. P. Singh, D. V. Avasthi, "Eddy current phenomena in laminated structures due to travelling electromagnetic fields", *Progress in Electromagnetics Research M* Vol. 18, pp. 159–169, 2011.

S. K. Mukerji, M. George, M. B. Ramamurthy, K. Asaduzzaman, "Eddy currents in solid rectangular cores", *Progress in Electromagnetics Research B* Vol. 7, pp. 117–131, 2008.

S. K. Mukerji, M. George, M. B. Ramamurthy, K. Asaduzzaman, "Eddy currents in laminated rectangular cores", *Progress in Electromagnetics Research* Vol. 83, pp. 435–445, 2008.

V. Subbarao, *Eddy Currents in Linear and Non-Linear Media*, Omega Scientific Publishers, New Delhi, 1991, pp. 36–39.

L. V. Bewley, *Two-Dimensional Fields in Electrical Engineering*, Dover Publication, New York, 1963.

22 Electromagnetic Compatibility

22.1 INTRODUCTION

The rapid advances in the areas of electrical and electronic engineering have immensely benefitted the humanity. These advances have made people's lives easy and comfortable. Simultaneously these advances have created problems to the engineers and researchers. Most of these problems are related to the mixing of electromagnetic signals of different devices operating in close proximity. Such a mixing is referred to as "electromagnetic interference", abbreviated "EMI". The smooth and efficient operation of all closely installed devices requires some effective measures to overcome the unwanted effects of electromagnetic interference. The quest of such measures has resulted in the development of a branch of electrical sciences which covers the unintentional generation, propagation and reception of electromagnetic energy with reference to the undesirable effects of EMI. This area is referred to as "electromagnetic compatibility", abbreviated "EMC". Its basic goal is to provide corrective measures to nullify the injurious effects of EMI in the electromagnetic environment of different equipment.

The problem of electromagnetic interference is the result of widespread use of different types of electromagnetic devices. Earlier the loads of most of the electrical appliances, used in the installations of conventional buildings, were linear in nature. These included ac and dc motors, resistive loads, filament lamps etc. These were resulting in almost insignificant interference between different devices or equipment. With the advances in technology the use of inverter driven ac-motors, discharge lamps, energy-saving lamps etc. have become quite common. All these produce narrow-band noise which is likely to spread all over the network. The root cause of this noise lies in switching of these devices at fixed frequencies above 9 kHz. The "switch-mode power supplies" operating in the range of 10 to 100 kHz produce this type of interference. Besides, the use of digital systems such as IT equipment for technical facility management, for industrial process automation systems, and for multimedia applications etc. has also increased many-fold. These have further aggravated the situation of electromagnetic interference.

The power supply systems are gradually becoming more powerful. As a result, the probability of electromagnetic interference is enhanced. Besides, the use of digital networks is increasing day by day and networks are becoming more and more sensitive. These networks are performing at higher data transfer rates and are increasingly used for safety-related tasks. In view of these developments the quality of electrical installations in all buildings has to be of high quality or in other words all the equipment involving electromagnetic phenomena has to be compatible. The electromagnetic incompatibility of such installations may lead to either higher costs or in the decrease of safety standards. Some of those involved in the development and installations of electromagnetic devices have tried to understand and address these problems. This chapter summarizes some of the developments in the area of EMC which are evolved in the recent past.

22.2 ROLE OF EMC AND SOURCES OF EMI

This section briefly presents the role of electromagnetic compatibility and the sources of electromagnetic interference.

22.2.1 Role of Electromagnetic Compatibility

Electromagnetic compatibility refers to the ability of an electrical or electronic system to operate properly in a disturbing electromagnetic environment and without disturbing the operation of other systems or components of other equipment. It basically addresses the issues of "emission" and "susceptibility" or "immunity". Emission refers to the unwanted generation of electromagnetic energy by various sources, and the required countermeasures. These measures are required to reduce such generation and to prevent the escape of energy into the external environment. Susceptibility (or immunity) refers to the correct operation of electrical equipment (the "victim") in the presence of electromagnetic disturbances. To achieve electromagnetic compatibility, both of these issues are to be properly addressed. Interfering sources must be suppressed and the potential victims fortified. In addition, the coupling paths between the sources and victims are also to be thoroughly studied and addressed so as to minimize electromagnetic interference.

22.2.2 Sources of Electromagnetic Interference

In general, all electrical conductive components of buildings and facilities therein may play a role in electromagnetic interference. These may act as a source (or EMI transmitter), as a drain (or EMI receiver), or as both. In buildings, the installed electrical conductors, metal pipelines, reinforcement bars, metal façades, and constructional steel work may become a source of EMI. Typical systems that may result in EMI include (i) power supply lines, (ii) measuring devices, (iii) control devices, (iv) alarm devices, and (v) computer installations including networks. All of these systems involve field phenomenon of one form or the other, and due to the flow of current and existence of potential between terminals, become the sources of EMI. At low frequencies, electric and magnetic fields act independently, whereas at high frequencies, propagating electromagnetic fields become important. As all of the fields in different frequency ranges are generated by electric charges and currents, the EMI due to these can be classified in accordance with the frequency ranges.

22.2.2.1 Low Frequency EMI

At low frequencies, the electric and magnetic fields have a relatively short range. This is mainly due to the reduction of intensity of these fields with inversely proportional to the distance from their sources. Thus, such fields remain confined mainly to the vicinity of the current carrying conductors.

An electrical field is proportional to the voltage of the electrical installation. Its contribution to EMI effects can only be visualized in the case of high-voltage installations. In general, electric fields do not play a major role in most installations located at farther distances. However, at short distances, electrical fields can become a potential source of EMI, e.g. in the case of cable trunks in which the cables run together.

Contrary to the electric field, which is proportional to the voltage, the magnetic field is proportional to the strength of current. Currents in many power supply systems may attain significantly higher values. This results in a strong magnetic field and thus in larger EMI effects. Such effects in TNC (Trans-National Corporation) types of installations are likely to be more prominent. Neutral (N) and protective earth (PE) conductors are combined to form a PEN conductor. Due to connections of these conductors to other parts of a building, currents may flow in every part of the building and may result in magnetic fields. The field may create an EMI environment almost everywhere. Furthermore, since part of the neutral return current flows in extraneous metal parts, it may add to the current in the TNC network. This creates an unbalanced condition and the net magnetic field of the TNC network may be enhanced.

The cathode ray tube (CRT)–type of computer terminal gets easily disturbed by magnetic fields of the order of 1.5 μT. These disturbances can be observed in the form of flickering on the screen. The magnetic field of this order can be generated by a single power line carrying a 10A 50 Hz current within a distance of 1.3 m. Larger CRT computer terminals (with screens larger than 17 inches) are even more sensitive to external magnetic fields. If the currents in the power line contain higher frequency components, the magnetic fields may even result in stronger EMI.

22.2.2.2 High-Frequency EMI

At high frequencies, electric and magnetic fields combine to form electromagnetic fields. These fields travel with the velocity of light through space in the form of electromagnetic waves. Thus, these have a greater potential for disturbing other devices at much larger distances. Radar, radio and tv transmitters, mobile phones, DECT (Digital European Cordless Telecommunication) telephones, wireless local area networks (WLAN), Bluetooth links, and industrial installations in the microwave range are some of the typical sources of electromagnetic fields. On occasion, power cables also act as antennas and propagate high-frequency signals. Some of these signals are intentionally used to travel, as in the case of power line communication. In other cases, these get unintentionally generated due to fast transients present on the network.

It can be concluded that electrical installations need proper immunization against unwanted electromagnetic fields. Thus, careful design and proper shielding of installations become necessities.

22.2.2.3 Elementary Coupling Model of EMI

As shown in Figure 22.1, the mechanism of EMI can be represented by a very simple model comprising a source, a coupling media or mechanism, and a disturbed device or victim.

The sources may include the lines of the electrical power system, antennas of wireless LAN systems, etc. Coupling is established through current if common conductors of different circuits are shared by electric, magnetic, or electromagnetic fields. The victim or disturbed drains may be an apparatus or a part of the electrical installation. The complete electromagnetic interaction of all the installations in a building or facility may be a complex combination of these elementary interactions. It needs to be noted that a drain or victim may also act as a potential source of EMI.

While planning or refurbishing a new installation, a matrix of all the possible sources, coupling paths, and disturbed objects is to be generated. This matrix may be used to assess the possible strength of mutual interference and to judge the occurrence and relevance of EMI disturbances. Such assessments will help in planning better preventive measures, rapid implementation, and cost-effective commissioning. The identified forms of EMI affecting drains of low- and high-voltage cables include (i) impedance or conductive coupling, (ii) inductive coupling, (iii) capacitive coupling, and (iv) radiative coupling. Inductive and capacitive couplings are the special cases of impedance coupling. The basic physical properties of these sources are summarized in Table 22.1. Each of these deserves due weight while assessing and planning a system.

The inductive coupling is the dominant disturbing phenomenon in buildings. It is followed by the capacitive and impedance coupling. In the case of radiative coupling, field strengths are usually well below the required limiting values. Thus, its impact has not yet assumed significance. In view of the increased use of wireless applications, radiative coupling may also become a potential source of EMI in the future.

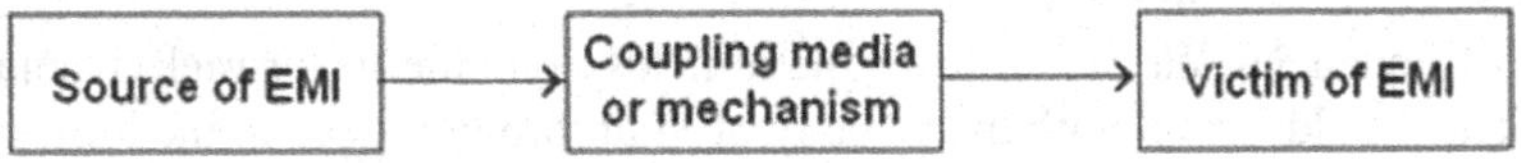

FIGURE 22.1 Elementary coupling model of EMI.

TABLE 22.1
Elementary Properties of EMI Coupling Types

Nature of Source Field	Interaction Domain	Nature of Coupling	Range of Operation
Electric	Low frequency	Capacitive	Short
Magnetic	Low frequency	Inductive	Short
Electromagnetic	High frequency	Radiative	Long

22.3 TYPES OF INTERFERENCE

Frequency-wise categorization is one way of looking at EMI. The other way of looking at the EMI is in accordance with the source and signal characteristics. The origin of noise can be manmade or natural.

22.3.1 Continuous Wave (CW) Interference

This type of interference arises where there is continuous emission from a source within a given frequency range. It can further be divided into the following subcategories in accordance with the frequencies, which may vary over a wide range, i.e. from dc to daylight.

22.3.1.1 Audio Frequency Interference (AFI)

This type of interference may be due to the sources operating in audio frequency range. In general, this range starts from a very low frequency and ranges up to 20 kHz. This range is sometimes extended up to 100 kHz. The AFI may be due to mains hum from the power supply units, nearby power supply wiring, transmission lines, or substations. Audio processing equipment (viz. audio power amplifiers and loudspeakers) and demodulation of a high-frequency carrier wave (viz. FM radio transmission) may also act as sources in this frequency range.

22.3.1.2 Radio Frequency Interference (RFI)

Sources in this range include wireless and radio frequency transmissions; television and radio receivers; industrial, scientific and medical equipment (ISM); and digital processing circuitry such as microcontrollers.

22.3.1.3 Broadband Interference

The sources in this category include solar activity, continuously operating spark gaps (e.g. arc welders), and CDMA (spread-spectrum) mobile telephony. This type of noise may spread across parts of either audio and radio frequency ranges, or both, with no particular frequency accentuated.

22.3.2 Pulse or Transient Interference

Some of the sources emit a short-duration pulse of energy. Such a pulse is called an electromagnetic pulse, abbreviated as EMP. The disturbance caused due to EMP is also referred to as transient disturbance. The energy contents of an EMP are usually broadband by nature. EMP often excites a relatively narrow-band damped sine wave response in the victim. The sources of EMP can be of isolated and repetitive nature.

22.3.2.1 Sources of Isolated EMP

The isolated pulses are generated by (i) switching action of electrical circuitry containing inductive loads such as relays, solenoids, and electric motors; (ii) electrostatic discharge when two charged objects come into close proximity or even contact; (iii) lightning electromagnetic pulse; (iv) nuclear electromagnetic pulse in the event of a nuclear explosion; (v) non-nuclear electromagnetic pulse weapons; and (vi) power line surges/pulses.

22.3.2.2 Sources of Repetitive EMP

The sources which may generate regular (or repetitive) pulse trains include (i) electric motors, (ii) gasoline engine ignition systems, and (iii) switching actions of digital electronic circuitry.

22.4 COUPLING MECHANISMS

Section 22.2 described various sources that can affect a victim. These are also illustrated in Figure 22.2, which shows the basic arrangement of noise source, coupling path, and victim (receptor or sink). The source and victim are usually electronic hardware devices. The source may also be a natural phenomenon such as lightning or electrostatic discharge (ESD).

This section deals with the four basic coupling mechanisms. These include conductive, capacitive, inductive (or magnetic), and radiative mechanisms. A victim may be affected through any of these mechanisms or through a combination of more coupling mechanisms working together. As illustrated in Figure 22.2, the lower path in the diagram involves inductive, conductive, and capacitive modes.

22.4.1 Conductive Coupling

The conductive coupling is also referred to as impedance or galvanic coupling. This occurs when different circuits use common lines and/or coupling impedances. This situation may be noticed when different circuits use the same voltage source in their circuit. It also occurs when the coupling path between the source and the receptor is formed by direct contact with a conducting body. Such a path may establish in the case of a transmission line, wire, cable, PCB trace, or in a metal enclosure.

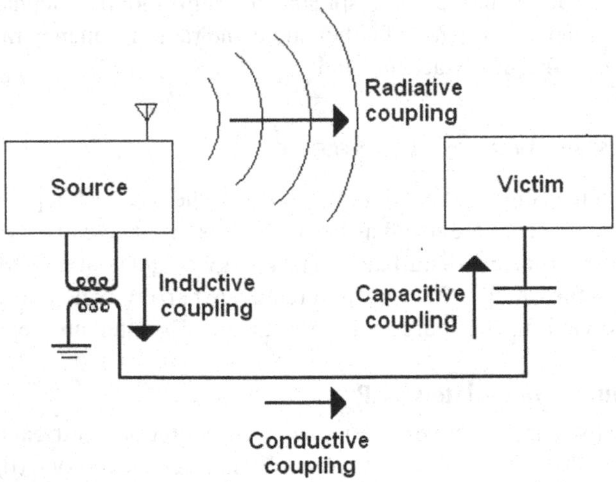

FIGURE 22.2 Four EMI coupling modes.

Conducted noise can also be characterized in accordance with the way of its appearance on different conductors. In the case of common mode or common impedance coupling, the noise appears in phase or in the same direction on two conductors. In differential mode, the noise on two conductors appears out of phase or in opposite directions.

At higher frequencies, self-inductance of the lines plays a significant role. This remains dominant even if the increase in resistance of the line due to the skin effect is accounted for, which is generally not negligible for fast transients and digital signals.

In view of Kirchhoff's laws, disturbing signals, through this coupling mode, may spread over the installation of an entire facility and may even affect the neighboring installations. Such coupling can be minimized by avoiding connections between independent systems. In cases where such connections become a necessity, self-inductance may be kept as low as possible.

Generally, galvanic decoupling of electrical power supply circuits can be more easily achieved by using TN-S system rather than a TN-C system.

22.4.2 Inductive Coupling

Inductive coupling is also called magnetic coupling. It occurs when a time-varying magnetic field exists between two parallel conductors (or source and receiver) separated by a short distance typically less than a wavelength. A time-varying external current generates a magnetic field, which induces a disturbing voltage in a neighboring circuit. The strength of this coupling depends mainly on (i) the strength of disturbing current, (ii) the distance between source and drain, and (iii) the frequency of disturbing field. Thus, magnetic coupling is a geometry and frequency dependent phenomenon. The disturbing signal becomes significant when (i) the currents of the external circuits become large, (ii) the currents of a go-and-return line become unbalanced, (iii) the circuits are closely located and encompass a large area, and (iv) when the signals of external circuit vary rapidly in time and therefore possess large high-frequency contents.

In some cases, inductive coupling may become useful in controlling the disturbances. This aspect can be explored by proper installation of cable trays and coaxial cables or when these are reliably connected with short paths with low impedance and at high frequencies. In such cases, inductive coupling provides shielding to the cables against external magnetic fields, especially at higher frequencies.

Studies have revealed that the effect of inductive coupling can be reduced (if not altogether neutralized) by having all connections of shielding facilities like cable trunks, cable channels, cabinets etc. with low resistance at high frequencies. Since the skin effect increases the resistance of electric conductors and this effect becomes more significant at higher frequencies, the geometries of conductors must be chosen so as to minimize this apparent resistance. This indicates that standard circular conductors are not the better choice and must be replaced by flat strip or braid conductor geometries, since in these the surface area is large and not thick. A short circuit loop may effectively work as a shielding device provided the protecting current flows and there is no disconnection in the short circuit loop. The shields have to be connected to ground at both ends to enable an unhindered flow of shielding current.

The intensity of magnetic coupling can be reduced by (i) keeping the area of any electrical installation as small as possible, (ii) maximizing the distances to the lines having high currents, (iii) separating power lines from data lines, and (iv) by using an EMC-friendly TN-S-type of networks. The unbalanced current of the TN-C network generates double the magnetic fields that a TN-S network generates.

22.4.3 Capacitive Coupling

This coupling occurs when a time-varying electrical field exists between two adjacent conductors separated by a distance typically less than a wavelength. Such a field of an external system produces time-varying charges in the disturbed system. The flow of the displacement currents connects the two systems and cause the disturbing voltages across the gap between the source and the receiver. Similar to inductive coupling, capacitive coupling becomes larger when (i) the two circuits are nearer, (ii) voltage difference of two circuits is large, and (iii) signals in the external circuit are rapidly changing with time and possess large high-frequency contents. With proper planning of cabling and shielding and careful installation, this type of disturbance can be avoided or at least minimized to the level of allowable tolerance.

The behavior of capacitive coupling is similar to that of inductive coupling. The disturbing voltage at low frequencies increases linearly with the frequency of disturbing signal and at high frequencies attains saturation. Again, fast disturbing signals containing large high-frequency components severely influence the disturbed circuit. Capacitive coupling can be reduced by using shielded cables.

22.4.4 Radiative Coupling

This is also called electromagnetic coupling. It occurs when source and victim are separated by a large distance, typically more than a wavelength. In this coupling, both the source and the victim act as radio antennas. The source radiates an electromagnetic signal, which in the form of a wave travels through the free space with the velocity of light. This signal may influence electrical installations in the near or far surroundings. The sources of such signals may include radio or TV transmitters, mobile telephones, or any other kind of wireless devices. The high-frequency components of fast signals or of fast transients (ESD, surge, burst lightning, etc.) may also cause radiations. These fields may creep-in by cables or through any other conductive parts of the electrical installation and may cause disturbances in the electrical systems in any part of the building. Disturbances on the power supply or data network containing high-frequency components may also contribute towards radiations. To estimate the magnitudes of radiated fields, the "Hertzian dipole" may be studied as an elementary model. The conductive parts of the electrical installation which may serve as antennas include (i) cables, (ii) openings and slots of cases, cubicles, etc., and (iii) printed board strips. Openings and slots of the cases of equipment radiate disturbances into the surroundings or the housing. Such signals may disturb other objects in the environment. Conducting elements (viz. cables and slots) start radiating when their linear dimension exceeds approximately half of the wavelength.

In practice it is not possible to completely close the housings. Some openings are to be provided for the entry of cables. Ventilation slots and gaps around the doors are also unavoidable. With such openings, the effectiveness of shielding of any housing is bound to be reduced. An acceptable level of shielding, however, can be achieved by judicious construction of the housing. The amount of leakage due to discontinuities in the shielding depends mainly on (i) the maximum linear dimension of the opening, (ii) the wave impedance, and (iii) the frequency of the source. In all practical installations, the maximal length of slots should be smaller than $\lambda/20$ to ensure a shielding effectiveness to the tune of at least 20 dB.

22.4.5 Complex EMI Scenario

Figure 22.3 illustrates the possibility of coexistence of all types of EMI couplings in a system. The proper functioning of such a system requires the generation and evaluation of a so-called EMC matrix in the planning process for both new and refurbished buildings.

22.5 EMC CONTROL PLAN

In view of the damaging effects of EMI, it is necessary to control such interference and reduce the risks to acceptable levels. A number of aspects should be addressed in order to control EMI and ensure EMC. These include (i) threat characterization, (ii) standard setting for emission and susceptibility levels, (iii) standard design compliance, and (iv) standard testing compliance. For complex equipment, EMC may require a dedicated EMC control plan involving all of these aspects.

22.5.1 Threat Characterization

Threat characterization requires understanding of (i) interfering sources and signals, (ii) coupling paths to the victim, and (iii) the electrical nature of the victim. The risk posed by the threat is usually of a statistical nature. Thus, the threat characterization and standard settings aim to reduce the probability of disruptive EMI to an acceptable level, rather than aiming for its total elimination. The other measures are discussed in the following subsections.

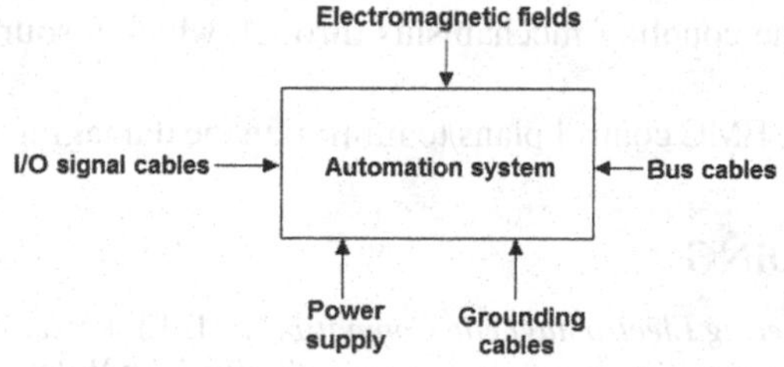

FIGURE 22.3 Coupling paths of an automation system.

22.5.1.1 Breaking a Coupling Path

This is equally effective at the start or the end of the path(s).

22.5.1.2 Grounding and Shielding

This aims to reduce emissions or divert EMI away from the victim by providing an alternative, low-impedance path. It may include (i) shielded housings, (ii) shielded cables, and (iii) earthing schemes such as star earthing for AF equipment and ground planes for RF equipment.

22.5.1.3 Other Measures

These may include (i) decoupling or filtering at critical points such as cable entries and high-speed switches, using RF chokes and/or RC elements; (ii) transmission line techniques for cables and wiring, such as balanced differential signal and return paths, and impedance matching; and (iii) avoidance of antenna structures, such as loops of circulating current, resonant mechanical structures, unbalanced cable impedances, or poorly grounded shielding.

22.5.2 Emissions Suppression Techniques

These include (i) use of spread-spectrum techniques for reducing EMC peaks, (ii) avoidance of unnecessary switching, (iii) avoidance of noisy circuits, (iv) employing harmonic wave filters, and (v) operation at lower signal levels.

22.5.3 Susceptibility Hardening

The measures that may be used for susceptibility hardening may include (i) use of fuses, trip switches, and circuit breakers; (ii) transient absorber devices; and (iii) design for operation at higher signal levels and reducing relative noise level.

DESCRIPTIVE QUESTIONS

Q22.1 Explain the meaning of EMI and EMC.
Q22.2 Discuss the role of electromagnetic compatibility.
Q22.3 Describe the sources of electromagnetic interference in different ranges of frequency.
Q22.4 Discuss different types of sources of interference.
Q22.5 Describe the coupling mechanisms through which a source can affect a victim.
Q22.6 Discuss the EMC control plans to minimize the damaging effects of EMI.

FURTHER READING

V. P. Kodali, *Engineering Electromagnetic Compatibility*, IEEE Press, 1996.
C. R. Paul, *Introduction to Electromagnetic Compatibility*, John Wiley, 1992.
H. W. Ott, *Noise Reduction Techniques in Electronic Systems*, A Wiley, 1988.
B. Keiser, *Principles of EMC*, Artech House, 1987.

Index

A

B

C

D

G

H

I

J

K

L

M

N

O

P

R

S

T

U

V

W

Z

Printed in the United States
by Baker & Taylor Publisher Services